Ammonoid
Paleobiology

TOPICS IN GEOBIOLOGY

Series Editors: **F. G. Stehli,** DOSECC, Inc., Gainesville, Florida
D. S. Jones, University of Florida, Gainesville, Florida

Ammonoid Paleobiology

Edited by

Neil H. Landman
American Museum of Natural History
New York, New York

Kazushige Tanabe
University of Tokyo
Tokyo, Japan

and

Richard Arnold Davis
College of Mount St. Joseph
Cincinnati, Ohio

Plenum Press • New York and London

Library of Congress Cataloging-in-Publication Data

Ammonoid paleobiology / edited by Neil H. Landman, Kazushige Tanabe,
and Richard Arnold Davis.
        p.   cm. -- (Topics in geobiology ; v. 13)
    Includes bibliographical references (p.   -   ) and index.
    ISBN 0-306-45222-7
    1. Ammonoidea.   I. Landman, Neil H.   II. Tanabe, Kazushige.
III. Davis, R. A. (Richard Arnold), 1942-   . IV. Series.
QE807.A5A43 1996
 564'.5--dc20   96-17035
                                                          CIP

ISBN 0-306-45222-7

©1996 Plenum Press, New York
A Division of Plenum Publishing Corporation
233 Spring Street, New York, N.Y. 10013

10 9 8 7 6 5 4 3 2 1

Printed in the United States of America

Macroconch of Late Cretaceous (Maastrichtian) *Discoscaphites gulosus* (Morton, 1834) from the Fox Hills Formation of South Dakota (x1). The original nacreous layer of the shell is preserved, which accounts for its colorful iridescence. (American Museum of Natural History 45282. Gift of Mrs. Helen Ross.) Photo: J. Beckett. © Copyright American Museum of Natural History.

To the Memory of Jost Wiedmann
(1931–1993)

Contributors

R. Thomas Becker Institut für Paläontologie, Humboldt-Universität Berlin, D-10155 Berlin, Germany

Hugo Bucher Centre des Sciences de la Terre, Univerité Claude-Bernard, Lyon I, 69622 Villeurbanne Cedex, France

John A. Chamberlain, Jr. Department of Geology, Brooklyn College, City University of New York, Brooklyn, New York 11210

Antonio G. Checa Departamento de Estratigrafía y Paleontología, and Instituto Andaluz de Ciencias de la Tierra, Universidad de Granada, Facultad de Ciencias, Granada 18071, Spain

Richard Arnold Davis Department of Chemistry and Physical Sciences, College of Mount St. Joseph, Cincinnati, Ohio 45233–1670

Larisa Doguzhaeva Paleontological Institute, Russian Academy of Sciences, 117647 Moscow, Russia

Jean-Louis Dommergues URA CNRS 157, Centre des Sciences de la Terre, Université de Bourgogne, F-21100 Dijon, France

Theo Engeser Geologisch-Paläontologisches Institut und Museum, Universität Hamburg, D-21046 Hamburg, Germany

Juan M. Garcia-Ruiz Instituto Andaluz de Ciencias de la Tierra, Consejo Superior de Investigaciones Científicas, Granada 18071, Spain

Jean Guex Institut de Géologie, Université de Lausanne, BFSH-2, CH-1015 Lausanne, Switzerland

Rainer Hengsbach Perwang 95, A-5163 Mattsee, Austria

Roger A. Hewitt Department of Geology, McMaster University, Hamilton, Ontario L8S 4M1, Canada

David K. Jacobs Department of Biology, University of California at Los Angeles, Los Angeles, California 90024-1606

Susan M. Klofak Department of Invertebrates, American Museum of Natural History, New York, New York 10024

Cyprian Kulicki Polish Academy of Sciences, Institute of Paleobiology, 02-089 Warsaw, Poland

Jürgen Kullmann Geologisch-Paläontologisches Institut, Universität Tübingen, D-72076 Tübingen, Germany

Neil H. Landman Department of Invertebrates, American Museum of Natural History, New York, New York 10024

Haruyoshi Maeda Department of Geology and Mineralogy, Faculty of Science, Kyoto University, Kyoto 606-01, Japan

Royal H. Mapes Department of Geological Sciences, Ohio University, Athens, Ohio 45701

Didier Marchand URA CNRS 157, Centre des Sciences de la Terre, Université de Bourgogne, F-21100 Dijon, France

Harry Mutvei Department of Paleozoology, Swedish Museum of Natural History, S-104 05 Stockholm, Sweden

Marion Nixon Research School of Geological and Geophysical Sciences, Birkbeck College and University College London, London WC1E 6BT, England

Takashi Okamoto Department of Earth Sciences, Ehime University, Matsuyama 790, Japan

Kevin N. Page Earth Science Branch, English Nature, Peterborough PE1 1UA, England

Adolf Seilacher Kline Geology Laboratory, Yale University, New Haven, Connecticut 06511-8161

Yasunari Shigeta Department of Paleontology, National Science Museum, Tokyo 160, Japan

Kazushige Tanabe Geological Institute, University of Tokyo, Tokyo 113, Japan

Peter Ward Department of Geological Sciences, University of Washington, Seattle, Washington 98195

Gerd E. G. Westermann Department of Geology, McMaster University, Hamilton, Ontario L8S 4M1, Canada

Jost Wiedmann (deceased) Geologisch-Paläontologisches Institut, Universität Tübingen, D-72076 Tübingen, Germany

Foreword: Ammonoids Do It All

Ammonoids are *the* quintessential fossils, seemingly covering all the major themes of paleontology. Method and theory of stratigraphic correlation using fossils? Albert Oppel, whose concepts of zonation were explicated and applied by W. J. Arkell exhaustively in his monumental works on the Jurassic System, immediately spring to mind—works based virtually exclusively on the stratigraphic distributions of ammonoid species. Evolution? W. Waagen leaps to mind, applying the term "mutation" to his ammonoid lineages, and thus introducing the word to the scientific literature well before geneticists co-opted "mutation" for their own, starkly different, use.

Extinction? Cretaceous heteromorphs were type examples of "racial senescence"—if now wholly discredited, nonetheless an important part of earlier discourse on what is one of the most compelling issues that paleobiology brings to general biological theory. I was myself stunned, when compiling data on the end-Cretaceous mass extinction in the late 1960s for a seminar conducted by Norman D. Newell, to find that the scaphitids—far from dwindling to a precious few as Cretaceous time was running out—were actually in the midst of an evolutionary radiation, an expansion of diversity cut abruptly short by whatever it was that disrupted things so badly 65 million years ago.

Indeed, though of course much remains to be learned about ammonoid phylogeny, every chart that I have seen published in the last 30 years showing the basic outlines of ammonoid evolution against the backdrop of Silurian–Cretaceous geologic time constitutes a stark object lesson on the resonance between evolution and extinction. The theme of early "experimentation" shows up amidst Devonian ammonoid diversity: the clymeniids constitute an arch example, with their siphuncle on the opposite side of the body from what proved to be the "normal" ammonoid condition—an experiment that failed to survive the late Devonian biotic crisis, thus forever depleting ammonoid morphological diversity. And are the goniatites, ceratites, and ammonites mere grades, as nearly everyone suspected back in the parallel-evolution-mad 1960s? Or are they, as now seems evident, genealogically coherent, monophyletic clades that represent radiations consequent to major biotic crises of the Permo–Triassic and Triassic–Jurassic boundaries? That grade-like patterns can come from evolutionary radiations following severe extinction bottlenecks is an aspect of evolutionary theory yet to be fully expounded. And it is the ammonoids that show such patterns best.

Biostratigraphy, evolution, extinction—not to mention biogeography, paleoecology, and functional morphology: of all major taxa in the fossil record, the ammonoids arguably do it best. But there is something more to them, a certain allure that makes them deserved rivals of trilobites as the most ardently desired and sought-after relics of the deep past. Ammonoids are at once exotic yet familiarly organic. Though nearly always simply the empty shells of long-dead animals, they nonetheless seem complete. They are almost always beautiful—and sometimes even colorful. It's probably the (nearly always planispiral) logarithmic spiral that, in spite of its mathematical precision, nonetheless casts an aura of intrigue and mystery to what is otherwise just another fossil. A few years back I published a lavishly illustrated book on fossils, using photographs of many of the finest specimens of all taxa from the rich paleontological collections of the American Museum of Natural History. And though I had skulls of a male and female Tertiary artiodactyl on the front cover, it is the photo on the back—of a pretty little pyritized specimen of the Jurassic ammonoid *Hecticoceras*—that attracted the most attention, and that has been subsequently reproduced over and over again.

I can only conclude that, over and above the prodigious intellectual contributions that continue to come from contemplation of these marvelous animals (as this present volume amply demonstrates), ammonoids also have that certain *je ne sais quoi* that will always keep them at the forefront of the paleontological realm. Ammonoids really do seem to have it all.

Niles Eldredge

The American Museum of Natural History
New York, New York

Preface

The ammonoids are a group of externally shelled cephalopods that appeared in the Late Silurian and became extinct at the end of the Cretaceous. They are generally acknowledged to be superb index fossils. In other words, their distribution in time and space permits precise biostratigraphic correlations across wide regions. Most studies of ammonoids focus on this biostratigraphic utility and tend to lose sight of, or are not concerned with, ammonoids as biological entities. Clearly, however, it is only through the study of the biology of these animals or, more properly, their paleobiology that we can begin to appreciate them as once-living organisms and try to understand why they are, in fact, such excellent index fossils. What was it about their life history and ecology, for example, that caused them to have a high incidence of species origination?

Of course, there have been paleobiological studies over the years, but these have been in the decided minority. Some of the earliest investigations focused on the functional significance of septa and sutures; E. Pfaff, H. H. Swinnerton, and, in the present day, notably Gerd E. G. Westermann have pursued these studies. Research on the early ontogeny of ammonoids was pioneered by W. Branco in 1879–1880 and has been continued by such workers as H. K. Erben, Tove Birkelund, and V. V. Druschits. Another outstanding contribution was the publication of A. E. Trueman in 1941 on ammonoid buoyancy and living orientation.

The last two decades have seen a growing number of paleobiological papers concentrating on ammonoid locomotion, buoyancy control, and embryonic development. Some of these studies have paralleled similar investigations on present-day *Nautilus*. Because this is the only externally shelled cephalopod alive today, many ammonoid researchers have studied it for the insights it provides into ammonoid paleobiology. Such interdisciplinary studies have been very successful and are evident in the proceedings of the several symposia on extinct and extant cephalopods in York (1979), Tübingen (1985), Lyon (1990), London (1991), Kyoto (1992), and Granada (1996).

This book is viewed as a synopsis of our current knowledge on ammonoid paleobiology. It is by no means complete. It contains both reviews of previous work and new material. Each of the editors of this book was involved in some way with the book previously published by Plenum Press on the biology of *Nautilus*. We regard this volume as a companion piece on ammonoids. The fact that ammonoids are extinct organisms, known only from their fossils, does impose some constraints, of course, on our paleobiological insights. Nevertheless, as the contents of this book demonstrate, it is surprising how much we have learned about these extinct animals.

However, *Ammonoid Paleobiology* also demonstrates how much is still unknown about these animals. The phylogeny of the Ammonoidea is still unclear, and many questions about life history, rate of growth, and locomotion are still unanswered. The potential for ammonoids to contribute to evolutionary studies is also still largely untapped. We hope this book will promote further research in these areas which will add to our understanding of ammonoid paleobiology.

As is evident, this book is not the work of a single author. Thus, it is not marked by a single vision, other than an interest in paleobiology. The different authors present a diversity of viewpoints. Indeed, some chapters even contain contradictory hypotheses and theories, demonstrating that many paleobiological issues are far from resolved. However, because this is a multiauthored work, experts in various fields cover specific subjects to an extent not possible in a single-authored volume. Moreover, the list of the different authors and their addresses provides some sense of the number of scientists who are currently studying ammonoid paleobiology and the variety of institutions with which they are associated.

Ammonoid Paleobiology is intended for a variety of readers. Certainly, workers on ammonoids will want to read it to gain familiarity with fields other than their own. Another major audience for whom this book is intended consists of graduate students in paleontology who want an up-to-date source on ammonoid paleobiology. We hope this book will serve that purpose and complement both the primary literature as well as the few synopses available—notably, the introductory material in the ammonoid volume of the *Treatise on Invertebrate Paleontology* (1957), the article on ammonoid paleobiology by W. J. Kennedy and W. A. Cobban (1976), and the textbook *The Ammonites—Their Life and Their World* by Ulrich Lehmann, recently republished in a second edition in 1990. We hope that the third major group of readers will be amateurs. Ammonoids, partly because they are such beautiful fossils, are prized by many collectors. This book will add depth to the meaning of these fossils and provide a context for their collection. Perhaps, also, it will encourage some amateurs to pursue more formal studies; indeed, many research scientists start out as amateur collectors.

The contents of this book are divided into seven major sections, in each of which there are one or more chapters. The first chapter by Theo Engeser discusses the systematic position of the Ammonoidea within the Cephalopoda. This discussion provides the phylogenetic framework for the rest of the book. Engeser argues, on the basis of several lines of evidence, that the Ammonoidea are more closely related to the Coleoidea than they are to the Nautiloidea and that, together, the Ammonoidea and Coleoidea form the taxon Angusteradulata.

The next section deals with the soft- and hard-part morphology of ammonoids. Marion Nixon describes the jaws and radula of ammonoids, which,

interestingly, furnish some of the most important data for Engeser's phylogenetic analysis. The muscle system is next described by Larisa Doguzhaeva and
Harry Mutvei. Although the only evidence of muscles consists of the scars left
on the inside surface of the shell, these allow a reconstruction of the muscular
system and how it may have functioned. The next two chapters, on shell
microstructure, by Cyprian Kulicki, and on color patterns, by Royal H. Mapes
and Richard Arnold Davis, focus on the microstructure and morphology of
the shell. The final chapter in this section, by Kazushige Tanabe and Neil H.
Landman, describes the septal neck–siphuncular complex in ammonoids. It
includes a description of the different kinds of septal necks and the various
ammonoid taxa in which they occur.

Part III deals closely with issues that have long figured in investigations of
ammonoid paleobiology. David K. Jacobs and John A. Chamberlain, Jr., discuss
buoyancy and the factors involved in ammonoid swimming. Takashi Okamoto
explains the construction of heteromorph ammonoids, those forms with
irregularly coiled shells. Antonio G. Checa and Juan M. Garcia-Ruiz address
the question of how ammonoid septa were constructed, reviewing many of
the hypotheses to date and introducing a new concept based on fractal models.
In the final chapter in this section, Roger A. Hewitt explains how measurements of the mechanical strength of the shell, septa, and siphuncle can be
used to generate accurate estimates of the depths at which ammonoids lived.
He includes a long table listing this information for many species.

Part IV is concerned with questions about ontogeny and the rate of growth.
The initial chapter (by Neil H. Landman, Kazushige Tanabe, and Yasunari
Shigeta) focuses on embryonic development. It summarizes data on the
embryonic shells of ammonoids, which average about 2 mm or less in
diameter. Next there is a chapter on the mode and rate of ammonoid growth,
by Hugo Bucher, Neil H. Landman, Susan M. Klofak, and Jean Guex. These
authors review the evidence bearing on the rate of growth and conclude that
most shallow-water forms reached maturity in about 5 years. Another multiauthored chapter (by Richard Arnold Davis, Neil H. Landman, Jean-Louis
Dommergues, Didier Marchand, and Hugo Bucher) completes this section and
discusses sexual dimorphism and the morphological modifications associated
with maturity.

The next two sections cover the topics of taphonomy and ecology, the two
being linked in that the first provides some of the data used to infer the second.
Haruyoshi Maeda and Adolf Seilacher present an overview of taphonomic
processes as well as particular examples from the Paleozoic and Mesozoic.
Rainer Hengsbach discusses ammonoid pathologies, not only those resulting
from injuries but also those related to possible parasites. Gerd E. G. Westermann concludes this section with a masterful account of ammonoid ecology
that brings into play many of the concepts treated in other chapters. He
advances the general theme that a great many ammonoids were nektic or
planktic rather than nektobenthic (see his use of the term "demersal").

The final section reviews the distribution of ammonoids in time and space, including their extinction at the end of the Cretaceous. In companion chapters, R. Thomas Becker and Jürgen Kullmann, on the one hand, and Kevin Page, on the other, treat Paleozoic and Mesozoic ammonoids, respectively. They trace the appearance and disappearance of biogeographic provinces and provide a taxonomic reference point for much of the book. The last two chapters explore the patterns and causes of ammonoid extinction. Jost Wiedmann and Jürgen Kullmann present an overview emphasizing the role of sea-level changes. Peter Ward concentrates on the pattern of ammonoid diversity during the Late Cretaceous and argues against a gradual decline over several geologic stages and in favor of an abrupt disappearance at the end of the Maastrichtian (the last stage of the Cretaceous).

The Glossary at the end of the book includes various technical terms used by the authors. Many of these terms are from disparate fields, ranging from physiology to cell structure to fluid dynamics, and demonstrates how work on ammonoids is truly multidisciplinary. The reader is referred to the glossary in the ammonoid volume of the *Treatise on Invertebrate Paleontology* for additional terms.

There were some practical matters that had to be confronted in putting this book together. One involved the choice of tense. Ammonoids are extinct; hence, strictly speaking, anything said about them, beyond pure morphology, should be in one of the past tenses. Unfortunately, use of the present perfect and past perfect tenses sometimes results in a more complicated syntax without a genuine improvement in understanding. Consequently, we have encouraged authors to use whichever tense seemed to work best in a given context, provided that the meaning was clear and that the usage within a given chapter was more or less consistent.

A second concern was the citation of genus and species names. General recommendation E10 of the *International Code of Zoological Nomenclature* suggests that the author and date of every genus and taxon of lower rank be cited at least once in each publication, and recommendation 51C further suggests the citation of original authors and dates as well as any revisers and their dates. Such citations for each of the many taxa that are mentioned in this book, some almost in passing, would have significantly interfered with the flow of the text. Hence, we did not strictly adhere to the *ICZN* recommendations. We left the decision up to individual authors. Some of them have chosen to add appendices at the ends of their chapters to provide this information.

A third editorial dilemma centered on whether a taxon, or even an informal group of animals, should be treated as singular or plural. For example, ammonoids each had a conch, but, strictly speaking, saying "ammonoids have a conch" makes no sense. Recognizing that American English is not very successful in dealing with collective nouns, we have let each author follow his or her own grammatical conscience, again, so long as the meaning was clear.

A final point was the choice of ammonoid classification used in this book. Most authors have chosen to follow the classification presented in the 1981 volume *The Ammonoidea*, edited by M. R. House and J. R. Senior, and in the two chapters on ammonoids, one by R. A. Hewitt *et al.* and the other by K. N. Page, in the 1993 compendium *The Fossil Record 2*, edited by M. J. Benton. However, as in other groups, the classification of ammonoids is always in a mild state of flux as a result of ongoing systematic revisions, and, as a consequence, there is some variation in taxonomic usage among the authors.

Many people have helped in the production of this volume. Individual authors have added acknowledgments at the ends of their chapters. The editors thank the authors themselves for supplying definitions of the terms listed in the Glossary. All of the chapters were reviewed by the three editors and were also sent out for additional reviews. The editors thank the following individuals, for reviewing all or parts of various chapters: Tomasz K. Baumiller, Klaus Bandel, John A. Chamberlain, Jr., Antonio G. Checa, Roger A. Hewitt, David K. Jacobs, Susan M. Kidwell, Susan M. Klofak, Ulrich Lehmann, Royal H. Mapes, E. T. Tozer, Janet Voight, Peter Ward, Wolfgang Weitschat, and Gerd E. G. Westermann.

Two people have provided truly outstanding assistance in the production of this book: Kathleen B. Sarg and Barbara P. Worcester, both of the American Museum of Natural History. Together, these individuals proofread, copyedited, and "word-processed" manuscripts, reformatted bibliographies, tracked down bibliographic information, arranged for permission to reproduce figures, corresponded with authors and the publisher, and drafted and revised many illustrations. They deserve enormous credit for the completion of this book. They were ably assisted in many of these tasks by Stephanie Crooms, Susan M. Klofak, and Andrew S. Modell, all of the American Museum of Natural History.

Finally, the editors thank the publisher, in the persons of Amelia M. McNamara and her assistant Kenneth Howell, for their patience and cooperation in seeing this project through, especially in the final stages of production.

During the preparation of this volume, one of the authors, Jost Wiedmann, died. Jost was a valued colleague, an expert ammonitologist, and a friend. In the last years of his life, Jost was working on the record of ammonoid extinction, which is summarized in this volume in a chapter with his coauthor Jürgen Kullmann. All of us mourn his passing, and, as a small tribute, we dedicate this volume to him.

Neil H. Landman
New York, New York

Kazushige Tanabe
Tokyo, Japan

Richard Arnold Davis
Cincinnati, Ohio

Contents

Chapter 4 • Ammonoid Shell Microstructure

Cyprian Kulicki

Chapter 5 • Color Patterns in Ammonoids

Royal H. Mapes and Richard Arnold Davis

Chapter 6 • Septal Neck–Siphuncular Complex of Ammonoids

Kazushige Tanabe and Neil H. Landman

Part III • Buoyancy, Swimming, and Biomechanics

Chapter 7 • Buoyancy and Hydrodynamics in Ammonoids

David K. Jacobs and John A. Chamberlain, Jr.

Chapter 8 • Theoretical Modeling of Ammonoid Morphology

Takashi Okamoto

Chapter 9 • Morphogenesis of the Septum in Ammonoids

Antonio G. Checa and Juan M. Garcia-Ruiz

Chapter 10 • Architecture and Strength of the Ammonoid Shell

Roger A. Hewitt

Part IV • Growth

Chapter 11 • Ammonoid Embryonic Development

Neil H. Landman, Kazushige Tanabe, and Yasunari Shigeta

Chapter 12 • Mode and Rate of Growth in Ammonoids

Hugo Bucher, Neil H. Landman, Susan M. Klofak, and Jean Guex

Chapter 13 • Mature Modifications and Dimorphism in Ammonoid Cephalopods

Richard Arnold Davis, Neil H. Landman, Jean-Louis Dommergues, Didier Marchand, and Hugo Bucher

Part V • Taphonomy

Chapter 14 • Ammonoid Taphonomy

Haruyoshi Maeda and Adolf Seilacher

Part VI • Ecology

Chapter 15 • Ammonoid Pathology

Rainer Hengsbach

Chapter 16 • Ammonoid Life and Habitat

Gerd E. G. Westermann

Part VII • Biostratigraphy and Biogeography

Chapter 17 • Paleozoic Ammonoids in Space and Time

R. Thomas Becker and Jürgen Kullmann

Chapter 18 • Mesozoic Ammonoids in Space and Time

Kevin N. Page

Chapter 19 • Crises in Ammonoid Evolution

Jost Wiedmann and Jürgen Kullmann

Chapter 20 • Ammonoid Extinction

Peter Ward

I

Phylogenetic Perspective

Chapter 1

The Position of the Ammonoidea within the Cephalopoda

THEO ENGESER

1. Introduction

The position of the Ammonoidea within the Cephalopoda is no longer in much dispute. Almost all those who study cephalopods agree that the Ammonoidea are more closely related to the Coleoidea than to the other ectocochleate cephalopods, i.e., the Nautiloidea (Jacobs and Landman, 1994). There also can be no serious doubts that the Ammonoidea were derived from bactritids (Jacobs and Landman, 1994). However, it is still worth the effort to state more precisely the position of the Ammonoidea within the Coleoidea clade (= Neocephalopoda) using a cladistic approach. Berthold and Engeser (1987) and Engeser (1990b) discussed the position of the Ammonoidea only in general.

A cladistic method also allows the reconstruction of the Bauplan of the Ammonoidea to some degree. This does not mean that this Bauplan is valid for all Ammonoidea, because some groups might have modified it to some

THEO ENGESER • Geologisch-Paläontologisches Institut und Museum, Universität Hamburg, D-20146 Hamburg, Germany.

Ammonoid Paleobiology, Volume 13 of *Topics in Geobiology*, edited by Neil Landman *et al.*, Plenum Press, New York, 1996.

extent. But it is valid for the "hypothetical ancestral ammonoid cephalopod" if we agree on a monophyletic origin of the Ammonoidea.

2. Reconstruction of the Ancestral Cephalopod

Phylogenetic study of a group using the cladistic approach starts with the study of Recent forms. These forms have many more characters than fossil forms, and a phylogenetic framework can usually be easily established. Fossil forms can later be placed within this framework.

The Recent Cephalopoda are certainly monophyletic, as can be demonstrated by several synapomorphies (see Berthold and Engeser, 1987; Engeser, 1990a,b). Their latest common ancestor has been called the hypothetical ancestral siphonopodean cephalopod (HASC) (Engeser, 1990a). This is not, however, the oldest cephalopod. It is only the latest common ancestor of Recent *Nautilus* and the Coleoidea lineage (Fig.1). Whereas the Coleoidea are still a comparatively diverse group, the genus *Nautilus* is the sole survivor of a formerly much larger group. Moreover, the species of *Nautilus* seem to have split from each other only a few million or even only a few hundred thousand years ago (see Woodruff *et al.*, 1987).

In order to understand the reconstruction of the Bauplan of the Ammonoidea, the Bauplan of HASC is repeated here briefly. However, no reasons are given for evaluating the polarity of character states. These reasons can be found elsewhere (Engeser and Berthold, 1987; Engeser, 1990a,b).

In the stem lineage of the Cephalopoda, at least 16 autapomorphies evolved that were already present in the latest common ancestor of the cephalopodean taxa with living representatives ("crown-group" Cephalopoda, or Siphonopoda of Lancester, 1877). The autapomorphies and plesiomorphies that were inherited from older ancestors allow a relatively detailed reconstruction of HASC (Fig. 1).

The autapomorphies include the following:

1. Development of a chambered shell (phragmocone) for buoyancy control; the phragmocone is slightly cyrtoconic.
2. Development of "many" arms.
3. Development of a funnel apparatus.
4. Transformation of simple, chitinous jaws into chitinous, beak-like jaws.
5. Further concentration and complication of the nervous system (development of a "brain").
6. Internal fertilization and copulatory organs.
7. Direct development of a yolk-rich germ without true larval stages.
8. Large embryonic conch (compared with other molluscs).
9. Relatively large coelomic cavity (compared with other molluscs).
10. Carnivorous life style.
11. Development of a crop.

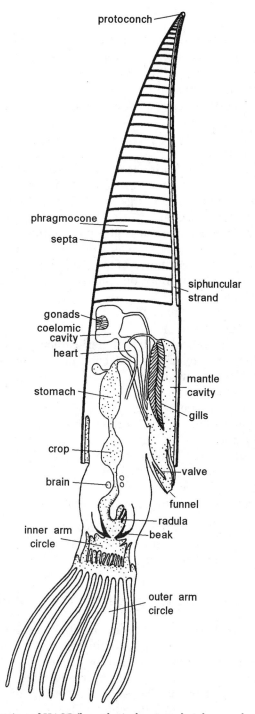

FIGURE 1. Reconstruction of HASC (hypothetical ancestral siphonopodean cephalopod). The arm number is highly questionable because it cannot be deduced from in- and out-group comparisons.

12. Development of nidamental glands.
13. Development of pericardial glands.
14. Development of the Needham's sac.
15. Reduction of the crystalline style.
16. Partially closed blood circulatory system.

Plesiomorphies include:

1. Marine living.
2. Radula (possibly with nine teeth in one row, four marginalia) [Comment: *Campitius titanicus* from the Lower Cambrian of the Westgard Pass area, California, is a large isolated radula with 13 elements per row (Firby and Durham, 1974). Although its former "owner" is unknown, it demonstrates that a group of molluscs with this character lived in the Early Cambrian seas. This radula might have belonged to a stem lineage representative of the Cephalopoda].
3. Two gills in a pallial cavity, one pair of kidneys, and a heart with one pair of auricles.
4. One pair of retractor muscles.
5. Simple pinhole eyes.
6. A single, high, conical shell with periostracum, prismatic, and nacreous layers; shell covering the visceral mass; mineralized parts of the shell consisting of aragonite.
7. A pair of statocysts.
8. Body bilaterally symmetrical.
9. Sexes separate and of roughly equal size.
10. Salivary glands.
11. Two oviducts, two spermiducts.
12. (?) r-selected reproductive strategy.
13. (?) Planktic early life phase.
14. Blood pigment consisting of hemocyanin.

The form and number of arms are unclear because it cannot be deduced from in- and outgroup comparisons. The arms might have been arranged in two circles as in *Nautilus* and many Coleoidea.

3. Phylogenetic Relationships within the Cephalopoda

Recent cephalopods can easily be distributed into two lineages: Recent *Nautilus* and the Coleoidea (Fig. 2). Both lineages are probably monophyletic. However, Recent cephalopods represent only a small fraction of the former diversity of their class, and it is certainly not advisable to enlarge these two taxa by inclusion of stem-lineage representatives. Thus, Lehmann (1967) proposed Angusteradulata and Lateradulata. The Angusteradulata are characterized by seven teeth and two marginalia, the Lateradulata by nine teeth and four marginalia. The latter condition probably represents the plesiomor-

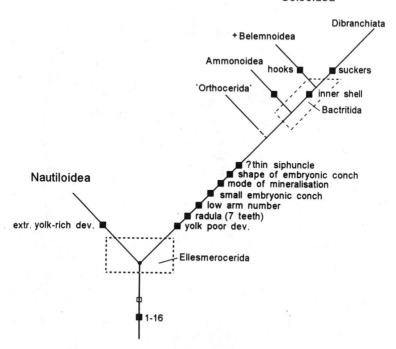

FIGURE 2. The phylogenetic tree of the Coleoidea lineage (= Angusteradulata). The black squares refer to apomorphic, and the white squares to plesiomorphic character states. Numbers 1–16 refer to the apomorphies defining HASC, as listed in the text. The positions of groups with hatched lines are uncertain.

phic character state (Berthold and Engeser, 1987). This division of the Cephalopoda is certainly somewhat problematic because the character "number of radula elements" and its distribution among the Cephalopoda are poorly known. However, fortunately, there are other characters that also demonstrate the monophyly of the Angusteradulata. Instead of Lateradulata, the older term Nautiloidea is used in this chapter. The latter term is, however, somewhat ambiguous. It has been used to describe all nautiloids (including Actinocerida and Endocerida) as well as in a more restricted sense.

3.1. Nautiloidea

Although the phylogeny of the Nautiloidea is not considered in this chapter, their supposed apomorphies are briefly outlined. This is done mainly to show the differences in early development and mode of mineralization of the embryonic conchs of Angusteradulata and Nautiloidea. The data are

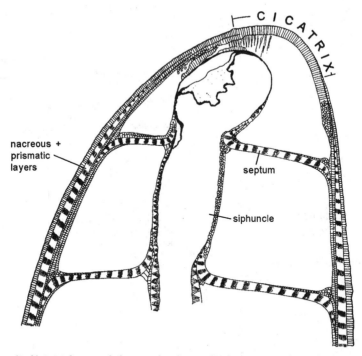

FIGURE 3. (Left) Initial part of the conch of *Pseudorthoceras* sp. from the Pennsylvanian Buckhorn Asphalt of Oklahoma. The first organic shell (area of cicatrix in this section) lacks nacreous and inner prismatic layers. (Right) Initial part of the conch of *Kionoceras* sp. The mode of mineralization and attachment of the siphuncle differ significantly from that in *Pseudorthoceras* sp. Slightly modified from Blind (1987).

largely taken from the studies of Arnold (1987), Arnold *et al.* (1987), and Tanabe *et al.* (1991).

The egg of Recent *Nautilus* is relatively large for such an animal (approximately 4 cm long; the diameter of the conch of the adult female is about 20 to 25 cm, depending on the species). The egg consists of an outer capsule and an inner capsule. The outer egg capsule consists of tough proteinaceous material. This capsule is sac-shaped and is attached to the substrate at its blunt end. There are a number of fine slits that allow sea water to circulate between the outer and the inner egg capsules. The inner egg capsule encloses the zygote and the egg jelly. The zygote measures 20 × 14 mm and contains a large amount of yolk. The cleavage of the germ is incomplete, as it is in all cephalopods ("discoidal cleavage"). The embryo develops on a disk-like cell plate on top of the yolk mass.

The shell gland secretes an initial shell that is entirely organic. The primary shell probably detaches from the mantle when the embryo is about 2 mm long. During reorganization of the yolk mass, this initial organic shell is stretched and pulled inward along a central line where cell tissue attaches to the shell before mineralization occurs. This process of attachment and deformation

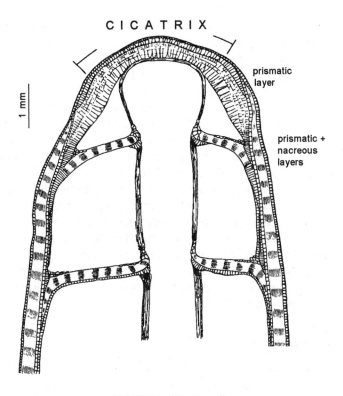

FIGURE 3. (Continued)

results in a scar-like feature called the cicatrix (Blind, 1987). In the course of this deformation, the primary shell is stretched and becomes radially wrinkled (called "low undulations" *sensu* Arnold *et al.* 1987). Then, the primary organic shell is mineralized by black organic deposits, needle-shaped aragonitic crystallites, and wheat-grain-like crystal aggregates (Blind, 1987; Arnold *et al.*, 1987; Landman *et al.*, 1989). The beginning of mineralization and the change from a wrinkled primary and early embryonic shell to a regularly sculptured secondary embryonic conch is usually visible in thin section as well as in the outer morphology (Erben and Flajs, 1975).

In *Nautilus*, a nacreous layer is present below the cicatrix. However, an inner prismatic layer is absent. In other nautiloids, a nacreous layer and an inner prismatic layer begin with the growth of the embryonic shell at the mantle edge, as in, e.g., *Pseudorthoceras* (Fig. 3) (Blind, 1987). *Nautilus* hatches with approximately seven completed septa. Other nautiloid groups usually hatched with fewer septa but not less than three (unpublished observations). The embryonic conch (i.e., the shell formed inside the egg capsule) of *Nautilus* is very large among cephalopods (up to 3.1 cm). This is very big even by nautiloid standards, where the average diameter (or length) of the embryonic conch is about 2 cm or less. The reproductive strategy of *Nautilus* (and probably of all fossil nautiloids) is K-selected, with few but well-developed large offspring. The adults can spawn several times and may live up to 20 years. The development time inside the egg is more than 1 year (Tanabe *et al.*, 1991), and maturity is reached in about 10 years. The development time in fossil forms with smaller embryonic conchs may have been shorter.

The type of early development in *Nautilus* has already been demonstrated in many other nautiloids, although there is a lot of variation, which may serve to differentiate groups at higher rank. For example, an illustration of a specimen of *Kionoceras* sp. from the Buckhorn Asphalt is included to demonstrate the utility of early shell mineralization data for systematics (Fig. 3). The Orthocerida show two types of early ontogenetic development and mineralization (see also below), which clearly demonstrates that the Orthocerida, in the classic sense, is a polyphyletic group. I have investigated several "nautiloid" orthocerids whose early ontogenetic development and mineralization are more similar to those of coiled nautiloids than to other orthocerids (e.g., the Carboniferous orthocerid family Kionoceratidae and the Triassic orthocerid family Trematoceratidae). These problems of orthocerid classification need to be resolved in the future.

3.2. Angusteradulata

The Angusteradulata comprise the Coleoidea, the Ammonoidea, the Bactritoidea, and at least some orthoceratids (Michelinoceratidae, Sphaerorthoceratidae, Arionoceratidae and probably other groups as well). The content of the taxon has not yet been fully investigated, mainly because of lack of data from the initial part of the shell.

As already stated, the "main" apomorphic character of the Angusteradulata was believed to be the character "number of radula elements" (Lehmann, 1967). However, in the Bactritida and in the angusteradulate orthocerids, we do not have any radulae preserved. Nevertheless, there are other strong arguments that the Angusteradulata is a monophyletic group, as explained below.

First, radula have seven teeth and two marginalia per row (see comments above).

Second, the general shape of the embryonic conch, its mode of mineralization, and its approximate size are consistent (see Chapter 11, this volume). The embryonic conch of the Angusteradulata is generally about 1.5 to 3 mm long. It is characteristically formed of a more or less spherical first chamber, which is separated from the remaining phragmocone by a more or less pronounced constriction, and a conical body chamber, which is modified into a spiral in more advanced groups.

The embryonic conch, excluding any septa, was initially entirely organic and was later rapidly mineralized. Therefore, the embryonic conch does not show any trace of growth lines. This type of mineralization was also different

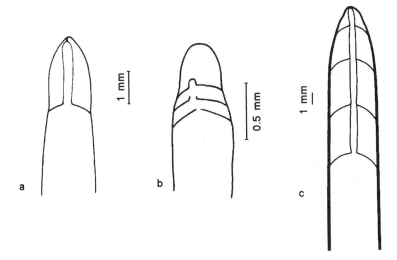

FIGURE 4. Embryonal conchs of three "orthocerids." (a) *Arionoceras* from the Silurian (modified from Serpagli and Gnoli, 1977). (b) An indeterminate sphaeorthocerid from the Silurian of Sardinia (please note the different scales). (c) *Trematoceras* from the Triassic of Northern Italy, a true nautiloid.

from that of the later conch. The initial organic shell was mineralized on the inside and outside by thin prismatic layers of aragonite. The outer organic periostracum, the nacreous layer, and the inner prismatic layer are absent.

After mineralization, the embryo partially retracted its posterior body mass from the first chamber, and the first septum was formed. The structure of the first septum is different from that of all following septa ("nacrosepta") because it is composed of only prismatic layers. In many groups, the posterior portion of the body mass was still in contact with the first chamber (formation of the cecum and prosiphon: see Kulicki, 1974; Bandel, 1986). In some orthocerids, e.g., the Arionoceratidae, the first chamber is enlarged (up to 3 mm), and, correspondingly, the cecum inside the first chamber is very long (Fig. 4; Serpagli and Gnoli, 1977). In the Belemnoidea, the posterior portion of the body mass was completely removed from the first chamber, which was sealed by an organic membrane (so-called "closing membrane" just below the first mineralized septum), and the siphuncular strand was attached to this membrane (Bandel *et al.*, 1984).

Before the animal hatched from the egg, generally one or several septa ("nacrosepta") were formed. The first chamber could be emptied of cameral liquid. After the first chamber was emptied, the small hatchling could have left the egg capsule. Hatching is generally indicated by a so-called "nepionic" constriction at the edge of the embryonic conch (Fig. 5). The postembryonic shell shows the typical four-layered structure of organic periostracum, outer prismatic layer, nacreous layer, and inner prismatic layer. We can assume that

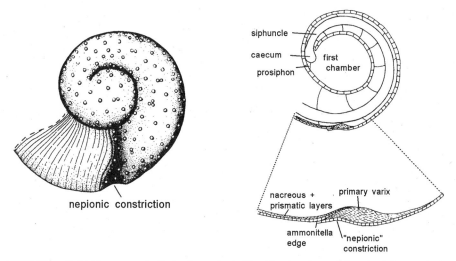

FIGURE 5. (Left) The ammonitella (embryonic shell) with tubercles and "nepionic" constriction. (Right) Median cross section of the ammonitella to demonstrate the differences in mineralization between the embryonic and postembryonic conch (slightly modified from Landman, 1988).

newly hatched animals lived in the plankton for some time. The reproductive strategy was r-selected with many small offspring.

Third, the Angusteradulata may already have had a hatching gland. Several orthocerids and ammonoids have structures that might be interpreted as devices to ease hatching from the egg. The tubercles on the ammonitella and the "primordial guard" on the outside of the embryonic conch of many sphaerorthoceratids (e.g., *Parasphaerorthoceras* Ristedt, 1969) might have borne glands to dissolve the egg capsule. They resemble comparable structures in Recent coleoid cephalopods (see Boletzky, 1982). However, the tubercles of the ammonitella and the "primordial guard" of some orthocerids do not indicate an endocochleate condition (see below).

Fourth, the Angusteradulata also probably had ten arms (see discussion below).

Fifth, the phragmocone was straight, and the siphuncle was situated at or near the center. The ancestral angusteradulate probably still had a subcentral siphuncle, although it shifted toward the venter in Bactritida, the Ammonoidea, and the Belemnitida. This condition can be interpreted as an apomorphy of these groups.

Sixth, the siphuncle was relatively thin and extended through the first septum into the first chamber, forming a cecum and prosiphon. The siphuncle is always located at some distance from the apex of the first chamber (see, e.g., the Belemnoidea). In contrast to the Nautiloidea, it is never attached directly to the apex of the first chamber. This feature might also be interpreted as apomorphic. However, the ancestral condition within the Cephalopoda is unknown.

4. Bauplan of the Ammonoidea

The following character compilation of the Bauplan of the Ammonoidea is based on theoretical considerations. This discussion overlaps with other chapters in the book that deal with specific problems of the paleobiology of ammonoids in greater detail. However, the Ammonoidea are cephalopods, and, as such, they inherited many characters from the latest common ancestor of all cephalopods. Therefore, it is worthwhile first to figure out the Bauplan of the Ammonoidea in order to recognize specific modifications of it in certain ammonoid groups. Several of the characters can easily be deduced from in- and out-group comparisons. The Bactritida are included here in the Ammonoidea because their separation is arbitrary. This group is paraphyletic or even polyphyletic. In the classical sense, the Bactritida include the ancestors of the Coleoidea and may even be true coleoids themselves (see Mapes, 1979: plate 18, Fig. 8, remains of a closing membrane?).

We begin first with plesiomorphic characters because they are more easily established.

1. There is a phragmocone for buoyancy. All ammonoids had a conch with phragmocone and living chamber. Probably all ammonoids used this phragmocone to establish neutral buoyancy. A creeping mode of life with a negatively buoyant phragmocone as proposed by Ebel (1992) certainly can be dismissed. If such a highly complicated structure were no longer needed, it would have quickly been reduced. Ebel (1992) miscalculated the weight of the phragmocone (see Chapters 7, 16, this volume). In subordinate groups, the phragmocone may even have been used for active upward and downward movement in the water column (Weitschat and Bandel, 1991) (see below).

2. The siphuncle is ventromarginal and very thin. In all ammonoids the siphuncle is ventromarginal, and its diameter is very small compared to that of nautiloids. This character stems from the latest angusteradulate ancestor. In some groups, the position of the siphuncle is not fixed in very early ontogenetic stages, and members of the Clymeniida had a dorsal siphuncle throughout life. Both character states are apomorphies of subordinate ammonoid groups.

3. The embryonic conch (ammonitella) is straight and has two septa, a cecum, and a prosiphon. The embryonic conch of the ancestral ammonoid comprises a more or less spherical first chamber that is sealed with a first septum. The siphuncle ends in a cecum that is attached to the bottom of the first chamber by the so-called prosiphon. The length of the embryonic conch is about 1.5 mm. The first chamber has a diameter of about 600 to 800 μm. A coiled ammonitella is an apomorphy of higher ammonoids (see below). The ammonitella was initially organic and then rapidly mineralized. Therefore, no growth lines are visible on the ammonitella (see discussion above). Growth lines start after the nepionic constriction, i.e., after the animal hatched from the egg.

4. The septal necks are retrochoanitic. The more primitive groups of Ammonoidea all have retrochoanitic septal necks. All "higher" Ammonoidea have modified septal necks. Retrochoanitic septal necks were inherited from HASC (compare Engeser, 1990a,b).

5. The shell is external. The Bauplan of the Ammonoidea includes an external shell. This is undisputed by almost all students of ammonoids. Some recent papers discussed the possibility that a subordinate group of the Ammonitida were endocochleates (Doguzhaeva, 1992). Even if the interpretation of an internal shell in this particular case is correct, which is doubtful, these forms would certainly not have been true endocochleates. It is possible that the mantle covered the surface of the shell as in some gastropods. The convergent evolution of a true endocochleate shell seems most unlikely. It would require permanent overgrowth of the shell gland during early ontogenesis.

6. The shell consists of periostracum, nacreous, and prismatic layers. This character was inherited from HASC. The nacreous layer is absent in the outer walls of the embryonic conch and in the first septum.

7. There is a hatching gland. The hatching gland has been considered an apomorphy of the Coleoidea by Berthold and Engeser (1987) and Engeser and Bandel (1988). However, the tubercles on the ammonitella and the "primordial guard" of many sphaerorthoceratids may be interpreted as hatching glands. Tanabe (1989) further proposed an endocochleate model for the Ammonoidea during early embryogenesis. The hatching gland could be a synapomorphy of all angusteradulate groups (see above).

8. All ammonoids had a hyponome. All cephalopods have a hyponome or funnel for locomotion, and this was already present in the latest common ancestor of all siphonopodean cephalopods. Because the ammonoids belong to the crown-group cephalopods, they used the funnel for locomotion. The funnel shows two modifications within the Cephalopoda, the closed funnel pipe of the Coleoidea and the funnel consisting of two flaps that occurs in *Nautilus*. Theoretical considerations do not allow us to determine the particular funnel modification of the Ammonoidea. In "higher" ammonoids, it was probably different from that in nautiloids. All nautiloids have a more or less well developed hyponomic sinus, which allows the animal to turn the funnel in all directions. This was certainly not possible in "higher" Goniatitida, Ceratitida, and Ammonitida because these forms developed a ventral salient or even a ventral keel in place of the hyponomic sinus. However, the Bauplan of the Ammonoidea still includes a hyponomic sinus.

9. All ammonoids had arms. All cephalopods have arms for catching and manipulating prey and carrion. The derivation of the arms from the anterior part of the foot or from epipodial tasters is still disputed. The pro and cons of these two hypotheses are beyond the scope of this discussion. Within the Cephalopoda, we basically know of two character states, the "many" arms of *Nautilus* and the ten arms of the Coleoidea (reduced to eight arms in the Octopoda). The number of arms in the Ammonoidea is an open question

because we still do not have a single ammonoid specimen with preserved arms. The number also cannot be deduced from in- and out-group comparisons, but I briefly discuss the results of theoretical considerations.

In my first papers on the phylogenetic classification of the Coleoidea, the ten-arm condition was considered apomorphic. This is certainly not the case, because the character state "ten arms" is already present in at least one ectocochleate group, the sphaerorthocerids (see Stürmer, 1985; Engeser, 1990b). The sphaerorthocerids seem to be closely related to the Bactritida. At least, theoretically, it is probable that the ammonoids had ten arms like the coleoids. Whether this was actually the case can only be shown by future finds.

Within Recent Cephalopoda, two arm types occur: the retractable, thin arms of *Nautilus* and the muscular arms equipped with suckers of the Dibranchiata. The suckers are probably an apomorphy of the Dibranchiata because the Belemnoidea (at least the "higher" groups) developed two rows of arm hooks on each arm. The arms of ammonoids certainly were more like those of *Nautilus*, although we do not know whether they were retractable or not. This might be the reason why we have not yet found ammonoids with clear impressions of arms.

10. They possessed both radula and beak. It probably was never disputed that ammonoids had a radula and a beak. They inherited these characters from the latest common ancestor of all siphonopodean cephalopods (HASC). Radulae have been found in several species and genera of ammonoids. The number of radular elements is seven to nine, generally seven teeth and, sometimes, two marginalia per row. Radulae differ considerably among ammonoids, and it might well turn out in the future that different kinds of radulae are characteristic of different subordinate groups. The Bauplan of the Ammonoidea also includes an organic beak with almost equal-sized lower and upper jaws. A similar beak must already have been present in HASC. The lower jaw was strongly modified and calcified in some major clades.

11. Development was direct with no larval stage; the reproductive strategy was r-selected. All cephalopods have direct development with no true larval forms (Boletzky, 1974), and therefore one can safely assume that ammonoids had a direct development too. This has been questioned by Erben *et al.* (1967), but in more recent papers it is accepted almost unequivocally (Chapter 11, this volume; Kulicki, 1974; Bandel, 1982, 1986).

Among Recent and fossil cephalopods, we can distinguish two different reproductive strategies, the K-selected strategy of *Nautilus* and the more or less r-selected reproductive strategy of most coleoids (this can be modified in some subordinate groups; cf. Landman, 1988). *Nautilus* lays few (less than a dozen) extremely large and yolk-rich eggs, and the embryo takes more than 1 year to hatch. *Nautilus* probably matures in about 10 years (see Ward, 1987) and can lay eggs for many years (see above). Most coleoid cephalopods grow rapidly and mature within 1 to 2 years. They reproduce once and die afterwards. The spawn may consist of several tens of thousands of small eggs (see Boletzky, 1989). The ammonoids clearly showed this second type of reproductive strategy. This is certainly the reason why ammonoids are usually common fossils in Late Paleozoic and Mesozoic sediments, whereas nautiloids are always rare. It is not clear which reproductive strategy is apomorphic and which is plesiomorphic. The extremely large and yolk-rich egg of *Nautilus* is certainly apomorphic, but it might well be that having large numbers of small eggs is apomorphic, too. This is indicated by a trend in the diameter of ammonoid eggs, as deduced from the size of embryonic conchs, which shows a decrease in size during evolution (Tanabe and Ohtsuka, 1985). The primitive condition would then be halfway between the two extremes.

12. These forms have two gills and a heart with two atria. The Bauplan of the Cephalopoda probably includes two gills, two gonoducts, two pericardioducts, and a heart with two atria (see Berthold and Engeser, 1987, for discussion). The doubling of several of these organs in Recent *Nautilus* can probably be regarded as an apomorphy. These basic characters stem from the latest common ancestor of all molluscs.

13. They have two eyes. All Recent cephalopods have two eyes. However, there are two modifications, the pinhole eye of *Nautilus* and the lens eye of the Coleoidea. We can surely assume that HASC was equipped with a pinhole eye similar to that of *Nautilus*. Theoretical considerations do not give a hint which type of eye the ammonoids had. However, it might be argued that the highly advanced lens eye of the Coleoidea was an adaptation to rapid swimming. In this respect, that is, swimming, the ammonoids were certainly more like *Nautilus*. Thus, it is preferable to assume a pinhole eye for the Ammonoidea.

14. The blood contained the pigment hemocyanin. All Recent Cephalopoda possess the blood pigment hemocyanin. It belongs to the Bauplan of the Cephalopoda. Thus, I assume that the Ammonoidea had "blue" blood, too. It has only about 20% the efficiency of hemoglobin.

15. There was one pair of strong retractor muscles. *Nautilus*, as the only extant representative of the ectocochleate cephalopods, possesses two strong retractor muscles; these are situated on the dorsolateral parts of the posterior portion of the living chamber. These strong muscles are used to force water through the funnel, thereby moving the animal by jet propulsion. In some Ammonoidea, a similar pair of lateral retractor muscles is also present (Sharikadzé *et al.*, 1990; Doguzhaeva and Mutvei, 1991). The contradictory conclusion of Weitschat and Bandel (1991) that lateral rectractor muscles were absent in all ammonoids has been proved to be erroneous. They investigated the organic sheets within ammonoid chambers. These sheets hid the muscle scars, with the exception of one small muscle scar on the dorsum. In many ammonoid genera, scars of retractor muscles are clearly visible (Sharikadzé *et al.*, 1990; Doguzhaeva and Mutvei, 1991; unpublished observations). This character was inherited from the HASC and, further back, from the latest conchiferan ancestor.

16. Reproduction was sexual with separate sexes. All Recent cephalopods reproduce sexually, and the sexes are separate. This certainly was also the case in fossil ammonoids. The character stems from HASC and, further back, from the hypothetical ancestral mollusc (HAM). In subordinate groups we can observe a strong dimorphism of the sexes (see Chapter 13, this volume).

17. Ammonoids probably followed a carnivorous life style. All Recent cephalopods are carnivorous, and it is most likely that HASC was already a predatory animal. It follows that ammonoids were carnivorous too. They may have developed several specialized forms of gathering food in different subordinate groups.

18. They possessed a crop as a prestorage and predigestion organ for food. The Bauplan of HASC also includes a crop. We can assume that ammonoids had a crop, too. The presence of a crop can be demonstrated by fossil finds (Lehmann, 1985, 1990).

19. Ammonoids were probably marine living. All Recent cephalopods are marine animals that rarely swim into brackish water. It follows that HASC and ammonoids were marine-living animals too.

The apomorphic characters of the Ammonoidea plus the Bactritida are listed below.

1. A suture line formed with a ventral lobe. The "orthocerid" ancestors of the Ammonoidea (plus Bactritida) had straight suture lines. The development of a folded suture can therefore be considered apomorphic. Folding of the suture line increased the internal surface of a chamber, which was covered by a hydrophilic mucus layer. In "higher" ammonoids, the folding progressed even further to attain the intricate suture lines characteristic of ammonites. Such complex septa probably served to provide storage for cameral water, which could then be removed quickly on demand. In contrast to *Nautilus*, which uses only jet propulsion for up and down movement in the water column, it seems plausible that "higher" ammonoids could have removed cameral liquid much faster and used this ability to move up and down in the water column (Weitschat and Bandel, 1991).

2. There was a prosuture with a slight dorsal saddle. In Ammonoidea, the suture of the first septum is different from that of all subsequent septa. It has a slight dorsal saddle (see Lehmann, 1990, and references therein). Compared with the angusteradulate orthocerids, this character can be considered apomorphic.

3. They possessed a marginal siphuncle. Ammonoidea and Bactritida share the apomorphic character "marginal siphuncle." This feature can be considered apomorphic, although in some "orthocerid" angusteradulates, the siphuncle is not strictly central but shifting slightly to the venter. The shift of the siphuncle from a subcentral to a marginal position occurred several times in cephalopod evolution.

Other characters usually regarded as typical ammonoid features (e.g., coiling of the embryonic and adult shell) are apomorphies of "higher" ammonoid groups.

5. Conclusions

The Bauplan of the Ammonoidea as presented here is based mainly on theoretical considerations and comparisons with presently available data. It must be stressed again that this is the Bauplan of the hypothetical ancestral ammonoid cephalopod (HAAC). Certainly, many ammonoid groups varied and changed this Bauplan significantly. However, all modifications must have been derived from this basic Bauplan. It can be expected that, with a more detailed reconstruction of HASC and with more data from the closest relatives of the Ammonoidea, the Bauplan of the hypothetical ancestral ammonoid will become better known. A cladistic analysis of the Ammonoidea is hampered by the gradual changes of morphology (problem of "grade groups") as well as numerous convergent developments. Nevertheless, the author hopes to present a cladistic classification of the Ammonoidea in the near future.

ACKNOWLEDGMENTS. I would like to thank the editors, especially Neil Landman, for correcting the English and patiently waiting so long for the final version of this paper.

References

Arnold, J. M., 1987, Reproduction and embryology of *Nautilus*, in: *Nautilus, The Biology and Paleobiology of a Living Fossil* (W. B. Saunders and N. H. Landman, eds.), Plenum Press, New York, pp. 353–372.

Arnold, J. M., Landman, N. H., and Mutvei, H., 1987, Development of the embryonic shell of *Nautilus*, in: *Nautilus, The Biology and Paleobiology of a Living Fossil* (W. B. Saunders and N. H. Landman, eds.), Plenum Press, New York, pp. 353–372.

Bandel, K., 1982, Morphologie und Bildung der frühontogenetischen Gehäuse von conchiferen Mollusken, *Facies* **8**:1–154.

Bandel, K., 1986, The ammonitella: A model of formation with the aid of the embryonic shell of archaeogastropods, *Lethaia* **19**:171–180.

Bandel, K., Engeser, T., and Reitner, J., 1984, Die Embryonalentwicklung von *Hibolithes* (Belemnitida, Cephalopoda), *N. Jb. Geol. Paläont. Abh.* **167**:275–303.

Berthold, T., and Engeser, T., 1987, Phylogenetic analysis and systematization of the Cephalopoda (Mollusca), *Verh. Naturwiss. Ver. Hamb. N.F.* **29**:187–220.

Blind, W., 1987, Vergleichend morphologische und schalenstrukturelle Untersuchungen an Gehäusen von *Nautilus pompilius*, *Orthoceras* sp., *Pseudorthoceras* sp. und *Kionoceras* sp., *Palaeontogr. Abt. A* **198**:101–128.

Boletzky, S. von, 1974, The "larvae" of Cephalopoda: A review, *Thalassia Jugosl.* **10**:45–76.

Boletzky, S. von, 1982, Structure tégumentaire de l'embryon et mode d'éclosion chez les céphalopodes, *Bull. Soc. Zool. Fr.* **197**:475–482.

Boletzky, S. von, 1989, Early ontogeny and evolution: The cephalopod model viewed from the point of developmental morphology, *Geobios Mém. Spec.* **12**:67–78.

Doguzhaeva, L., and Mutvei, H., 1991, Organization of the soft body in *Aconeceras* (Ammonitina), interpreted on the basis of shell morphology and muscle-scars, *Palaeontogr. Abt. A* **218**:17–33.

Ebel, K., 1992, Mode of life and soft body shape of heteromorph ammonites, *Lethaia* **25**:179–193.

Engeser, T., 1990a, Major events in cephalopod evolution, in: *Major Evolutionary Radiations*, Vol. 42 (P. D. Taylor and G. P. Larwood, eds.), Clarendon Press, Oxford, pp. 119–138.

Engeser, T., 1990b, Phylogeny of the fossil coleoid Cephalopoda (Mollusca), *Berl. Geowiss. [A]* **124**:123–191.

Erben, H. K., and Flajs, G., 1975, Über die Cicatrix der Nautiloideen, *Mitt. Geol.-Paläont. Inst. Univ. Hamburg* **44**:59–68.

Firby, J. B., and Durham, J. W., 1974, Molluscan radula from earliest Cambrian, *J. Paleontol.* **48**:1109–1119.

House, M. R., 1988, Major features of cephalopod evolution, in: *Cephalopods—Present and Past* (J. Wiedmann and J. Kullmann, eds.), Schweizerbart'sche Verlagsbuchhandlung, Stuttgart, pp. 1–16.

Kulicki, C., 1974, Remarks on the embryogeny and postembryonal development of ammonites, *Acta Palaeontol. Pol.* **19**:201–204.

Landman, N. H., 1988, Early ontogeny of Mesozoic ammonites and nautilids, in: *Cephalopods—Present and Past* (J. Wiedmann and J. Kullmann, eds.), Schweizerbart'sche Verlagsbuchhandlung, Stuttgart, pp. 215–228.

Landman, N. H., Arnold, J. M., and Mutvei, H., 1989, Description of the embryonic shell of *Nautilus belauensis* (Cephalopoda), *Am. Mus. Novit.* **2960**:1–16.

Lehmann, U., 1967, Ammoniten mit Kieferapparat und Radula aus Lias-Geschieben, *Paläontol. Z.* **41**:38–45.

Lehmann, U., 1985, Zur Anatomie der Ammoniten: Tintenbeutel, Kiemen, Augen, *Paläontol. Z.* **59**:99–108.

Lehmann, U., 1990, *Ammonoideen—Leben zwischen Skylla und Charybdis*, Vol. 2, Ferdinand Enke Verlag, Stuttgart.

Mapes, R. H., 1979, Carboniferous and Permian Bactritoidea (Cephalopoda) in North America, *Univ. Kans. Paleontol. Contrib.* **64**:1–75.

Serpagli, E., and Gnoli, M., 1977, Silurian cephalopods of the Meneghini collection (1857) with reproduction of the original plate, *Boll. Soc. Paleontol. Ital.* **16**:137–142.

Sharikadzé, M. Z., Lominadzé, T. A., and Kvantaliani, I. V., 1990, Systematische Bedeutung von Muskelabdrücken spätjurassischer und frühkretazischer Ammonoideae, *Zentralb. Geol. Wiss.* **18**:1031–1039.

Stürmer, W., 1985, A small coleoid cephalopod with soft parts from the lower Devonian discovered using radiography, *Nature* **318**:53–55.

Tanabe, K., 1989, Endocochliate embryo model in the Mesozoic Ammonitida, *Hist. Biol.* **2**:183–196.

Tanabe, K., and Ohtsuka, Y., 1985, Ammonoid early internal shell structure: Its bearing on early life history, *Paleobiology* **11**:310–322.

Tanabe, K., Tsukahara, J., Fukuda, Y., and Taya, Y., 1991, Histology of a living *Nautilus* embryo: Preliminary observations, *J. Ceph. Biol.* **2**:13–22.

Weitschat, W., and Bandel, K., 1991, Organic components in phragmocones of Boreal Triassic ammonoids: Implications for ammonoid biology, *Paläontol. Z.* **65**:269–303.

Woodruff, D. S., Carpenter, M. P., Saunders, W. B., and Ward, P. D., 1987, Genetic variation and phylogeny in *Nautilus*, in: *Nautilus, The Biology and Paleobiology of a Living Fossil* (W. B. Saunders and N. H. Landman, eds.), Plenum Press, New York, pp. 65–83.

II

Structure of Hard and Soft Tissues

Chapter 2

Morphology of the Jaws and Radula in Ammonoids

MARION NIXON

1. Introduction

Some ammonoids have been found with structures still in the living chamber that are now recognized as jaws (beaks or mandibles) because small teeth, characteristic of those of the radula of extant Cephalopoda, lie between upper and lower jaws. The radulae so far found in ammonoids have the same number of teeth in each transverse row as do those of living Coleoidea, and the teeth are of similar shape and form. The first discovery was of just a few teeth in *Eoasianites* (Closs and Gordon, 1966), a member of the taxon Goniatitina. A year later, eight specimens of the same species, in which the radula and jaws were preserved together, were reported (Closs, 1967). The jaws and radulae of two adult microconchs of *Eleganticeras elegantulum,* a Liassic ammonite, were found by Lehmann (1967); other descriptions have followed, mostly of

MARION NIXON • Research School of Geological and Geophysical Sciences, Birkbeck College and University College London, London WC1E 6BT, England.

Ammonoid Paleobiology, Volume 13 of *Topics in Geobiology,* edited by Neil Landman *et al.,* Plenum Press, New York, 1996.

jaws but also of radulae (for reviews see Lehmann, 1971a, 1981a,b, 1988, 1990; Dagys *et al.*, 1989).

Fossils known as aptychi have been recorded for more than two centuries (see Moore and Sylvester-Bradley, 1957), and *Aptychus* was originally a generic name but has come to be used as a morphological term rather than a zoological name (Arkell, 1957). An aptychus is slightly curved, calcitic, and shaped roughly like the valves of a bivalve; they may be found either singly or in pairs, when they lie side by side. An anaptychus is a single structure of dark horny material. The form of the aptychus is indicated by a prefix such as cornaptychus or laevaptychus (Trauth, 1938; Lehmann, 1981a). Since the Devonian, aptychi occur as univalved horny plates. In the Jurassic, plates of this type became covered by two symmetrical calcareous plates, and, together, the whole structure is termed aptychus (Lehmann, 1990). Many aptychi are found in isolation, but others remain in the living chamber and, rarely, in the aperture of ammonoids. Various suggestions have been proposed for the function of apthychi, and as early as 1864, Meek and Hayden considered the possibility that they represented lower jaws. However, the interpretation of the aptychus as a lid, or operculum, received most support (see Trauth, 1938; Schindewolf, 1958), although these paleontologists followed different paths to reach the same conclusion (see Lehmann, 1981a).

The anatomic position and function of these structures have been a source of controversy for a considerable time (see Lehmann, 1981a,b, 1988; Morton, 1981; Dagys *et al.*, 1989). The interpretation of anaptychi and aptychi of Jurassic and Cretaceous ammonites as jaws (Lehmann, 1970, 1972) has been largely accepted during the past two decades. However, one report suggested that jaws and an operculum were present in the body chamber of an ammonite (Bandel, 1988), but this has since been found to be erroneous (Dagys *et al.*, 1989; Lehmann and Kulicki, 1990; Seilacher, 1993).

The review that follows includes a brief description of the buccal mass of living *Nautilus* and *Octopus,* as their anatomy is better known, for comparison with that of fossil forms. The ammonoids included are those in which the jaws and a radula are present in the living chamber and have been reconstructed or described from sections.

2. The Mouthparts

2.1. Nautiloidea

The upper jaw of modern *Nautilus* has a hard, sharp, calcareous tip, the rhyncholite, shaped like an arrowhead (see Teichert *et al.*, 1964; Saunders *et al.*, 1978; Nixon, 1988a), that fits within the lower jaw (Fig. 1a). The calcareous tip of the lower jaw, the conchorhynch, generally bears several rows of denticles on the oral surface. The rhyncholite and conchorhynch, formed of calcite and a little magnesite, are white and in marked contrast to the dark,

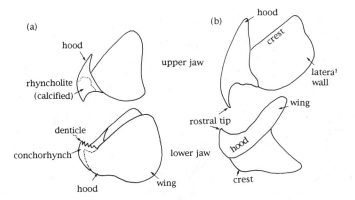

FIGURE 1. The upper and lower jaws of *Nautilus* (a) and a modern teuthoid (b).

almost black, organic material forming the major portion of the jaws; this is a chitin–protein complex (Saunders *et al.*, 1978; Lowenstam *et al.*, 1984; Crick *et al.*, 1985).

The radula of *Nautilus* is wide and has nine teeth and four plates in each transverse row (Fig. 2a; see Nixon, 1988a, 1995a). The central rhachidian tooth and lateral teeth 1 and 2 are relatively small. Marginal teeth 1 and 2 predominate the teeth in each transverse row; both are unicuspid, tall, and curved, and each has an associated marginal plate.

2.2. Coleoidea

The buccal mass of *Octopus* is shown diagramatically in Fig. 3. Mandibular muscles encase the upper and lower jaws to which they are attached via a

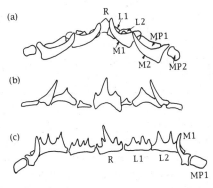

FIGURE 2. Radulae. Outline drawings of one transverse row of teeth from *Nautilus* (a), *Octopus* (b), and a member of the Bolitaenidae (c). L1, lateral tooth 1; L2, lateral tooth 2; M1, marginal tooth 1; MP1, marginal plate 1; M2, marginal tooth 2; MP2, marginal plate 2; R1, rhachidian tooth.

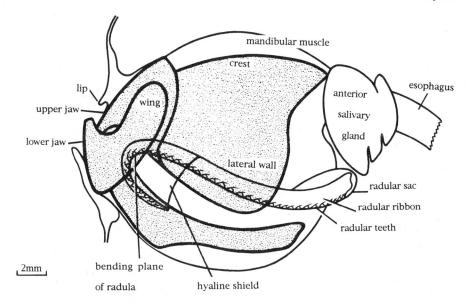

FIGURE 3. The buccal mass of a modern coleoid, *Octopus vulgaris*. The major structures are shown diagramatically.

layer of tall columnar cells. This layer is formed mainly of three types of cells. The most numerous are beccublast cells; these contain cell-long fibrils, which may have contractile and/or tensile properties. Complex trabeculae extend from the beccublast cells into the matrix of the beak. The cell-long fibrils are attached to the trabeculae, and at the opposite end of the cells they are anchored to the beccublast–muscle interface in close association with the muscles that move the beak. The second cell type is probably the major source of the hard tissue of the beak, and the third cell type contains a mixture of fibrils and secretory tissue (Dilly and Nixon, 1976).

The jaws of modern coleoids consist of a chitin–protein complex (Hunt and Nixon, 1981). The rostrum of the upper jaw fits within that of the lower jaw (Fig. 3). The lower jaw has characteristic features, and these allow identification to the generic, and even species, level in some cases (Fig. 1b; see Clarke, 1962, 1980, 1986).

The radula of a present-day coleoid generally has seven teeth and two marginal plates in each transverse row (Fig. 2b,c; see Nixon, 1988a, 1995a). The nomenclature of the teeth remains the same as for *Nautilus* (Fig. 2a) except for the omission of the two outer elements, marginal tooth 2 and marginal plate 2, on either side. There are a few exceptions to the presence of seven teeth in each transverse row in coleoids; for example, *Gonatus* has five teeth, the cirrate octopod *Cirrothauma* only vestigial teeth, and *Spirula* has no radula at all (see Nixon, 1988b). The central, or rhachidian, tooth has a mesocone, and lateral cusps may be present or absent. Each of the lateral teeth generally has one cusp, but there may be more (Fig. 2c). The marginal tooth is a single cusp, generally as tall as, or taller than, the rhachidian tooth. A marginal plate is present in many coleoids.

2.3. Ammonoidea

2.3.1. Anarcestina

Six specimens of *Gyroceratites* (Mimoceratidae), Lower to Middle Devonian (Miller *et al.*, 1957), were X-rayed by W. Stürmer, and when Bandel (1988) reexamined these films, he considered that parts of the buccal apparatus could be discerned in their body chambers.

2.3.2. Goniatitina

Jaws and radula of *Cravenoceras* (Cravenoceratidae), Lower to Upper Carboniferous (Miller *et al.*, 1957; Kullmann, 1981), have been found preserved in the anterior part of the living chamber of one specimen (Tanabe and Mapes, 1995). The jaws are of black chitinous material, apparently near their original position, as the anterior rostral portion of the lower beak is directed anteriorly and the radula lies in the front of the buccal cavity. The lower jaw has survived well; the outer lamella is wide and extended posteriorly on either side to form wings, the inner lamella is short, and the rostral tip is beak-like. The upper beak, also chitinous, is poorly preserved. The radula has seven teeth and a marginal plate on either side in each transverse row. The rhachidian tooth has a major cone with small lateral cusps, and it is of similar height to lateral tooth 1; lateral tooth 2 is unusual as it is taller than the marginal tooth.

One specimen of *Wiedeyoceras* (Wiedeyoceratidae), Middle Pennsylvanian (Miller *et al.*, 1957; Kullmann, 1981), with the buccal apparatus in its body chamber (Fig. 4b) was found among the Mazon Creek fauna. There was no indication of calcification associated with the jaws (Saunders and Richardson, 1979).

It was in a specimen of *Eoasianites* (Neoicoceratidae), Lower Carboniferous-Lower Permian (Fig. 4a; Arkell *et al.*, 1957), from a marine transgressive series of the Itararé Formation of the Late Carboniferous to Early Permian, that a radula was found first *in situ* between the jaws (Closs and Gordon, 1966). Eight additional specimens (shell diameters 75–100 mm) were found, each with the buccal apparatus in the living chamber (Closs, 1967). The radular teeth are in excellent condition, and their shape and form easy to discern. Seven teeth are present in each transverse row of the ribbon, namely, a central rhachidian tooth together with two lateral teeth and a marginal tooth on either side (Fig. 5a). The rhachidian tooth has a distinct mesocone and a relatively wide base; lateral tooth 1 also has a wide base and a single cusp, which is slightly taller than the mesocone of the central tooth. Lateral tooth 2 is similar to, but shorter than, marginal tooth 1, and both are similar in shape and form.

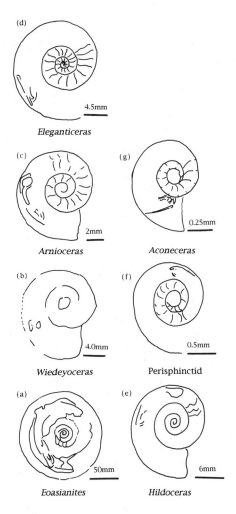

FIGURE 4. Ammonoids seen in median cross section with the feeding apparatus in the living chamber. (a and b) Lower Carboniferous-Lower Permian; (c–e) Lower Jurassic; (f) Upper Jurassic; (g) Lower Cretaceous. Redrawn from (a) Bandel (1988); (b) Saunders and Richardson (1979); (c and e) Lehmann (1981a); (d) Lehmann (1967); (f) Kulicki and Wierzbowski (1983); (g) Doguzhaeva and Mutvei (1992).

The radula is 5.5 mm wide, the rhachidian tooth 1.0 mm, and the lateral teeth 1.5–2.0 mm in height (Closs, 1967). The lower jaw has an inner plate with a smooth surface that is somewhat thicker than the outer plate. The biting edge appears strong but lacks a pointed rostrum. The upper jaw has an outer plate with striations extending concentrically around the edge. It has a length of 20 mm but is considered originally to have been about 30 mm long.

Bandel (1988) examined the specimens described by Closs in 1967 and reconstructed the upper and lower jaws (Fig. 5a). The upper jaw has an inner plate less than half the length of the outer plate. The lower jaw is larger, and the inner and outer plates are of similar size. The length of the lower jaw is 37 mm and 16 mm in specimens with shell diameters of 150 mm and 110 mm, respectively. The shell of the living chamber of the four specimens examined had been largely dissolved before fossilization, leaving wrinkled sheets representing the inner layers of organic material of the shells (Closs, 1967; Bandel, 1988). No calcareous material associated with the jaws has been recorded, and it is perhaps possible that this could have been lost at the time of demineralization of the shells.

2.3.3. Phylloceratina and Lytoceratina

A lower jaw of *Neophylloceras* (Phylloceratidae), Hauterivian to Maastrichtian (Arkell *et al.*, 1957), with a calcified tip, like a conchorhynch, was found in a specimen attributed to this genus (Lehmann *et al.*, 1980a; Tanabe *et al.*, 1980; Kanie, 1982), but its orientation is uncertain (Dagys *et al.*, 1989).

A calcareous tip is present on a structure described as a lower jaw of *Tetragonites* (Tetragonitidae), Upper Cretaceous (Arkell *et al.*, 1957) (Lehmann *et al.*, 1980; Tanabe *et al.*, 1980a; Kanie, 1982), but there is confusion as to its orientation (Dagys *et al.*, 1989). *Gaudryceras,* another tetragonitid, Turonian to Maastrichtian (Arkell *et al.*, 1957), was found with upper and lower jaws. These were reconstructed, and both had calcified tips (Lehmann *et al.*, 1980; Tanabe *et al.*, 1980a; Kanie, 1982); X-ray diffraction analysis showed the tips to be mostly of wilkeite with some calcite. As in *Tetragonites,* there is uncertainty about the orientation of the jaws (Dagys *et al.*, 1989).

2.3.4. Ancyloceratina

The buccal apparatus of *Scalarites mihoensis* (Diplomoceratidae), Turonian to Coniacian (Arkell *et al.*, 1957; Wright, 1981) has been reconstructed, and the lower and upper jaws are shown in Fig. 5i (Tanabe *et al.*, 1980b). Both jaws are of black horny material. The upper jaw has lateral walls joined anteriorly by the hood and rostrum; the latter is thick and sharp, and there is a thin calcareous layer on the inner lamellar surface. This jaw is slightly shorter than the lower. The lower jaw has a relatively large outer lamella and a very short inner lamella; two sharply marked ridges extend from near the rostral tip to the edge of the hood. There is a calcareous layer on the posterior region of the outer lamella.

The lower jaw of *Scaphites* (Scaphitidae), upper Albian to Campanian (Arkell *et al.*, 1957) has an inner organic layer with an outer calcitic layer on its flanks. The upper beak is two-layered; it has an anterior projection, although this lacks a cutting edge. Posteriorly, there is a lateral wall on either side; there is no associated calcareous deposit (Lehmann, 1972, 1981b). The jaws are similar to those of *Hildoceras* and *Eleganticeras* (Lehmann, 1972). The form and composition of the jaws of three other members of the family, *Discoscaphites, Hoploscaphites,* and *Jeletzkytes,* are the same; the lower jaw

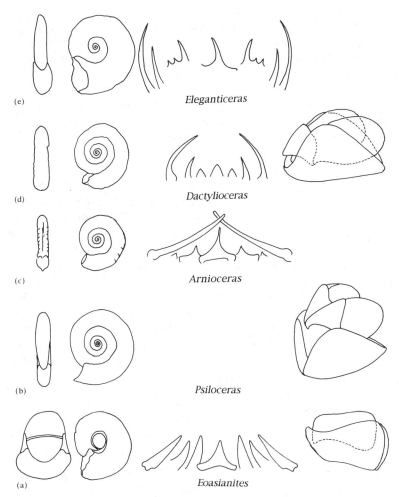

(e) *Eleganticeras*

(d) *Dactylioceras*

(c) *Arnioceras*

(b) *Psiloceras*

(a) *Eoasianites*

FIGURE 5. The feeding apparatus of Ammonoidea. The outline drawings show an example of the radula and the upper and lower jaws of the genus. (a) Lower Carboniferous-Lower Permian, (b–f) Lower Jurassic, (g) Lower Cretaceous, (h) Upper Cretaceous, (i) Upper Cretaceous. Ca, calcified region; L1, lateral tooth 1; L2, lateral tooth 2; M1, marginal tooth 1; MP1, marginal plate 1; R, rhachidian tooth. Redrawn from Bandel (1988), Doguzhaeva and Mutvei (1992), Lehmann (1979, 1981a), Tanabe (1983), and Tanabe *et al.* (1980b).

is larger and thicker than the upper, and the former is reinforced with an outer calcified layer (Landman and Waage, 1993).

2.3.5. Ammonitina

The upper and lower jaws of *Psiloceras* (Psiloceratidae), Hettangian (Arkell *et al.*, 1957) were reconstructed by Lehmann (1970, 1981b) and are shown in

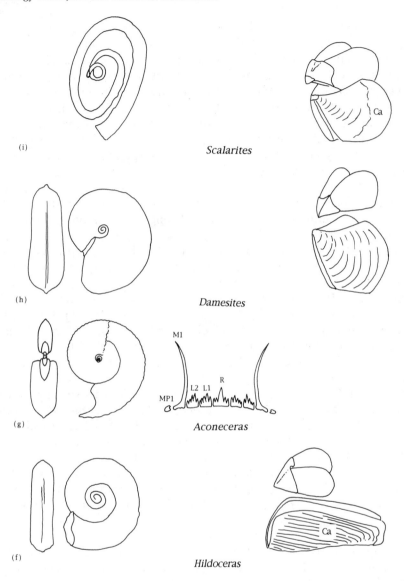

(i)

Scalarites

(h)

Damesites

(g)

Aconeceras

(f)

Hildoceras

FIGURE 5. (Continued)

Fig. 5b. The lower jaw consists of an outer plate with a shallow anterior fold forming an inner portion. The upper jaw consists of two anterior layers; the outer forms a hood, and the inner has two lateral walls.

The upper and lower jaws, radula, part of the esophagus, and crop/stomach of *Arnioceras* (Arietitidae), Sinemurian (Arkell *et al.*, 1957) have been found in the living chambers of three specimens (Fig. 4c; Lehmann, 1971b, 1981b; Lehmann and Weitschat, 1973). In one specimen, shell diameter 13.6 mm, the lower jaw is 4.0 mm long, and the upper 1.6 mm long; the radula lies between the jaws and is 2.4 mm long. Another conch, shell diameter about 21 mm, has a lower jaw 4.7 mm long and an upper 3.0 mm long; the radula is 0.72 mm wide and has seven teeth in each transverse row. The jaws are of a black horny material, and the lower is covered externally with a thin layer of calcite. The upper jaw is a single structure with a pointed rostrum; its hood rises slightly above the inner plate. In the lower jaw the outer plate is folded anteriorly to form a narrow rim. The radula is well preserved in one specimen, and the details of the teeth are discernible (Lehmann, 1971b). Seven teeth are present in each transverse row. The tall, slender marginal tooth is more than twice the height of the rhachidian tooth (Fig. 5c). A single cusp is present on each of the lateral teeth. The rhachidian tooth has a wide base and a tapering mesocone. Approximately eight transverse rows of teeth are present on 1 mm of radular ribbon. Another specimen has a tube extending backward from the radula; this enters a sac-like structure, which has rather thick, somewhat "spongy" walls (Lehmann, 1971b, 1981b). In the smallest and best-preserved specimen, a piece of esophagus leads to the crop/stomach (Fig. 4c). The latter is 2 mm in diameter, and within its walls are the remains of foraminiferans and ostracod shells, 0.15–0.20 mm in size (Lehmann, 1971b).

Figure 5d shows a reconstruction of the jaws of *Dactylioceras* (Dactylioceratidae), lower Toarcian (Arkell *et al.*, 1957), made from serial sections of one specimen in which the buccal apparatus was still present in the living chamber (Lehmann, 1979). Both jaws are of similar length; the lower is a single structure with a short rim anteriorly; the upper jaw has lateral walls and a distinct rostrum and hood anteriorly. No calcareous material was found associated with the jaws. The radula is incompletely preserved between the jaws, but individual teeth could be discerned, including a tall, slender marginal tooth, 1.2 mm in height, the lateral and rhachidian teeth being taller. The arrangement and number of teeth could not be ascertained positively, however; the probable form of one transverse row of teeth is shown in Fig. 5d (Lehmann, 1979).

Eleganticeras elegantulum (Hildoceratidae), lower Toarcian (Arkell *et al.*, 1957), was the second ammonoid in which a radula and jaws were found in the living chamber (Fig. 4d; Lehmann, 1967). The inner layer of the lower jaw is continuous; it is of horny material with a thin outer coat of calcite. The upper jaw is smaller than the lower jaw and lacks any associated mineralization. Hemisections of two microconchs of different shell sizes indicate the growth of the buccal apparatus. The radula has seven teeth in each transverse row (Fig. 5e). The slender, tapering marginal tooth is the tallest in each row. The rhachidian tooth has a slim mesocone and small lateral cusps. The first lateral tooth has three cusps, the innermost being almost as tall as the mesocone of the rhachidian tooth. Lateral tooth 2 is similar to, but smaller than, the marginal tooth. Lehmann (1967) found seven and nine transverse rows of teeth on 1-mm lengths of radular ribbon in specimens of 31 mm and 17.5 mm shell diameter, respectively. Thus, the size of the teeth increases with the growth of the animal, as they do in living cephalopods (Nixon, 1969).

A specimen of *Hildoceras* (Hildoceratidae), lower Toarcian (Arkell *et al.*, 1957), with buccal apparatus and crop/stomach present in the living chamber (Fig. 4e) was described by Lehmann and Weitschat (1973). A reconstruction of the jaws was made using serial sections of an adult macroconch, 167 mm shell diameter. The lower jaw has an inner organic layer and an outer, smaller, calcareous layer, and it is longer than the upper jaw (Fig. 5f; Lehmann, 1981b). The radula, lying between the jaws, has seven teeth in each transverse row (Lehmann, 1981b). In one macroconch, 181 mm shell diameter, a length of esophagus extends from the buccal apparatus to a crop/stomach, 8 mm diameter, lying further back in the living chamber. This sac contained an upper and lower jaw from a smaller ammonite (Lehmann and Weitschat, 1973).

A large number of fragmented but small aptychi of the same species lies together in one part of the living chamber of one specimen (Michael, 1894), then referred to as *Oppelia sterapsis* but since identified as probably *Neochetoceras* (Oppeliidae), Kimmeridgian (Arkell *et al.*, 1957), by Lehmann and Weitschat (1973). Another specimen, 55 mm shell diameter, has a few small, broken aptychi in a sac 7 mm in diameter lying close to the leading septa; these aptychi are similar in shape and form to that at the front of the living chamber (Lehmann and Weitschat, 1973).

Twenty juvenile specimens of *Aconeceras*, (Oppeliidae), upper Barremian to Lower Albian (Arkell *et al.*, 1957), with radula and jaws present in their body chambers have been investigated by Doguzhaeva and Mutvei (1990, 1993; Fig. 4g). The valves of aptychi, and in some cases counteraptychi, surround the radula. The valves of the aptychi are calcified along their anterior edges; the inner layer is short and is mainly calcified; the outer layer is mostly organic. The counteraptychi are calcified and consist of two calcified layers; these layers are much thinner than those of the aptychi (Doguzhaeva and Mutvei, 1993). The radula, 0.45 mm wide, has seven teeth and two marginal plates in each transverse row (Fig. 5g). The slender, tapering marginal tooth is notable for its extreme height. All of the remaining teeth in each row are multicuspidate. The mesocone of the rhachidian tooth is distinct, and each lateral tooth has five cusps, the second cusp being the tallest in both lateral tooth 1 and 2. The teeth consist of organic material with a thick and probably mineralized wall. The best-preserved tooth has an inner layer with numerous parallel, rod-shaped elements set at right angles to the surface, whereas the outer layer is denser in structure. Energy-dispersive analyses reveal the presence of silica and calcium in the teeth (Doguzhaeva and Mutvei, 1992); the same minerals are present in the radula of modern *Nautilus* (Tanabe and Fukuda, 1987a).

An ammonitella, shell diameter 0.96 mm, possibly of *Aconeceras trautscholdi*, was found with the buccal apparatus one-third of the way into the body chamber (Kulicki *et al.*, 1988). Both the upper and lower jaws have inner and outer lamellae and calcified apical tips.

The lower jaw of a microconch described as *Normannites* (Lehmann, 1972) but since recognized as *Stephanoceras* (Stephanoceratidae), middle Bajocian (Arkell *et al.*, 1957; Donovan *et al.*, 1981) (G.E.G. Westermann, personal communication) has an inner organic layer with an outer calcareous coat on its flanks. The upper jaw is two-layered anteriorly, with a beak-like tip, although it lacks a cutting edge; lateral walls are present, but no associated calcareous material (Lehmann, 1972, 1981a). *Parkinsonia* is another stephano-ceratid, but of the upper Bajocian to lower Bathonian (Arkell *et al.*, 1957). A pair of aptychi were found with an upper jaw in a fragment of a living chamber. The aptychi consist of a continuous layer, 13 mm in length and 9 mm wide, that is bent inwards anteriorly to form an inner lamella. The upper jaw is smaller, 8 mm in length and 3.7 mm wide, and possesses an inner lamella. Both jaws are of horny material, and the lower has associated calcareous tissue (Lehmann, 1978).

The jaws of *Quenstedtoceras* (Cardioceratidae), upper Callovian (Arkell *et al.*, 1957), are similar to those of *Hildoceras* and *Eleganticeras*; the lower jaw is of organic material and has a calcareous coat on its flanks (Lehmann, 1972). The upper jaw has two lobes; it is double-layered anteriorly to form a beak-like tip, but it lacks a cutting edge; there is no calcareous material associated with this jaw (Lehmann, 1981b).

The buccal apparatus of Perisphinctidae, Middle Jurassic–Upper Jurassic (Arkell *et al.*, 1957), was found in specimens of juvenile perisphinctids possessing shells with only 2.25 whorls (Fig. 4f; Kulicki and Wierzbowski, 1983).

The jaws of *Physodoceras* (Aspidoceratidae), Kimmeridgian (Arkell *et al.*, 1957), are like those of *Hildoceras* and *Eleganticeras* (Lehmann, 1972). The lower jaw has an organic inner layer, which is bent inward to form an anterior inner lamella. There is a calcareous coat on the sides, and this is thick at the front edge. The upper jaw has two lobes; the anterior part has two layers forming a beak-like tip, but it lacks a cutting edge (Lehmann, 1972, 1981b). Besides the buccal apparatus there is a region in the living chamber, presum-ably crop/stomach, containing pieces from the arms and calyx of the stalkless crinoid *Saccocoma* (Lehmann and Weitschat, 1973).

Specimens of two species of *Damesites* (Desmoceratidae), Cenomanian to Campanian (Arkell *et al.*, 1957), were found with the buccal apparatus still in the body chamber (Tanabe, 1983). In *D. ainuanus,* only an upper jaw is present, but in *D. semicostatus*, both jaws are intact. A three-dimensional reconstruc-tion of the jaws is shown in Fig. 5h. They are composed of a dark-colored homogeneous material. The upper jaw lies in the middle of the living chamber in an immature specimen, 26.6 mm shell diameter. The second species, *Tragodesmoceroides subcostatus*, shell diameter 269 mm, contains a lower jaw near the aperture, and, after reconstruction, it was seen to be similar to that of *Damesites semicostatus*.

In *Reesidites* (Collignoniceratidae), Turonian (Arkell *et al.*, 1957), the buccal apparatus lies in the body chamber. The structure of the jaws was reconstructed from thin serial sections (Tanabe and Fukuda, 1987b). The upper jaw is of dark, homogeneous, layered chitinous material, entirely without associated calcification. Anteriorly, where the jaw is thickest, there is a beak-like projection. The lower jaw is wide and curved with a median depression; it is thickest anteriorly. In one specimen the rostral region has rows of ridges and grooves; the tops of ridges are serrated and sharp. The lower jaw is chitinous with an outer calcareous layer, which has a prismatic structure; this jaw is larger than the upper jaw.

3. The Buccal Mass

The mouthparts described after being found in living chambers of ammonoids include the upper and lower jaws and radulae; all are hard-tissue structures composed of organic material, and in some there is associated mineralization.

The earliest jaws to be reconstructed belong to an ammonoid of the Middle Pennsylvanian Period; they are long, and there is a shallow but distinct blunt rostrum (Fig. 5a). The ammonoids from the Lower Jurassic have a lower jaw with a shallow anterior fold and lateral walls with a covering of calcareous material and an upper beak with lateral walls joined anteriorly by a fold that forms a blunt rostrum (Fig. 5b,d,f). Jaws from Cretaceous ammonoids differ, as there is a calcified rostral tip on both the upper and lower jaw as well as some calcareous material on the outer surface of the lateral wall of the lower jaw (Fig. 5h,i).

Ammonoid jaws differ from those of modern cephalopods, and there are differences in the arrangement and sites of attachment of their muscles. In some, the muscles are attached to the inner surface of the lower jaw (aptychus type; Kaiser and Lehmann, 1971). In other ammonoids, the muscles may be attached to areas of the outer surface that are devoid of calcification. The beccublast cells of modern coleoids provide the mechanism of attachment of muscles to the jaws (Dilly and Nixon, 1976). These cells make an imprint on the surface of the jaw, and similar imprints have been described on the outer lamella of the lower jaw of *Gaudryceras* (Tanabe and Fukuda, 1987a).

The jaws are of horny, organic material, and the lower jaw of some has a deposit of calcium on the outer surface of the lateral wall (Table I). In Cretaceous forms the rostral tip of both jaws is calcified; this region on the upper jaw resembles a rhyncholite and, on the lower jaw, a conchorhynch (Fig. 5h,i; Tanabe *et al.*, 1980a). Calcification has not been detected in association with the jaws of goniatites from the Upper Carboniferous.

The radulae discovered so far in ammonoids all have seven teeth in each transverse row (Fig. 5). So too does the radula of *Jeletzkya*, an Upper Carboniferous fossil with ten arms (Johnson and Richardson, 1968; Saunders and

TABLE I. Presence of Calcium Deposited on Jaws of Ammonoids[a]

Genus	Period	Upper jaw	Lower jaw
Scalarites	U. Cretaceous	Ca	Ca
Reesidites	U. Cretaceous	-	Ca
Scaphites	Cretaceous	-	Ca
Neophylloceras	Cretaceous	Ca-tip	Ca-tip
Tetragonites	U. Cretaceous	Ca-tip	Ca-tip
Gaudryceras	U. Cretaceous	Ca-tip	Ca-tip
Aconeceras	L. Cretaceous	Ca-tip	Ca-tip
Physodoceras	U. Jurassic		Ca
Quenstedtoceras	M. Jurassic	—	Ca
Stephanoceras	M. Jurassic	—	Ca
Hildoceras	L. Jurassic		Ca
Eleganticeras	L. Jurassic	—	Ca
Dactylioceras	L. Jurassic	—	Ca
Arnioceras	L. Jurassic		Ca
Eoasianites	L. Carboniferous–L. Permian	—	—
Wiedeyoceras	U. Carboniferous	—	—

[a]Ca, mineral is present, generally on lateral walls; Ca-tip, calcium is present on the rostral tip of the jaw.

Richardson, 1979). This number is characteristic for modern Coleoidea (Fig. 2b,c) and suggests a close ancestral relationship between ammonoids and coleoids. This was recognized by Lehmann (1967), as he placed them together in the subclass Angusteradulata, narrow radula, and placed the Nautiloidea, with nine teeth, in the subclass Lateradulata, wide radula. The seven-toothed radula separates the ammonoids and the coleoids from the nautiloids, in which nine teeth are present in each transverse row (Fig. 2a). *Paleocadmus*, a fossil from the Mazon Creek fauna (Saunders and Richardson, 1979) has nine teeth, and they are strikingly similar in shape and form to those of living *Nautilus* (see Nixon, 1988a).

Each transverse row of teeth on the radular ribbon of ammonoids has a central, rhachidian tooth with three teeth on either side, each side being a mirror image of the other (Fig. 5a,c–e,g). The rhachidian tooth generally has a mesocone, and in some ammonoids, such as *Eleganticeras* and *Aconeceras*, two cusps are present. Lateral tooth 1 is of similar height to or shorter than the central tooth; in *Eoasianites* and *Dactylioceras*, it is simple, but it is multicuspid in the others. Lateral tooth 2 is more variable in that it is a simple cusp in *Eoasianites*, *Dactylioceras*, and *Eleganticeras*, and possibly also in *Arnioceras*, whereas in *Aconeceras*, it is multicuspid and similar to lateral tooth 1. Marginal tooth 1 is generally long, especially in *Arnioceras* and *Aconeceras* (Fig. 5c,g).

The shape and form of the radular teeth in these ammonoids do show some differences, but the variation is no greater than is apparent among modern coleoids (see Nixon, 1995b). Teuthoid squids generally have simple teeth, as does the ammonoid *Eoasianites*. The rhachidian and lateral teeth of *Arnioceras* resemble those of the small Sepiolidae, except that in the latter the marginal teeth are not especially long. The teeth of *Aconeceras* (Doguzhaeva and Mutvei, 1990, 1992) show considerable differentiation, and all but the marginal teeth are multicuspidate; this radula resembles that of the Bolitaenidae (Thore, 1949), a group of gelatinous, meso- to bathypelagic Octopoda (Fig. 2c).

4. Food and Feeding

The sacs in the living chambers of four genera of Jurassic ammonites each contained whole organisms or fragments, indicating predatory or scavenging habits (Table II; see Lehmann, 1988). The debris is generally in a discrete, sac-like structure, which in some of the specimens, appears to be joined by a tube to the buccal apparatus (Lehmann and Weitschat, 1973). The remains in these sacs were examined, and the material present identified as animals belonging to the same phyla as the prey eaten by modern cephalopods (see Nixon 1985, 1987, 1988a,b). *Nautilus* has been observed in the sea feeding on the shed exuviae of large crustaceans, and it ingests other skeletal material (Ward and Wicksten, 1980). *Sepia* consumes some skeletal tissue, as its prey can be identified from their stomach contents (Guerra, 1985; Guerra *et al.*, 1988). The remains of the animals could have entered the ammonites during the taphonomic process. This seems unlikely, as the debris present is in one area of the living chamber in each of the specimens, generally behind and well separated from the buccal apparatus, and even at the far end of the living chamber in one specimen.

Fragmented jaws from smaller ammonites were identified in *Hildoceras* (Table II). Two specimens of *Neochetoceras* had consumed ammonites; in one, the fragments belonged to the same species as the predator. The devoured ammonites are generally one tenth the size of the ammonite predator (Lehmann, 1981b). The jaws could have been ingested during scavenging, but

TABLE II. The Identity of Remains Exposed *in Situ* in the Crop/Stomach in the Living Chamber of Ammonites

Genus[a]	Contents of crop/stomach
(a) *Neochetoceras*	Mass of fragmented cephalopod jaws in two specimens
(b) *Physodoceras*	Arms and calyx segments of stalkless crinoid, *Saccocoma*
(c) *Hildoceras*	Jaws, upper and lower, of smaller ammonites
(d) *Arnioceras*	Ostracods and foraminiferans
(e) *Svalbardiceras*	Ostracod valves broken into pieces

[a]Source: (a–c) Lehmann and Weitschat (1973), (d) Lehmann (1971b), (e) Lehmann (1985).

cannibalism is known among modern coleoids (see Nixon, 1987). Whole or broken ostracods were found in *Arnioceras* and *Svalbardiceras*; the former ammonoid is considered to have inhabited depths of 30–50 m. *Physodoceras* had consumed a free-living crinoid. Assuming that the esophagus passes through the central nervous system as it does in modern cephalopods, then whatever is ingested has to be small enough to pass along the esophagus without damage to the brain.

The jaws of the ammonoids described so far are somewhat variable in shape and form. The lower jaw is generally a little longer than the upper jaw but is notably longer in *Hildoceras* (Fig. 5f). The lower jaw folds anteriorly, leaving a space internal to the major part of the jaw. In this it differs from modern cephalopods, where the space is external to the larger part of the jaw. The upper jaws are similar to their modern counterparts except that the lateral walls are separate and not joined by the crest as in living forms. In several of the ammonoids, the apical portion of the lower jaw does not form a rostrum; the exceptions to this are the Lower Cretaceous forms in which the rostrum is more distinct and its tip is calcified (Table I). The upper jaw has an apical point forming a blunt rostrum. This resembles the jaw of modern octopods rather than that of the squids, whose rostra are generally sharper (see Clarke 1962, 1980, 1986; Nixon 1988a). The lower jaw of the Jurassic ammonite is shovel-like, and there is a substantial calcified covering (Fig. 5f). The calcification presumably had a significant function in the life of this animal. The increased weight of the lower jaw would have kept the ammonoid close to, but not necessarily in contact with, the substratum and protected the buccal apparatus when foraging (see Lehmann, 1981b). Any opening–closing movements of the jaws would have disturbed bottom-dwelling organisms. These could be taken into the buccal cavity but would have been accompanied by considerable volumes of water. The calcareous coat of the lower jaw would provide rigidity and facilitate the action of muscles attached to the jaw and of the buccal complex itself, all within the buccal cavity, to expel excess water taken in with food organisms (Morton and Nixon, 1987). Westermann (1990) has tentatively proposed that the thinly calcified shovel-like aptychus of *Praestriaptychus* of the Stephanoceratidae has a supplementary function in which the two valves of the aptychus would, by their movements, produce water currents. The jaws of the heteromorph *Scalarites* appear little different from those found in other Cretaceous ammonites (Fig. 5). Nesis (1986) has proposed that heteromorphs possessing a crook-shaped bend at the apertural end of the shell may have fed with the aid of a mucus net to trap small, slow-moving organisms.

Little is known of the function of the radula in living cephalopods. The radula of *Octopus vulgaris* is concerned mainly with the passage of food toward the entrance of the esophagus (Altman and Nixon, 1970). It is involved in the very early stages of drilling the shells of molluscs (Nixon, 1979) and the exoskeletons of crustaceans (Nixon and Boyle, 1982); the later stages of drilling involve only the toothed regions of the posterior salivary gland papilla and duct (see Nixon 1988a; Nixon and Maconnachie, 1988). The role of the radular teeth in the life of modern cephalopods is otherwise not known, and as yet none has been found to possess such tall and seemingly delicate marginal teeth as those found in some of the ammonites (Fig. 5c–g; see Nixon, 1995b).

5. Summary

Clearly, cephalopods have been predatory and/or scavenging in habit since at least the Permian, as is evident from the crop/stomach contents found in some ammonoids, and the buccal apparatus has retained its major features throughout this period. The horny jaws have shown changes in this time, but their main characters have remained. Mineralization of the lower jaws is seen in the majority of ammonoids, together with the lateral walls or the rostral tip and also in the rostrum of the upper jaws of Cretaceous forms. The upper and lower jaws are calcified in both fossil and living nautiloids. The radulae of ammonoids have seven teeth in each transverse row, as do those of living Coleoidea and a ten-armed fossil of the Upper Carboniferous. Thus, the Ammonoidea and the Coleoidea with only seven teeth differ from fossil and modern Nautiloidea, which have nine teeth in each transverse row of the radula. Several other features also separate the nautiloids from the ammonoids and coleoids (Jacobs and Landman, 1993).

ACKNOWLEDGMENTS. I should like to thank Professor D. T. Donovan, Drs. L. A. Doguzhaeva, U. Lehmann, and H. Mutvei, and the three editors for the many helpful criticisms and comments on the chapter.

References

Altman, J. S., and Nixon, M., 1970, Use of the beaks and radula by *Octopus vulgaris*, *J. Zool. (Lond.)* **161**:25–38.

Arkell, W. J., 1957, Aptychi, in: *Treatise on Invertebrate Paleontology*, Part L, *Mollusca* 4 (R. C. Moore, ed.), Geological Society of America and University of Kansas Press, Lawrence, KS, pp. 437–440.

Arkell, W. J., Kummel, B., and Wright, C. W., 1957, Systematic descriptions, in: *Treatise on Invertebrate Paleontology*, Part L, *Mollusca 4* (R. C. Moore, ed.), Geological Society of America, and University of Kansas Press, Lawrence, KS, pp. 129–464.

Bandel, K., 1988, Operculum and buccal mass of ammonites, in: *Cephalopods—Present and Past* (J. Wiedmann and J. Kullman, eds.), Schweizerbart'sche Verlagsbuchhandlung, Stuttgart, pp. 653–678.

Clarke, M. R., 1962, The identification of cephalopod "beaks" and the relationship between beak size and total body weight, *Bull. Br. Mus. (Nat. Hist.) Zool.* **8**(10):421–480.

Clarke, M. R., 1980, Cephalopoda in the diet of sperm whales of the Southern Hemisphere and their bearing on sperm whale biology, *Discov. Rep.* **37**:1–324.

Clarke, M. R., ed., 1986, *A Handbook for the Identification of Cephalopod Beaks*, Clarendon Press, Oxford.

Closs, D., 1967, Goniatiten mit Radula und Kieferapparat in der Itararé Formation von Uruguay, *Paläont. Z.* **41**:19–37.

Closs, D., and Gordon, M., Jr., 1966, An Upper Paleozoic radula, *Notas Estud.* **1**:73–75.

Crick, R. E., Burkart, B., Chamberlain, J. A., Jr., and Mann, K. O., 1985, Chemistry of calcified portions of *Nautilus pompilius, J. Mar. Biol. Assoc. U.K.* **65**:415–420.

Dagys, A. S., Lehmann, U., Bandel, K., Tanabe, K., and Weitschat, W., 1989, The jaw apparati of ectocochleate cephalopods, *Paläont. Z.* **63**:41–53.

Dilly, P. N., and Nixon, M., 1976, The cells that secrete the beaks in octopods and squids (Mollusca, Cephalopoda), *Cell Tissue Res.* **167**:229–241.

Doguzhaeva, L., and Mutvei, H., 1990, Radulae, aptychi and counteraptychi in Cretaceous ammonites (Mollusca: Cephalopoda), *Dokl. Akad. Nauk SSSR* **313**:192–195 (in Russian).

Doguzhaeva, L., and Mutvei, H., 1992, Radula of the Early Cretaceous ammonite *Aconeceras* (Mollusca: Cephalopoda), *Palaeontogr. Abt. A* **223**:167–177.

Doguzhaeva, L. A., and Mutvei, H., 1993, Shell ultrastructure, muscle-scars, and buccal apparatus in ammonoids, *Geobios Mém. Spéc.* **15**:111–119.

Donovan, D. T., Calloman, J. H., and Howarth, M. K., 1981, Classification of the Jurassic Ammonitina, in: *The Ammonoidea*, Systematics Association Special Volume 18 (M. R. House and J. R. Senior, eds.), Academic Press, London, pp. 101–155.

Guerra, A., 1985, Food of the cuttlefish *Sepia officinalis* and *S. elegans* in the Ria de Vigo (NW Spain) (Mollusca: Cephalopoda), *J. Zool. (Lond.)* **207**:511–519.

Guerra, A., Nixon, M., and Castro, B. G., 1988, Initial stages of food ingestion by *Sepia officinalis* (Mollusca: Cephalopoda), *J. Zool. (Lond.)* **214**:189–197.

Hunt, S., and Nixon, M., 1981, A comparative study of protein composition in the chitin–protein complexes of the beak, pen, sucker disc, radula and oesophageal cuticle of cephalopods, *Comp. Biochem. Physiol.* **68B**:535–546.

Jacobs, D. K., and Landman, N. H., 1993, *Nautilus*—a poor model for the function and behavior of ammonoids? *Lethaia* **26**:101–111.

Johnson, R. G., and Richardson, E. S., Jr., 1968, Ten-armed fossil cephalopod from the Pennsylvanian in Illinois, *Science* **159**:526–528.

Kaiser, P., and Lehmann, U., 1971, Vergleichende Studien zur Evolution des Kiefferapparates rezenter und fossiler Cephalopoden, *Paläont. Z.* **45**:18–32.

Kanie, Y., 1982, Cretaceous tetragonatid ammonite jaws: A comparison with modern *Nautilus* jaws, *Trans. Proc. Palaeont. Soc. Jpn. N.S.* **125**:239–258.

Kulicki, C., and Wierzbowski, A., 1983, The Jurassic ammonites of the Jagua Formation, Cuba, *Acta Palaeontol. Pol.* **28**:369–384.

Kulicki, C., Doguzhaeva, L. A., and Kabanov, G. K., 1988, *Nautilus*-like jaw elements of a juvenile ammonite, in: *Cephalopods—Present and Past* (J. Wiedmann and J. Kullmann, eds.), Schweitzerbart'sche Verlagsbuchhandlung, Stuttgart, pp. 679–686.

Kullmann, J., 1981, Carboniferous goniatites, in: *The Ammonoidea*, Systematics Association Special Volume 18 (M. R. House and J. R. Senior, eds.), Academic Press, London, pp. 37–48.

Landman, N. H., and Waage, K. M., 1993, Scaphitid ammonites of the Upper Cretaceous (Maastrichtian) Fox Hills Formation in South Dakota and Wyoming, *Bull. Am. Mus. Nat. Hist.* **215**:1–257.

Lehmann, U., 1967, Ammoniten mit Kieferapparat und Radula aus Lias-Geschieben, *Paläont. Z.* **41**:38–45.

Lehmann, U., 1970, Lias-Anaptychen als Kieferelemente (Ammonoidea), *Paläont. Z.* **44**:25–31.

Lehmann, U., 1971a, New aspects in ammonite biology, in: *Proceedings North American Paleontological Convention, September 1969,* Part I, 1251–1269.

Lehmann, U., 1971b, Jaws, radula, and crop of *Arnioceras* (Ammonoidea), *Palaeontology (Lond.)* **14**:338–341.

Lehmann, U., 1972, Aptychen als Kieferelemente der Ammoniten, *Paläont. Z.* **46**:34–48.

Lehmann, U., 1978, Über den Kieferapparat von Ammoniten der Gattung *Parkinsonia, Mitt. Geol.-Paläont. Inst. Univ. Hamburg* **48**:79–84.

Lehmann, U., 1979, The jaws and radula of the Jurassic ammonite *Dactylioceras*, *Palaeontology (Lond.)* **22**:265–271.

Lehmann, U., 1981a, *The Ammonites: Their Life and Their World*, Cambridge University Press, Cambridge.

Lehmann, U., 1981b, Ammonite jaw apparatus and soft parts, in: *The Ammonoidea*, Systematics Association Special Volume 18 (M. R. House and J. R. Senior, eds.), Academic Press, London, pp. 275–287.

Lehmann, U., 1985, Zur Anatomie der Ammoniten: Tintenbeutel, Kiemen, Augen, *Paläont. Z.* **59**:99–108.

Lehmann, U., 1988, On the dietary habits and locomotion of fossil cephalopods, in: *Cephalopods—Present and Past* (J. Kullmann and J. Wiedmann, eds.), Schweitzerbart'sche Verlagsbuchhandlung, Stuttgart, pp. 633–640.

Lehmann, U., 1990, *Ammonoideen: Leben zwischen Skylla und Charybdis*, Enke, Stuttgart.

Lehmann, U., and Kulicki, C., 1990, Double function of aptychi (Ammonoidea) as jaw elements and opercula, *Lethaia* **23**:325–331.

Lehmann, U., and Weitschat, W., 1973, Zur Anatomie und Okologie von Ammoniten: Funde von Kropf und Kiemen, *Paläont. Z.* **47**:69–76.

Lehmann, U., Tanabe, K., Kanie, Y., and Fukuda, Y., 1980, Über den Kieferapparat der Lytoceratacea (Ammonoidea), *Paläont. Z.* **54**:319–329.

Lowenstam, H. A., Traub, W., and Weiner, S., 1984, *Nautilus* hard parts: A study of the mineral and organic constituents, *Paleobiology* **10**:268–279.

Meek, A. K., and Hayden, F. V., 1864, Palaeontology of the upper Missouri; Invertebrates, *Smithson. Contrib. Know.* **14**:118–121.

Michael, R., 1894, Über Ammoniten-Brut mit Aptychen in der Wohnkammer von *Oppelia steraspis*, *Z. Dtsch. Geol. Ges.* **46**:697–702.

Miller, A. K., Furnish, W. M., and Schindewolf, O. H., 1957, Paleozoic Ammonoidea, in: *Treatise on Invertebrate Paleontology*, Part L, *Mollusca 4* (R. C. Moore, ed.), Geological Society of America and University of Kansas Press, Lawrence, KS, pp. 11–79.

Moore, R. C., and Sylvester-Bradley, P. C., 1957, Taxonomy and nomenclature of aptychi, in: *Treatise on Invertebrate Paleontology*, Part L, *Mollusca 4* (R. C. Moore, ed.), Geological Society of America and University of Kansas Press, Lawrence, KS, pp. 465–471.

Morton, N., 1981, Aptychi: The myth of the ammonite operculum, *Lethaia* **14**:57–61.

Morton, N., and Nixon, M., 1987, Size and function of ammonite aptychi in comparison with buccal masses of modern cephalopods, *Lethaia* **20**:231–238.

Nesis, K. N., 1986, On the feeding and the causes of extinction of certain heteromorph ammonites, *Paleont. Zh.* **1**:8–15 (in Russian).

Nixon, M., 1969, Growth of the beak and radula of *Octopus vulgaris*, *J. Zool. (Lond.)* **159**:363–379.

Nixon, M., 1979, Hole-boring in shells by *Octopus vulgaris* in the Mediterranean, *Malacologia* **18**:431–443.

Nixon, M., 1985, Capture of prey, diet, and feeding of *Sepia officinalis* and *Octopus vulgaris* (Mollusca: Cephalopoda) from hatchling to adult, *Vie Milieu* **35**:255–261.

Nixon, M., 1987, Cephalopod diets, in: *Cephalopod Life Cycles*, Vol. II (P. R. Boyle, ed.), Academic Press, London, pp. 201–219.

Nixon, M., 1988a, The buccal mass of fossil and recent Cephalopoda, in: *The Mollusca, Paleontology and Neontology of Cephalopods*, Vol. 12 (M. R. Clarke, and E. R. Trueman, eds.), Academic Press, San Diego, pp. 103–122.

Nixon, M., 1988b, The feeding mechanisms and diets of cephalopods—living and fossil, in: *Cephalopods—Present and Past* (J. Kullmann and J. Wiedmann, eds.), Schweitzerbart'sche Verlagsbuchhandlung, Stuttgart, pp. 641–652.

Nixon, M., 1995a, A nomenclature for the radula of the Cephalopoda (Mollusca)—living and fossil, *J. Zool. (Lond.)* **236**:73–81.

Nixon, M., 1995b, The radulae of Cephalopoda, *Smithson. Contrib. Zool.* (in press).

Nixon, M., and Boyle, P. R., 1982, Hole-drilling in crustaceans by *Eledone cirrhosa* (Mollusca: Cephalopoda), *J. Zool. (Lond.)* **196**:439–444.

Nixon, M., and Maconnachie, E., 1988, Drilling by *Octopus vulgaris* in the Mediterranean, *J. Zool. (Lond.)* **216**:687–716.

Saunders, W. B., and Richardson, E. S., Jr., 1979, Middle Pennsylvanian (Desmoinesian) Cephalopoda of the Mazon Creek Fauna, Northeastern Illinois, in: *Mazon Creek Fossils* (M. H. Nitecki, ed.), Academic Press, New York, pp. 333–359.

Saunders, W. B., Spinosa, C., Teichert, C., and Banks, R. C., 1978, The jaw apparatus of Recent *Nautilus* and its palaeontological implications, *Palaeontology* **21**:129–141.

Schindewolf, O. H., 1958, Über Aptychen (Ammonoidea), *Palaeontogr. Abt. A* **111**:1–46.

Seilacher, A., 1993, Ammonite aptychi: How to transform a jaw into an operculum, *Am. J. Sci.* **293A**:20–32.

Tanabe, K., 1983, The jaw apparatuses of Cretaceous desmoceratid ammonites, *Palaeontology (Lond.)* **26**:677–686.

Tanabe, K., and Fukuda, Y., 1987a, Mouth part histology and morphology, in: *Nautilus—The Biology and Paleobiology of a Living Fossil* (W. B. Saunders and N. H. Landman, eds.), Plenum Press, New York, pp. 313–322.

Tanabe K., and Fukuda, Y., 1987b, The jaw apparatus of the Cretaceous ammonite *Reesidites*, *Lethaia* **20**:41–48.

Tanabe, K., and Mapes, R., 1995, Jaws and radula of the Carboniferous ammonoid Ammonoidea, *Cravenoceras*, *J. Paleont.* **69**:703–707.

Tanabe, K., Fukuda, Y., Kanie, Y., and Lehmann, U., 1980a, Rhyncolites and conchorhynchs as calcified jaw elements in some Late Cretaceous ammonites, *Lethaia* **13**:157–168.

Tanabe, K., Hirano, H., and Kanie, Y., 1980b, The jaw apparatus of *Scalarites mihoensis*, a late Cretaceous ammonite, in: *Professor Saburo Kanno Memorial Volume*, Institute of Geosciences, University of Tsukuba, Tsukuba, pp. 159–165.

Teichert, C., Moore, R. C., and Zeller, D. E. N., 1964, "Rhyncholites" in: *Treatise on Invertebrate Paleontology*, Part K, *Mollusca 3* (R. C. Moore, ed.), Geological Society of America and University of Kansas Press, Lawrence, KS, pp. 467–484.

Thore, S., 1949, Investigations on the "Dana" Octopoda. Part I. Bolitaenidae, Amphitretidae, Vitreledonellidae, and Alloposidae, *Dana Rep.* **6**(33):1–85.

Trauth, F., 1938, Die Lamellaptychi des Oberjura und der Unterkriede, *Palaeontogr. Abt. A* **88**:115–229.

Ward, P., and Wicksten, M. K., 1980, Food sources and feeding behavior of *Nautilus macromphalus*, *Veliger* **23**:119–124.

Westermann, G. E. G., 1990, New developments in ecology of Jurassic—Cretaceous ammonoids, in: *Atti del Secondo Convegno Internazionale, Fossili, Evoluzione, Ambiente, Pergola 1987* (G. Pallini, F. Cecca, S. Cresta, and M. Santantonio, eds.), Technostampa, Ostra Vetere, Italy, pp. 459–478.

Wright, C. W., 1981, Cretaceous Ammonoidea, in: *The Ammonoidea*, Systematics Association Special Volume 18 (M. R. House and J. R. Senior, eds.), Academic Press, London, pp. 37–48.

Chapter 3

Attachment of the Body to the Shell in Ammonoids

LARISA DOGUZHAEVA and HARRY MUTVEI

LARISA DOGUZHAEVA • Paleonotological Institute, Russian Academy of Sciences, 117647 Moscow, Russia.   HARRY MUTVEI • Department of Paleozoology, Swedish Museum of Natural History, S-104 05 Stockholm, Sweden.

Ammonoid Paleobiology, Volume 13 of *Topics in Geobiology*, edited by Neil Landman *et al.*, Plenum Press, New York, 1996.

1. Introduction

One of the most intriguing paleobiological problems in ammonoids is to interpret the organization of their muscular system in order to obtain a better understanding of their locomotion and, ultimately, their mode of life. Despite the effects of diagenesis, many ammonoid shells have surprisingly retained visible muscle, ligament, and mantle attachment scars. These scars have been extensively investigated over the last 30 years; in fact, during the quarter of a century that has passed since the classical paper by Jordan (1968), the number of genera exhibiting preserved attachment scars has doubled and is now approximately 80.

The question of the attachment of the mantle to the shell has also received attention. Ultrastructural studies have revealed pore canals that strengthened the mantle attachment to the shell (Doguzhaeva and Mutvei, 1986, 1993).

Finally, indications that shells in certain ammonoids could have been semiinternal or internal (Druschits *et al.*, 1978; Doguzhaeva and Mutvei, 1986, 1993) have given new impetus for further investigations of the mantle–shell attachment problem.

2. Previous Studies on Attachment Scars

Crick (1898) described paired dorsal attachment scars in two Paleozoic ammonoid genera and 11 Mesozoic ammonoid genera, including both hetero-morphs and normally coiled ammonoids (monomorphs). Jones (1961) de-scribed an unpaired ventral attachment scar in addition to the paired dorsal scars in a Cretaceous heteromorph. Jordan (1968) restudied the ammonoids dealt with by Crick (1898) and found that only about 40% of the attachment scars described by Crick could be recognized with certainty. Jordan described attachment scars in 22 additional Mesozoic genera. He was the first to point out that several genera have a sinus line ("indentation," "*Einbuchtung*") with an adoral opening on each side of the body chamber. Doguzhaeva and Kabanov (1987) and Doguzhaeva and Mutvei (1991) described a pair of previously unknown lateral, adorally directed lobes associated with attachment scars in two Mesozoic genera. Bandel (1982), Weitschat (1986), Weitschat and Bandel (1991), Landman and Bandel (1985), and Sarikadze *et al.* (1990) demonstrated the occurrence of an unpaired middorsal scar in the internal (dorsal) sutural lobe in several ammonoids.

In Recent *Nautilus*, attachment of the body to the shell occurs at several sites. The mantle edge is attached at the shell aperture by periostracum, which is secreted in the periostracal groove and is very thin in all species except in *N. scrobiculatus* (see Ward, 1987). Numerous minute epithelial extensions from the mantle margin probably project into vertical pore canals at the shell aperture in order to strengthen the weak periostracal attachment (Doguzhaeva and Mutvei, 1986). In the posterior portion of the body, the powerful cephalic retractor muscles, longitudinal mantle muscles, palliovisceral ligament, and septal myoadhesive band have their attachment sites on the wall of the body chamber (Mutvei *et al.*, 1993).

3. Terminology

The following terms are used to describe attachment scars in ammonoids:

1. Paired dorsal scars. These scars are situated immediately in front of the last suture on dorsal or dorsolateral sides of the body chamber (= "*paarige Muskelansatz-Strukturen*," Jordan, 1968; "*Abdruck des Lateralmuskels*," Sarikadze *et al.*, 1990). Most writers have interpreted these scars as attachment sites of the retractor muscles (dp, Figs. 1A,B, 2B–D, 3B–D).

2. Unpaired middorsal scar. This scar is situated in the internal (= dorsal, according to the Russian terminology) lobe of the suture (Bandel, 1982; Landman and Bandel, 1985; Weitschat, 1986; Sarikadze *et al.*, 1990; Weitschat and Bandel, 1991). As demonstrated by Weitschat and Bandel (1991), this scar is paired in the first chamber but becomes unpaired

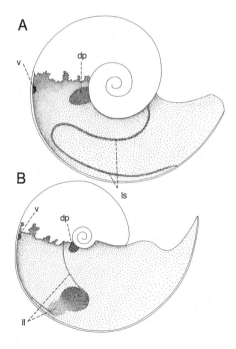

FIGURE 1. (A) Drawing of *Melchiorites* with lateral sinus (ls), paired dorsal scars (dp), and unpaired ventral scar (v). (B) Drawing of *Aconeceras* with lateral lobe and paired lateral attachment scars (ll), paired dorsal scars (dp), and unpaired ventral scar (v). (B modified from Doguzhaeva and Mutvei, 1991, text fig. 8.)

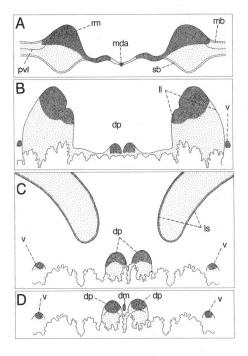

FIGURE 2. Attachment scars projected in one plane. (A) *Nautilus* with scars of paired cephalic retractor muscles (rm), unpaired middorsal scar (mda), scar of mantle myoadhesive band (mb), scar of septal myoadhesive band (sb), and scar of pallioviseral ligament (pvl). (B) *Aconeceras* with lateral lobe (ll), paired dorsal scars (dp), and unpaired ventral scar (v). (C) *Melchiorites* with lateral sinus (ls), paired dorsal scars (dp), and unpaired ventral scar (v). (D) Ammonoid with unpaired dorsal scar (dm), paired dorsal scars (dp), and unpaired ventral scar (v). The apertural direction is toward the top. (A and B modified from Doguzhaeva and Mutvei, 1991, text fig. 5A,B; D modified from Jordan, 1968, Fig. 7.)

in subsequent chambers. It probably was the attachment site of the pallioviseral ligament (dm, Figs. 2D, 3G).

3. Unpaired ventral scar. This scar is situated in front of the ventral lobe of the suture (Vogel, 1959; Jones, 1961; Doguzhaeva and Mutvei, 1991; "*Sipho-Struktur*," "*dunkeles Sipho-Band*," Jordan, 1968; "*Abdruck des Ventralmuskels*," Sarikadze *et al.*, 1990). Its probable function was to support the circumsiphonal invagination in the posterior portion of the body (v, Figs. 1A,B, 2B–D, 3A,F).

4. Paired lateral scars. These scars are in the shape of an adorally directed lobe on each side of the body chamber (Doguzhaeva and Kabanov, 1988; Doguzhaeva and Mutvei, 1991). This lobe extends from the last suture to about the midpoint of the body chamber (ll, Figs. 1B, 2B).

5. Lateral sinus. A sinus with an adoral opening is situated on the side of the body chamber (= "indentation," "*Einbuchtung*," Jordan, 1968; "*Abdruck des vorderen Lateralmuskels*," Sarikadze *et al.*, 1990). It extends from the posterior portion of the body chamber to the shell aperture (ls, Figs. 1A, 2C, 3E).

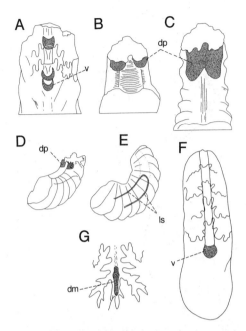

FIGURE 3. (A) Unpaired ventral attachment scar (v) in *Amaltheus*. (B) Paired dorsal attachment scars (dp) in *Pachylytoceras*. (C) Paired dorsal attachment scars (dp) in *Amaltheus*. (D,E) Paired dorsal attachment scars (dp) and lateral sinus (ls) in *Pleurolytoceras*. (F) Unpaired ventral attachment scar (v) in *Sonninia*. (G) Unpaired middorsal attachment scar (dm) in *Pseudocrioceratites*. (A–F modified from Jordan, 1968, Figs. 15, 5, 6, 4, 28, 17: G modified from Sarikadze *et al.* 1990.)

6. Annular elevation. This elevation forms a narrow zone in front of the last suture in some ammonoid genera ("*Haftband Struktur*," "*Annulus Struktur*").

4. Attachment Scars in Recent *Nautilus*

The following muscles are attached to the wall of the body chamber in *Nautilus* (see Mutvei, 1957; Mutvei *et al.*, 1993):

1. A pair of powerful cephalic retractor muscles originate from the lateral sides of the body (rm, Figs. 4A,B, 5B). Their attachment areas are large and crescent-shaped. These areas occur on the lateral and dorsolateral sides of the body chamber (rm, Fig. 6A,B).
2. A pair of small hyponome retractor muscles occur on the ventral surface of the cephalic retractor muscles (hr, Fig. 4A,B). The attachment sites of these muscles on the shell wall cannot be distinguished from those of the cephalic retractors.

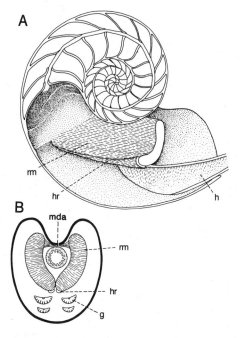

FIGURE 4. (A) Diagrammatic median cross section through a specimen of *Nautilus* showing the cephalic retractor muscle (rm), hyponome retractor muscle (hr), and hyponome (h). (B) Transverse cross section through the body chamber of a specimen of *Nautilus* showing the cephalic retractor muscles (rm), hyponome retractor muscles (hr), middorsal attachment site (mda), and gills (g). Note that the powerful cephalic retractors form the roof above the ventral mantle cavity. (Modified from Mutvei, 1964, Fig. 2.)

3. Longitudinal mantle muscles originate from the mantle (anterior) myoadhesive band (mb, Fig. 5B). This band is attached to the shell wall along an annular zone immediately in front of the scars of the cephalic retractor muscles (mb, Fig. 6A,B).

4. A narrow septal myoadhesive band (= "septal aponeurosis," Willey, 1902) of the body epithelium (sb, Fig. 5B) is attached to the shell wall in front of the last septum (sb, Fig. 6A,B).

5. An unpaired middorsal scar occurs in front of the last septum (mda, Figs. 2A, 6B). This scar probably represents the principal attachment site of the palliovisceral ligament (pvl, mda, Figs. 2A, 4B, 5B).

The individual attachment sites of the cephalic retractor muscles and myoadhesive bands are usually difficult to distinguish from each other on the inner surface of the body chamber wall. Instead, these areas appear together as an annular zone in front of the last septum. This annular zone, called the annulus by Owen (1832) and the annular elevation by Mutvei (1957), is broad on the lateral sides but narrow on the dorsal side of the body chamber.

The annular elevation consists of a thin inner prismatic layer of the shell wall, termed the myostracal layer by Doguzhaeva and Mutvei (1986). The

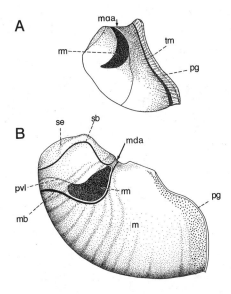

FIGURE 5. Diagrammatic lateral views of the mantle in *Nautilus*. Early embryonic preseptal stage (A) and adult stage (B). (Modified from Mutvei et al., 1993, Fig. 10B,C.) Abbreviations: m, mantle; mb, mantle myoadhesive band; pg, periostracal groove; pvl, palliovisceral ligament; rm, cephalic retractor muscle; sb, septal myoadhesive band; se, septal epithelium; tm, terminus of longitudinal mantle muscles.

mantle epithelium, which is attached to this layer, is composed of "palisade-like cells" (Bandel and Spaeth, 1983). In adult growth stages, the myostracal layer forms a narrow ridge on the anterior border of the annular elevation. In fresh shells the annular elevation is covered by a comparatively soft, brownish, organic layer that, after drying, shrinks and remains as a thin sheet. Crick (1898) demonstrated that only the anterior ridge of the annular elevation can usually be recognized in fossils. This writer filled the body chamber of a specimen of Recent *Nautilus* with paraffin and then dissolved the shell wall. The imprint of the anterior ridge formed an incised line on the internal artificial paraffin mold of the body chamber, but the individual sites of the attachment areas could not be distinguished.

5. Preservation and Morphology of Attachment Scars in Ammonoids

5.1. Paired Dorsal (Umbilical) Attachment Scars

Crick (1898) pointed out that the dorsal attachment scars in adult ammonoid shells almost always form a narrow ridge along their periphery, similar to that in Recent *Nautilus*. This ridge appears on the surface of the

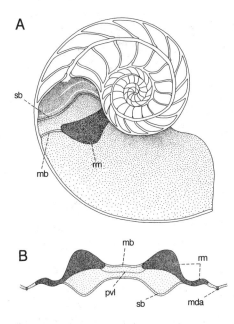

FIGURE 6. (A) Diagrammatic median cross section of a *Nautilus* shell showing the position of attachment scars of the cephalic retractor muscles (rm), mantle myoadhesive band (mb), and septal myoadhesive band (sb) in the posterior portion of the body chamber. (B) Attachment scars projected in one plane. The adapertural direction is toward the top. Abbreviations: pvl, palliovisceral attachment scar; mda, middorsal scar. (Modified from Mutvei, 1957, Figs. 2A, 3.)

internal mold of the body chamber as an incised line. In some ammonoids, the surface of the scar "is very slightly roughened" (Crick, 1898, p. 74).

Jordan (1968) described several modes of preservation of the attachment scars. In addition to the preservation noted by Crick (1898), Jordan found that the imprint of these scars on the internal mold of the body chamber commonly can be distinguished by a black color caused by a thin layer of fine crystalline pyrite. Previous sites in the phragmocone appear as a continuous dorsal black band (= "*dunkeles Band*" of Jordan). In other cases, the dorsal attachment scars and their previous positions are indicated by a continuous light dorsal band (= "*helles Schalen Band*" of Jordan) that extends from the phragmocone to the body chamber. This light band consists of an additional calcareous shell layer that was secreted on the attachment site on the inner surface of the body chamber.

The paired dorsal scars are situated immediately in front of the last suture. They are small in comparison to the size of the body chamber. In monomorphic shells, their position is related to the shape of the whorl cross section; in shells with broad cross section, the scars have a dorsal position, whereas in shells with narrow cross section, the scars also extend to the side (Jordan, 1968, p. 22). The dorsal scars are broad or narrow; they are separated or united middorsally (dp, Fig. 3B,C). The outline of some attachment scars is distinctly marked (dp, Fig. 1A,B), whereas in other scars the posterior boundary cannot be clearly observed (dp, Fig. 3B–D). Rakus (1978) designated the first type the "*type fermé*" and the second type the "*type ouvert.*"

Most writers have interpreted the dorsal attachment scars as the areas of origin of the cephalic retractor muscles (e.g., Crick, 1898; Jordan, 1968; Mutvei, 1964; Doguzhaeva and Mutvei, 1991).

5.2. Unpaired Middorsal Attachment Scar

The middorsal attachment scar is small and situated immediately inside the internal (dorsal) lobe of the last suture. Its surface is typically rugose (Landman and Bandel, 1985; Weitschat, 1986; Weitschat and Bandel, 1991; Sarikadze *et al.*, 1990). Its position in earlier growth stages is commonly clearly visible between successive septa. In Triassic ammonoids, this scar has a long slender outline, whereas in Jurassic and Cretaceous ammonoids, it has a round to oval outline. During ontogeny this scar is paired just adoral of the first septum (proseptum) and then becomes unpaired on the adoral sides of succeeding septa (Bandel, 1982; Landman and Bandel, 1985; Weitschat and Bandel, 1991).

The middorsal scar seems to correspond to the middorsal area in Recent *Nautilus*. This area is distinct and has a rugose surface in the preseptal embryonic stage. It is situated here at the dorsal margin of the protoseptum (Arnold *et al.*, 1987, Figs. 4,10A; Landman *et al.*, 1989, Fig. 2B–D). In somewhat later ontogenetic growth stages, a prominent dorsal septal depression appears in front of the middorsal area (Arnold *et al.*, 1987, Fig. 10B,C). In adult shells, the dorsal septal depression is replaced by a septal furrow. In these shells, the surface of the middorsal area shows a highly variable ornamentation (Arnold *et al.*, 1987, Fig. 10D). By dissecting fresh adult animals one can observe that the palliovisceral ligament is attached at the middorsal area (Mutvei *et al.*, 1993, mspv, Fig. 7B). We still do not know what tissues or organs are housed within the dorsal septal depression.

In several well-preserved ammonoids, only the middorsal attachment scar could be clearly distinguished, but not the paired dorsal and unpaired ventral scars (e.g., Weitschat and Bandel, 1991). Therefore, some writers believe that the unpaired middorsal scar is, in fact, identical to the paired dorsal scars. However, this is not the case. In *Ptychoceras* and *Dorsoplanites*, for example, distinct paired dorsal scars and an unpaired middorsal scar occur in one and the same specimen.

5.3. Unpaired Ventral Attachment Scar

This scar is situated at a variable distance in front of the ventral lobe of the suture. Its shape is round, oval, or crescent (v, Figs. 1A,B, 2B–D, 3A,F, 7A). In

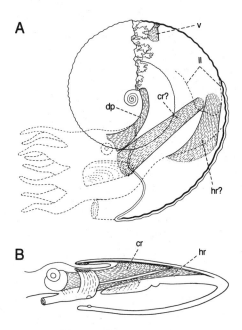

FIGURE 7. (A) A reconstruction of the muscle system in the Cretaceous ammonoid *Aconeceras*. Abbreviations: cr?, paired cephalic retractor muscles; dp, paired dorsal muscles; hr?, paired hyponome retractor muscles; ll, ventrolateral lobe; v, unpaired ventral muscle or ligament. (Modified from Doguzhaeva and Mutvei, 1991, text fig. 8B.) (B) Paired cephalic retractor muscles (cr) and paired hyponome retractor muscles (hr) in a squid. (Modified from Wells, 1988, Fig. 1B.)

a manner similar to the paired dorsal scars, the ventral scar is preserved on the internal mold of the body chamber as a surface impression or as a peripheral ridge-like thickening. In some shells it is colored black by pyrite (Jordan, 1968). In earlier growth stages, the sites of the ventral scars sometimes appear as a black band on the phragmocone ("*dunkles Sipho-Band*" of Jordan, 1968), whereas other times they appear as a row of individual scars (v, Fig. 3A).

Jordan (1968, p. 28) pointed out that the position of the ventral scar is functionally related to the projection of the last prochoanitic septal neck into the body chamber. He also said that this scar could have been an attachment site for a ligament or a muscle. Doguzhaeva and Mutvei (1991, p. 31) showed that the ventral scar in *Aconeceras* is situated at a distance of about one to three chamber lengths in front of the last suture. This distance corresponds to the length of the primary membrane of the last connecting ring, which is found occasionally in the body chamber (Druschits and Doguzhaeva, 1974; Kulicki, 1979; Westermann, 1982; Doguzhaeva, 1988). The length of this membrane indicates the depth of the circumsiphonal invagination in the posterior portion of the body. It is, therefore, probable that the muscle or ligament that was attached to the ventral scar was used to maintain the shape and position of the circumsiphonal invagination during the growth and forward migration of the body in the body chamber.

5.4. Lateral Sinuses

In some ammonoid genera, the surface of the internal mold of the body chamber shows an incised line on each side. This line forms a posteriorly directed sinus ("indentation," "*Einbuchtung*," Jordan, 1968; "*Abdruck des vorderen Lateralmuskelns*," Sarikadze *et al.*, 1990) that extends from the posterior portion of the body chamber to the shell aperture (ls, Figs. 1A, 2C, 3E). In some cases, the area surrounded by the sinus forms an imprint on the surface of the internal mold of the body chamber, but in other cases, this area is colored black by pyrite or manganese hydroxide (Jordan, 1968, Fig. 21, Pl. 10, Fig. 1). Jordan (1968) compared the lateral sinuses in ammonoids with the pallial sinus in recent bivalves.

5.5. Lateral Attachment Scars

These scars form adaperturally directed lobes on each side of the body chamber (ll, Figs. 1B, 2B; Doguzhaeva and Mutvei, 1991). Here the shell wall is thickened by additional nacreous and inner prismatic layers. The outline of the lateral lobe is usually marked by a low, narrow rim of shell material. The lateral lobe extends from the last suture to about the middle of the body chamber. Its adapertural portion is subdivided into two minor lobes, i.e., a ventral and dorsal lobe. To date, about 50 specimens of *Aconeceras* and several specimens of *Quenstedtoceras* are known to have this type of attachment scar.

Doguzhaeva and Mutvei (1991) postulated that the cephalic retractor muscles originated from the dorsal minor lobe and the hyponome retractors from the ventral minor lobe (cr?, hr?, Fig. 7A).

5.6. Annular Elevation

The imprint of a narrow zone immediately in front of the last suture has been observed on the surface of the internal cast of the body chamber in a limited number of ammonoid genera. This imprint also follows the outline of the last suture. Jordan (1968, p. 4) pointed out that in larger ammonoids the annular elevation appears as a row of fine tubercles on the inner surface of the shell wall. In smaller ammonoids, the annular elevation "appears as a dark line." Jordan interpreted the annular elevation as the attachment site of septal and mantle myoadhesive bands, similar to those in Recent *Nautilus*.

6. Taxonomic Occurrence of Attachment Scars

Jordan (1968) listed about 30 genera in which one or several types of attachment scars were known. In the present paper we have listed between 70 and 80 genera (Table I). However, it should be emphasized that our knowledge of attachment scars in ammonoids is still very limited. The main reason for this is the state of preservation; even in extremely well preserved shells, attachment scars are rare.

TABLE I. List of Genera with Known Attachment Scars[a]

Genera	Age	Type of attachment scars	Authors
Goniatitina			
Goniatites	C	dp,v,an	Crick, 1898; Jordan, 1968; K. Tanabe, personal communication, 1994
Muensteroceras	C1	dp	Crick, 1898; Jordan, 1968
Ceratitida			
Anagymnotoceras	Tr2Ans	dm	Weitschat, 1986
Amphipopanoceras	Tr2Ans	dm	Weitschat, 1986; Weitschat and Bandel, 1991
Czekanowskites	Tr2Ans	dm	Weitschat and Bandel, 1991
Indigirites	Tr2Lad	dm	Weitschat and Bandel, 1991
Nathorstites	Tr2Lad	dm	Weitschat and Bandel, 1991
Aristoptychites	Tr2Lad	dm	Weitschat and Bandel, 1991
Arctoptychites	Tr2	dm	Weitschat and Bandel, 1991
Sphaerocladiscites	Tr2Lad	dm	Weitschat and Bandel, 1991
Stolleites	Tr3Crn	dm	Weitschat and Bandel, 1991
Paracladiscites	Tr3Crn	dm	Weitschat and Bandel, 1991
Ceratites	Tr2Lad	dp,v	Unpublished observations
Phylloceratina			
Trogophylloceras	J1	dp	Jordan, 1968
Holcophylloceras	J2Baj	v	Unpublished observations
Phyllopachyceras	K1Apt	v	Sarikadze *et al.*, 1990; Unpublished observations
Euphylloceras	K1Apt	v	Sarikadze *et al.*, 1990; Unpublished observations
Salfeldiella	K1Apt	ls	Sarikadze *et al.*, 1990
Lytoceratina			
Lytoceras	J1–2	dp,v,ls,an	Crick 1898; Jordan, 1968; Rakus, 1978
Derolytoceras	J1Toa	dp	Rakus, 1978
Pachylytoceras	J1Toa	dp	Jordan, 1968
Pleurolytoceras	J1Toa	dp, ls	Jordan, 1968
Pictetia	K1Apt	dm	Sarikadze *et al.*, 1990
Tetragonites	K1Apt	v	Sarikadze *et al.*, 1990; Unpublished observations
Hemitetragonites	K1Apt	dp, v	Sarikadze *et al.*, 1990; Unpublished observations
Ptychoceras	K1Apt	dp,dm,v	Sarikadze *et al.*, 1990; Unpublished observations
Hamites	K1Hau	dp	Crick, 1898; Jordan, 1968

TABLE I. Continued

Genera	Age	Type of attachment scars	Authors
Lytoceratina (continued)			
Baculites	K2Tur	dp	Crick, 1898; Jordan, 1968; Unpublished observations
Diplomoceras	K2Cmp	dp, v	Crick, 1898; Jordan, 1968
Turrilites	K2Cen	dp,an(?)	Crick, 1898; Jordan, 1968
Hoploscaphites	K2Mas	dp,v	Landman and Waage, 1993
Ammonitida			
Psilocerataceae			
Alsatites	J1Het	dp	Crick, 1898; Jordan, 1968
Arietites	J1Sin	dp,v(?),an(?)	Crick, 1898; Jordan, 1968
Asteroceras	J1Sin	dp	Jordan, 1968
Paroxynoticeras	J1Sin	dp,v	Rakus, 1978
Eoderocerataceae			
Eoderoceras	J1Sin	dm	Rakus, 1978
Amaltheus	J1Plb	dp,v,an,ls	Crick, 1898; Jordan, 1968
Androgynoceras	J1Plb	dp,v(?)	Crick, 1898
Amauroceras	J1Plb	dm	Lehmann, 1990
Hildocerataceae			
Arieticeras	J1Plb	dp	Jordan, 1968
Grammoceras	J1Toa	ls	Jordan, 1968
Dorsetensia	J2Baj	dp,v	Jordan, 1968; Unpublished observations
Ludwigia	J2Baj	dp	Jordan, 1968
Sonninia	J2Baj	dp,v	Crick, 1898; Jordan, 1968
Staufenia	J2Baj	dp,an	Jordan, 1968
Leioceras	J2Baj	dp	Jordan, 1968
Haplocerataceae			
Distichoceras	J2–3	ls	Crick, 1898; Jordan, 1968
Paroecotraustes	J2	dp	Jordan, 1968
Hecticoceras	J2	dp,ls,an(?)	Crick, 1898; Jordan, 1968
Creniceras	J3	dp,an	Crick, 1898; Jordan, 1968
Stephanocerataceae			
Clydoniceras	J2Bth	dp, an(?)	Crick, 1898; Jordan, 1968
Bullatimorphites	J2Bth	v	Jordan, 1968
Quenstedtoceras	J2Clv	dm,v,ll	Jordan, 1968; Landman and Bandel, 1985; Sarikadze *et al.*, 1990; Lehmann, 1990; Doguzhaeva and Mutvei, 1991; Unpublished observations
Cardioceras	J3Oxf	dp,v(?),an	Crick, 1898; Jordan, 1968
Amoeboceras	J3Oxf	dp,an	Crick, 1898; Jordan, 1968
Aconeceras	K1Apt	dp,v,ll	Doguzhaeva and Kabanov, 1988; Sarikadze *et al.*, 1990; Doguzhaeva and Mutvei, 1991
Perisphinctaceae			
Grossouvria	J1Ckv	v	Jordan, 1968
Siemiradzkia	J2Bth	v	Jordan, 1968
Aspidoceras	J3Kim	dp	Crick, 1898; Jordan, 1968
Peltoceras	J2Clv	dp	Crick, 1898; Jordan, 1968
Virgatites	J3Kim	v	Unpublished observations
Indosphinctes	J2Clv	dp,v	Sarikadze *et al.*, 1990
Elatmites	J2Clv	v,an	Unpublished observations

TABLE I. Continued

Genera	Age	Type of attachment scars	Authors
Perisphinctaceae (continued)			
Dorsoplanites	J3Kim	dp, dm	Unpublished observations
Polyptychites	K1Vlg	v	Vogel, 1959
Olcostephanus	K1Vlg	v	Unpublished observations
Desmocerataceae			
Beudanticeras	K1Apt	v	Unpublished observations
Melchiorites	K1Apt	dp,v, ls	Sarikadze *et al.*, 1990; Unpublished observations
Zurcherella	K1Apt	ls	Sarikadze *et al.*, 1990
Pseudosilesites	K1Apt	dp,v	Sarikadze *et al.*, 1990
Hauericeras	K2Sant.	dm	Obata *et al.*, 1978
Hoplitaceae			
Epicheloniceras	K1Apt	dp	Sarikadze *et al.*, 1990
Deshayesites	K1Apt	dp,v,ls	Doguzhaeva and Kabanov, 1988; Sarikadze *et al.*, 1990
Parahoplites	K1Apt	dp,v	Landman and Bandel, 1985
Colombiceras	K1Apt	dp,v	Sarikadze *et al.*, 1990
Acanthohoplites	K1Apt	dp,v	Sarikadze *et al.*, 1990
Hypacanthoplites	K1Apt	dm,v	Landman and Bandel, 1985; Sarikadze *et al.*, 1990
Diadochoceras	K1Apt	v	Unpublished observations
Euhoplites	K1Alb	dm	Landman and Bandel, 1985
Ancylocerataceae			
Aegocrioceras	K1Hau	dp	Crick, 1898; Jordan, 1968
Ancyloceras	K1Apt	dp	Crick, 1898; Jordan, 1968; Doguzhaeva and Mikhailova, 1982
Pseudocrioceratites	K1Apt	dm	Sarikadze *et al.*, 1990
Acanthocerataceae			
Tissotia	K2Kon	dp(?)	Crick, 1898; Jordan, 1968

[a]Abbreviations: dp, dorsal paired attachment scars; dm, middorsal unpaired attachment scar; v, ventral unpaired attachment scar; ls, lateral sinus; ll, lateral lobe; an, annular elevation. Geologic stages are abbreviated according to Harland *et al.* (1989). Genera are arranged into superfamilies according to Arkell *et al.* (1957) with the exception of the heteromorphs, which are arranged according to Mikhailova (1983).

Ammonoids belonging to goniatitids, ceratitids, phylloceratids, lytocera-tids, and ammonitids all seem to possess paired dorsal attachment scars and probably also an unpaired ventral attachment scar. Goniatitids also have paired dorsal attachment scars, but the occurrence of an unpaired ventral attachment scar is still somewhat doubtful (Jordan, 1968). The unpaired middorsal attachment scar is at present known only in ceratitids, lytoceratids, and five superfamilies of Ammonitina. Thus, it is probable that the Ammonoidea as a whole have paired dorsal scars, unpaired middorsal scars, and unpaired ventral scars. In addition to these three types of common attachment scars, phylloceratids, lytoceratids, and four superfamilies of Ammonitina also have paired lateral sinuses. Finally, paired lateral attachment scars have so far been found in only two genera of the Stephanocerataceae (Ammonitina). These genera also have paired dorsal and unpaired ventral attachment scars.

7. Functional Interpretation of Attachment Scars

7.1. Recent *Nautilus*

On the basis of anatomic features, Griffin (1900) and Mutvei (1957, 1964) concluded that the powerful cephalic retractor muscles in *Nautilus* are necessary for swimming by jet propulsion. These muscles form a roof above the spacious mantle cavity (rm, Fig. 4A,B). Their contraction pulls the head into the body chamber and presses the roof of the mantle cavity down, expelling water from the mantle cavity through the hyponome. This interpretation of the swimming mechanism was later confirmed by direct observations (Ward, 1987; Chamberlain, 1987; Wells, 1987, 1988). Thus, the position and course of the cephalic retractor muscles and their relationship to the ventral mantle cavity are essential in understanding the swimming mechanism in *Nautilus*. This mechanism also requires that the body chamber and body be comparatively short and that the cephalic retractors be able to extend straight across to the head (rm, Fig. 4A).

7.2. Ammonoids

Interpretation of the function of attachment scars in ammonoids is still highly speculative because of the difficulties of comparing these scars with those in Recent *Nautilus*. The length of the body chamber in ammonoids varies from slightly less than one-half whorl to two whorls. For example, in Cretaceous genera, the body chamber length usually is about three-quarters of the whorl; in some Permian and Triassic genera, the length of the body chamber equals one and one-half to two whorls (Fig. 8A; Doguzhaeva and Mutvei, 1986). Thus, in general, the body chamber in ammonoids is longer than that in *Nautilus* (Fig. 6A) and in most fossil nautiloids. This indicates that the mantle cavity in most ammonoids was narrower and longer than that in *Nautilus*. Unlike the condition in *Nautilus* (rm, Fig. 4A,B), the paired dorsal muscles, usually considered to correspond to the cephalic retractors, therefore could not have formed a roof above the mantle cavity because they were situated close to the dorsal side of the body chamber in monomorphic shells (dp, Figs. 8A,B, 9A,B; Mutvei, 1964; Mutvei and Reyment, 1973; Doguzhaeva and Mutvei, 1986).

In contrast, in *Aconeceras* and *Quenstedtoceras*, the body was comparatively short, and the lateral attachment scars extended to the middle of the body chamber (Fig. 1B). Therefore, the two paired muscles that originated from these scars were able to extend straight across to the head and to the base of the hyponome (Fig. 7A). These muscles are, therefore, interpreted as

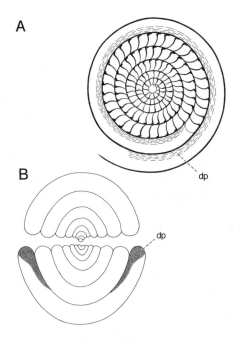

FIGURE 8. Median (A) and transverse (B) cross section of the shell of the Triassic ceratitid *Proarcestes* showing the probable position of the paired dorsal muscles (dp) in the long body chamber. (Modified from Doguzhaeva and Mutvei, 1986, Fig. 10A,B.)

cephalic retractors and hyponome retractors (cr?, hr?, Fig. 7A). The hyponome retractors are very small in *Nautilus* (hr, Fig. 4A,B) but well developed in Recent coleoids (hr, Fig. 7B). It is still unknown whether these lateral attachment scars occur in other ammonoids or whether they are confined to the superfamily Stephanocerataceae of the Ammonitina.

Another comparatively widespread feature in many ammonoids consists of paired lateral sinuses. Jordan (1968) pointed out the similarity between these sinuses and the pallial sinus in bivalves. The pallial sinus in bivalves marks the site of the muscular attachment of the funnel, which develops from the mantle. If the lateral sinuses in ammonoids correspond to the pallial sinus in bivalves, we have to assume that ammonoids also had mantle extensions or organs that required a sinus-shaped attachment scar.

As is well known, the shape of the apertural margin in many ammonoids differs considerably from that in Recent *Nautilus* and the majority of fossil nautiloids. Doguzhaeva and Mutvei (1991) emphasized that in some ammonoids, at least, the shells may have been semiinternal or internal. If the latter were true, we do not have enough data to determine the locomotory and swimming mechanism, the relationship of the soft body to the body chamber, the buoyancy system, or the mode of life in these ammonoids.

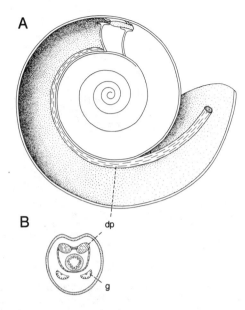

FIGURE 9. (A) Diagrammatic median cross section of an ammonoid shell showing the probable course of the paired dorsal muscles (dp) along the dorsal wall. (B) Transverse cross section of the body chamber showing the probable position of the paired dorsal muscles (dp) and ventral mantle cavity with gills (g). Note that the dorsal muscles probably did not form a roof above the mantle cavity as in *Nautilus*. Modified from Mutvei (1964, Figs. 3E, 7A).

8. Wrinkle Layer

In the shell wall of ammonoids, two inner prismatic layers can be distinguished, the myostracal layer, which is secreted in the posterior portion of the body chamber on the attachment scars, and the wrinkle layer, which is secreted at the shell aperture.

The wrinkle layer occurs in all Paleozoic and Mesozoic ammonoids with smooth or slightly ornamented shells (for references, see Doguzhaeva and Mutvei, 1986). This layer forms sharp-edged ridges on the dorsal and lateral sides of the shell wall (w, dw, Fig. 10B). The sharp edges point obliquely in the apertural direction. In the ventral portion of the shell wall, the ridges are usually absent, and the prismatic wrinkle layer has a smooth surface.

Doguzhaeva and Mutvei (1986) compared the wrinkle layer with the oblique prismatic layer, which covers the inner shell surface in many gastropods with narrow shells, and with a similar layer in Recent *Nautilus*. As in gastropods (Gainey and Wise, 1975), the sharp ridges of the wrinkle layer (w) seem to have provided mechanical support for the mantle (m) to temporarily maintain a fixed position in the living chamber (Fig. 10B). This might have

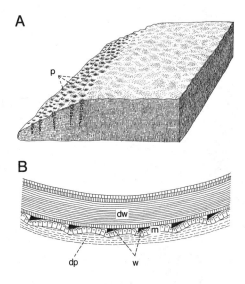

FIGURE 10. (A) Block diagram of the shell wall in *Nautilus* showing the pore canals (p) at the apertural margin. (B) Median cross section of the dorsal shell wall (dw) in an ammonoid showing the ridges of the wrinkle layer (w) and their relationship to the mantle epithelium (m) and dorsal muscle (dp). (Modified from Doguzhaeva and Mutvei, 1986, Fig. 10F.)

facilitated the forward movement and extension of the body in the body chamber.

9. Pore Canals

In the shells of the Triassic genus *Phyllocladiscites*, distinct pore canals are observed in the inner myostracal shell layer (Doguzhaeva and Mutvei, 1986). Extensions of the myoadhesive epithelium were probably inserted in these canals in order to strengthen the attachment of this epithelium to the shell at the origin of the retractor and mantle muscles. Batten (1984) found pores (tubules) in the myostracal layer of a deep-sea limpet and interpreted them as possible sites for muscle cell insertion. In *Aconeceras*, small pore canals also occur in the inner prismatic layer of the shell wall in early ontogenetic stages. In addition, exceptionally large pore canals are found in the keel of *Aconeceras* (see Doguzhaeva and Mutvei, 1993).

In Recent *Nautilus*, pore canals occur in the inner prismatic, mantle attachment layer (Doguzhaeva and Mutvei, 1986, Figs. 8D, 9B). The canals are well developed in fully grown shells at the apertural margin where the mantle attachment layer is considerably thickened (Fig. 10A). Pore canals from previous growth stages are no longer visible, probably because they become filled in by shell material. This may also explain why pore canals are rare in ammonoid shells.

10. Septal Attachment

Henderson (1984) described complex septa in the Cretaceous lytoceratid *Pseudophyllites*. Here, each septum is curved backward to form a pronounced depression, termed the septal recess. The septal recess extends to the surface of the previous septum. The successive septal recesses link one septum to the next, forming a tunnel. The tunnel itself is septate. The ventral median wall of each recess forms an inner septum which has a fluted periphery, similar to that of the septum proper. Henderson believed that muscles in *Pseudophyllites* were attached to the periphery of the septum proper and inner septum and not, as in Recent *Nautilus*, to the wall of the body chamber.

References

Arkell, W. J., Furnish, W. M., Kummel, B., Miller, A. K., Moore, R. C., Schindewolf, O. H., Sylvester-Bradley, P. C., and Wright, C. W., 1957, *Treatise on Invertebrate Paleontology,* Part L, *Mollusca 4,* Geological Society of America and University of Kansas Press, Lawrence, KS.

Arnold, J. M., Landman, N. H., and Mutvei, H., 1987, Development of the embryonic shell of *Nautilus,* in: *Nautilus: The Biology and Paleobiology of a Living Fossil* (W. B. Saunders and N. H. Landman, eds.), Plenum Press, New York, pp. 373–400.

Bandel, K., 1982, Morphologie und Bildung der frühontogenetischen Gehäuse bei conchiferen Mollusken, *Facies* **7:**1–198.

Bandel, K., and Spaeth, C., 1983, Beobachtungen am rezenten *Nautilus, Mitt. Geol. Paläont. Inst. Univ. Hamburg* **53:**9–26.

Batten, R. L., 1984, Shell structure of the Galapagos rift limpet *Neomphalus fretterae* McLean, 1981, with notes on muscle scars and insertions, *Am. Mus. Novit.* **2776:**1–13.

Chamberlain, J. A., 1987, Locomotion of *Nautilus,* in: *Nautilus: The Biology and Paleobiology of a Living Fossil* (W. B. Saunders and N. H. Landman, eds.), Plenum Press, New York, pp. 489–525.

Crick, G. C., 1898, On the muscular attachment of the animal to its shell in some Cephalopoda (Ammonoidea), *Trans. Linn. Soc. Lond. 2nd Ser., Zoology* **7:**71–113.

Doguzhaeva, L. A., 1988, Siphuncular tube and septal necks in ammonoid evolution, in: *Cephalopods—Present and Past* (J. Wiedmann and J. Kullmann, eds.), Schweizerbart'sche Verlagsbuchhandlung, Stuttgart, pp. 291–302.

Doguzhaeva, L. A., and Kabanov, G. K., 1987, Muscle scars and assumed mode of life in Aptian ammonoid *Aconeceras,* in: *Molluscs, Results and Perspectives of Research,* Nauka, Leningrad, (in Russian).

Doguzhaeva, L. A., and Kabanov, G. K., 1988, Muscle scars in ammonoids, *Doklady Akademii Nauk SSSR* **301:**210–212.

Doguzhaeva, L. A., and Mutvei, H., 1986, Functional interpretation of inner shell layers in Triassic ceratid ammonites, *Lethaia* **19:**195–209.

Doguzhaeva, L. A., and Mutvei, H., 1991, Organization of the soft body in *Aconeceras* (Ammonitina), interpreted on the basis of shell morphology and muscle scars, *Palaeontogr. A* **218:**17–33.

Doguzhaeva, L. A., and Mutvei, H., 1993, Structural features in Cretaceous ammonoids indicative of semi-internal or internal shells, in: *The Ammonoidea: Environment, Ecology, and Evolutionary Change,* Systematics Association Special Volume 47 (M. R. House, ed.), Clarendon Press, London, pp. 99–104.

Druschits, V. V., and Doguzhaeva, L. A., 1974, On some features of the morphogenesis of phylloceratids and lytoceratids (Ammonoidea), *Paleontol. Zh.* **1:**42–53 (in Russian).

Druschits, V. V., Doguzhaeva, L. A., and Mikhailova, I. A., 1978, Unusual coating layers of ammonites, *Paleontol. Zh.* **2**:36–44 (in Russian).

Gainey, L. F., and Wise, S. W., 1975, Archaeogastropod (Mollusca) shell: Functional morphology of the oblique prismatic layer, *Trans. Am. Microsc. Soc.* **94**:411–413.

Griffin, L. E., 1900, The anatomy of *Nautilus pompilius*, *Mem. Natl. Acad. Sci.* **8**:101–230.

Harland, W. B., Armstrong, R. L., Cox, A. V., Craig, L. E., Smith, A. G., and Smith, D. G., 1989, *Geologic Time Scale*, Cambridge University Press, Cambridge.

Henderson, R. A., 1984, A muscular attachment proposal for septal function in Mesozoic ammonites, *Palaeontology* **27**:461–486.

Jones, D. L., 1961, Muscle attachment impressions in a Cretaceous ammonite. *J. Paleontol.* **35**:502–504.

Jordan, R., 1968, Zur Anatomie mesozoischer Ammoniten nach den Strukturelementen der Gehäuse-Innenwand, *Beih. Geol. Jahrb.* **77**:1–64.

Kulicki, C., 1979, The ammonite shell: Its structure, development and biological significance, *Palaeontol. Pol.* **39**:97–142.

Landman, N. H., and Bandel, K., 1985, Internal structures in the early whorls of Mesozoic ammonites, *Am. Mus. Novit.* **2823**:1–21.

Landman, N. H., and Waage, K. M., 1993, Scaphitid ammonites of the Upper Cretaceous (Maastrichtian) Fox Hills Formation in South Dakota and Wyoming, *Bull. Am. Mus. Nat. Hist.* **215**:1–257.

Landman, N. H., Arnold, J. M., and Mutvei, H., 1989, Description of the embryonic shell of *Nautilus balauensis* (Cephalopoda), *Am. Mus. Novit.* **2960**:1–16.

Lehmann, U., 1990, *Ammonoideen*, F. Enke Verlag, Stuttgart.

Mikhailova, I. A., 1983, *Taxonomy and Phylogeny of Cretaceous Ammonoids*, Nauka, Moscow (in Russian).

Mutvei, H., 1957, On the relations of the principal muscles to the shell in *Nautilus* and some fossil nautiloids, *Ark. Mineral. Geol.* **2**:219–254.

Mutvei, H., 1964, Remarks on the anatomy of Recent and fossil Cephalopoda, *Stockholm Contrib. Geol.* **11**:79–102.

Mutvei, H., and Reyment, R. A., 1973, Buoyancy control and siphuncle function in ammonoids, *Palaeontology* **16**:623–636.

Mutvei, H., Arnold, J. M., and Landman, N. H., 1993, Muscles and attachment of the body to the shell in embryos and adults of *Nautilus balauensis* (Cephalopoda), *Am. Mus. Novit.* **3059**:1–15.

Obata, I., Futakami, M., Kawashita, Y., and Takahashi, T., 1978, Apertural features in some Cretaceous ammonites from Hokkaido, *Bull. Nat. Sci. Mus. Ser. C* **4**:139–155.

Owen, R., 1832, *Memoir on the Pearly Nautilus (Nautilus pompilius, Linn.) with Illustrations of its External Form and Internal Structure*, Richard Taylor, London.

Rakus, M., 1978, Sur l'existence de deux types distincts d'empreintes de muscles rétracteurs chez les ammonites, *Bull. Soc. Vaudoise Sci. Nat.* **354**:74, 139–145.

Sarikadze, M. Z., Lominadze, T. A., and Kvantaliani, I. V., 1990, Systematische Bedeutung von Muskelabdrücken spätjurassischer Ammonoideae, *Z. Geol. Wiss. Berl.* **18**:1031–1039.

Vogel, K. P., 1959, Zwergwuchs bei Polyptychiten (Ammonoidea), *Geol. Jahrb.* **76**:469–540.

Ward, P. D., 1987, *The Natural History of Nautilus*, Allen and Unwin Ltd., London.

Weitschat, W., 1986, Phosphatisierte Ammonoideen aus der Mittleren Trias von Central-Spitzbergen, *Mitt. Geol. Paläont. Inst. Univ. Hamburg* **61**:249–279.

Weitschat, W., and Bandel, K., 1991, Organic components in phragmocones of Boreal Triassic ammonoids: Implications for ammonoid biology, *Paläontol. Z.* **65**:269–303.

Wells, M. J., 1987, Ventilation and oxygen extraction by *Nautilus*, in: *Nautilus: The Biology and Paleobiology of a Living Fossil* (W. B. Saunders and N. H. Landman, eds.), Plenum Press, New York, pp. 339–350.

Wells, M. J., 1988, The mantle muscle and mantle cavity in cephalopods, in: *The Mollusca, Form and Function*, Vol. 11 (E. R. Truman and M. R. Clarke, eds.), Academic Press, London, pp. 287–300.

Westermann, G. E. G., 1982, The connecting rings of *Nautilus* and Mesozoic ammonoids: Implications for ammonoid bathymetry, *Lethaia* **15**:373–384.

Willey, A., 1902, Contribution to the natural history of the Pearly *Nautilus*, in: *Zoological Results Based on Material from New Britain, New Guinea, Loyalty Islands and Elsewhere, Collected During the Years 1895, 1896 and 1897*, Cambridge University Press, Cambridge, pp. 691–830.

Chapter 4

Ammonoid Shell Microstructure

CYPRIAN KULICKI

1. Introduction

This chapter is not devoted to shell microstructure alone. In addition to presenting a description of the structure of the individual layers that compose the ammonoid shell, we also discuss the distribution and relationships of these layers to one another as well as their ultrastructure where possible.

Because aragonite, the chief mineral that makes up the ammonoid shell, is metastable and transforms into calcite as a function of time, pressure, and temperature (Dullo and Bandel, 1988), it is difficult to obtain specimens that

CYPRIAN KULICKI • Polish Academy of Sciences, Institute of Paleobiology, 02-089 Warsaw, Poland.

Ammonoid Paleobiology, Volume 13 of *Topics in Geobiology*, edited by Neil Landman *et al.*, Plenum Press, New York, 1996.

are well enough preserved for study. This explains why all the major micro- and ultrastructural studies of ammonoids have been conducted on Mesozoic material from platform deposits.

2. Embryonic Stage

2.1. Existing Structural Models

The term "ammonitella" was introduced by Druschits and Khiami (1969) to denote the initial chamber and the first whorl to the primary constriction, together with the proseptum, prosiphon, and cecum; thus, for many paleontologists this term denotes the embryonic shell. Earlier, Ruzhentsev and Shimanskij (1954) and Makowski (1962, 1971) used the name "protoconch" for the initial chamber plus the first whorl to the primary constriction, and in their understanding, the term protoconch is a synonym for the term ammonitella as used by Druschits and Khiami (1969).

Because of controversies on the structure and relationships of the layers in the wall of the ammonitella, and because of diverse opinions on the embryogenesis of ammonoids, the problem of the microstructure of the ammonitella is of especial interest.

The distinct morphological and microstructural character of the ammonitella had already been observed in the 19th century (Hyatt, 1872; Branco, 1880), but knowledge on the subject rapidly improved with the use of electron microscopy. Birkelund (1967) and Birkelund and Hansen (1968) were the first to apply transmission electron microscopy (TEM) in studies of the microstructure of ammonitellae of Upper Cretaceous *Saghalinites* and *Scaphites*. The preparation method was described by Hansen (1967) and depended generally on slight etching of the surfaces of polished cross-sections in EDTA and on removing the collodial replicas and sputtering them with carbon. By these methods, the following was determined (Fig. 1B):

1. The wall of the initial chamber is built of two sublayers without a distinct boundary between them. The inner sublayer consists of crystals perpendicular to the inner shell surface, whereas crystals in the external sublayer are distributed without a distinct orientation. Both sublayers of the initial chamber wedge out in the vicinity of the base of the proseptum.

2. The wall of the first whorl appears as a prismatic layer on the inner surface of the initial chamber. After the wall of the initial chamber wedges out, the wall of the first whorl continues without much change to the primary constriction. It is similar in construction to the wall of the initial chamber; i.e., it consists of two sublayers, an inner sublayer having more regular crystals perpendicular to the inner shell surface and an external, thinner sublayer with less regularly oriented crystals.

The apertural part of the ammonitella has a constriction known as the primary constriction (see Chapter 11). The prismatic and subprismatic sublayers of the first whorl become much thinner, and the nacreous primary varix develops beneath them. The proseptum is constructed of the same crystalline matter as that of the internal prismatic layer of the initial chamber and the wall of the first whorl.

Erben *et al.* (1968, 1969) were the first to use scanning electron microscopy (SEM) to study the microstructure of ammonoids. The method of preparing specimens for SEM is much simpler than that for TEM. In their monograph, these authors described a model for the structure and development of the ammonitella (Fig. 1A). According to this model, the wall of the initial chamber initially consists of two subprismatic layers. These wedge out, and only some time later do fully prismatic sublayers appear on the inner surface of the initial chamber. Two of these sublayers also wedge out, and only the sublayer beyond the base of the proseptum continues until the end of the primary constriction. The dorsal part of the proseptum represents an additional layer on the inner

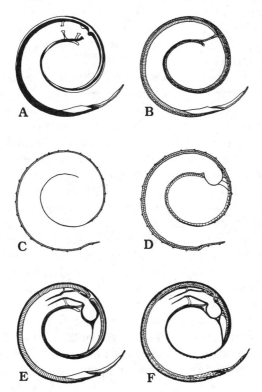

FIGURE 1. Models of the structure of the ammonitella shell in longitudinal cross section. (A) Modified from Erben *et al.*, 1969, Fig. 5. (B) Modified from Birkelund, 1981, Fig. 2. (C,D) Modified from Bandel, 1982, Figs. 47, 48. (E) Modified from Kulicki, 1979, Fig. 6. (F) Modified from Kulicki and Doguzhaeva, 1994, Fig. 14.

surface of the initial chamber, whereas the ventral part of the proseptum represents a continuation of the innermost layer of the initial chamber and first whorl. The flange is composed of a prismatic layer and is separated from the proseptum and wall of the initial chamber by a discontinuity surface, suggesting that it is formed later, according to this model (Fig. 1A).

Kulicki (1979), on the basis of excellently preserved material of *Quenstedtoceras* and *Kosmoceras* from Luków, presented a model of the formation of the ammonitella (Fig. 1E). In this model, the wall of the initial chamber has two layers, best seen in its dorsal and apical parts. The inner layer has a regular prismatic structure and represents the wall proper of the initial chamber, continuing into the outer prismatic layer of the first whorl up to the ammonitella edge. The outer layer of the wall of the initial chamber is a continuation of the mural part of the proseptum and represents the dorsal wall of the first whorl of the ammonitella. This layer is thickest opposite the primary varix and is subprismatic. Crystallites in this layer are oriented parallel, diagonal, or perpendicular to the shell surface. The boundary between this layer and the wall proper of the initial chamber is not distinct. A distinct boundary between the dorsal wall and outer prismatic layer of the first whorl develops at the appearance of the tuberculate sculpture characteristic of the ammonitella.

In medial and paramedial cross sections through the outer saddle of the proseptum, on the inner surface of the venter of the initial chamber, there is an inner prismatic layer of regular structure linked to the base of the proseptum. This layer is separated from the wall proper of the initial chamber by a thin layer of microcrystalline structure. The maximum thickness of this thin layer occurs in the middle part of the base of the proseptum. The prismatic layer of the proseptum commonly continues as one of the main components of the wall of the first whorl ("medial prismatic layer" of Kulicki, 1979).

All of the above models assume simultaneous secretion of the organic phase of the shell together with mineralization. This is what occurs in the formation of the postembryonic shell in Recent molluscs. In contrast, Bandel (1982, 1986) presented a model based on shell development in some Archaeogastropoda in which the larval shell is formed in two phases. In the first phase, the shell consists of only elastic, organic matter, and in the second phase, the organic primary shell is calcified. The direction of calcification may not have been consistent with the direction of secretion of the organic shell. In the case of the ammonitella, the wall of the first whorl and umbilical walls of the initial chamber would have been calcified first. Only later would the remaining wall of the initial chamber and proseptum have been mineralized (Fig. 1C,D).

Bandel's (1982, 1986) interpretation has been confirmed by well-preserved ammonitellae of *Aconeceras* representing different calcification stages (Kulicki, 1989; Kulicki and Doguzhaeva, 1994). Four stages of calcification of the ammonitella have been recognized.

Stage 1 is represented by specimens in which the wall of the first whorl, including the primary constriction, and the lateral walls of the initial chamber are calcified.

Stage 2 is represented by specimens in which, in addition, the part of the wall of the initial chamber that separates the interior of the initial chamber from the lumen of the first whorl is calcified.

Stage 3 is represented by specimens that have a calcified first whorl, initial chamber, proseptum, and nacreous primary varix. In *Quenstedtoceras* ammonitellae from Łuków, there is another septum adapertural of the proseptum (the first nacroseptum).

Stage 4 is represented by ammonitellae of larger, postembryonic specimens. This stage is characterized by a distinct thickening from the inside of the inner prismatic layer and, commonly, by the addition of extra prismatic layers from the inside (Fig. 1F).

2.2. Structure of the Ventral and Lateral Walls of the Ammonitella

The wall of the first whorl of ammonitellae of *Aconeceras* that represent the first calcification stage ends on the side of the initial chamber at a so-called "calcification front" (Figs. 2B, 3). According to Kulicki and Doguzhaeva (1994), the calcification front is the zone where the appearance and growth of mineral components calcifying the primary organic shell are observed. During calcification, this zone moved from the apertural part of the shell toward the initial chamber and proseptum.

The calcified wall of the first whorl consists of three layers. The outer and inner layers are made of closely packed plates that, in turn, consist of closely adjoining aragonitic needles perpendicular to the outer and inner shell surfaces. The plates of the outer and inner layers are joined in the medial layer by aragonitic needles that are a continuation of those of the overlying and underlying layers (Figs. 4E, 5A,B). These needles are not so closely packed as they are in the outer and inner layers, and there seem to be gaps filled with secondary calcite.

According to the model of Kulicki and Doguzhaeva (1994), the outer and medial layers correspond to the shell at the first stage of calcification, whereas the inner layer is a later thickening of the shell from the inside (Fig. 3). In *Aconeceras*, the estimated thickness of the wall of the primary organic shell is 3–4 μm, similar to that in the cooccurring genus *Deshayesites*; in *Quenstedtoceras* this thickness is about 11 μm. Such differences in the thickness of the primary organic shell affect the structure of the calcified shell wall.

A cross section through the wall of the first whorl in *Quenstedtoceras*, some 30° adapical of the primary varix, is shown in Fig. 6A. The outer layer has a regular prismatic structure and a thickness of about 1 μm. Prisms consist of aragonitic needles 0.1–0.2 μm in diameter oriented perpendicular to the

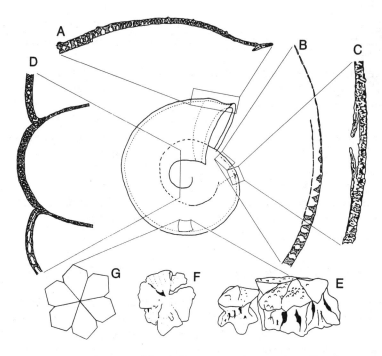

FIGURE 2. Diagrammatic drawing of the shell structure of an ammonitella of *Aconeceras*. (A) Longitudinal cross section through the apertural margin. (B) Longitudinal cross section through the calcification front at the beginning of the second calcification stage. (C) Longitudinal cross section through the wall of the initial chamber in the basal region of the future proseptum. Same stage as in B. (D) Transverse cross section through the umbilical wall of the initial chamber and first whorl. (E) Oblique view of aragonitic pseudohexagonal trilling with a single sector moved slightly aside. (F) Pseudohexagonal trilling from the medial layer of the delaminated shell wall. (G) View of idealized pseudohexagonal trilling of aragonite. (A–G after Kulicki and Doguzhaeva, 1994).

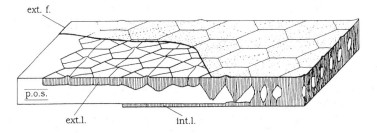

FIGURE 3. Diagram showing the calcification front of the originally organic shell wall of an ammonitella. Adapertural direction is toward the right. Abbreviations: ext.f., external organic film; ext.l., layer of external plates of the calcified shell wall of the ammonitella; int.l., layer of internal plates of the calcified shell wall of the ammonitella in the process of thickening from the inside; p.o.s., primary organic shell wall. (After Kulicki and Doguzhaeva, 1994.)

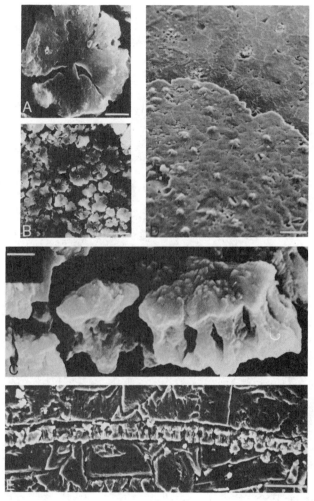

FIGURE 4. *Aconeceras* sp., Lower Cretaceous, Aptian, Symbirsk region, Russia. (A) View of pseudohexagonal trilling on the surface of the delaminated shell wall of the ammonitella. Treated in H_2O_2. Scale bar, 1 μm. (B) Numerous pseudohexagonal trillings on the surface of the delaminated shell wall of the ammonitella. Treated in H_2O_2. Scale bar, 5 μm. (C) Pseudohexagonal trilling in oblique view. Single sector is moved slightly aside. Empty interspaces are visible only in the medial part. External shell surface is on top. Treated in H_2O_2. Scale bar, 1 μm. (D) Ventral view of the external surface of the shell of the ammonitella with tubercles located in the central parts of the pseudohexagonal trillings. The surface of the internal mold is visible above, with marked boundaries between pseudohexagonal trillings. Scale bar, 5 μm. (E) Longitudinal cross section through the well-calcified shell wall of the ammonitella with aragonitic needles and a three-layered structure clearly visible. Treated in H_2O_2. Scale bar, 10 μm. (A–E after Kulicki and Doguzhaeva, 1994).

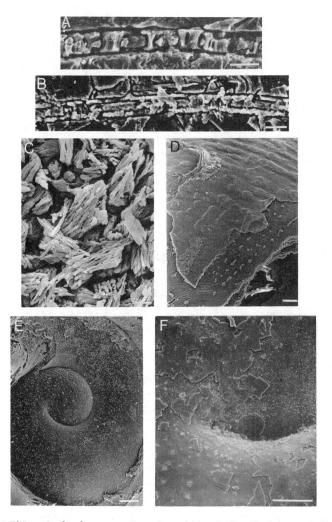

FIGURE 5. (A,B) Longitudinal cross sections through the shell wall of ammonitellae of *Acone-ceras* in various calcification stages. (A) Medial layer not fully developed, showing broad interspaces filled with secondary calcite. Treated in H_2O_2. Scale bar, 5 μm. (B) Medial layer showing only sporadic connections. Treated in H_2O_2. Scale bar, 5 μm. (C) Spherulitic–prismatic layer in the shell wall of Recent *Nautilus pompilius* L. Acicular crystallites of this layer are connected by numerous thin, trabecular elements. Treated in H_2O_2. Scale bar, 2 μm. (D) View of the internal surface of the dorsal wall in the third whorl of *Quenstedtoceras*. The wrinkle pattern is visible. Scale bar, 20 μm. (E) Tuberculate microornamentation on an ammonitella of *Quenstedtoceras*. Scale bar, 50 μm. (F) Tuberculate microornamentation on an ammonitella of *Quenstedtoceras* masked by a thin external covering, presumably diagenetic in origin. Scale bar, 15 μm. (A,B, after Kulicki and Doguzhaeva, 1994.)

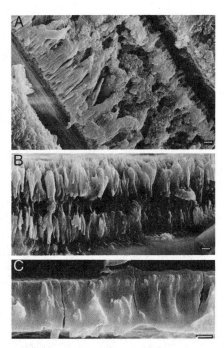

FIGURE 6. Longitudinal cross sections through the shell wall of an ammonitella of *Quenstedto-ceras*. Treated in H_2O_2. (A) Longitudinal cross section through the first whorl close to the primary varix. External surface is toward the right. Scale bar, 1 μm. (B) Longitudinal cross section through the initial chamber, close to the base of the proseptum. External surface is toward the top. Scale bar, 5 μm. (C) Longitudinal cross section through the apical part of the initial chamber. External surface is toward the top. Scale bar, 1 μm. (A–C after Kulicki and Doguzhaeva, 1994.)

outer surface. The inner prismatic layer is 2.5–3.0 μm thick. It also has a regular prismatic structure and needle-like crystallites oriented perpendicular to the outer surface. The diameter of the crystallites is 0.2 μm.

Between the outer and inner prismatic layers is a medial layer approximately 10 μm thick. In this layer, regular crystalline elements occur rarely and are oriented differently from those of the inner and outer layers. These elements are bigger and better developed in the vicinity of the inner prismatic layer. Elsewhere, the medial layer has an irregular, grainy structure and is a mixture of aragonite and organic matter or kerogen. According to Kulicki and Doguzhaeva (1994), the irregular, grainy structure of the medial layer results from the limiting effect of organic matter on the growth of aragonitic crystals. In transverse cross sections through the initial chamber and first whorl, the outer prismatic layer and the medial layer wedge out on the umbilical seam, and only the inner prismatic layer continues across the dorsum. This confirms earlier presentations by Bandel (1982, 1986) and Kulicki (1979).

In *Aconeceras*, the needle-like crystallites of the inner, outer, and medial layers in the wall of the first whorl of the ammonitella become aggregated into structural elements of a higher order (Kulicki and Doguzhaeva, 1994). In the outer and inner layers, there are closely adjoining plates. On the outer surface of the ammonitella, one can see plates forming aggregates of an even higher order, comparable to aragonitic pseudohexagonal trillings consisting of six triangular plates (Fig. 2E–G). The specimen in Fig. 4D shows a fragment of the outer surface of the first whorl with pseudohexagonal trillings; each of these has, in its central part, a tubercle of the exterior ornamentation of the shell. On the internal mold of the first whorl, shown in the same figure, the boundaries between pseudohexagonal trillings are distinct and indicate that what is shown on the outer surface is consistent with what is imprinted on the mold. The surface of the delaminated wall shows pseudohexagonal trillings in which the triangular elements are less closely packed (Fig. 4A–C). In the medial part there are distinct gaps, which were originally filled with organic matter but now are empty or filled with kerogen or secondary calcite.

In *Quenstedtoceras*, in which the medial layer is thicker than it is in *Aconeceras*, aragonitic needles are aggregated into larger prisms; however, aggregations into structures of a still higher order, such as the pseudohexagonal trillings observed in *Aconeceras*, do not occur, nor are there direct junctions between the outer and inner surfaces (Fig. 6A,B). Differences in the structure of the ventral and lateral walls of the first whorl of *Aconeceras* and *Quenstedtoceras* were interpreted by Kulicki and Doguzhaeva (1994) as reflecting variations in thickness and structure of the primary organic wall undergoing calcification.

2.3. Apertural Zone of the Ammonitella

The terminology used here to describe different structural and morphological elements is derived from Druschits *et al.* (1977) and Landman and Waage (1982). The morphological–structural distinctness of the apertural zone of the ammonitella was first recorded by Hyatt (1872), but the significance of this zone in terms of the ontogeny of ammonoids is not the subject of the present chapter.

In the first calcification stage of the ammonitella of *Aconeceras*, a primary constriction is already present in the apertural zone (Fig. 2A). The nacreous layer is not present at this stage. In the adapical part of the apertural zone, the layer of inner plates ends abruptly, forming a small step. The medial layer decreases in thickness in the ad apertural direction and wedges out in the middle of this zone. The layer of outer plates continues past the end of the medial layer to the ammonitella edge, where it forms a short return section directed toward the shell interior (Fig. 2A). A similar situation has been illustrated by Erben *et al.* (1969) and Kulicki (1974, 1979).

The nacreous primary varix is formed only in the third stage of calcification of the ammonitella. The lamellae of the nacreous layer form a characteristic arrangement, described for the first time by Kulicki (1974). The most external lamellae of the nacreous layer are relatively short and, on the posterior side, end on the inner surface beneath the outer prismatic layer. However, on the anterior side, they reach the return section of the outer prismatic layer. The inner lamellae reach further backward and are generally longer. The plates of the nacreous layer of the primary varix are arranged in vertical stacks, similar in structure to those of the nacreous layer of the postembryonic shell. The vertical stacks in the nacreous layer of the primary varix have been illustrated by Birkelund and Hansen (1974), Drushits *et al.* (1977), Druschits and Doguzhaeva (1981), and Ohtsuka (1986).

Kulicki (1974, 1979) indicated that the character of the outer prismatic layer of the ammonitella and that of the outer prismatic layer of the postembryonic shell are identical. The breadth of the thinning zone of the prismatic and subprismatic layers of the ammonitella should correspond to the breadth of the secretory zone. During the formation of the ammonitella, this zone already was divided into two subzones. The nacreous layer of the primary varix formed after a change in secretion of the posterior subzone during the withdrawal of the mantle edge. This interpretation explains the characteristic arrangement of the lamellae in the primary varix and the formation of the return section of the outer prismatic layer.

Consistent with the interpretation of Kulicki and Doguzhaeva (1994), the breadth of the thinning zone of the two outer layers of the ammonitella corresponds to the breadth of the secretory zone of the primary organic shell. The short return section of the outer prismatic layer in specimens without any trace of a nacreous layer points to the presence of a periostracal groove. During calcification of the primary organic shell, the primary constriction had already formed. The secretory tissue at the apertural edge stopped its secretory activity but remained closely connected to the shell. A similar suggestion has been made by Bandel (1982). Beyond the zone of close contact with the secretory tissue, the epithelium was free, as evidenced by the step formed in the proximal part of the thinning apertural edge.

2.4. Structure of the Wall of the Initial Chamber and Proseptum

In ammonitellae of *Aconeceras*, the umbilical walls of the initial chamber are calcified at the same stage as the ventral and lateral walls of the first whorl. In all known ammonitellae, the umbilical walls of the initial chamber are relatively thick, and their thickness is comparable to that of the wall of the first whorl. With respect to their structure, they also resemble the wall of the first whorl; i.e., they have the same three layers that characterize the wall of the first whorl (Fig. 2D).

Transverse cross sections through the first whorl of ammonitellae of *Aconeceras* (see Kulicki and Doguzhaeva, 1994) and *Quenstedtoceras* (see Kulicki, 1979; Bandel, 1982, 1986) show that the most external components of the umbilical wall of the initial chamber and of the lateral wall of the first whorl wedge out at the umbilical seam. Only the internal regular prismatic layer of the umbilical wall goes under the umbilical seam as the primary wall of the initial chamber (Fig. 2D). In the apical part of the initial chamber of the ammonitella of *Quenstedtoceras*, the 3.0- to 3.5-μm wall made of organic matter is penetrated through its entire thickness by needle-like aragonitic crystallites (Fig. 6C). In longitudinal cross section, this layer appears to reach the end of the flange.

In medial and paramedial cross sections through the ventral wall of the initial chamber, one or two prismatic layers are visible under the wall proper of the initial chamber (Fig. 6B; Kulicki, 1979). Kulicki (1979) and Kulicki and Doguzhaeva (1994) claimed that the wall proper of the initial chamber continues as the outer prismatic layer of the first whorl. Birkelund (1967, 1981), Birkelund and Hansen (1968), Erben *et al.* (1969), Tanabe *et al.* (1980, 1993b), and Tanabe and Ohtsuka (1985) concluded that the wall of the initial chamber wedges out in the vicinity of the base of the proseptum. According to Bandel (1982, 1990), in median section, the inner prismatic layer of the initial chamber extends from the base of the proseptum and continues as the primary wall of the initial chamber, ending in the flange. These observations of Bandel are contradictory to those of Kulicki (1979) conducted on the same material, namely, *Quenstedtoceras* from Łuków (see Fig. 1).

Opinions concerning the relationship between the proseptum and the shell wall are not consistent either. Some authors, such as Grandjean (1910), Miller and Unklesbay (1943), Arkell (1957), Birkelund and Hansen (1968), Erben *et al.* (1969), Kulicki (1975, 1979), Bandel (1982), and Landman and Bandel (1985) have considered the proseptum to be continuous with the internal layers of the shell wall. But Hyatt (1872), House (1965), Erben (1962, 1966), Druschits and Khiami (1970), and Druschits and Doguzhaeva (1981) have suggested that the relationship of the proseptum to the shell wall is the same as that of all other septa.

Studies on the incompletely calcified ammonitellae of *Aconeceras* (see Kulicki and Doguzhaeva, 1994) have shown that the proseptum is not yet calcified, even though the adjacent shell wall is partially calcified and even thickened from the inside. During further calcification, the organic wall of the proseptum, including its broad base on the ventral side, was calcified, and the shell wall became thicker on the inside. Thus, ventrally and laterally, the inner prismatic layer covered the proseptal surface (Fig. 2C). In medial cross section, a similar situation can be observed (Erben *et al.*, 1969; Kulicki, 1975, 1979). In the dorsal region, the proseptum has a very long mural part that covers a large area of the wall proper of the initial chamber. Bandel (1982, 1990) interpreted the mural part of the proseptum as continuing to the aperture of the ammonitella; this is consistent with the occurrence of a dorsal wall in the ammonitella. As shown in paramedial cross sections distant from the plane of bilateral symmetry, the ventral part of the proseptum continues as the inner prismatic layer of the first whorl (Birkelund and Hansen, 1974, Fig. 2; Kulicki, 1979, Fig. 10C).

Most authors consider the proseptum to be a one- or two-layered structure made of aragonitic prisms oriented perpendicular to the outer surface. Only

Landman and Bandel (1985, Fig. 33) illustrated the three-layered structure of the proseptum in *Euhoplites* sp., in which the irregularly prismatic proseptal layer is sandwiched between two layers of a more homogeneous prismatic material that originally may have been organic. These authors (1985, Fig. 21) also noted distinct wrinkles on the surface of the proseptum of *Baculites* sp.

2.5. Dorsal Wall of the Ammonitella

Birkelund (1967, 1981), Birkelund and Hansen (1968, 1974), and Erben *et al.* (1969) have denied the existence of a dorsal wall in the ammonitella. However, Kulicki (1979) has shown that the outermost layer overlying the wall proper of the initial chamber as seen in medial and paramedial cross sections is the dorsal wall of the first whorl (Fig. 7A). Its structure is subprismatic, and it is not separated from the wall proper of the initial chamber by a sharp boundary. In the apertural region of the ammonitella, the dorsal wall is reduced in thickness, and, with the appearance of tuberculate sculpture, a distinct boundary appears between the dorsal wall and the ventral wall of the previous whorl.

In the third calcification stage in the ammonitella of *Quenstedtoceras* there are two septa and a nacreous primary varix (Fig. 7F). Kulicki and Doguzhaeva (1994) have shown that in this developmental stage the outer surface of the initial chamber in the apertural region is covered by distinct structures resembling those of the wrinkle layer (Fig. 7C,D). Inside the living chamber these structures reach about 90° from the aperture. In longitudinal cross section, these structures are triangular, with the gentle slope pointing adapically and the much steeper slope facing adaperturally. These triangular elements have a prismatic structure with the long axes of the prisms perpendicular to the overall surface of the wall of the ammonitella rather than to the surfaces of the gentle adapical slopes of the triangles.

Kulicki (1979) and Doguzhaeva and Mutvei (1986b) recognized the complex structure of the dorsal wall. The first author distinguished two components, an outer and an inner. The wrinkle-like layer of the ammonitella is the outer component. It was produced by the anterior part of the mantle and, later in ontogeny, was covered by more internal components that were produced by the posterior part of the body.

Generally, the ammonitellae available for study are in the fourth calcification stage; i.e., they are the apical parts of large specimens. Thus, the dorsal wall at what had been the edge of the ammonitella has the thickness and structure characteristic of the posterior part of the living chamber (Fig. 7A). On the basis of such specimens, Kulicki (1979) stated that the dorsal wall of the ammonitella covers the outer surface of the wall proper of the initial chamber, that the dorsal wall has a subprismatic structure, and that the boundary between these two layers is not too distinct. In the dorsal wall of a specimen illustrated by Kulicki (1979, Pl. 44, Fig. 3) and shown here (Fig. 7A), diagonal wrinkle-like elements are visible. (These were not noticed at the time by the author.) Figure 7 also shows that there is a distinct boundary between the dorsal and ventral walls of the two whorls; this is a result of the appearance of the characteristic tuberculate microornamentation on the surface of the ammonitella (Fig. 7B). Kulicki and Doguzhaeva (1994) believed that this distinct boundary between the above-mentioned walls was caused by hatching, which also resulted in a decrease in the thickness of the dorsal wall. These differences in the dorsal wall between the embryonic and postembryonic stages may explain why so many authors have incorrectly presented the wedging out of the wall proper of the initial chamber.

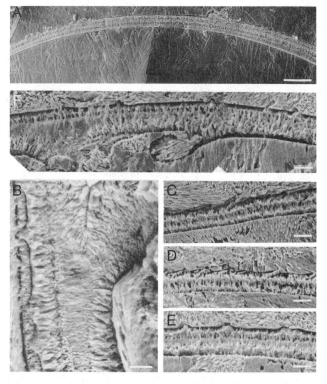

FIGURE 7. Longitudinal cross sections through the shell wall of the ammonitella of *Quenstedtoceras* sp., Middle Jurassic, Callovian, Łuków, Poland. (A) Medial cross section through the ventral part of the initial chamber opposite the primary varix. Oblique elements of the wrinkle layer are visible in the external, relatively thick layer. Treated in EDTA. Scale bar, 20 μm. (B) Medial cross section through the base of the proseptum. Note that tubercles in the outer layer are formed of simple extensions of aragonitic needles. The layer occurring between the tuberculate elements represents the dorsal wall of the next whorl. Treated in EDTA. Scale bar, 3 μm. (C–F) Medial cross sections through the ventral wall of the initial chamber of an ammonitella of *Quenstedtoceras* in the third calcification stage. Specimen treated in HCl. (C,D) Wrinkle layer of the ammonitella shell wall. Scale bar, 5 μm. (E) Medial cross section close to the proseptum. Scale bar, 5 μm. (F) Medial cross section through the base of the proseptum and second septum. Scale bar, 5 μm. (A,B after Kulicki, 1979; C–F after Kulicki and Doguzhaeva, 1994).

2.6. Ornamentation of the Ammonitella

The presence of a tuberculate microornamentation in Mesozoic ammonitellae has been documented in the past (Brown, 1892; J. P. Smith, 1901; W. D. Smith, 1905). More detailed investigations were possible with the use of SEM (Kulicki, 1974, 1979; Bandel, 1982; Bandel *et al.*, 1982; Landman, 1985, 1987, 1988; Tanabe, 1989).

In general, tuberculate microornamentation is characteristic of Jurassic and Cretaceous ammonitellae. However, there are differences in the density of tubercles, in their size, and in their distribution on the surface of the shell. Bandel (1982) showed that tuberculate microornamentation does not occur on the wall dividing the interior of the initial chamber from the lumen of the first whorl. In addition, Tanabe (1989) showed that tubercles do not occur on some exposed surfaces of the ammonitella. For example, tubercles are connected into a continuous layer on the umbilical walls of the initial chamber of *Anapachydiscus*. In *Phyllopachyceras* they occur only on the lateral wall of the first whorl in the vicinity of the umbilical seam.

Bandel *et al.* (1982, p. 387) found that "some smaller tubercles appear to be emergent ends of single large prisms, but the larger ones show the complex spherulitic structure." On the basis of cross sections through the walls of ammonitellae of *Quenstedtoceras*, Kulicki (1979) determined that tubercles are a continuation of the prisms of the outer prismatic layer and that they are not separated from the outer shell surface by any discontinuity (Fig. 7B,E,F). It also has been mentioned that larger tubercles have a spherulitic character (Bandel *et al.*, 1982), but this is not obvious in the specimen shown in Fig. 7B. Here, the tubercles consist of needles approximately 0.2 μm in diameter that parallel one another and are perpendicular to the outer shell surface. This is similar to the situation in the outer prismatic layer.

Tanabe (1989, Fig. 5) presented the tubercles as separate elements on the surface of the outer prismatic layer or, as in the umbilical wall of *Anapachydiscus*, sometimes fused into a continuous layer, forming a smooth outer surface. On this basis, Tanabe formulated an endocochliate embryo model in which the tuberculate microornamentation, or the layer corresponding to it, was formed by a fold of the mantle covering the outside of the ammonitella.

In *Aconeceras trautscholdi*, the tubercles on the ammonitella surface invariably are connected with the central parts of the pseudohexagonal trillings (Kulicki and Doguzhaeva, 1994). Figure 5E,F illustrates the surface of an ammonitella of *Quenstedtoceras*. Tubercles are visible between the wall of the initial chamber and that of the first whorl. In Figure 5F, there are patches of a thin layer that mask the tuberculate microornamentation. The origin of this layer is not clear, but it is most probably of diagenetic origin. It is possible that Tanabe (1989, Figs. 1A,2C) observed such a layer on the umbilical walls of *Anapachydiscus* and *Phyllopachyceras ezoense*. In any case, the diversity of interpretations about the structure of the tuberculate microornamentation by different authors leaves the subject open for further investigation.

3. Postembryonic Stage

3.1. Products of the Anterior Mantle Edge

Ammonoids resemble Recent *Nautilus* and other Recent molluscs, e.g., bivalves, gastropods, and monoplacophorans, both in the products of secretion and in the morphology of the apertural edge of the shell. This similarity makes it possible to argue that these groups are almost identical with respect to the processes of biocalcification taking place at the mantle edge.

Biocalcification in *Nautilus* has been described by Crick and Mann (1987) and Grégoire (1987) and is briefly summarized below. In *Nautilus* the epithelium of the mantle edge is typically folded. The outer fold is separated from the rest of the mantle by a periostracal groove, where the outermost layer of the molluscan shell, the periostracum, is secreted. The inner wall of the periostracal groove adheres strongly to the outer surface of the periostracum, thus protecting the space filled with extrapallial fluid from the external environment. In this way, all the products of secretion participating in the construction of the outer shell pass through the underlying epithelium before being exuded into the extrapallial fluid. The periostracum insures the isolation of the secretory environment and is, at the same time, a substrate for the nucleation of mineral components. The first zone lying immediately behind the periostracal groove is where the outer prismatic layer is produced, and the deposition of the nacreous layer is associated with another zone, located farther away. Production of $CaCO_3$ is controlled by the disequilibrium of the extrapallial fluid as the CO_2 content varies. The organic matrix accounts for the crystal orientation. The deposition of the inner prismatic layer is confined to the zone of the myoadhesive epithelium (Mutvei, 1964).

The anterior mantle edge is responsible for the production of sculpture in ammonoids. Some sculptural elements, e.g., ribs and nodes, are formed from both the outer prismatic and nacreous layers, so that convex shapes on the outer surface correspond to concave ones on the inner surface. Both layers generally retain the same thickness throughout the extent of the rib or node. In the posterior part of the body chamber and in the phragmocone, where the inner prismatic layer is present, a certain smoothing out of the interior surface of the sculpture may occur.

Occasionally, as in *Hypophylloceras*, sculptural elements are composed mostly of the outer prismatic layer so that the inner surface of the shell wall remains flat (Birkelund and Hansen, 1974, 1975; Birkelund, 1981). Protruding sculptural elements—spines or a sharp keel—may be totally cut off from the cavity of the body chamber by the inner prismatic layer; as a result, the space within such elements remains free (Hölder, 1952a,b; Erben, 1972b). Numerous Paleozoic and Mesozoic ammonoids feature regularly spaced varices, which are thickenings of the nacreous layer. Such varices appear as constrictions on inner molds. Constrictions as outer sculptural elements may occur by themselves or be accompanied by varices. Mature growth stages of micro- and macroconchs may show modifications connected with thickenings of the apertural edge; these thickenings are chiefly composed of the nacreous layer.

3.1.1. Periostracum

This is the most external layer of the shell wall. It is organic and was produced in the periostracal groove of the mantle. In the ontogeny of ammonoids, the periostracal groove would have already been formed in the final stage of embryonic development, as indicated by the return section of the outer prismatic layer and lack of the nacreous primary varix at the apertural edge of the ammonitella (Kulicki and Doguzhaeva, 1994).

The periostracum in extant molluscs may have a complex, multilayered structure as in, e.g., *Mytilus edulis* (see Dunachie, 1963) or be double-layered as in *Nautilus macromphalus* (see Ward, 1987). The periostracum of ammonoids has not been investigated yet.

3.1.2. Outer Prismatic Layer

The mantle zone directly behind the periostracal groove secretes the outer prismatic layer directly on the inner surface of the periostracum. In all normally coiled ammonoids (Erben *et al.*, 1969; Kulicki, 1979), this layer occurs in the ventral, lateral, and umbilical portions of the wall but wedges out at the umbilical seam. Earlier reports of the total lack of an outer prismatic layer in some normally coiled ammonoids (for example, Bøggild, 1930; Mutvei, 1967; Birkelund and Hansen, 1974) are based on incompletely preserved material (Birkelund and Hansen, 1975; Erben *et al.*, 1969). In heteromorphs in which a portion of the shell has no contact with the shell of earlier developmental stages, the outer prismatic layer occurs not only ventrally and laterally but also in the dorsal part (Doguzhaeva and Mikhailova, 1982; Doguzhaeva and Mutvei, 1989).

In early postembryonic stages, the outer prismatic layer comprises a considerable percentage of the total thickness of the shell wall: over 75% of the total shell wall thickness in *Aconeceras*, and about 50% in *Quenstedtoceras*. During later development, the thickness of the outer prismatic layer decreases considerably to only a small fraction of the total shell wall thickness. Along the second shaft in *Ptychoceras*, the outer prismatic layer wedges out completely, and the nacreous layer becomes the outermost layer in the wall (Doguzhaeva and Mutvei, 1989). Despite many years of investigation, I have not found any outer prismatic layer in Early Cretaceous *Bochianites*, although this layer is well preserved in cooccurring normally coiled ammonoids.

From the microstructural point of view, the outer prismatic layer consists of regular prisms, each with needle-like crystallites 0.2–0.5 μm in diameter; these are perpendicular to the outer shell surface. This regularity of the prisms in the outer prismatic layer of ammonoids distinguishes this layer from the outer layer of present-day *Nautilus*, which consists of two sublayers (Mutvei,

1964). The outer sublayer in *Nautilus* has a grainy structure; each grain is made of crystallites that are mutually parallel in arrangement or slightly spherulitic (Fig. 5C).

3.1.3. Nacreous Layer

The nacreous layer of ammonoids, as in *Nautilus*, was produced in two secretory zones, that of the shell wall and that of the septa. The first of these has a belt-like shape, encircling the inside of the aperture with the exception of the dorsal part in normally coiled ammonoids. The second zone of nacreous secretion is a large area at the posterior part of the body. Although the secretory product is the same in both zones, the relationship of the surface of secretion to the elements of the nacreous layer (lamellae) varies. Generally, the growth direction in the zone of septal secretion is perpendicular to the surface of the septum and to the interlamellar membranes. On the other hand, in the shell wall, the interlamellar membranes occur in very long sections parallel to the inner surface of the outer prismatic layer, and the growth zone cuts them diagonally.

The basic elements of the nacreous layer are aragonitic hexagonal plates, which are arranged in layers separated by interlamellar membranes. These plates are placed one on top of another, forming vertical stacks. Such a columnar arrangement of nacre is characteristic of ectocochliate cephalopods and of gastropods. The growth surfaces of columnar nacre described by Wise (1970) and Erben (1972a) show that the nuclei of newly formed plates are always deposited in the central part of subordinate plates and afterwards accreted. The lateral walls of neighboring hexagonal plates are separated from each other by intercrystalline membranes of conchiolin. According to Mutvei (1980), the hexagonal tablets of stack nacre are composed of a variable number of crystalline sectors, which are separated from each other by vertical radial organic membranes. These sectors represent contact and interpenetrant twins. The central portion of each tablet in the stack is occupied and connected vertically by a central organic accumulation (Mutvei, 1980, 1983); these accumulations commonly occur in ammonoid septa (Fig. 8A,B).

Ultrastructural studies of the conchiolin of the nacreous layer of molluscs have been conducted by many authors, but only Grégoire (1966, 1980) included ammonoids in his investigations. Experimental studies on the conchiolin of the nacreous layer of molluscs show that it is thermoresistant. Even after having been heated to 900°C for 5 hr, it is still biuret-positive (Grégoire, 1966, 1968, 1972, 1980; Voss-Foucart and Grégoire, 1971). In diagenesis, the chemical composition of molluscan conchiolin changes rapidly but afterwards can remain stable and unchanged over millions of years, even under metamorphic conditions (Voss-Foucart and Grégoire, 1971; Grégoire, 1980; Weiner *et al.*, 1979). The mineral component of the hexagonal plates of nacre has been examined since the 19th century, but even present-day authors offer differing opinions on its structure.

FIGURE 8. Longitudinal cross section through the nacreous layer of Middle Jurassic *Quenstedto-ceras*. Treated in H_2O_2. (A) Longitudinal cross section through the shell wall of the living chamber of an adult macroconch of *Quenstedtoceras*. Note the numerous central organic accumulations. Scale bar, 10 μm. (B) The same view as in A, but enlarged. The complex structure of the central organic accumulations is visible. Scale bar, 1 μm. (C) Longitudinal cross section through the nacreous layer of the second whorl. The interlamellar conchiolin membranes were removed during preparation. The crystalline elements are similar to the angular plates of nacre in *Mytilus*. Scale bar, 1 μm. (D) Longitudinal cross section through the inner prismatic layer in the ventral wall of the second whorl of *Quenstedtoceras*. Scale bar, 2 μm.

In addition to being present in ectocochliate cephalopods, stack nacre is characteristic of gastropods, monoplacophorans, and primitive pelecypods such as *Nucula* and *Trigonacea* (see Erben *et al.*, 1968). Orientation of axes *a* and *b*, respectively, in different plates is parallel in cephalopods (Grégoire, 1962; Wise, 1970) but random in gastropods (Wise, 1970).

On the basis of his work on gastropods, Erben (1974) concluded that hexagonal plates of nacre represent compact crystals rather than polycrystal-line aggregates, as suggested by Mutvei (1969, 1970, 1979). From study of chromium-sulfate-treated preparations, Erben (1974) observed, apart from interlamellar and intercrystalline membranes, intracrystalline organic elements such as diagonal sheets or vertical pillars. Mutvei (1969, 1970, 1979), on the other hand, interpreted these elements as acicular crystallites.

Chromium sulfate preparations of the nacreous layer of *Quenstedtoceras* show structural characters such as those seen in *Nautilus* (see Mutvei, 1972a) and *Haliotis* (see Erben, 1974, Pl. 4, Figs. 1–5), the only difference being that in *Quenstedtoceras* the crystalline elements between interlamellar membranes are perpendicular to those membranes. The same *Quenstedtoceras* specimens prepared in H_2O_2 (Fig. 8C; method used by Kulicki and Doguzhaeva, 1994) display a great similarity in their nacre to that of present-day *Mytilus edulis* treated with sodium hypochlorite solution and etched with 25% glutaraldehyde solution for 2 days (Mutvei, 1979). The crystalline elements in these specimens of *Quenstedtoceras* correspond in size and shape to the angular plates reported by Mutvei (1979). They are in layers corresponding to the lamellae of nacreous plates. Organic elements have been totally removed during the preparation.

The thickness of the lamellae of the nacreous layer of ammonoids is constant in different parts of the shell. They are generally 0.25 μm thick, but they can be several times thicker in the transition zone to the outer prismatic layer.

Doguzhaeva and Mutvei (1989) described pores in the nacreous layer of the heteromorph *Ptychoceras*. They consider these to be an inherent characteristic of these ammonoids. In the nacreous layer of the Early Cretaceous heteromorph *Bochianites,* similar pores also are observed, but it is difficult to eliminate the possibility that they were made by boring organisms in spite of their "rowlike distribution" (Doguzhaeva and Mutvei, 1989).

3.2. Inner Prismatic Layer

This layer is situated on the inner surface of the nacreous layer at some distance from the apertural edge of the shell. In specimens in the final stages of ontogeny, the inner prismatic layer comes closer to the apertural zone (Doguzhaeva and Mutvei, 1986b). This layer results from secretory activity of the myoadhesive epithelium (Mutvei, 1964). Although the literature provides various data on the occurrence of the inner prismatic layer in the ontogeny of different ammonoids, all ammonoids that have been examined by the present author had it developed starting as far adapically as the second or third septum.

In transverse cross sections of normally coiled ammonoids, it is the only shell layer that extends across the venter and dorsum; this is in contrast to the outer prismatic and nacreous layers, which wedge out at the umbilical seam and do not extend across the dorsum. The thickness of this layer varies; it is generally thickest near the umbilical shoulders, i.e., in the place where the majority of authors have reconstructed the attachments of the retractor muscles. Makowski (1962, Fig. 11) described extremely well-developed muscle attachments in adult (gerontic) specimens of *Quenstedtoceras*, as shown by a substantial thickening of the inner prismatic layer causing a narrowing of the inside diameter of the whorl in the posterior part of the body chamber. An especially strong development of the inner prismatic layer has also been observed in some Phylloceratina; on the dorsum, this layer is more than twice as thick as the whole ventral wall of the previous whorl (Birkelund and Hansen, 1974, 1975; Birkelund, 1981).

There is not always a close contact between the inner prismatic layer and the layers it covers or lines. The general tendency is to smooth out all unevenness or hollows, e.g., as in the floored hollow spines of *Kosmoceras* (see Erben, 1972b), or to smooth out the fine ribbing on the exterior of the previous whorl (Birkelund and Hansen, 1974, 1975; Howarth, 1975).

With respect to microstructure, this layer generally consists of regular prisms with visible needle-like crystallites (Fig. 8D). On the dorsal side, a spherulitic–prismatic arrangement is sometimes observed. In Triassic *Phyllocladiscites*, Doguzhaeva and Mutvei (1986b) have described well-developed pores in the inner prismatic layer, filled with crystalline matrix. These pores are not visible in cross sections of the wall but can be seen only on the surface.

3.3. Modifications of the Shell Wall

3.3.1. Inner Shell Wall of the Dactylioceratidae

Howarth (1975) described the so-called inner shell wall in the Dactylioceratidae, a modification unknown in other groups of ammonoids. The main shell wall in representatives of the Dactylioceratidae consists of layers with a normal sequence: outer prismatic layer, nacreous layer, and inner prismatic layer. The inner shell wall consists of an outer prismatic layer and an inner nacreous layer. On the inside, the inner shell wall is covered by a septal prismatic layer.

The appearance of the inner shell wall coincides with the appearance of ornamentation. It is not closely attached to the inner surface of the main shell wall but smoothes out this wall from the inside, leaving empty spaces between it and the concave surfaces of ribs. The outer prismatic layer of the inner shell wall is not continuous—it is missing in some places. In ontogenetic development, the nacreous layer of the inner shell wall first appears as an insert in the inner prismatic layer of the main shell wall. With respect to its microstructure, the nacreous layer of the inner shell wall does not differ from the nacreous layer of the main shell wall (Howarth, 1975). Although the inner prismatic layer is not split by a nacreous layer in other groups of ammonoids, there is an analogy to the splitting of the inner prismatic layer by the conspicuous organic membranes in *Nautilus* (see Doguzhaeva and Mutvei, 1986b).

3.3.2. Dorsal Wall

In heteromorphs, where the shell walls of adjacent "whorls" do not join, the dorsal wall of the "free whorls" usually has the same structure as do other shell parts. In normally coiled ammonoids, on the other hand, the microstructure of the dorsal wall changes, and the dorsal wall is a wall covering the previous whorl and laterally limited by the umbilical seams.

According to Kulicki (1979), the dorsal wall of ammonoids generally consists of two components: the outer wrinkle layer and the inner prismatic layer. The latter may have a complex microstructure, as discussed in the previous section.

The term *"Runzelschicht"* (wrinkle layer) was introduced by Sandberger and Sandberger (1850) to denote a thin layer superimposed on the test of Devonian goniatites. This layer had been recognized earlier by Keyserling (1846). The descriptions of these authors leave no doubt that the layer described is connected with the dorsal wall. The Sandbergers also used another expression, *"Ritzstreifen,"* to denote the markings preserved on internal molds of the inside of the lateral and ventral parts of the ammonoid whorl. Subsequent authors have used the term *Runzelschicht* for both categories of elements. The problem of the wrinkle layer in Paleozoic goniatites was discussed by House (1971), who insisted that the terms *"Runzelschicht"* and *"Ritzstreifen"* denote the same structure and suggested that this structure be referred to as the dorsal or ventral wrinkle layer, respectively. According to House (1971), all Paleozoic ammonoids possessing such structures are smooth and ribless: the Anarcestaceae and Pharciceratceae display both kinds of wrinkle layers, the Cheiloceratceae only the dorsal one, and the Clymeniina has just a narrow belt on the dorsal side. Tozer (1972) verified and precisely defined the meaning of these expressions.

The wrinkle layer occurs on the dorsal side of normally coiled ammonoids in the vicinity of the aperture; in some instances, it has been observed beyond the aperture (Senior, 1971; Tozer, 1972). The secretory zone of this layer is connected with the mantle edge, and, therefore, it is comparable to the black film of *Nautilus* with which it shares a similar microornamentation. *Ritzstreifen*, as defined by the Sandbergers, are situated at the back of the body chamber, and, therefore, the zone of their formation lies behind the secretory zone of the nacreous layer.

Kulicki (1979) described the development of the wrinkle layer in the body chamber of *Quenstedtoceras* (Figs. 5D, 9). A similar wrinkle layer and developmental sequence has been presented in Triassic *Proarcestes* by Doguzhaeva and Mutvei (1986b). Walliser (1970) described the variability in ornamentation and occurrence of the *Runzelschicht* and *Ritzstreifen* without differentiating these two categories.

The main elements of the wrinkle layer of *Quenstedtoceras* are triangular in longitudinal cross section and are composed of prisms perpendicular to the surfaces of the steep adapertural sides of the triangles. Each triangle is filled with smaller prisms, arranged more chaotically. In addition to the prisms, the triangular elements contain abundant organic matter. Nassichuk (1967) assumed that the wrinkle layer of *Clistoceras* may have been organic. Zakharov and Grabovskaya (1984), however, assigned the wrinkle layer in *Zelandites japonicus* to the nacreous layer, a fact that is not clearly shown in their figure

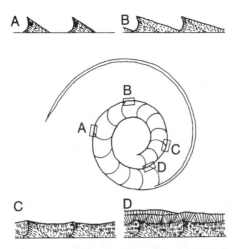

FIGURE 9. Diagrammatic sketch showing the development of the dorsal wall in the living chamber in *Quenstedtoceras*. (A) Vicinity of apertural margin. (B,C) Inside the living chamber. (D) Close to the last septum. (A–D after Kulicki, 1979, Fig. 9.)

(Zakharov and Grabovskaya, 1984, Fig. 3). Bayer (1974) described the wrinkle layer of Mesozoic ammonoids as occurring only on the dorsal side. He argued that the wrinkle layer did not constitute a separate layer but was merely part of the inner prismatic layer.

3.3.3. Umbilical Plugs and Encrusting Layers

Umbilical plugs comparable to those in *Nautilus pompilius* are observed in some Paleozoic and Triassic ammonoids, especially in strongly involute forms such as *Clistoceras* (see Nassichuk, 1967) and *Nathorstites* (see Tozer, 1972).

In present-day *Nautilus pompilius*, the surface of the umbilicus is partly covered by black film on which lies the material of the umbilical plug. This umbilical callus consists of spherulitic prismatic deposits and elements of the nacreous layer (Grégoire, 1966; unpublished observations). In shells of adult individuals the umbilical plug does not fill the whole umbilical space but only the outer part.

Nassichuk (1967) called the formations described in *Clistoceras* as "helicolateral deposits" but did not confirm their homology to umbilical plugs in *Nathorstites* and *Nautilus pompilius*. Tozer (1972), in contrast, confirmed the close homology between the helicolateral deposits of *Clistoceras* and the umbilical plugs of *Nathorstites* and *N. pompilius*. There is no detailed description of the structure of the abovementioned deposits apart from the statement by Tozer (1972) that, in the formation of the umbilical callus in *Nathorstites*, a secondary nacreous material is included. Bogoslovsky (1969) illustrated specimens of *Prolobites* in which the umbilicus is covered by a calcium carbonate wall.

Druschits *et al.* (1978) described multilayered structures encrusting the umbilical surface in the Late Cretaceous genus *Gaudryceras*. These do not cover the entire shell surface in that some spaces are left out, especially at the umbilical seams. Starting from the third whorl, the encrusting layer separates the ventral wall of the third whorl from the inner prismatic layer of the fourth whorl. Birkelund (1981) has confirmed the occurrence of such encrusting layers in specimens of *Gaudryceras* from Japan. The thickness of the encrusting layers may be several times that of the shell wall itself. Both Druschits *et al.* (1978) and Birkelund (1981) have observed the occurrence of nacreous structures in these layers. I have had an opportunity to examine specimens of *Gaudryceras* from Russia and have observed that the encrusting layers are enriched in organic matter.

3.4. Septa

According to the commonly accepted view, the first two septa have been called prosepta (but see Chapter 11). They were supposed to differ from other septa in their relationship to the wall of the outer shell (Grandjean, 1910; Böhmers, 1936; Voorthuysen, 1940; Miller and Unklesbay, 1943; Miller *et al.*, 1957; Arkell, 1957; Druschits and Doguzhaeva, 1974; Tanabe *et al.*, 1980). Miller and Unklesbay (1943) pointed out that prosepta are adaperturally concave in medial cross section, whereas all remaining septa are convex toward the aperture.

On the basis of SEM observations, Erben *et al.* (1969) confirmed the distinct microstructural character of prosepta in relation to all other septa. They also observed that the second septum (primary septum, "*Primärseptum*"), like all subsequent septa, is separated from the shell wall by a conchiolin layer.

A nacreous layer in the second septum has been observed by Birkelund and Hansen (1974), Kulicki (1979), Bandel (1982), Landman and Bandel (1985), and Landman (1987). Kulicki (1979), on the basis of variously oriented cross sections, observed that in planes far from the midline, the nacreous layer in the second septum is much thicker than the prismatic layer, whereas in medial cross sections, the reverse is true. Erben *et al.* (1969), using strictly medial cross sections, may not have recorded the occurrence of a nacreous layer in the building of the second septum, although they examined the same genera as did Kulicki (1979).

The main component in the construction of septa, with the exception of the proseptum, is the nacreous layer. Sporadically, the adapertural surface and, less commonly, the adapical one may be covered by a prismatic layer and, more rarely, a spherulitic prismatic one (Birkelund and Hansen, 1974; Howarth, 1975; Kulicki, 1979; Kulicki and Mutvei, 1982). The connection between the septum and septal neck on the adapertural surface in *Quenstedtoceras* illustrates the prismatic layer passing collaterally and continuously into the lamellae of the nacreous layer (Kulicki and Mutvei, 1982, Text Fig. 3, Pl. 5, Fig. 1).

In the connections of the septa to the shell wall, a distinct border separating the two is visible. Lamellae of the nacreous layer bend strongly adaperturally and pass continuously into the prismatic layer, the so-called septal prismatic layer of Howarth (1975), which may be developed to different degrees.

3.5. Septal Neck–Siphuncular Complex

The expression "siphuncular complex," introduced by Tanabe *et al.* (1993a), includes the following elements: cecum and prosiphon, siphuncular tube (= connecting ring), siphuncular membranes, auxiliary deposit (= auxiliary anterior deposit), cuff (= auxiliary posterior deposit), and angular deposit.

3.5.1. Cecum and Siphuncular Tube

The cecum is the spherical ending of the siphuncular tube in the initial chamber. Originally composed of organic matter, conchiolin, the cecum generally is phosphatized in the fossil state. The problem of preservation of the cecum applies equally well to the other organic elements of the siphuncular complex, i.e., the siphuncular tube and the siphuncular membranes.

The main mineral in these elements is francolite (Andalib, 1972; Hewitt and Westermann, 1983). The siphuncular tube is very rarely calcified (Joly, 1976). Nautiloids of the same age have nonphosphatized siphuncular tubes (Fukuda in Obata *et al.*, 1980; unpublished observations). Hewitt and Westermann (1983) presented a tentative interpretation of the ammonoid siphuncular tube as having been originally chitinous, but stiffened internally with phosphate crystals.

In its anterior part, the cecum in Mesozoic forms, e.g., *Kosmoceras* and *Quenstedtoceras*, is connected directly or by a calcareous deposit to the achoanitic proseptum (Kulicki, 1979). In *Tornoceras* the relationship between the proseptum and the cecum was illustrated by House (1965, Fig. 2) and also by Bogoslovsky (1971, Fig. 6), and, although these two authors interpreted the described elements quite differently, they both indicated a retrochoanitic septal neck in the proseptum. Thus, it can be expected that the relationship between the conchiolin wall of the cecum and the proseptum is the same as that between the siphuncular tube and the retrochoanitic septal neck in *Nautilus*. The outer part of the cecum in many genera is covered by a calcareous deposit (Druschits and Doguzhaeva, 1974, 1981; Druschits *et al.*, 1983; Tanabe *et al.*, 1979, 1980).

The cecum is connected to the inner surface of the initial chamber by means of the so-called prosiphon, which is an organic structure of complex construction. Prosiphons, even in the same species, may display considerable morphological variability. Apart from tube-like elements, which vary as to the degree of flattening and bending, there is another very variable element: membranes that join the tubular elements to the inner surface of the initial chamber (Kulicki, 1979; Bandel, 1982; Ohtsuka, 1986; and citations above).

The cecum in cross section may show a multilayered concentric structure (Ohtsuka, 1986). The cecum and prosiphon commonly exhibit wrinkles or tension lines, proving their former elasticity.

The siphuncular tube in ammonoids, as opposed to that in *Nautilus*, is constructed only of organic matter, i.e., multilayered concentric membranes of conchiolin, each of which consists of lace-like, tuberculate microfibrils 1–2 μm in diameter with elongate pores (Obata *et al.*, 1980; Grégoire, 1984).

In the retrochoanitic condition, comparable to that in present-day *Nautilus*, the adapertural end of each segment of the siphuncular tube is a continuation of the organic membranes of the nacreous layer of the septal neck. The adapical end of each segment of the siphuncular tube terminates in an auxiliary deposit attached to the inner surface of the previous septal neck. In the prochoanitic condition, the only difference is the relationship between the siphuncular tube and the septal neck at the adapertural end of the tube segment. In this case, the organic segment of the siphuncular tube is connected with the cuff and not with the distal part of the septal neck.

According to Mutvei (1972a), in present-day *Nautilus* there are no organic layers in common joining the adjacent segments of the siphuncular tube. Therefore, Grégoire's (1987, Fig. 12) schematic drawing, in which the auxiliary ridge is connected to both the adapertural and adapical segments of the siphuncular tube, seems to be misleading. Ammonoids in the retrochoanitic condition do not exhibit such connections either. As for ammonoids in the prochoanitic condition, Kulicki and Mutvei (1982) have observed that there are connections in common between adjacent segments of the siphuncular tube via the basal lamella of the cuff. This phenomenon is easy to explain because the siphuncular tube protruding into the body chamber was longer than the distance between septa. In this case, after the soft body moved forward into a position to form a new septum, because a partially secreted siphuncular tube was already present, the basal lamella undergoing calcification became connected to both the adapical and adapertural segments of the siphuncular tube. An additional connection of the outer layer of the siphuncular tube to the adapertural tip of the prochoanitic septal neck has also been reported by Kulicki and Mutvei (1982). In Kulicki's (1979) opinion, the siphuncular tube in the prochoanitic condition may have been secreted both from the inside, by the epithelium of the siphuncular cord, and from the outside, by the epithelium of the circumsiphonal invagination. The outermost layer of the tube segment would have been secreted by the epithelium of the circumsiphonal invagination.

Birkelund and Hansen (1974) have described a double-layered siphuncular tube in a specimen of the Maastrichtian ammonoid *Saghalinites wrighti*. The inner, initially organic layer of this siphuncular tube displays a fibrous structure, whereas the outer layer is composed of calcium carbonate with a prismatic structure. The state of preservation of the specimen prevented these authors from determining whether the mineral of the outer layer was calcite or aragonite. But the fact that the strontium content of this layer differed from that in the surrounding rock was, in their opinion, a confirmation of the primary nature of this calcium carbonate layer and, hence, evidence of a similarity between the siphuncular structure in ammonoids and that in present-day *Nautilus*.

3.5.2. Septal Necks

There is a special structure called the septal neck around the siphuncular perforation of the septum. The septal neck is made of a nacreous layer, which is a continuation of the nacreous layer of the septum and, thus, an integral part of the septum.

In ammonoids, three types of septal necks are distinguished: retrochoanitic, intermediate, and prochoanitic, and in each type the relationship of the siphuncular tube to the septal neck is different. Because these are discussed in Chapter 6, they are only briefly treated here.

3.5.2.a. Retrochoanitic Condition. The relationship of the siphuncular tube to the septal neck in the retrochoanitic condition in ammonoids is completely comparable to the condition existing in present-day *Nautilus*. In *Nautilus*, the siphuncular tube is a noncalcified continuation of the retrochoanitic septal neck. The nacreous layer of the septal neck is connected to the siphuncular tube by a modified nacreous layer. This layer is enriched in conchiolin and has a spherulitic–prismatic structure (Mutvei, 1972a; Doguzhaeva and Mutvei, 1986a). The adapical end of the siphuncular tube segment is connected to the inner surface of the previous septal neck by means of an auxiliary deposit.

According to Doguzhaeva and Mutvei (1986a, p. 13), these formations "seem to have a prismatic structure . . . probably similar to that in corresponding deposits in the *Quenstedtoceras*." Expressions such as "seem to have" and "probably" express a degree of uncertainty. The problem is that the majority of post-Triassic ammonoids have retrochoanitic septal necks only in very early whorls, whereas older ammonoids with well-developed retrochoanitic necks are characterized by extensive recrystallization, which blurs the original microstructure.

The nature of the auxiliary deposits and their position on the inner surface of retrochoanitic septal necks may differ in various ammonoids and may vary from one ontogenetic stage to another (Miller and Unklesbay, 1943).

3.5.2.b. Prochoanitic Condition. Prochoanitic septal necks, by definition, are directed adaperturally. The wall of the siphuncular tube is not a continuation of the nacreous layer of the septal neck. Commonly, the distal parts of these necks are not in contact with the siphuncular tube (Voorthuysen, 1940; Kulicki, 1979). In longitudinal cross section, it can be observed that the lamellae of the nacreous layer wedge out on the inner surface of the neck in its distal part (Kulicki and Mutvei, 1982).

Siphuncular deposits in the Ammonitina are best known in *Quenstedtoceras* (see Kulicki, 1979; Kulicki and Mutvei, 1982). Generally, the inner surface of the septal neck is covered by a cuff, which is the adapertural

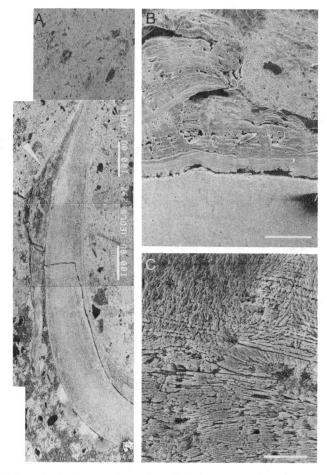

FIGURE 10. (A) Longitudinal cross section through the last septum of a fully grown macroconch of *Quenstedtoceras*. A well–developed cuff is visible, and there is a large space between it and the proximal part of the neck, which is filled with sediment. The arrow points to the basal lamella of the auxiliary deposit. Treated in EDTA. Scale bar, 100 μm. (B) Longitudinal cross section through one of the last septa in a macroconch of *Quenstedtoceras*. The nacreous layer of the septal neck (below) is covered by the cuff. The adapical termination of the auxiliary deposit is visible above. Treated in EDTA. Scale bar, 100 μm. (C) Longitudinal cross section through one of the last septa in a macroconch of *Quenstedtoceras* (different specimen than that in B). The nacreous layer of the septal neck (below) is covered by the cuff (part, above), which shows a distinct fine lamination and a spherulitic–prismatic arrangement of mineral components. The basal lamella of the cuff, with distinct spherulites, lies directly on the surface of the nacre. Treated in EDTA. Scale bar, 10 μm. (A–C, from H. Mutvei and C. Kulicki, unpublished data).

termination of the siphuncular tube segment. An auxiliary deposit, which is the adapical termination of the next segment of the siphuncular tube, frequently covers the cuff from the inside.

In middle and late ontogenetic stages in *Quenstedtoceras*, the outermost part of the cuff, called the basal lamella, is distinguished as a separate element (Fig. 10C). It remains in direct contact with the nacreous layer of the septal neck. Kulicki and Mutvei (1982) distinguished it because of the difference in its microstructure with respect to the rest of the septal neck and because it is connected to the wall of the siphuncular tube both adaperturally and adapically. These authors also have distinguished the basal lamella in the auxiliary deposit, which is connected at its distal end to the wall of the siphuncular tube.

The cuff and auxiliary deposit are identical in microstructure (Fig. 10B). They both show fine lamellae (about 1 μm thick) and a spherulitic-prismatic system of mineral components similar to the modified nacreous layer in the septal neck of *Nautilus* (see Kulicki and Mutvei, 1982). The basal lamellae have a stronger spherulitic–prismatic calcification than the rest of the deposit, and, thus, the lamellar structure is less visible there (Kulicki and Mutvei, 1982). These studies did not confirm the interpretation of Bandel and Boletzky (1979) and Bandel (1981, 1982) regarding an original porous zone in siphuncular deposits in the Ammonitina.

In a newly formed septum, one commonly can observe that the basal lamella of the cuff is not in contact with the inner surface of the proximal part of the septal neck and that there is an empty space between these two elements (Fig. 10A; H. Mutvei and C. Kulicki, unpublished data). In more adapical septa, this space is filled by sparry calcareous material and forms a so-called "angular deposit" (Kulicki, 1979). Angular deposits are formed as a result of inorganic calcification; they are almost always present between the cuff and the adapical part of the prochoanitic septal neck, with the exception of the last septum. This suggests rapid formation of angular deposits during the lifetime of the animal (Kulicki, 1979).

3.5.2.c. Intermediate Condition. Septal necks characteristic of this condition represent an intermediate form between typical retro- and prochoanitic necks. However, such necks do not necessarily develop into prochoanitic necks during ontogeny and therefore do not represent a transition in a morphogenetic sense (Chap. 6). Such necks have been described by Doguzhaeva (1973), Doguzhaeva and Mutvei (1986a), Druschits and Doguzhaeva (1974), Kulicki (1979), and Tanabe *et al.* (1993a).

Generally, it can be said that during the transition from retrochoanitic septal necks to fully prochoanitic necks, the secretion of the siphuncular complex became more independent of the secretion of the septum and septal neck than in the retrochoanitic condition.

3.6. Intracameral Membranes

The inner surfaces of chamber walls and the adapical and adoral surfaces of septa are covered with phosphatized membranes that were originally organic. These are called, in general, intracameral, conchiolin, or cameral membranes. Those that remain in contact with the siphuncular tube are specifically known as siphuncular membranes. Siphuncular membranes have been described by Grandjean (1910), Schulga-Nesterenko (1926), Erben and Reid (1971), Westermann (1971), Bayer (1975), Kulicki (1979), Bandel and Boletzky (1979), Bandel (1981, 1982), Tanabe *et al.* (1982), Weitschat (1986), and Weitschat and Bandel (1991).

The ultrastructure of intracameral membranes was described by Tanabe *et al.* (1982) and Grégoire (1984). According to these authors, these membranes are composed of thin (0.015–0.15 µm) fibers that lack a distinct orientation. These membranes are comparable to the pellicle ("brown membrane") in *Nautilus* and *Spirula*. However, they differ considerably in having finer fibers that have no preferred orientation (see also Chapter 6).

Siphuncular membranes display great variability in different ammonoid groups. Although Permian and Triassic genera seem to have especially well-developed and numerous siphuncular membranes (Schulga-Nesterenko, 1926; Weitschat and Bandel, 1991), they are less numerous and not so well-developed in post-Triassic forms. These membranes connect the siphuncular tube to both the adapical and adapertural surfaces of the septa. They may also form connections to the ventral or ventrolateral shell wall. Especially complex systems of siphuncular membranes have been described by Weitschat and Bandel (1991) in such Triassic genera as *Anagymnotoceras* and *Czekanowskites*.

Horizontal membranes, as noted by Weitschat and Bandel (1991), represent a case of extremely well-developed siphuncular membranes. These horizontal membranes divide the phragmocone chamber into two main compartments of nearly equal size, a ventral and a dorsal compartment. These membranes are connected to the siphuncular tube, to the septa at their midheight, and to the inner surfaces of the flanks of the outer shell wall. These membranes commonly are perforated.

There is another kind of intracameral membrane known as a transverse membrane. Such membranes have also been described by Weitschat and Bandel (1991). These membranes are not always connected to the siphuncle but generally separate the chamber into compartments. Transverse membranes also show perforations.

Zaborski (1986) and Hewitt *et al.* (1991) have described membranous structures, the so-called "pseudosepta," in vascoceratids from Nigeria. Pseudosutures, i.e., the lines of attachment of pseudosepta to the shell walls, have been reported from Mesozoic ammonoids (John, 1909; Hölder, 1954; Vogel, 1959; Schindewolf, 1968; Bayer, 1977) as well as from Paleozoic goniatites (Landman *et al.*, 1993).

There are two interpretations for the formation of intracameral membranes. The first, accepted by the majority of researchers, posits that organic membranes are secreted in close contact with the mantle surface, not unlike the periostracum or other mineral–organic layers of the shell, e.g., the nacreous layer. Weitschat and Bandel (1991) have recently discussed this point of view. According to the second interpretation, although the organic matter of the intracameral membranes is secreted by the mantle surface, the membranes themselves are produced by dehydration of a jelly-like cameral fluid (Bandel and Boletzky, 1979; Hewitt *et al.*, 1991; Westermann, 1992).

Functionally, hydrophilic intracameral membranes could have served either as pathways for transport of cameral fluid or as structures helping to maintain the fluid within certain reservoirs and preventing any overflow. This may have made it possible to keep cameral and circumsiphonal fluids in a state of decoupling, which may have been beneficial from a physiological point of view (Mutvei, 1967; Kulicki, 1979; Kulicki and Mutvei, 1988; Ward, 1987; Weitschat and Bandel, 1991).

4. Summary

The major limitation to micro- and ultrastructural studies of ammonoids is diagenesis, i.e., the state of preservation of the specimens used. This explains why Jurassic and Cretaceous ammonoids are among the best studied, while the data on Paleozoic and Triassic forms are generally inadequate. It is only under exceptionally favorable conditions of preservation, such as those in the Buckhorn asphalts of Pennsylvanian age, that the study of the ultrastructure of orthoconic cephalopods was possible, revealing a great similarity between these forms and present-day *Nautilus* (see Mutvei, 1972b).

At the ultrastructural level, the three main layers of the postembryonic shell of ammonoids do not differ significantly from those of the shell of present-day *Nautilus*. The same is also true of the septa. However, a clearly distinctive element is the ammonitella, which is now the focus of attention of many ammonitologists. Despite intensive study, the number and distribution of the layers in the ammonitella, as well as the relationships among them, are still unclear. In addition, the new model of calcification of the ammonitella, described on the basis of specimens of *Aconeceras* (see Kulicki and Doguzhaeva, 1994), requires further verification in other ammonoid groups.

Knowledge of the ultrastructure of both the mineral and organic components in ammonoids is meager compared with what is known about other Recent molluscs. Some results obtained in the course of ultrastructural studies by Grégoire (1984) and Obata *et al.* (1980) seem to show that such research is not impossible.

ACKNOWLEDGMENTS. The author is deeply grateful to Neil Landman (American Museum of Natural History) and Richard Davis (College of Mount St. Joseph) for reviewing this manuscript, correcting its language, and also for valuable comments and remarks.

References

Andalib, F., 1972, Mineralogy and preservation of siphuncles in Jurassic cephalopods, *N. Jb. Geol. Paläont. Abh.* **140**:33–48.

Arkell, W. J., 1957, Introduction to Mesozoic Ammonoidea, in: *Treatise on Invertebrate Paleontology*, Part L, *Mollusca 4* (R. C. Moore, ed.), Geological Society of America and University Kansas Press, Lawrence, KS, pp. 81–129.

Bandel, K., 1981, The structure and formation of the siphuncular tube of *Quenstedtoceras* compared with that of *Nautilus* (Cephalopoda), *N. Jb. Geol. Paläont. Abh.* **161**:153–171.

Bandel, K., 1982, Morphologie und Bildung der frühontogenetischen Gehäuse bei conchiferen Mollusken, *Facies* **7**:1–198.

Bandel, K., 1986, The ammonitella: A model of formation with the aid of the embryonic shell of archaeogastropods, *Lethaia* **19**:171–180.

Bandel, K., 1990, Cephalopod shell structure and general mechanisms of shell formation, in: *Skeletal Biomineralization: Patterns, Processes and Evolutionary Trends*, Vol. I (J. G. Carter, ed.), Van Nostrand Reinhold, New York, pp. 97–115.

Bandel, K., and Boletzky, S. v., 1979, A comparative study of the structure, development, and morphological relationships of chambered cephalopod shells, *Veliger* **21**:313–354.

Bandel, K., Landman, N., and Waage, K. M., 1982, Micro-ornament on early whorls of Mesozoic ammonites: Implications for early ontogeny, *J. Paleontol.* **56**(2):386–391.

Bayer, U., 1974, Die Runzelschicht—ein Leichtbauelement der Ammonitenschale, *Paläontol. Z.* **48**(1–2):6–15.

Bayer, U., 1975, Organische Tapeten im Ammoniten—Phragmocon und ihr Einfluß auf die Fossilisation, *N. Jb. Geol. Paläont. Mh.* **1975**(1):12–25.

Bayer, U., 1977, Cephalopoden Septen. Teil 1. Konstruktionsmorphologie des Ammoniten-Septums, *N. Jb. Geol. Paläont. Abh.* **154**:290–366.

Birkelund, T., 1967, Submicroscopic shell structures in early growth stage of Maastrichtian ammonites (*Saghalinites* and *Scaphites*), *Medd. Dan. Geol. Foren.* **17**(1):95–101.

Birkelund, T., 1981, Ammonoid shell structure, in: *The Ammonoidea*, Systematics Association Special Volume 18 (M. R. House and J. R. Senior, eds.), Academic Press, London, pp. 177–214.

Birkelund, T., and Hansen, H. J., 1968, Early shell growth and structures of the septa and the siphunculer tube in some Maastrichtian ammonites, *Medd. Dan. Geol. Foren.* **18**:71–78.

Birkelund, T., and Hansen, H. J., 1974, Shell ultrastructures of some Maastrichtian Ammonoidea and Coleoidea and their taxonomic implications, *K. Dan. Vidensk. Selsk. Biol. Skr.* **20**(6):2–34.

Birkelund, T., and Hansen, H. J., 1975, Further remarks on the post-embryonic Hypophylloceras shell, *Bull. Geol. Soc. Den.* **24**:87–92.

Bøggild, O. B., 1930, The shell structure of the molluscs, *K. Dan. Vidensk. Selsk. Skr. Raekke 9* **2**(2):233–326.

Bogoslovsky, B. I., 1969, Devonskie Ammonoidei, I, Agoniatity, *Trans. Paleont. Inst. Akad. Nauk SSSR* **124**:1–341 (in Russian).

Bogoslovsky, B. I., 1971, Devonskie Ammonoidei, II, Goniatity, *Trans. Paleont. Inst. Akad. Nauk SSSR* **127**:1–216 (in Russian).

Böhmers, J. C. A., 1936, *Bau und Struktur von Schale und Sipho bei permischen Ammonoidea*, Dissertation, Drukkerij University, Amsterdam, Apeldoorn.

Branco, W., 1880, Beiträge zur Entwicklungsgeschichte der fossilen Cephalopoden, *Palaeontographica* **27**:17–81.

Brown, A., 1892, The development of the shell in the coiled stage of *Baculites compressus* Say, *Proc. Acad. Nat. Sci. Phila.* **44**:136–142.

Crick, R. E., and Mann, K. O., 1987, Biomineralization and systematic implications, in: *Nautilus—The Biology and Paleobiology of a Living Fossil* (W. B. Saunders and N. H. Landman, eds.), Plenum Press, New York, pp. 115–134.

Doguzhaeva, L. A., 1973, Vnutriene stroienie rakoviny roda *Megaphyllites*, *Byull. Mosk. Ova. Ispyt. Prir. Otd. Geol.* **48**(6):161.

Doguzhaeva, L. A., and Mikhailova, I. A., 1982, The genus *Luppovia* and the phylogeny of Cretaceous heteromorph ammonoids, *Lethaia* **15**:55–65.

Doguzhaeva, L. A., and Mutvei, H., 1986a, Retro and prochoanitic septal necks in ammonoids, and transition between them, *Palaeontogr. Abt. A* **195**:1–18.

Doguzhaeva, L. A., and Mutvei, H., 1986b, Functional interpretation of inner shell layers in Triassic ceratid ammonites, *Lethaia* **19**:195–209.

Doguzhaeva, L. A., and Mutvei, H., 1989, Ptychoceras—a heteromorphic lytoceratid with truncated shell and modified ultrastructure, *Palaeontogr. Abt. A* **208**:91–121.

Druschits, V. V., and Doguzhaeva, L. A., 1974, Some morphogenetic characteristics of phylloceratids and lytoceratids (Ammonoidea), *Paleontol. J.* **8**(1):37–48.

Druschits, V. V., and Doguzhaeva, L. A., 1981, *Ammonity pod elektronnym mikroskopom*, Moskva University Press (in Russian).

Druschits, V. V., and Khiami, N., 1969, O niekotorykh voprosakh izuchenia rannikh stadii ontogeneza ammonitov, in Tez. Dokl. na sveshch. po probl. *Puti i zakonomiernosti istoritscheskogo rozvitia zivotnykh i rostitelnykh organizmov*, Moscow, pp. 26–30 (in Russian).

Druschits, V. V., and Khiami, N., 1970, Stroienie sept, stenki protokonkha i natchalnykh oborotov rakoviny nekotorykh ranniemelovykh ammonitov, *Paleontol. Zh.* **1970**(1):35–47 (in Russian).

Druschits, V. V., Doguzhaeva, L. A., and Mikhailova, I. A., 1977, The structure of the ammonitella and the direct development of ammonites, *Paleontol. J.* **11**(2):188–199.

Druschits, V. V., Doguzhaeva, L. A., and Mikhailova, I. A., 1978, Neobytchnye oblekayusche sloi ammonitov, *Paleontol. Zh.* **1978**(2):36–44 (in Russian).

Druschits, V. V., Muravin, E. S., and Baranov, V. N., 1983, Morfogenez rakovin srednevolzhskikh ammonitov roda *Virgatites, Lomonosovella, Epivirgatites*, *Vestn. Mosk. Univ. Ser. 4 Geol.* **1983**:35–44 (in Russian).

Dullo, W. C., and Bandel, K., 1988, Diagenesis of molluscan shells: A case study, in: *Cephalopods—Present and Past* (J. Wiedmann and J. Kullmann, eds.), Schweizerbart'sche Verlagsbuchhandlung, Stuttgart, pp. 719–729.

Dunachie, J. F., 1963, The periostracum of *Mytilus edulis*, *Trans. R. Soc. Edinb.* **65**(15):383–411.

Erben, H. K., 1962, Über den Prosipho, die Prosutur und die Ontogenie der Ammonoidea, *Paläontol. Z.* **36**:99–108.

Erben, H. K., 1966, Über den Ursprung der Ammonoidea, *Biol. Rev.* **41**:631–658.

Erben, H. K., 1972a, Über die Bildung und das Wachstum von Perlmut, *Biomineralisation* **4**:15–46.

Erben, H. K., 1972b, Die Mikro- und Ultrastruktur abgedeckter Hohlelemente und die Conellen des Ammoniten–Gehäuses, *Paläontol. Z.* **46**:6–19.

Erben, H. K., 1974, On the structure and growth of the nacreous tablets in gastropods, *Biomineralisation* **7**:14–22.

Erben, H. K., and Reid, R. E. H., 1971, Ultrastructure of shell, origin of conellae and siphuncular membranes in an ammonite, *Biomineralisation* **3**:22–31.

Erben, H. K., Flajs, G., and Siehl, A., 1968, Ammonoids: Early ontogeny of ultramicroscopical shell structure, *Nature* **219**:396–398.

Erben, H. K., Flajs, G., and Siehl, A., 1969, Die frühontogenetische Entwicklung der Schalenstruktur ectocochleater Cephalopoden, *Palaeontogr. Abt. A* **132**:1–54.

Grandjean, F., 1910, Le siphon des ammonites et des belémnites, *Bull. Soc. Geol. Fr. Ser. 4* **10**:496–519.

Grégoire, C., 1962, On submicroscopic structure of the *Nautilus* shell, *Bull. Inst. R. Sci. Nat. Belg.* **38**(49):1–71.

Grégoire, C., 1966, On organic remains of Paleozoic and Mesozoic cephalopods (nautiloids and ammonoids), *Bull. Inst. R. Sci. Nat. Belg.* **42**(39):1–36.

Grégoire, C., 1968, Experimental alteration of the *Nautilus* shell by factors involved in diagenesis and metamorphism, Part I, Thermal changes in conchiolin matrix of mother-of-pearl, *Bull. Inst. R. Sci. Nat. Belg.* **44**(25):1–69.

Grégoire, C., 1972, Experimental alteration of the *Nautilus* shell by factors involved in diagenesis and in metamorphism, Part III, Thermal and hydrothermal changes in the mineral and organic components of the mural mother-of-pearl, *Bull. Inst. R. Sci. Nat. Belg.* **48**(6):1–85.

Grégoire, C., 1980, The conchiolin matrices in nacreous layers of ammonoids and fossil nautiloids: A survey, *Akad. Wiss. Lit. Abh. Math. Naturwiss. Kl. Mainz* **1980**(2):1–128.

Grégoire, C., 1984, Remains of organic components in the siphonal tube and in the brown membrane of ammonoids and fossil nautiloids. Hydrothermal simulation of their diagenetic alterations, *Akad. Wiss. Lit. Abh. Math. Naturwiss. Kl. Mainz* **1984**(5):5–56.

Grégoire, C., 1987, Ultrastructure of the *Nautilus* shell, in: *Nautilus—The Biology and Paleobiology of a Living Fossil* (W. B. Saunders and N. H. Landman, eds.), Plenum Press, New York, pp. 463–486.

Hansen, H. J., 1967, A technique for depiction of grind sections of foraminifera by aid of compiled electronmicrographs, *Medd. Dan. Geol. Foren.* **17**:128.

Hewitt, R. A., and Westermann, G. E. G., 1983, Mineralogy, structure and homology of ammonoid siphuncles, *N. Jb. Geol. Paläont. Abh.* **165**(3)378–396.

Hewitt, R. A., Checa, A., Westermann, G. E. G., and Zaborski, P. M., 1991, Chamber growth in ammonites inferred from colour markings and naturally etched surfaces of Cretaceous vascoceratids from Nigeria, *Lethaia* **24**:271–287.

Hölder, H., 1952a, Über Gehäusebau, insbesondere Hohlkiel jurassischer Ammoniten, *Palaeontogr. Abt. A* **102**:18–48.

Hölder, H., 1952b, Der Hohlkiel der Ammoniten und seine Entdeckung durch F. A. Quenstedt, *Jb. Vaterl. Naturk. Wurtt.* **1952**:37–50.

Hölder, H., 1954, Über die Sipho-Anheftung bei Ammoniten, *N. Jb. Geol. Paläont. Mh.* **1954**(8):372–379.

House, M. R., 1965, A study in the Tornoceratidae: The succession of *Tornoceras* and related genera in the North American Devonian, *Phil. Trans. R. Soc. Lond. B* **250**(763):79–130.

House, M. R., 1971, The goniatite wrinkle-layer, *Smithson. Contrib. Paleobiol.* **3**:23–32.

Howarth, M. K., 1975, The shell structure of the Liassic ammonite family Dactylioceratidae, *Bull. Br. Mus. (Nat. Hist.) Geol.* **26**:45–67.

Hyatt, A., 1872, Fossil cephalopods of the Museum of Comparative Zoology: Embryology, *Bull. Mus. Comp. Zool. Harv.* **3**:59–111.

John, R., 1909, *Über die Lebensweise und Organisation des Ammoniten*, Inaugural Dissertation, University of Tübingen, Stuttgart.

Joly, B., 1976, Les Phylloceratidae malgaches au Jurassique. Generalités sur la Phylloceratidae et quelques Juraphyllitidae, *Doc. Lab. Géol. Fac. Sci. Lyon* **67**:1–471.

Keyserling, A., 1846, *Wissenschaftliche Beobachtungen auf einer Reise in das Petschora-Land im Jahre 1843*, St. Petersburg.

Kulicki, C., 1974, Remarks on the embryogeny and postembryonal development of ammonites, *Acta Palaeontol. Pol.* **19**(2):201–224.

Kulicki, C., 1975, Structure and mode of origin of the ammonite proseptum, *Acta Palaeontol. Pol.* **20**(4):535–542.

Kulicki, C., 1979, The ammonite shell: Its structure, development and biological significance, *Palaeontol. Pol.* **39**:97–142.

Kulicki, C., 1989, Archaeogastropod model of mineralization of ammonitella shell, in: *Skeletal Biomineralization: Patterns, Processes and Evolutionary Trends, Short Course in Geology*, Vol. 5, Part 2 (J. G. Carter, ed.), American Geophysical Union, Washington, DC, p. 324.

Kulicki, C., and Doguzhaeva, L. A., 1994, Development and calcification of the ammonitella shell, *Acta Palaeontol. Pol.* **39**:17–44.

Kulicki, C., and Mutvei, H., 1982, Ultrastructure of the siphonal tube in *Quenstedtoceras* (Ammonitina), *Stockholm Contrib. Geol.* **37**:129–138.

Kulicki, C., and Mutvei, H., 1988, Functional interpretation of ammonoid septa, in: *Cephalopods—Present and Past* (J. Wiedmann and J. Kullmann, eds.), Schweizerbart'sche Verlagsbuchhandlung, Stuttgart, pp. 719–729.

Landman, N., 1985, Preserved ammonitellas of *Scaphites* (Ammonoidea, Ancyloceratina), *Am. Mus. Novit.* **2815**:1–21.

Landman, N., 1987, Ontogeny of Upper Cretaceous (Turonian–Santonian) scaphitid ammonites from the Western interior of North America: Systematics, developmental patterns, and life history, *Bull. Amer. Mus. Nat. Hist.* **185**:117–241.

Landman, N., 1988, Early ontogeny of Mesozoic ammonites and nautilids, in: *Cephalopods—Present and Past* (J. Wiedmann and J. Kullman, eds.) Schweizerbart'sche Verlagsbuchhandlung, Stuttgart, pp. 215–228.

Landman, N., and Bandel, K., 1985, Internal structures in the early whorls of Mesozoic ammonites, *Am. Mus. Novit.* **2823**:1–21.

Landman, N., and Waage, K., 1982, Terminology of structures in embryonic shells of Mesozoic ammonites, *J. Paleontol.* **56**:1293–1295.

Landman, N. H., Tanabe, K., Mapes, R. H., Klofak, S. M., and Whitehill, J., 1993, Pseudosutures in Paleozoic ammonoids, *Lethaia* **26**:99–100.

Makowski, H., 1962, Problem of sexual dimorphism in ammonites, *Palaeontol. Pol.* **12**:1–92.

Makowski, H., 1971, Some remarks on the ontogenetic development and sexual dimorphism in the Ammonoidea, *Acta Geol. Pol.* **21**(3):321–340.

Miller, A. K., and Unklesbay, A. G., 1943, The siphuncle of late Paleozoic ammonoids, *J. Paleontol.* **17**:1–25.

Miller, A. K., Furnish, W. M., and Schindewolf, O. H., 1957, Paleozoic Ammonoidea, in: *Treatise on Invertebrate Paleontology*, Part L, *Mollusca 4* (R. C. Moore, ed.), Geological Society of America and University Kansas Press, Lawrence, KS, pp. 11–79.

Mutvei, H., 1964, On the shells of *Nautilus* and *Spirula* with notes on the shell secretion in non-cephalopod molluscs, *Ark. Zool.* **16**(14):221–278.

Mutvei, H., 1967, On the microscopic shell structure in some Jurassic ammonoids, *N. Jb. Geol. Paläont. Abh.* **129**(2):157–166.

Mutvei, H., 1969, On the micro and ultrastructure of the conchiolin in the nacreous layer of some recent and fossil Molluscs, *Stockholm Contrib. Geol.* **20**(1):1–17.

Mutvei, H., 1970, Ultrastructure of the mineral and organic components of molluscan nacreous layers, *Biomineralisation* **2**:48–61.

Mutvei, H., 1972a, Ultrastructural studies on cephalopod shells, Part 1, The septa and siphonal tube in *Nautilus, Bull. Geol. Inst. Univ. Upps.* **3**:237–261.

Mutvei, H., 1972b, Ultrastructural studies on cephalopod shells, Part 2, Orthoconic cephalopods from the Pennsylvanian Buckhorn Asphalt, *Bull. Geol. Inst. Univ. Upps.* **3**:263–272.

Mutvei, H., 1979, On the internal structures of the nacreous tablets in molluscan shells, *Scanning Electron Microsc.* **1979**(II):451–462.

Mutvei, H., 1980, The nacreous layer in molluscan shells, in: *The Mechanisms of Biomineralization in Animals and Plants, Proceedings 3rd International Biomineralization Symposium*, (M. Omori and N. Watabe, eds.), Tokai University Press, Tokyo, pp. 49–56.

Mutvei, H., 1983, Flexible nacre in the nautiloid *Isorthoceras*, with remarks on the evolution of cephalopod nacre, *Lethaia* **16**:233–240.

Nassichuk, W. W., 1967, A morphological character new to ammonoids portrayed by *Clistoceras* gen. nov. from Pennsylvanian of Arctic Canada, *J. Paleontol.* **41**:237–242.

Obata, I., Tanabe, K., and Fukuda, Y., 1980, The ammonite siphuncular wall: Its microstructure and functional significance, *Bull. Natl. Sci. Mus. (Tokyo) Ser. C (Geol.)* **6**(2):59–72.

Ohtsuka, Y., 1986, Early internal shell microstructure of some Mesozoic Ammonoidea: Implications for higher taxonomy, *Trans. Proc. Palaeontol. Soc. Jpn. New Ser.* **141**:275–288.

Ruzhentsev, V. E., and Shimanskij, V. N., 1954, Nizhnepermskye svernutye i sognutye nautiloidei juzhnogo Urala, *Trans. Paleontol. Inst. Akad. Nauk SSSR* **50**:1–150.

Sandberger, G., and Sandberger, F., 1850–1856, *Die Versteinerungen des rheinischen Schichtensystems in Nassau*, Kreidel & Nieder, Wiesbaden, pp. 1–564.

Schindewolf, O. H., 1968, Analyse einen Ammonites-Gehäuses, *Akad. Wiss. Lit. Abh. Math. Naturwiss. Kl. Mainz* **1968**(8):139–188.

Schulga-Nesterenko, M. J., 1926, Internal structure of the shell in Artinskian ammonites, *Byull. Mosk. Ova. Ispyt. Prir. Otd. Geol.* **4**(1–2):81–100 (in Russian).

Senior, J. R., 1971, Wrinkle-layer structures in Jurassic ammonites, *Palaeontology* **14**:107–113.

Smith, J. P., 1901, The larval coil of *Baculites*, *Am. Nat.* **35**(409):39–49.

Smith, W. D., 1905, The development of *Scaphites*, *J. Geol.* **13**:635–654.

Tanabe, K., 1989, Endocochliate embryo model in the Mesozoic Ammonoidea, *Hist. Biol.* **2**:183–196.

Tanabe, K., and Ohtsuka, Y., 1985, Ammonoid early internal shell structure: Its bearing on early life history, *Paleobiology* **11**(3):310–322.

Tanabe, K., Obata, I., Fukuda, Y., and Futakami, M., 1979, Early shell growth in some Upper Cretaceous ammonites and its implications to major taxonomy, *Bull. Natl. Sci. Mus. (Tokyo), Ser. C (Geol.)* **5**(4):153–176.

Tanabe, K., Fukuda, Y., and Obata, I., 1980, Ontogenetic development and functional morphology in the early growth stages of three Cretaceous ammonites, *Bull. Natl. Sci. Mus. (Tokyo), Ser. C (Geol.)* **6**(1):9–26.

Tanabe, K., Fukuda, Y., and Obata, I., 1982, Formation and function of the siphuncle-septal neck structures in two Mesozoic ammonites, *Trans. Proc. Palaeontol. Soc. Jpn. New Ser.* **128**:433–443.

Tanabe, K., Landman, N., and Weitschat, W., 1993a, Septal necks in Mesozoic Ammonoidea: Structure, ontogenetic development, and evolution, in: *The Ammonoidea: Environment, Ecology, and Evolutionary Change* (M. R. House, ed.), Clarendon Press, Oxford, pp. 57–84.

Tanabe, K., Landman, N., Mapes, R., and Faulkner, C., 1993b, Analysis of a Carboniferous embryonic ammonoid assemblage from Kansas, U.S.A.—Implications for ammonoid embryology, *Lethaia* **26**:215–224.

Tozer, E. T., 1972, Observations on the shell structure of Triassic ammonoids, *Palaeontology (Lond.)* **15**(4):637–654.

Vogel, K. P., 1959, Zwergwuchs bei Polyptychiten (Ammonoidea), *Geol. Jahrb.* **76**:469–540.

Voorthuysen, J. H., 1940, Beitrag zur Kenntnis des inneren Baus von Schale und Sipho bei Triadischen Ammoniten, Dissertation, Amsterdam University, Van Gorcum & Co., Assen.

Voss-Foucart, M. F., and Grégoire, C., 1971, Biochemical composition and submicroscopic structure of matrices of nacreous conchiolin in fossil cephalopods (nautiloids and ammonoids), *Bull. Inst. R. Sci. Nat. Belg.* **47**(41):1–42.

Walliser, O. H., 1970, Über die Runzelschicht bei Ammonoidea, *Göttinger Arb. Geol. Paläont.* **5**:115–126.

Ward, P. D., 1987, *The Natural History of Nautilus*, Allen & Unwin, Boston.

Weiner, S., Lowenstam, H. A., Taborek, B., and Hood, L., 1979, Fossil mollusk shell organic matrix components preserved for 80 million years, *Paleobiology* **5**(2):144–150.

Weitschat, W., 1986, Phosphatisierte Ammonoideen aus der Mittleren Trias von Central-Spitzbergen, *Mitt. Geol. Paläont. Inst. Univ. Hamburg* **61**:249–279.

Weitschat, W., and Bandel, K., 1991, Organic components in phragmocones of Boreal Triassic ammonoids: Implications for ammonoid biology, *Paläontol. Z.* **65**(3–4):269–303.

Westermann, G. E. G., 1971, Form, structure and function of shell and siphuncle in coiled Mesozoic ammonoids, *Life Sci. Contrib. R. Ont. Mus.* **78**:1–39.

Westermann, G. E. G., 1992, Formation and function of suspended organic cameral sheets in Triassic ammonoids—discussion, *Paläontol. Z.* **66**(3–4):437–441.

Wise, S. W., 1970, Microarchitecture and mode of formation of nacre (mother-of-pearl) in pelecypods, gastropods and cephalopods, *Eclogae Geol. Helv.* **63**:775–797.

Zaborski, P. M. P., 1986, Internal mould markings in a Cretaceous ammonite from Nigeria, *Palaeontology (Lond.)* **29**(4):725–738.

Zakharov, Yu. D., and Grabovskaia, B. S., 1984, Stroienie rakoviny roda *Zelandites* (Lytoceratida), *Paläontol. Zh.* **1984**(1):19–29.

Chapter 5

Color Patterns in Ammonoids

ROYAL H. MAPES and RICHARD ARNOLD DAVIS

ROYAL H. MAPES • Department of Geological Sciences, Ohio University, Athens, Ohio 45701.
RICHARD ARNOLD DAVIS • Department of Chemistry and Physical Sciences, College of
Mount St. Joseph, Cincinnati, Ohio 45233-1670.

Ammonoid Paleobiology, Volume 13 of *Topics in Geobiology*, edited by Neil Landman *et al.*,
Plenum Press, New York, 1996.

1. Introduction

A color pattern in an ammonoid was first reported and illustrated by d'Orbigny in 1842 (p. 185, Pl. 45, Fig. 4). He recognized that the pattern preserved on the shell of a specimen of *Asteroceras stellare* (at that time known as *Ammonites stellaris*) from the Lower Jurassic was a remnant of a biologically produced color pattern emplaced by the animal when it was alive. Furthermore, he recognized that these kinds of patterns were observable only on well-preserved specimens. These brief observations laid the groundwork for the now generally recognized conclusions that color patterns are scarce on ammonoids and that they have paleobiological significance. Since d'Orbigny's time, there have been periodic reports of ammonoid color patterns (see Table I), and some workers have attempted to integrate this information into a greater understanding of the biology of ammonoids.

Inherent within all analyses of, and critical to an understanding of, ammonoid color patterns is the biology of present-day cephalopods. Coloration in the soft tissues of living coleoids includes bioluminescence, permanent

TABLE I. Source, Description, and Geographic Data on Ammonoids with Well-Documented Color Patterns

Taxon	Source	Pattern[a]	Pattern description[c]	Country
Cretaceous				
Tetragonites glabrus (Jimbo)	Tanabe and Kanie, 1978	L	D on L	Japan
Tetragonites sp.	Tanabe and Kanie, 1978	R	D on L	Japan
Protexanites (Protexanites) botanti shimizui Matsumoto	Matsumoto and Hirano, 1976	L	D on L	Japan
Protexanites (Anatexanites) fukazawai (Yabe and Shimizu)	Matsumoto and Hirano, 1976	L	D on L	Japan
Paratexanites (Parabevahites) serratomarginatus (Redtenbacher)	Matsumoto and Hirano, 1976	L	D on L	Japan
Libycoceras afikpoense Reyment	Reyment, 1957	S	D on L	Nigeria
Submortoniceras woodsi (Spath)	Kennedy *et al.*, 1981	L	D on L	South Africa
Calliphylloceras sp. aff. *C. demedoffi*	Bardhan *et al.*, 1993	R	D on L	India

Table I. (Continued)

Taxon	Source	Pattern[a]	Pattern description[c]	Country
Jurassic				
Amaltheus stokesi (Sowerby)	Spath, 1935	L[b]	D on L	England
Amaltheus subnodosus (Young and Bird)	Pinna, 1972	L	D on L	Germany
Amaltheus gibbosus (Schlotheim)	Pinna, 1972	L	D on L	Germany
Amaltheus margaritatus de Montfort	This report	L	D on L	England
Androgynoceras lataecosta (Sowerby)	Spath, 1935; Arkell, 1957	L	D on L	England
Tragophylloceras loscombi Spath	Pinna, 1972	L	D on L	England
Leioceras sp.	Greppin, 1898; Arkell, 1957	L	L on D	Switzerland
Asteroceras stellare (Sowerby)	d'Orbigny, 1842; Arkell, 1957	L	L on D	France
Asteroceras stellare (Sowerby)	Manley, 1977	S	D on L (internal mold)	England
Pleuroceras spinatum (Bruguiere)	Schindewolf, 1928, 1931; Arkell, 1957	R	D on L	Germany
Pleuroceras salebrosum (Hyatt)	Pinna, 1972; Lehmann, 1990	S	D on L	Germany
Pleuroceras transiens (Frentzen)	Pinna, 1972	S	D on L	Germany
Pleuroceras sp. aff. *solare* (Phillips)	Heller, 1977	L	D on L	Germany
Triassic				
Owenites koeneni Hyatt and Smith	Tozer, 1972; Mapes and Sneck, 1987	R	D on L	United States
Owenites sp. cf. *O. koeneni*	Mapes and Sneck, 1987	R	D on L	United States
Owenites sp.	Mapes and Sneck, 1987	C	D on L	United States
Dieneroceras spathi Kummel and Steele	Mapes and Sneck, 1987	L	D on L	United States
Dieneroceras subquadratum (Smith)	This report	L	L on D	United States
Prosphingites slossi Kummel and Steele	Mapes and Sneck, 1987	R	D on L	United States

[a]Pattern groups include the following: S, spots; L, longitudinal; R, radial/transverse; and C, combined.

[b]The specimen of Jurassic *Amaltheus stokesi* described by Spath (1935) is listed with a longitudinal pattern (L). This pattern is confined to an area of sublethal damage; the original pattern adapical of the sublethal damage consists of transverse bands (see Fig. 2C).

[c]Pattern description provides insight as to whether the pattern is dark on light (D on L) or light on dark (L on D). With the exception of that in *Libycoceras afikpoense*, the patterns described by Reyment (1957) are not listed because none of the patterns he described has been confirmed and because some of them are clearly iridescence rather than color patterns in the sense described herein. Also, it should be noted that different species names have been utilized in different references for the same specimen and that the names used herein are, perhaps, not the most current designations.

tissue coloration, and the remarkable chromatophores that allow nearly instantaneous color changes that can result in a chameleon-like background
blending. Unfortunately, with the limitations of the fossil record, there are no
preserved fossils that indicate ammonoids had any of these characteristics or
abilities.

Nautilus, with reddish brown transverse bands in the shell, is the only
present-day ectocochliate cephalopod. Thus, perforce, fossil-cephalopod
workers long have utilized *Nautilus* as the "working model" to explain color
patterns in ammonoids. This single taxon, of course, is not the only possible
model. The shells of fossil nautiloids are known to have a variety of color
patterns, including longitudinal bands, transverse bands, wavy bands, chev-
rons, etc. (Teichert, 1964, pp. K23–K25; see also Foerste, 1930; Hoare, 1978;
Mapes and Hoare, 1987). Ammonoids are known to have some but not all of
the types of color patterns observed in nautiloids.

The subject matter of this chapter is those colors and color patterns that
result from pigments. The substances that characterize color patterns in
present-day molluscs are deposited in the outer part of the shell or in the
periostracum (Hollingworth and Barker, 1991). Presumably the same held true
for ammonoids. There are many different kinds of pigments (biochromes) in
various present-day organisms. However, only two sorts are likely to be
preserved in fossils, melanin and members of the porphyrin group of tetrapyr-
roles (Hollingworth and Barker, 1991). Porphyrins fluoresce under ultraviolet
light, and melanin is a black pigment that can appear as black, brown, reddish
brown, or even yellow, depending on its concentration. However, little is
known of the identity of pigments in fossil ammonoids, at least in part because
paleontologists have been reluctant to destroy those precious few specimens
that exhibit color patterns.

Color can result from sources other than pigments. For example, unaltered
aragonitic shell material in molluscs sometimes exhibits multicolored iridescence. This is one of the structurally produced chromatic effects included
under the term schemochromy. This particular one results from interference
of incident and reflected light waves in very fine laminations of calcium
carbonate separated from one another by similarly thin spaces filled with
another material, for example, water (Fox, 1972). Examples of this kind of
coloration are well known in ammonoids from the Late Cretaceous of the
Western Interior of North America and in other cephalopods from Mesozoic
exposures around the world. There are rare occurrences in rocks from the Late
Paleozoic of North America (D. Work, personal communication, 1989). Some
of the specimens described but not illustrated by Reyment (1957) probably
are examples of structurally produced coloration.

As indicated previously, this chapter deals only with colors and color
patterns that result from actual pigments. It does not cover structurally

produced coloration, nor does it deal with the many postmortem effects on colors and color patterns.

Specimens referred to in this chapter are reposited in the British Museum (Natural History) [BM(NH)] and the Ohio University Zoological Collection (OUZC), as indicated.

2. Possible Ammonoid Colors and Color Patterns

The occurrence of present-day molluscs (including gastropods, pelecypods, and *Nautilus*) with patterns of pigments in their shells provides a basic framework for the recognition of patterns that can be expected to occur in the ammonoids and for the evaluation of putative patterns that have been reported. It must be recognized that there are no known specimens of ammonoids that have been demonstrated to have preserved the true, original colors they had in life. Rather, what one sees are *patterns* of black, grays, browns, creams, and white in or on the walls of the conchs.

Strictly speaking, to have a color pattern, a shell either must have areas of no pigment interspersed with areas in which one or more pigments are present or must have areas of pigment of other colors. The following is a generalized list and brief discussion of potential color schemes that have been or might be encountered in ammonoids.

2.1. Transparency

In organisms inhabiting aquatic environments, there are various adaptations in numerous phyla that allow the transmission of light through part or all of the organism's body. Packard (1988) speculated about the intriguing possibility that some ammonoids might have had a transparent shell. Dynamic color changes like those produced chromatophorically in extant coleoids might have been visible through such shells.

Of course, numerous ammonoid conchs are known in which the shell material has been recrystallized into translucent or even transparent calcite. However, there is no evidence that any ammonoids had transparent or even translucent shells in life. (Nor, for that matter, have chromatophores been documented in fossil cephalopods.)

2.2. Achromatism

Some molluscs do not impregnate their shells with any organic pigments. This results in a white, porcelaneous conch of aragonite. Recognition of this condition in ammonoids may be impossible because their shells tend to absorb the coloration of the surrounding sediment during diagenesis.

2.3. Monochromatism

Some molluscs impregnate their shells uniformly with organic pigments. Again, as with the achromatic condition, recognition of this kind of biological coloration will be difficult. However, what may be a possible example was discussed briefly but not illustrated by Matsumoto and Hirano (1976). They noted some specimens of *Gaudryceras* that each have a uniform brown color and no pattern. It is possible that what Matsumoto and Hirano observed was periostracum. Another possibility, as summarized by Doguzhaeva and Mutvei (1993), is that this brown shell layer was deposited in the late stages of ontogeny, when the conch was covered by mantle.

2.4. Irregular and Regular Patterns of Spots

Irregular and regular patterns of spots are known to occur in present-day molluscs. Recognition of a biologically produced, irregular pattern of spots in a fossil shell is extremely difficult because blotching is a common phenomenon produced by diagenesis and weathering. On the other hand, patterns of spots that are symmetrically arranged on either side of the medial plane of a planispiral shell almost certainly were emplaced when the animal was alive.

FIGURE 1. Color pattern of longitudinal bands of semiconnected dark spots on a light background in a *Pleuroceras salebrosum* (Hyatt) from the Jurassic at Marloffstein near Erlanger, Germany, ×1.8. (Photograph by Sig. Engelhart; copy provided by Ulrich Lehmann.)

Examples of approximately equal-sized spots in longitudinal rows (that is, following the coiling) on the outside of the shell include *Pleuroceras salebrosum* (Fig. 1; see Pinna, 1972; Lehmann, 1990) and *P. transiens* (see Pinna, 1972). The significance of Manley's (1977) report of longitudinal rows of spots in *Asteroceras stellare* is unclear in that the spots are not in the outer part of the shell, but, rather, on a thin coating on the internal mold of the body chamber (Fig. 2D).

2.5. Transverse Zig-Zags and Chevrons

Examples of these patterns are known to occur in extant molluscs and in fossil orthoconic and cyrtoconic nautiloids (see Foerste, 1930; Teichert, 1964). No examples of these patterns have been documented in ammonoids.

2.6. Transverse and Radial Stripes

Transverse stripes and bands are common in extant molluscs, including *Nautilus*. A number of occurrences are known in the Ammonoidea (Figs. 2C, 3A–F, 4B,C). Several examples of this type of pattern have been described from the Triassic of Nevada (Figs. 3A–F, 4B,C) (Tozer, 1972; Mapes and Sneck, 1987), including *Owenites koeneni* (Figs. 3A–C) and *O.* sp. cf. *O. koeneni* (Fig. 3E,F). Also, *Prosphingites slossi* from the same locality has transverse bands (Fig. 3D). Jurassic reports are more problematic. Schindewolf (1928, 1931) described transverse stripes on *Pleuroceras spinatum*. In addition, Reyment (1957) reported, but did not illustrate, a specimen he identified as *Macrocephalites* from the Jurassic of the former Soviet Union that he said had colored ribs near the umbilicus. Also, Spath (1935) described a longitudinal pattern on *Amaltheus stokesi*; however, reanalysis of this specimen (Fig. 2C) strongly suggests that the longitudinal pattern on the orad end of the shell occurs in an area of sublethal damage and that the apical end of the shell preserves a transverse band pattern. Cretaceous examples include four specimens of *Calliphylloceras* sp. aff. *C. demedoffi* (see Bardhan *et al.*, 1993) and *Tetragonites* sp. (see Tanabe and Kanie, 1978). Reyment (1957) also reported but did not illustrate a specimen he identified as *Speetoniceras versicolor* from the Early Cretaceous (Neocomian) of the former Soviet Union that he said had light brown ribs on the body chamber.

2.7. Longitudinal Stripes or Bands

Extant molluscs exhibiting this kind of pattern are common, and longitudinal stripes and bands are well known in fossil cephalopods (Figs. 1, 2A,B,D,E, 4D). Ammonoids show considerable variation in the number of stripes or bands, their widths, and their placement on the conch (ventral, lateral, and umbilical).

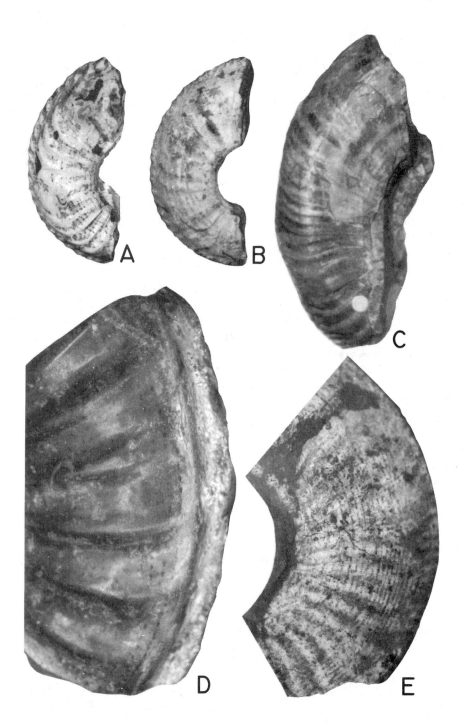

FIGURE 2. (A,B) Left lateral views of *Amaltheus gibbosus* (Schlotheim) and *A. margaritatus* de Montfort [BM(NH) C67340], respectively, with narrow dark longitudinal bands on a white background. Both specimens are from the same Jurassic locality in Germany, both at ×1.0. (C) Left lateral view of *Amaltheus stokesi* (Sowerby) with transverse dark bands on a dark gray-brown background. A large repaired wound with longitudinal dark bands related to the shell damage and repair and not to the original color pattern is also present, ×0.6. (D) Right lateral view of the steinkern described by Manley (1977) of a Jurassic *Asteroceras stellare* (Sowerby) showing longitudinal rows of black spots on a dark brown background. This spot pattern is provisionally accepted as a color pattern; however, the process of preservation remains unknown, ×0.65. (E) Right lateral view of *Amaltheus stokesi* (Sowerby) from the Jurassic of England, showing longitudinal thin dark bands on a white background, ×0.7. (We are aware that C and E are identified as the same taxon but have different color patterns. Additional study of this problem will be necessary to resolve this conflict.)

◄——

One of the two Triassic ammonoids with longitudinal stripes (both ventral and umbilical) is identified as *Dieneroceras spathi* (see Mapes and Sneck, 1987). The other that was recently discovered with a similar pattern is *D. subquadratum* (Fig. 4D). Jurassic examples include two species of *Amaltheus* (see Pinna, 1972), *Androgynoceras lataecosta* (see Spath, 1935; Arkell, 1957), *Asteroceras stellare* (see d'Orbigny, 1842; Arkell, 1957), *Leioceras* sp. (see Greppin, 1898; Arkell, 1957, p. L94), several species of *Pleuroceras* (see Schindewolf, 1928, 1931; Pinna, 1972; Arkell, 1957; Heller, 1977; Lehmann, 1990), and *Tragophylloceras loscombi* (see Pinna, 1972). The Jurassic example described by Spath (1935) of *Amaltheus stokesi* is probably the result of sublethal damage. Cretaceous examples were described for two species of *Protexanites* and one species of *Paratexanites* (see Matsumoto and Hirano, 1976) and for *Tetragonites glabrus* (see Tanabe and Kanie, 1978). The report by Reyment (1957) of *Libycoceras afikpoense* is accepted with reservation; however, the description by Kennedy *et al.* (1981) of *Submortoniceras* based on two specimens from South Africa is well documented.

2.8. Combination Patterns

Combinations of two or more of the previously described solid band patterns are also common in extant molluscs. Only a single instance is known among ammonoids, namely, one specimen identified by Mapes and Sneck (1987) as *Owenites* sp. This individual, from the Triassic of Nevada, has a combined longitudinal and transverse pattern (Fig. 3G).

In each complete mature specimen of present-day *Nautilus*, the ventral portion of the body chamber lacks the reddish brown bands that occur on the dorsal two-thirds of the conch; this is a combination of achromatism with a color pattern consisting of transverse stripes. There also are known fossil nautiloid shells that each have a color pattern on part of the conch but not on the rest of the shell (see, for example, Teichert, 1964, p. K23, Fig. 9D). Reyment (1957, Fig. 1) reported pigmented specimens of *Libycoceras afikpoense* from

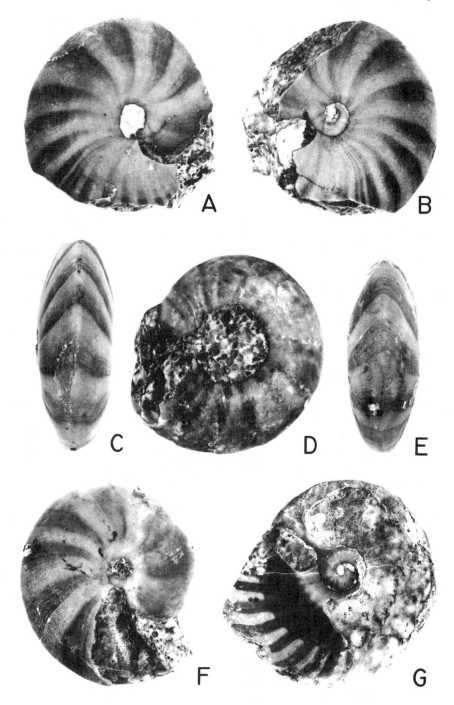

FIGURE 3. Examples of transverse and combined color patterns from the Triassic of Nevada, U.S.A. (A,B,C) Right, left, and ventral views, respectively, of a nearly mature *Owenites koeneni* Hyatt and Smith with pronounced dark transverse bands on a medium-gray background, ×2.0. (D) Lateral view of *Prosphingites slossi* Kummel and Steele with dark transverse bands on a medium-gray background, ×2.2. (E,F) Ventral and right lateral views, respectively, of a juvenile *Owenites* sp. cf. *O. koeneni* with dark transverse bands on a medium-gray background, ×2.0. (G) Left lateral view of an *Owenites* sp. with combined dark transverse and longitudinal bands, ×1.2.

FIGURE 4. (A) Left lateral view of a combined false color pattern consisting of transverse dark bands and a longitudinal light band in the umbilical region of the phragmocone of a Pennsylvanian goniatite *Gatherites morrowensis* Miller and Moore from the Bloyd Formation at Gaither Mountain near Harrison in north-central Arkansas. The dark transverse bands are produced by periodic shell thickenings that coincide with constrictions in the internal mold, and the longitudinal band represents the exposed conch where the succeeding whorl, which has been removed by taphonomic processes, did not cover, ×2.7. (B,C) Left lateral and dorsal views, respectively, of a Triassic *Prosphingites slossi* Kummel and Steele from Nevada with dark transverse bands on a medium gray background, ×3.0. (D) Right lateral view of a Triassic *Dieneroceras subquadratum* (Smith) (OUZC 1300) from Nevada with a longitudinal white band on a dark-gray background.

the Late Cretaceous (early Maastrichtian) of Nigeria. He interpreted the brown spots he observed to be remnants of a broad longitudinal band on the midline of the conch and of a narrower longitudinal stripe on each flank. His drawn reconstruction shows the last one-half volution to be pattern-free, like that of mature *Nautilus*. His argument that the spots represent remnants of longitudinal stripes and that the pattern ended on the body chamber would have been stronger had he provided some photographic documentation. (Attempts to reexamine his specimens have met with failure—the relevant material in the British Museum of National History does not exhibit these patterns, and the material registered in Sweden cannot be located.)

3. Patterns through Time and Space

Far too few recorded occurrences of color patterns on ammonoids are available for detailed analyses of their geographic and evolutionary patterns through time and space (Table II). Indeed, the fact that most of the well-reported occurrences in the Cretaceous are from Japan (Matsumoto and Hirano, 1976; Tanabe and Kanie, 1978) may indicate that Japanese invertebrate paleontologists have been particularly active in reporting color patterns. The same holds true with European and British investigators in reporting longitudinal lines of spots in ammonoids from the Jurassic of England (Manley, 1977) and Germany (Pinna, 1972; Heller, 1977).

In a general sense, however, it is immediately striking that despite hundreds of thousands (millions?) of Paleozoic ammonoid specimens collected, there is not a single instance of a confirmed color pattern. This is particularly surprising in light of the relatively high incidence of color patterns in other

TABLE II. Distribution of Reported Ammonoid Color Patterns by Pattern Type, Geologic Age, and Geographic Occurrence[a]

Age	Spots	Radial bands	Longitudinal bands	Combined (radial and longitudinal)
K	Nigeria (1?)	India (1)	Japan (4)	NR
		Japan (1)	South Africa (1)	
Jr	England (1)	Germany (1)	England (2)	
	Germany (2)	England (1)	Germany (4)	NR
			Switzerland (1)	
			France (1)	
Tr	NR	Nevada (3)	Nevada (2)	Nevada (1)

[a]NR indicates no report. Numbers in parentheses indicate the number of taxa and/or patterns described or illustrated. The report by Reyment (1957) of the color pattern in *Libycoceras afikpoense* from the Cretaceous of Nigeria is included with strong reservation; all the remainder of his descriptions are omitted. The description of *Amaltheus stokesi* appears to be based on an abnormal specimen with longitudinal bands in an area of sublethal damage and radial bands on the undamaged shell (see Fig. 2C); because of the possibility of confusion, this specimen is omitted from the table.

molluscs associated with these ammonoids in well-preserved faunas from the North American Midcontinent. We provide two examples below.

The famous *Konservat–Lagerstätte* called the Buckhorn asphalt (Late Carboniferous, Oklahoma USA) has yielded a moderately diverse cephalopod fauna. There are several fossil-bearing beds at this locality, and moderate lithologic variation in each bed has been observed. Detailed sedimentological studies of this site have not been published; however, the deposit is, in part, a carbonate sand and essentially unaltered (aragonite and primary calcite) shells glued together by a natural asphalt (R. Crick, personal communication, 1995). Since the initial work by Smith (1938), extensive studies have been done on the shell structures of the diverse and plentiful orthoconic nautiloids and ammonoids. Despite the extraordinary fossil preservation, not a single cephalopod with a color pattern has been observed, even though at least six specimens of the cooccurring gastropod *Naticopsis* have been reported with a color pattern of dark bands on a white background (Squires, 1976).

Another Carboniferous occurrence with a plentiful and diverse cephalopod fauna is known from the Imo Formation in north-central Arkansas USA (Gordon, 1965; Saunders, 1973; Saunders *et al.*, 1977; Mapes, 1979). Color patterns on numerous kinds of cephalopods (Table III) and other invertebrates from this site have been reported (Jeffery *et al.*, 1994). The cephalopods with color patterns include bactritoids and coiled, cyrtoconic, and orthoconic nautiloids. However, despite the thousands of ammonoids recovered from the same beds and in the same exposures as the bactritoids and nautiloids, not a single specimen with a color pattern has been observed or reported.

TABLE III. Cephalopoda from the Imo Formation (Upper Mississippian, Arkansas) with Observed Color Patterns

Occurrence	Source	Pattern type
Bactritoidea		
?*Rugobactrites*	Mapes, 1979	Black patches on dorsum
Angustobactrites saundersi	Mapes, 1979	Dark longitudinal bands on sides
Coiled nautiloids		
Liroceras bicostatum	R. H. Mapes, this report	White transverse bands
Undescribed new genus related to *Stroboceras*	R. H. Mapes, this report	Light longitudinal bands
Orthoconic nautiloids		
?*Tripteroceroides* sp.	Gordon, 1965	Dark chevrons on dorsum
Euloxoceras sp.	R. H. Mapes, this report	Dark longitudinal bands on sides
?*Mooreoceras* sp.	R. H. Mapes, this report	Dark dorsum and light venter
Cyrtoconic nautiloids		
?*Mariceras* sp. A	R. H. Mapes, this report	Dark longitudinal bands
?*Mariceras* sp. B	Windle, 1973; R. H. Mapes, this report	Dark chevrons

Based on these empirical observations, it appears that Paleozoic ammonoids were either achromatic or monochromatic. A piece of supporting evidence is provided by the way that present-day *Nautilus* lays down its shell layers as compared to the occurrence of *Runzelschicht* (wrinkle layer) in Paleozoic ammonoids. *Nautilus* deposits its "black layer" of melanin-rich organic material on top of the previously developed whorl, which contains the color pattern in the outermost shell layer. *Nautilus* then deposits nacreous shell on top of the "black layer." Thus, beneath the dorsal shell of the outermost whorl is the "black layer," and beneath that is the color pattern of the previous whorl. The *Runzelschicht* occupies the same position in Paleozoic (House, 1971) and Mesozoic (Tozer, 1972) ammonoids as does the combined nacreous layer and "black layer" in *Nautilus*. In other words, it covers the outermost shell layer of the previous whorl and, hence, any color pattern in the wall of the previous whorl. Ammonoid color patterns have been observed through the thin wrinkle layer in Triassic (Mapes and Sneck, 1987) and Cretaceous specimens (Tanabe and Kanie, 1978; Bardhan *et al.*, 1993). If color patterns were present in Paleozoic ammonoids, they should be detectable in at least some specimens when the conchs are extracted from fine-grained limestone matrices, concretions, or shales. We have observed tens of thousands of Late Paleozoic ammonoids, but not a single instance where such a color pattern has been preserved. Thus, we are forced to conclude that Paleozoic ammonoids may not have secreted pigments in patterns in their shells.

4. Recognition of Fossil Color Patterns

The patterns that are described herein are presumed to represent the original, biologically deposited color pattern that was present when the animal was alive. When compared to those of *Nautilus* and other externally shelled molluscs (for example, pelecypods and gastropods), the color patterns in ammonoids have characteristics that support a biological origin.

4.1. Symmetry

Overall bilaterally symmetrical color patterns on a bilaterally symmetrical conch would seem to indicate that these patterns were originally present on the shell of the living animal.

4.2. Confinement of the Pattern to the Outer Shell Layer

According to Grégoire (1987), Valenciennes (1839) was the first person to point out that the pigment that gives rise to the coloration in *Nautilus* is confined to the outermost portion of the porcelaneous layer of the shell wall. In this chapter also, the color patterns accepted as being of primary biological origin seem to be in the outermost layer of the shell (for examples see Pinna, 1972; Mapes and Sneck, 1987).

An exception to this is the report by Manley (1977), who described a pattern of elongated dark spots in a thin coating on the steinkern of a body chamber of *Asteroceras stellare* (Sowerby). This set of dark spots almost certainly constitutes a legitimate, biologically derived color pattern; however, the mechanism by which this pattern was preserved is an unexplained geochemical mystery. Another set of dark color markings were reported to occur on the steinkerns of Cretaceous ammonoid phragmocones by Hewitt *et al.* (1991); these are related to internal organic membranes in the camerae and do not constitute a color pattern in the sense used in this chapter.

4.3. False Color Patterns

Recognition of false (= nonbiological) color patterns as such can be a problem. Probably the most common occurrence is the condition of shell thickening that occurs at constrictions, pseudoconstrictions, and varices (see Chapter 12, this volume). Under the right circumstances, the semitranslucent, recrystallized ammonoid test will be a uniform gray color, but, where a shell thickening occurs, the thickened test appears to be a darker gray color band. Thus, a pattern of periodic dark transverse bands is produced; these give the impression of a biologically developed color pattern. Another false pattern is the longitudinal band that is sometimes observed on a phragmocone where the umbilical seam of a former whorl once rested. Foerste (1930, p. 143) observed such a false color pattern on a specimen of the Paleozoic goniatite *Homoceras striolatum* (Phillips), in which the pattern consisted of dark "brown, and rather thick markings along the suture lines as seen through the translucent shell." Both of these kinds of false transverse and longitudinal color patterns are exhibited on a specimen of the goniatite *Gatherites morrowensis* (Miller and Moore) (Fig. 4A) from the Lower Pennsylvanian of Arkansas.

This kind of false color pattern can be detected by noting the coincidence of darker bands and shell thickenings, by determining whether the pigmentation of the pattern is developed throughout the thickness of the shell or is confined to the outer layer of the shell, and by determining whether a part of the conch is missing that would be coincident with the position of the umbilical seam. Some of these criteria were applied by Bardhan *et al.* (1993) when they correctly reported biologically emplaced radial color bands that were present only in the outer shell layer at each constriction on four specimens of *Calliphylloceras* sp. aff. *C. demedoffi* from the Cretaceous of India. (The fact that color bands correspond in position to constrictions in these specimens may well result from growth allometry. If constrictions and related features represent a slowdown in the rate the peristome of the conch moved forward, but pigment was deposited at a constant rate over time, then relatively more pigment would have been deposited at a constriction than between constrictions. This could result in darker bands at constrictions.)

4.4. Ultraviolet Light Detection of Relict Color Patterns

Several workers have discovered relict color patterns in molluscs by using ultraviolet light (for examples see Neuffer, 1971; Wilson, 1975). Several techniques to enhance the patterns under ultraviolet light include exposure to sunlight and soaking in household bleach (Wilson, 1975). However, at this time there are no reports of color patterns in ammonoids having been revealed by fluorescence under ultraviolet light.

4.5. Chemical Detection of Color Patterns

Detection of color patterns in ammonoid shells by chemical analysis has yet to be reported. In part, this is because of the limited numbers of specimens available for study, together with the fact that most chemical analyses that have been used to test hard parts of other invertebrate taxa required destruction of the specimen (Blumer, 1965).

5. Taphonomy of Color Patterns

Based on the available evidence, it is apparent that at least some ammonoid taxa had well-developed color patterns. Yet, when large collections of ammonoids are examined in repositories, virtually all of the recovered ammonoids show no patterns. This holds true even for conspecific specimens from the same exposures and layers that have yielded the specimens with color patterns enumerated in Table I. The taphonomic processes by which these biologically derived color patterns in ammonoids and in other organisms are destroyed are nowhere near well enough studied or understood. It is fair to say, however, that preservation of color patterns depends on the chemistry and stability of the pigments involved, on the original mineralogy of the shells, and on the taphonomic processes to which the specimens have been subjected (Hollingworth and Barker, 1991). Some of the more obvious mechanisms of color pattern destruction are briefly outlined in the following.

5.1. Mechanical Destruction

Because in most molluscan shells the color pattern consists of one or more organic pigments in the outermost part of the shell, it can be expected that any appreciable amount of abrasion will remove the color pattern. Such abrasion might occur in life or be postmortem. However, when mechanical fragmentation processes cause shells to be broken into pieces, it is logical that some of the pieces should be recovered with some part of the color patterns intact—like fragments of painted pottery at an archaeological site. This has not been observed in the fossil record, which forces the conclusion that other, more significant taphonomic forces were operating to destroy the organic pigments that form the color patterns. An additional consideration must include the fact that most ammonoid conchs are relatively thin sided and that the color patterns are confined to the outermost layer of that relatively thin shell.

5.2. Sun "Bleaching"

Present-day molluscs characteristically lose their color patterns and become white in a relatively short time (days to months) when exposed to extended periods of sunlight while lying on the beach or even during flotation. The presumed process by which this happens is the breakdown of pigments as a result of exposure to the ultraviolet light in sunlight. It may well be also that chemical changes are enhanced by the range of temperature variations associated with the diurnal cycle of exposure to direct sunlight.

5.3. "Bleaching" in Submarine Environments

Sun "bleaching" is probably most important in fossil shells preserved in beach or shoreline deposits; however, a large number of the cephalopods in many collections are not from such environments. Indeed, there is a growing body of evidence that many ammonoid conchs did not float long distances to become beached on strand lines (Boardman *et al.*, 1984). Rather, they were negatively buoyant for a variety of reasons, and the conchs sank to the bottom and eventually were buried by sediment essentially where they lived. These specimens undoubtedly could not have lost their color patterns to sun "bleaching".

Present-day molluscs that are dead and remain in marine environments, unburied by sediment, lose their color patterns. Degradation of pigments can take place by the action of bacteria and through boring by endolithic algae. It may well be that some organic pigments are unstable with long-term exposure to marine waters themselves. Even after burial, the interstitial water in the sediment almost certainly is deleterious to pigments under certain circumstances.

5.4. Recrystallization

The conversion of molluscan shell from original aragonite to secondary calcite is very common and probably is responsible for much of the color pattern loss in molluscs. Recrystallization probably not only has a negative impact on the chemistry of the organic pigments but also destroys the organic membranes that acted as templates for biomineralization as the ammonoid secreted its shell.

5.5. Diagenesis

Based on the reported occurrences of ammonoid color patterns (Table I), the best chances for color-pattern preservation are in fine-grained carbonate concretions, in limestone beds that formed very early in the process of diagenesis, or in rapidly dewatered clays, where the conchs were buried and sealed from pore waters relatively rapidly. There are no reported occurrences of ammonoid color patterns being preserved in sandstone or coarse-grained sediments, even though turbidity currents and storm deposits could have been expected to bury conchs rapidly, including those with color patterns.

Detailed analysis of the diagenetic history of matrices that contain color-bearing ammonoid conchs almost certainly would yield very important insights; however, none has been so analyzed. Comparison of rocks of nearby units that do not yield such specimens probably would reveal slight lithologic variations that are extremely important in color-pattern preservation.

6. Possible Functions of Ammonoid Color Patterns

A number of workers have attempted to define the paleobiological function or functions of colors and color patterns produced by molluscs as well as of those of other invertebrates (see, for example, the discussions and the references cited in Fox, 1972; Kennedy and Cobban, 1976; Hollingworth and Barker, 1991; Bardhan et al., 1993; and, especially, Kobluk and Mapes, 1989). The many potential functions and reasons that have been proposed may be subdivided into five categories: those related to metabolism, to vision, to shell strength, to light screening, and to thermoregulation.

6.1. Metabolism

It commonly is stated that pigments are by-products of metabolic processes, and the relationship of pigments to molluscan diet and waste disposal has been commented on by numerous authors (see Kobluk and Mapes, 1989). In some gastropods and pelecypods, for example, metabolic by-products or dietary wastes are disposed of by being secreted into the shell as pigments (Comfort, 1951). In some gastropods, shell color may be altered by changes in diet, whereas in other snail taxa, no such relationship can be established (Hollingworth and Barker, 1991). However, as pointed out by Boucot (1990, p. 447), it is unlikely that metabolic storage is the whole story.

Regardless of whether pigments in ammonoid shells are metabolic by-products, they are actual and potential sources of information about the metabolism of these animals. For example, the composition of pigments could tell us about the chemical processes that occurred in the living ammonoid. Patterns that are not continuous longitudinally on the shell show that the secretory activities of the pigment-producing cells at a given place on the mantle were able to be turned on and off during ontogeny. Longitudinal bands of pigment separated by stripes without pigment demonstrate that different loci on the mantle edge were able to secrete different materials simultaneously. Stripes that are neither longitudinal nor parallel to growth lines indicate that a locus for the deposition of pigment moved laterally along the edge of the mantle during ontogeny. Thus, not surprisingly, ammonoid cephalopods were like many other groups of molluscs with respect to the processes of color-pattern formation.

6.2. Vision

An ammonoid might avoid becoming a meal of a would-be predator if it were invisible to that predator. On the other hand, a hungry ammonoid might have a better chance of catching another animal if the ammonoid were invisible to potential prey. Thus, an obvious possible function for ammonoid color patterns was camouflage, whereby the ammonoid looked like its background. Such adaptive concealment might involve the actual matching of the colors and patterns of the background against which a given animal might find or place itself. This is the way that chameleons are portrayed in cinema cartoons. At a less mimetic level, a given color-pattern might countershade an organism or break up its outline sufficiently that it might not be noticed by potential predators or prey. Stenzel's (1964, p. K71) description of *Nautilus* might well have applied to some ammonoids: "The brown bands disrupt the contour of the animal, like the bands on zebras and warships; *Nautilus* is extremely difficult to see even in clear shallow water, because the color markings resemble shadows cast by sea fans and seaweeds and simulate the effect of the play of sunlight on surface ripples of the sea."

According to Packard (1988), some present-day octopi have white spots that tend to attract the retinal foveas of would-be predators and, thus, divert attention away from other features that might reveal to the predator the presence of the octopus as such. There is, however, no evidence of this phenomenon, which Packard termed *epistrepsis*, in ammonoids.

In certain circumstances, concealment may not be to an animal's best advantage. An organism may need to display the fact that it belongs to a particular species, either to attract potential mates or to warn off potential competitors. In the former case, it may need to advertise its gender and the fact that it is in a sexually mature condition. Mapes and Sneck (1987) reported two different color patterns within the Triassic ammonoid *Owenites* from the same locality and suggested that they represented sexual antidimorphs (compare Fig. 3A–C and 3G).

Likewise, within the realm of potential advertising are territorial displays and coloration warning that an organism is dangerous or distasteful and should be left alone (aposematic coloration). Of course, we have no evidence that such phenomena occurred in the Ammonoidea. Nor has anyone suggested that ammonoids resorted to mimicry in the strict sense of that word.

As pointed out by Packard (1988) and Kobluk and Mapes (1989), among others, the effectiveness of a particular color pattern as camouflage depends on the nature and quality of the visual apparatus and nervous system of the would-be predator or potential prey. Not all organisms are equipped to differentiate colors, shades of gray, shapes, and patterns to the same extent and in the same detail. According to Kobluk and Mapes (1989, p. 71), coleoid cephalopods are capable of forming images and differentiating colors, whereas *Nautilus* is unable to perceive color or to distinguish different shades of gray. We do not know, of course, how ammonoids perceived the colors and color patterns of their brethren (or anything else, for that matter).

Moreover, the properties of the water in which the ammonoid lived cannot be ignored; depth and turbidity both would have affected the wavelength (color) and the intensity of the light available for vision in both predators and prey. At the very least, it is a reasonable assumption that colors and color patterns were important visually only to ammonoids in the photic zone. However, in darker environments, pigments may have served a visual function in that they would have absorbed light, thereby reducing the albedo (reflectance) of the pigmented animal, which would have helped the creature disappear into the semidarkness.

Not all shells of ammonoids are smooth, of course. Sculpture can produce shadows and break up the outline of an animal and thus be functionally equivalent to color patterns (Cowen *et al.*, 1973). Moreover, ribs, plications, and so on, when combined with color patterns, might have been doubly effective as agents of camouflage as well as of other phenomena related to vision.

It must be pointed out that some present-day gastropods that have shells bearing color patterns also have a periostracum that is so thick and opaque that the color pattern cannot be seen in the living animal (William K. Emerson, personal communication, 1995). In such cases, the color pattern can serve no function that depends on its visibility. It is known that some ammonoids had a thick periostracum (for example, *Gaudryceras* from the Upper Cretaceous; see Doguzhaeva and Mutvei, 1993). Even though color patterns have not been reported in this genus, there is at least the possibility that the color patterns of some ammonoids may not have been visible in life.

6.3. Shell Strength

Melanins and some other pigments may have a function unrelated to color patterns. These substances add strength and hardness to materials (Burtt,

1979); this would make them more resistant to abrasion and, perhaps, to breakage. If this were so in ammonoid shell material, it is possible that the beneficial effects of color patterns were not necessarily restricted to photic environments.

6.4. Light Screening

In some invertebrates, pigments seem to serve a function of filtering out harmful amounts and wavelengths of light and other electromagnetic radiation (Kobluk and Mapes, 1989). It is not impossible that color and color patterns in ammonoids may have served some such function.

6.5. Thermoregulation

It is a commonplace that pigmentation in terrestrial animals can be important in temperature control. Kobluk and Mapes (1989, p. 77) argued that a light ventral surface might be similarly important for aquatic organisms in "very shallow tropical waters where brightly lit sands can be significantly warmer than the surrounding waters."

6.6. Possibilities versus Probabilities

Of the various possible functions for color and color patterns enumerated in the preceding paragraphs, three seem likely for ammonoids: camouflage (in the broad sense of that word), displays, and storage of metabolic by-products ("waste disposal").

Bardhan *et al.* (1993), following Seilacher (1972) and others, argued that low variability of color patterns within an animal population indicates that the color pattern under consideration has adaptive significance. On the other hand, Packard (1988), in his discussion of the pattern-recognition system of a predator versus the pattern-generating system of a prey animal, posited that a predator is less likely to recognize an individual prey animal if the prey image of the given species varies widely. In Packard's words (p. 100), ". . . in visual matters, high variability may be a positive advantage." (As an aside, this argument is equally applicable to visual effects produced by sculpture, peristome configuration, and so on.)

At the risk of "stating the obvious," it must be pointed out that a given color or color pattern may have had no one, single function. It may have served a multitude of functions, perhaps under different ecological or ethological circumstances. It also is possible that the occasional particular coloration in a particular taxon may have served no adaptive function at all.

7. Future Studies of Ammonoid Color Patterns

As is patently obvious in the preceding parts of this chapter, there is much to be learned with respect to ammonoid color and color patterns. Moreover, even what we do know about these subjects has the real, mostly unrealized potential to throw light on other aspects of the paleobiology of the Ammonoidea.

More material is needed. The limited number of specimens of ammonoids bearing color patterns currently available simply is not adequate for extensive, in-depth studies. In order to gain greater understanding of paleobiology and the paleoenvironmental conditions of these animals, cephalopod workers must build a more extensive data base. The way to start is to search available collections and report color pattern occurrences. Moreover, ammonoid workers in the field need to "keep their eyes open" for color patterns. So many questions might be answered with the aid of more specimens. How much variability in color pattern is there within a species? Among the species of a genus? What about color patterns on steinkerns (Manley, 1977)?

Present-day ammonoid systematics are established on the basis of suture pattern, conch form, and sculpture. What would be the impact if color pattern data were integrated? Bardhan *et al.* (1993) suggested that color patterns varied somewhat on the four specimens of what they considered to be a single species of *Calliphylloceras*, and different species of *Pleuroceras* have been reported with different color patterns. Boucot's comment on taxonomy (1990, p. 569), although written in reference to brachiopods, is applicable to ammonoids: "I would not be shocked to find that a species . . . was equivalent to two or three living species, in view of our ignorance about such things as shell color patterns . . . ."

What pigments are present in ammonoids? Some of the new (to ammonoid studies) methods of detection of color patterns and of microchemical analysis should yield new and exciting results.

Did ammonoids have chromatophores? Packard (1988, p. 91) pointed out that nickel is known to be concentrated in chromatophores. He suggested minute scanning of belemnoids for tiny point concentrations of this element. The same perhaps could be done on specimens of ammonoids that seem to have remnants of soft parts preserved. If ammonoids did have chromatophores, fixed color patterns in shells are reduced to but a small part of a much larger and potentially dynamic chromatic picture.

Analysis of the taphonomy of color patterns—how they are preserved, altered, and destroyed—not only would allow a better assessment of the taphonomic history of ammonoid conchs with color patterns but also would permit a better understanding of color patterns in the once-living animals.

Paleoecological studies need to be done to relate ammonoid color patterns to the conditions that existed when and where the animals lived. Did particular patterns predominate in particular environments, for example, depths? Is there a correlation between color patterns and the incidence and nature of predation events?

There is no question that future functional analyses of ammonoid color patterns in relation to shell shape and sculpture, subaqueous optics, paleoenvironmental and taphonomic conditions, and so on will give real insights into ammonoid paleobiology.

ACKNOWLEDGMENTS. We wish to thank the Ohio University Research Committee and the National Geographic Society for providing support of the field work that allowed the collecting of some specimens with color patterns in Nevada. Also, we wish to thank Drs. David Work, Rex Crick, and William K. Emerson for providing unpublished information for this project. We are grateful to Drs. Ulrich Lehmann and K. Tanabe for providing pictures and information regarding ammonoids with color patterns. Additionally, part of this research was supported by a grant from the National Science Foundation (EAR-9117700) to R. H. Mapes and G. Mapes. Acknowledgment is also made by R. H. Mapes to the donors of the American Petroleum Research Fund, administered by the American Chemical Society, for partial support of this research (PRF No. 15821-AC2 and PRF No. 20742-B8-C).

References

Arkell, W. J., 1957, Introduction to Mesozoic Ammonoidea, in: *Treatise on Invertebrate Paleontology*, Part L, *Mollusca 4* (R. C. Moore, ed.), Geological Society of America and University of Kansas Press, Lawrence, KS, pp. L80–L129.

Bardhan, S., Jana, S. K., and Datta, K., 1993, Preserved color pattern of a phylloceratid ammonoid from the Jurassic Chari Formation, Kutch, India, and its functional significance, *J. Paleontol.* **67**(1):140–143.

Blumer, M., 1965, Organic pigments; their long-term fate, *Science* **149**:722–726.

Boardman, D. R. III, Mapes, R. H., Yancy, T. E., and Malinky, J. M., 1984, A new model for the depth-related allogenic community succession within North American Pennsylvanian cyclothems and implications on the black shale problem, in: *Limestones of the Mid-Continent* (N. J. Hyne, ed.), *Tulsa Geol. Soc. Spec. Pub.* **2**:144–182.

Boucot, A. J., 1990, *Evolutionary Paleobiology of Behavior and Coevolution*, Elsevier, Amsterdam.

Burtt, E. H., Jr., 1979, Tips on wings and other things, in: *The Behavioral Significance of Color* (E. H. Burtt, Jr., ed.), Garland STPM Press, New York, pp. 75–110.

Comfort, A., 1951, The pigmentation of molluscan shells, *Biol. Rev. Camb. Phil. Soc.* **26**(3):285–301.

Cowen, R., Gertman, R., and Wiggett, G., 1973, Camouflage patterns in *Nautilus* and their implications for cephalopod paleobiology, *Lethaia* **6**(2):201–213.

Doguzhaeva, L., and Mutvei, H., 1993, Structural features in Cretaceous ammonoids indicative of semi-internal or internal shells, in: *The Ammonoidea: Environment, Ecology and Evolutionary Change*, Systematics Association Special Volume 47 (M. R. House, ed.), Clarendon Press, Oxford, pp. 99–114.

d'Orbigny, A., 1842, *Paléontologie Française. Terraines Oolitiques ou Jurassiques. I. Céphalopodes*, Masson, Paris.

Foerste, A. F., 1930, The color-patterns of fossil cephalopods and brachiopods, with notes on gastropods and pelecypods, *Contrib. Mus. Paleont. Univ. Mich.* **3**(6):109–150.

Fox, D. L., 1972, Chromatology of animal skeletons, *Am. Sci.* **60**:436–447.

Gordon, M., Jr., 1965, *Carboniferous cephalopods of Arkansas*, *U.S. Geological Survey Professional Paper 460*, U.S. Geological Survey, Washington, DC.

Grégoire, C., 1987, Ultrastructure of the *Nautilus* shell, in: *Nautilus—The Biology and Paleobiology of a Living Fossil* (W. B. Saunders and N. H. Landman, eds.), Plenum Press, New York, pp. 463–486.

Greppin, E., 1898, Description des fossiles du Bajocien Supérieur des environs de Bâle, *Mém. Soc. Paléont. Suisse* **25**:1–52.

Heller, F., 1977, Ein *Pleuroceras* aff. *solare* (PHILL.) mit gut erhaltener Farbzeichnung aus den Amaltheentonen Frankens, *Geol. Bl. Nordost-Bayern Angrenzende Geb.* **27**(3–4):161–168.

Hewitt, R. A., Checa, A., Westermann, G. E. G., and Zaborski, P. M., 1991, Chamber growth in ammonites inferred from colour markings and naturally etched surfaces of Cretaceous vascoceratids from Nigeria, *Lethaia* **24**(3):271–287.

Hoare, R. D., 1978, Annotated bibliography on preservation of color patterns on invertebrate fossils, *Comp. Sig. Gam. Eps.* **55**(3):39–63.

Hollingworth, N. T. J., and Barker, M. J., 1991, Colour pattern preservation in the fossil record: Taphonomy and diagenetic significance, in: *The Processes of Fossilization* (S. K. Donovan, ed.), Belhaven Press, London, pp. 105–119.

House, M. R., 1971, The goniatite wrinkle-layer, *Smithson. Contrib. Paleobiol.* **3**:23–32.

Jeffery, D. L., Hoare, R. D., Mapes, R. H., and Brown, C. J., 1994, Gastropods (Mollusca) from the Imo Formation (Mississippian, Chesterian) of North-Central Arkansas, *J. Paleontol.* **68**(1):58–79.

Kennedy, W. J., and Cobban, W. A., 1976, Aspects of ammonite biology, biogeography, and biostratigraphy, *Palaeontol. Assoc. Spec. Pap. Palaeontol.* **17**.

Kennedy, W. J., Klinger, H. C., and Summesberg, H., 1981, Cretaceous faunas from Zululand and Natal, South Africa. Additional observations on the ammonite subfamily Texanitinae Collignon, 1948, *Ann. S. Afr. Mus.* **86**(4):115–155.

Kobluk, D. R., and Mapes, R. H., 1989, The fossil record, function, and possible origins of shell color-patterns in Paleozoic marine invertebrates, *Palaios* **4**(1):63–85.

Lehmann, U., 1990, *Ammonoideen—Leben zwischen Skylla und Charybdis*, Ferdinand Enke Verlag, Stuttgart.

Manley, E. C., 1977, Unusual pattern preservation in a Liassic ammonite from Dorset, *Palaeontology (Lond.)* **20**(4):913–916.

Mapes, R. H., 1979, Carboniferous and Permian Bactritoidea (Cephalopoda) in North America, *Univ. Kans. Paleontol. Contrib.* Article 64.

Mapes, R. H., and Hoare, R. D., 1987, Annotated bibliography for preservation of color-patterns on invertebrate fossils, *Comp. Sig. Gam. Eps.* **65**:12–17.

Mapes, R. H., and Sneck, D. A., 1987, The oldest ammonoid "colour" patterns: Description, comparison with *Nautilus*, and implications, *Palaeontology (Lond.)* **30**(2):299–309.

Matsumoto, T., and Hirano, H., 1976, Colour patterns in some Cretaceous ammonites, *Palaeontol. Soc. Jpn. Trans. Proc. N. S.* **102**:334–342.

Neuffer, O., 1971, Nachweis von Farbungsmustern an tertiaren Bivalven unter UV–Licht, *Heinz-Tobien Festschrift Hess. Landesamt Bodenforsch. Abh.* **60**:121–129.

Packard, A., 1988, Visual tactics and evolutionary strategies, in: *Cephalopods—Present and Past* (J. Wiedmann and J. Kullmann, eds.), Schweizerbart'sche Verlagsbuchhandlung, Stuttgart, pp. 89–103.

Pinna, G., 1972, Presenza di tracce di colore sul guscio di alcune ammoniti della famiglia Amaltheidae Hyatt, 1867, *Soc. Ital. Sci. Nat. Mus. Civ. Stor. Nat. Milano Atti.* **113**(2):193–200.

Reyment, R. A., 1957, Über Farbspuren bei einigen Ammoniten, *N. Jb. Geol. Paläont. Mh.* **7–8**:343–351.

Saunders, W. B., 1973, Upper Mississippian ammonoids from Arkansas and Oklahoma, *Geol. Soc. Am. Spec. Pap.* **145**.

Saunders, W. B., Manger, W. L., and Gordon, M., Jr., 1977, Upper Mississippian and lower and middle Pennsylvanian ammonoid biostratigraphy of northern Arkansas, *Okla. Geol. Surv. Gdbk.* **18**:117–137.

Schindewolf, O. H., 1928, Über Farbstreifen bei *Amaltheus (Paltopleuroceras) spinatum* (Brug.), *Paläontol. Z.* **10**:136–143.

Schindewolf, O. H., 1931, Nochmals über Farbstreifen bei *Amaltheus (Paltopleuroceras) spinatus* (Brug.), *Paläontol. Z.* **13**(4):284–287.

Seilacher, A., 1972, Divaricate patterns in pelecypod shells, *Lethaia* **5**:325–343.

Smith, H. J., 1938, *The Cephalopod Fauna of the Buckhorn Asphalt*, private edition, distributed by University of Chicago Libraries, Chicago.

Spath, L. F., 1935, On colour-markings in ammonites, *Ann. Mag. Nat. Hist. Ser. 10* **15**(87):395–398.

Squires, R. L., 1976, Color-pattern of *Naticopsis (Naticopsis) wortheniana*, Buckhorn Asphalt deposit, Oklahoma, *J. Paleontol.* **50**(2):349–350.

Stenzel, H. B., 1964, Living *Nautilus*, in: *Treatise on Invertebrate Paleontology*, Part K, *Mollusca 3* (R. C. Moore, ed.), Geological Society of America and University of Kansas Press, Lawrence, KS, pp. K59–K93.

Tanabe, K., and Kanie, Y., 1978, Colour markings in two species of tetragonitid ammonites from the Upper Cretaceous of Hokkaido, Japan, *Sci. Rep. Yokosuka City Mus.* **25**:1–6.

Teichert, C., 1964, Morphology of hard parts, in: *Treatise on Invertebrate Paleontology*, Part K, *Mollusca 3* (R. C. Moore, ed.), Geological Society of America and University of Kansas Press, Lawrence, KS, pp. K13–K53.

Tozer, E. T., 1972, Observations on the shell structure of Triassic ammonoids, *Palaeontology* (*Lond.*) **15**(4):637–654.

Valenciennes, M. A., 1839, Nouvelles recherches sur le Nautile flambé (*Nautilus pompilius*, Lam.), *Arch. Mus. Nat. Hist.* **2**:257–314. (The date for this paper is commonly given as 1841; however, van der Hoeven's copy is designated "Exemplaire d'auteur" and is dated 1839.)

Wilson, E. C., 1975, Light show from beyond the grave, *Terra* **13**(3):10–13.

Windle, D. L., Jr., 1973, *Studies in Carboniferous nautiloids: Cyrtocones and annulate orthocones*, Ph.D. Dissertation, University of Iowa, Iowa City, Iowa.

Chapter 6

Septal Neck–Siphuncular Complex of Ammonoids

KAZUSHIGE TANABE and NEIL H. LANDMAN

1. Introduction

One of the characteristic features of virtually all of the fossil Cephalopoda is the presence of a chambered shell with a siphuncle. Among extant cephalo-

KAZUSHIGE TANABE • Geological Institute, University of Tokyo, Tokyo 113, Japan. NEIL H. LANDMAN • Department of Invertebrates, American Museum of Natural History, New York, New York 10024.

Ammonoid Paleobiology, Volume 13 of *Topics in Geobiology*, edited by Neil Landman *et al.*, Plenum Press, New York, 1996.

pods, chambered shells have been completely lost or reduced, for example, to chitinous gladii. Only three genera, *Nautilus, Sepia,* and *Spirula,* preserve chambered shells in the modern fauna. Denton and Gilpin-Brown (1973) demonstrated experimentally that the chambered shells of these three genera function as hydrostatic floats to adjust the density of the living animal to that of seawater by means of low-pressure gas within the chambers. They have also suggested that this basic function of chambered shells has remained the same in a great variety of forms and over a geologically long period of time.

In most extant and fossil chambered cephalopods, the siphuncular system consists of a tube-like organ that extends from the apex of the shell to the base of the living chamber. It passes through an opening in each septum around which there is a collar-like projection called the septal neck.

The shape and microstructure of septal necks differ markedly at higher taxonomic levels (Appellöf, 1893; Naef, 1922; Bandel and Boletzky, 1979). For example, in most of the Nautiloidea and Coleoidea excluding some Aulaco-cerida and *Naefia,* septal necks always are directed adapically (retrochoanitic), whereas in the Ammonoidea, the direction of septal necks commonly changes during ontogeny from adapical (retrochoanitic) to adoral (prochoanitic). Branco (1879–1880) first documented these changes in his study of the internal structures of ammonoid shells.

Since the work of Branco, many papers on ammonoid septal necks have been published describing their microstructure (Grandjean, 1910; Böhmers, 1936; Birkelund and Hansen, 1968, 1974; Kulicki, 1979; Tanabe *et al.*, 1979; Birkelund, 1981; Bandel, 1981, 1982; Kulicki and Mutvei, 1982; Doguzhaeva and Mutvei, 1986; Zakharov, 1989), ontogenetic development (Spath, 1933; Miller and Unklesbay, 1943; Bogoslovskaya, 1959; Mutvei, 1967; Bogoslovsky, 1969; Druschits and Khiami, 1970; Zakharov, 1971, 1974; Druschits and Doguzhaeva, 1974; Druschits *et al.*, 1976; Tanabe *et al.*, 1982; Doguzhaeva and Mutvei, 1986), and evolution (Schindewolf, 1950; Druschits *et al.*, 1976; Doguzhaeva, 1988). These papers have clarified our understanding of the structure of septal necks and their development in ontogeny and phylogeny. However, there still is no comprehensive treatment of septal necks among all ammonoid taxa.

We have made an extensive survey of septal necks in Paleozoic and Mesozoic Ammonoidea based on well-preserved specimens from many regions throughout the world (see Appendix). Our data base includes material from 107 species belonging to 27 superfamilies distributed among eight suborders. This chapter synthesizes our observations on these septal necks and includes a discussion of their evolution and morphogenesis.

Illustrated specimens are reposited in the following institutions: American Museum of Natural History (AMNH), Yale Peabody Museum (YPM), University of Iowa (SUI), Hamburg Universität (SGPIMH), Kyushu University (GK), and University Museum of the University of Tokyo (UMUT).

2. Septal Necks and Associated Structures

2.1. Septal Neck Architecture

As in other chambered cephalopods, the septal necks of ammonoids are continuous with the rest of the septum and are primarily composed of nacre (Tanabe *et al.*, 1993). They are, therefore, elements of the septa, not of the siphuncular tube. The septal necks of ammonoids can be classified into four types on the basis of their overall shape (Fig. 1): (1) *retrochoanitic necks* (Miller *et al.*, 1957), which are projected only adapically; (2) *modified retrochoanitic necks* (Tanabe *et al.*, 1993), which are projected adapically on the ventral side but both adapically and adorally (type A) or only adorally (type B) on the dorsal side; (3) *amphichoanitic necks* (Druschits and Khiami, 1970; = transitional necks of Kulicki, 1979), which are projected both adorally and adapically; and (4) *prochoanitic necks* (Miller *et al.*, 1957), which are projected only adorally.

Modified retrochoanitic necks have previously been classified as prochoanitic necks because of their adoral projection on the dorsal side (Druschits *et al.*, 1976; Kulicki, 1979; Doguzhaeva and Mutvei, 1986; Doguzhaeva, 1988). Amphichoanitic necks have also been considered special cases of prochoanitic necks (Druschits *et al.*, 1976; Birkelund, 1981; Doguzhaeva, 1988). However, both modified retrochoanitic and amphichoanitic necks are clearly

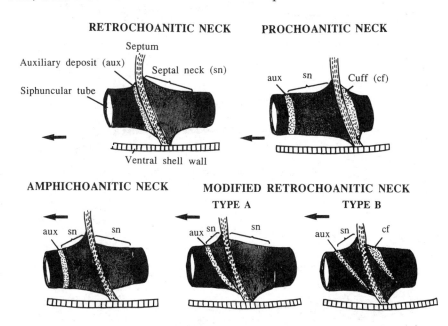

FIGURE 1. Three-dimensional views of retro-, modified retro-, pro-, and amphichoanitic septal necks in the Ammonoidea. Arrows indicate adoral direction. Septa are simplified. Modified from Tanabe *et al.* (1993, Figs. 4.1, 4.2).

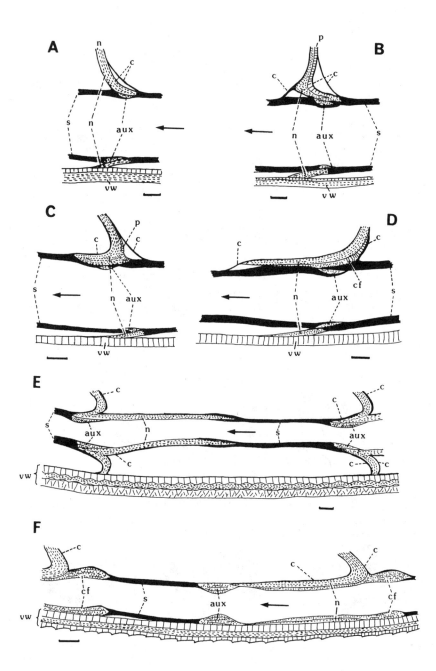

FIGURE 2. Diagrammatic median cross-sections through the septal necks of some Mesozoic Ammonoidea. (A-C) Ontogenetic transformation from retro- to Type A modified retrochoanitic necks in the fourth whorl of *Stolleyites tenuis* (Stolley) (Ceratitina: Nathorstitaceae). From the Carnian of Spitsbergen, AMNH 44354. (A) 42nd septum, (B) 43rd septum, (C) 46th septum. (D) Type B modified retrochoanitic neck in the fifth whorl of *Parapopanoceras paniculatum* Popow (Ceratitina: Megaphyllitaceae). From the Anisian of Taimyr, Arctic Siberia, AMNH 44352. (E) Amphichoanitic neck in the third whorl of *Hypophylloceras subramosum* (Shimizu) (Phylloceratina: Phyllocerataceae). From the Coniacian of the Haboro area, Hokkaido, AMNH 44364. (F) Prochoanitic neck in the third whorl of *Gaudryceras denseplicatum* (Jimbo) (Lytoceratina: Lytocerataceae). From the Coniacian of the Haboro area, Hokkaido, AMNH 44371. n, nacreous layer of the septum; p, spherulitic–prismatic layer of the septum; s, siphuncular tube; c, thin organic membrane (cameral membrane); aux, auxiliary deposit; cf, cuff; vw, ventral shell wall. Arrows indicate the adoral direction. Scale bars, 50 μm.

◄ ───

distinguished from prochoanitic necks on the basis of shape and associated structures (Tanabe *et al.*, 1993; Figs. 4.1, 4.2).

2.2. Associated Structures

Septal necks are associated with a number of organic and calcified tissues that altogether constitute the septal neck–siphuncular complex (Tanabe *et al.*, 1993; = siphuncular system of Bandel and Boletzky, 1979), a tube of alternating rigid and flexible rings (Hewitt *et al.*, 1993; Fig. 5.2). The associated tissues consist of (1) the siphuncular tube, (2) auxiliary deposit, (3) cuff, and (4) cameral membranes.

The siphuncular tube (= siphonal tube of Mutvei, 1967; s, Figs. 2–6, 11) is a tube that extends from the adoral side of one septal neck to the adapical side of the next. Because this tube connects two consecutive septa, it is also called a connecting ring (Miller *et al.*, 1957; Birkelund, 1981). The siphuncular tube consists of multilayered concentric membranes of conchiolin, each of which is constructed from lace-like tuberculate microfibrils like those of the inner horny layer of the siphuncular tube of Recent *Nautilus* (Erben and Reid, 1971; Obata *et al.*, 1980; Westermann, 1982; Hewitt and Westermann, 1983; Grégoire, 1984; Hasenmueller and Hattin, 1985; Fig. 6).

The auxiliary deposit is a calcified deposit that appears on the adoral side of all kinds of septal necks (aux, Figs. 1, 2, 4, 11). This term originally was introduced by Miller and Unklesbay (1943) to denote calcified deposits on the septal necks of Paleozoic ammonoids; it later was redefined by Kulicki (1979). The auxiliary deposit covers the inner surface of the septal neck and is replaced adorally by the conchiolin membranes of the siphuncular tube. In exceptionally well-preserved specimens, this deposit shows an annular layering consisting of loosely packed, aggregated spherulites (Kulicki and Mutvei, 1982, Pl. 2; Tanabe *et al.*, 1993, Figs. 4.8, 4.9; Fig. 4A).

Another spherulitic–prismatic deposit called the cuff by Kulicki and Mutvei (1982) commonly occurs on the adapical side of prochoanitic necks

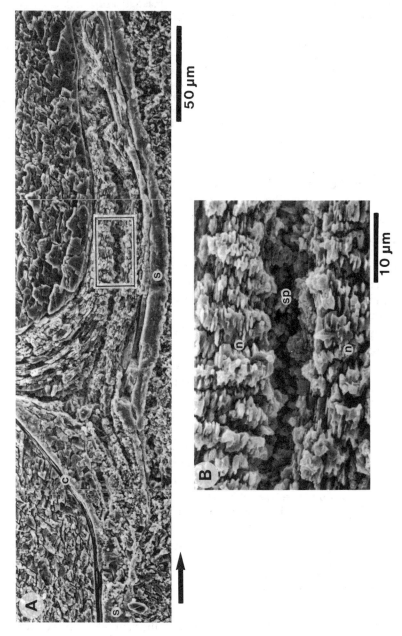

FIGURE 3. Median cross section through the dorsal portion of a type A modified retrochoanitic neck in the fourth whorl of *Arctohungarites ventroplana* (Popow) (Ceratitina: Danubitaceae). (A) Nacreous lamellae form a prominent fold directed both adorally and adapically. (B) Close-up of A, showing modified nacre with a spherulitic prismatic structure (sp) in the axial portion of the fold. From the Middle Anisian of East Taimyr, Arctic Siberia. UMUT MM 19061. n, nacreous lamellae; s, siphuncular tube. The arrow indicates the adoral direction.

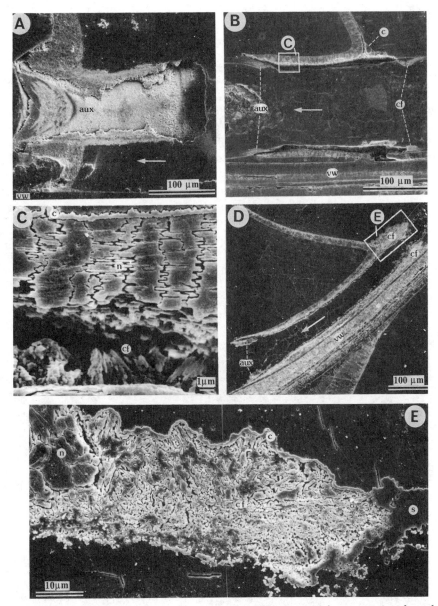

FIGURE 4. Septal necks in Cretaceous ammonoids. (A) Parasagittal cross section through an amphichoanitic neck in the fourth whorl of *Phyllopachyceras ezoense* (Yokoyama) (Phyllocerat-ina: Phyllocerataceae). From the Campanian of the Abeshinai area, Hokkaido, AMNH 44366 (same specimen reported by Tanabe *et al.*, 1993, Fig. 4.7C). (B,C) *Beudanticeras* sp. (Ammonitina: Desmocerataceae). From the Albian of southern Alaska, AMNH 44375 (same specimen described by Tanabe *et al.*, 1993, Fig. 4.10). (B) Median cross section through a long prochoanitic neck with well-developed cuff and auxiliary deposit in the third whorl. (C) Midportion of the septal neck on its dorsal side showing the cuff inside the nacreous layer. (D,E) *Anagaudryceras limatum* (Yabe) (Lytoceratina: Tetragonitaceae). From the Coniacian of the Haboro area, Hokkaido, AMNH 44374 (same specimen reported by Tanabe *et al.*, 1993, Fig. 4.9A,B). (D) Median cross section through a long prochoanitic neck with well-developed cuff and auxiliary deposit in the fourth whorl. (E) Adapical end of the septal neck on its dorsal side showing a well-developed cuff. Abbreviations are the same as in Fig. 2. Arrows indicate the adoral direction.

(Druschits and Doguzhaeva, 1974; Bandel, 1981; Kulicki and Mutvei, 1982; cuff or cf, Figs. 1, 2F, 4B–E, 11) and of type B modified retrochoanitic necks (Tanabe *et al.*, 1993, Fig. 4.5; Fig. 2D). Other septal necks lack cuffs; the conchiolin membranes of the siphuncular tube are attached directly to the nacreous layer of the septal neck on its adapical side (Figs. 2A–C, 3A, 11).

The adapical and adoral surfaces of the septa and outer surfaces of the septal necks and siphuncular tube are covered by thin (several microns thick) organic membranes, called cameral membranes (c, Figs. 2–4). These organic membranes are sometimes detached from the siphuncular tube and septum in the adapical region of a chamber, where they form one or two narrow spaces called "decoupling room" by Bandel and Boletzky (1979; Figs. 2A–C, 3A, 4B). Such organic membranes have been observed in well-preserved specimens of various ammonoid taxa (Grandjean, 1910; Shoulga-Nesterenko, 1926; Schindewolf, 1968; Erben and Reid, 1971; Bayer, 1975).

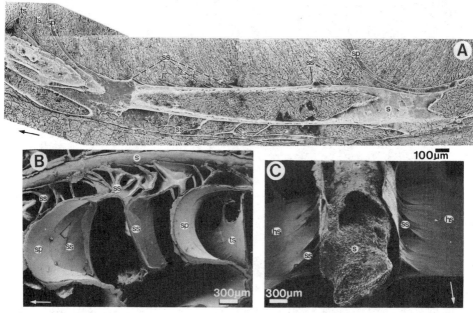

FIGURE 5. Well-preserved conchiolin cameral membranes in Permian and Triassic ammonoids. (A) *Artinskia electraensis* (Plummer and Scott) (Prolecanitina: Medlicottiaceae). Median cross section through the siphuncular tube showing siphuncular sheets (ss) and transverse sheets (ts), both of which are arranged parallel or subparallel to the septal surface. From the Middle Permian of Beck Mountain, Nevada, UMUT PM 19040-2. (B) *Czekanowskites hayesi* (McLearn) (Ceratitina: Danubitaceae), view into hollow chambers showing siphuncular sheets and transverse sheets. From the middle Anisian of Spitsbergen, SGPIMH 3593 (same specimen reported by Weitschat and Bandel, 1991, Fig. 10). (C) *Aristoptychites kolymensis* Kiparisova (Ceratitina: Pinacocerataceae). Frontal view of the ventral portion of a hollow chamber showing horizontal sheets (hs) and siphuncular sheets (ss). From the upper Ladinian of Spitsbergen, SGPIMH 3598 (same specimen as figured by Weitschat and Bandel, 1991, Fig. 18). sp, septum; s, siphuncular tube. Arrows indicate the adoral direction.

Recently, Weitschat and Bandel (1991) described cameral membranes preserved as phosphatized sheets in Triassic ammonoids from Spitsbergen and Siberia. According to their observations, the cameral membranes in these ammonoids form a complex network and can be grouped into three types, based on their distribution and orientation: (1) siphuncular sheets, (2) transverse sheets, and (3) horizontal (longitudinal) sheets. Siphuncular sheets are arranged parallel or subparallel to the septal surface and, together with horizontal sheets, subdivide a chamber into many compartments (Fig. 5).

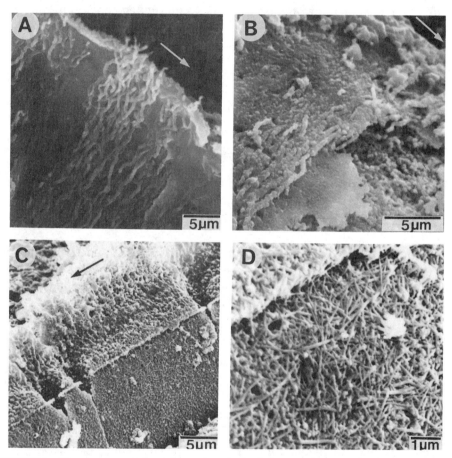

FIGURE 6. Ultrastructure of the siphuncular tube in Cretaceous ammonoids. (A,B) *Damesites semicostatus* Matsumoto (Ammonitina: Desmocerataceae). Inner (A) and outer (B) surfaces of the siphuncular tube consist of multilayered membranes of conchiolin. Each membrane is composed of reticulate, lace-like microfibrils, which are arranged perpendicular to the axis of the siphuncular tube. From the lower Santonian of the Obira area, Hokkaido, GK. H 8085 (same specimen reported by Obata *et al.,* 1980, Pl. 1, Fig. 5; Pl. 3, Fig. 7). (C,D) *Subprionocyclus (Reesidites) minimus* (Hayasaka and Fukada) (Ammonitina: Acanthocerataceae). A thin membrane consisting of unoriented microfibrils covers the outer surface of the siphuncular tube. From the upper Turonian of the Manji area, Hokkaido, UMUT MM 19344 (same specimen reported by Tanabe *et al.,* 1982, Pl. 67, Figs. 2,3). Arrows indicate the adoral direction.

Siphuncular sheets originate from the outer organic layer of the siphuncular tube and are attached dorsally to horizontal sheets and ventrally to the inner surface of the shell wall. Transverse sheets extend from the inner surface of the dorsal wall in a direction subparallel to the septal surface (ts, Fig. 5B). Commonly, these sheets stretch across the fluted margins of septa, forming small enclosures in front of the lobes. As suggested by Weitschat and Bandel (1991), cameral membranes may have been present in all ammonoids, but, because of their delicate structure, they have been mostly destroyed some time after death, possibly through bioerosion by fungi or bacteria. The ultrastructure of these cameral membranes consists of unoriented conchiolin microfibrils (less than 0.2 μm in diameter) (Tanabe *et al.*, 1982; Grégoire, 1984; Hasenmueller and Hattin, 1985; Figs. 6C, D), which are comparable to the pellicles (brown membranes) within the camerae of *Nautilus, Spirula,* and *Sepia* (Grégoire, 1962, 1973; Denton and Gilpin-Brown, 1966, 1973; Bandel and Boletzky, 1979). The pattern of distribution of cameral membranes in ammonoids is similar to that in *Sepia*. The cameral membranes might have served for cameral liquid transport (Bandel and Boletzky, 1979; Tanabe *et al.*, 1982; Weitschat and Bandel, 1991).

3. Distribution of Septal Necks among Suborders

3.1. Bactritina

The taxonomic position of the Bactritina has long been uncertain, but this group currently is regarded as the ancestral stock of the Ammonoidea (House, 1985). Septal necks are short and always retrochoanitic in this long-ranging (Devonian to Triassic) suborder (Erben, 1964a,b; Mapes, 1979).

3.2. Anarcestina, Agoniatitina, Gephuroceratina, and Clymeniina

Available data indicate that septal necks are retrochoanitic throughout ontogeny in the Anarcestina (e.g., *Anarcestes, Latanarcestes*: Spath, 1933; Bogoslovsky, 1969; Druschits *et al.*, 1976), Agoniatitina (e.g., *Agoniatites, Mimagoniatites*: Schindewolf, 1935; Bogoslovsky, 1969), Gephuroceratina (e.g., *Manticoceras, Timanites*: Bogoslovsky, 1969; Druschits *et al.*, 1976), and Clymeniina (e.g., *Clymenia, Platyclymenia*: Druschits *et al.*, 1976).

3.3. Prolecanitina

Septal necks were observed in two Permian species, *Pronorites praepermicus* (Pronoritidae) and *Artinskia electraensis* (Medlicottiidae). Both species have relatively long retrochoanitic necks throughout ontogeny (Fig. 5A), a condition similar to that in other genera previously studied (e.g., *Daraelites,*

Propinacoceras: Böhmers, 1936; *Sakmarites*: Bogoslovskaya, 1959; *Prolecanites*: Druschits *et al.*, 1976).

3.4. Goniatitina

Both retrochoanitic and modified retrochoanitic septal necks have been recognized in this suborder (Table I). All septal necks are initially retrochoanitic. Retrochoanitic necks are commonly replaced during ontogeny by modified retrochoanitic necks, first type A, then type B. In most species of the Goniatitaceae, Dimorphoceratacea, and Cyclolobaceae, this transformation occurs between four and seven whorls. In species of the Adrianitaceae and Agathicerataceae, modified retrochoanitic necks develop between two and three whorls; their appearance is relatively earlier than that in species of other superfamilies.

The transformation from retro- to type A modified retrochoanitic necks in the Goniatitina is essentially similar to that in the Ceratitina, documented in detail by Kulicki (1979, Fig. 1), Doguzhaeva and Mutvei (1986, Figs. 1–3), and Tanabe *et al.* (1993, Fig. 4.3).

Septal necks of *Goniatites* (Goniatitidae), *Owenoceras* (Gastrioceratidae), and *Homoceras* (Homoceratidae) are retrochoanitic in all of the ontogenetic stages we observed (up to eight whorls). It is unclear, however, whether these genera developed modified retrochoanitic necks in later ontogeny.

3.5. Ceratitina

As in the Goniatitina, both retrochoanitic and modified retrochoanitic necks occur in the Ceratitina (Table II). In every species studied septal necks are initially retrochoanitic. The retrochoanitic condition persists throughout ontogeny in *Olenikites, Svalbardiceras, Wasatchites, Owenites* (Noritaceae), and *Karangatites* (Danubitaceae), at least in all of the ontogenetic stages we observed. In other species retrochoanitic necks are transformed during ontogeny into modified retrochoanitic necks, first type A, then type B. The transformation from retro- to modified retrochoanitic necks has been documented carefully in two nathorstitaceans *Nathorstites gibbosus* (see Kulicki, 1979, Fig. 1) and *Stolleyites tenuis* (see Tanabe *et al.*, 1993, Fig. 4.3; Fig. 2A–C) and in the megaphyllitacean *Megaphyllites prometheus* (see Doguzhaeva and Mutvei, 1986, Text figs. 1–3). Examination of well-preserved specimens of *Arctohungarites ventroplana* (Danubitaceae) from Arctic Siberia reveals that the dorsal sides of modified retrochoanitic necks consist of a prominent fold of nacreous lamellae directed both adorally and adapically (Fig. 3A). The axial portion of the fold is occupied by modified nacre with a spherulitic prismatic structure (Fig. 3B; see also Tanabe *et al.*, 1993, Fig. 4.3D).

The timing of the ontogenetic transformation from retro- to modified retrochoanitic necks is markedly variable even among species within the same family. For example, in *N. gibbosus* and *S. tenuis* of the Nathorstitidae, type A modified retrochoanitic necks develop between four and five whorls, and this condition persists for the rest of ontogeny, whereas in *Indigirites tozeri* of the same family, type A modified retrochoanitic necks are themselves replaced by type B modified retrochoanitic necks (Table II).

3.6. Phylloceratina

There are two kinds of septal necks in this suborder, retrochoanitic and amphichoanitic (Table III). In every species examined, septal necks are initially retrochoanitic and become amphichoanitic in early to middle ontogeny. This change occurs at the end of the second whorl in the Middle Triassic (Ladinian) genus *Indigirophyllites*, whereas it occurs at some point within the first whorl in Early Jurassic through Cretaceous genera. In the Early Jurassic species *Tragophylloceras loscombei*, amphichoanitic necks are further modified at the beginning of the fourth whorl so that the dorsal side of the septal neck becomes completely projected adorally; here we provisionally describe them as modified amphichoanitic necks (labeled A' in Table III).

During the transformation from retro- to amphichoanitic necks, a small inflection appears on the adoral side of the septal neck. This inflection becomes more prominent in subsequent septa and eventually develops into an adorally directed fold. In Late Cretaceous species of *Hypophylloceras*, the adapical portion of this fold is relatively long in later ontogeny. The adoral and adapical portions of this fold appear to be separated by a sharp boundary, which led Birkelund and Hansen (1974) to interpret the long adapical portion of the fold as a calcified element of the siphuncular tube and to call it a "false septal neck." Birkelund (1981) further noted that this "false septal neck" is surrounded by organic tissue of the siphuncular tube. Our SEM observations of well-preserved specimens of *H. subramosum* from Japan, however, reveal that the axial portion of the septal fold lacks any organic membrane and consists, instead, of modified nacre with a spherulitic prismatic structure (Tanabe *et al.*, 1993, Fig. 4.7B). In another species, *Phyllopachyceras ezoense*, the nacreous lamellae of the septal fold can be traced without any break in microstructure to the adapical end of the septal neck (Fig. 4A).

3.7. Lytoceratina

Retro- and prochoanitic septal necks have been recognized in this suborder (Table III). In all Cretaceous species examined, retrochoanitic necks occur only in the first whorl and are replaced by prochoanitic necks in the second whorl. There is, however, some variation among species in the timing of appearance of prochoanitic necks. In *Gaudryceras denseplicatum*, for example, septal necks change abruptly from retro- to prochoanitic between septa 2 and 3, whereas in *Anagaudryceras limatum*, retrochoanitic necks gradually change from retrochoanitic through a transitional phase (r′, Table III) to prochoanitic over the course of the first whorl (septa 1–9).

In most species, prochoanitic necks gradually lengthen in middle to later ontogeny and become attached ventrally to the inner surface of the shell wall (Figs. 2F, 4D). This change occurs simultaneously with a shift of the siphuncular tube toward the venter. As a result, the nacreous lamellae on the ventral sides of the septal necks are very thin in *A. limatum* and *G. denseplicatum* (Tanabe *et al.*, 1993; Fig. 2F) and may be absent in *G. tenuiliratum* (Doguzhaeva and Mutvei, 1986). The long prochoanitic necks of the Lytoceratina commonly are associated with well-developed cuffs and auxiliary deposits (Figs. 2F, 4D,E).

3.8. Ammonitina

Retro- and prochoanitic septal necks have been identified in this suborder (Table III). They are short in most species except for Cretaceous Desmoceratidae. Of 17 species examined, retrochoanitic necks occur in only four species, an Early Jurassic hildoceratid, *Eleganticeras elegantulum*, and three Cretaceous desmoceratids, *Beudanticeras* sp., *Desmoceras japonicum*, and *Damesites damesi*. In *E. elegantulum*, the transformation from retro- to prochoanitic necks occurs in the middle of the second whorl between septa 11 and 12 (Ohtsuka, 1986, Pl. 46, Fig. 2). The three desmoceratids each have a short retrochoanitic neck at septum 2, but the septal necks of all subsequent septa are prochoanitic. Retrochoanitic necks also are known to occur in the first whorl of several Jurassic genera such as *Promicroceras* (see Erben *et al.*, 1969, Pl. 1) and *Arietites* (see Grandjean, 1910, Fig. 7). In other species, septal necks are prochoanitic throughout ontogeny.

The prochoanitic necks of some genera are associated with well-developed cuffs and auxiliary deposits in middle to later ontogeny (e.g., *Quenstedtoceras*, see Kulicki, 1979; Bandel, 1981; Kulicki and Mutvei, 1982; *Beudanticeras*, see Tanabe *et al.*, 1993; Fig. 4B,C).

3.9. Ancyloceratina

In five species of the Scaphitidae and Baculitidae examined, only prochoanitic necks occur throughout ontogeny (Table III). Cuffs and auxiliary deposits are poorly developed in these Cretaceous heteromorphs.

4. Discussion

4.1. Evolutionary Patterns

The type of septal neck varies among suborders (Tables I–III). In all Paleozoic suborders except the Goniatitina, only retrochoanitic necks are present. In the Goniatitina and Ceratitina, retrochoanitic necks either persist

TABLE I. Ontogenetic Changes in the Type of Septal Neck in the Goniatitina[a]

Superfamily	Family	Species	Number of whorls
Dimorphocerataceae	Dimorphoceratidae	Dimorphoceras politum (Shumard)	
	Girtyoceratidae	Girtyoceras meslerianum (Girty)	
		Eumorphoceras plummeri Miller & Youngquist	
		Gatheries morrowensis (Miller & Moore)	mrA
Goniatitaceae	Goniatitidae	Goniatites multiliratus Gordon	
		Goniatites sp. aff. G. crenistria Phillips	
		Goniatites kentuckiensis Miller	
		Goniatites choctawensis Shumard	
	Agathiceratidae	Agathiceras applini Plummer & Scott	
Neoglyphiocerataceae	Cravenoceratidae	Cravenoceras richardsonianum (Girty)	mrA mrB
		Cravenoceras incisum (Hyatt)	mrA mrB
Schistocerataceae	Schistoceratidae	Schistoceras missouriense (Miller & Faber)	mrA mrB
Gastriocerataceae	Gastrioceratidae	Pseudogastrioceras simulator (Girty)	mrB
		Owenoceras bellilineatum (Miller & Owen)	mrA
	Reticuloceratidae	Retites semiretia McCaleb	
		Arkanites relictus (Quinn, McCaleb & Webb)	mrA
	Glaphyritidae	Glaphyrites hyattianus (Girty)	mrA mrB
		Glaphyrites warei (Miller & Owen)	mrA mrB
		Glaphyrites jonesi (Miller & Owen)	mrA
		Glaphyrites clinei (Miller & Owen)	mrA
		Glaphyrites welleri (Smith)	mrA mrB
	Homoceratidae	Homoceras subglobosum (Bisat)	

Superfamily	Family	Species
Shumarditaceae	Shumarditidae	*Shumardites cuyleri* Plummer & Scott
	Perrinitidae	*Properrinites bakeri* (Plummer & Scott)
		Perrinites sp.
Adrianitaceae	Adrianitidae	*Texoceras* sp.
		Adrianites dunbari Miller & Furnish
		Pseudagathiceras difuntense Miller
		Crimites elkoensis Miller, Furnish & Clarke
Popanocerataceae	Popanoceratidae	*Popanoceras annae* Ruzhencev
		Peritrochia erebus Girty
Metalegocerataceae	Metalegoceratidae	*Metalegoceras bakeri* (Miller & Parizek)
		Metalegoceras baylorense White
Cyclolobaceae	Cyclolobidae	*Mexicoceras guadalupense* (Girty)
Thalassocerataceae	Thalassoceratidae	*Eothalassoceras inexpectans* (Miller & Owen)
	Bisatoceratidae	*Bisatoceras greenei* Miller & Owen
		Bisatoceras n. sp.

[a] r, retrochoanitic; mrA, type A modified retrochoanitic; mrB, type B modified retrochoanitic. Classification after Kullmann (1981) and Glenister and Furnish (1981).

TABLE II. Ontogenetic Changes in the Type of Septal Neck in the Ceratitina[a]

Superfamily	Family	Species	Septal neck ontogeny (by number of whorls, 1–8)
	Olenikitidae	Olenikites spiniplicatus (Mojsisovics)	r (~4) → mrA (~5)
		Subolenekites altus (Mojsisovics)	mrB (~6)
		Svalbardiceras spitzbergensis Frebold	r (~4)
Noritaceae	Ophiceratidae	Nordophiceras schmidti (Mojsisovics)	r (~2) → mrA (~5)
	Prionitidae	Wasatchites tardus (McLearn)	r (~3)
	Meekoceratidae	Arctoceras blomstrandi (Oeberg)	r (~2)
		Boreomeekoceras keyserlingi (Mojsisovics)	r (~2) → mrA (~4) → mrB (~6)
		Bajarunia euomphala (Gabb)	r (~2) → mrA (~5)
	Melagathiceratidae	Juvenites septentrionalis Smith	r (~3)
	Paranannitidae	Owenites koeneri Hyatt & Smith	r (~1) → mrA (~2)
		Prosphingites czekanowskii (Mojsisovics)	r (~4) → mrB (~5)
Xenodiscaceae	Xenodiscidae	Xenoceltites subevolutus Spath	r (~2) → mrA (~5)
		Kelteroceras bellulum Ermakova	r (~3)
	Paraceltitidae	Paraceltites elegans Girty	r (~1) → mrA (~2) → mrB (~3)

Superfamily	Family	Species					
		Stenopopanoceras mirabile Popow	r	mrA	mrA		mrB
		Megaphyllites prometheus Shevyrev[1]	r				
		Amphipopanoceras asseretoi Dagys & Konstantinov	r	mrA			mrB
Nathorstitaceae	Nathorstitidae	*Indigirites tozeri* Weitschat & Lehmann	r	mrA	mrA		mrB
		Stolleyites tenuis (Stolley)	r				mrA
		Nathorstites gibbosus Stolley[2]	r				mrA
Ceratitaceae	Ceratitidae	*Frechites laqueatus* (Lindstroem)	r	mrA	mrB		
		Gymnotoceras blakei (Gabb)	r				
	Sibiritidae	*Siberites eichwaldi* (Mojsisovics)	r		mrA		
	Keyserlingitidae	*Olenekoceras nikitini* (Mojsisovics)	r	mrA			mrB
Danubitaceae	Danubitidae	*Czekanowskites rieberi* Dagys & Weitschat	r	mrA			mrB
		Arctohungarites ventroplana Popow	r	mrA	mrA		
	Aplococeratidae	*Karangatites evolutus* Popow	r				
	Longobarditidae	*Lenotropites caurus* (McLearn)	r	mrA			
Pinacocerataceae	Ptychitidae	*Aristoptychites kolymensis* Kiparisova	r	mrA			mrB

ᵃr, retrochoanitic; mrA, type A modified retrochoanitic; mrB, type B modified retrochoanitic. Sources: 1, Doguzhaeva and Mutvei (1986); 2, Kulicki (1979). Classification after Tozer (1981).

TABLE III. Ontogenetic Changes in the Type of Septal Neck in the Phylloceratina, Lytoceratina, Ammonitina, and Ancyloceratina[a]

Sub-order	Superfamily	Family	Species	Number of whorls (1–7)
Phylloceratina	Phyllocerataceae	Ussuritidae	Indigirophyllites spitsbergensis (Oeberg)	r ——— a
Phylloceratina	Phyllocerataceae	Juraphyllitidae	Tragophylloceras lascombei (J. Sowerby)	r — a — a'
Phylloceratina	Phyllocerataceae	Phylloceratidae	Phylloceras nilssoni (La Verpilliere)	r — a
Phylloceratina	Phyllocerataceae	Phylloceratidae	Calliphylloceras velledae (Michelin)[1]	r — a
Phylloceratina	Phyllocerataceae	Phylloceratidae	Hypophylloceras subramosum (Spath)	r — a
Phylloceratina	Phyllocerataceae	Phylloceratidae	Phyllopachyceras ezoense (Yokoyama)	r — a
Lytoceratina	Lytocerataceae	Lytoceratidae	Lytoceras batesi Trask	r —?— p
Lytoceratina	Tetragonitaceae	Tetragonitidae	Eotetragonites balmensis (Breistroffer)	r — p
Lytoceratina	Tetragonitaceae	Tetragonitidae	Tetragonites glabrus (Jimbo)	r — p
Lytoceratina	Tetragonitaceae	Tetragonitidae	Gabbioceras angulatum Anderson	r — p
Lytoceratina	Tetragonitaceae	Gaudryceratidae	Eogaudryceras aureum (Anderson)	r —?— p
Lytoceratina	Tetragonitaceae	Gaudryceratidae	Gaudryceras denseplicatum (Jimbo)	r — p
Lytoceratina	Tetragonitaceae	Gaudryceratidae	Gaudryceras tenuiliratum Yabe	r — p
Lytoceratina	Tetragonitaceae	Gaudryceratidae	Gaudryceras striatum (Jimbo)	r — p
Lytoceratina	Tetragonitaceae	Gaudryceratidae	Anagaudryceras limatum (Yabe)	r — r' — p

Suborder	Superfamily	Family	Species	Siphuncle type[a]
Ammonitina	Hildocerataceae	Hildoceratidae	Eleganticeras elegantulum (Young & Bird)[2]	r … p
	Eoderocerataceae	Amaltheidae	Pleuroceras sp.	p
		Dactylioceratidae	Peronoceras fibulatum (J. Sowerby)	p
		Eoderoceratidae	Promicroceras sp.	p
	Psilocerataceae	Arietitidae	Coronioceras reynsei (Spath)	p
	Stephanocerataceae	Cardioceratidae	Quenstedoceras sp.[3]	p
		Kosmoceratidae	Kosmoceras sp.[3]	p
	Desmocerataceae	Desmoceratidae	Beudanticeras sp.	r … p
			Desmoceras japonicum (Yabe)	r … p
			Damesites damesi (Jimbo)	r … p
			Damesites sugata (Forbes)	p
		Pachydiscidae	Menuites sp.	p
	Acanthocerataceae	Collignoniceratidae	Collignoniceras woollgari (Mantell)	p
			Subprionocylcus neptuni (Geinitz)	p
			Subprionocyclus minimus (Hayasaka & Fukada)	p
	Hoplitaceae	Placenticeratidae	Metaplacenticeras subtilistriatum (Jimbo)	p
	Spirocerataceae	Spiroceratidae	Spiroceras calloviense Morris	? … p
Ancyloceratina	Scaphitaceae	Scaphitidae	Scaphites planus (Yabe)	p
			Scaphites preventricosus Cobban[5]	p
			Otoscaphites puerculus (Jimbo)	p
			Clioscaphites vermiformis (Meek & Hayden)[5]	p
	Turrilitaceae	Baculitidae	Baculites sp.[4]	p

[a] r, retrochoanitic; p, prochoanitic; a, amphichoanitic; a, modified amphichoanitic (adorally projected on the dorsal side and both adorally and adapically projected on the ventral side). Sources: 1, Druschits et al. (1976); 2, Ohtsuka (1986); 3, Kulicki (1979); 4, Landman (1982); 5, Landman (1987).

throughout ontogeny or develop into modified retrochoanitic necks, first type A, then type B. In the Phylloceratina, retrochoanitic necks are present within the first two whorls and later develop into amphichoanitic necks. In the three other Mesozoic suborders, the Lytoceratina, Ammonitina, and Ancyloceratina, prochoanitic necks occur throughout most of ontogeny. Retrochoanitic necks appear only on the first few septa of the Lytoceratina and on the first septum in some species of the Ammonitina.

Retrochoanitic necks may represent the phylogenetically primitive (plesiomorphous) condition. This is the only type of septal neck that occurs in the outgroup taxon the Orthoceratida, and it is the first kind of septal neck to appear in the ontogenetic development of most ammonoids.

Previous authors have suggested that the ontogenetic appearance of prochoanitic necks has accelerated gradually over the course of evolutionary history within and between suborders starting with the Goniatitina through the Ceratitina and Phylloceratina and ending with the Lytoceratina (Schindewolf, 1950; Zakharov, 1974; Druschits and Doguzhaeva, 1974; Druschits et al., 1976; Doguzhaeva, 1988). According to these authors, the pattern of acceleration within suborders represents a continuation of that between suborders.

This hypothesis is provocative and raises an immediate question. How can we compare the timing of the ontogenetic appearance of prochoanitic necks among suborders when this kind of septal neck is not present in all suborders? The Goniatitina and Ceratitina lack prochoanitic necks and instead develop modified retrochoanitic necks. The Phylloceratina also lack prochoanitic necks and develop amphichoanitic ones. Neither modified retrochoanitic nor amphichoanitic necks appear in the early ontogenetic development of the Lytoceratina or Ammonitina.

We still can pursue this line of inquiry by comparing instead the timing of the ontogenetic transformation itself, whether it be from retrochoanitic to modified retrochoanitic, retrochoanitic to amphichoanitic, or retrochoanitic to prochoanitic. The timing of the ontogenetic transformation is equivalent to the ontogenetic duration of the retrochoanitic condition. So, to rephrase the original hypothesis, does the ontogenetic duration of the retrochoanitic condition systematically decrease over the course of evolutionary history among and within suborders?

To answer the first part of this question, we measured maximum ontogenetic duration of the retrochoanitic condition, expressed in number of shell whorls, within each suborder. In the Goniatitina and Ceratitina, retrochoanitic necks persist throughout ontogeny or develop into modified retrochoanitic necks at some point between the first and seventh whorls. In the Phylloceratina and Ammonitina, the retrochoanitic condition persists for a maximum of about two whorls. In the Lytoceratina and Ancyloceratina, the retrochoanitic condition persists for a maximum of less than one-half whorl. Clearly, the maximum ontogenetic duration of the retrochoanitic condition is shorter in phylogenetically more derived suborders.

Does a similar pattern occur within suborders? In other words, is the ontogenetic duration of the retrochoanitic condition shorter in phylogeneti-

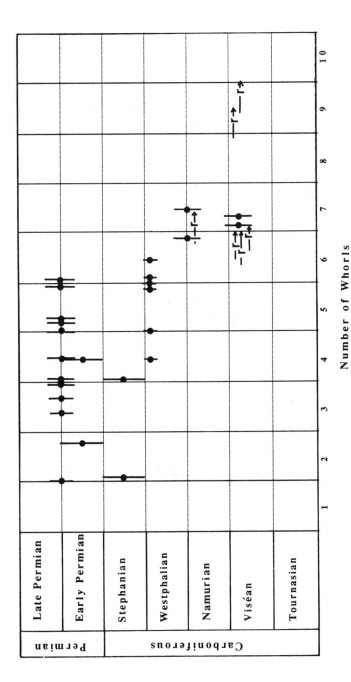

FIGURE 7. Ontogenetic duration of the retrochoanitic condition measured in number of shell whorls, plotted against geologic age, indicated by vertical bars, for 37 species of the Goniatitina. Each black dot indicates the average value for species represented by more than two specimens. Arrows with the letter "r" indicate species in which the retrochoanitic condition persists throughout ontogeny, at least in all of the ontogenetic stages we observed.

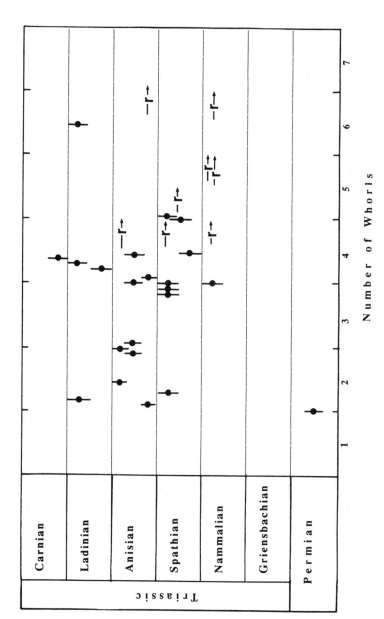

FIGURE 8. Ontogenetic duration of the retrochoanitic condition, measured in number of shell whorls, plotted against geologic age, indicated by vertical bars, for 31 species of the Ceratitina. Each black dot indicates the average value for species represented by more than two specimens. Arrows with the letter "r" indicate species in which the retrochoanitic condition persists throughout ontogeny, at least in all of the ontogenetic stages we observed.

cally more derived species within the same suborder? As a first approach to answering this question, we compared the ontogenetic duration of the retrochoanitic condition among species within suborders versus geologic time for three suborders (Figs. 7–9).

In the Goniatitina, the retrochoanitic condition is present over approximately seven whorls or throughout ontogeny in geologically older species (Fig. 7). In geologically younger species, the retrochoanitic condition does not persist throughout ontogeny. However, it is present over as few as one whorl to as many as six whorls. In the Ceratitina, the ontogenetic duration of the retrochoanitic condition varies broadly among species irrespective of geologic time (Fig. 8). It is present for as little as one whorl or may persist throughout ontogeny. In the Phylloceratina, the ontogenetic duration of the retrochoanitic condition is generally longer but more variable in geologically older species (Fig. 9). In Triassic and Jurassic species, the retrochoanitic condition is present from 0.8 to 2.7 whorls, whereas in Cretaceous species, it is present from 0.8 to 1.3 whorls. On the basis of these data, it appears that the duration of the retrochoanitic condition is highly variable among species within suborders.

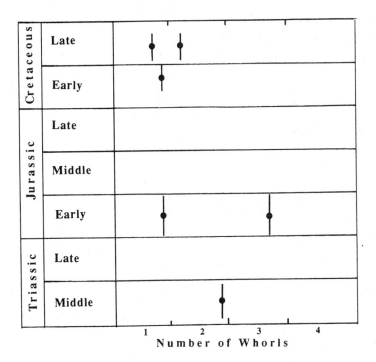

FIGURE 9. Ontogenetic duration of the retrochoanitic condition, measured in number of shell whorls, plotted against geologic age, indicated by vertical bars, for six species of the Phylloceratina. Each black dot indicates the average value for species represented by more than two specimens.

4.2. Morphogenesis of Septal Necks

The morphogenesis of septal necks and their associated structures must be considered in the context of chamber formation. The formation of septal necks has been discussed by a number of workers (e.g., Branco, 1879–1880; Mutvei, 1967; Mutvei and Reyment, 1973; Kulicki, 1979; Bandel, 1981; Tanabe et al., 1982; Kulicki and Mutvei, 1982; Henderson, 1984; Doguzhaeva and Mutvei, 1986; Doguzhaeva, 1988). All of these authors assumed that the rear mantle moved rapidly forward prior to septal formation, as is presumed to occur in present-day Nautilus (Ward et al., 1981; Ward and Chamberlain, 1983).

However, recent discoveries of pseudosutures (Zaborski, 1986; Hewitt et al., 1991; Landman et al., 1993; Lominadzé et al., 1993) and organic cameral membranes (Weitschat and Bandel, 1991; Fig. 5) in Paleozoic and Mesozoic ammonoids have suggested that mantle translocation in ammonoids may have occurred gradually. On the basis of careful observations of pseudosutures preserved in Cretaceous vascoceratids and Jurassic lytoceratids, Hewitt et al. (1991, p. 285) concluded that "the posterior aponeurosis crept forward by gradual secretion of an aqueous gel containing fluted chitinous pseudosepta." These pseudosepta may correspond to the siphuncular sheets illustrated in Fig. 5 and described by Weitschat and Bandel (1991). The presence of pseudosutures on the adoral side of the last septum in a Carboniferous goniatite clearly indicates that formation of these cameral membranes occurred prior to the secretion of the next mineralized septum (Fig. 10).

During the development of modified retro-, amphi-, and prochoanitic necks, there probably was invagination of the septal epithelium, as first suggested by Branco (1879–1880) in his discussion of the formation of prochoanitic necks. There are currently two views to explain the developmental sequence of this invagination during septal formation. The first view, proposed by Kulicki (1979, Fig. 3), Tanabe et al. (1982, Text-fig. 1), and Lominadzé et al. (1993), is that the invagination of the septal epithelium was initially absent and developed only in the course of septal secretion.

Other authors (Mutvei, 1967, Text-fig. 4; Mutvei and Reyment, 1973, Text-fig. 6; Henderson, 1984, Text-fig. 10g,h; Doguzhaeva and Mutvei, 1986, Text-fig. 8; Doguzhaeva, 1988, Text-fig. 5B,C) have argued that the invagination of the septal epithelium was already present in the initial stage of chamber formation. According to this interpretation, the formation of the septal neck and siphuncular tube involved the following two steps: (1) the invaginated septal epithelium secreted a relatively long and thin organic membrane ("primary membrane") that extended adorally to the site of the future septal neck, and (2) the siphuncular epithelium secreted the organic membranes of the siphuncular tube beneath the primary membrane at the same time as the nacreous layers of the septal neck formed on both sides of the epithelial invagination. Doguzhaeva (1988, Text-fig. 5C) further postulated that secretion of a new septum may have been considerably delayed relative to that of the primary membrane because in some ammonoids this membrane extends two chamber lengths in front of the last formed septum.

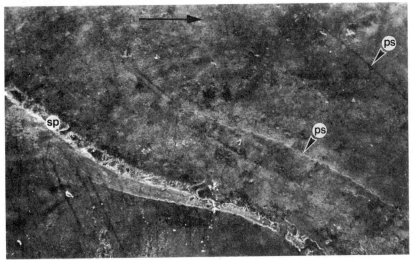

100μm

FIGURE 10. Pseudosutures (ps) preserved at the base of the body chamber in *Goniatites multiliratus* Gordon (Goniatitina: Goniatitaceae). From the Upper Mississippian (Chesterian) of Oklahoma (UMUT PM 19018-6). sp, last septum. The arrow indicates the adoral direction.

The primary membrane described by these authors clearly is equivalent to the cameral membrane. If the above model is correct, the "primary membrane" should be present in the axial portion of the fold of modified retro- and amphichoanitic necks and between the cuff and neck of prochoanitic necks. Such an organic membrane has never been observed in these regions of the septal neck (Tanabe *et al.*, 1993). As already described, the organic membrane (= pellicle) on the adapical sides of these septal necks is not strongly projected adorally and is sometimes detached from the siphuncular tube and septum (c, Figs. 2B,C, 4B). These observations are consistent with the interpretation of Kulicki (1979) and Tanabe *et al.* (1982).

We propose the following morphogenetic model, which emphasizes slow translocation of the posterior mantle and gradual deepening of the epithelial invagination during the formation of amphi- and prochoanitic septal necks (Fig. 11).

1. The rear mantle initially was attached at the site of the lobes and saddles on the adoral side of the septum; then it detached and crept forward (Fig. 11A-1,B-1,C-1).
2. As the rear mantle gradually moved forward, it repeatedly secreted organic siphuncular sheets, which left traces of their attachment along the inside surfaces of the chamber walls (pseudosutures; Fig. 11A-2~3,B-2~3,C-2~3). In contrast, other cameral membranes, namely, the transverse and horizontal organic sheets, do not conform to the configuration of the septal epithelium and presumably did not form in direct contact with it. These organic sheets may have been further modified as a conse-

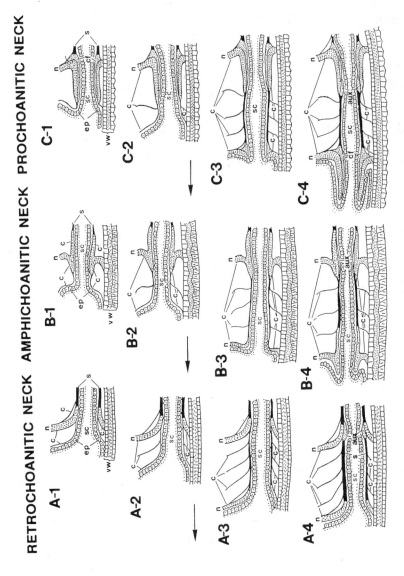

FIGURE 11. Morphogenetic model showing successive development of retro- (A-1→4), amphi- (B-1→4), and prochoanitic (C-1→4) necks in the Ammonoidea. See text for details. c, cameral membrane; ep, septal epithelium; sc, siphuncular cord; s, siphuncular tube; cf, cuff; aux, auxiliary deposit; vw, ventral shell wall. Outlines of the soft tissues are dotted. Arrows indicate the adoral direction.

quence of a contraction episode of the rear mantle after membrane secretion.

3. When the mantle ceased to move forward, the siphuncular epithelium secreted the rest of the siphuncular tube (Fig. 11A-3,B-3,C-3) and the auxiliary deposits on the adoral side of the already-formed septal neck (Fig. 11A-4,B-4,C-4).

4. During the formation of amphi- and prochoanitic necks, in contrast to that of retrochoanitic necks, there was a gradual invagination of a part of the rear mantle epithelium, followed by secretion of the nacreous layers of the septal neck on both the inner and outer sides of this invagination (Fig. 11B-4,C-4). In the formation of modified retrochoanitic necks, as stressed by Doguzhaeva and Mutvei (1986), the epithelial invagination occurred only on the dorsal side of the septal neck. The depth of this invagination increased during the formation of modified retrochoanitic, amphichoanitic, and prochoanitic septal necks.

The cuff, which occurs on the adapical side of prochoanitic necks and type B modified retrochoanitic necks, may have developed after completion of the epithelial invagination (Fig. 11C-4). The highly porous microstructure of the cuff is similar to that of the pillar zone within the long septal necks of *Spirula*. This suggests that the cuff may have formed by remote biomineralization in a blister connected to the organic pellicle during osmotic pumping of the cameral liquid (Seilacher and Chinzei, 1993).

5. Summary and Conclusions

Septal necks in the Ammonoidea are classified into four types: (1) retrochoanitic (projected only adapically), (2) modified retrochoanitic [projected adapically on the ventral side but both adapically and adorally (type A) or only adorally (type B) on the dorsal side], (3) amphichoanitic (projected both adorally and adapically), and (4) prochoanitic (projected only adorally).

The septal neck is continuous with the rest of the septum and is composed primarily of nacre. A spherulitic–prismatic deposit, called the auxiliary deposit, commonly occurs on the adoral side of all kinds of septal necks. Another spherulitic–prismatic deposit (the cuff) appears on the adapical side of type B modified retrochoanitic and prochoanitic necks.

The kinds of septal necks differ among suborders. The ontogenetic transformation from retrochoanitic to other kinds of septal necks occurs earlier in phylogenetically more derived suborders but does not show any clear-cut trend among species within suborders.

The morphogenesis of septal necks and their associated structures must be considered in the context of chamber formation. We propose a model that emphasizes slow translocation of the septal mantle prior to secretion of a mineralized septum and a gradual invagination of the epithelial tissue during formation of amphi- and prochoanitic septal necks.

Appendix: List of

Suborder & Family	Species	Horizon
Bactritina Bactritidae	*Bactrites* sp.	U. Miss. (Chesterian)
Goniatitina Dimorphoceratidae	*Dimorphoceras politum* (Shumard)	M. Pennsylvanian
Girtyoceratidae	*Girtyoceras meslerianum* (Girty) *Eumorphoceras plummeri* Miller & Youngquist *Gatherites morrowensis* (Miller & Moore)	U. Miss. (Chesterian) U. Miss. (Chesterian) L. Penn. (Morrowan)
Goniatitidae	*Goniatites multiliratus* Gordon *Goniatites* sp. aff. *G. crenistria* Phillips *Goniatites kentuckiensis* Miller *Goniatites choctawensis* Shumard	U. Miss. (Chesterian) U. Miss. (Chesterian) U. Miss. (Chesterian) M. Miss. (Meramecian)
Agathiceratidae	*Agathiceras applini* Plummer & Scott	L. Pennsylvanian
Cravenoceratidae	*Cravenoceras richardsonianum* (Girty) *Cravenoceras incisum* (Hyatt)	U. Miss. (Chesterian) U. Miss. (Chesterian)
Schistoceratidae	*Schistoceras missouriense* (Miller & Faber)	U. Penn. (Missourian-Virgilian)
Gastrioceratidae	*Pseudogastrioceras simulator* (Girty) *Owenoceras bellilineatum* (Miller & Owen)	L. - M. Permian U. Miss. (Chesterian)
Reticuloceratidae	*Retites semiretia* McCaleb *Arkanites relictus* (Quinn, McCaleb & Webb)	L. Penn. (Morrowan) L. Penn. (Morrowan)
Glaphyritidae	*Glaphyrites hyattianus* (Girty) *Glaphyrites warei* (Miller & Owen) *Glaphyrites jonesi* (Miller & Owen) *Glaphyrites clinei* (Miller & Owen) *Glaphyrites welleri* (Smith)	M. Penn. (Desmoinesian) M. Penn. (Desmoinesian) M. Penn. (Desmoinesian) M. Penn. (Desmoinesian) M. Penn. (Desmoinesian)
Homoceratidae	*Homoceras subglobosum* (Bisat)	M. Carbon. (L. Namurian)
Shumarditidae	*Shumardites cuyleri* Plummer & Scott	U. Penn. (Missourian-Virgilian
Perrinitidae	*Properrinites bakeri* (Smith) *Perrinites* sp.	L. Permian M. Permian
Adrianitidae	*Texoceras* sp. *Adrianites dunbari* Miller & Furnish *Pseudagathiceras difuntense* Miller *Crimites elkoensis* Miller, Furnish & Clarke	M. Permian M. Permian M. Permian M. Permian
Popanoceratidae	*Popanoceras annae* Ruzhencev *Peritrochia erebus* Girty	M. Permian M. Permian
Metalegoceratidae	*Metalegoceras bakeri* (Miller & Parizek) *Metalegoceras baylorense* White	M. Permian M. Permian
Cyclolobidae	*Mexicoceras guadalupense* (Girty)	M. Permian
Thalassoceratidae	*Eothalassoceras inexpectans* (Miller & Owen)	M. Penn. (Desmoinesian)
Bisatoceratidae	*Bisatoceras greenei* Miller & Owen *Bisatoceras* n. sp.	L. Penn. (Morrowan) M. Penn. (Desmoinesian)
Prolecanitina Pronoritidae	*Pronorites praepermicus* Karpinsky	M. Permian (Artinskian)
Medlicottiidae	*Artinskia electraensis* (Plummer & Scott)	M. Permian

Ammonoid Species Studied

Locality	Sample
Jack Fork Creek, Pontotoc Co., Oklahoma	N=2 (UMUT PM 19017-1, 2)
Henry Co., Missouri	N=1 (SUI 1755)
Jack Fork Creek, Pontotoc Co., Oklahoma	N=5 (UMUT PM 19023-1~5)
San Saba, Texas	N=1 (UMUT PM 19030)
Gather Mt., Arkansas	N=1 (UMUT PM 19032)
Jack Fork Creek, Pontotoc Co., Oklahoma	N=5 (UMUT PM 19018-1~6)
Ahloso, Pontotoc Co., Oklahoma	N=2 (UMUT PM 19019-1, 2)
Ada, Oklahoma	N=1 (SUI 1718)
Clarita, Coal Co., Oklahoma	N=2 (UMUT PM 19020-1, 2)
Coleman, Texas	N=1 (SUI 1766)
Wapanucka, Johnston Co., Oklahoma	N=1 (UMUT PM 19021)
San Saba, Texas	N=3 (UMUT PM 19022-1~3)
Jacksboro, Texas	N=1 (SUI 1746)
Sublette Range, Wyoming	N=2 (SUI 1738, 1740)
Henry Co., Missouri	N=1 (SUI 1713)
Thompson, Arkansas	N=1 (UMUT PM 19031)
Bradshaw Mt., Carroll Co., Arkansas	N=1 (UMUT PM 19029)
Okmulgee, Oklahoma	N=3 (UMUT PM 19025-1~3)
Collinsville, Tulsa, Oklahoma	N=6 (UMUT PM 19026-1~6)
Collinsville, Tulsa, Oklahoma	N=1 (UMUT PM 19027)
Collinsville, Tulsa, Oklahoma	N=1 (UMUT PM 19028)
Henry Co., Missouri	N=1 (SUI 1726)
Stonehead Beck, Yorkshire, England	N=2 (UMUT PM 19024-1, 2)
Jacksboro, Texas	N=1 (SUI 1791)
Dugout Mt., Brewster Co., Texas	N=1 (SUI 1790)
Clayslide, Iron Mt., Texas	N=1 (AMNH 41183-1)
El Capitan, Texas	N=2 (UMUT PM 19037-1, 2)
La Difunta, Coahuila, Mexico	N=1 (SUI 1764)
Unknown	N=1 (SUI 1790)
Buck Mt., Nevada	N=1 (UMUT PM 19038)
Aktubinsk, Russia	N=1 (SUI 1777)
Bone Springs, Texas	N=2 (UMUT PM 19039-1, 2)
Buck Mt., Nevada	N=1 (UMUT PM 19034)
Buck Mt., Nevada	N=1 (UMUT PM 19035)
South Wells, Culberson Co., Texas	N=1 (SUI 1782)
Okmulgee, Oklahoma	N=3 (UMUT PM 19036-1~3)
Henry Co., Missouri	N=1 (SUI 1719)
Okmulgee, Oklahoma	N=2 (UMUT PM 19033-1, 2)
Middle Urals, Russia	N=1 (SUI 1686)
Buck Mt., Nevada	N=3 (UMUT PM 19040-1~3)

Suborder & Family	Species	Horizon
Ceratitina		
Olenikitidae	*Olenikites spiniplicatus* (Mojsisovics)	Spathian
	Subolenekites altus (Mojsisovics)	Spathian
	Svalbardiceras spitzbergensis Frebold	U. Spathian
Ophiceratidae	*Nordophiceras schmidti* (Mojsisovics)	U. Spathian
Prionitidae	*Wasatchites tardus* (McLearn)	Smithian
Meekoceratidae	*Arctoceras blomstrandi* (Oeberg)	Smithian
	Boreomeekoceras keyserlingi (Mojsisovics)	U. Spathian
	Bajarunia euomphala (Keyserling)	L. Spathian
Melagathiceratidae	*Juvenites septentrionalis* Smith	Smithian
Paranannitidae	*Owenites koeneri* Hyatt & Smith	Smithian
	Prosphingites czekanowskii (Mojsisovics)	U. Spathian
Xenodiscidae	*Xenoceltites subevolutus* Spath	Smithian
	Kelteroceras bellulum Ermakova	Smithian
Paraceltitidae	*Paraceltites elegans* Girty	M. Permian
Parapopanoceratidae	*Parapopanoceras paniculatum* Popow	M. Anisian
	Stenopopanoceras mirabile Popow	L. Anisian
	Amphipopanoceras asseretoi Dagys & Ermakova	M. Anisian
Nathorstitidae	*Indigirites tozeri* Weitschat & Lehmann	U. Ladinian
	Stolleyites tenuis (Stolley)	L. Carnian
Ceratitidae	*Frechites laqueatus* (Lindstroem)	U. Anisian
	Gymnotoceras blakei (Gabb)	U. Anisian
Sibiritidae	*Siberites eichwaldi* (Mojsisovics)	U. Spathian
Keyserlingitidae	*Olenekoceras nikitini* (Mojsisovics)	U. Spathian
Danubitidae	*Czekanowskites rieberi* Dagys & Weitschat	M. Anisian
	Arctohungarites ventroplana Popow	M. Anisian
Aplococeratidae	*Karangatites evolutus* Popow	L. Anisian
Longobarditidae	*Lenotropites caurus* (McLearn)	L. Anisian
Ptychitidae	*Aristoptychites kolymensis* Kiparisova	U. Ladinian

Locality	Sample
Mouth of Oelenek River, East Siberia	N=2 (AMNH 44347, 44348)
East Taimyr, Arctic Siberia	N=1 (UMUT MM 19041)
Wallenbergfjellet, Spitsbergen	N=3 (AMNH 44349, UMUT MM 19042-1, 2)
Oelenek River, Arctic Siberia	N=1 (UMUT MM 19043)
Botneheia, Spitsbergen	N=2 (AMNH 44350, UMUT MM 19044)
Stensiö-Fjellet, Spitsbergen	N=1 (AMNH 44351)
Oelenek River, Arctic Siberia	N=2 (UMUT MM 19045-1, 2)
Lena River, Arctic Siberia	N=1 (UMUT MM 19046)
Crittenden Springs, Nevada	N=1 (UMUT MM 19048)
Crittenden Springs, Nevada	N=1 (UMUT MM 19049)
Oelenek River, Arctic Siberia	N=1 (UMUT MM 19047)
Wallenbergfjellet, Spitsbergen	N=2 (AMNH 44346, UMUT MM 19050)
Buur River, Arctic Siberia	N=1 (UMUT MM 19051)
Bone Springs, Texas	N=1 (UMUT PM 19052)
Cape Tsvetkov, Taimyr, Arctic Siberia	N=2 (AMNH 44352, UMUT MM 19053)
Taimyr, Arctic Siberia	N=1 (UMUT MM 19054)
Lena River, Arctic Siberia	N=1 (UMUT MM 19055)
Botneheia, Spitsbergen	N=2 (AMNH 44353, UMUT MM 19056)
Kongressfjellet, Spitsbergen	N=3 (AMNH 44354, 44355, 44356)
Wallenbergfjellet, Spitsbergen	N=2 (AMNH 44357, UMUT MM 19057)
Crittenden Springs, Nevada	N=1 (UMUT MM 19059)
Oelenek River, Arctic Siberia	N=2 (UMUT MM 19058-1, 2)
Oelenek River, Arctic Siberia	N=1 (UMUT MM 19060)
Mt. Tuara-Khayata, Oelenek, Arctic Sib.	N=2 (AMNH 44358, 44359)
East Taimyr, Arctic Siberia	N=1 (UMUT MM 19061)
Oelenek Bay, Arctic Siberia	N=1 (UMUT MM 19062)
Oelenek Bay, Arctic Siberia	N=1 (UMUT MM 19063)
Botneheia, Spitsbergen	N=2 (AMNH 44360, 44361)

Suborder & Familly	Species
Phylloceratina	
Ussuritidae	*Indigirophyllites spitsbergensis* (Oeberg)
Juraphyllitidae	*Tragophylloceras lascombei* (J. Sowerby)
Phylloceratidae	*Phylloceras nilssoni* (La Verpilliere)
	Hypophylloceras subramosum (Spath)
	Phyllopachyceras ezoense (Yokoyama)
Lytoceratina	
Lytoceratidae	*Lytoceras batesi* Trask
Tetragonitidae	*Eotetragonites balmensis* (Breistroffer)
	Tetragonites glabrus (Jimbo)
	Gabbioceras angulatum Anderson
Gaudryceratidae	*Eogaudryceras aureum* (Anderson)
	Gaudryceras denseplicatum (Jimbo)
	Gaudryceras tenuiliratum Yabe
	Gaudryceras striatum (Jimbo)
	Anagaudryceras limatum (Yabe)
Ammonitina	
Hildoceratidae	*Eleganticeras elegantulum* (Young & Bird)
Amaltheidae	*Pleuroceras* sp.
Dactylioceratidae	*Peronoceras fibulatum* (J. Sowerby)
Eoderoceratidae	*Promicroceras* sp.
Arietitidae	*Coronioceras reynsei* (Spath)
Desmoceratidae	*Beudanticeras* sp.
	Desmoceras japonicum (Yabe)
	Damesites damesi (Jimbo)
	Damesites sugata (Forbes)
Pachydiscidae	*Menuites* sp.
Collignoniceratidae	*Collignoniceras woollgari* (Mantell)
	Subprionocyclus neptuni (Geinitz)
	Subprionocyclus minimus (Hayasaka & Fukada)
Placenticeratidae	*Metaplacenticeras subtilistriatum* (Jimbo)
Spiroceratidae	*Spiroceras calloviense* Morris
Ancyloceratina	
Scaphitidae	*Scaphites planus* (Yabe)
	Otoscaphites puerculus (Jimbo)

Horizon	Locality	Sample
M. Triassic (Anisian)	Botneheia, Spitsbergen	N=2 (AMNH 44362, 44363)
L. Jurassic (Pliensbachian)	Charmouth, Dorset, England	N=1 (AMNH 41348)
L. Jurassic	Monte Catria, Apenin, Italy	N=1 (YPM 1831)
U. Cretaceous (Coniacian)	Haboro, Hokkaido, Japan	N=1 (AMNH 44364)
U. Cretaceous (L. Campanian)	Saku, Hokkaido, Japan	N=1 (AMNH 44366)
L. Cretaceous	California	N=1 (AMNH 44367)
L. Cretaceous (M. Albian)	California	N=1 (AMNH 44368)
U. Cretaceous (L. Campanian)	Saku, Hokkaido, Japan	N=1 (UMUT MM 19064)
L. Cretaceous (U. Aptian)	California	N=1 (AMNH 44370)
L. Cretaceous (L. Albian)	California	N=1 (AMNH 44369)
U. Cretaceous (Coniacian)	Haboro, Hokkaido, Japan	N=1 (AMNH 44371)
U. Cretaceous (U. Santonian)	Embetsu, Hokkaido, Japan	N=1 (AMNH 44372)
U. Cretaceous (L. Campanian)	Saku, Hokkaido, Japan	N=1 (UMUT MM 19065)
U. Cretaceous (Coniacian)	Haboro, Hokkaido, Japan	N=2 (AMNH 44373, 44374)
L. Jurassic (L. Toarcian)	Near Hamburg, Germany	N=2 (UMUT MM 19066-1, 2)
L. Jurassic (U. Pliensbachian)	Unterstürmig, Germany	N=1 (UMUT MM 19067)
L. Jurassic (L. Toarcian)	Whitby, Yorkshire, England	N=1 (UMUT MM 19068)
L. Jurassic (L. Sinemurian)	Dorset, England	N=2 (UMUT MM 19069-1, 2)
L. Jurassic (L. Sinemurian)	Schwabisch Gmünd, Germany	N=1 (UMUT MM 19070)
L. Cretaceous (Albian)	Southern Alaska	N=1 (AMNH 44375)
U. Cretaceous (Cenominian)	Obira, Hokkaido, Japan	N=1 (UMUT MM 19071)
U. Cretaceous (L. Santonian)	Haboro, Hokkaido, Japan	N=1 (AMNH 44376)
U. Cretaceous (L. Campanian)	Saku, Hokkaido, Japan	N=2 (UMUT MM 19072-1, 2)
U. Cretaceous (L. Campanian)	Saku, Hokkaido, Japan	N=1 (UMUT MM 19073)
U. Cretaceous (M. Turonian)	Black Hills, South Dakota	N=1 (UMUT MM 19074)
U. Cretaceous (M. Turonian)	Urakawa, Hokkaido, Japan	N=1 (UMUT MM 19075)
U. Cretaceous (U. Turonian)	Manji, Hokkaido, Japan	N=3 (UMUT MM 19076-1~3)
U. Cretaceous (U. Campanian)	Embetsu, Hokkaido, Japan	N=1 (UMUT MM 19077)
M. Jurassic (Callovian)	Wiltshire, England	N=1 (YPM 01854)
U. Cretaceous (M. Turonian)	Obira, Hokkaido, Japan	N=1 (UMUT MM 19078)
U. Cretaceous (M. Turonian)	Obira, Hokkaido, Japan	N=1 (UMUT MM 19079)

ACKNOWLEDGMENTS. We thank Royal Mapes (Ohio University), Wolfgang Weit-schat (Hamburg Universität), Haruyoshi Maeda (Kyoto University), Yasunari Shigeta (National Science Museum, Tokyo), and Norm Brown (San Diego, CA) for providing us with excellent specimens of ammonoids of Carboniferous to Cretaceous age; Copeland MacClintock (Yale Peabody Museum), Tim White (YPM), and Julia Golden (University of Iowa) for arranging loans of museum specimens; and Dolf Seilacher (Yale University), John Chamberlain, Jr. (Brooklyn College), Roger Hewitt (Leigh-on-Sea, Essex, England), Harry Mutvei (Swedish Museum of Natural History), Richard Davis (College of Mount St. Joseph), and Antonio Checa (Universidad de Granada) for reviewing an early draft of this manuscript. W. Weitschat kindly provided SEM photos of Triassic ammonoids with cameral membranes from Spitsbergen. Peling Fong assisted in SEM work, and Susan Klofak and Jane Whitehill assisted in specimen preparation (all AMNH). This work was supported by a Grant-In-Aid from the Japanese Ministry of Education, Science, and Culture for 1990–91 (No. 01460062).

References

Appellöf, A., 1893, Die Schalen von *Sepia, Spirula* und *Nautilus.* Studien über den Bau und das Wachstum, *K. Svensk. Vetenskapsakad. Handl.* **25**:1–106.

Bandel, K., 1981, The structure and formation of the siphuncular tube of *Quenstedtoceras* compared with that of *Nautilus* (Cephalopoda), *N. Jb. Geol. Paläont. Abh.* **161**:153–171.

Bandel, K., 1982, Morphologie und Bildung der frühontogenetischen Gehäuse bei conchiferen Mollusken, *Facies* **7**:1–198.

Bandel, K., and Boletzky, S. v., 1979, A comparative study of the structure, development and morphological relationships of chambered cephalopod shells, *Veliger* **21**:313–354.

Bayer, U., 1975, Organische Tapeten im Ammoniten-Phragmokon und ihr Einfluss auf die Fossilisation, *N. Jb. Geol. Paläont. Mh.* **1975**(1):12–25.

Birkelund, T., 1981, Ammonoid shell structure, in: *The Ammonoidea*, The Systematics Association Special Volume 18 (M. R. House and J. R. Senior, eds.), Academic Press, London, pp. 177–214.

Birkelund, T., and Hansen. H. J., 1968, Early shell growth and structures of the septa and the siphuncular tube in some Maastrichtian ammonites, *Medd. Dan. Geol. Foren.* **18**:71–78.

Birkelund, T., and Hansen, H. J., 1974, Shell ultrastructures of some Maastrichitian Ammonoidea and Coleoidea and their taxonomic implications, *K. Dan. Vidensk. Selsk. Biol. Skr.* **20**:1–34.

Bogoslovskaya, M. F., 1959, Internal structure of the shells of some Artinskian ammonoids, *Paleont. Zh.* **1959**(1):49–57 (in Russian).

Bogoslovsky, B. I., 1969, Devonian Ammonoidea. I. *Agoniatites, Trans. Paleont. Inst. Akad. Nauk SSSR* **124**:1–341 (in Russian).

Böhmers, J. C. A., 1936, Bau und Struktur von Schale und Sipho bei permischen Ammonoidea, Dissertation, Drukkerij University, Apeldoorn.

Branco, W., 1879–1880, Beiträge zur Entwickelungsgeschichte der fossilen Cephalopoden, *Palaeontographica* **26**(1879):15–50; **27**(1880):17–81.

Denton, E. J., and Gilpin-Brown, J. B., 1966, On the buoyancy of the pearly *Nautilus, J. Mar. Biol. Assoc. U.K.* **46**:723–759.

Denton, E. J., and Gilpin-Brown, J. B., 1973, Floatation mechanisms in modern and fossil cephalopods, in: *Advances in Marine Biology*, Volume 11 (F. S. Russell and M. Yonge, eds.), Academic Press, London, pp. 197–268.

Doguzhaeva, L., 1988, Siphuncular tube and septal necks in ammonoid evolution, in: *Cephalopods—Present and Past* (J. Wiedmann and J. Kullmann, eds.), Schweizerbart'sche Verlagsbuchhandlung, Stuttgart, pp. 291–301.

Doguzhaeva, L., and Mutvei, H., 1986, Retro- and prochoanitic septal necks in ammonoids, and transition between them, *Palaeontogr. Abt. A* 195:1–18.

Druschits, V. V., and Doguzhaeva, L. A., 1974, Some morphogenetic characteristics of phylloceratids and lytoceratids (Ammonoidea), *Paleontol. J.* 8(1):37–48.

Druschits, V. V., and Khiami, N., 1970, Structure of the septa, protoconch walls and initial whorls in early Cretaceous ammonites, *Paleontol. J.* 4(1):26–38.

Druschits, V. V., Bogslovskaya, M. F., and Doguzhaeva, L. A., 1976, Evolution of septal necks in the Ammonoidea, *Paleontol. J.* 10(1):37–50.

Erben, H. K., 1964a, Bactritoidea, in: *Treatise on Invertebrate Paleontology*, Part K, *Mollusca 3* (R. C. Moore, ed.), Geological Society of America and University of Kansas Press, Lawrence, KS, pp. 491–505.

Erben, H. K., 1964b, Die Evolution der ältesten Ammonoidea, *N. Jb. Geol. Paläont. Abh.* 120:107–212.

Erben, H. K., and Reid, R. E. H., 1971, Ultrastructure of shell, origin of conellae and siphuncular membranes in an ammonite, *Biomineralisation* 3:22–31.

Erben, H. K., Flajs, G., and Siehl, A., 1969, Die frühontogenetische Entwicklung der Schalenstruktur ectocochleater Cephalopoden, *Palaeontogr. Abt. A* 132:1–54.

Glenister, B. F., and Furnish, W. M., 1981, Permian ammonoids, in: *The Ammonoidea*, The Systematics Association Special Volume (M. R. House and J. R. Senior, eds.), Academic Press, London, pp. 49–64.

Grandjean, F., 1910, Le siphon des ammonites et des belémnites, *Bull. Soc. Géol. France, Ser. 4,* 10:496–519.

Grégoire, C., 1962, On submicroscopic structure of the *Nautilus* shell, *Bull. Inst. R. Sci. Nat. Belg.* 38:1–71.

Grégoire, C., 1973, On the submicroscopic structure of the organic components of the siphon in the *Nautilus* shell, *Arch. Int. Physiol. Biochim.* 81:299–316.

Grégoire, C., 1984, Remains of organic components in the siphonal tube and in the brown membrane of ammonoids and fossil nautiloids. Hydrothermal simulation of their diagenetic alterations, *Akad. Wiss. Lit. Abh. Math. Naturwiss. Kl. (Mainz)* 1984(5):1–56.

Hasenmueller, W. A., and Hattin, D. E., 1985, Apatitic connecting rings in moulds of *Baculites* sp. from the middle part of the Smoky Hill Member, Niobrara Chalk (Santonian) of western Kansas, *Cretaceous Res.* 6:317–330.

Henderson, R. A., 1984, A muscle attachment proposal for septal function in Mesozoic ammonites, *Palaeontology* 27:461–486.

Hewitt, R. A., and Westermann, G. E. G., 1983, Mineralogy, structure and homology of ammonoid siphuncles, *N. Jb. Geol. Paläont. Abh.* 165:378–396.

Hewitt, R. A., Checa, A., Westermann, G. E. G., and Zaborski, P. M., 1991, Chamber growth in ammonites inferred from colour markings and naturally etched surfaces of Cretaceous vascoceratids from Nigeria, *Lethaia* 24:271–287.

Hewitt, R. A., Abdelsalam, U. A., Dokainish, M. A., and Westermann, G. E. G., 1993, Comparison of the relative strength of siphuncles with prochoanitic and retrochoanitic septal necks by finite-element analysis, in: *The Ammonoidea: Environment, Ecology and Evolutionary Change*, The Systematics Association Special Volume 47 (M. R. House, ed.), Clarendon Press, Oxford, pp. 85–98.

House, M. R., 1985, Correlation of mid-Palaeozoic ammonoid evolutionary events with global sedimentary pertubations, *Nature* 313:17–22.

Kulicki, C., 1979, The ammonite shell: Its structure, development and biological significance, *Palaeont. Pol.* 39:97–142.

Kulicki, C., and Mutvei, H., 1982, Ultrastructure of the siphonal tube in *Quenstedtoceras* (Ammonitina), *Stockholm Contrib. Geol.* 37:129–138.

Kullmann, J., 1981, Carboniferous Goniatites, in: *The Ammonoidea*, The Systematics Association Special Volume 18 (M. R. House and J. R. Senior, eds.), Academic Press, London, pp. 37–48.

Landman, N. H., 1982, Embryonic shells of *Baculites*, *J. Paleontol.* **56**:1235–1241.

Landman, N. H., 1987, Ontogeny of Upper Cretaceous (Turonian–Santonian) sacphitid ammonites from the Western Interior of North America: Systematics, developmental patterns, and life history, *Bull. Am. Mus. Nat. Hist.* **185**(2):117–241.

Landman, N. H., Tanabe, K., Mapes, R. H., Klofak, S. M., and Whitehill, J., 1993, Pseudosutures in Paleozoic ammonoids, *Lethaia* **26**:99–100.

Lominadzé, T. A., Sharikadzé, M. Z., and Kvantaliani, I. V., 1993, On mechanism of soft body movement within body chamber in ammonites, *Geobios Mém. Spéc.* **15**:267–273.

Mapes, R. H., 1979, Carboniferous and Permian Bactritoidea (Cephalopoda) in North America, *Univ. Kans. Paleontol. Contrib. Artic.* **64**:1–75.

Miller, A. K., and Unklesbay, A. G., 1943, The siphuncle of Late Paleozoic ammonoids, *J. Paleont.* **17**:1–25.

Miller, A. K., Furnish, W. M., and Schindewolf, O. H., 1957, Paleozoic Ammonoidea, in: *Treatise on Invertebrate Paleontology*, Part L, *Mollusca 4* (R. C. Moore, ed.), Geological Society of America and University of Kansas Press, Lawrence, KS, pp. 11–79.

Mutvei, H., 1967, On the microscopic shell structure in some Jurassic ammonoids, *N. Jb. Geol. Paläont. Abh.* **129**:157–166.

Mutvei, H., and Reyment, R. A., 1973, Buoyancy control and siphuncle function in ammonoids, *Palaeontology* **16**:623–636.

Naef, A., 1922, *Die Fossilen Tintenfische*, Gustav Fischer, Jena.

Obata, I., Tanabe, K., and Fukuda, Y., 1980, The ammonite siphuncular wall: Its microstructure and functional significance, *Bull. Natl. Sci. Mus. (Tokyo), Ser. C* **6**:59–72.

Ohtsuka, Y., 1986, Early internal shell microstructure of some Mesozoic Ammonoidea: Implications for higher taxonomy, *Trans. Proc. Palaeont. Soc. Jpn. New Ser.* **141**:275–288.

Schindewolf, O. H., 1935, Zur Stammesgeschichte der Cephalopoden, *Jb. Preuss. Geol. Landesanst.* **55**:258–283.

Schindewolf, O. H., 1950, *Grundfragen der Paläontologie*, Schweizerbart'sche Verlagsbuchhandlung, Stuttgart.

Schindewolf, O. H., 1968, Analyse eines Ammoniten-Gehäuses, *Akad. Wiss. Lit. Abh. Math. Naturwiss. Kl. (Mainz)* **8**:139–188.

Schoulga-Nesterenko, M., 1926, Nouvelles données sur l'organisation intérieure des conques des ammonites de l'étage d'Artinsk, *Bull. Soc. Nat. Moscou Sec. Géol.* **4**(1–2):81–100.

Seilacher, A., and Chinzei, K., 1993, Remote biomineralization II: Fill skeletons controlling buoyancy in shelled cephalopods, *N. Jb. Geol. Paläont. Abh.* **190**:363–373.

Spath, L. F., 1933, The evolution of the Cephalopoda, *Biol. Rev.* **8**:418–462.

Tanabe, K., Obata, I., Fukuda, Y., and Futakami, M., 1979, Early shell growth in some Upper Cretaceous ammonites and its implications to major taxonomy, *Bull. Nat. Sci. Mus. (Tokyo), Ser. C* **5**:153–176.

Tanabe, K., Fukuda, Y., and Obata, I., 1982, Formation and function of the siphuncle–septal neck structures in two Mesozoic ammonites, *Trans. Proc. Palaeont. Soc. Jpn. New Ser.* **128**:433–443.

Tanabe, K., Landman, N. H., and Weitschat, W., 1993, Septal necks in Mesozoic Ammonoidea: Structure, ontogenetic development and evolution, in: *The Ammonoidea: Environment, Ecology and Evolutionary Change*, The Systematics Association Special Volume 47 (M. R. House, ed.), Clarendon Press, Oxford, pp. 57–84.

Tozer, E. T., 1981, Triassic Ammonoidea: Classification, evolution and relationship with Permian and Jurassic forms, in: *The Ammonoidea*, The Systematics Association Special Volume 18 (M. R. House and J. R. Senior, eds.), Academic Press, London, pp. 66–100.

Ward, P. D., and Chamberlain, J. A., Jr., 1983, Radiographic observation of chamber formation in *Nautilus pompilius*, *Nature* **304**:57–59.

Ward, P. D., Greenwald, L., and Magnier, Y., 1981, The chamber formation cycle in *Nautilus macromphalus*, *Paleobiology* **7**:481–493.

Weitschat, W., and Bandel, K., 1991, Organic components in phragmocones of Boreal Triassic ammonoids: Implications for ammonoid biology, *Paläontol. Z.* **65**:269–303.

Westermann, G. E. G., 1982, The connecting rings of *Nautilus* and Mesozoic ammonoids: Implications for ammonoid bathymetry, *Lethaia* **15**:373–384.

Zaborski, P. M. P., 1986, Internal mould markings in a Cretaceous ammonite from Nigeria, *Palaeontology* **29**:725–738.

Zakharov, Yu. D., 1971, Some features of the development of the hydrostatic apparatus in early Mesozoic ammonoids, *Paleontol. J.* **5**(1):24–33.

Zakharov, Yu. D., 1974, New data on internal shell structure in Carboniferous, Triassic and Cretaceous ammonoids, *Paleontol. J.* **8**(1):25–36.

Zakharov, Yu. D., 1989, New data on biomineralization of the Ammonoidea, in: *Skeletal Biomineralization: Patterns, Processes and Evolutionary Trends, Short Course in Geology*, Vol. 5, Part II (J. G. Carter, ed.), American Geophysical Union, Washington, DC, p. 325.

III

Buoyancy, Swimming, and Biomechanics

Chapter 7

Buoyancy and Hydrodynamics in Ammonoids

DAVID K. JACOBS and JOHN A. CHAMBERLAIN, JR.

DAVID K. JACOBS • Department of Biology, University of California at Los Angeles, Los Angeles, California 90024-1606.   JOHN A. CHAMBERLAIN, JR. • Department of Geology, Brooklyn College, City University of New York, Brooklyn, New York 11210.

Ammonoid Paleobiology, Volume 13 of *Topics in Geobiology*, edited by Neil Landman *et al.*, Plenum Press, New York, 1996.

1. Introduction

Information pertaining to the function of ammonoid shells is generated by analogy to living cephalopods, by measurement or experiment designed to elucidate the properties of the ammonoid shell in life situations, and by examination of the distribution and sedimentary environments in which ammonoid fossils are preserved. Virtually all discussions of ammonoid shell function implicitly or explicitly incorporate more than one of these approaches. The combination of analogy with empirical work in the field and laboratory makes the reconstruction of the function of ammonoid shells and interpretation of ammonoid life habits and mode of life particularly intriguing. These interpretations have led to many lively debates among paleobiologists. In this chapter, we examine ammonoid buoyancy and locomotion. We evaluate arguments that have been used to reconstruct the buoyancy and locomotor properties of these extinct cephalopods, discuss recent advances in the understanding of ammonoid locomotion, and suggest directions in which the study of these aspects of ammonoid paleobiology may proceed in the future. Other chapters of this book explore aspects of the structural issues pertaining to the implosion strength of ammonoid shells (Hewitt, Chapter 10, this volume) as well as the environmental information that can be brought to bear on the subject (Westermann, Chapter 16, this volume).

Bone, shell, and muscle tissue, the functional tissues of active marine animals, are denser than water. Two different solutions to this problem have evolved in animals that spend large amounts of time in the water column. Some expend energy through swimming to keep their dense bodies "up" in the water column, e.g., sharks, tuna, and a few squid. Others contain structures or tissues that are less dense than sea water and offset the dense tissues of the body. The ammonia-containing tissues of some squid are lighter than seawater, and many teleost fish employ a swim bladder to attain neutral buoyancy. The chambered cephalopod shell provides another solution to this buoyancy problem. In addition to amonoids and some coleoids, a number of now extinct Paleozoic taxa and the Mesozoic and Cenozoic nautilid clade, including modern *Nautilus*, have such chambered shells.

In modern cephalopods with chambered shells (Fig. 1), it has been demonstrated that the siphuncle empties water from the chambers, rendering the shell less dense than sea water. The buoyant shell then compensates for the dense soft parts of the organism, resulting in neutral buoyancy. Neutral buoyancy allows cephalopods with chambered shells to swim without having to expend muscular energy just to stay in the water column. However, the shell itself results in extra drag, resisting movement and rendering swimming less efficient. This leads to the expectation that shell shape should be modified in such a way as to minimize this energy loss during locomotion. Experimentation with flow tanks designed to ascertain which shell shapes optimize swimming efficiency has been one of the major directions of ammonoid

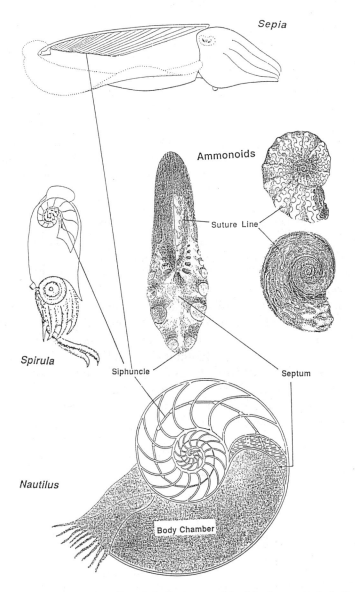

FIGURE 1. Cephalopods with chambered shells discussed in this chapter include the modern *Sepia, Spirula,* and *Nautilus* as well as extinct ammonoids. The chambers are emptied osmotically by the siphuncle and are passively filled with gas through diffusion to a final pressure less than 1 atmosphere. As a consequence of this low internal pressure, the shell structure must support water pressure when these organisms are submerged. The tubular siphuncle of the externally shelled *Nautilus* and ammonoids and the internally shelled *Spirula* traverses the chambers within the shell. In the internal shell of *Sepia,* the cuttlefish, the siphuncle forms the ventral/posterior surface of the chambered shell. All these chambered shells are lighter than sea water when empty and compensate for the muscle mass of the organism, which is denser than sea water. Of these forms, only *Sepia* can adjust its buoyancy rapidly so as to go up and down in the water column on a daily basis. In the other forms the shells serve to maintain neutral buoyancy. Squids without chambered shells, such as *Loligo* and *Illex,* are not shown. (From Jacobs, 1992a, Fig. 1.)

research. We explore issues pertaining to ammonoid locomotion after reviewing arguments relating to buoyancy.

2. Buoyancy

The chambered shells of cephalopods have inspired functional interpretations of shell form for hundreds of years. By the time of Hooke (Derham, 1726),
it was recognized that these chambered shells function in buoyancy. However,
it was not until the work of Denton and Gilpin-Brown (1961a,b,c, 1966; Denton
et al., 1961, 1967) that siphuncular emptying of the shell was understood.
These studies demonstrated that the chambers of cephalopod shells are
osmotically emptied of cameral liquid, that the pressure within the shell is at
one atmosphere or less, and that the shells of cephalopods supported the
hydrostatic load of the overlying water column. The realization that shelled
cephalopods support hydrostatic load had profound consequences for the
interpretation of the structural function of cephalopods (see Jacobs, 1992a).
This understanding of siphuncular function that developed in the 1960s and
1970s also has a number of implications for the understanding of buoyancy.
In this section we discuss experimental work on living cephalopods and how
this work has led to a better understanding of the limitations of the shell-emp-
tying mechanism in ammonoids. In particular, we review the arguments
indicating that "decoupling" of fluid from the siphuncle does not play an
important role in cephalopod shell design and that because of the small size
of the ammonoid siphuncle, ammonoids could not change density on a daily
basis. We then briefly discuss the implications of the many calculations of
buoyancy and static stability of the shell.

2.1. History

Prior to 1960 it was not clear how the chambered shells of cephalopods
were emptied of water during chamber formation and growth. Many 19th and
early 20th century workers thought that pressurized gas played a role in
emptying the chambers. This idea was first advocated in 1695 by Hooke
(Derham, 1726), who also argued that gas generation could be used to fill and
empty the shell, permitting passive, that is, buoyancy-driven, vertical migration. In contrast, Owen (1832) observed that the vascularization required for
gas generation in fish swim bladders was not present in *Nautilus*. He felt that
a gas pressure mechanism was not likely. Only a few workers took Owen's
lead; Buckland (1836), Pfaff (1911), Spath (1919), and Westermann (1956)
inferred that there was little gas pressure in the chambers of cephalopods and
interpreted shell function with this in mind. However, none of these authors
correctly interpreted the means of chamber evacuation, which involves an
osmotic pump (Fig. 2).

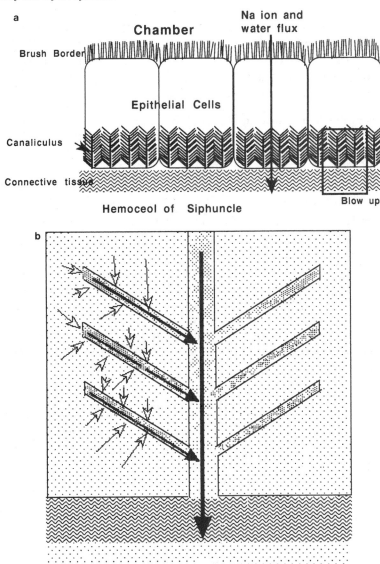

FIGURE 2. Details of the "standing-gradient osmotic flow" structure and mechanism (after Diamond and Bossert, 1967). A similar mechanism is known to function in the siphuncular epithelium of cephalopods (Greenwald *et al.*, 1982). Figure 2a shows the general arrangement of epithelial cells, siphuncle, and chamber. Figure 2b is a blowup of the intracellular ultrastructure. In the minute tubes depicted, a high concentration of sodium ions is maintained by active transport. The high standing gradient of ions between the inside and outside of these small tubes combined with their large surface area relative to cross-sectional area generates a diffusive flux of water into the tube. The water is subsequently forced out of the tube by this osmotically generated pressure, and a flux of water then proceeds out of the tube, ultimately into the hemocoel of the siphuncle. Pressure generated by the pump depends on the ionic difference that can be maintained across the small tubes as well as the geometry of the tubes.

The first person to consider an osmotic mechanism in cephalopods was Bruun (1943), who worked on *Spirula* (Fig. 1). He was attracted to the osmotic mechanism because he had not observed gas emanating from *Spirula* that were rapidly retrieved from considerable depth and felt that an osmotic pump would provide a simple means of emptying the shell:

> Undoubtedly, the most simple explanation [of chamber emptying] could be given if it could be proved, that the pressure inside the shell is always about 1 atm . . . In this case we need only assume that a chamber . . . is first filled with fluid, which is absorbed and replaced by air, by a fairly simple osmotic mechanism. (Bruun, 1943, p. 11)

However, Bruun was in a quandary. *Spirula* was frequently retrieved from depths greater than 1,000 m. The osmotic difference between fresh and salt water produces an osmotic pressure of about 25 atmospheres, a pressure equivalent of a 250-m column of water. Consequently, this difference in salt concentration can pump water out of a chamber only at depths of 250 m or less. As a consequence of this depth discrepancy, Bruun (1943, 1950) did not accept the osmotic pumping mechanism. This same conflict between the observed habitat depth of modern chambered cephalopods and the apparent or presumed depth limitation on osmotic pumping continued to trouble cephalopod workers even after the osmotic mechanism itself had been adequately demonstrated.

2.2. Discovery of the Osmotic Mechanism

The cuttlebone of cuttlefish consists of a series of fine chambers. The siphuncle of cuttlefish is not tubular, as it is in other chambered cephalopods; it is sheet-like, forming the large ventral surface of the cuttlebone where it abuts all the chambers (Fig. 1). When Denton and Gilpin-Brown set out to explore the buoyancy mechanism of the cuttlefish, *Sepia officinalis,* they expected to find pressurized gas within the chambers of freshly caught specimens. What they found was not what they expected.

Denton and Gilpin-Brown (1961a,b,c) and Denton *et al.* (1961) noted a number of features of cuttlefish biology that allowed them to infer the mechanism by which water is removed from the chambers. (1) Gas does not come out of cuttlefish when they are rapidly retrieved from depth in the sea. (2) Cuttlefish vary in density between day and night; during the day they are negatively buoyant, and during the night neutral or positively buoyant; the change in density results from a change in the density of the cuttlebone, as opposed to other tissue. (3) After puncture of the chambers of a cuttlebone submerged in a fluid, fluid flows into the chambers rather than gas flowing out. (4) More fluid flows into recently formed chambers than into chambers formed earlier in ontogeny; the older chambers fill about 20% when punctured; the gas in these chambers was primarily nitrogen with much less oxygen than in the atmosphere but substantially more carbon dioxide. (5) The fluid remaining in partially emptied chambers is less salty than sea water.

From their observations, these authors (Denton and Gilpin Brown, 1961a,b,c; Denton *et al.*, 1961) inferred that the cameral liquid in shelled cephalopods is pumped out osmotically. An active transport mechanism moves ions across the siphuncular membrane, and water follows as a consequence of the difference in salt concentration across the membrane. The low-oxygen, high-carbon-dioxide composition of the gas recovered from the chambers indicates passive diffusion of gas from the blood of the cuttlefish into the chambers. A passive diffusion mechanism is further supported by the pattern of increasing gas pressure with age of the chambers. Thus, gas composed primarily of nitrogen diffuses into the chambered shells of cephalopods after they have been emptied by the osmotic mechanism. The pressure in the chambers does not exceed 1 atmosphere; contrary to the arguments of earlier workers, gas pressure plays no role in emptying the chambers of cephalopods. (Some might suspect that the pressure of gas dissolved at depth in the sea, and hence in the blood of cephalopods, would be above the pressure of gas in the atmosphere. However, the pressure of a gas dissolved in a fluid is largely determined by the pressure of that gas at the surface of the fluid. If this were not the case, the pressure of gas in water at depth would be supersaturated relative to water near the surface. Such conditions, when they exist, are unstable. Upward movement of the water at depth then results in supersaturation and rapid or even explosive degassing.)

Denton and Gilpin-Brown (1966) and Denton *et al.* (1967) then went on to perform similar studies on *Nautilus* and *Spirula* (Fig. 1). The results they obtained were very similar to their observation on *Sepia*, with the exception that there were no diurnal changes in density. Unlike *Sepia*, both *Nautilus* and *Spirula* were close to neutral buoyancy throughout the period during which they were observed. Denton and Gilpin-Brown (1973) then went on to interpret their results in the context of fossil cephalopods, including ammonoids, and argued that a similar mechanism of shell emptying was indicated in fossil forms with chambered shells and siphuncles.

2.3. Decoupling

Denton and Gilpin-Brown were troubled by one aspect of their results. They observed, as Bruun (1943) had before them, that *Spirula* as well as *Nautilus* spent much of their time below 250 m. This figure is the approximate depth equivalent of the osmotic pressure produced by the difference in salinity between fresh and salt water. To Bruun, as well as to Denton and Gilpin-Brown, the salt concentration in sea water appeared to set a depth limit below which an osmotic pumping mechanism could not work. *Nautilus* had often been recovered below this depth. *Spirula* was virtually always recovered from depths much greater than 250 m. So how did these organisms succeed in osmotically emptying their chambered shells if they were found at depths greater than 250 m? An osmotic mechanism was clearly indicated by the low ionic concentration of the fluid that remained in the chambers and by the low gas pressures observed in the shell. Denton and Gilpin-Brown were at an impasse. To reconcile their observations on pumping mechanism with the depth difference between fresh and salt water, Denton and Gilpin-Brown invented "decoupling" (see Jacobs, 1992a, for review).

In their decoupling argument, Denton and Gilpin-Brown (1966) and Denton *et al.* (1967) suggested that *Nautilus* and *Spirula* pumped out their shells at depths shallower than 250 m and made only temporary descents into deeper water. Furthermore, it was argued that, while these animals were at depths in excess of 250 m, fluid in the chambers was "decoupled"; that is, it was not in contact with the siphuncle, and consequently, the pumping mechanism was not operative. In *Spirula*, orientation was thought to result in decoupling, ". . . when *Spirula* is in its normal [head-down, see Fig. 1] swimming position, this liquid [in the chambers] will be almost completely decoupled from the permeable region [of the siphuncle] . . ." (Denton *et al.*, 1967, p. 188). Bruun (1950) also reinterpreted his depth estimates for *Spirula*, rendering them shallower and more consistent with the osmotic pumping mechanism. Subsequently, additional workers have interpreted chamber, septal, and sutural design in ammonoids as indicative of, or adaptations for, decoupling or preventing the liquid retained in the chambers from contacting the siphuncle, at least in certain shell orientations.

2.4. Lack of Support for the Decoupling Argument

Unfortunately, the decoupling argument appears to have little merit. Lack of fluid contacting the siphuncle from the inside would not prevent fluid from flowing across the siphuncular membrane back into the chambers when the hydrostatic pressure imparted by the overlying water exceeded the osmotic pressure difference between fresh and salt water. Thus, there seems to be no physical basis for decoupling. In fact, organisms can concentrate salts in intracellular structures (Diamond and Bossert, 1967); as a consequence, osmotic pumping need not be directly limited by the salt concentration of the environment (Fig. 2b).

In their later work, Denton and Gilpin-Brown (1973) cited a study of hyperosmotic pumping employed by sea birds to remove salt from their body fluids and suggested that a similar system played a role in emptying cephalopod shells at depths greater than the osmotic pressure difference between fresh and salt water (Diamond and Bossert, 1967). Clearly, Denton and Gilpin-Brown themselves were not entirely satisfied with the explanatory value of the decoupling mechanism. Subsequently, Greenwald *et al.* (1982) studied the *Nautilus* siphuncle at the ultrastructural level and documented the presence of structures associated with a hyperosmotic pump (Fig. 2). In such a pump,

Na$^+$ ions are actively transported across the cell membrane into microscopic intracellular channels; negatively charged Cl$^-$ ions follow passively to maintain the charge balance; water then diffuses into the channel in response to the high ionic concentration. As a result, high pressures in the small channels exceed the hydrostatic pressure, and water flows down them into the siphuncle. The geometry of the small channels is important. Water diffuses readily across the relatively large surface area of the tube, but flow down the tube and into the lumen of the siphuncle is much more rapid than diffusion. Thus, intracellular channels or canaliculi can transport fluid against very high pressures as long as ions are present to be pumped. There is, of course, an energetic expense associated with continuous transportation of Na$^+$ across membranes to maintain the high osmotic gradient.

It was the appearance of a conflict between an osmotic pumping mechanism and observed habitat depth that precipitated the decoupling hypothesis. It is now known that there is no precise depth limit to osmotic pumping (Greenwald et al., 1982). In addition, observations of Nautilus fail to document decoupled water retained in the older chambers (Ward, 1979). When the cameral liquid falls below the level of the siphuncle in the last formed chamber of Nautilus, transport of cameral liquid to the siphuncle and out of the chamber continues unimpeded (Ward, 1979). Capillary action in the wetted organic sheet, the pellicle, that covers the septal surface undoubtedly facilitates this final phase of chamber emptying. Given the potential for fluid transport on the wetted surfaces of the chamber, as well as the potential for transport of water in the vapor phase in the partial vacuum of the forming chamber, it seems highly unlikely that fluid can be successfully isolated, that is, decoupled, from the siphuncle. In the absence of theoretical or empirical support, there seems to be little justification for continued application of the concept of decoupling to the interpretation of cephalopod shell function. However, decoupling continues to be invoked in the paleontological literature as an adaptive explanation for various aspects of ammonoid shell morphology (e.g., Bandel and Boletzky, 1979; Weitschat and Bandel, 1991; Saunders, 1995).

2.5. Diurnal Density Changes and Vertical Movements?

Many authors from the 17th century onward have assumed that all chambered cephalopods, including ammonoids, could adjust their buoyancy at will, permitting them to rise or sink in the water column on a daily basis. However, more recent data suggest that such density-driven migrations are not likely to occur in the majority of cephalopods with a narrow siphuncular tube. Because of the osmotic nature of the siphuncular pump, the pumping rate is likely to be proportional to the siphuncular area and inversely related to the depth, that is, the pressure that must be pumped against. Ward (1982) has obtained empirical data on the emptying rates of Nautilus. Based on these data, Ward calibrated the per-area pumping rate of siphuncular tissue. He then argued convincingly that the small tubular siphuncles of many cephalopods preclude diurnal changes in buoyancy at any appreciable depth. The siphuncular area is sufficiently small, relative to the size of the chambers and the size of the organism as a whole, that any change in buoyancy generated in a diurnal cycle would be trivial. These calculations suggest that virtually all chambered cephalopods, including ammonoids and nautilids, are not capable of daily migration based on changes in buoyancy. In these forms with narrow siphuncles, vertical movements can be accommodated through active locomotion. *Nautilus* is observed to swim vertically relatively frequently (e.g., O'Dor *et al.*, 1993).

Ward (1982) argued that only two groups of cephalopods have siphuncles large enough to permit the rapid changes of density required for passive vertical movements. These are the living sepiids, or cuttlefish, and the fossil endoceratids. Thus, the vision first promoted by Hooke (Derham, 1726), that the chambered cephalopods can rapidly change their density and migrate vertically, is the exception rather than the rule. Ward (1982) also argued that in forms living at sufficient depth, pumping rate and chamber emptying are so slow that they limit not only diurnal migration but growth rate as well.

So what prevented ammonoids, or other cephalopods, from evolving larger siphuncles permissive of more rapid chamber emptying? There appear to be two responses to this question, one structural and the other energetic. In tubular siphuncles, strength is inversely related to diameter. Tubular siphuncles of large diameter are necessarily weaker and fail at shallower depths than smaller siphuncles of comparable wall thickness. Thus, a large tubular siphuncle permits rapid density changes but constrains the animal to relatively shallow water. Sepiids have avoided the structural constraints imposed by a tubular siphuncle; the siphuncular surface of sepiids forms a sheet on the outside of the shell. This siphuncular sheet is supported by very closely spaced septa (Fig. 1). However, Ward and Boletzky (1984) argue that even in these cuttlefish, there is still a structural trade-off. Shallow-water forms have relatively flat shells with large surface areas, permitting the most rapid changes in buoyancy. Those sepiids that live in deeper water have more cylindrical shells. This cylindrical shell form is stronger, but it also has a smaller ratio of siphuncular surface area to shell volume and less rapid pumping rates.

In addition to mechanical limitations, there are energy costs to pumping and to maintaining empty chambers. These costs should be proportional to siphuncular area and increase with depth. When a siphuncle is exposed to fluid under pressure, it must pump constantly to overcome diffusion of water into the chamber. The greater the depth, and the greater the area of the siphuncle, the greater the energetic costs of chamber emptying and subsequent maintenance. It may be that septal necks play an important role in limiting siphuncular area and minimize the energetic costs of maintaining empty chambers at depth. (See Tanabe and Landman, Chapter 6, this volume, for a discussion of septal neck evolution in ammonoids.) In any case, life in deep water is likely to constrain the amount of siphuncular area that can be maintained as well as limit the maximum diameter of tubular siphuncles.

2.6. Buoyancy Calculations

The chambered shells of living cephalopods displace water and are positively buoyant. The chambered portion of the shell balances the negative buoyancy of the soft parts and body chamber, which are denser than sea water. Shell and soft parts together are then at, or close to, neutral buoyancy. These density and volume relationships can be conceived of as a set of equations and unknowns constrained by the requirement of neutral buoyancy. Thus, if density and volumes of some of the variables can be measured or inferred, others can be calculated in fossil forms when they cannot be observed. In addition, a stable orientation and the forces required to rotate the shell from

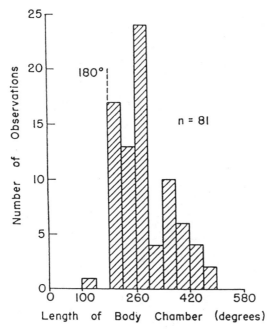

FIGURE 3. Histogram of Trueman's (1941) body-chamber length data (from Raup, 1967, Text-Fig. 17). Note that body-chamber lengths are longer than those of *Nautilus* (124° to 167°: Saunders and Shapiro, 1986). Also note that the peak in the distribution is near 300°. A body-chamber length of 300° produces a stable orientation with the aperture well above the location of the aperture in *Nautilus* (see Saunders and Shapiro, 1986). The position of the peak of the histogram suggests that the Jurassic ammonoids Trueman (1941) studied are not well designed to maximize static stability but, instead, are designed so that a horizontally produced thrust would not induce rotation requiring stabilization.

this stable orientation can be determined if one knows the density and geometry of the phragmocone, body chamber, and soft parts (Fig. 3).

2.6.1. Calculations of Body Chamber Length and Static Stability

A large number of workers (Trueman, 1941; Raup, 1967; Chamberlain, 1981; Saunders and Shapiro, 1986; Swan and Saunders, 1987; Ebel, 1990; Shigeta, 1993) have performed calculations to determine the density and static stability of ammonoids. The assumptions and intentions of these workers have varied. Trueman (1941) set out to determine the body chamber lengths, relative stability, and stable orientations of various ammonoid shell forms. Subsequent workers have employed computers to generate models of ammonoids using coiling equations such as those of Raup (1967; Fig. 4). The algorithms involved propagate an expanding whorl section using the isometric relationship of spirally coiled shells first discussed by Mosely (1838).

These calculations require many simplifying assumptions. Accretionary growth at the aperture generates the coiled form of the molluscan shell. Coiling is not dictated by mathematical parameters; the coiling parameters of am-

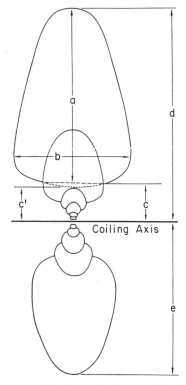

FIGURE 4. Geometric parameters of shell morphology used in this chapter (from Raup, 1967, Text-Fig. 1). Expansion rate (Raup's W) = $(d/e)^2$. Umbilical ratio (Raup's D) = (c/d). Whorl cross-sectional shape (Raup's S) = (b/a). Thickness ratio = $[b/(d+e)]$.

monoids are derived from the shell after the fact. Many ammonoids clearly do not retain the same coiling parameter during the entire course of their growth. This is most evident in the heteromorphs. Some workers have tried to confront the issue of changing coiling parameters by calculating shell growth as a differential deposition of material at the aperture (Okamoto, 1988; Ackerly, 1989). In addition, the shapes used as whorl sections, or generating curves, in calculations performed to date are simple geometric forms such as circles and polygons. These are, of course, not identical with the whorl cross sections of actual ammonoids.

The complete body chamber is preserved in only a small fraction of ammonoids. One application of coiling equations has been to calculate the body chamber length of a range of ammonoids, assuming that the living animals were neutrally buoyant. Such calculations require information about shell thickness, shell density, and the volume and density of the negatively buoyant soft parts. The actual distribution of the shell material in a phragmo-cone is not easy to assess in detail. Terms for the shell material in the septa and ornament are often incorporated in these calculations (Raup and Chamberlain, 1967; Westermann, 1971; Saunders and Shapiro, 1986). Generally, soft parts are assumed to have the same density as those of *Nautilus*, and to exactly fill the ammonoid body chamber.

Raup (1967) calculated body chamber length and static stability for a range of coiled forms with circular whorl sections. He then compared these theoretical results with the distribution of ammonoid forms in the space defined by a set of morphological variables pertaining to planispiral shells, i.e., an ammonoid shell morphospace. The coiling parameters W (whorl expansion rate), D (distance of the generating curve from the coiling axis), and S (shape of the aperture) define the axes of this morphospace (Fig. 4). Raup then observed that the distribution of actual ammonoid shapes occupies only a fraction of the possible morphospace defined by the variables and axes and compared this distribution to the results of buoyancy and static stability calculations. Raup found that planispiral ammonoid morphology was concentrated in a general region of W from 1.5 to 3, smaller than that for modern *Nautilus*. Expansion of the coil was constrained by the relationship where the whorls touch, $W = 1/D$. Adult goniatites are more involute (smaller D) than most other ammonoids. Raup concluded that shell locomotion and static stability favored planispiral forms and that some degree of whorl overlap was advantageous to conserve shell material and strengthen the shell. However, large amounts of whorl overlap or very small D were not favored, possibly because of negative effects on shell buoyancy and static stability. Raup also indicated a general relationship between degree of compression of the whorl (S) and the other variables. Involute forms (low D) with large whorl expansion rates (W) tended to have laterally compressed whorls (small S), whereas more evolute forms tended to have less compressed whorls (S approaching 1 or greater). We will return to this relationship among S, W, and D when we discuss the effects of shell shape on swimming and its relationship to ontogeny (Sections 3.7 and 3.8).

Raup and Chamberlain (1967) developed a computer model to calculate body chamber length, relative static stability, and orientation of ammonoid shells from the coiling parameters devised by Raup (1967). Saunders and Shapiro performed additional calculations involving apertural areas and shapes (Saunders and Shapiro, 1986; Shapiro and Saunders, 1987). They argued that these factors, not considered by Raup (1967), may constrain ammonoid shell design. In addition, Saunders and Shapiro's examinations of Mississippian ammonoids suggest that shell thickness greatly influences the calculated body chamber lengths and orientations. Previous workers used a single relative shell thickness to compare all ammonoids. Saunders and Shapiro (1986) employed thickness terms for the shell and septum derived from digitized sections of the specimens as well as densities of shell material that more closely approximate those of the nacreous shell. Of the five forms they examined, all had much longer body chambers than *Nautilus*. However, four of the forms had stable orientations that placed the aperture at low angles from the vertical in a similar position to that of *Nautilus* (Fig. 5). Saunders and Shapiro argued that this orientation, found in many Paleozoic forms, is associated with benthic feeding as it is in *Nautilus*. In some Paleozoic ammonoids a ventral sulcus is also present, suggesting that these forms were able to direct the jet downward and, perhaps, backward under the shell to accommodate forward swimming, as is the case in *Nautilus*.

In all these exercises assumptions are made. Saunders and Shapiro (1986), Raup (1967), and Raup and Chamberlain (1967) assumed a circular whorl section. This may not be an important departure from the observed morphology in the Paleozoic forms examined. However, in more compressed shells, an assumption of a circular whorl section is likely to produce an overestimate of shell volume relative to the mass of the shell material, and, consequently, an overestimate of body chamber length. In addition, retraction or extension of the body in or out of the body chamber would change the center of mass. This would alter the static orientation of the shell. Trueman (1941) suggested that this gave ammonoids some control over orientation.

Swan and Saunders (1987) built on their earlier morphometric and hydrostatic work (Saunders and Shapiro, 1986; Saunders and Swan, 1984) in an analysis of the body chamber length and apertural angle of Namurian ammonoids. They explored the significance of changing whorl expansion, W, departure of the generating curve from the center of coiling, D, and shell thickness, TH, on apertural angle and body chamber length. They found that, given an assumption of neutral buoyancy, whorl expansion and shell thickness had large effects on body chamber length and apertural angle. They then plotted these variables as a function of W and TH and found that about two-thirds of Namurian forms occupied a region of morphospace that generated body chamber lengths on the order of 300° to 500°. The model suggests that these forms also tended to have low apertural angles. This contrasts with

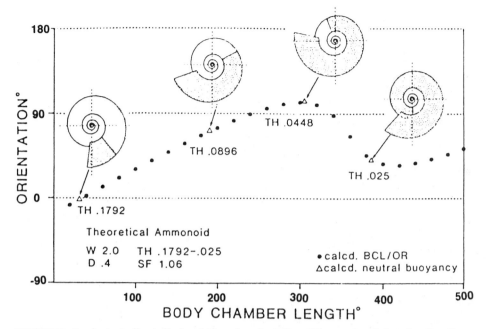

FIGURE 5. In planispirally shelled cephalopods, orientation of the aperture is largely a function of body-chamber length, as shown (from Saunders and Shapiro, 1986, Fig. 8). If one assumes uniform density of soft parts and uniform distribution of water in the ammonoid body chamber, then shell thickness and whorl expansion rate become the most important variables controlling body chamber length and orientation. Mesozoic (primarily Jurassic) ammonoids studied by Trueman (1941; Raup, 1967) appear to occupy the peaks in the orientation curve near body chamber lengths of 300° and 500°. These peaks should produce aperture orientations near 90° (see Fig. 3). Namurian ammonoids (Saunders and Shapiro, 1986; Swan and Saunders, 1987) tend to have body chamber lengths slightly longer than one whorl in length, near 400°. This should stabilize the aperture below the shell in a position similar to that of modern *Nautilus*; however, *Nautilus* has a much shorter body chamber of approximately 150° (Shapiro and Saunders, 1987). *TH*, relative thickness of the shell wall; *OR*, orientation measured in degrees from vertical. *BCL*, body-chamber length; *SF*, septal–siphuncle factor. See Fig. 4 for other abbreviations.

Trueman's (1941) data. Derived primarily from Jurassic ammonites, these data show a bimodal distribution of body chamber lengths (Raup, 1967) with a paucity of ammonoids having body chambers around 360°. A body chamber length of 360° is thought to generate minimum static stability in these open-coiled, low-*W*–high-*D* Jurassic morphs. This instability is associated with apertures oriented either well below or well above the middle of the coiled shell.

2.6.2. Limits, Assumptions, and Tests of Neutral Buoyancy

As we have discussed, a number of workers have performed calculations based on an assumption of neutral buoyancy. Other workers have attempted to determine whether fossil forms departed from neutral buoyancy. Ebel

(1983) made calculations of shell density based on shell thickness measurements reported by Westermann (1971). He concluded that all ammonoids were slightly negatively buoyant benthic organisms. Other workers have argued that all such departures from neutral buoyancy fall within the measurement errors of the calculations (Westermann, Chapter 16, this volume). They base this assessment on the difficulties in measuring shell and septal thickness in a sufficiently precise and comprehensive way, even in a single well-preserved specimen. Determining the length of the body chamber can often be difficult except in exceptionally well-preserved material. Simplifying assumptions, such as the geometric form of the whorl shape, inherent in calculations of volume from coiling equations also appear to preclude precise calculation of buoyancy in fossil forms. For example, Ebel (1983) used triangular and trapezoidal cross sections to calculate shell displacement and the volume of shell material. These geometric forms will, necessarily, generate an overestimate of shell wall volume to phragmocone volume if there is any roundness to the whorl sections of the actual specimens. Neutral buoyancy is observed in all modern chambered cephalopods. It is difficult to imagine that a cephalopod would retain the siphuncle and chambered shell intact over long periods of evolutionary time if they no longer functioned to generate neutral buoyancy. Nevertheless, Ebel (1992) has persisted in his argument, inferring a benthic mode of life of a range of ammonoids.

There are additional uncertainties in the other assumptions required to calculate the buoyancy of fossil cephalopods. *Nautilus* soft-part density is used as a proxy for the body density of ammonoids, and these soft parts are presumed to fill the entire body chamber. However, ammonoids are only distantly related to modern *Nautilus* (see Lehmann, 1967; Jacobs and Landman, 1993); they are much more closely related to coleoids. All coleoids have larger mantle cavities than *Nautilus* (Chamberlain, 1992). A large mantle imparts greater swimming efficiency, as discussed in the following section. If ammonoids had mantle cavities comparable in size to coleoid mantle cavities, the body chamber might not be as completely filled with dense proteinaceous material as is envisioned in these calculations. The amount of dense propulsive muscular tissue varies in cephalopods, ranging from 30% in *Sepia* to 40% in *Illex*, as opposed to *Nautilus,* which is 4.5% by weight propulsive muscle (O'Dor and Webber, 1991). Other workers have noted that the density of the nacreous material that makes up the ammonoid shell is subject to variation as a consequence of the relative proportions of aragonite and protein matrix (Mutvei, 1983). Thus, there are a host of assumptions required to calculate the density in life of an ammonoid.

Declining density through ontogenetic size increase is expected of the soft tissues of swimming organisms. Drag is a function of area; it is most limiting on small organisms, which necessarily have large area-to-volume ratios. To compensate for their relatively large drag, small, early ontogenetic stages often have a higher ratio of dense muscle to other soft tissue than adults. Adults also often have large amounts of positively buoyant fat required for reproduction. Thus, soft-tissue density should decline with increasing size in ontogeny. (Similarly, giant squids are less dense than smaller squids because they can afford to devote a smaller proportion of their tissue to propulsive muscle and use some of their volume to accommodate urea-filled tissues that are osmotically equivalent to, but lighter than, sea water.) Such an ontogenetic decrease in density would be sufficient to explain the results of Shigeta (1993), who examined the density of *Tetragonites glabrus* and two other Lower Turonian forms through their ontogeny. He concluded that early in ontogeny, they were positively buoyant, and later in ontogeny, they became neutral or negatively buoyant. These results are interpreted as indicating a planktonic juvenile and benthic adult stage. If Shigeta's results are sufficiently reliable to overcome problems in shell volume measurements, they may record responses of the shell to ontogenetic changes in density of the soft parts rather than changes in buoyancy. Earlier in ontogeny, a proportionally larger phragmocone-to-body-chamber ratio would be required if neutral buoyancy is to be maintained with denser soft parts. The majority of Ebel and Shigeta's determinations of density are within 10% of neutral buoyancy. With the large number of variables in the analysis and the uncertainty of those variables, it is surprising that the results of Ebel (1983), Shigeta (1993), and Tanabe *et al.* (1995) are as close to neutral buoyancy as they are. These results could just as easily be interpreted as a confirmation of neutral buoyancy. Much more dramatic results would be needed to demonstrate convincingly that a chambered cephalopod could not attain neutral buoyancy.

2.7. Evolution of the Buoyancy Apparatus

The functional complex of chambers and siphuncle is a unique derived feature occurring only in cephalopods. Hence, it is difficult to envision the evolution of these structures from other molluscs that lack them. Most workers depict the evolution of the chambered cephalopod shell as proceeding from a simple, cap-shaped, monoplacophoran-like shell. Subsequent addition of chambers in the apex of such a shell is thought to have generated the first Cambrian cephalopod, presumably much like the fossil *Plectronoceras* (Yochelson *et al.*, 1973). Surprisingly, formation of apical chambers occurs in some gastropods and bivalves (Owen, 1878). In elongate gastropods with small whorl expansion, such as vermetids, septa seal off the unused, early ontogenetic portions of the shell far from the aperture. In some cemented bivalves, chambers appear to form as a response to rapid growth, with intermittent carbonate deposition inside the pallial line. Thus, chambers produced for other reasons in a mollusc ancestral to the cephalopods could have been exaptively incorporated into the chambered buoyant shell of a cephalopod progenitor.

The evolution of the siphuncle appears to be more difficult to account for than that of the chambers. However, on this issue as well, Owen provides us with interesting observations and speculation. Owen (1878) observed a *Spon-*

dylus that had formed a shell containing a series of chambers. In this particular cemented bivalve, the muscle attachment penetrated a series of chambers. As ontogeny proceeded, the septa formed around the preexisting muscle, which remained attached to the underlying layers of the shell. Owen speculated that a muscle penetrating a series of chambers in this fashion might have provided an intermediate form between cephalopods and other molluscs. Stated in modern terms, Owen's concept implies that the siphuncle of cephalopods evolved from a muscle that fortuitously penetrated a set of chambers that formed around it. To further elaborate on this speculation, we note that the muscles and nerves operate by setting up an ionic gradient across their cell membranes. A capacity for pumping ions across a membrane would be an essential preadaptive or exaptive function of any tissue that was to evolve the osmotic pumping capacity of the siphuncle.

3. Swimming

The swimming ability of ammonoids is controversial. Some workers argue that ammonoids as a whole swam relatively fast. Others maintain that they did not swim at all. From the outset of this discussion, we would like to make it clear that an external chambered shell, which most if not all ammonoids are thought to have possessed, is an impediment to swimming. (Some have argued that ammonoid shells were internal, e.g., Gray, 1845; Doguzhaeva and Mutvei, 1989.) The shell, as a consequence of its large volume, produces significant drag forces. Thus, ectocochliate cephalopods are not likely to swim with the same efficiency as fish or squid (e.g., Chamberlain, 1981, 1992; Jacobs, 1992b). On the other hand, all modern cephalopods, with the possible exception of the cranchiid squid, are capable of reasonably active swimming. Even *Octopus* can swim. In addition, some shell shapes in ammonoids suggest adaptation to locomotion. It is precisely because the cephalopod shell is an impediment to locomotion that one would expect selective forces to mold its shape to reduce drag and acceleration forces that resist movement.

One way to place limits on the locomotor abilities of ammonoids is to compare the force required by, or energy dissipated in, a particular swimming behavior with estimates of the amount of force, or energy, that the organism had available to perform that activity. Which physical variables (e.g., energy, force, power) are appropriate for analysis will depend on, among other factors, the duration of the activity involved. The supply of the relevant variables, such as energy or force, can then be compared to the demand, the amount required for a particular set of swimming behaviors. In this discussion, we divide swimming behavior into three distinct categories based on the time scale of the behavior. At short time scales (seconds) that pertain to acceleration, or short bursts of activity involved in prey capture or escape, the potential for thrust generation can be compared with the forces of acceleration and drag that resist such rapid movement. Over longer time periods (minutes to hours), the ability to metabolize and generate a constant stream of aerobic energy will limit swimming performance. By comparing the sustainable supply of aerobically generated power available, with the power required to maintain constant locomotion, the range of velocities that ammonites could sustain for such periods of time can be constrained (Jacobs, 1992b). Over yet longer time periods, the limits on behavior will depend on energy reserves. By estimating the amount of energy stored, and determining the energetic costs of transportation, one can establish the distance the organism can travel between fill-ups (meals). One can also determine the velocity that confers the most efficient transportation, permitting travel between two points with the least energetic cost. Calculations of some aspects of cost of transportation have recently been performed on squid and *Nautilus* (O'Dor, 1988a; O'Dor and Webber, 1991; Chamberlain, 1990; O'Dor *et al.*, 1993). Similar calculations may place important constraints on the reconstruction of ammonoid modes of life.

These three different levels, anaerobic "burst" performance, aerobically sustainable activity, and long-term cost of transportation, on which aquatic locomotion can be analyzed pertain to selective influences in nature that are likely to be optimally resolved by different shell shapes. This suggests that the relationship between ammonoid form and hydrodynamics may not have been simple and that a single hydrodynamically optimal shell form should not be expected. On the other hand, consideration of an appropriate range of factors may permit insights into the adaptive constraints on shell shape in ammonoids.

Various approaches have been taken in assessing the resources available for propulsion (Mutvei and Reyment, 1973; Reyment, 1973; Mutvei, 1975; Chamberlain, 1981; Ebel, 1990; Jacobs, 1992b). Chamberlain (1981) and Jacobs (1992b) attempted to assess swimming velocity across the range of planispiral forms using estimates of force and energy available for propulsion. However, with the exception of Jacobs (1992b), all attempts at examining the resources available for propulsion in ammonoids have relied heavily on analogy to swimming in *Nautilus*. Analyses that refer directly to the swimming mechanics of *Nautilus* may not always be relevant to the swimming of ammonoids because of the great phylogenetic distance between ammonoids and *Nautilus* (Jacobs and Landman, 1993).

In addition to difficulties in analysis of the "supply side," that is, the resources available for locomotion, there are also many issues yet to be assessed that relate to the "demand side," or resources required for movement of an ammonoid. To date, analyses of locomotion in fossil cephalopods have focused heavily on drag forces (Schmidt, 1930; Kummel and Lloyd, 1955; Chamberlain, 1976, 1980, 1981; Jacobs, 1992b). As a result, these analyses assume, and necessarily pertain only to, steady or continuous locomotion. Drag is not an adequate proxy for the costs associated with acceleration from rest. As a consequence, the influence of standing-start locomotion such as lunging predation on ammonoid shell form has not been addressed experimentally.

Drag forces most clearly relate to the costs of continuous swimming; however, even in this case, some of the costs of locomotion are not accounted for by drag alone. Locomotion in all marine animals is, necessarily, unsteady; in cases involving jet propulsion, locomotion is likely to be particularly unsteady and to require additional energy inputs to overcome repeated acceleration (Daniel, 1984). Despite these limitations, analyses of drag provide a point of departure in that they identify an important component of the energetic costs of locomotion. Interestingly, even when this one variable is considered, no single shell morphology is uniquely advantageous; a range of morphologies can be identified that have advantageously low drag under different conditions (Jacobs, 1992b). In addition, some aspects of shell form impart large costs in terms of drag and are clearly disadvantageous in terms of locomotion. This, in itself, is an interesting conclusion. In other cases, the absence of an advantage to a particular form in terms of drag alone may suggest that the form was adapted to minimize forces associated with acceleration rather than drag. We argue that this is the case for oxycones.

In the remainder of the chapter, we (1) highlight a number of uncertainties inherent in assessing locomotion in ammonoids in which direct knowledge of the soft parts is limited; (2) argue that analogy to modern *Nautilus* may not always be appropriate in interpreting ammonoid hydrodynamics; (3) review approaches to the supply-side issues that may provide reasonable constraints on aspects of ammonoid locomotion; (4) discuss the body of work dealing with drag forces and the conclusions that can be drawn from that approach, including comparisons of the power available with the power dissipated in drag; (5) explore additional issues related to acceleration and acceleration-re-action forces; (6) consider the ecological and evolutionary implications for ammonoids of studies that have examined the costs of transportation in living cephalopods; (7) examine how these hydrodynamic phenomena relate to particular aspects of ammonoid morphology such as shell shape and orna-ment; and (8) discuss how these factors relate to broader aspects of ammonoid paleobiology, such as ontogeny, reproductive strategy, paleoenvironment, and mode of life.

3.1. Investigation of Ammonoid Swimming: Uncertainties and Assumptions

Arguments regarding form and function often pertain to structures whose functions can be directly observed. In the case of fossils, evidence of function is usually indirect at best, and the structures themselves may only be partly preserved. As regards ammonoids, a plethora of well-preserved ammonoid shells are available, but direct evidence of ammonoid soft part anatomy is limited. The shell contributes a large fraction to the energy dissipated during locomotion. By direct experimental examination of the shell, we can gain some knowledge of the demand side of the locomotor equation. However, the orientation of the shell during swimming, the means of propulsion, as well as the shape of the soft parts can have important influences on the results (Chamberlain, 1980). Consequently, assumptions about the nature of ammonoid locomotion are necessary before even the simplest exercise in gathering data about drag or other aspects of locomotion is performed. The supply side is necessarily even more difficult; propulsion is directly linked to the rarely preserved soft tissues.

In virtually all examinations of drag in ammonoids, drag measurements have been conducted on shells facing backward, that is, with the aperture and any extended soft parts on the trailing side of the shell. This is despite the fact that both *Nautilus*, and coleoids are observed to swim both backwards and forwards. The rationale for this approach is rarely expressly stated. It has become in some ways a tradition of the discipline. Presumably, the justification for this approach is based on the fact that the rapid escape swimming of *Nautilus* is backward. However, even *Nautilus* spends the majority of its time swimming slowly (O'Dor *et al.*, 1993) and, presumably, in a forward direction. In addition to the issue of forward versus backward swimming, differences in orientation of the shell aperture influence drag (Chamberlain, 1976). Thus, assumptions of direction and orientation are likely to affect measurements of the energy dissipated in drag and acceleration.

Other assumptions pertain to the mechanism of propulsion. Most cephalopods are capable of jet propulsion, and jet propulsion has been assumed in every study of ammonoid locomotion to date. However, virtually all coleoids also propel themselves with fins, which are used to stabilize and maneuver.

3.2. Swimming in Ammonoids and Analogy to Modern Cephalopods

Several authors have advanced arguments suggesting that ammonoids could not swim, or could not swim well, as ammonoids lack features that are required for swimming in *Nautilus*. In particular, ammonoids appear to lack the large paired retractor muscles and shell stability important in *Nautilus* swimming. However, the assumption that ammonoids and *Nautilus* require the same set of features for locomotion is difficult to support (Jacobs and Landman, 1993). As noted, both stratigraphic and cladistic analyses of fossil cephalopod phylogeny indicate a much closer relationship between ammonoids and modern coleoids than between either of these groups and *Nautilus*. Thus, a coleoid analogue may be more informative as regards the ammonoid mode of life.

Coleoids use a compressible mantle to generate the jet; *Nautilus* uses the shell as a containment vessel, drawing the body back into the shell while compressing the water in the mantle as though the body were a piston in a cylinder (Chamberlain, 1987, 1990). Thus, coleoids swim with a very different mechanism than *Nautilus*. Jacobs and Landman (1993) argued that a coleoid mantle could have functioned in ammonoids and that a coleoid mantle may have evolved in the lineage that gave rise to both ammonoids and coleoids. Coleoids lost their body chamber early in their evolution. Without a body chamber, coleoids could not have swum as *Nautilus* does via retraction into the body chamber. Consequently, coleoids ought to have evolved a contracting mantle prior to, or coincident with, the evolutionary loss of the body chamber.

Several lines of evidence suggest that coleoids are the sister taxon of ammonoids (Lehmann, 1967; Berthold and Engeser, 1987; Jacobs and Landman, 1993). The close relationship of ammonoids and coleoids and the impossibility of a *Nautilus*-like swimming mechanism in forms lacking a body chamber suggest that ammonoids possessed a coleoid-like contractile mantle. This argument is supported by observations of muscle scars. Many ammonoids have medial dorsal and ventral muscle scars (Crick, 1898; Jordan, 1968; Doguzhaeva and Mutvei, 1991; Chapter 3, this volume). These medial scars are much like the attachment of the head and funnel retractor muscles to the gladius, the much reduced shell found in many squid (Jacobs and Landman, 1993; see Fig. 10, Chapter 3, this volume). These muscles in squid do not play a direct role in propulsion. A contractile mantle compresses the water to form the jet. If the small medial muscle scars observed in many ammonoids are homologues of the head and funnel retractor muscles of coleoids, and if ammonoids had a squid-like mantle, the small size of these muscles would not limit propulsive power. In addition, an attachment scar with a lateral reentrant is present in the body chambers of some ammonoids (Jordan, 1968; Doguzhaeva and Mutvei, 1991; Chapter 3, this volume). This scar and reentrant have been analogized to the siphonal sinus in the pallial line, the line of mantle attachment in bivalves.[*] In bivalves, this sinus is associated with the ability to evert and extend the siphon. If this structure serves an analogous function in ammonoids, it suggests that the siphon and portions of the mantle could be everted out of the shell, perhaps permitting the function of coleoid-like mantle contractions (Jacobs and Landman, 1993).

To continue the analogy to coleoid swimming, it should be observed that many coleoids have fins that they use for low-velocity locomotion. It may be that some ammonoids had fins similar to those that occur in coleoids. The possibility of fins in ammonoids has yet to be examined by cephalopod workers.

A number of authors have presented trends in the evolution of cephalopod locomotor design based on the assumption that the condition found in living *Nautilus* is ancestral (O'Dor and Wells, 1990; O'Dor and Webber, 1991; Wells

---

[*]Doguzhaeva has published material that shows large lateral muscle scars. However, other authors (K. Tanabe, personal communication, 1992) have argued that these marks indicate the attachment of the mantle, analogous to the pallial line of bivalves. Other authors have presented some equivocal evidence of lateral musculature in a few other forms (e.g., Crick, 1898). Bandel (1986) reported an ontogenetic series in *Quenstedtoceras*, where the lateral muscle scars shift during ontogeny to a more medial position. This could relate to an evolutionary change from forms with lateral to forms with more medial musculature.

and O'Dor, 1991; O'Dor *et al.*, 1993). However, to date this assumption has not been supported by argument, and it is not clear that the locomotor design of *Nautilus* is the ancestral condition in cephalopods. The last shared ancestor of *Nautilus* and the living coleoids is thought to have been in the Ellesmerocerida of Ordovician age (House, 1981). [The Ellesmerocerida are a group of Cambrian and Ordovician age that gave rise to all other cephalopods. Thus, it is a massively paraphyletic grouping that, in inclusive phylogenetic terms, is equivalent to the Cephalopoda as a whole (Jacobs and Landman, 1993)]. The Ellesmerocerida are straight shelled; the traditional phylogeny suggests that the lineage leading to *Nautilus* contains a succession of coiled tarphycerids and uncoiled oncocerids prior to the final coiling in the nautilids. If this evolutionary succession of coiling and uncoiling is correct, it is likely to have influenced locomotor design. Thus, evolution of swimming in the *Nautilus* lineage may have been complex. Similar complexities may also have occurred in the evolutionary succession leading from the ellesmeroceratids to the coleoids and ammonoids.

Recent work on *Nautilus* documents two distinct mechanisms of propulsion. At low swimming speed *Nautilus* uses the same mechanism for propulsion as is used to ventilate the gills, i.e., oscillation of the wings of the funnel and collar folds (Wells and Wells, 1985). This mechanism produces a constant stream of water. During more rapid escape swimming the retractor muscles come into play, producing the pulsed jet thought of as typical of *Nautilus* swimming (Chamberlain, 1987, 1990). Because there are two mechanisms of propulsion in *Nautilus*, it seems fair to ask whether one or the other of them represents the ancestral condition. A facile answer would be that the mechanism associated with respiration is more likely to be ancestral; after all, molluscs outside the cephalopods produce respiratory currents. On the other hand, if the retractor muscle mechanism of propulsion found in *Nautilus* is ancestral in the cephalopods, then those ellesmeroceratids, tarphyceratids, oncocerids, and nautilids that comprise the *Nautilus* lineage ought to have retractor muscle scars that resemble those of *Nautilus*. Examination of muscle scars in appropriate fossil forms could serve to substantiate inferences regarding the evolution of locomotor design in cephalopods.

Chamberlain (1981) and Ebel (1990) have provided distinctly different arguments as to why static stability limits swimming velocity in ectocochliate cephalopods. These arguments are based on the idea that the force of the cephalopod jet opposes the gravitiational force stabilizing the shell (Fig. 6). This premise is based on observations of *Nautilus*, where the jet exits near the venter of the shell and acts to rotate the shell. Consequently, *Nautilus* swims with a rocking motion. First the jet rotates the shell, and then the static forces rotate the shell back to an upright orientation (Chamberlain, 1981, 1987). It has long been observed that many ammonites, especially those with long body chambers, are much less stable than *Nautilus*. This results from the smaller distance between the center of mass and center of buoyancy in these forms (Trueman, 1941; Raup, 1967). This lack of stability has been invoked as a

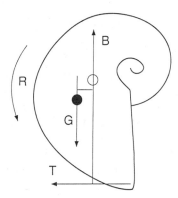

FIGURE 6. Forces operating on *Nautilus* during rapid swimming (after Chamberlain, 1981, Fig. 4, and from Jacobs and Landman, 1993, Fig. 2). The filled circle is the center of mass on which gravity, the arrow marked *G*, exerts a downward force; the open circle is the center of buoyancy on which the buoyancy force, *B*, imparts an upward force. The center of buoyancy and mass are offset. This imparts dynamic stability to the shell by generating a restorative moment, *R*, which is here shown opposing the thrust force of the jet marked *T*.

limitation on swimming. However, the long body chamber of many ammonites places the aperture higher, nearer the midline of the shell (Fig. 3). In this position, a jet directed aft would not act on a long lever arm, and would not impart a large rotation to the shell. Thus, although long-body-chambered ammonoids may lack static stability, such stability may not be required in these forms. Some, especially late Paleozoic ammonoids, are relatively stable (Swan and Saunders, 1987). Many of these forms also have a hyponomic sinus, suggesting that the jet was produced near the venter in these stable forms. The hyponomic sinus also suggests that the hyponome could bend back underneath the shell. Perhaps, these ammonoids could swim forward with their jet directed back under the shell. *Nautilus* swims in this fashion when searching for food. Thus, static stability and a hyponomic sinus may be a complex of characters related to forward swimming or possibly benthic feeding. The absence of these characters in other ammonoids does not necessarily imply that they could not swim. (See Jacobs and Landman, 1993, for a more detailed argument; also see Westermann (Chapter 16, this volume) for a slightly different perspective.) In addition, most other swimming organisms, including coleoids, use dynamic forces associated with the swimming process to maintain or change orientation.

3.3. Supply-Side Issues

As was observed in the previous discussion, direct analogies to the swimming mechanism of *Nautilus* can artificially constrain interpretation of swimming in ammonoids, in part because ammonoids are so distantly related to *Nautilus* that there may be few similarities in their actual swimming mechanics. On the other hand, the coleoids, although much more closely related to ammonoids, are very diverse. It is difficult to generate a single number for force or power available from such a diverse group (O'Dor and Webber, 1991). However, ammonoids are also a diverse group, and we would argue that the range of values for important locomotory parameters found in coleoids should be considered, in addition to those found in *Nautilus*, in any reconstruction of ammonoid locomotion. Forces produced by the jet, pressure produced by the jet, and oxygen consumption have been measured in *Nautilus*, cuttlefish, octopus, and the squids *Loligo* and *Illex* (see O'Dor and Webber, 1991, for a recent review). A number of additional observations on jet propulsion in other invertebrates such as salps (Madin, 1990), medusae (Daniel, 1985; Demont and Gosline, 1988), and scallops (Dadswell and Weihs, 1990) may also have implications for cephalopod locomotion.

Oxygen consumption measurements in living cephalopods can be used in estimating maximum sustainable swimming velocities in ammonoids. This approach was employed by Jacobs (1992b), who used an estimate of the power available for swimming based on assessments of metabolic scope in living cephalopods. Such studies were first used extensively to examine swimming in fish (e.g., Brett, 1965). The approach involves taking measurements of oxygen consumption at rest and then forcing the animal to exercise at its aerobic maximum. (Aerobic maximum is often operationally defined as the maximum performance sustainable for one hour.) Usually, swimming velocity is manipulated, and oxygen concentration measured, in a closed tank with recirculating water. Often a stimulus to swim, such as an electrified baffle, is also provided. Metabolic scope is then obtained by subtracting resting oxygen consumption from the oxygen consumption observed during maximal sustained activity. Such measurements have been obtained in *Loligo* (O'Dor, 1982), *Illex* (Webber and O'Dor, 1986), *Nautilus* (Wells and Wells, 1985; O'Dor *et al.*, 1990), and *Sepia* (O'Dor and Webber, 1991).

To apply these data to ammonoid swimming, values of metabolic scope inferred from studies of oxygen consumption in living organisms have to be converted to power available for propulsion. This requires the assumption of a number of efficiency terms such as the conversion of oxygen consumption to muscular force as well as the efficiency of conversion of muscular force to propulsive force. This last exercise requires the consideration of Froude efficiency.

The energetic efficiency of jet propulsion is determined, in part, by Froude efficiency. Energy consumed in swimming includes a term for the energy imparted to the water to produce thrust. This has the form typical of kinetic energy, $E = \frac{1}{2}mU^2$, where E is energy, m is the mass of the water, and U is the velocity. However, the thrust or momentum imparted by the jet is simply mU, or mass times velocity. Because energy increases as a square of velocity, the most economical way to produce momentum is to maximize the mass and minimize the velocity of the water accelerated to produce thrust. Undulatory or finned swimming accelerates larger masses of water than jets. As a consequence, such swimming has considerable advantage in Froude efficiency over jet propulsion, especially for continuous swimming. In addition, the larger the mantle cavity, the greater the amount of water that can be accelerated by the jet of the cephalopod; consequently, a larger mantle cavity should confer greater efficiency. It is noteworthy that *Nautilus* has one of the smallest mantle cavities of any cephalopod. Coleoids have much larger mantle cavities. Uncertainty as to the mantle size and swimming mechanism in ammonoids introduces uncertainties in assessing the Froude efficiency of ammonoid swimming.

When data derived from squid are compared to ectocochliate forms, some provision also has to be made for the proportion of the shell volume that does not contribute to propulsion. Jacobs (1992b) has used this approach to infer maximum sustainable swimming speeds in ammonoids by comparing power available, to power required to overcome drag forces. These results are discussed further in the following sections on drag and power.

In addition to approaches based on oxygen consumption, power output has been determined from observations of jet pressure, jet velocity, and changes in mantle cavity volume determined from high-speed film. Trueman and Packard (1968) pioneered the recording of mantle pressures in swimming cephalopods. They inserted cannulae into the mantle cavities of tethered squid and octopus. With this approach, pressures in the mantle could be related to the thrust measured via the forces on the tether. With these initial data in hand, equations treating the range of forces produced in swimming squid and *Nautilus* were developed (Johansen *et al.*, 1972; Bone *et al.*, 1981; Chamberlain, 1987; O'Dor, 1988b). Subsequently, jet pressures were determined in free-swimming squid (*Illex*) and *Nautilus* as well as in *Nautilus* in the wild, using various transmitter arrangements (Webber and O'Dor, 1986; O'Dor *et al.*, 1990, 1993).

Jet pressure averaged over a period of time, in combination with knowledge of the funnel orifice area or changes in mantle volume, can be used to calculate power consumption. Such calculations have been performed for aerobically sustainable high-speed swimming as well as for higher speeds produced during anaerobic respiration. O'Dor and Webber (1991) compiled data from a number of sources, performed these calculations, and tabulated the results for a number of cephalopods as well as other jet-propelled invertebrates. These calculations entail a different set of assumptions and sources of error than are inherent to assessments of power based on oxygen consumption. They can also be used to assess sustainable swimming velocities in ammonoids, as is discussed in the following section on drag. Power measurements, derived from jet pressure during the most active periods of swimming observed, can also be used to assess the potential for more rapid bursts of activity supported by anaerobic respiration. In addition, instantaneous pressures produced by the mantle cavities of various cephalopods are even higher than the average mantle pressures associated with maximum swimming speeds. These peak pressures, in combination with the forces resisting acceleration, provide some idea of the acceleration that can achieved by ammonoids. This issue is touched on in the discussion of acceleration.

3.4. Drag and Analyses of Constant Velocity

Drag is the primary force resisting the steady movement of an object through a fluid. Hydrodynamic work on ammonoids conducted to date has primarily involved an examination of drag. All such analyses contain the simplifying assumption that acceleration, and the forces associated with acceleration such as acceleration reaction, are not a large factor in ammonoid swimming. This is unlikely to be uniformly true in that jet propulsion in aquatic organisms is inherently unsteady. Another assumption of such analyses is that jet propulsion is the means by which ammonoids swam. We discuss some of the constraints likely to be imposed by acceleration and the unsteady nature of aquatic locomotion after reviewing the work that has been done measuring drag and analyzing various steady-state properties from it.

3.4.1. Reynolds Number and Coefficient of Drag

Reynolds number (Re) and coefficient of drag (Cd) are dimensionless numbers used to characterize flow regimes around, and to compare drag forces on, swimming organisms. The Reynolds number can be thought of as a ratio of inertial forces divided by viscous forces resisting flow. In swimming organisms, Reynolds numbers are typically calculated as a function of the length of the organism in the direction of flow times the velocity of the organism divided by kinematic viscosity. In water at room temperature, where drag experiments are conducted, or in Mesozoic seas, the Reynolds number of an ammonoid can be approximated by its diameter, in centimeters, times its swimming velocity, in centimeters per second, times 100. Under these conditions, an ammonoid 5 cm in diameter swimming at 20 cm/sec will be operating at Re of 10^4; an ammonoid 20 cm in diameter swimming at 5 cm/sec will also be operating at Re of 10^4. If these ammonoids, of different size, are identical in shape, then the flow pattern around them should be similar as long as the Reynolds number is the same. As a result of this concept of flow similarity, observations taken at one combination of velocity, size, and viscosity will pertain to all combinations that yield the same Reynolds number. This permits considerable flexibility in experimental design and interpretation of results (see Vogel, 1981, for a more detailed explanation).

Drag can be conceived of as having two components, "pressure," or inertial drag, and "skin friction," or viscous drag. Inertial drag relates to the displacement of the mass of fluid out of the path of a moving object. Viscous drag results from the stickiness or adhesion of the fluid to the surface of the object. Roughly speaking, pressure drag is a function of the cross-sectional area of an object perpendicular to flow or movement, whereas the viscous component of drag is a function of surface area rather than cross-sectional area. At high

Reynolds numbers, the inertial drag will predominate, whereas at low Reynolds numbers, the viscous component of drag will predominate. Inertial drag generally varies with the square of velocity; viscous drag is a first-order function of velocity.

Coefficient of drag (*Cd*) is a nondimensional number. It consists of drag force corrected for the expected change in inertial drag with velocity:

$$Cd = \frac{\text{drag force}}{0.5(\text{density of the fluid})(\text{Area})(\text{velocity})^2}$$

Coefficient-of-drag calculations are devised to facilitate comparisons of drag by accounting for differences in velocity and size. To balance the force term in the numerator and generate a nondimensional coefficient, an area term must be incorporated. In studies on cephalopods, an area term devised from volume [(volume)$^{2/3}$] is used. It is argued that volume is the appropriate comparative metric in analyses of swimming organisms (Vogel, 1981). In this typical formulation of coefficient of drag, there is no compensation for skin friction drag, just for pressure drag. Skin friction may be important at low

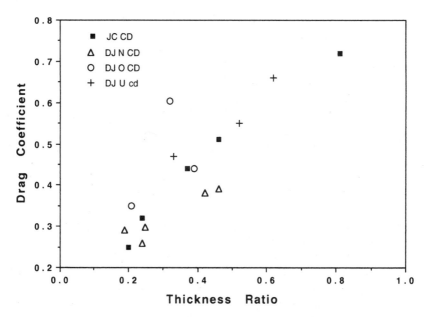

FIGURE 7. Effect of shell morphology for *Re* above 15,000. ■, shell models differing in degree of whorl compression (*S*) without bodies, data from Chamberlain (1976) recalculated to thickness ratio. △, +, ○, data from Jacobs (1992b), all with body prostheses attached during drag measurement. △, involute, narrowly umbilicate forms; ○, ornamented forms; +, openly umbilicate forms. These data indicate a general increase in drag coefficient with thickness ratio at higher *Re*. The opposite tends to be the case at lower *Re* (≤8000; Fig. 9). The data also indicate that ornament and large umbilici increase drag and that the presence of a body reduces drag.

Reynolds numbers, and lower Reynolds numbers may have characterized the swimming of many ammonoids. For comparisons based on coefficient of drag to be relevant, they must be made in a region of Reynolds numbers comparable to that experienced by the ammonoids in question.

3.4.2. Studies of Drag on Ammonoid Shells

Schmidt (1930), Kummel and Lloyd (1955), Chamberlain (1976), and, more recently, Jacobs (1992b) measured drag forces on models of ammonoids and calculated their coefficients of drag. Schmidt and Kummel and Lloyd directed flows past models of ammonoids in a relatively crude fashion and determined that compressed forms, specifically oxycones, narrow involute forms that taper toward the shell margin, had the lowest drag coefficients. Chamberlain (1976) examined a suite of models based on variation in the Raupian coiling parameters W and D in a naval test tank. He also examined a few models with compressed whorl sections. Forms with narrow whorl sections had substantially lower drag, as had been observed by the earlier workers. Chamberlain also found that variation in W and D produced systematic

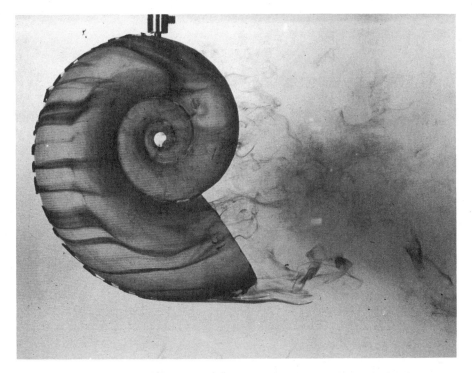

FIGURE 8. Flow structure at $Re = 3000$ for a lytoceratid shell model. Crystals of potassium permanganate attached to the model's venter generate the dye. Boundary-layer separation occurs at the shoulder of the umbilicus, where the dye streams deflect. Also note vortices of dye trailing the shell and in the umbilicus. (From Chamberlain, 1976, Pl. 84, Fig. 1.)

variation in drag, primarily as a result of change in the size of the umbilicus and thickness ratio of the shell (Fig. 7). Flow visualization studies showed that these results were induced by alteration of flow structure, particularly wake size and vortex features, induced by different aspects of shell morphology (Fig. 8).

The studies of Schmidt (1930), Kummel and Lloyd (1955), and Chamberlain (1976) were conducted at relatively high Reynolds numbers, well in excess of 10^4. These values are at the upper limit of Reynolds numbers that would be expected for ammonoid swimming. Jacobs (1992b) explored a range of lower Reynolds numbers (2000–25,000) and found that skin-friction drag had an important influence at these lower values. Thin forms with larger surface areas were less efficient at low Reynolds numbers. It was only at Reynolds numbers above 7000–8000 that compressed ammonoids had lower coefficients of drag. Thus, it was only at higher Reynolds numbers that thin

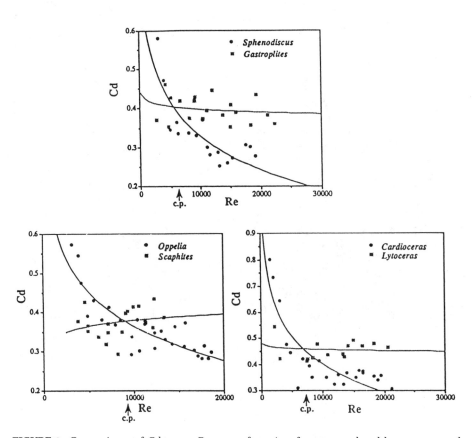

FIGURE 9. Comparisons of Cd versus Re curves for pairs of compressed and less-compressed ammonoids of similar form reveal a crossing point (shown by arrow on Re axes) at Re from 5000 to 8000. At Re below the crossing point, the less-compressed form has a lower Cd than the more "streamlined" compressed form. (After Jacobs, 1992b, Fig. 5.)

shell shapes were advantageous. This result suggests that there was no single optimal morphology for swimming in ammonoids. Which particular shell shape was advantageous depended on size and swimming velocity (Fig. 9). That different shell forms may be optimal under different conditions may permit interpretation of some aspects of the modes of life of different ammonoids. It also leads to important predictions: for example, if ammonoids required shells optimally shaped to reduce drag throughout their lives, then (1) ammonoid shell shape should become progressively more compressed during ontogeny, and (2) evolutionary change in shell shape in ammonoid lineages should relate to changes in the hydrodynamic regime as documented by facies changes. However, before examining these issues, we consider comparisons based on power, rather than drag coefficient, and additional forces associated with acceleration.

3.4.3. Power

Examining swimming in the context of a dimensional variable, such as power, has some advantages over looking at coefficient of drag alone. Power available for propulsion will increase as a function of volume, a cube of linear dimension. However, drag scales to an area term, so the consumption of power by drag should increase only as a square of the linear dimension. As a consequence of these scaling relationships, swimming organisms can swim faster with increasing size. The limitations of this relationship on swimming are not intuitively obvious from the Cd alone. In addition, the power required for swimming can be compared to the amount of power available, yielding estimates of swimming velocity.

It is relatively simple to calculate power required for swimming at a given velocity. Power is work per unit time; work is a force operating through a distance. By examining the units, it is apparent that power is simply the drag force times the velocity. One can also use Cd versus Re plots to calculate power required for swimming at a range of sizes and velocities (e.g., Jacobs, 1992b). Power required per unit volume can then be compared between different forms at a range of sizes and velocities. Such a comparison is made in Fig. 10 between *Gastroplites*, a moderately thick, involute form, and *Sphenodiscus*, a thin oxycone. (Relative thickness or degree of compression of the coil as a whole is best compared using the thickness ratio, *T.R.*, which is calculated by dividing the maximum width by the diameter of the shell. The *Gastroplites* specimen examined by Jacobs, 1992b, had a *T.R.* of 0.42, and the *Sphenodiscus* had a *T.R.* of 0.19.) This comparison illustrates the range of size and swimming behaviors in which one form will have an energetic advantage relative to the other. The results suggest that there is a broad range of combinations of small size and low velocity where thicker forms are substantially more efficient. There is also a large region of high velocity and moderate size where thinner forms are markedly more efficient than thicker forms in terms of power consumption per unit volume. For large forms traveling slowly, the power differences are small, and the advantage conferred on either morph is probably

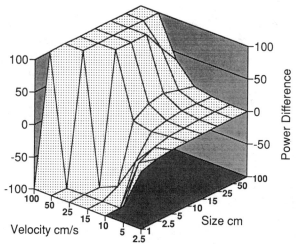

FIGURE 10. A three-dimensional representation of the difference in power consumption per unit volume for a thick form, *Gastroplites*, and a thin form, *Sphenodiscus*. To calculate power difference, power required per unit volume for *Sphenodiscus* was subtracted from that for *Gastroplites*. (From Jacobs, 1992b, Fig. 6.) When values are negative, as they are at low multiples of velocity and size, then *Gastroplites* has a lower power required per unit volume; at high multiples of size and velocity, values are positive, and *Sphenodiscus* is preferred. At combinations of large size and low velocity, power required per unit volume is small, and so are power differences. At combinations of small sizes and high velocity, power differences are substantial and likely to be biologically significant. The surface was truncated at a power difference of 100 ergs/sec per cm^3.

trivial. Thus, size and swimming velocity are likely to bear on which shell shapes will be preferred in terms of lower power consumption in steady swimming. Examination of the power required per unit volume indicates that thicker forms have an advantage up to a higher *Re* than if one considers *Cd* alone (Fig. 9).

To determine a swimming velocity from the power-required data, one needs some measure of the power available for propulsion. It is difficult to know what limited the power available to ammonoids. However, comparison to living organisms suggests that the limits of power available depend strongly on the duration of the activity. On very short time scales, large amounts of power may be mustered by anaerobic muscle contraction. Brief, dramatic pulses of power output are important in lunging predation and in escape from predation. Analyses of maximal thrust forces produced in living cephalopods could be useful in assessing the limits of acceleration in ammonoids. Alternatively, peak propulsive power available in moderately energetic cuttlefish or highly energetic squid can be used to estimate maximum power production in ammonoid soft tissues.

On intermediate time scales, power generated for propulsion is limited by respiration. Sustainable or aerobic respiratory potential can be constrained by analogy to the maximal oxygen consumption of living cephalopods or by estimates based on average jet pressure as discussed previously. These figures can then be used to calculate maximum sustainable swimming velocities. One of the difficulties encountered in this approach is that different cephalopods are able to mobilize different amounts of aerobic or sustained metabolic energy. O'Dor and Webber (1991) tabulated information indicating that *Nautilus* has an extremely small metabolic scope, whereas *Illex*, in particular, has a very large metabolic scope.* The difference is only partially explained by the large shell and relatively small retractor muscles of *Nautilus*. It seems inappropriate to constrain ammonoids to low, *Nautilus*-based, power production values, in part because ammonoids may have had a much higher ratio of soft parts and propulsive muscle to the chambered portion of the shell (Trueman, 1941; Swan and Saunders, 1987), and because ammonoids are more closely related to the energetically diverse coleoids than to *Nautilus* (Jacobs and Landman, 1993).

Jacobs (1992b) employed an intermediate metabolic scope of 200 ml O_2/kg per hr in his analysis of maximum sustainable swimming speeds in ammonoids. This figure is roughly twice that for *Nautilus* and about half that of some of the figures available at the time for *Loligo*, although, recently, higher figures have been reported for *Loligo* and *Illex* (see O'Dor and Webber, 1991). The figure given above is very near that for recently reported oxygen consumption values for *Sepia* (O'Dor and Webber, 1991). Jacobs (1992b) used values from the literature on Froude efficiency and muscular efficiency and the relationship of body volume to total volume to generate a power-available figure of 400 ergs/sec per cm^3. This upper limit in power available was then compared with data on power required, derived from flow tank observations of drag on ammonoid models, to determine maximum swimming velocities. The most striking aspect of these results is the increase in velocity that can be attained with increasing size (Fig. 11). At small sizes, thicker forms could have attained slightly higher maximum velocities than thinner forms. At larger size, thinner forms could attain higher swimming speeds.

Despite the multiple assumptions required to estimate power available, the resulting maximum sustainable swimming velocities are relatively robust.

---

*These exceptional qualities of *Illex* do not result from evolution of novel functional features. *Illex* simply has more muscle, and more mitochondria, leading to a life style that requires even more prey than other squid to feed this energy-intensive system. Parallels can be drawn to terrestrial locomotion. Pronghorn antelopes have more muscle, respiratory tissue, and mitochondria than other running animals and achieve higher speeds, in part by achieving a greater flux of energy than other runners (Linstedt *et al.*, 1992). What these parallels suggest is that many of the organisms thought of as runners and swimmers are not adapted to attain peak velocities. Apparently, very few organisms are precisely adapted for peak performance of a narrowly defined locomotory variable. Trade-offs, both with other locomotory parameters and perhaps with other nonlocomotory features that confer selective advantage, compromise perfect adaptation to a particular attribute.

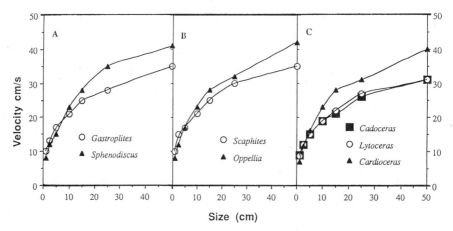

FIGURE 11. Maximum sustainable swimming velocity (MSV) is presented as a function of size for pairs of less-compressed and compressed ammonoids: *Gastroplites* and *Sphenodiscus*; and *Scaphites* and *Oppelia*. The umbilicate forms *Cadoceras*, *Lytoceras*, and *Cardioceras* are also graphed together. The MSV was determined by interpolation of the power-required data using a maximum power available figure of 400 ergs/sec per cm³. Note that differences between the velocities of all the forms are not large and that curves for compressed and less compressed forms cross between 5 and 10 cm in size. Multiples of size and velocity above 300 represent extrapolations beyond the *Re* range of drag measurements made. (From Jacobs, 1992b. Fig. 7.)

Power-required figures increase dramatically with velocity. As a consequence, large variations in the amount of power available produce much smaller variations in the velocity calculated. However, it appears from the calculations of O'Dor and Webber (1991), based in part on alternative means of calculating power output from jet pressure, that active squid such as *Illex* have metabolic scopes as much as four times higher than the *Sepia*-like figure assumed in the analysis. In addition, *Illex* incorporates added muscular efficiencies that render their overall power output as much as an order of magnitude higher than that of a modest performer such as *Sepia*. These power outputs are comparable to those that can be generated via anaerobic respiration in *Sepia*. We take this figure as an upper bound on power available for swimming associated with anaerobic respiration in ammonoids.

In *Sepia*, maximum swimming velocities of approximately 65 cm/sec are attained for brief periods of time. These short periods of activity supported by anaerobic respiration in *Sepia* are associated with a power output of 10,000 ergs/sec per cm³ (O'Dor and Webber, 1991). If one assumes that the inert shell of an ammonoid is 40% of the total volume and could not contribute to propulsion, one is still left with a figure of 6000 ergs/sec per cm³ potentially available for burst performance in ammonoids. With this figure, comparisons can then be made with the power available tables in Jacobs' (1992b) work. Resulting velocities assume energy loss through drag alone. Given these assumptions, a 1-cm diameter *Sphenodiscus* could attain velocities of just over 20 cm/sec, and a *Gastroplites* could have briefly attained speeds of a little over 25 cm/sec. This compares to a result of 8 and 10 cm/sec, respectively, for these two forms operating in an aerobically sustainable fashion. At 25 cm in length, a *Sphenodiscus* could attain velocities on the order of 100 cm/sec, faster than the maximum velocity observed in *Sepia*. At this size, *Gastroplites* could have swum at about 70 cm/sec, similar to maximum burst swimming speed of *Sepia*.

The above result, suggesting that *Sphenodiscus* could have swum faster than *Sepia* does today, may be a consequence of lower drag in *Sphenodiscus*. At least the model used in assessing the drag of *Sphenodiscus* lacked the fins that may contribute to drag in *Sepia*, and, by all accounts, *Sepia* is not designed for maximal velocity among modern cephalopods. However, the thrust in *Sepia* is not entirely devoted to overcoming drag. Some of it must accelerate the water that fills the mantle cavity during each propulsive cycle, and there may be additional costs associated with acceleration reaction. Because of these extra costs, which are not accounted for in *Sphenodiscus*, it seems that the above maximum swimming speeds calculated for *Sphenodiscus* and *Gastroplites* represent an upper bound on estimates of maximum anaerobic or burst performance.

A lower limit on the potential burst performance of ammonoids may be ascertained by looking at the power required for burst swimming in *Nautilus*. For reasons mentioned previously, including the larger ratio of body to chambered portion of the shell in ammonoids and their phylogenetic proximity to the more energetic squid, ammonoids are likely to have more power available per unit volume than *Nautilus*. *Nautilus* mobilizes about 1000 ergs/sec per cm^3 of power during burst swimming. With this amount of energy, the power-required data suggest that *Sphenodiscus* could have attained swimming speeds of 12 cm/sec at 1 cm in diameter and on the order of 55 cm/sec at 25 cm in diameter. *Gastroplites* could have attained speeds close to 15 cm/sec at 1 cm in diameter and speeds of close to 40 cm/sec at 25 cm in diameter.

So what can we conclude from this exercise of using power consumption in living cephalopods to constrain swimming speeds in ammonoids? First, the results are dependent on the assumptions made, such as which modern analogues to use, whether aerobically sustainable swimming or anaerobic bursts of activity are of interest, and whether it can be assumed that power consumed by drag alone adequately represents total power consumption. Despite the uncertainties, some patterns emerge. There is a clear relationship between size and swimming velocity. Swimming velocity in small ammonoids will be constrained relative to that of large ones. Consistent with the observations of power required, thick forms have advantages in terms of maximum velocity at small sizes. With increasing size, thinner ammonoids gain an increasing relative advantage. Depending on the assumptions, estimates of the maximum anaerobically supported swimming speed of a 25-cm diameter *Sphenodiscus* range from 55 cm/sec to 100 cm/sec. Anaerobically supported jet propulsion results in speeds of 30 cm/sec for *Nautilus*, 65 cm/sec for *Sepia* of about this same size, and up to 170 cm/sec for *Illex* of slightly greater length (O'Dor and Webber, 1991). Scallops can attain relatively remarkable speeds. *Placopecten magellanicus*, a mere 5 cm in shell height, can attain velocities of 55 cm/s (Dadswell and Wiehs, 1990). This relatively high velocity is generated by a power production per unit mass nearly three times that of *Sepia* and almost on par with *Loligo* (O'Dor and Webber, 1991). This dramatic result is possible perhaps because anaerobic "burst" swimming is the only kind of swimming a scallop does. It need not invest in additional respiratory structures required for aerobically sustained locomotion. This achievement is all the more remarkable in that only 5.4% by weight of a scallop is muscle. If a *Sphenodiscus* were to generate power similar to that of a scallop, a 5-cm individual could have overcome drag forces equivalent to 90 cm/sec and a 25-cm individual could have overcome drag forces equivalent to over a meter per second. However, in such bursts of activity, acceleration forces may make an important contribution to the power required for movement.

3.5. Acceleration

In addition to overcoming the drag forces resisting steady movement, a swimming organism must overcome acceleration forces. These include forces that are typically thought of as acceleration, *i.e.*, forces that are simply a function of mass and rate of velocity change. However, swimming also results in transport of fluid in the organism's boundary layer and fluid entrainment in its wake. During steady movement, the mass and momentum of this fluid remain relatively constant. However, during acceleration, this mass of fluid must be accelerated along with the organism. The additional mass associated with this fluid is referred to as "added mass," and the force necessary to accelerate it is referred to as "acceleration reaction." Acceleration reaction has had an adaptive influence on ammonoid shell shape. A coefficient that is a function of shape, referred to as the added mass coefficient, can be defined. Thus, we can infer some aspects of how acceleration reaction forces will behave in shells of a variety of shapes.

A typical formulation for the acceleration reaction force is given by Daniel (1984):

$$G = -arV(du/dt)$$

where G is the acceleration reaction force, a is the added mass coefficient, r is the density of the fluid, V is volume, and du/dt is acceleration. As can be seen in Fig. 12, the added mass coefficient, a, is a function of the thickness ratio. Laterally compressed, low-thickness-ratio shells have lower added mass coefficients than broader shells that necessarily accelerate a proportionately larger volume of associated water. Thus, slender shells will have an additional hydrodynamic advantage in acceleration beyond that associated with drag

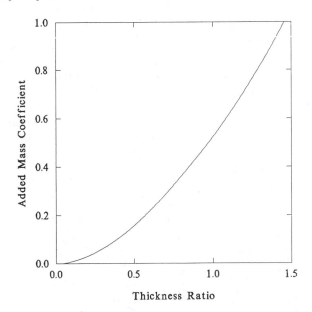

FIGURE 12. Added-mass coefficient versus thickness ratio (after Daniel, 1984). Additional resistance to acceleration results from the necessity of accelerating an "added mass" of water in the boundary layer and entrained wake of an object changing velocity in a fluid. This acceleration reaction is a linear function of velocity change as well as a function of the added-mass coefficient. This dimensionless coefficient relates to the shape of the object and its orientation relative to the direction of acceleration. Thin forms have substantially less added mass than thick forms.

alone. Acceleration reaction is different from drag in that it resists deceleration as well as acceleration. Therefore, in swimming that has cycles of acceleration and deceleration, the momentum imparted to the associated water will retard deceleration as well as slowing acceleration. In fact, in an ideal fluid, one that has mass but no viscosity, the momentum imparted during acceleration should return during deceleration. In the real world, the efficiency with which the momentum is retained with the swimming organism and limits deceleration between power strokes is likely to depend on a number of factors. These include the symmetry of the form, the symmetry of the acceleration–deceleration cycle, and the loss of momentum from the formation of vortices. Consequently, there is likely to be an energetic cost associated with acceleration and deceleration in swimming cephalopods, although it is difficult to assess it on theoretical grounds alone. However, this cost is likely to be substantially smaller than the acceleration reaction forces associated with acceleration from rest.

During acceleration from a standing start, costs are initially related to acceleration alone. As speed increases, energy costs linked to drag increase, usually as the square of velocity. In addition, the force leading to acceleration, such as the pulse of the jet, is not uniform; it goes up to a peak and then declines. This multiplicity of variables make it difficult to use force measurements recorded in swimming squid to calculate the expected acceleration history and final velocity resulting from a single jet pulse. However, costs attributable to added mass increase rapidly with thickness ratio. Figure 12 suggests that at a thickness ratio of 0.4, the added mass coefficient is at least twice that of a form with a thickness ratio of 0.2, a doubling of the cost of added mass with a doubling of thickness ratio. Daniel (1984) presented some data for the ratio of drag costs to acceleration costs in standing-start locomotion in a variety of swimming organisms. The results range from 48% of the total for a small squid accelerating at 2000 cm/sec^2 to 62% for a medusa accelerating at 700 cm/sec^2 to 92% for a salp accelerating at 23 cm/sec^2. Paradoxically, forms that accelerate rapidly to high velocity will have proportionally less energy loss from added mass and proportionally more energy loss from drag than forms that accelerate more slowly. Added mass is most important, relative to drag, in the initial phases of acceleration, before high speeds and hence high drag costs are achieved. So added mass will be particularly costly to forms that engage in standing-start locomotion and forms in which speed is reduced almost to zero between successive pulses of the jet. This may be the case in small forms, such as the early ontogenetic stages of ammonoids (discussed in Section 3.8).

Clearly, the primary shape variable controlling the added mass coefficient, and hence, the cost borne by ammonoids, is the thickness ratio of the form. In addition, in ammonoids, the umbilicus traps a pool of water that must be accelerated when the organism changes velocity. Quadrate forms may also have additional costs related to added mass relative to shell forms with tapered whorl sections. These observations suggest that it may be possible to differentiate between forms that minimize added mass and those that are designed primarily to minimize drag forces. Thus, acceleration reaction is interesting in the context of those aspects of ammonoid shell form that cannot be easily rationalized in terms of reducing drag. As discussed previously, Jacobs (1992b) observed that the form that confers the least drag is a function of *Re*. At lower Reynolds numbers, a range of relatively thick forms had lower drag and power requirements; at higher Reynolds numbers, thinner forms were more advantageous. However, drag did not discriminate strongly between thin forms that vary in other aspects of whorl shape. For example, compressed oxyconic forms with a sharp venter and closed umbilicus do not have significantly lower drag than other ammonoids with comparable compression but a different whorl shape. Thus, it appears that oxycones did not have a large advantage, in terms of drag alone, in the range of Reynolds numbers likely to be relevant to swimming in ammonoids. However, it is exactly this greater degree of compression and loss of the umbilicus that would be likely to reduce added mass in a coiled cephalopod shell. In this case, an analogy can be drawn between the teardrop shape of low-drag continuously swimming fish and the even more elongate form associated with fish such as barracudas, pike, and gars. These fish are too long to be streamlined. All these fish are ambush predators; they lie passively waiting for prey and then accelerate rapidly to grab them. Similarly, in the fossil record of aquatic vertebrates, forms "too long" to be "streamlined" are interpreted as ambush predators (Massare, 1988). In a planispiral shell, the increased radius associated with narrow oxyconic shell form is analogous to the elongate form of vertebrate ambush predators. Further analogy can perhaps be drawn to the barracuda, which sits in the water column concealed by countershading, waiting for prey to approach. This kind of passive "sit and wait" behavior would be effective in externally shelled cephalopods. Neutral buoyancy, conferred by the cephalopod shell, lowers the energetic cost of waiting in the water column.

3.6. Cost of Transportation

Cost of transportation is determined by taking the total amount of energy consumed (per unit mass or volume) during swimming and dividing it by the velocity, yielding an energetic cost per unit distance (Schmidt-Nielsen, 1972;

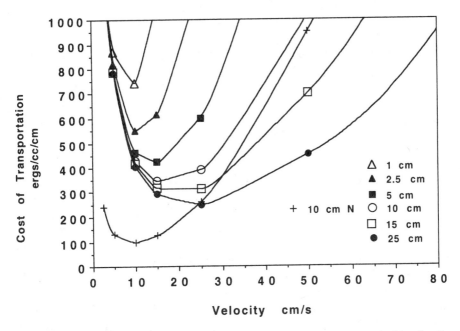

FIGURE 13. Cost of transportation calculated over a range of swimming velocities for *Gastroplites* ranging in diameter from 1 to 25 cm. These calculations employ power consumption associated with drag (Jacobs, 1992b). Efficiency of conversion of metabolic energy to propulsive force is assumed to be 10%. A resting metabolic rate of *Sepia* (3900 ergs/sec per cm^3 from O'Dor and Webber, 1991) is incorporated. This total power consumption is then divided by velocity, yielding the energy required to travel a unit distance. These plots illustrate the declining energy requirements for locomotion with increasing size. The 10 cm N curve incorporates a *Nautilus* resting metabolic rate (1/7th that of *Sepia*).

O'Dor, 1988a; O'Dor and Webber, 1991). Trade-off between the energy required for locomotion and the energetic cost of maintaining other life functions determines the velocity that is lowest in cost per unit of distance traveled (Fig. 13). If life functions can be maintained at low energetic cost, extremely slow swimming speeds may be advantageous, long fasting periods can be sustained, and long slow searches for intermittent food resources become possible. This may be the case in *Nautilus* (see Wells, 1987; Chamberlain, 1990). *Nautilus* requires very low amounts of energy to sustain its metabolic needs and little energy to propel itself at low velocities (O'Dor *et al.*, 1990). This permits a scavenging mode of life in the forereef environment, where *Nautilus* is thought to subsist on crab molts and other scavenged low-energy resources. Without this ability to search for food for long periods of time, it would not be possible for *Nautilus* to subsist on such ephemeral resources. This analysis of *Nautilus* suggests that a chambered shell permits maintenance in the water column with minimal energy; consequently, externally shelled cephalopods have opportunities for low-energy modes of life.

Wells and O'Dor (1991) argued that a low-energy mode of life similar to that found in *Nautilus* pertained to fossil cephalopods including ammonoids. In addition, they argued that fish displaced shelled cephalopods from high-energy niches throughout the Late Paleozoic and Mesozoic.[*] This argument seems overly simplistic in that the major radiation of teleost fish did not occur until the Cretaceous. In addition, this interpretation does not acknowledge the close phylogenetic relationship between ammonoids and coleoids. Studies on squid suggest that they have some of the highest aerobic capacities of, and expend energy faster than other, swimming organisms. Many squid, especially those of small size, have to feed very often to maintain their high-energy life style, the opposite of the situation in *Nautilus*.

Cost-of-transportation curves for ammonoids can be generated using the power dissipated in drag and a value for the resting metabolic rate derived from modern analogues. The power consumption can then be back-calculated to the metabolic scope using efficiency assumptions. We used the power-required data and the efficiency assumptions of Jacobs (1992b) to reconstruct cost of transportation for ammonoids. In this analysis it became clear that assumptions regarding resting metabolic rate have profound consequences for the cost of transportation curves generated (Figs. 13 and 14). The low resting rates of *Nautilus,* when applied to ammonoid morphs, result in very low transportation costs at low swimming velocities. If we assume the higher resting metabolic rates of *Sepia*, higher swimming velocities generate the

---

[*]It is interesting to note that "living fossil" relicts of the Paleozoic fauna engage in lower-energy modes of life than successive faunas. This is true of brachiopods relative to more recently evolved filter-feeding groups, and Thayer (1992) has argued that it is a general trend; a trend suggestive of the Red Queen hypothesis of Van Valen (1973), where every participant has to run faster and faster, and presumably respire more, to maintain the status quo.

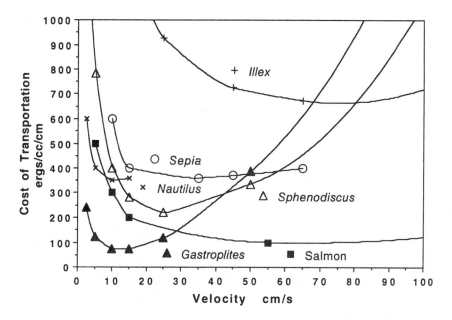

FIGURE 14. Cost of transportation based on the drag data of Jacobs (1992b) for two ammonoids are compared to observed costs of transportation for other swimmers (O'Dor and Webber, 1991). Efficiency of conversion of metabolic energy to propulsive force is assumed to be 10%. For a 25-cm *Gastroplites*, a *Nautilus* resting metabolic rate is assumed, and for a 25-cm *Sphenodiscus*, a *Sepia* resting metabolic rate is assumed. Costs of transportation for the ammonoids are relatively low at low velocities, especially when one assumes a *Nautilus* resting metabolic rate. However, at higher velocities, transportation costs rise rapidly in comparison to nonshelled swimmers. This general result is likely to be robust. Potential sources of error include the uncertainty of appropriate resting metabolic rates, the variable size of the modern taxa from which metabolic rates were derived, and the possibility that the added mass and other costs increase inefficiency of the ammonoids beyond the factor of 10 used in the calculation. Additional sources of inaccuracy for the data derived from modern taxa include the role of finned propulsion in *Sepia* (see O'Dor and Webber, 1991 for discussion).

lowest-cost transportation. Note also that cost of transportation is highly dependent on the size of the organisms. As size increases, per-unit costs decline, and optimal speeds occur at slightly higher velocities. We can also compare these calculations of ammonoid cost of transportation, based on various assumptions, to published data on cost of transportation for living cephalopods and fish (Fig. 14). The cost-of-transportation curves for ammonoids, although they are low at relatively low velocities, rise more rapidly with increasing velocity than the curves for cephalopods that lack external shells. This is a consequence of the drag associated with the shell. (Presumably, the cost of transportation for *Nautilus* also rises rapidly at speeds above those plotted, although this is not evident in the published figure of cost of transportation for *Nautilus*: O'Dor and Webber, 1991.)

The comparison of data from living taxa with the reconstructed curves for ammonoids suggests that, if one assumes the resting metabolic rate of *Nautilus*, the *Gastroplites* model has the lowest cost of transportation. The coefficient of drag relative to Reynolds number is likely to be very similar in *Gastroplites* and *Nautilus*; thus, the lower value determined for *Gastroplites* may be a consequence of the efficiency assumptions in the analysis. *Nautilus* combines a relatively inefficient swimming mechanism with very low resting and active metabolic rates. The swimming mechanism of ammonoids was probably not identical to that of *Nautilus*. Combining the propulsive efficiency of *Sepia* with the resting metabolic rate of *Nautilus* generates a cost of transportation similar to that reconstructed for *Gastroplites*. However, it is not clear that high swimming efficiency or, alternatively, high metabolic scope can cooccur with low resting metabolic rates. High metabolic scope seems to be correlated with high resting metabolic rate. This makes some sense. Respiratory structures and a high density of mitochondria in muscle are necessary to mobilize large amounts of energy, but any residual activity in these structures while the organism is at rest will raise the resting metabolic rate. Thus, it is not clear whether a low resting metabolic rate can accompany higher swimming efficiency, or a mode of life that incorporates periods of very active aerobic metabolism.

So did ammonoids have low energy mobilization and a low-key life comparable to *Nautilus*, or the high-energy life style exhibited by some squid? Probably no generalization is possible. Some variables may have been correlated, as they are among coleoids; some ammonoids may have had lower resting metabolic rates as well as a limited amount of energy available for locomotion; other forms may have been capable of more active locomotion at the cost of higher resting metabolic rates and lower slow-speed swimming efficiency. In addition, some ammonoids may have had greater anaerobic capacity and burst performance than others. Coleoids range in their activities from very high-energy, high-speed squid to forms like *Vampyroteuthis* and cranchid squid, which lead relatively low-energy life styles. Diversity in energy use seems plausible in ammonoids as well.

As noted, the chambered shell of cephalopods permits suspension in the water column without expenditure of energy. This is an advantage in low-energy life styles. At the same time, because the shell contributes additional drag, it is disadvantageous for continuous rapid swimming. Thus, ammonoids may not have been pursuit predators, comparable to tuna or some squid, that spend long periods of time chasing down prey at high speed. This would deny the utility of the neutrally buoyant shell in limiting energetic expenditure. However, life styles that require only intermittent bursts of energy, such as ambush predation, seem possible, and oxyconic shell shape, in particular, may have been conducive to such a mode of life.

3.7. Hydrodynamic Interpretation of Shell Shape

There have been many and varied interpretations of the relationship between locomotion and shell shape in ammonoids. We have already discussed several of the factors that are likely to influence shell shape in planispiral ammonoids, including drag, power consumption, and acceleration reaction. As was observed earlier, coefficient of drag and power consumption for steady swimming vary with Reynolds number (Jacobs, 1992b) such that different shell shapes have lower drag at different *Re*. Smaller forms that swim slowly will benefit from only a moderate degree of compression or thickness ratio from 0.3 to 0.4. At higher Reynolds numbers pertaining to larger and faster-swimming ammonoids, more compressed shell shapes will enjoy hydrodynamic advantages. This relationship between shell shape and swimming velocity is evident in Figs. 9 and 10. As was also discussed earlier, acceleration produces additional forces that may constrain shell shape.

In regard to particular morphologies, most authors agree that heteromorphic forms must have been poorer swimmers than closely coiled planispiral forms. Our data (Jacobs *et al.*, 1994) indicate that uncoiling in *Scaphites* is accompanied by an increase in *Cd* and a decrease in hydrodynamic efficiency. It also seems likely that torted forms might have had particular difficulty in locomotion. However, orthocones such as *Baculites* may have been capable of rapid acceleration with the body chamber forward as well as vertical movement. A ventral sulcus is evident in some well-preserved body chambers of larger specimens of *B. compressus*, suggesting that the hyponome was bent back under the shell, permitting forward locomotion. In rapid forward swimming, drag and acceleration forces may have rotated the shell to a subhorizontal orientation. In such a position, the elongate form would reduce acceleration reaction forces. Thus, one could envision *Baculites* spending much of their time suspended vertically in the water column, intermittently darting after prey. Many midwater coleoids may lead similar existences today.

There are a range of possible ammonoid modes of life. They will necessarily involve different trade-offs in locomotor design. In Table I, three compo-

TABLE I. Swimming Behavior

Shape	Slow, continuous swimming	Fast, continuous swimming	Acceleration
Compressed involute			
Oxyconic—*Sphenodiscus*	Poor	Good	Excellent
Round venter—*Oppelia*	Moderate	Excellent	Good
Quadrate—*Anahoplites*	Good?	Good	Moderate
Compressed umbilicate	Moderate	Good	Moderate
Less compressed			
Involute–Juvenile—*Scaphites*	Excellent	Moderate	Moderate
Evolute round whorl—*Lytoceras*	Moderate	Moderate	Poor
Evolute compressed whorl	Good	Moderate	Moderate
Orthoconic—*Baculites*	Poor	Poor	Excellent?

nents of locomotion—slow continuous swimming, fast continuous swimming, and acceleration—are compared in a number of ammonoid shell morphs. This table is intended to summarize the general implications for adult ammonoid shell morphology of a range of behavioral possibilities and hydrodynamic factors.

3.8. Ornament

Chamberlain and Westermann (1976) suggested that shell ornament may lower the Reynolds number at which the transition to boundary layer turbulence occurs, thereby reducing drag. Fine ornament can reduce drag in bluff bodies traveling at particular velocities. This is why golf balls are made with dimples. These small surface features trigger the onset of turbulent flow in the boundary layer at lower Reynolds numbers than occurs in smooth objects. This in turn results in a fully turbulent wake that is smaller and dissipates less energy than the larger series of vortices normally generated in the wake of an object traveling at the Reynolds numbers in question. For ornament to produce this effect, it must be relatively small, usually less than 1% of the shell diameter, otherwise the ornament itself extends beyond the boundary layer and generates additional drag forces. Chamberlain (1981) measured drag and calculated Cds for forms with the appropriate ornament and found the predicted reduction in coefficient of drag. However, this effect will be important only at Re well in excess of 40,000, that is, for large ammonoids swimming fast. Chamberlain (1981, Fig. 3) presented evidence suggesting that at least some larger ammonoids have the appropriate ornament to produce this kind of drag reduction, but further research is needed to establish whether this is a pervasive pattern among larger ammonoids.

The distribution of ornament on the ammonoid shell could have influenced drag and stability. For example, globose forms with deep umbilici often have a series of nodes on the umbilical shoulder. We speculate that such ornament may have generated a particular set of vortices that spanned the umbilicus and reattached the flow on the downstream side, thereby reducing the drag produced by the umbilicus. Presumably such an effect could operate over only a limited range of Re. Quadrate, especially compressed quadrate, forms often have ornament situated on the ventrolateral shoulder. Perhaps such ornament may have kept flow around the edge (venter and dorsum) of the whorl from interacting with the flow passing by the flanks. This division of the flow into two fields could aid in maintaining flow attachment, preventing the early onset of vortices in the wake. Presumably, such ornament would be advantageous only under a restricted set of swimming conditions.

In addition to umbilical and ventrolateral ornament, many laterally compressed forms, such as *Cardioceras,* have falciform ribbing on the flanks that increases in size through ontogeny. The largest ribs, those nearest the aperture, are oriented subparallel to flow in the backward swimming position, where the aperture and siphon trail behind the shell. Rotation from such a posture would orient these ribs perpendicular to flow, potentially generating drag forces that would restore the animal to the "normal" swimming posture. Thus, we argue that such ornament may play a role in dynamic stabilization of the shell. This observation may also relate to "Buckman's law of covariation" (Westermann, 1966).

Observations of drag on, and of flow patterns around, ammonoid shells have been conducted on models subjected to steady flow (Fig. 8). However, even if an ammonoid were to swim at a reasonably constant rate, the surrounding flow would be unsteady because of the filling of the mantle cavity and pulsation of the jet. In salps (swimming tunicates), pulsing of the jet entrains vortices, adding to the mass of water accelerated (Madin, 1990). This increases the Froude efficiency of the system. Whether the jet can function in a similar manner in living cephalopods has yet to be addressed. In fossil shelled forms, interactions between the jet and the wake produced by the shell and shell ornament may have been complex.

Despite the above arguments suggesting the beneficial role of some ornament in swimming, many ammonoids possess large ornament. Chamberlain (1981) and Jacobs (1992b) found that large ornament significantly increased drag (Fig. 7). Therefore, it is difficult to interpret such ornament as advantageous in terms of locomotion. A number of other adaptive explanations have been invoked for large ornament. For example, analyses showing increasing ornamentation in ammonoids through the Mesozoic have been interpreted as an adaptive response to increased predation pressure (Ward, 1981).

3.9. Hydrodynamics and Ammonoid Life History

Ammonoids hatched at a very small size for cephalopods (Landman *et al.*, 1983; Chapter 11, this volume). An ammonoid 10 cm in diameter is two orders of magnitude larger than a hatchling. As mentioned earlier, smaller forms will have relatively less power relative to drag than larger forms. Thus, ammonoids could swim between one and two orders of magnitude faster as adults than at hatching. Both dimension and velocity contribute to the Reynolds number; therefore, adult ammonoids may have operated at Reynolds numbers three to four orders of magnitude greater than hatchlings. This increase in Reynolds numbers during ontogeny suggests that the shell shape that will confer minimal drag ought to change as well. A 1-mm diameter hatchling ammonoid could have swum at perhaps 1 cm/sec, a Reynolds number of around 10. At this low Reynolds number, skin friction drag will predominate. Consequently, hatchlings with nearly spherical shell shapes that minimize surface area and skin friction drag would be expected. Later in ontogeny more compressed shell shapes would be expected. Such changes in thickness ratio are observed in most ammonoids during their ontogeny (Fig. 15).

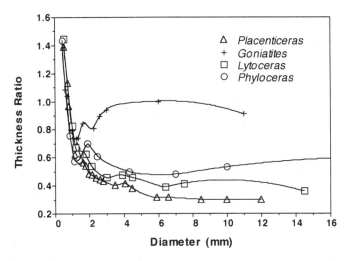

FIGURE 15. Ontogenetic change in thickness ratio for *Lytoceras, Phyloceras, Placenticeras,* and *Goniatites* (data from Smith, 1897, 1898, 1900). Ammonoids hatch at about 1 mm in size. Consequently, immediately after hatching, they must have swum at low Reynolds numbers, where the optimal thickness ratio approaches 1.0. With growth to 10 mm, the optimal thickness ratio will decline to 0.3–0.5. These plots of thickness ratio in early ontogeny suggest that ammonoids responded to this changing optimum during their posthatching growth. *Goniatites* show a pattern of early ontogenetic compression and then reexpansion. This may indicate a declining influence of hydrodynamic factors on shell form late in ontogeny.

In addition to drag, added mass could play a role in the evolution of shell form in early ontogeny. Small forms have difficulty maintaining momentum against large drag forces, and their swimming may be characterized by more frequent acceleration than larger, later ontogenetic, stages. Thus, acceleration reaction may be especially important in hatchling ammonoids. The added energetic cost of acceleration reaction and the lower additional mass coefficient of compressed shells (Fig. 12) suggest that laterally compressed shells would be advantageous at smaller sizes and earlier ontogenetic stages than if shells were optimized to minimize drag alone.

It is also clear from the cost-of-transportation analysis that hatchling ammonoids will not have the energy resources necessary to move more than very modest distances. This effectively consigns the early ontogenetic stages of ammonoids to a passive or planktonic mode of life in the water column. Alternatively, juveniles may have sought the protection of the benthic boundary layer, the region of low velocity next to the bottom. However, in some deposits, bottom conditions were continuously anoxic, and early ontogenetic stages of ammonoids are present, e.g., the Sharon Springs member of the Pierre Shale (Landman, 1988). Preservation of juveniles in these circumstances suggests that the early ontogenetic stages were in the water column, not adjacent to the bottom.

Because there is such a high cost to locomotion in early ontogenetic stages, it is surprising that ammonoids do not have larger embryos (Jacobs, 1992b). *Nautilus* embryos hatch at sufficient size that much of the ontogenetic change in Reynolds numbers encountered by ammonoids is avoided. Jacobs (1992b) argued that the serpenticonic forms so typical of ammonoids (Raup, 1967) permit a change in thickness ratio that conforms with a changing optimal shape with increasing Reynolds number during ontogeny. Few nautilids are serpenticonic, perhaps because they hatch at a sufficiently large size that the need for shape change in early ontogeny is precluded.

A number of coleoids, especially octopods, brood their young. *Argonauta,* for example, builds a shell to serve, in part, as a brood chamber. Because the small embryos of ammonoids swam poorly, it is tempting to speculate that some ammonoids might also have brooded their young until they reached sizes capable of more adequate locomotion. The frequent occurrence of dimorphism in ammonoids is suggestive of brooding. Dimorphism is usually recognized in the last whorl suggesting that it might be related to the onset of sexual maturity. Mature modifications typically expand the body chamber of one of the dimorphs, suggesting that one of the sexes might have sufficient space for a brood chamber. One example of this behavior is the hook-like expansion of the body chamber in *Scaphites.*

3.10. Hydrodynamics and Habitat

As noted, ammonoids are not well designed for continuous rapid swimming. As a consequence, one would expect them to have occupied habitats that do not require such swimming. These might involve nektobenthic behavior in habitats with moderate or low current velocities as well as modes of life divorced from the bottom in epeiric seas or contained within oceanic gyres. Some oceanic populations of squid are contained within gyres, and their reproductive cycle is often closely associated with the circulation pattern. If ammonoids depended on benthic resources, they must have swum, at least intermittently, against the ambient current. Otherwise, they would not have been able to stay near spatially limited resources. This requirement, in combination with the relationship of optimal shell shape to Reynolds number and swimming speed, suggests that there may be instances in which shell shape is closely related to, and perhaps controlled in an evolutionary sense, by ambient current velocity (Jacobs, 1992b; Jacobs *et al.*, 1994).

If ammonoids living in higher-energy environments were required to swim faster, they would be expected to have more compressed shells than those ammonoids that lived in quieter low-energy environments. Thus, there is an expected relationship between ambient current velocity and shell shape. The testable expectation is that other indicators of current velocity, such as grain size, should be correlated with degree of compression of the shell. A number of workers have related ammonoid faunas to environment (Batt, 1989; Westermann, Chapter 16, this volume). It is true that highly compressed forms such as *Sphenodiscus* and *Placenticeras* are often an important component of near-shore faunas. However, other aspects of the mode of life, i.e., lunging predation versus scavenging, are likely to affect swimming behavior and influence shell shape. In order to limit this overprint of behavior, closely related forms should be compared between different environments.

Lineage studies in basins undergoing environmental change may be the best way to substantiate relationships between shell shape and environment. Bayer and McGhee (1984; McGhee *et al.*, 1991) examined ammonoid lineages in shallowing upward cycles in the Middle Jurassic of the German basin and suggested that several lineages, including Leioceratinae and Graphoceratinae, became larger, more involute, and more compressed during the regressive higher-energy portion of these "Klupfel cycles." Similarly, two independent scaphite lineages, *Hoploscaphites,* and *Jeletzkytes,* are thought to extend from the shaley facies of the Pierre Shale to the overlying, shallower, sandier Fox Hills Formation. In each lineage the forms in the higher-energy Fox Hills Formation are more compressed (Landman and Waage, 1993). Similar, correlated changes in shell compression and environment are evident in *Scaphites* lineages that cross successive facies boundaries in the Niobrara Cyclothem (D. K. Jacobs, N. H. Landman, and J. A. Chamberlain, unpublished observation). Correlated shell shape and sedimentary features are even evident within

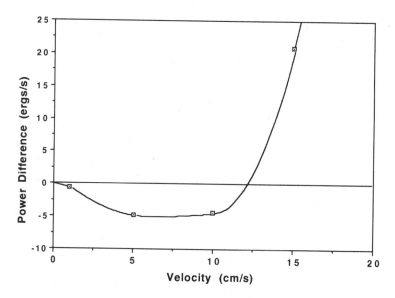

FIGURE 16. Difference in power requirements at various velocities for two forms of juvenile *Scaphites whitfieldi.* Power required for swimming in a thin form found in near-shore sandy environments has been subtracted from that for a thick form found in quieter, shaley sediment. Where the power difference is negative, at low velocities the thick form requires less energy for locomotion. At higher velocities the thin form is at an advantage.

species. In Upper Cretaceous *Scaphites whitfieldi*, specimens collected from sandy proximal portions of the Turner Sandy member of the Carlile Shale are more compressed and could have swum more efficiently at somewhat higher velocities than thicker specimens collected from more distal offshore facies of the Turner (Jacobs *et al.*, 1994; Fig. 16). In this case, other adaptive factors were constrained by examination of closely related ammonoids, permitting the documentation of a relationship between environmental energy and ammonoid shell shape.

4. Summary

As in other chambered cephalopods, the chambered portion of the ammonoid shell, the phragmocone, generated the buoyant force needed to offset the weight of the shell and muscular tissues which are denser than sea water. This buoyancy mechanism requires a rigid shell. In this, and other respects, the buoyancy mechanism of cephalopods is distinct from other buoyancy adjustment mechanisms found in marine organisms such as the swim bladders of fish. The rigid shell, especially when it is external, as it is thought to have been in ammonoids, produces constraints on locomotion. Thus the two topics of this chapter, buoyancy and hydrodynamics interact with each other, and are also clearly linked to other aspects of ammonoid paleobiology, such as the hydrostatic forces borne by the phragmocone, and the paleocology, or mode of life, of ammonoids discussed elsewhere in this volume.

4.1. Buoyancy

The chambered shells of cephalopods have long inspired the imagination of natural historians. In the 17th century, Hooke was the first to articulate a vision of the buoyancy function of the chambered shell (see Jacobs, 1992 for historical review). He felt that the shell went up and down in the water column as pressurized gas filled and emptied the chambered shell. Although some prescient 19th and earlier 20th century workers took exception to this view, it was not until the 1960s that Hooke's gas-pressure mechanism was shown to be incorrect. Experimental work by Denton and Gilpin-Brown (1961a,b,c, 1966; Denton *et al.*, 1961, 1967) demonstrated that chambered cephalopod shells were pumped out osmotically and actually generated a partial vacuum in the chambers. The chambered shells of cephalopods never contain gas above atmospheric pressure, consequently, there is a pressure difference across the phragmocone wall. Thus, the shell must support large hydrostatic forces when the animal is submerged. Hooke's argument that cephalopods changed their buoyancy on a frequent or regular basis is not likely to have been true of ammonoids. Ward (1982) argued cogently that the rate of fluid removal from the chambers and hence the rate of buoyancy change was proportional to siphuncular area. In ammonoids, as opposed to modern *Sepia*, which does change density on a daily basis, the siphuncular area was too small to support rapid or daily change in buoyancy. Thus ammonoids probably maintained neutral buoyancy throughout life. Despite the evidence for an osmotic pump and low pressures in the chambers, textbooks continue to refer to gas pressure in cephalopod shells.

Decoupling, the separation of fluid in the chambers from the siphuncle, was thought to shut off the osmotic pump and permit cephalopods to descend below the 240m osmotic pressure difference between fresh and salt water. This concept of decoupling was advanced because Denton and Gilpin-Brown did not know that the siphuncle could concentrate salts permitting pumping well beyond the osmotic difference between fresh and salt water (Jacobs, 1992a). It is now clear that the decoupling explanation was not necessary, it would not have shut off the osmotic pump in the fashion envisioned, and there is no evidence of decoupled water in modern cephalopods. However, despite the absence of any supporting evidence, workers continue to invoke decoupling to explain aspects of ammonoid morphology that they find inexplicable on other functional grounds (e.g., Bandel and Boletzky, 1979; Weischat and Bandel, 1991; Saunders, 1995).

Given simple coiled shells, and assuming neutral buoyancy as well as the densities of soft parts and shell material, shell geometry will necessarily relate to static stability, resting orientation, and body chamber length. Trueman (1941) was the first to informally explore this set of relationships. Starting with Raup these relationships have been formalized and computerized. More recently a number of workers have tried to document departures from neutral buoyancy in particular fossil forms (Ebel, 1990; Shigeta, 1993). However, given the small departures from neutral buoyancy calculated and the range of assumptions regarding soft part density as well as simplifications of morphology inherent in the calculations, these arguments are not convincing. In the absence of stronger evidence, it would seem prudent to assume that fossil chambered cephalopods, including ammonoids, were capable of achieving neutral buoyancy given that this is the primary function of the chambered shell and siphuncle in modern cephalopods.

Some gastropods and bivalves produce chambers in the apical portion of the shell. Thus, the unique evolutionary aspect of the chambered cephalopod shell is not the septum, but the siphuncle and its osmotic pump. We follow Owen (1878) in speculating that the siphuncle may have evolved from regions of muscle attachment that penetrated the chambers in an ancestral cephalopod.

4.2. Swimming

Until recently studies of ammonoid locomotion have focused on calculation of coefficients of drag on various models usually at Reynolds numbers not representative of the whole range of possible swimming modes. Thus, earlier work focused on determining *the* optimal shell form for swimming in planispiral ammonoids. More recent studies and our discussion here argue for a more complex adaptive topology with a variety of optimal shapes depending on size and hypothesized swimming behavior. In this work we subdivided the ammonoid locomotion problem into two components, a demand side and a supply side.

The demand side is the amount of a physical property, such as force or power, dissipated in overcoming drag or acceleration. Experimentation with model forms or theoretical calculation can be used to determine these properties. Knowledge of the supply of power or propulsive force is then required to determine what kinds of locomotory performance could actually have been achieved. This supply side requires reference to biological models or modern analogues.

In the past, analogy to *Nautilus* has been used to constrain the supply side of the equation. Given the closer relationship of coleoids to ammonoids, we argue that it is important to consider coleoid features in assessing the swimming ability of ammonoids and to consider the potential implications of a coleoid-like mantle and fins in ammonoids. Studies on coleoids document a range of behavior and locomotor performance; these studies can help define the possible range of locomotion behavior in ammonoids which may have been similarly diverse.

Depending on the time-scale of the activity envisioned, three different kinds of data derived from modern analogues can constrain the supply side and can be used to assess maximal or optimal performance: 1) short bursts of activity usually involve a large acceleration component and are primarily limited by the propulsive force or thrust that can be generated; we use studies of the propulsive force generated in modern organisms to assess the possible limits of acceleration and rapid movement in ammonoids; 2) over longer time periods, of minutes to hours, limits on aerobic metabolism constrain performance, and oxygen consumption information from studies of modern swimming organisms can be used to assess the maximum aerobically sustainable levels of activity; 3) at still longer time periods, energetic costs of behavior, such as the cost of transportation, will prove important. If resting metabolic rates are low, as they are in *Nautilus*, low speed transpiration can be very efficient and slow speed searches for ephemeral or intermittent resources can be conducted efficiently. In addition, the neutral buoyancy of the chambered shell provides a low cost means of remaining on station in the water column.

Considering both the demand side and supply side of the equation, a number of ammonoid modes of life can be envisioned. Ammonoids of intermediate compression may have been relatively efficient slow speed swimmers, more compressed forms may have been able to swim somewhat more rapidly. Due to the extra drag associated with the shell, ammonoids would have been poor high speed swimmers or pursuit predators. Oxycones, narrow involute keeled forms, appear to be designed not for continuous swimming but to minimize added mass in acceleration. They may have been ambush predators, waiting in the water column and lunging at prey. The relationship between shell shape and hydrodynamic function has additional implications for changes in ammonoid shell shape that occur during ontogeney, as well as changes in shell shape associated with changes in facies.

ACKNOWLEDGMENTS. We would like to thank N. Landman, R. Davis, G. Westermann, B. Saunders, K. Sarg, B. Worcester, R. O'Dor, and T. Baumiller for their comments on the manuscript. This work was supported by NSF grant #EAR-9104888.

References

Ackerly, S. C., 1989, Kinematics of accretionary shell growth, with examples from brachiopods and molluscs, *Paleobiology* **15**:157–164.

Bandel, K., 1986, The ammonitella: A model of formation with the aid of the embryonic shell of archaeogastropods, *Lethaia* **19**:171–180.

Bandel, K., and Boletzky, S. v., 1979, A comparative study of the structure, development, and morphologic relationships of chambered cephalopod shells, *Veliger* **21**:313–354.

Batt, R. J., 1989, Ammonite shell morphospace distribution in the Western Interior Greenhorn Sea and some paleoecological implications, *Palaios* **4**:32–43.

Bayer, U., and McGhee, G. R., Jr., 1984, Iterative evolution of Middle Jurassic ammonite faunas, *Lethaia* **17**:1–16.

Berthold, T., and Engeser, T., 1987, Phylogenetic analysis and systematization of the Cephalopoda, *Verh. Naturwiss. Ver. Hamb. N.F.* **29**:187–220.

Bone, Q., Pulsford, A., and Chubb, A. D., 1981, Squid mantle muscle, *J. Mar. Biol. Assoc. U.K.* **61**:327–342.

Brett, J. R., 1965, The relation of size to rate of oxygen consumption and sustained swimming speed of sockeye salmon (*Oncorhynchus nerka*), *J. Fish Res. Board Can.* **22**:1491–1501.

Bruun, A. F., 1943, The biology of *Spirula spirula* (L.), *Dana Rep.* **4**:1–46.

Bruun, A. F., 1950, New light on the biology of *Spirula*, a mesopelagic cephalopod, in: *Essay on the Natural Sciences in Honor of Captain Allan Hancock*, University of Southern California Press, Los Angeles, pp. 61–72.

Buckland, W., 1836, *Geology and Mineralogy Considered with Reference to Natural Theology*, Vol. 1, William Pickering, London.

Chamberlain, J. A., Jr., 1976, Flow patterns and drag coefficients of cephalopod shells, *Palaeontology (Lond.)* **19**:539–563.

Chamberlain, J. A., Jr., 1980, The role of body extension in cephalopod locomotion, *Palaeontology (Lond.)* **23**:445–461.

Chamberlain, J. A., Jr., 1981, Hydromechanical design of fossil cephalopods, in: *The Ammonoidea*, Systematics Association Special Volume 18 (M. R. House and J. R. Senior, eds.), Academic Press, London, pp. 289–336.

Chamberlain, J. A., Jr., 1987, Locomotion of *Nautilus*, in: *Nautilus—The Biology and Paleobiology of a Living Fossil* (W. B. Saunders and N. H. Landman, eds.), Plenum Press, New York, pp. 489–525.

Chamberlain, J. A., Jr., 1990, Jet propulsion of *Nautilus*: A surviving example of early Paleozoic locomotor design, *Can. J. Zool.* **68**:806–814.

Chamberlain, J. A., Jr., 1992, Cephalopod locomotor design and evolution: The constraints of jet propulsion, in: *Biomechanics and Evolution* (J. M. V. Rayner and R. J. Wootton, eds.), Cambridge University Press, Cambridge, pp. 57–98.

Chamberlain, J. A., Jr., and Westermann, G. E. G., 1976, Hydrodynamic properties of cephalopod shell ornament, *Paleobiology* **2**:316–331.

Crick, G. S., 1898, On the muscular attachment of the animal to the shell in some fossil Cephalopoda (Ammonoidea), *Trans. Linn. Soc. N.Y.* **7**:71–113.

Dadswell, M. J., and Weihs, D., 1990, Size related hydrodynamic characteristics of the giant scallop, *Placopecten magellanicus* (Bivalvia: Pectinidae), *Can. J. Zool.* **68**:778–785.

Daniel, T. L., 1984, The unsteady aspects of locomotion, *Am. Zool.* **24**:121–134.

Daniel, T. L., 1985, Cost of locomotion: Unsteady medusan swimming, *J. Exp. Biol.* **119**:149–164.

DeMont, M. E., and Gosline, J. M., 1988, Mechanics of jet propulsion in the hydromedusan jellyfish, *Polyorchis penicillatus*. II. Energetics of the jet cycle, *J. Exp. Biol.* **134**:333–345.

Denton, E. J., and Gilpin-Brown, J. B., 1961a, The buoyancy of the cuttlefish *Sepia officinalis* (L.), *J. Mar. Biol. Assoc. U.K.* **41**:319–342.

Denton, E. J., and Gilpin-Brown, J. B., 1961b, The distribution of gas and liquid within the cuttlebone, *J. Mar. Biol. Assoc. U.K.* **41**:365–381.

Denton, E. J., and Gilpin-Brown, J. B., 1961c, The effect of light on the buoyancy of the cuttlefish, *J. Mar. Biol. Assoc. U.K.* **41**:343–350.

Denton, E. J., and Gilpin-Brown, J. B., 1966, On the buoyancy of pearly *Nautilus, J. Mar. Biol. Assoc. U.K.* **46**:365–381.

Denton, E. J., and Gilpin-Brown, J. B., 1973, Floatation mechanisms in modern and fossil cephalopods, *Adv. Mar. Biol.* **11**:197–268.

Denton, E. J., Gilpin-Brown, J. B., and Howarth, J. V., 1961, The osmotic mechanism of the cuttlebone, *J. Mar. Biol. Assoc. U.K.* **41**:351–364.

Denton, E. J., Gilpin-Brown, J. B., and Howarth, J. V., 1967, On the buoyancy of *Spirula spirula, J. Mar. Biol. Assoc. U.K.* **47**:181–191.

Derham, W., 1726, *Philosophical Experiments and Observations of the late Eminent Dr. Robert Hooke*, Derham, London.

Diamond, J. M., and Bossert, W. H., 1967, Standing-gradient osmotic flow—A mechanism for coupling water and solute transport in epithelia, *J. Gen. Physiol.* **50**:2061–2083.

Doguzhaeva, L., and Mutvei, H., 1989, *Ptychoceras*—A heteromorph lytoceratid with truncated shell and modified ultrastructure (Mollusca: Ammonoidea), *Palaeontogr. Abt. A* **208**:91–121.

Doguzhaeva, L., and Mutvei, H., 1991, Organization of the soft body in *Aconeceras* (Ammonitina), interpreted on the basis of shell morphology and muscle scars, *Palaeontogr. Abt. A* **218**:17–33.

Ebel, K., 1983, Berechnungen zur Schwebefähigkeit von Ammoniten, *N. Jb. Geol. Paläont. Mh.* **1983**:614–640.

Ebel, K., 1990, Swimming abilities of ammonites and limitations, *Paläontol. Z.* **64**:25–37.

Ebel, K., 1992, Mode of life and soft body shape of heteromorph ammonites, *Lethaia* **25**:179–194.

Gray, J. E., 1845, On the animal of *Spirula, Ann. Nat. Hist.* **15**:257–261.

Greenwald, K. P., Cook, C. B., and Ward, P., 1982, The structure of the chambered *Nautilus* siphuncle: The siphuncular epithelium, *J. Morphol.* **172**:5–22.

House, M. R., 1981, On the origin, classification and evolution of the early Ammonoidea, in: *The Ammonoidea*, Systematics Association Special Volume 18 (M. R. House and J. R. Senior, eds.), Academic Press, London, pp. 3–36.

Jacobs, D. K., 1992a, The support of hydrostatic load in cephalopod shells—adaptive and ontogenetic explanations of shell form and evolution from Hooke 1695 to the present, in: *Evolutionary Biology*, Vol. 26 (M. K. Hecht, B. Wallace, and R. J. MacIntyre, eds.), Plenum Press, New York, pp. 287–349.

Jacobs, D. K., 1992b, Shape, drag, and power in ammonoid swimming, *Paleobiology* **18**:203–220.

Jacobs, D. K., and Landman, N. H., 1993, Is *Nautilus* a good model for the function and behavior of ammonoids? *Lethaia* **26**:101–110.

Jacobs, D. K, Landman, N. H., and Chamberlain, J. A., Jr., 1994, Ammonite shell shape covaries with facies and hydrodynamics: Iterative evolution as a response to changes in basinal environment, *Geology* **22**:905–908.

Johansen, W., Soden, P. D., and Trueman, E. R., 1972, A study in jet propulsion: An analysis of the motion of the squid, *Loligo vulgaris, J. Exp. Biol.* **56**:155–156.

Jordan, R., 1968, Zur Anatomie mesozoischer Ammoniten nach den Strukturelementen der Gehäuse-innenwand, *Geol. Jahrb. Beih.* **77**:1–64.

Kummel, B., and Lloyd, R. M., 1955, Experiments on the relative streamlining of coiled cephalopod shells, *J. Paleontol.* **29**:159–170.

Landman, N. H., 1988, Early ontogeny of Mesozoic ammonites and nautilids, in: *Cephalopods— Present and Past* (J. Wiedmann and J. Kullmann, eds.), Schweizerbart'sche Verlagsbuchhandlung, Stuttgart, pp. 215–228.

Landman, N. H., and Waage, K. M., 1993, Scaphitid ammonites of the Upper Creatceous (Maastrichtian) Fox Hills Formation in South Dakota and Wyoming, *Bull. Am. Mus. Nat. Hist.* **215**:1–257.

Landman, N. H., Rye, D. M., and Shelton, K. L., 1983, Early ontogeny of *Eutrephoceras* compared to recent *Nautilus* and Mesozoic ammonites: Evidence from shell morphology and light stable isotopes, *Paleobiology* **9**:269–279.

Lehmann, U., 1967, Ammoniten mit Kieferapparat und Radula aus Lias-Geschieben, *Paläontol. Z.* **41**:38–45.

Lindsedt, S. L., Hokanson, J. F., Wells, D. J., Swain, S. D., Hopper, H., and Navarro, V., 1992, Running energetics in pronghorn antelope, *Nature* **353**:748–750.

Madin, L. P., 1990, Aspects of jet propulsion in salps, *Can. J. Zool.* **68**:765–777.

Massare, J. A., 1988, Swimming capabilities of Mesozoic marine reptiles: Implications for method of predation, *Paleobiology* **14**:187–205.

McGhee, G. C., Bayer, U., and Seilacher, A., 1991, Biological and evolutionary responses to transgressive-regressive cycles, in: *Cycles and Events in Stratigraphy* (G. Einsele, W. Ricken, and A. Seilacher, eds.), Springer-Verlag, Berlin, pp. 696–708.

Moseley, H., 1838, On the geometrical form of turbinated and discoid shells, *Phil. Trans. R. Soc. Lond.* **128**:351–370.

Mutvei, H., 1975, The mode of life in ammonoids, *Paläontol. Z.* **49**:196–206.

Mutvei, H., 1983, Flexible nacre in the nautiloid *Isorthoceras*, with remarks on the evolution of cephalopod nacre, *Lethaia* **16**:223–240.

Mutvei, H., and Reyment, R. A., 1973, Buoyancy control and siphuncle function in ammonoids, *Palaeontology* (*Lond.*) **16**:623–636.

O'Dor, R. K., 1982, Respiratory metabolism and swimming performance of the squid, *Loligo opalescens, Can. J. Fish. Aquat. Sci.* **39**:580–587.

O'Dor, R. K., 1988a, The energetic limits on squid distributions, *Malacologia* **29**:113–119.

O'Dor, R. K., 1988b, The forces acting on swimming squid, *J. Exp. Biol.* **137**:421–442.

O'Dor, R. K., and Webber, D. M., 1991, Invertebrate athletes: Trade-offs between transport efficiency and power density in cephalopod evolution, *J. Exp. Biol.* **160**:93–112.

O'Dor, R. K., and Wells, M. J., 1990, Performance limits of "antique" and "state-of-the-art" cephalopods, *Nautilus* and squid, *Am. Malacol. Union Prog. Abstr. 56th Ann. Meeting,* p. 52.

O'Dor, R. K., Wells, M. J., and Wells, J., 1990, Speed jet pressure and oxygen consumption relationships in free-swimming *Nautilus, J. Exp. Biol.* **154**:383–396.

O'Dor, R. K., Forsythe, J., Webber, D. M., Wells, J., and Wells, M. J., 1993, Activity levels of *Nautilus* in the wild, *Nature* **362**:626–627.

Okamoto, T., 1988, Analysis of heteromorph ammonoids by differential geometry, *Palaeontology* (*Lond.*) **31**:35–52.

Owen, R., 1832, *Memoir on the Pearly Nautilus,* Royal College of Surgeons, London.

Owen, R., 1878, On the relative positions to their construction of the chambered shells of cephalopods, *Proc. Zool. Soc. Lond.* **1878**:955–975.

Pfaff, E., 1911, Über Form und Bau der Ammonitensepten und ihre Beziehungen zur Suturlinie, *Jahr. Nieder. Geol. Ver. Hann.* **1911**:207–223.

Raup, D. M., 1967, Geometric analysis of shell coiling: Coiling in ammonoids, *J. Paleontol.* **41**:43–65.

Raup, D. M., and Chamberlain, J., 1967, Equations for volume and center of gravity in ammonoid shells, *J. Paleontol.* **41**:566–574.

Reyment, R. A., 1973, Factors in the distribution of fossil cephalopods. Part 3: Experiments with exact models of certain shell types, *Bull. Geol. Inst. Univ. Uppsala N.S.* **4**:7–41.

Saunders, W. B., 1995, The ammonoid suture problem: Relationship between shell and septal thickness and sutural complexity in Paleozoic ammonoids, *Paleobiology*, 21:343–355.

Saunders, W. B., and Shapiro, E. A., 1986, Calculation and simulation of ammonoid hydrostatics, *Paleobiology* 12:64–79.

Saunders, W. B., and Swan, R. H., 1984, Morphology and morphologic diversity of Mid-Carboniferous (Namurian) ammonoids in time and space, *Paleobiology* 10:195–228.

Schmidt, H., 1930, Über die Bewegungsweise der Schalencephalopoden, *Paläontol. Z.* 12:194–208.

Schmidt-Nielsen, K., 1972, Locomotion: Energy cost of swimming, flying and running, *Science* 177:222–228.

Shapiro, E. A., and Saunders, W. B., 1987, *Nautilus* shell hydrostatics, in: *Nautilus—The Biology and Paleobiology of a Living Fossil* (W. B. Saunders and N. H. Landman, eds.), Plenum Press, New York, pp. 527–545.

Shigeta, Y., 1993, Post-hatching life history of Cretaceous Ammonoidea, *Lethaia* 26:133–146.

Smith, J. P., 1897, The development of *Glyphioceras* and the phylogeny of the Glyphioceratidae, *Proc. Calif. Acad. Sci.* 1:105–128.

Smith, J. P., 1898, The development of *Lytoceras* and *Phylloceras*, *Proc. Calif. Acad. Sci.* 1:129–152.

Smith, J. P., 1900, The development and phylogeny of *Placenticeras*, *Proc. Calif. Acad. Sci.* 3:181–232.

Spath, L. F., 1919, Notes on ammonites, *Geol. Mag.* 56:26–58, 65–74, 115–122, 170–177, 220–225.

Swan, R. T. H., and Saunders, W. B., 1987, Function and shape in Late Paleozoic (Mid-Carboniferous) ammonoids, *Paleobiology* 13:297–311.

Tanabe, K., Shigeta, Y., and Mapes, R. H., 1995, Early life history of Carboniferous ammonoids inferred from analysis of shell hydrostatics and fossil assemblages. *Palaios*, 10:80–86.

Thayer, C. W., 1992, Escalating energy budgets and oligotrophic refugia: Winners and drop-outs in the Red Queen's race, Fifth North American Paleontology Conference Abstracts and Program, *Paleontol. Soc. Spec. Pub.* 6:290.

Trueman, A. E., 1941, The ammonite body chamber, with special reference to the buoyancy and mode of life of the living ammonite, *Q. J. Geol. Soc. (Lond.)* 96:339–383.

Trueman, E. R., and Packard, A., 1968, Motor performances of some cephalopods, *J. Exp. Biol.* 49:495–507.

Van Valen, L., 1973, A new evolutionary law, *J. Evol. Theory* 1:1–30.

Vogel, S., 1981, *Life in Moving Fluids: The Physical Biology of Flow*, Princeton University Press, Princeton, NJ.

Ward, P. D., 1979, Cameral liquid in *Nautilus* and ammonites, *Paleobiology* 5:40–49.

Ward, P. D., 1981, Shell sculpture as a defensive adaptation in ammonoids, *Paleobiology* 7:96–100.

Ward, P. D., 1982, The relationship of siphuncle size to emptying rates in chambered cephalopods: Implications for cephalopod paleobiology, *Paleobiology* 8:426–433.

Ward, P. D., and Boletzky, S. von, 1984, Shell implosion depth and implosion morphologies in three species of *Sepia* (Cephalopoda) from the Mediterranean Sea, *J. Mar. Biol. Assoc. U.K.* 64:955–966.

Webber, D. M., and O'Dor, R. K., 1986, Monitoring the metabolic rate and activity of free-swimming squid with telemetered jet pressure, *J. Exp. Biol.* 126:205–224.

Weitschat, W., and Bandel, K., 1991, Organic components in phragmocones of Boreal Triassic ammonoids: Implications for ammonoid biology, *Paläontol. Z.* 65:269–303.

Wells, M. J., 1987, Ventilation and oxygen extraction by *Nautilus*, in: *Nautilus—The Biology and Paleobiology of a Living Fossil* (W. B. Saunders and N. H. Landman, eds.), Plenum Press, New York, pp. 339–348.

Wells, M. J., and O'Dor, R. K. 1991, Jet propulsion and the evolution of cephalopods, *Bull. Mar. Sci.* 49:419–432.

Wells, M. J., and Wells, J., 1985, Ventilation and oxygen uptake by *Nautilus*, *J. Exp. Biol.* 118:297–312.

Westermann, G. E. G., 1956, Phylogenie der Stephanocerataceae und Perisphinctaceae des Dogger, *N. Jb. Geol. Paläont. Abh.* **103**:233–279.

Westermann, G. E. G., 1966, Covariation and taxonomy of the Jurassic ammonite *Sonninia adicra* (Waagen), *N. Jb. Geol. Paläont. Abh.*, **124**:19–312.

Westermann, G. E. G., 1971, Form, structure and function of shell and siphuncle in coiled Mesozoic ammonoids, *Life Sci. Contrib. R. Ont. Mus.* **78**:1–39.

Yochelson, E. L., Flower, R. H., and Webers, G. F., 1973, The bearing of the new Late Cambrian monoplacophoran genus *Knightoconus* upon the origin of the Cephalopoda, *Lethaia* **6**:275–310.

Chapter 8

Theoretical Modeling of Ammonoid Morphology

TAKASHI OKAMOTO

1. Introduction

Theoretical morphology, which was first developed by Raup and Michelson (1965), is a means of describing the morphological spectra of extant and fossil organisms using a mathematical growth model. Raup (1966, 1967) simulated the three-dimensional morphology and growth pattern of marginally growing

TAKASHI OKAMOTO • Department of Earth Sciences, Ehime University, Matsuyama 790, Japan.

Ammonoid Paleobiology, Volume 13 of *Topics in Geobiology*, edited by Neil Landman *et al.*, Plenum Press, New York, 1996.

molluscan shells by several simple parameters and reproduced these shell shapes with the aid of computer graphics. His approach can be applied not only to interpret the functional and adaptive constraints of morphology but also to analyze morphogenesis. With the recent development of the computer and its graphic techniques, the theoretical morphological approach becomes useful for understanding the morphology of extant and extinct animals including ammonoids.

In this chapter, I discuss the rules and mechanisms of shell formation in ammonoids and determine the most appropriate theoretical model for simulating the morphology of ammonoids. If a shell maintains a gnomic mode of growth, the resultant shell morphology must be either orthoconic, planispiral, or helicoid. The majority of ammonoids have a planispiral shell morphology, and such isomorphic shells do not clearly reveal the mechanism of shell formation. In order to determine the general growth mechanism of ammonoid shells, I have focused mainly on heteromorph ammonoids rather than "normally" coiled (i.e., planispiral) ones, because some of the coiling anomalies observed during the growth of heteromorph ammonoids provide clues for understanding the rules governing morphogenesis.

2. Theoretical Modeling of Shell Coiling

The coiling geometry of gastropod and cephalopod shells often gives an impression of plastic natural beauty. Many workers have tried to model the coiling geometry of these shells from the viewpoint of theoretical morphology. I review several theoretical models that describe the shell morphology of ammonoids and some other animals. Simple models are generally less accurate, even if they are widely applicable. On the other hand, improvement of models for more accuracy inevitably requires more parameters and sometimes decreases their general applicability. Because any theoretical model should be regarded as a tool for interpreting something, the quality of a model must depend on whether workers use it.

As Savazzi (1990) pointed out, theoretical models for analyzing the growth dynamics of coiled shells can be classified into fixed-frame methods and moving-frame methods. Most models developed or adopted by previous authors relied on the former method, and only two (the growing tube model and Ackerly's model, mentioned below) are based on the latter. One of the main purposes of theoretically modeling the morphology of coiled shells, including ammonoids, is to clarify the functional constraints in the morphology of actual shells. Logarithmic models, such as Raup's (1966) model, based on fixed-frame methods, are still useful in a functional approach to theoretical morphology. However, my concern in this chapter is to identify the morphogenetic program controlling the actual morphology of ammonoids. The hypothetical shell-generating process with a moving frame is closer to actual shell growth than the Raupian and other traditional shell growth models, and thus, a model based on the moving-frame method is necessary to begin to under-stand the morphogenetic program of shell coiling.

2.1. Logarithmic Model

Mosley (1838) first described the shell morphology of gastropods and nautiloids using a geometric model based on the equiangular spiral and developed an equation for shell volume. Many other classical works have been carried out subsequently, describing the form and growth of coiled shells, and Thompson (1942) comprehensively reviewed these works in his discussion of the equiangular spiral. In order to estimate the shell volume of ectocochliate cephalopods, Trueman (1941) also described the ammonoid shell form using equations based on the equiangular spiral. Raup (1966) developed a logarith-mic growth model (Raup's model) in which a generating curve revolves around a coiling axis within a cylindrical coordinate system. The hypothetical shell morphology is expressed by the trace of the generating curve. Four simple ratios that determine the attributes of the generating curve are required: the expansion rate of the generating curve per revolution (W), the distance between the generating curve and the coiling axis (D), the translation rate of the generating curve along the axis per revolution (T), and the shape of the generating curve (S). This model is widely applicable to tubular shells showing accretionary growth and can approximate not only "typical" (plan-ispiral or helicoid) shells of ammonoids and gastropods but also the "de-pressed" shells of bivalves and brachiopods with a high W. Using computer graphics, Raup (1966) illustrated the wide variation in hypothetical shell morphology produced by this model. Later, Raup (1967) mathematically analyzed the shell morphology of ammonoids, discussing the geometric constraints involved from the viewpoint of shell function. Raup's model has been applied subsequently in many works on the theoretical and functional morphology of ammonoids and other molluscan shells (i.e., Rex and Boss, 1976; McGhee, 1978, 1980; Tanabe *et al.*, 1981; Saunders and Swan, 1984; Saunders and Shapiro, 1986).

2.2. Improvement of the Logarithmic Model

Although Raup's model is simple and very useful in general, the resultant hypothetical form does not accurately simulate actual shell morphology. The main incongruities are that (1) the Raupian generating curve does not coincide with the biological generating curve (i.e., the actual shell aperture or commis-sure), and (2) the actual shell coiling does not always conform to a logarithmic spiral. In order to achieve a more detailed approximation that is more widely applicable, several workers have attempted to improve on the logarithmic model.

The generating curve in Raup's model is generally elliptic in shape and shares the same plane as the coiling axis. The difference between the theoretical and biological generating curve is conspicuous in bivalve shells, whereas it may be negligible in typically coiled gastropod and cephalopod shells. Thus, an additional parameter that defines the orientation of the generating curve relative to the coiling axis is sometimes required for approximating the apertural posture. McGhee (1980) analyzed the shell geometry of articulate brachiopods by introducing an additional parameter B, defined as the angle between the geometric and biological generating curves.

Even if the plane of the geometric generating curve coincides with that of the biological generating curve, the shape of the actual shell aperture (or commissure) is not a simple elliptic curve. The actual shell aperture commonly occupies a three-dimensional space curve and may change during growth. When a "modified generating curve" is modeled as a looped space curve, which is more similar in shape to that of an actual aperture, the most accurate representation of the shell is produced. Savazzi (1985, 1987) drew several shapes of hypothetical bivalves and gastropods in this way. Furthermore, because the modified generating curve is usually defined as a sequential set of points, each point can be revolved and expanded at a different speed. The disproportional growth model suggested by Bayer (1978) can allometrically change the shape of the generating curve during growth. Using a similar growth model, McGhee (1978) illustrated torsion of the commissure plane in some bivalves, and Checa (1991) analyzed heteromorphic growth of some gastropods and ammonoids.

The most accurate approximation of the shell aperture, however, inevitably requires a number of additional parameters and, consequently, a modification of the simple model to a more complicated and less convenient one. Even if the theoretical generating curve perfectly matches the actual shell aperture, the range of variation observed in the patterns of molluscan shell coiling cannot be described adequately by the gnomic coiling model. It is quite common that the shells of ammonoids and gastropods more or less change their mode of coiling during growth. Raup (1966) tried to model the ontogenetic change in the coiling geometry of molluscan shells by changing his parameters during growth. Burnaby (1966) developed an allometric coiling model in which whorl radius (the length between the coiling axis and the external margin of the shell) allometrically increases with respect to the spiral length of the whorls. He applied this model to an arcestid ammonoid showing a decreasing whorl expansion rate during growth. The models of Bayer and McGhee previously mentioned can also potentially express allometric growth in whorl expansion and translation. Because the definition of allometry includes, of course, isometric growth, a model of allometric coiling provides a better approximation of the ontogenetic development of various morphotypes than a simple isometric model.

However, some shells (e.g., nostoceratid ammonoids and vermetid gastropods) that show tremendous ontogenetic variation in the mode of coiling can

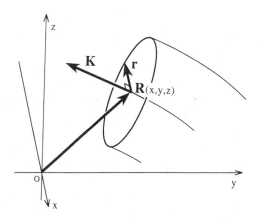

FIGURE 1. Schematic diagram showing a tubular shell surface. The shell surface is expressed as $R + r$. The trajectory of the center of the generating curve $R(x,y,z)$ and the corresponding tube radius $|r|$ are given by an equation in the fixed coordinate system. The differential of R indicates the direction of shell growth (K), which intersects r perpendicularly.

never be approximated, even by the allometrically defined coiling models. For describing such kinds of irregular coiling, a consistent coiling axis is invalid. Previously, I proposed a heteromorphic coiling model that is applicable to more complicated tubular objects (Okamoto, 1984). Using this model, I described the coiling of the heteromorph ammonoid *Nipponites* using computer-generated illustrations. In this model, the coiling shell, which is regarded as a smooth tubular object having a circular cross section, is expressed by equations describing the centerline of the tube and the corresponding length of the radius within a fixed coordinate system. Thus, the generating curve is always perpendicular to the growth direction of the tube (Fig. 1). Illert (1987, 1989) independently proposed a model that is almost identical with my model (Okamoto, 1984). Thus, I refer to this method as the Okamoto–Illert model. Although the Okamoto–Illert model is sometimes confused with the "growing-tube model" of Okamoto (1988a), mentioned later, based on differential geometry, the two are totally different. Illert (1989) introduced the concept of the Frenet frame merely for defining the generating curve relative to the direction of growth. By adopting the Okamoto–Illert method, any coiling shell geometry can be described. However, this model requires specific equations to express the centerline of the tube for every coiling geometry. Usually, these equations are defined quite artificially within a fixed coordinate system, and it is often difficult to determine their biological meaning.

2.3. Growing-Tube Model

The morphology of molluscan shells can be understood as a result of infinitesimal growth at every growth stage. The mode of growth is controlled by certain morphogenetic rules, which must be defined *ad hoc* within a local framework near the aperture and not in reference to an artificially defined, fixed coordinate system. In order to simulate the accretionary growth of molluscan shells realistically, it is necessary to reconstruct the local mode of growth near the aperture (or commissure) without the traditional concept of fixed coordinates. The "growing-tube model" (Okamoto, 1986, 1988a), which is based on the concept of differential geometry using a moving frame (a local coordinate system called the Frenet frame), is totally different from previously proposed models in defining the mode of growth without reference to any fixed coordinate system. The Frenet frame is always situated at the closest point to the generating curve and shifts with growth. This frame is composed of three unit vectors, the tangent (*t*), the principal normal (*n*), and the binormal vectors (*b*), which all intersect each other perpendicularly (Fig. 2). The tangent vector indicates the direction of shell growth. The principal normal vector indicates the direction from the center of the generating curve to the maximum growth point (MGP), which signifies the point of maximum growth at this growth stage. In this model, the morphology of the shell is regarded as a growing tube and is expressed as the locus of a circular generating curve. The mode of growth at every stage is expressed as the motion of the frame itself. Although Okamoto (1988a) used the Frenet frame for explaining his three differential parameters (*E*, *C*, and *T*), the definition of these parameters is rather complicated. Savazzi (1990) suggested that Okamoto's (1988a) parameters are easier to explain by a "standardized moving frame" instead, in which the norm of the frame (the unit length of the tangent, principal normal, and binormal vectors) coincides with the tube radius at every growth stage and increases throughout growth (Fig. 2). The standardized moving frame thus

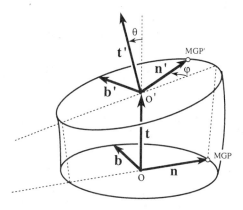

FIGURE 2. Schematic diagram showing the growing tube model (Okamoto, 1988a) using the standardized moving frame (*t*, *n*, and *b*). After one unit of growth ($\Delta s = 1$), the frame shifts, revolves, and extends with the movement of the generating curve, and is shown as *t'*, *n'*, and *b'*. In this figure, the three differential parameters *E*, *C*, and *T* are equivalent to $|n'| / |n|$, θ, and φ respectively.

shifts, revolves, and extends with the movement of the generating curve, and the differential parameters express the infinitesimal motion of the standardized moving frame at any given stage. E, C, and T are defined as the rate of increase of a unit vector length, the rate of revolution of the tangent vector within the tangential plane, and the rate of revolution of the binormal vector within the normal plane, respectively. Each differential parameter is usually expressed as a function of a growth parameter s:

$$E = E(s), \quad C = C(s), \quad T = T(s)$$

where s is the integral of the infinitesimal movement of the standardized moving frame relative to its unit length. In other words, s may be regarded as the growing tube length measured by an "extending yardstick." To reproduce

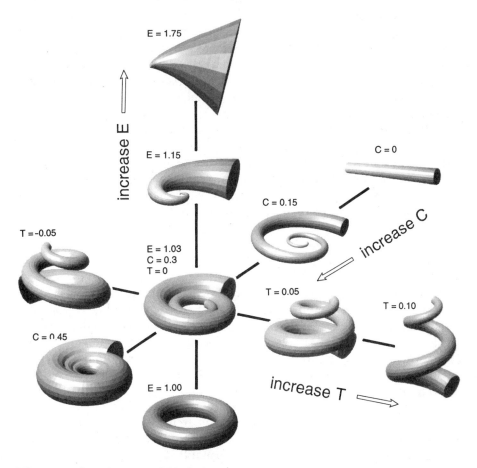

FIGURE 3. Three-dimensional block diagram showing the spectrum of hypothetical shells produced by the growing-tube model.

the shell morphology, a sequential set of data defining each of the parameters is input into a computer as a "script." When the three differential parameters maintain constant values with increasing s, then isomorphic shell forms are produced (Fig. 3). Normally coiled ammonoids are characterized by having low E, constant C, and zero T, whereas in heteromorph ammonites, C and T commonly change during growth.

One of the great advantages in describing shell morphology using the growing-tube model is the wide applicability and ease with which diverse coiled shells can be compared. Even when the shells are irregularly coiled heteromorph ammonoids, their coiling geometries can easily be reproduced graphically (Okamoto, 1988a, Text Figs. 10 and 11), and we can compare the difference and similarity of coiling mode using geometrically consistent parameters. Another advantage of this model may be ease of theoretical modeling and computer simulation because of the perfect similitude among the generating curve, the standardized moving frame, and the fact that every parameter is retained in this model. On the other hand, there is a disadvantage in this model in that the differential parameters, especially shell torsion, are difficult to estimate from actual specimens. Okamoto (1988a,b) indirectly determined the three differential parameters for actual specimens from the shell form equation in the Okamoto–Illert model and from some other simplified equations for curves.

Shell morphology should be regarded as the result of the integration of infinitesimal growth throughout ontogeny. Because any fixed coordinate system is artificially introduced to describe the position and shape of objects, it never represents a meaningful framework for growing organisms. The growing-tube model, in which the modes of growth are expressed without a fixed framework, can be regarded as a more accurate model of the actual mode of growth.

2.4. Ackerly's Model

Independent of the development of the growing-tube model, Ackerly (1987, 1989) developed another moving-frame model, which I call "Ackerly's model" (Fig. 4). His model is, however, somewhat different from the growing-tube model, although both models are based on an almost identical concept. Ackerly attempted to focus on the motion of an actual aperture rather than a hypothetical generating curve that is always perpendicular to the growth direction in the growing-tube model. With the introduction of a rotation axis and a translation axis, two parameters, α (rotation) and γ (downward plunge of the translation axis), are defined to trace the movement of the center of the aperture (centroid; Ackerly, 1989). The composition of the two transformations expressed by α and γ roughly corresponds to that of C and T of the growing-tube model. The rate of enlargement of the aperture is expressed by δ (dilation). Two more parameters are necessary to define the orientation of

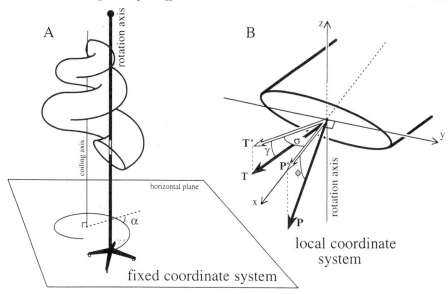

FIGURE 4. Schematic diagram showing Ackerly's model. (A) Relationship among the shell, the rotation axis, and the fixed coordinate system. The coiling axis of the shell is usually oriented in the vertical direction of the fixed coordinate system. The rotation axis, which is always perpendicular to the horizontal plane, moves with growth so as to contain the center of the aperture. (B) Relationship between the rotation axis and the local coordinate system. Because the z-axis coincides with the rotation axis, the x–y plane is parallel to the horizontal plane of the fixed coordinate system. The apertural pole (P) is contained in the x–y plane. The motion of the aperture with growth is expressed by five angular parameters: α, γ, δ, σ, and ϕ. P, apertural pole; P', x–y projection of the apertural pole; T, translation axis; T', x–y projection of the translation axis; α, rotation angle; γ, downward plunge of the translation axis; σ, angle between the apertural pole and the translation axis, measured in the x–y plane; ϕ, downward plunge of the apertural pole. (δ, dilation, which is equivalent to one-half of the apical angle of the cone, is not shown in this figure.)

the aperture: i.e., ϕ, the angle between the apertural plane and the rotation axis (equivalent to the downward plunge of the apertural pole), and σ, the angle between the apertural pole and the translation axis in the projection of the X–Y plane. Although this model might be useful to analyze the morphology of gastropod-like shells in which a consistent direction of the coiling axis can be defined throughout growth, it is not sufficiently applicable to the general morphology of many heteromorph ammonoids. Ackerly's model is theoretically inadequate for the following two reasons.

First, his model is discordant with the concept of a "moving reference system fixed to the shell aperture." He stressed the importance of local growth at the aperture. The local coordinate system suggested by Ackerly (1989, Fig. 5) is, however, a frame best fitted neither to the aperture nor to the trajectory of the curve; the rotation axis always coincides with the z-axis, and the apertural pole shifts in the x–z plane of the local coordinate system. The morphological significance of the parameters based on this coordinate system (e.g., σ and γ) is rather obscure.

Second, Ackerly's model is not free from the constraint of a fixed direction of the coiling axis, which is a traditional concept. In his model, although the rotation axis (of the local coordinate system) changes its position according to the progression of the aperture, it is always parallel to the z-axis of the fixed coordinate system. Although a gastropod-like shell can usually be situated in the fixed coordinate system so that the coiling axis is vertical (that is, the coiling axis and the rotation axis have the same direction), any other orientation in that system is also possible. However, the changing pattern of these parameters depends completely on the relationship between the coiling axis of the shell and the rotation axis. This problem is fatal in trying to model heteromorph ammonoids because the direction of the coiling axis cannot be determined uniquely in most of these forms.

3. Hydrostatics

The most distinct characteristic of ectocochlian cephalopods in relation to theoretical morphology is probably the existence of a septate phragmocone. This organ must have developed as an apparatus to maintain neutral buoyancy of animals in the water column, as observed in Recent *Nautilus* (Denton and Gilpin-Brown, 1966). The shell, which consists of a septate phragmocone and a living chamber, is hydrostatically very important. The phragmocone reduces the total density of the organism, affects life orientation, and influences other hydrostatic properties of the animal. In any simulation of ammonoid morphogenesis, we need to consider the growth of the phragmocone and its hydrostatic effects.

3.1. Estimation of Total Density

Nautilus maintains neutral buoyancy throughout growth (Denton and Gilpin-Brown, 1966). Because the body chamber filled with the soft body is heavier than sea water, this load must be counterbalanced by a phragmocone filled with gas. If ammonoids had a nektoplanktic mode of life similar to that of extant *Nautilus*, the phragmocone of ammonoids would also have had a large enough capacity to achieve neutral buoyancy. If the volume and corresponding density of each part are known (phragmocone and body chamber), we can test whether the animal could have attained neutral buoyancy by calculating the total average density. Based on the abovementioned hydrostatic conditions, Trueman (1941) estimated the total average density of ammonoids during life. The volumes of the phragmocone and body chamber were calculated using equations based on the logarithmic model, and the densities of the shell and soft body were estimated as 2.94 g/cm^3 and 1.13 g/cm^3, respectively. Trueman calculated that the total densities of ammonoids were usually close to that of sea water (1.026 g/cm^3), and, consequently, he presumed that ammonoids had a nektoplanktic mode of life similar to that of *Nautilus*. Although these densities have subsequently been revised [2.62 g/cm^3 for the shell (Reyment, 1958) and 1.062 g/cm^3 for the soft body (Raup and Chamberlain, 1967)], the hydrostatic concept and Trueman's method have been followed by many other workers (Raup, 1967; Heptonstall, 1970; Tanabe, 1975, 1977; Ward and Westermann, 1977; Ebel, 1983; Okamoto, 1988b). Most of these workers except for Ebel (1983) accepted that ammonoids, including heteromorphs, had a sufficient volume of phragmocone to achieve neutral buoyancy. However, little attention has been paid to the errors caused by inaccurate measurements or the difference between actual and theoretical models. Shell thickness and the relative size of the phragmocone to the body chamber are especially difficult to measure with high accuracy, although both parameters strongly influence the calculation of total density.

3.2. Hydrostatic Variables

The posture of a neutrally buoyant animal in sea water is determined by the distribution of its centers of buoyancy and gravity. The center of buoyancy, that is, the theoretical point at which the force of buoyancy acts upward, coincides with the center of volume. The gravitational force acts downward at the center of gravity and lies below the center of buoyancy. Under such conditions, the distance between the centers of buoyancy and gravity indicates the stability of the animal.

If ammonoids were neutrally buoyant in sea water, their life orientation and other hydrostatic properties can be estimated from the relative distribution of the centers of buoyancy and gravity. Trueman (1941) estimated the life orientation of ammonoids under the assumption of neutral buoyancy. As he pointed out, if ammonoids were somewhat heavier than sea water, their life orientation would not have changed markedly. Although Ebel (1985, 1992) reconstructed a gastropod-like orientation of some ammonoid shells with a benthic mode of life, such a life orientation would have been hydrostatically unstable even if these ammonoids were slightly negatively buoyant. Trueman empirically determined the centers of buoyancy and mass from actual specimens. He used the center of mass (= center of the body chamber) instead of the center of gravity because the latter is difficult to determine by empirical methods. Any errors in estimating the center of gravity affect the resultant life orientation, especially if the center of gravity is close to the center of buoyancy. Some additional assumptions are also required in this method: each chamber within the phragmocone that was filled with gas and the body chamber that was filled with the soft body must be regarded as consisting of homogeneous material. Raup and Chamberlain (1967) and Raup (1967) calculated the life orientation of normally coiled ammonoids with the aid of a computer. They assumed a constant shell thickness relative to the whorl radius of the shell

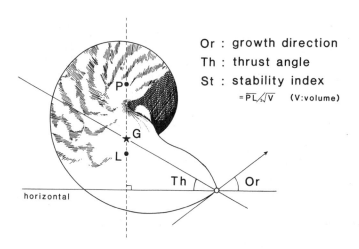

FIGURE 5. Schematic diagram showing the definitions of three hydrostatic parameters in ectocochliate cephalopods. P, center of phragmocone; L, center of body chamber; G, center of gravity. The center of buoyancy, which is not shown in this figure, must be situated between P and G.

and a constant density of the shell and soft body. Therefore, if the four Raup parameters are given (usually $T = 0$ and $S = 1$), one can determine how the length and shape of the body chamber affect neutral buoyancy and the life orientation of the animal in the water column.

Body-chamber length (BCL) is usually defined as the angular length of the body chamber around the coiling axis. Although BCL is somewhat convenient for expressing the relationship between the size of the body chamber and the resultant life orientation in normally coiled ammonoids, this index can hardly be applied to heteromorphs. In order to achieve wider applicability, I introduce another parameter (PT) defined as the ratio of the length of the phragmocone to the total shell length. This value must be more or less constant among ammonoids to maintain neutral buoyancy, because the density of the phragmocone and the body chamber can be regarded as nearly constant in most ammonoids. The actual value of PT ranges from 0.55 to 0.60. The following three indices, discussed below, are useful in discussing the hydrostatic properties of the ammonoid shell. I modified these indices slightly so as to achieve wider applicability because they are primarily defined only for normally coiled ammonoids (Fig. 5).

3.2.1. Growth Direction

In this chapter, the life orientation of an ammonoid is expressed by the growth direction (Or), which is defined as the angle measured in degrees between the direction of shell growth and the horizontal plane (see also Okamoto, 1988c). This value is positive when the ammonoid has an upward direction of shell growth. In previous papers, life orientation was expressed by the apertural orientation, defined as the angle between the apertural margin and the vertical plane, including the coiling axis (e.g., Raup, 1967; Saunders and Shapiro, 1986). However, this angle not only is influenced by the life orientation in sea water but also depends on the obliquity of the aperture, which is defined as the angle between the apertural margin and the plane normal to the direction of shell growth. The apertural orientation can be expressed as the sum of the growth direction and the obliquity of the aperture and is identical to the growth direction when the apertural plane is perpendicular to the direction of shell growth.

3.2.2. Shell Stability Index

The stability of ammonoid shells suspended in sea water is affected by the distance between the centers of buoyancy and gravity. When an ammonoid shell is tilted in sea water, the restoring moment acting on the center of gravity is expressed as the product of the force of buoyancy and the horizontal component of the distance between the centers of gravity and buoyancy. Therefore, a shell with a large distance between the two centers experiences a large restoring moment and is more stable than one with a small distance. However, the center of gravity of the whole animal cannot be estimated without determining the density, volume, and center for each part. Trueman (1941) discussed shell stability in sea water in terms of the distance between the center of buoyancy and the center of mass of the body chamber instead of the center of gravity of the whole animal. This alteration is acceptable because the average density of the phragmocone and of the body chamber can be regarded as nearly constant. Following Trueman's logic, Raup (1967) and Raup and Chamberlain (1967) defined a stability index as the ratio of the distance between the centers of buoyancy and mass (of the body chamber) to the total shell diameter. However, the total shell diameter is difficult to define in heteromorph ammonoids. In order to apply this index to heteromorph ammonoids, I chose the cube root of the volume of the whole shell instead of the shell diameter. Thus, the redefined shell stability index (St) equals the ratio of the distance between the centers of buoyancy and mass to the cube root of the shell's total volume. Using this definition, shells with a similar shape but different size have the same value of the shell stability index. Note that the hydrostatic relationship does not actually remain constant with increasing shell size even in isomorphic growth, because buoyancy and gravity increase proportionally with the cube of shell size, while the distance between the two centers maintains a direct proportion to shell size.

3.2.3. Thrust Angle

It is difficult to determine the swimming ability of ammonoids from shell morphology. Chamberlain (1976, 1981), Chamberlain and Westermann (1976), and Jacobs (1992) estimated swimming ability by empirically measuring the hydrodynamic properties of ammonoid shells. However, theoretical calculations of the hydrodynamic properties of coiled shells are fairly difficult to do. Here I use the thrust angle, which is a hydrostatic index introduced by Ebel (1990), to estimate the swimming ability of an ammonoid. Thrust angle (*Th*) is defined as the angle between the horizontal plane and the vector from the ventral margin of the aperture to the center of gravity. Propulsion of ammonoids was presumably produced by a water jet from the hyponome situated near the ventral side of the aperture, as observed in extant *Nautilus*. If the resisting force of the water is ignored, thrust should act in the direction of the center of gravity for the most effective motion; otherwise, a portion of that force will be wasted on shell rotation. For horizontal locomotion, therefore, it is desirable that the hyponome be situated at the same level as the center of gravity (i.e., *Th* = 0).

3.3. Hydrostatic Properties of Normally Coiled Ammonoids

The hydrostatic difference between *Nautilus* and normally coiled ammonoids has been thoroughly analyzed and discussed by Trueman (1941), Raup (1967), Saunders and Shapiro (1986), Ebel (1990), and others. Although there is variation in the interpretation of the mode of life of ammonoids, the results of the calculations of hydrostatic properties are quite similar among all authors: (1) if neutral buoyancy is presumed, a negative correlation between whorl expansion rate and body-chamber length is expected; (2) life orientation depends strongly on body-chamber length (in degrees), and this orientation reaches a maximum value (usually 90° to 110°) when the body-chamber length attains 300°; (3) shell stability is influenced by body-chamber length (or by the whorl expansion rate, which correlates negatively with body-chamber length)—a shell with a long body chamber has relatively poor stability; and (4) thrust angle, which depends on the body-chamber length, is 0° (horizontal) when the body chamber length equals approximately 300°.

I examine here the shell morphology and hydrostatic properties of 12 Late Cretaceous ammonoids (Fig. 6, Table I). The method of calculation is based on Raup and Chamberlain (1967). Although ammonoids exhibit wide variation in shell morphology, many of them have a comparatively smaller whorl expansion rate and a longer body chamber than extant *Nautilus*. The negative correlation between these two parameters is observed in Cretaceous ammonoids (Fig. 6); serpenticonic ammonoids with a small whorl expansion rate have a long body chamber, commonly more than one whorl in length (e.g., *Gaudryceras tenuiliratum*, W = 1.7, BCL = 365°), whereas many oxyconic ammonoids with a larger whorl expansion rate have a shorter body chamber, usually less than 300° (e.g., *Metaplacenticeras subtilistriatum*, W = 2.7, BCL = 190°). The shell morphology of normally coiled ammonoids can be classified into three types according to the length of the body chamber. Hydrostatic properties and the presumed behavior in sea water are different among these types.

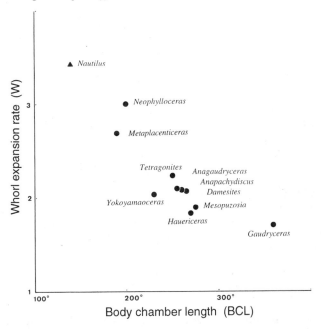

FIGURE 6. Relationship between whorl expansion rate (*W*) and body-chamber length (*BCL*) in several Late Cretaceous ammonoids and present-day *Nautilus*. Species are referred to in Table I. Data for *Neophylloceras* and *Gaudryceras* are from juvenile specimens.

TABLE I. Measurements and Hydrostatic Properties of *Nautilus pompilius* and Several Late Cretaceous Ammonites[a]

Species	W	E	UD	BH	BCL	PT	Or	St	Th
Nautilus pompilius	3.43	1.180	0.05	0.77	140	0.61	24	0.75	−37
Neophylloceras subramosum	3.01	1.154	0.08	0.53	200	0.54	52	0.70	−14
Metaplacenticeras subtilistriatum	2.69	1.123	0.15	0.46	190	0.59	54	0.78	−15
Tetragonites glabrus	2.24	1.086	0.24	1.09	250	0.57	79	0.41	9
Anagaudryceras limatum	2.10	1.070	0.30	0.93	255	0.59	88	0.42	13
Anapachydiscus fascicostatus	2.09	1.081	0.23	1.12	260	0.58	85	0.36	14
Damesites damesi	2.08	1.105	0.07	0.75	265	0.58	88	0.35	15
Yokoyamaoceras jimboi	2.04	1.065	0.32	0.65	230	0.63	80	0.57	13
Mesopuzosia pacifica	1.90	1.058	0.33	0.73	275	0.61	96	0.36	20
Hauericeras angustum	1.84	1.044	0.42	0.54	270	0.63	64	0.45	23
Gaudryceras tenuiliratum	1.72	1.041	0.41	1.09	365	0.58	65	0.11	−12
Didymoceras awajiense	—	1.025	—	—	—	0.58	67	0.87	−11
Eubostrychoceras japonicum	—	1.014	—	—	—	0.58	52	1.38	2

[a]*W*, whorl expansion rate; *E*, rate of whorl enlargement; *UD*, ratio of umbilical diameter to shell diameter; *BH*, ratio of whorl breadth to height; *BCL*, angular length of the body chamber; *PT*, ratio of the phragmocone length to the total shell length; *Or*, growth direction; *St*, shell stability index; *Th*, thrust angle. See Figs. 2 and 5 for definitions of *E*, *Or*, *St*, and *Th*.

The short-body-chamber type (BCL = 120–240°) is characterized by having a large whorl expansion rate, a low growth direction, a high shell stability, and a high thrust angle. Because the hydrostatic properties of this type are comparable to those of modern *Nautilus*, a similar mode of life and behavior in sea water are presumed. In *Nautilus*, the fastest locomotion without rocking is observed when the animal is moving in a backward ascending direction; shell rocking is inevitable when motion is in any other direction (Chamberlain, 1981). Although this type is rarely found in ammonoid species, among the Cretaceous ammonoids, *Neophylloceras* and *Metaplacenticeras* may be included in this type. The behavior of these species in sea water was probably comparable to modern *Nautilus*.

The moderate-body-chamber type (BCL = 240–360°) is characterized by having a moderate whorl expansion rate, a high growth direction, a relatively low shell stability, and a low thrust angle. Many species of Cretaceous ammonoids belong to this type. Chamberlain (1981) suggested that ectocochliates having a lower shell stability might have been weak swimmers, relying on the restoring moment. As Saunders and Shapiro (1986) pointed out, however, the rotational component in propulsive thrust is minimized or eliminated when the propulsion is directed through the center of gravity. If ammonoids could have controlled their shell orientation by tentacles during locomotion, their swimming ability might not have been limited by static shell stability. The nearly horizontal thrust angle of this type suggests a relatively high capacity for horizontal locomotion.

The long-body-chamber type ($BCL \cong 360°$) is characterized by having a small whorl expansion rate, a low growth direction, an extremely low shell stability, and a high thrust angle. A serpenticone is a typical shape of this type. Four of the five case studies of Saunders and Shapiro (1986, Fig. 9 a–d) are also included in this type. It is less common in Cretaceous ammonoids, and only *Gaudryceras tenuiliratum* belongs to this type among the species examined. Although the growth direction and the thrust angle of this type are similar to those of *Nautilus*, the mode of locomotion was probably different; the extremely low stability of the shell would by no means have produced enough restoring moment. As Chamberlain (1981) suggested, the hydrostatic properties of this type seem to be disadvantageous for swimming.

3.4. Hydrostatic Properties of Heteromorph Ammonoids

Trueman (1941) empirically estimated the life orientation of some heteromorph ammonoids. Using the aforementioned growing-tube model, Okamoto (1988b) also calculated the ontogenetic change of life orientation and shell stability in some nostoceratid heteromorphs. However, a detailed analysis including calculation of other hydrostatic variables has never been carried out. Here, I show two case studies that analyze the ontogenetic changes of hydrostatic variables in Cretaceous nostoceratid ammonoids (Figs. 7 and 8).

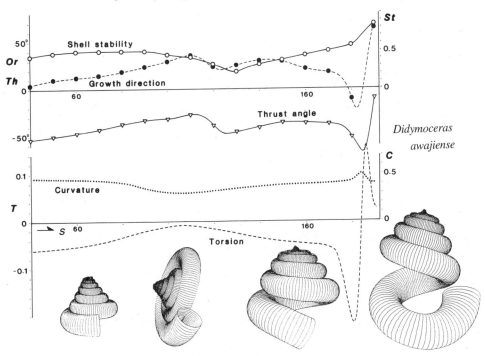

FIGURE 7. Ontogenetic changes of hydrostatic properties in a Late Cretaceous nostoceratid *Didymoceras awajiense*. See Figs. 2 and 5 for definitions of the parameters. Reconstructed life orientations for some growth stages are shown on the bottom.

In this simulation, I follow the same technique adopted in Okamoto (1988b) for calculating the center of buoyancy of the whole animal, the center of mass of the body chamber, and other physical quantities. However, the center of gravity, which is necessary to estimate the thrust angle, can not be calculated without knowing the densities of the materials involved. Therefore, I assume that the average density of the phragmocone was 0.5 g/cm^3. The average density of the body chamber is determined so as to achieve neutral buoyancy of the whole animal.

The mode of coiling in *Didymoceras awajiense* is roughly divided into three stages (Fig. 7). The shell in the early stage is turreted, and the apical angle of the helix is approximately 70°. Both the degree of curvature and the degree of torsion decrease with growth, and the coiling is nearly planispiral at the end of this stage. The degree of torsion increases again in the middle stage. Whorls gradually become turreted again, and nearly isomorphic growth is maintained throughout this stage. Using computer simulations, we observe that the hydrostatic properties change through ontogeny. This species has a relatively high shell stability, low growth direction, and high thrust angle; a fluctuation caused by a decrease of shell torsion occurs during the transition from the early to middle stages. The parameters of shell coiling fluctuate in the late stage; shell torsion increases and then rapidly decreases. The shell first elongates along the coiling axis and then develops a hook-like body chamber, which suddenly breaks away from the previously formed spire. Shell stability increases throughout the late growth stage and attains a maximum value. The growth direction decreases with the increase in shell elongation, then rapidly increases, and attains a maximum value (65°) at the end of shell growth. The thrust angle becomes very small (less than 10°) at the end of ontogeny.

The coiling mode in *Eubostrychoceras japonicum* changes ontogenetically and can be divided into three growth stages (Fig. 8). The shell in the early stage consists of open and almost planispiral whorls without torsion. At this stage, the shell has a relatively high life orientation (40–60°), moderate shell stability, and small thrust angle. The shell in the middle growth stage is represented by an open and highly turreted torticone with a high shell torsion. The growth direction decreases rapidly and maintains a downward direction throughout this stage. The shell stability and the thrust angle both maintain fairly high values. In the late stage, unlike that of *D. awajiense*, the final hook is simply formed without shell elongation; shell torsion becomes small and

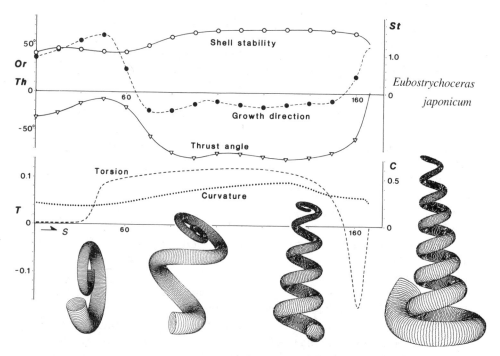

FIGURE 8. Ontogenetic changes of hydrostatic properties in a Late Cretaceous nostoceratid *Eubostrychoceras japonicum*. See Figs. 2 and 5 for definitions of the parameters. Reconstructed life orientations for some growth stages are shown on the bottom.

then reverses. Shell stability also maintains a high value in this stage. The growth direction increases to 50°, and the thrust angle becomes almost horizontal (0°).

Thus, the hydrostatic properties of heteromorph ammonoids may be quite variable and may differ among species and growth stages. Therefore, we cannot assume a uniform mode of life and swimming ability for these animals. The juvenile stages (early and middle stages) of many heteromorph ammonoids, including the two examples described above, are, however, closer to modern *Nautilus* than to normally coiled ammonoids in having a high shell stability, a low to moderate growth direction, and a high thrust angle. The aberrant shell morphology of the ammonoids at these stages suggests that they probably were not rapid swimmers. Probably they could neither have turned upside down nor tilted their shell laterally as can *Nautilus*. Also, they could not have moved horizontally without rocking their shell. The high stability and low angle of growth direction might have been favorable for scavenging or capturing food on the sea bottom. On the other hand, adult heteromorphs with a retroversal body chamber have different hydrostatic properties. These shells are characterized by having an extremely high shell stability, a high growth direction, and an almost horizontal thrust angle. This fact suggests a different life style from that of the juveniles, e.g., plankton feeding, because the highly elevated aperture of adult shells might have been poorly adapted to search for prey on the sea bottom, and because high shell stability would have limited their agility and flexibility. The horizontal thrust angle suggests that adult heteromorphs would have had a somewhat better horizontal mobility than juveniles.

4. Theoretical Modeling of Ammonoid Morphogenesis

Theoretical methods to model the morphology of molluscan shells have often been carried out to understand some of the functional constraints acting on the coiling organisms. The distribution of actual shell morphology is usually analyzed within the range of geometrically possible shell shapes (Raup, 1966, 1967; McGhee, 1980; Saunders and Swan, 1984; Savazzi, 1987; Swan and Saunders, 1987). In contrast, little attention has been paid to the underlying morphogenetic rules. In order to understand ammonoid shell morphology from the viewpoint of constructional morphology, we should study the morphogenetic programs and constraints that influence shell growth.

The shell growth models based on the moving-frame method trace the mode of coiling of tubular shells and reconstruct the shell morphologies. Although a tremendous variety of three-dimensionally coiled shells is theoretically produced by such models by inputting various combinations of parameters, the area occupied by actual heteromorph ammonoids in this hypothetical morphospace is very small and discontinuous. Traditional approaches to theoretical morphology are ineffective in interpreting such "aberrant" shell morphologies and their discontinuous variation, so that other models are required to approximate the actual ammonoid morphology.

4.1. Possibility of Self-Regulatory Ability

Modeling of shell growth using the "growing-tube model" of Okamoto (1988a) is the most appropriate method for analyzing and comparing the morphology of free tubular shells as in heteromorph ammonoids. The locally defined framework around the aperture is necessary to model the growth process, which proceeds according to some unknown morphogenetic rules. It should be pointed out, however, that the growing-tube model does not completely simulate the actual morphogenesis of shell coiling.

We can easily understand why this is so by analyzing the coiling patterns of *Polyptychoceras*. The parallel shafts connected with U-shaped semiwhorls observed in *Polyptychoceras* are rather difficult (although possible) to simulate using the growing-tube model. This is because shell growth in this model is carried out *ad hoc* by reading a script, which is usually a graph or a data file of parameters, and because small aberrations in each growth stage tend to accumulate during such hypothetical shell growth. Therefore, the shafts in the hypothetical *Polyptychoceras* sometimes do not run parallel to each other (Fig. 9). If organisms actually adopted a morphogenetic program similar to the growing-tube model, the shell morphologies shown in Fig. 9B,C would have been produced in actual specimens. The shaft in actual specimens, however, runs parallel to previously formed shafts. This suggests that the actual morphogenetic program could control coiling by reference to the aperture and the

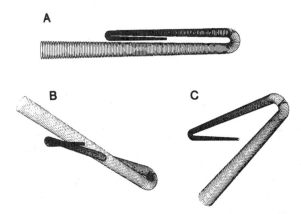

FIGURE 9. Hypothetical shell forms of *Polyptychoceras* (Late Cretaceous diplomoceratid) produced by the growing-tube model. Although several shafts connected with U-shaped semiwhorls run parallel to each other in the actual specimens of *Polyptychoceras* as shown in A, shell variations B and C are hypothetically produced by specifying 10% larger and smaller values of shell curvature, respectively.

previously formed shell so as to regulate shell morphology. Merkt (1966) described a curiously coiled ammonoid shell with an oyster encrusted on one side. As the oyster encrusted the living ammonoid, the hydrostatic balance of the ammonoid was disturbed by the additional weight of the growing oyster. The whorls of the ammonoid shell grew abnormally on the side opposite the side encrusted by the oyster. The host reaction of the ammonoid suggests the possibility that ammonoids had a self-regulatory ability to modify their shells.

The shell form of ammonoids, even when the coiling is as complicated as it is in some heteromorph ammonoids, can be understood to be the result of the integration of infinitesimal growth increments at the aperture. Although a local coordinate system on the shell is necessary to model ammonoid shell growth, the self-regulatory mechanism of shell coiling is difficult to describe using only a local coordinate system. The gravitational direction, which provides an absolute framework for ammonoid shell growth, may be an alternative to an artificial, fixed coordinate system. Assuming that ammonoids were neutrally buoyant, their life orientation can be determined by calculating the distance between the centers of buoyancy and gravity. The actual mode of growth in ammonoid shells is probably determined *a posteriori* by some morphogenetic rules defined with respect to the surrounding conditions, that is, the position of the aperture and the growth direction relative to the gravitational vector, the distribution of the centers of buoyancy and gravity, and the previously formed shell, among other things. No theoretical model can currently reconstruct the actual morphogenesis of ammonoid shells. However, the most accurate model to explain ammonoid morphogenesis must be a rule-defined simulation based on a local framework at the aperture and on a global framework based on the reconstructed life orientation.

4.2. A Model Explaining the Orientation of Ribs

Ribs on the ammonoid shell must indicate the shape of the apertural margin at every growth stage, because growth lines usually run parallel to ribs (Okamoto, 1988b). The rib direction of ammonoids was probably influenced by the life orientation. Because, in a normally coiled ammonoid, an almost constant life orientation was probably maintained throughout growth, a change in rib obliquity is rarely observed. However, remarkable changes in life orientation during growth are presumed to have occurred in some heteromorph ammonoids. Okamoto (1988b) previously analyzed the relationship between rib obliquity and the presumed life orientation in four species of Cretaceous heteromorphs and reached the following conclusions: (1) the ventral margin of the shell usually coincides with the bottom margin, that is, the lowest point on the generating curve, and (2) the apertural plane maintains a constant angle (apertural angle) to the vertical at every growth stage. Figure 10 shows the presumed ontogenetic changes in life orientation and rib pattern

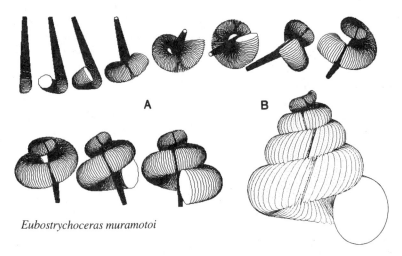

Eubostrychoceras muramotoi

FIGURE 10. Results of a computer simulation of *Eubostrychoceras muramotoi* (Late Cretaceous nostoceratid). (A) Inferred ontogenetic change of life orientation. A turnover of life orientation is expected at about one and one-half whorls after the beginning of helicospiral shell growth. (B) Theoretically produced ribbing pattern assuming a constant apertural angle. The developmental changes of rib obliquity in the actual specimen of this species are well approximated by this rule-defined simulation. (Modified from Okamoto, 1988b, Text Figs. 6 and 10.)

in *Eubostrychoceras muramotoi* with the aid of computer graphics, assuming a constant apertural angle of 40°. The theoretically produced profile of *E. muramotoi* closely simulates the ontogenetic changes of shell coiling in real specimens, in which the abrupt change of rib obliquity always occurs in the early helicoid stage.

On the basis of the concept of differential geometry, the rib-forming process could also be expressed by some additional parameters such as "post-torsion" as suggested by Savazzi (1990). However, the interpretation of a constant apertural angle is more appropriate; the simulation in which some rules of growth are specified (instead of a "script" expressed by some differential parameters) is probably much closer to the process of rib formation in the actual morphogenetic program.

4.3. A Model Explaining Meandering Whorls

A rule-defined simulation is also applicable to the mode of coiling. *Nipponites*, a Cretaceous nostoceratid, exhibits meandering shell growth during the middle growth stage. The mode of coiling of this ammonoid is so aberrant that it has often been regarded as an adaptation to a sessile mode of life like vermetid gastropods, which also have an irregular conch (Diener, 1912; Moore *et al.*, 1952; Tasch, 1973). However, the coiling of *Nipponites* is actually quite regular, as pointed out by Yabe (1904). On the basis of theoretical calculations

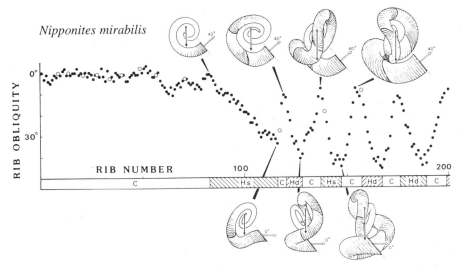

FIGURE 11. Diagrammatic figure showing the ontogenetic change of rib obliquity, mode of coiling, and reconstructed life orientation in a Late Cretaceous nostoceratid *Nipponites mirabilis*. There are three modes of coiling, Hs, Hd, and C, which represent sinistral helicoid, dextral helicoid, and crioceratoid (planispiral), respectively. (Modified from Okamoto, 1988c, Figs. 6 and 8.)

of the total average density of *N. occidentalis*, Ward and Westermann (1977) documented that there was sufficient phragmocone volume to achieve neutral buoyancy.

Okamoto (1988c) observed the relationship between shell coiling and rib obliquity in this genus. *Nipponites* regularly switches among three modes of coiling throughout the middle growth stage: dextral and sinistral helicoid coiling occur alternately with crioceratoid coiling in between. Rib obliquity also shows a regular oscillation corresponding to the switching of the coiling mode; when the rib obliquity approaches some maximum value, the mode of coiling switches from crioceratoid to dextral/sinistral helicoid, and when the rib obliquity approaches some minimum value, the mode of coiling returns to crioceratoid again. If the assumption of a constant apertural angle, mentioned above, applies to *Nipponites*, the change in rib obliquity suggests a change in life orientation. I described this regularity in coiling as a "growth direction regulatory model" (Okamoto, 1988c). In this model, three modes of coiling, crioceratoid, dextral, and sinistral helicoid, are mathematically defined, and the hypothetical ammonoid shell can switch its mode of coiling in order to maintain its growth direction within some appropriate range of elevation angles to the horizontal plane (Fig. 11). To simulate the meandering whorls, the upper and the lower limits of growth direction are initially specified. Crioceratoid coiling is initially adopted and continues as long as the growth direction is within some given range. When the life orientation exceeds a certain upper limit, the mode of coiling switches to dextral or sinistral helicoid, which acts to decrease the elevation angle of the growth

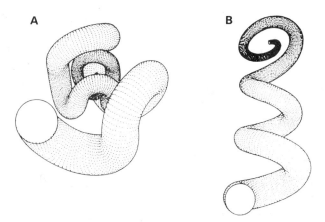

FIGURE 12. Hypothetical shell-coiling patterns constructed by computer simulation using the growth direction regulatory model. (A) Meandering coiling pattern as in *Nipponites* usually appears if moderate upper and lower limits of the growth direction are assumed. (B) Persistent helicoid coiling appears when a smaller value of the lower limit of the growth direction is specified. (Modified from Okamoto, 1988c, Fig. 11.)

direction. If the life orientation drops below a certain lower limit during helicoid coiling, the mode of coiling changes again to crioceratoid. A meandering shell like *Nipponites* is automatically produced by such a mechanism of coiling feedback without programming of the meandering itself (Fig. 11). Variation in the amplitude of the meandering that is actually observed in *Nipponites* can also be interpreted as the result of changing the acceptable limits of the growth direction.

The simulation outlined above suggests that an abrupt change of shell morphology may occur when the mechanism of coiling feedback fails during growth. If the lower limit of the growth direction is extremely low, for example, the helicoid mode of coiling does not return to crioceratoid. Thus, an entirely different shell form like *Eubostrychoceras japonicum*, a probable ancestor of *Nipponites*, is produced by the same program (Fig. 12). The regulatory mechanism of growth direction suggests that the change of coiling pattern from simple helicoid to meandering occurs abruptly without intermediate forms, although the change in the acceptable limits of growth direction is continuous. The actual fossil record also supports the abrupt appearance of *Nipponites* (Matsumoto, 1977; Okamoto, 1989).

5. Conclusions

The actual mode of coiling in ammonoids is probably determined by some morphogenetic rules interacting with the environment and some physical property of the previously formed part of the shell. The best theoretical model to explain ammonoid morphology and its developmental and phylogenetic change must be a rule-defined simulation in which a theoretical shell can control its mode of growth so as to regulate its morphology. I have shown two examples of such growth simulations based on a local framework at the aperture and a global framework with respect to life orientation. The adequacy of these models can be tested by observing the actual variation in shells in comparison with the variability in hypothetical shells. Although the shell morphology regulated by a feedback mechanism is rather resistant to occasional anomalies or deviations during growth, this mechanism has the potential to produce drastic changes in shell morphology. One of the most interesting features of this model is that the discontinuous change of shell morphology predicted under the rule-defined simulation correlates well with the actual variability and morphological saltation observed in the clade of nostoceratid ammonoids. The morphological transformation from *Eubostrychoceras* to *Nipponites* was probably accomplished by only a slight modification of the morphological program, even though it seems to be visually drastic.

References

Ackerly, S. C., 1987, Using "local" coordinates to analyze shell form in molluscs (abstract), *Geol. Soc. Am. Abstr. Progs.* **19**:566.

Ackerly, S. C., 1989, Kinematics of accretionary shell growth, with examples from brachiopods and molluscs, *Paleobiology* **15**:147–164.

Bayer, U., 1978, Morphologic programs, instabilities, and evolution: A theoretical study, *N. Jb. Geol. Paläont. Abh.* **156**:226–261.

Burnaby, T. P., 1966, Allometric growth of ammonoid shells: A generalization of the logarithmic spiral, *Nature* **209**:904–906.

Chamberlain, J. A., Jr., 1976, Flow patterns and drag coefficients of cephalopod shells, *Palaeontology (Lond.)* **19**:539–563.

Chamberlain, J. A., Jr., 1981, Hydromechanical design of fossil cephalopods, in: *The Ammonoidea*, Systematics Association Special Volume 18 (M. R. House and J. R. Senior, eds.), Academic Press, London, pp. 289–336.

Chamberlain, J. A., Jr., and Westermann, G. E. G., 1976, Hydrodynamic properties of cephalopod shells, *Paleobiology* **2**:316–331.

Checa, A., 1991, Sectorial expansion and shell morphogenesis in molluscs, *Lethaia* **24**:97–114.

Denton, E. F., and Gilpin-Brown, J. B., 1966, On the buoyancy of the pearly *Nautilus*, *J. Mar. Biol. Assoc. U.K.* **46**:723–759.

Diener, C., 1912, Lebensweise und Verbreitung der Ammoniten, *N. Jb. Mineral. Geol. Paläontol.* **2**:67–89.

Ebel, K., 1983, Berechnungen zur Schwebefähigkeit von Ammoniten, *N. Jb. Geol. Paläontol. Monatsh.* **1983**:614–640.

Ebel, K., 1985, Gehäusespirale und Septenform bei Ammoniten unter der Annahme vagil benthischer Lebensweise, *Paläontol. Z.* **59**:109–123.

Ebel, K., 1990, Swimming abilities in ammonites and limitations, *Paläontol. Z.* **64**:25–37.

Ebel, K., 1992, Mode of life and soft body shape of heteromorph ammonites, *Lethaia* **25**:179–193.

Heptonstall, W. B., 1970, Buoyancy control in ammonoids, *Lethaia* **3**:317–328.

Illert, C., 1987, Formulation and solution of the classical problem. I. Shell geometry, *Nuovo Cimento* **9**:791–813.

Illert, C., 1989, Formulation and solution of the classical problem. II. Tubular three-dimensional seashell surfaces, *Nuovo Cimento* **11**:761–780.

Jacobs, D. K., 1992, Shape, drag, and power in ammonoid swimming, *Paleobiology* **18**:203–220.

Matsumoto, T., 1977, Some heteromorph ammonites from the Cretaceous of Hokkaido, *Mem. Fac. Sci. Kyushu Univ. Ser. D, Geol.* **23**:303–366.

McGhee, G. R., 1978, Analysis of the shell torsion phenomenon in the Bivalvia, *Lethaia* **11**:315–329.

McGhee, G. R., 1980, Shell form in the biconvex articulate Brachiopoda: A geometric analysis, *Paleobiology* **6**:57–76.

Merkt, J., 1966, Über Austern und Serperln als Epöken auf Ammonitengehäusen, *N. Jb. Geol. Paläontol. Abh.* **125**:467–479.

Moore, R. C., Lalicker, O., and Fischer, A., 1952, *Invertebrate Fossils.* McGraw-Hill, New York.

Moseley, H., 1838, On the geometrical forms of turbinated and discoid shells, *Phil. Trans. R. Soc. Lond.* **1838**:351–370.

Okamoto, T., 1984, Theoretical morphology of *Nipponites* (a heteromorph ammonoid), *Fossils* (Kaseki), *Palaeontol. Soc. Jpn.* **36**:37–51 (in Japanese).

Okamoto, T., 1986, Analysis of morphology in heteromorph ammonites, *Abstr. Ann. Meet. Palaeontol. Soc. Jpn.* **1986**:34 (in Japanese).

Okamoto, T., 1988a, Analysis of heteromorph ammonoids by differential geometry, *Palaeontology (Lond.)* **31**:35–52.

Okamoto, T., 1988b, Changes in life orientation during the ontogeny of some heteromorph ammonoids, *Palaeontology (Lond.)* **31**:281–294.

Okamoto, T., 1988c, Developmental regulation and morphological saltation in the heteromorph ammonite *Nipponites*, *Paleobiology* **14**:272–286.

Okamoto, T., 1989, Comparative morphology of *Nipponites* and *Eubostrychoceras* (Cretaceous nostoceratids), *Trans. Proc. Palaeont. Soc. Jpn N.S.* **265**:117–139.

Raup, D. M., 1966, Geometric analysis of shell coiling: General problems, *J. Paleontol.* **40**:1178–1190.

Raup, D. M., 1967, Geometric analysis of shell coiling: coiling in ammonoids, *J. Paleontol.* **41**:43–65.

Raup, D. M., and Chamberlain, J. A., Jr., 1967, Equation for volume and center of gravity in ammonoid shells, *J. Paleontol.* **41**:566–574.

Raup, D. M., and Michelson, A., 1965, Theoretical morphology of the coiled shell, *Science* **147**:1294–1295.

Rex, M. A., and Boss, K. J., 1976, Open coiling in recent gastropods, *Malacologia* **15**:289–297.

Reyment, R. A., 1958, Some factors in the distribution of fossil cephalopods, *Stockholm Contrib. Geol.* **1**:97–184.

Saunders, W. B., and Shapiro, E. A., 1986, Calculation and simulation of ammonoid hydrostatics, *Paleobiology* **12**:64–79.

Saunders, W. B., and Swan, A. R. H., 1984, Morphology and morphologic diversity of mid-Carboniferous (Namurian) ammonoids in time and space, *Paleobiology* **10**:195–228.

Savazzi, E., 1985, SELLGEN a BASIC program for the modeling of molluscan shell ontogeny and morphogenesis, *Comput. Geosci.* **11**:521–530.

Savazzi, E., 1987, Geometric and functional constraints on bivalve shell morphology, *Lethaia* **23**:195–212.

Savazzi, E., 1990, Biological aspect of theoretical shell morphology, *Lethaia* **23**:195–212.

Swan, A. R. H., and Saunders, W. B., 1987, Function and shape in late Paleozoic (mid-Carboniferous) ammonoids, *Paleobiology* **13**:297–311.

Tanabe, K., 1975, Functional morphology of *Otoscaphites puerculus* (Jimbo), an Upper Cretaceous ammonite, *Trans. Proc. Palaeont. Soc. Jpn. N.S.* **99**:109–132.

Tanabe, K., 1977, Functional evolution of *Otoscaphites puerculus* (Jimbo) and *Scaphites planus* (Yabe), Upper Cretaceous ammonites, *Mem. Fac. Sci. Kyushu Univ. D. Geol.* **23**:367–407.

Tanabe, K., Obata, I., and Futakami, M., 1981, Early shell morphology in some Upper Cretaceous heteromorph ammonites, *Trans. Proc. Palaeontol. Soc. Jpn. N.S.* **124**:215–234.

Tasch, P., 1973, *Paleobiology of the Invertebrates*, John Wiley & Sons, New York.

Thompson, D. W., 1942, *On Growth and Form*, Cambridge University Press, Cambridge.

Trueman, A. E., 1941, The ammonite body-chamber, with special reference to the buoyancy and mode of life of the living ammonite, *Q. J. Geol. Soc. (Lond.)* **96**:339–383.

Ward, P. D., and Westermann, G. E. G., 1977, First occurrence, systematics, and the functional morphology of *Nipponites* (Cretaceous Lytoceratina) from the Americas, *J. Paleontol.* **51**:367–372.

Yabe, H., 1904, Cretaceous Cephalopoda from the Hokkaido, Part 2, *J. Coll. Sci. Imp. Univ. Tokyo* **20**:1–45.

Chapter 9

Morphogenesis of the Septum in Ammonoids

ANTONIO G. CHECA and JUAN M. GARCIA-RUIZ

1. Introduction

Ammonoid septa are aragonitic structures that divide the shell internally into a series of chambers, the most adoral of which (also the largest one) is the living chamber and is occupied by the ammonoid's soft body. The septal

ANTONIO G. CHECA • Departamento de Estratigrafía y Paleontología, and Instituto Andaluz de Ciencias de la Tierra, Universidad de Granada, Facultad de Ciencias, Granada 18071, Spain. JUAN M. GARCIA-RUIZ • Instituto Andaluz de Ciencias de la Tierra, Consejo Superior de Investigaciones Científicas, Granada 18071, Spain.

Ammonoid Paleobiology, Volume 13 of *Topics in Geobiology*, edited by Neil Landman *et al.*, Plenum Press, New York, 1996.

surface is roughly transverse to the shell tube. The most typical feature of the septum is its marginal corrugation. Individual folds are given different names according to their polarity. Adorally bulging major folds are called saddles, and apically directed folds are lobes. Minor elements of saddles and lobes are folioles and lobules, respectively. Marginal complication progressively decreases toward the septum center, which is a slightly undulated to flat surface. For each saddle or lobe, the fold or element placed opposite (i.e., linked by a minimal distance through the septum to the other side or to the other wall of the whorl cross section) is always of the same polarity (Figs. 9, 10). Therefore, the septum is adorally concave when sectioned across two opposite saddles and adorally convex when this is done across opposite lobes. On the basis of this property, the septum is an anticlastic surface.

The ammonoid septum long has been of interest to paleontologists. Complexly fluted septa are enigmatic for several reasons, principally because of the absence of close recent analogues, either in cephalopods or in any other biological group. The constructional hypotheses put forward to explain septal morphology (especially its marginal fluting) and variability are based mainly on adaptive rather than purely fabricational aspects. The distinction between these two aspects of morphogenetic approach is important. The adaptive aspect refers to the suitability of a structure to its function or functions. On the other hand, fabrication deals with those biophysical forces or processes leading to pattern formation (for a longer discussion, see, e.g., Seilacher, 1973; Reif *et al.*, 1985). In this chapter, we attempt to explain the shape of ammonoid septa using purely fabricational criteria. Thus, we are not concerned here with the possible functions the septa may have had for the ammonoid.

Early fabricational hypotheses were based on Owen's (1832) preseptal gas hypothesis for *Nautilus*, according to which chamber formation involved the creation of a gas-filled space afterwards closed by a calcareous septum (see Ward, 1987, p. 32). As far as ammonoids are concerned, the first hypotheses are to be found in Tate and Blake (1876) and Solger (1901), who suggested that the adoral convexity of the septum was evidence of pressure from behind the animal and that the saddles and their associated folioles were portions of an uncalcified septal surface inflated under internal gas pressure, whereas lobes and lobules represented muscular attachment sites of the membrane to the shell wall. This view was later supported by Schmidt (1925). Swinnerton and Trueman (1918) also thought that gas pressure was active during ontogeny and demonstrated that, in the adult septum of *Dactylioceras*, lobes have greater amplitudes than saddles; i.e., they extend from the septum center further apically than saddles do adorally, so that the average profile is adorally convex. These views were later accepted by most ammonitologists dealing with the subject (e.g., Spath, 1919; Buckman, 1919–1921).

Denton and Gilpin-Brown (1966, 1973) and Denton (1974) showed that during body translocation *Nautilus* eventually creates a (cameral) liquid-filled space between the body and the last fully mineralized septum. Once the new septum has mineralized to half its final thickness, the animal empties the last-formed chamber by siphuncular osmotic pumping. The now gas-filled chamber finally develops a partial pressure of nitrogen in equilibrium with the atmosphere, so that it is the shell that has to withstand hydrostatic pressure. A similar process was proposed for *Sepia*. If ammonoids followed a *Nautilus*-like mode of chamber emptying, they could not inflate their septal membrane by gas pressure.

A major breakthrough occurred in the 1970s with a series of papers that provided the basis for fabricational models that were considered valid until only recently. Seilacher (1973, 1975) compared the continuous anticlastic curvature of the septal surface to a soap bubble spread out along the contour of the suture line, with the difference that in ammonoids marginal undulations caused by the sutural lobes extend further toward the center. Thus, the septum is not a minimal-energy surface. According to his view the ammonoid septum could be understood as an elastic membrane subject to radial tension. Septal sutures would then consist of a series of defined tie points to the shell wall, connected by saddles bulging anteriorly, mainly resulting from muscular pull-off. This model also implied an initial planar adhesion of the mantle to the shell wall for each septum.

Westermann (1975), although agreeing with Seilacher on the tie-point concept, postulated that the posterior mantle had a stiff (flexible) margin (aponeurosis) that, on withdrawal from the septum, was distorted but largely maintained its shape during translocation, and in which reattachment would occur. This mantle was reinflated at the folioles between tie points as a result of a positive (adorally directed) pressure differential between the cameral liquid and the rear body. Subsequently, "the mantle will tighten along the flutes, deepening and extending them axially toward the center of the future septum" (Westermann, 1975). Therefore, the suture did not form just anywhere but along a preformed, stiff, but somewhat elastic aponeurosis that caused exact duplication of the septal (and sutural) shape. In his comprehensive study, Bayer (1977a,b, 1978a,b) envisioned a weakly elastic septal mantle surface that gradually became more complicated through ontogeny and that was attached not only to single tie points but all along its length. The concept of attachment along the entire sutural length gave rise to the classic discussion about "inverted" suture lines in ammonoids (Section 4.2).

Although differing in some respects, especially as to the degree of flexibility and homogeneity of the rear mantle and the role of cameral liquid, these three models all agreed that the hydrostatic pressure in the liquid of the last, incomplete chamber was subequal to or higher than that in the rear of the soft body. Here it is worth mentioning the detailed chamber-construction models in certain ammonoids by Zaborski (1986) and Hewitt *et al.* (1991), which were based for the first time on the paleontological evidence provided by pseudo-sutures (Section 4.1). These latter authors concluded that the shape of the ammonoid septum was largely the result of a gradual peripheral fluting of the posterior mantle epithelium; that is, the rear mantle did not fabricate the septal morphology *de novo* but rather from a permanently fluted body (see

Section 4.1), thus supporting Westermann's (1975) standpoint. This fact was later incorporated by Seilacher (1988) in his revised fabrication model. From the viewpoint of Hewitt *et al.* (1991), tie points did not represent fixed sites of mantle attachment but especially adhesive adapical margins of a fluted mantle.

2. The Viscous Fingering Model

2.1. Fractals as a Geometric Tool

Fractals are continuous but not differentiable geometric objects characterized by the property of dilation symmetry. Dilation is a type of symmetry operation similar to the rotations, translations, mirrors, and other symmetry operations that every geologist knows from his mineralogic and crystallographic background. Unlike the case of Euclidean objects, such invariance on scale is nontrivial in the case of fractals. We can illustrate this by considering a nonfractal (Euclidean) object, for instance, the square in Fig. 1 (left). The area $A(L)$ of any square grows with its linear size as $A(L) = L^2$, and it is also clear that the number of squares $N(l)$ of linear size l needed to cover a square of linear dimension $L > l$ increases as $N(l) = l^{-2}$. Obviously, the squares look the same whatever the length of their edges, and, therefore, we can say that they are self-similar objects showing dilation symmetry. There is nothing new in this but the basis of how to express the area of a square in terms of a typical length, the edge size. The trivial dilation symmetry leads to an exponent 2, which coincides with the topological dimension of the space in which the square is embedded. As a second example, we start by considering a simple line segment (Fig. 1, center). We can express the length of the segment in terms of a scale stick l, and we see that the length of the segment varies with l as $L(l) = nl^1$; i.e., we find again the equivalence between the exponent and the topological dimension of the space embedding the object and therefore a new example of trivial dilation symmetry. It is clear that the number N of pieces of length $l < L$ is related to the applied reduction factor r, by the formal expression:

$$N = \left|\frac{1}{r}\right|^D \text{ or } D = \frac{\log N}{\log 1/r}$$

Consider now the intricate line shown in Fig. 1 (right), which is known as a Koch curve. This curve is created from a segment of linear dimension l, defined by the closed interval [0,1]. The first stage of the construction is obtained by dividing [0,1] into three parts of equal length and replacing the inner third, [1/3,2/3], by two segments of equal length forming a "cape" with an internal angle of 60°. This procedure is iterated for each new linear segment created in the previous stage. After i iterations, we obtain the so-called Koch

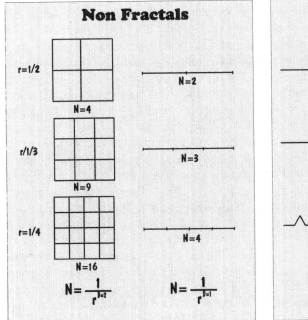

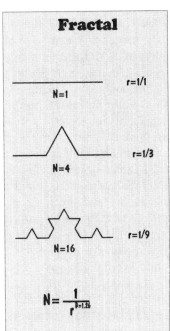

FIGURE 1. Non-fractal (Euclidean) and fractal (Koch curve) objects: r, reduction factor; N, number of partitions; D, fractal dimension. See Section 2.1.

curve. It is clear that each iteration increases the number of linear segments N by 4 when a reduction factor $r = 1/3$ is applied. Thus, according to the previous relation, we obtain:

$$D = \frac{\log 4^n}{\log 3^n} = \frac{\log 4}{\log 3} = 1.2619$$

Note that in this case, D is not an integer but a fractional number smaller than the topological dimension in which the object is embedded. From the singular *fractional* character of this exponent, Mandelbrot (1982) created the word *fractal* to refer to geometric objects having this particular nontrivial scale invariance. The main difference in nontrivial scale behavior of fractal objects can be obtained from an evaluation of their size in the d-1 dimension. For the square of Fig. 1, it is trivial that the length enclosing the area is a constant value. In contrast, for the case of the fractal Koch curve, such a length increases as we magnify the structure, i.e., the length is a function of the scale of measurement. There is no characteristic scale size for a fractal object, and the structures that they exhibit at a given scale return persistently at any other scale we try. It follows that the length of a fractal curve increases *ad infinitum* as we increase the resolution. This is true only for all the exact or deterministic fractals that are created by iterating a known generation procedure and whose fractal dimension can be analytically calculated. The structures exactly fulfilling such a definition are ideal or mathematical objects, like the Koch curve.

Unlike mathematical fractals, natural patterns arise from spatial or temporal phenomena exhibiting partial correlation over many scales. They are the result of iterations of random fluctuations, and therefore, natural fractal objects are not strictly self-similar in a mathematical sense, but they are scale-invariant structures over a range of magnification, and ability to manage them as fractal objects depends on the scale free interval. Unfortunately, the fractal dimension and the range of magnification showing fractal behavior cannot be derived but have to be measured. There are several methods to measure the fractal dimension of a physical object (Longley and Batty, 1989; Korvin, 1992; Peitgen *et al.*, 1992). The most popular are the covering methods, mainly the box-counting method. It consists in covering the object with a grid made of a number N of square boxes of edge ε. The number of boxes needed to cover the object depends on the linear size ε of each box in the grid, and the dependence of the number N of filled boxes on ε is $N \propto (\varepsilon)^d$. Varying ε generates a set of $N(\varepsilon)$ values, which are plotted versus ε using logarithmic axes. For fractal objects, the resulting graph will be a straight line, and its slope gives the fractal dimension.

Some cautions must be observed when carrying out such a simple measurement. To put the problem in terms of deterministic fractals, let us consider the fifth iteration of an exactly self-similar Koch curve that is known to have an exact D value of $\ln 4 / \ln 3 = 1.26 \dots$ Figure 2A shows the log–log plot obtained for the Koch curve measured with the box-counting method. The linear size of the Koch curve was 480 pixels, and the range of measurement used was from $\varepsilon = 2$ to $\varepsilon = 220$ pixels. The slope of the linear relationship obtained is 1.17, a value not close to the right value of 1.26 in spite of the fact that the correlation coefficient was very high. The reason for this difference is twofold. First, it arises from the fact that for the smaller values of the measuring unit ε, we start to obtain information from the range of scale showing Euclidean behavior. Then, it is convenient to remove the data corresponding to the values of $\varepsilon < 7$ pixels (a length that corresponds approximately to the length of the linear segments forming the fifth iteration of the Koch curve). Note also that for large values of the measuring unit, the log N values start to oscillate, showing a periodic behavior with a period related to the generation procedure used in the construction of the Koch curve. This behavior becomes clear when

---

FIGURE 2. Log–log plots for the measurement of the fractal dimension by the box-counting method. The slope of the linear regression coincides with the estimated fractal dimension (see Section 2.1 for details). (A) Data for the fifth iteration of the triadic Koch curve (see Fig. 1). (B) Filtered data with a fast Fourier transform algorithm. ε, size of box edge (in pixels); $N(\varepsilon)$, number of boxes needed to cover the object; sd, standard deviation of the slope value; R, regression coefficient; SD, standard deviation of the R value; N, number of data points.

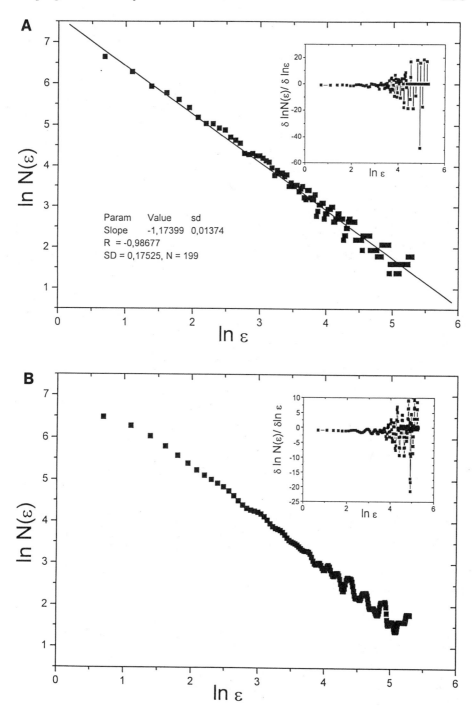

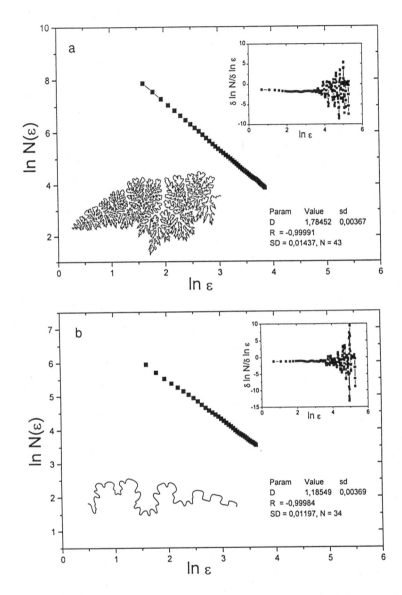

FIGURE 3. Log–log plots for the determination of the fractal dimension (*D*) of the septal sutures of three different ammonoid genera. (a) *Puzosia* (in Arkell *et al.*, 1957, Fig. 476). (b) *Cottreauites* (in Arkell *et al.*, 1957, Fig. 531). (c) *Prohysteroceras* (in Arkell *et al.*, 1957, Fig. 523). ε, size of box edge (in pixels); $N(\varepsilon)$, number of boxes needed to cover the object; *D*, fractal dimension; *sd*, standard deviation of the *D* value; *R*, regression coefficient; *SD*, standard deviation of the *R* value; *N*, number of data points.

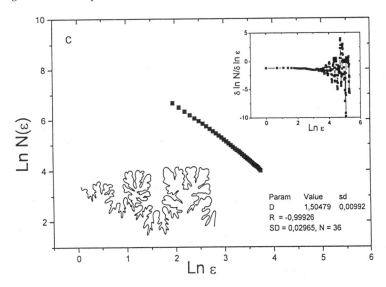

FIGURE 3. (Continued)

any smoothing procedure such as the fast Fourier transform is applied (Fig. 2B). This effect can be reduced by using values of $N(\varepsilon)$ corresponding to the minimum number of boxes of edge ε needed to cover the structure that are obtained by changing the origin of the measurement lattice for every ε computation.[*] Finally, it is clear that for values of the measuring unit ε close to or larger than the linear size (480 pixels) of the measured object (not shown in the plots), the slope will take a zero value because the number of squares will be kept constant for such large ε values. Therefore, the general trend in the middle range of ε values is characteristic of the covering dimension of the object, and we found in this particular case of the fifth iteration of the Koch curve a measured D value of 1.27, which is reasonably close to the theoretical fractal dimension of the Koch curve.

The wavy, tree-like (arborescent) structure displayed by suture lines suggests at first glance that they are fractal objects. In fact, in the last two decades a number of authors (Vicencio, 1973; Guex, 1981; Bayer, 1985; Damiani, 1986, 1990; Seilacher, 1988) realized that fractal geometry is a good tool for describing suture complexity in a quantitative way, as the fractal dimension is a measure of the extent to which the septal suture fills a two-dimensional space and varies between 1 and 2. When measuring D for a given ammonoid suture, we face the same problems found for the case of deterministic fractals, but

---

[*]Box-C (introductory level) and Ffract (research level) are software devoted to the measurement of the fractal dimension by the box-counting method. Both are available from F. Otálora and J. M. García-Ruiz upon request. Please use the following e-mail address: otalora@goliat.ugr.es.

now we lack the right analytically derived value of D. Figure 3 shows the value for three suture lines of different complexity. The suture lines were digitized at a resolution of 150 dpi, and the field dimension was 480×480 pixels. The measuring process was performed for boxes with edge values (ε) from 2 to 200 pixels. Data were filtered with a fast Fourier transform algorithm. After removing the points corresponding to smaller ε values and those corresponding to the region where $\delta \ln N / \delta \ln \varepsilon$ clearly oscillates, we obtain the experimental value of the fractal dimension. Note that D varies with the degree of complexity displayed by the suture lines. It follows from the above discussion that any comparative study based on the values of the fractal dimension of suture lines has to consider carefully the procedure used to estimate such D values.

2.2. Beyond Description: The Viscous Fingering Model

The fractal dimension of suture lines has been measured for several ammonoids. Long (1985) measured D values for the genera *Baculites* and *Placenticeras*, and García-Ruiz *et al.* (1990) and García-Ruiz and Checa (1993) measured the fractal dimension for ten suture lines displaying a wide range of complexity. They found that D values vary between 1.16 for embryonic sutures and 1.56 for adult ones. Boyajian and Lutz (1992) also measured the fractal dimension of sutures of more than 600 ammonoid genera and tried to use them in studies of evolution. In spite of the fact that fractal geometry can obviously be an excellent tool to quantify highly complicated patterns emerging from natural processes, it has to be stressed that neither the fractal character of a biological structure nor the measurement of its fractal dimension provides morphogenetic information by itself. A protocol for morphogenetic studies of biological scale-invariant structures should consist of (García-Ruiz, 1992; García-Ruiz and Otálora, 1994): (1) noting the self-similar character, (2) measuring its fractal dimension, (3) searching for the existence of archetypical models of physical or chemical instabilities eventually leading to the formation of growth patterns reminiscent qualitatively and quantitatively (D values) of the natural pattern under study, (4) ascertaining if the physical (or chemical) driving force developing the biological structure fits the physical (or chemical) constraints of the model suspected to generate it, and (5) testing the hypothesis with *ad hoc* experiments.

Following the above protocol, we start by considering that the complexity of suture lines remains constant or more generally increases during ontogeny; i.e., they are growth patterns. The second interesting feature is that this complexity increases by successive splitting of the lobes, which leads to their typical arborescent structure. We know that tree-like patterns, either anisotropic (e.g., crystalline dendrites) or random structures (e.g., suture lines), can be obtained from a growth mechanism fulfilling the Laplace equation. The problem arises from the fact that there are a wide range of chemical and physical processes that fulfill the Laplace condition. Stanley (1987) discussed the three main variants of the original Laplace equation. It seems reasonable to remove the cases of dielectric breakdown and dendritic solidification because it is difficult to imagine that suture formation could be triggered by their corresponding potential fields. The physics underlying the third mechanism, the so-called "viscous fingering" phenomenon, is closer to septal suture formation, as it deals with a fluid mechanics problem.

Viscous fingering (VF) refers to a type of growth pattern obtained when a fluid with viscosity μ_1 invades another one of viscosity $\mu_2 > \mu_1$. Experiments designed to study the phenomenon in the laboratory make use of the so-called Hele–Shaw cell, which is a simple device made of two parallel glass plates separated from each other by a gap typically of a millimeter (Fig. 4). The cell

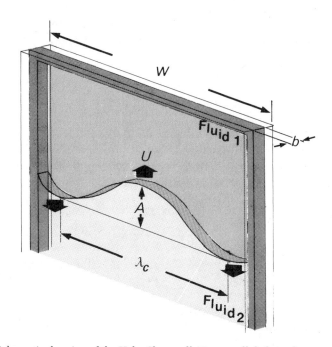

FIGURE 4. Schematic drawing of the Hele–Shaw cell. Two parallel glass plates are separated a distance b (cell thickness), and the space enclosed between them is sealed. Two fluids (1 and 2) of different viscosities are then injected within the cell. Initially, fluid 2, which is here assumed to be more viscous but less dense, occupies the upper side of the cell, while fluid 1 remains underneath. This is the situation with the systems of, e.g., water–glycerine or water–oil. At this stage, the interface between the two fluids is horizontal. When the cell is turned upside down, fluid 1 (the one with greater density) tries to get to the lower side of the cell and presses against fluid 2; therefore, the less viscous fluid (fluid 1) intrudes the more viscous one (fluid 2). When this condition is fulfilled, the interface between the two fluids becomes unstable and splits into a number of undulations (viscous fingers) with a predictable wavelength (λ_c) (equation 1 in text). In other variants of the Hele–Shaw cell, the less viscous fluid can be injected directly within the more viscous fluid. A, wave amplitude; W, width of the cell; U, velocity of displacement of the interface. (Adapted from Feder, 1988.)

contains the fluid of higher viscosity, and the less viscous fluid is injected into the other. The interface between the two liquids advances in the direction of the pressure gradient. It does not move parallel to itself but, rather, becomes wavy, creating some protruding parts and splitting into a number of fingers (viscous fingers), which grow with a volume (surface) scaling with the radius (L) of the pattern in the form $V \propto L^D$. In this Saffman–Taylor-type instability, the front is known to be unstable with respect to perturbations with a wavelength larger than a critical value (see, e.g., Feder, 1988):

$$\lambda_c = 2\,\pi \left[\frac{\sigma}{\left(\dfrac{\mu_2}{k_2} - \dfrac{\mu_1}{k_1} \right)(U - U_c)} \right]^{1/2} \tag{1}$$

where σ is surface tension, μ viscosity, k the permeability of the Hele–Shaw cell ($k = b^2/12$), and U_c the critical velocity of the invading fluid, which is given by

$$U_c = g[\rho_1 - \rho_2] \Big/ \left(\frac{\mu_2}{k_2} - \frac{\mu_1}{k_1} \right)$$

b being the cell thickness and ρ density. Note that for $\rho_2 > \rho_1$, we find $U_c < 0$ for the case $\mu_2 > \mu_1$, so that the system is unstable even at $U = 0$. All wavelengths are unstable, but those perturbations with a wavelength $\lambda_m = \sqrt{3}\,\lambda_c$ grow faster and dominate the pattern. Assuming that $\mu_1 << \mu_2$,

$$\lambda_m = \pi\,b \sqrt{\frac{\sigma}{U\mu}} = \frac{\pi\,b}{\sqrt{C_a}} \tag{2}$$

where $C_a = U\mu/\sigma$ is the dimensionless capillary number defined by the ratio of viscous to capillary forces. Note that surface tension opposes the instability, as it works to reduce the length of the interface between the fluids, and that the larger the capillary number the higher the complexity of the structure. Therefore, in order to reduce the length scale provoked by surface tension, liquids with zero (or very low) interfacial tension are currently used to create growth patterns that are fractal features with a scaling factor $D = 1.7$. Nevertheless, by varying the values of the physical parameters discussed above, a wide range of fractal patterns with different fractal dimensions can be created (see, e.g., Van Damme, 1989).

Our proposal to understand the formation of suture lines in the framework of viscous fingering phenomena relies on the biophysics of the problem. We have to remember that septa are just the mineral record of the interface between two fluids: the soft rear part of the mantle and the less viscous fluid in the last camera. It is currently accepted that ammonoid camerae worked as pressure vessels, and it has been also inferred that the rear part of the mantle was pushed by the cameral fluid in the last stages of body translocation during the process of chamber formation (Section 4). Therefore, it seems reasonable to conclude that we are facing a fluid displacement problem between two liquids with a finite viscosity ratio $\mu_1/\mu_2 \ll 1$. Moreover, we know that the displacement of the interface (as seen from its mineral record) created corrugated shapes that became, in some cases, very complex and display fractal features. Therefore, this displacement does nor occur in the stable regime but is characteristic of viscous fingering (Lenormand, 1985), in agreement also with the expected value of the viscosity ratio.

We can conclude that the "viscous fingering approach" emulates the patterns displayed by sutures either as single lines or as stages of an ontogenetic sequence. But we also realize that the original VF model has to be enhanced, introducing some variants in order to provide a full explanation of this intriguing morphogenetic problem. In particular, we have to cope with the difficulties arising from the three-dimensional geometry of the septa, namely, their higher marginal complexity compared to the center. Therefore, we need to constrain the development of the instability mainly to the contact of the cameral liquid–rear body interface with the outer shell. To introduce the problem, let us consider a fortuitous experiment recently described in the literature (García-Ruiz, 1992). The phenomenon occurs in an anticrack window made up of three crystal plates separated from each other by a polyvinyl thin film. For reasons suggested by García-Ruiz (1992), air invaded the gap between the plates and pushed the polyvinyl film. The pressure field created by the pumped air fluctuates and permits the development of a fingered growth pattern up to 70 cm in diameter (see figures in García-Ruiz, 1993) that propagates at a very low rate (≈ 0.1 mm/day). This large-scale fingering phenomenon displays other interesting features that are also relevant to ammonoid septal sutures, as in the existence of banding, which reveals a pulsating pressure field (Section 4). A detailed physical study of this case must also consider the action of adhesive forces, but the resulting growth pattern lets us suspect that it can be considered within the framework of Saffman–Taylor-type instabilities. We next present an *ad hoc* experimental study aiming to capture as far as possible the biophysical background of the problem and to discuss the results obtained in comparison with real ammonoid septal sutures.

2.3. The Shape of Ammonoid Septa: An Experimental Study

2.3.1. Viscous Fingers in Three Dimensions

In order to determine the extent to which the three-dimensional geometry of the interface created in viscous-fingering processes approximates that of actual septa, we have performed a set of experiments. Ordinary Hele–Shaw cells are not useful in this respect, for the proximity of the two plates (typically less than 2 mm) causes the marginal viscous fingers to deform the entire

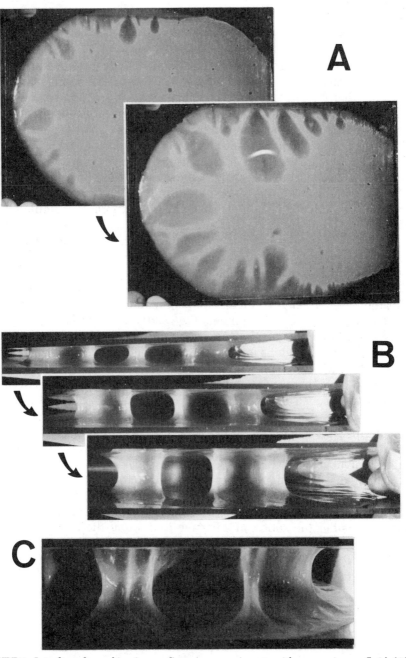

FIGURE 5. Interfaces formed in viscous-fingering experiments with a very viscous fluid. (A) Top view of the experiment, in which the viscous material is sandwiched between two glass plates, which are subsequently pulled apart at the left end; adhesion viscous fingers develop symmetrically on both plates. Two successive stages in a single experiment reveal small undulations growing in a dendritic fashion as both plates are separated. Undulation size decreases towards the contact end (right) of both plates. (B,C) Front view of the interface. Symmetrical lobe-like undulations are joined by concave ridges. As in ammonoids, the interface is not a minimum-area surface. (B) Three successive stages of a single test. (C) Final stage of a different run. Note "bifid" and "trifid" lobes.

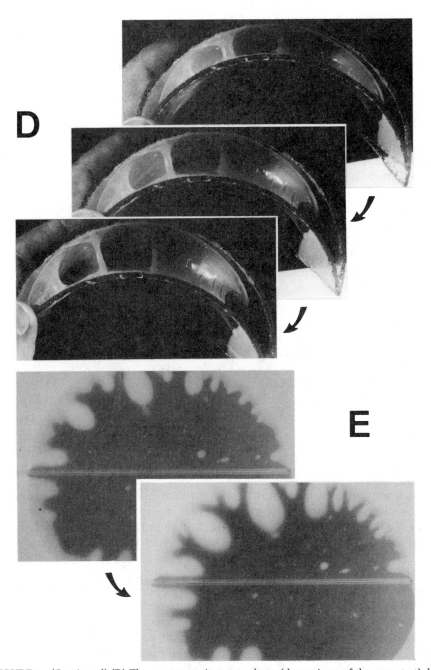

FIGURE 5. (Continued) (D) The same experiment as above (three views of the same test), but with two curved (concentric) plates, to simulate an ammonoid whorl cross section. Ridges are invariably perpendicular to plate surfaces and thus shortest in length. (E) Two successive views of a single experiment similar to A, in which the siphonal lobe is simulated by gluing a glass rod to the internal surface of the upper plate. The screening effect induced by the rod leads consistently to a central lobe that grows with plate separation. Arrows in A, B, D, and E indicate succession of stages.

interface drastically (Fig. 4). For more or less stable interfaces spanning distances on the order of centimeters, very viscous, slow-flowing materials are necessary. The only available material fitting these conditions is a children's toy marketed in Spain under the name Blandy Blub® (Mattel). Blandy Blub® is a very viscous liquid that behaves in a viscoelastic fashion for high deformation rates. Viscoelastic materials deform increasingly the longer a force is applied; once the stress is removed, recovery is also gradual, so that any measurement of strain is time dependent (see, e.g., Wainwright *et al.*, 1976). Blandy Blub® has a density somewhat above that of water and a high power of hydration, which prevents using water or water solutions as secondary liquids in our experiments. We tried to push Blandy Blub® with air from below, but its relatively high density made some experimental attempts unsuccessful. We therefore tested a variant of viscous-fingering pattern formation learned from an unplanned experiment described elsewhere (García-Ruiz, 1992): the viscoelastic fluid is sandwiched between two glass plates, which then are separated, forcing air to penetrate the fluid. Separation leads to the formation of adhesion figures dominated by viscous fingers (Fig. 5A). Note that, from equation 1, when the invading fluid is less dense and less viscous than the invaded fluid, the critical velocity (U_c) is negative. This implies that the Saffmann–Taylor instability can be obtained not only by injecting the less viscous fluid into the more viscous one (i.e., by increasing pressure) but also by suctioning the more viscous fluid.

Thompson (1942) and Damiani (1986, 1990) previously compared ammonoid septal sutures to adhesion figures, although from a purely morphological perspective.

2.3.2. The Anticlastic Morphology of the Septal Surface

In our experiments the plates are separated at only one end while remaining in contact at the opposite end (right end in Fig. 5A), at which time the whole viscous mass advances toward the contact end, and its moving interface begins to form "sutures" at the contacts with the two plates. The resultant two- and three-dimensional patterns display some geometric properties that also can be found in ammonoid septa.

Several types of sutures arise when the separation rate between the two plates is varied. At the beginning of the experiment, a goniatitic-like pattern is achieved. As the separation progresses, the previous pattern typically becomes ceratitic-like (Fig. 5A, advanced phase), with lobes that have two or three lobules, but saddles remain more or less smooth. Higher separation usually leads to the abrupt detachment of the viscous liquid from the upper plate, with the transitory formation of a complex, ammonitic-like pattern. These results agree with equation 2, in which λ_m varies inversely with U (plate separation rate). Of course, the factor controlling the pattern of the septal suture in ammonoids is likely to have been something other than the rate of forward motion of the rear soft body within the shell (see below).

A "suture" formed on one plate is symmetrical to that formed on the other plate, in the sense that any lobe or saddle has its counterpart on the opposite plate. Moreover, corresponding lobe tips on both plates are linked by a ridge that is always convex in transverse section (i.e., perpendicular to both plates) toward the viscous fluid (Fig. 5B,C). Saddles also are joined by strips, but these are more or less flat or even slightly concave, depending on the rate of motion. In summary, viscous fingers develop only at the contact with the two plates, and the pattern intrudes the center of the interface, although in a progressively deadened fashion. Deadening of the marginal pattern toward the center of the interface is expected to be faster as the cohesiveness of the material increases. Our results are consistent with the finding of Seilacher (1975) that the septum is not a minimal-area surface. The surface thus obtained can be called anticlastic and is organized (autoorganized) without the need for tie points, proposed by earlier authors (Section 1). Hewitt *et al.* (1991) postulated the existence of a cameral gel with the same density and viscosity of the rear soft body in order to explain the "apparently unstable, anticlastic topology of the uncalcified septal membrane and mantle." The viscous-fingering model interprets septal morphology as a true instability and requires measurable viscosity differences between the cameral liquid and rear mantle. A cameral gel is therefore unnecessary to our model.

In order to reproduce the ammonoid whorl cross section more accurately, an experiment similar to that described above was performed with two half-cylindrical glass surfaces, one fitting into the other. When the surfaces are pulled apart, with the viscous liquid in between, an interface forms that is similar to that described above. The ridges joining opposite elements are clearly perpendicular to the two glass surfaces (Fig. 5D). This suggests that opposite elements are linked across minimal distances, and this is exactly what can be observed in actual septa, as was pointed out by Westermann (1956, 1958) and Seilacher (1975, 1988). In some cases, where whorl sections change from depressed to compressed during ontogeny, the mutual relationship between lobes varies according to this "law" of minimal distances (see, e.g., Bayer, 1977a, Fig. 10). This fact is interesting because in ammonoids with depressed sections, either the number of dorsal and ventral lobes is the same or there is one more ventral lobe. The length of the dorsal section is obviously smaller than that of the ventral section; this means that dorsal elements must have a shorter wavelength than ventral elements. At the same time, the flat central area of the septum is not midway from the ventral and dorsal walls but nearer to the dorsal side (see Pfaff, 1911, Figs. 1, 8, 9; Swinnerton and Trueman, 1918, Figs. 4, 6; Westermann, 1975, Fig. 5). Therefore, we can state that, in ammonoids, λ_m (which controls lobe depth and width) is not a function of the distance between opposite elements but, rather, of the distance from the elements to the septum center; in the case of small unpaired lobes positioned on the umbilical seam, their size depends on the distance to the ridge joining adjacent flutes. Laterally compressed sections of planulate ammonoids differ from the above in that the internal lobe is connected to the closer lateral or umbilical elements (see, e.g., Bayer 1977a, Fig. 8; Seilacher, 1975, Fig. 5; Fig. 10A), and, from here toward the external lobe, elements on both sides of the section are mutually connected and are, thus, the same size.

A striking feature in our experiments is that the convexity of the transverse arch between corresponding lobe tips and, hence, the complexity of the "sutural" pattern increase with the rate of separation. In other words, for a given span between plates, goniatitic-like viscous fingers are only slightly indented, whereas the reverse is true for ammonitic-like ones. We have tried to determine whether this situation also occurred in ammonoids by calculating the ratio of the depth of the lateral lobe L and half the straight distance across the septum between L and its opposite (i.e., its symmetrical L in compressed sections or a dorsal element in depressed sections). This ratio is therefore a measure of the concavity of the local transverse section of the septum through L. When it is plotted for a set of ammonoids (Fig. 6), it can be seen how the three main types of sutures (goniatitic, ceratitic, and ammonitic) display similar ranges of values. Therefore, the transition from one type to the other(s) probably did not result from single variations in pressure differential across the rear body, as in our experiments (see above). Differences are more likely to be of a structural nature (Section 4). Within each group there is a certain homogeneity. Paleozoic ammonoids display a very narrow range of values. Triassic forms display a wider range, the higher values corresponding to sutures with bottleneck saddles, which could have suffered additional

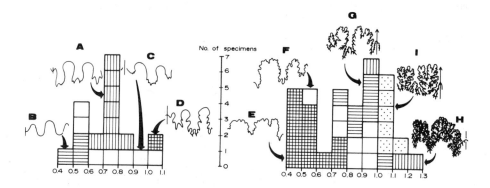

FIGURE 6. Frequency bar graphs of the ratio between lateral lobe depth (distance between lobe tip and the line joining tops of adjacent saddles, measured along the main axis of the lobe) and half its distance to the septum center (1/2 b in Fig. 7 or 8). This ratio is a measure of the indentation of the L lobe relative to the whole septum. The different classes of Ceratitina and Ammonitina are differentiated according to complication of the septal suture (see sutures for comparison). A, Goniatitina and Prolecanitina; B–D, Ceratitina; E–H Ammonitina; I, Lytoceratina and Phylloceratina. The different suborders cannot be differentiated according to their ranges of values. Ceratitina with the highest values (C) display bottleneck saddles. There is an overall relationship in the Ammonitina between lobe depth and complication. Measurements were taken from adult and subadult specimens in Haniel (1915), Miller *et al.* (1957), and at the Departamento de Estratigrafía y Paleontología of the University of Granada.

elastic expansion in a rubber-balloon-like manner (Seilacher, 1975; Section 4.1; Fig. 11D). Within Jurassic and Cretaceous ammonoids, there is a certain correlation between lateral lobe indentation and the complexity of the suture line. The process of increasing indentation with plate separation observed in experiments holds, nevertheless, for the ontogenetic complexity of the suture, judging by the findings of Westermann (1971, p. 21) that the minimum radii of curvature of the septal margin remain more or less similar during growth in *Sonninia* and *Leopoldia*.

2.3.3. The External (Ventral) Lobe and Septal Necks

The external lobe is present in all ammonoids except for some Paleozoic forms. It can easily be reproduced with the two-plate experiment by gluing a rod to the internal face of the upper plate and repeating the experiment (Fig. 5E). In all cases two lobes grow symmetrically, one on each side of the rod. This is caused by a "screening" effect, which in Hele–Shaw cells leads to the development of fingers at the contact with the cell's outer boundaries (see García-Ruiz *et al.*, 1990, Fig. 5). Lobes developed in this way never display a median saddle, which is associated with the development of prochoanitic septal necks.

Septal necks in some Ammonitina are prochoanitic beginning with the second septum, whereas in other ammonoids the transition from retro- to prochoanitic septal necks begins a varied number of whorls after the protoconch, depending on the group (Doguzhayeva and Mutvei, 1986; Landman, 1988; Tanabe *et al.*, 1993). Westermann (1975, p. 250) assumed that this transition reflected the inversion of pressures across the interface between the septum and the cameral fluid from apically to adorally positive. There is, however, evidence that septal suture differentiation into several-order flutes as a result of compression of the mantle from the rear began at least from the second septum on (Section 2.5). Additionally, the beginning of second-order frilling never coincided with proversion of the septal necks.

Retrochoanitic necks are characteristic of *Nautilus*, in which the horny tube of the siphuncle forms synchronously with the nacreous septum, the septal neck being continuous with the organic tube of the siphuncle (Mutvei, 1975; Bandel, 1981; Ward, 1987). Ammonoids, in contrast, displayed a different kind of siphuncle formation. Drushchits and Doguzhayeva (1974), Kulicki (1979), and Westermann (1982) presented convincing examples of sections of siphonal tube within the living chambers of ammonoids, indicating that these formed before body translocation began. This is confirmed by the mineralogic relationships between septal necks and connecting rings (Bandel, 1981; Kulicki and Mutvei, 1982; Henderson, 1984; Doguzhayeva and Mutvei, 1986; Tanabe *et al.*, 1993). According to Henderson (1984, p. 482), prochoanitic septal necks partly reproduce the invaginated shape of the mantle to form a sleeve, which is responsible for the precocious secretion of the connecting rings.

Ward and Westermann (1976) also showed that suture "inversion" in a Cretaceous ammonite (Section 4.2) did not affect the median saddle of the external lobe normally surrounding the siphuncle. This is understandable because this saddle probably formed part of the ventral sleeve, with different mechanical properties from the rest of the rear body.

2.4. Size, Number, and Position of the Elements of the Septal Margin

A suitable approach to the question of whether ammonoid shells somehow acted as Hele–Shaw cells to develop viscous fingers is to identify in ammonoids the parameters in equation 1 and, especially, equation 2. The two glass plates forming the Hele–Shaw cell are equivalent to the dorsal and ventral shell walls that define the ammonoid whorl section (see, e.g., Fig. 5D). Cell thickness b is equivalent to the distance between opposite elements of the septal suture, i.e., those linked by minimal distances across the septum. In planulate ammonoids, with laterally compressed whorl sections, b has to be measured between symmetrical elements except for those elements within the zone of overlap with the preceding whorl in which b is defined as the

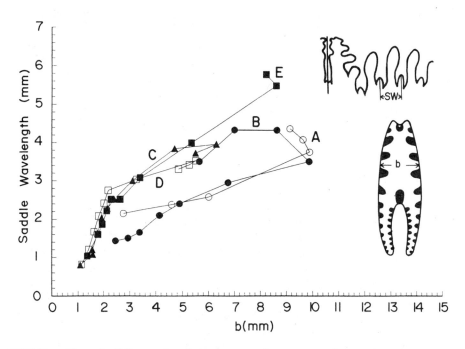

FIGURE 7. Plots of saddle wavelength (*SW*) against the transverse distance between opposite saddles (*b*) for the external suture in several Paleozoic Prolecanitida. (A) *Propinacoceras transitorium* (Haniel, 1915, Pl. 47, Fig. 5). (B) *Medlicottia orbygniana* (Handel, 1915, Pl. 47, Fig. 8). (C) *Artioceras*. (D) *Aktubinskia*. (E) *Artinskia* (C–E, Seilacher, 1975, Fig. 5).

distance between opposite elements of the ventral and dorsal walls (see Figs. 7, 10A). In ammonoids with depressed whorls, b is always the distance between opposite elements of the ventral and dorsal walls (Figs. 8, 10B). Within a given whorl cross section, b is, in general, variable, usually being low toward the umbilical seam and high toward the equatorial plane of the ammonoid spiral. A plot of saddle wavelength against the corresponding b in several Paleozoic ammonoids (Fig. 7) allows us to conclude that the relationship between the size of ammonoid saddles and b is direct, as predicted from equation 2. This accounts for the long-recognized fact that the relative sizes of sutural elements closely correspond to the distances to be spanned on both sides of the cross section (Pfaff, 1911; Spath, 1919; Arkell, 1957; Westermann, 1954, 1956, 1958, 1971; Seilacher, 1975, 1988). According to Westermann (1971, 1975), there is also a constant relationship between length (or depth) and radial width of sutural elements for a given suture; that is, sutural elements behave as fractal (self-similar) objects.

In Hele–Shaw cells, the number of lobes is proportional to the width of the cell (W) (Fig. 4), because the greater the value W the more fingers of a given λ_m that may be inset within the cell. A similar correlation is found in goniatitids, in which the ventral and dorsal walls of the whorl cross section are parallel, thus resembling curved Hele–Shaw cells. In all of the goniatitids measured, the ratio between cross-sectional width (measured along the sec-

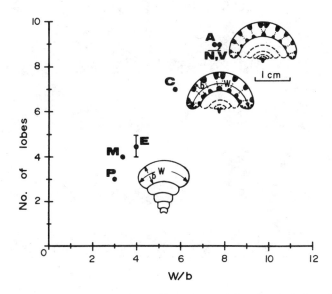

FIGURE 8. The relationship between the number of outer lobes (not including the umbilical area) and the ratio of W (width of the whorl cross section along the septum center) to b (mean height perpendicular to W) is linear in goniatitids with similar sutural patterns. A, *Adrianites*; C, *Crimites*; E, *Eoasianites*; M, *Metalegoceras*; N, *Neocrimites*; P, *Paralegoceras texanum*; V, *Vidrioceras*. A, E, M, P, and V from figures in Miller *et al.* (1957); C and N from Seilacher (1975, Fig. 5).

tion center, parallel to the ventral and dorsal walls) and its mean height (measured perpendicular to both walls) is proportional to the number of the major outer lobes (Fig. 8).

Nevertheless, the development of new lobes in ammonoids is more complicated and requires further explanation. As a general rule, the number of lobes increases with the addition of "narrow" spaces or corners to the whorl, such as ventral depressions and umbilical extensions, for these are usually zones of small b. Thus, in general, the number of lobes in ammonoids is not a function of absolute but of relative length of the cross-sectional perimeter (compared, for example, to the enclosed area or the mean value of b). Circular whorl sections display the lowest perimeter-to-area ratio, which accounts for the four-rayed symmetrical arrangement of the elements of the septal suture in the uncoiled straight *Bochianites* (Fig. 9) and *Spiroceras* (see Westermann, 1956, 1958). Six-lobed sutures are displayed by orthocones with subcircular to oval cross sections, such as *Baculites* (Fig. 9). Seilacher (1988, Fig. 1) also attributed to the six-lobed *Kabylites* a circular whorl cross section; the speci-

FIGURE 9. Relationship between shell shape and the number and size of septal suture elements. The straight heteromorph *Bochianites*, with a completely circular whorl cross section, displays only four lobes of similar size. *Kabylites* and *Baculites*, with oval sections, have hexalobed sutures. In normally coiled shells (*Lytoceras*, *Cadoceras*), septal elements vary in size and increase in number as the cross section departs from equidimensionality. (Slightly modified from Seilacher, 1988.)

men illustrated by this author was originally described by Durand-Delga (1954), who did not show the cross section, although it is oval rather than circular (height/width = 0.9, according to Durand-Delga's original description; see tentative reconstruction in Fig. 9). Most Cretaceous heteromorphs display six lobes, even some with near-circular cross sections, which might suggest some genetic fixing of the number of lobes (G. E. G. Westermann, personal communication).

In coiled ammonites, overlapping between the successive whorls leads to the development of narrow, acute-angled extensions of the whorl cross section toward the umbilicus. Many small lobes typically concentrate here because these are zones of small b. The rule of relative enclosed area also holds in this case. For instance, fast-expanding, involute shells bear cross-sections which are usually also highly overlapping on the preceding whorls (Fig. 10A). As is well known, shells of this kind bear many more lobes than do serpenticones. In fact, Checa (1986) found a direct relationship between the expansion rate of the whorl cross section and the number of umbilical lobes in phylogenetically related species. A similar trend is apparent in Tanabe (1977, see in particular Fig. 13 and plates). Hewitt and Westermann (1987) also demonstrated a significant relationship between the number of umbilical flutes and the degree of overlap (whorl-overlap-shape parameter of these authors) in several groups of Jurassic Ammonitina. Oxyconic shells, in which the whorl cross section typically bears large overlap extensions onto the preceding whorl, display the highest perimeter-to-area ratio and, therefore, the largest number of lobes among ammonites. On the opposite side, the least lobed forms are certain Lytocerataceae, with slightly overlapping whorls and rounded whorl sections.

The same criterion holds for the usual ontogenetic increase in the number of lobes in ammonoids. If an ammonoid section grew isometrically, the number of lobes would remain the same, because any value of b (and hence of λ_m) that could be measured in the whorl cross section would increase in the same proportion. Isometric growth is not usually the case in ammonoids, because development started with nearly nonoverlapping ammonitella-type sections; subsequent overlapping of the whorl cross section onto the preceding whorls leads to the ontogenetic development of new lobes (Fig. 10). The position in which new lobes appear seems to be determined by the pattern of expansion of the whorl cross section (Westermann, 1956; Checa, 1991). This could be the case with the development of adventitious (A) lobes of the Paleozoic Goniatitida as compared to umbilical (U) lobes characteristic of Prolecanitida and Mesozoic ammonites (see Wiedmann and Kullmann, 1980). In Paleozoic goniatites, the appearance of new lobes can be related to the ontogenetic modification of the cross-sectional shape. In these, the ventral wall of the section expands suddenly compared to the dorsal wall (Fig. 10B), resulting in an involute, overlapping shell, but this process differs markedly from that of ammonoids with U lobes (Fig. 10A).

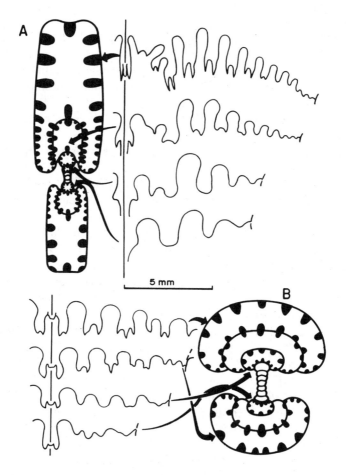

FIGURE 10. (A) Paleozoic ammonoids with U-type lobe development (*Artioceras riphaeum*) lead to many small lobes at the umbilical area. (B) In ammonoids with A lobes (*Prostacheoceras juresanense*), relative enlargement of the whorl cross section is achieved in the ventrolateral area. Both types of shell development lead to involute morphologies with very different shapes. (Synthetic figure after Ruzhentshev, 1963; Seilacher, 1975. Scale valid for sections.)

The same processes account for the secondary fluting of the external saddle in the Medlicottiidae (Prolecanitina), which develops simultaneously with a ventral depression (Seilacher, 1988; Fig. 10A). This creates two well-defined ventrolateral shoulders (symmetrically disposed about the midline), which are thus zones of small *b*, and on both sides of which sutural elements may span.

2.5. Early Postembryonic Sutures

The contrast between proseptal suture and second suture (some authors' primary suture) has always been a controversial subject. Proseptal morphology can be explained easily according to the model of embryonic development of ammonoids proposed by Bandel (1982, 1986) and Landman and Bandel (1985). When the visceral mass withdrew from the protoconch, it still was attached to it by retractor muscles (to the inner side of the flange) and siphuncular tissue (to the walls of the protoconch). The visceral mass formed an organic sheet with a central opening (later to accommodate the cecum and siphuncle) that was firmly attached to the inner walls of the protoconch. Attachment of the organic precursor of the proseptum to the ammonitella occurred along a prismatic ridge that is especially well developed along the bases of the lateral lobes. The remainder of the organic precursor of the proseptum was then free to bulge adorally to form a median saddle, although it is not possible to determine whether this was done by purely mechanical pull-off or by positive hydrostatic pressure from inside the protoconch. This sheet later mineralized to form the prismatic proseptum. After formation of the proseptum, the retractor muscles reattached in two bundles, adoral of the proseptum. Subsequently, hatching occurred, and the cecum and prosiphon formed.

The second septum (primary septum of other authors) is the first nacreous septum; it can be distinguished morphologically from the proseptum by the occurrence of a lobe formed in the place of the proseptal saddle. At the same time, sutural morphology became completely reorganized. This is explained as a consequence of "screening," resulting from the existence of a siphuncular tube that was extruded while the soft body translocated (Section 2.3.3). The siphuncle initially occupied much of the chamber height, so that it also may have directly influenced the formation of a dorsal lobe, adoral of which the retractor muscle was inserted. The formation of an external lobe in the living animal meant that, on either side of the external lobe, an external saddle had to have formed, the size of which depended on the transverse diameter of the whorl cross section of the ammonitella at that point (b in equation 2). The next elements were the lateral lobes, also with predetermined λ_m, and so on. The good fit of the morphology of the second septum to the viscous-fingering model, together with its adoral convexity (Landman and Bandel, 1985, p. 15), suggests that adorally directed pressures very probably began to shape the rear body at this point in ontogeny and continued to do so throughout growth.

Early postembryonic sutures contained within the ammonitella did not usually frill. In ammonites, frilling often began when the cross section of the tube suddenly expanded after the ammonitella stage or after the tightly coiled neanic whorls (Kulicki, 1974; Hewitt, 1985; Checa, 1991, p. 107), and, consequently, individual elements of the septal suture began to exceed the critical value of λ. Nevertheless, the smooth elements of initial sutures maintained a rank higher than that of subsequent lobules and folioles throughout the ontogeny and constituted the main lobes and saddles of septal sutures contained within the ammonitella tube. This is consistent with viscous-fingering processes, in which initial undulations do not recede or disappear (provided that the pressure system is maintained) but rather amplify and frill as the two plates are separated (Section 2.3), according to equation 2.

2.6. Conclusions

In summary, the main morphological features of the septum and its suture can easily be explained within the framework provided by the viscous-fingering model. Saffman–Taylor instabilities generated in Hele–Shaw cells are stochastic processes; that is, the resulting fluted morphology is never exactly the same when the experiment is repeated, although the topological properties remain. Nevertheless, in ammonoids the conch constrained the randomness of the pattern. As we have seen, the position of the siphuncle and its mode of secretion determined the formation of an external lobe and adjacent undulations, whereas the cross-sectional shape and its expansion rate determined the size, number, and position of the septal folds (Fig. 17). These constrained "natural" viscous-finger experiments, repeated under the same initial conditions (i.e., individuals of the same species), would have led to indistinguishable patterns. This accounts for the similarity of septal sutures within taxonomically related ammonoids, most of the variability probably being linked to differences in conch morphology.

The viscous-fingering model also requires that the less viscous cameral liquid somehow pushed the soft rear body (which had to have viscoelastic properties). Additional evidence concerning the pressure system within the ammonoid chambers and the mechanical properties of the rear body is explored below.

3. Pseudosutures, Pseudosepta, and Cameral Membranes

In phragmocone steinkerns of the Triassic ceratitids *Koninckites* and *Clypeoceras*, John (1909) noted some enigmatic markings that partly reproduced the pattern of the septal sutures. To these he gave the name *"Pseudolobenlinie."* These pseudosutures apparently were caused by cameral membranes closing some lobes of the rear septum of the camera (Bayer, 1977a, Fig. 15). Later, Vogel (1959), Jordan (1968), and Schindewolf (1965) described spiral imprints (so-called drag bands), which they interpreted as striae formed during movement of the body inside the chamber. These are different from the siphuncle-associated structures of Hölder (1954) and Henderson (1984), which were, rather, residual siphuncular membranes. Zaborski (1986) also recognized well-preserved pseudosutures in Nigerian vascoceratids, interestingly associated with drag bands. Because the latter were apparently the continuation onto the chamber walls of the lobes and lobules, this author interpreted them as traces left by tie-point attachments (see also Seilacher, 1988). A closer inspection reveals that vascoceratid drag bands generally coincide better with lobule flanks than they do with lobule tips. They usually are paired for a single lobule, and, when better preserved, are merely fused telescoped pseudolobules (Hewitt *et al.*, 1991). Very recently, Lominadze *et al.* (1993) and Landman *et al.* (1993) reported interesting sets of pseudosutures from Devonian to Aptian ammonoids.

The term "cameral membranes" is applied to curtain-like phosphatic (originally organic) structures found within the camerae, with a complex arrangement and mutual relationships. In addition to those reported by John (1909), cameral membranes have been described and illustrated by Westermann (1971), Erben and Reid (1971), Bayer (1975), Weitschat (1986), and, especially, by Weitschat and Bandel (1991). This last paper deals exclusively with Triassic ammonoids but contains the most complete and beautifully preserved inventory of phosphatic membranes ever seen. Cameral membranes were differentiated by these authors into (1) siphuncular sheets that diverge from the sides of the siphuncle towards the ventral shell wall in a longitudinal arrangement, (2) transverse sheets that close off septal flutes, and (3) horizontal sheets that extend between the centers of consecutive septa in contact with the siphuncle, thus dividing the chamber into dorsal and ventral halves.

Weitschat and Bandel (1991) claimed that all of these sheets were produced by direct replication of the rear mantle through a series of complex movements and deformations during the process of chamber formation. This mode of formation, particularly for the first two types of cameral membranes, was opposed by Westermann (1992), who stated that they were products of desiccation of a cameral hydrogel, shaped by surface-tension processes. The hypothesis that ammonoids secreted a cameral hydrogel instead of saline water was put forward by Hewitt et al. (1991). Whether or not the desiccation hypothesis is accepted, extreme deformations of the rear body in order to secrete the various sheets cannot be reconciled with the coexistence of pseudosutures (which we have observed extensively in Triassic ammonoids) and their three-dimensional equivalents, pseudosepta, which suggest an almost constant, slightly deflated shape of the rear epithelium during the translocation process. One of the authors (A.G.C.) recently restudied the original Triassic material in Hamburg and advanced the following conclusions: (1) Some pseudosutures might be independent of any kind of membranes intruding the camerae, but others might be the marginal imprint of some "siphuncular sheets connected to additional vertically directed membranes" (see Weitschat and Bandel, 1991, Figs. 8–10), which are interpreted here as remains of true pseudosepta. (2) All other kinds of membranes reported from Boreal Triassic ammonoids might well be desiccation products of a cameral hydrogel or mucus-enriched water, as proposed by Westermann (1992). He explained horizontal sheets as chamber linings separate from the walls, although other alternatives can be proposed.

The membranes forming the pseudosepta and some other precursors of the pseudosutures (probably peripheral concentrations of mucus or slimy material) thus were replicated by the posterior mantle during its forward movement. Once the septum was completed, drainage of the cameral liquid gave rise to reservoirs that remained inert long enough to produce mucus-enriched surfaces; these later dried and hardened, producing cameral membranes. The two kinds of structures, pseudosepta and the remaining cameral membranes, must then be separated from the genetic point of view. As will be seen below, information provided by pseudosutures is essential for understanding the processes that took place during soft-body translocation.

4. Differentiation and Mechanical Properties of the Septal Mantle

4.1. Interpretation of Pseudosutures

Pseudosuture successions, which are no doubt imprints of the rear mantle margin, provide essential information about mantle differentiation. Pseudosuture density generally increases toward the true septum, thus reflecting the decreasing rate of translocation, provided that they were secreted periodically. As a whole, pseudosutures, in addition to other structures, reveal that the rear mantle was not intruded on by the cameral fluid during most of the translocation process and that pressurization and differentiation into several-order flutes took place shortly before or after definitive emplacement of the rear body. Evidence of this includes the following:

1. Drag bands in vascoceratids (Zaborski, 1986, Text-fig. 2; Hewitt *et al.*, 1991) and in *Acanthohoplites nolani* (Fig. 11G) diverge adorally within the saddles (this is particularly true for the external saddles), which suggests expansion of the septal mantle during about the last third of chamber length.

2. The pattern of pseudosutures in specimens of the Lytoceratida and Phylloceratida suggest more drastic changes in shape of the septal mantle than in vascoceratids (Fig. 11F,H,I), because the saddles of the basal pseudosuture consist of a series of dendrites, implying that the septal mantle changed from an almost completely invaginated shape to the definitive septal suture. This change in shape apparently occurred after translocation had finished. In viscous-fingering experiments performed with high-viscosity fluids, the interface does not revert to a flat shape when the pressure is released but forms a complex pattern dominated by dendrites of the kind noticed in phyllo- and lytoceratids (García-Ruiz, 1992). On the other hand, Lominadze *et al.* (1993) considered these dendritic imprints different from pseudosutures, interpreting them as lines of longitudinal contact between the rear body and the inner surface of the outer shell.

3. The apical margin of the septal recess leading to the dorsal lobe of *Pseudophyllites* (see Henderson, 1984) is fluted in a fashion similar to the true septal margin. The same applies to the inner septum apically closing the dorsal lobe. Throughout translocation, the septal recess, including its complexly fluted margin, had to contract and slide along the walls of the tunnel formed by the septal recess of the previous septum, so that its increase in complexity would have been possible only after the translocation stage. Saddle-lobe polarities, both of the apical septal recess and of the inner septum, are consistent with the existence of pressures from within the chamber.

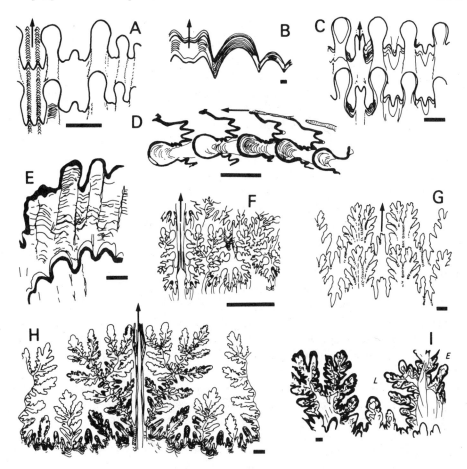

FIGURE 11. Pseudosutures in ammonoids. Suture complexity grows from the pseudosutures to the definitive septal suture with the intricacy of the latter. (A) *Neopronorites permicus*, Lower Permian. (B) *Timanites acutus*, Upper Devonian. (C) *Kargalites typicus*, Lower Permian. (D) *Amphipopanoceras* sp. cf. *A. medium*, Anisian, Central Spitzbergen, DSPUG.AC.Hb1., Departamento de Estratigrafía y Paleontología, University of Granada, kindly provided by Dr. W. Weitschat (Geologische-Paläontologisches Institut, Hamburg University); pseudosutures reveal a rubber-like behavior (elastic inflation) of the external saddle. (E) *Vascoceras* sp., part of right external saddle, Lower Turonian. (F) *Phyllopachyceras baborense*, Upper Aptian. (G) *Acanthohoplites nolani*, Upper Aptian. (H) *Hemitetragonites* sp., Upper Aptian. (I) *Lytoceras jurense*, Upper Toarcian, external (*E*) and lateral (*L*) lobes indicated. (A, B, C, F, G, and H reproduced from Lominadze *et al.*, 1993, with the permission of *Geobios*. E and I from Hewitt *et al.*, 1991, with the permission of the author and The Lethaia Foundation.) Scale bars, 2 mm. Arrows indicate growth direction and position of the siphuncle (except for E and I).

As stated above, there are marked differences between members of the Lytoceratina/Phylloceratina and other forms with simple sutures (such as vascoceratids, *Acanthohoplites*, and some Paleozoic ammonoids) regarding the degree of deformation of the rear mantle in the last stages of the translocation process (Fig. 11). In vascoceratids there is evidence of a 1.25-fold increase in length from the pseudosutures to the septal sutures, whereas *Lytoceras* increased this length some 1.5- to 2.5-fold (Hewitt *et al.*, 1991). In the case of individuals of the Phylloceratina and Lytoceratina, the main saddles of the pseudosutures were invaginated completely; this suggests that, after detachment from the last-formed septum, the walls of the soft body saddles converged and stuck together, producing a linear dendritic pattern (see 2 above). At the same time, the soft lobes expanded to "fill" the extra space left by the saddles. This process was reversed when the position of the subsequent septum was reached: saddle walls detached and began to differentiate into many low-order flutes, whereas the lobes contracted transversely (judging by oblique drag bands merging into lobules; Fig. 11F). A somewhat less drastic figure is implied for *Acanthohoplites* (Fig. 11G). The behavior revealed by this deformation pattern is clearly nonelastic. We therefore no longer refer to an "inflation" process, because this implies rubber-like behavior (see Seilacher, 1975). The same applies to vascoceratids, though to a lesser extent. The only case in which pseudosutures reveal a characteristic elastic behavior is that of the smooth "bottleneck" saddles of the Triassic ammonoid *Amphipopanoceras* (Fig. 11D; see also Seilacher, 1975, Fig. 3).

The loose but clear correlation between the complexity of the definitive septal suture and the deformation of the rear mantle (as revealed by pseudosutures), mentioned above, was also noticed by Lominadze *et al.* (1993). In our opinion, this correlation suggests that a major factor controlling the degree of complexity and, thus, the type of septal suture was the extent to which the mantle was able to deform in a nonlinear fashion under cameral fluid pressure, provided that other parameters were held constant. In this way, highly viscoelastic rear mantles led to the characteristically complex sutures of the Phylloceratina and Lytoceratina, whereas progressively stiffer rear mantles produced simple ammonitic, ceratitic/pseudoceratitic, or goniatitic septal sutures. This fits with the findings of Bayer (1977a) that the rear mantle of some Lower Jurassic forms with rather simple ammonitic sutures (see his Figs. 11–13) were stiff and produced small folds and wrinkles when the organic "blueprint" of the septum crossed prominent ribs.

Additional evidence for differences in mantle viscoelasticity between the various groups of ammonoids is the correlation between the number of lobes in the primary suture and the type of subsequent sutural frilling. Paleozoic ammonoids with goniatitic sutures display three-lobed primary sutures. Most Triassic "mesoammonoids" acquired quadrilobate primary sutures, whereas from the Jurassic onwards ammonites achieved quinquelobate and even hexalobate (Upper Cretaceous tetragonitids) primary sutures (see Wiedmann and Kullmann, 1980).

In summary, the analysis of pseudosutures offers two conclusions: (1) the rear mantle (at least for ammonites) behaved as a viscoelastic material and (2) was pressed by or against (see below) the cameral liquid in the last stages of or after translocation. Note that both conclusions closely match the predictions made by the viscous-fingering model.

4.2. Interpretation of Rare Pseudoinverted Sutures

Several other data allow us to ascertain that differentiation of the rear mantle into flutes of various orders was the result of positive pressures created within the cameral liquid. The case of sutural "inversion" (or, better, pseudoinversion) reported by Ward and Westermann (1976) in a specimen of *Glyptoxoceras subcompressum* provides good evidence of this (Fig. 12A). This rare case was explained by the authors as being caused by a reversion of the pressure system such that the cameral fluid failed to push (or support) the rear body. They also stated that pressure reversion would have changed the concavity of the flutes subtended between consecutive tie points; i.e., true inversion of the lobulations took place. Bayer (1978a) rejected this explanation on the basis of a comparison with a pseudoinverted external saddle in *Bifericeras* (his Fig. 8); he attributed the inversion to a radial compression of the septal mantle associated with a reduction of the septal distance. However, there is no evidence of mechanical compression of the rear mantle in *Glyptoxoceras*. Ward and Westermann's (1976) original hypothesis of pressure reversion is more likely. We think, nevertheless, that true reversion of flute polarities did not occur: adapically directed pressure would inflate posteriorly bulging lobulations and deflate anteriorly bulging ones. Even though with a modified meaning, we adopt here Bayer's term pseudoinversion. Following this process, reconstruction of pseudoinverted sutures leads to figures that are closer to normal sutures than those produced if we assume flute-polarity reversion (Fig. 12B). Therefore, the existence of tie points, assumed by Ward and Westermann (1976), does not need to be invoked to explain the case of *Glyptoxoceras*.

A simple quantification demonstrates that there is a change in complexity of the septal suture associated with pressure-gradient reversion. For a measure of complexity, we counted the number of curvature changes or minor lobulations along each septal suture ("tie points" of Ward and Westermann, which are considered here more a morphological than a physiological feature; see also Hewitt *et al.*, 1991, p. 286). When this is plotted against the septum number (Fig. 12C), it can be seen that the first pseudoinverted suture (no. 3) is less complex than the preceding one and that there is no appreciable increase in complexity during the pseudoinverted stage. The next subsequent normal suture (no. 8) is disproportionately complex as compared to the preceding ones. Rear mantle pressurization was therefore responsible for septal suture fluting.

In laboratory experiments with viscous fingers, pressure reversion (i.e., squeezing of the plates back together) leads to the disappearance of the

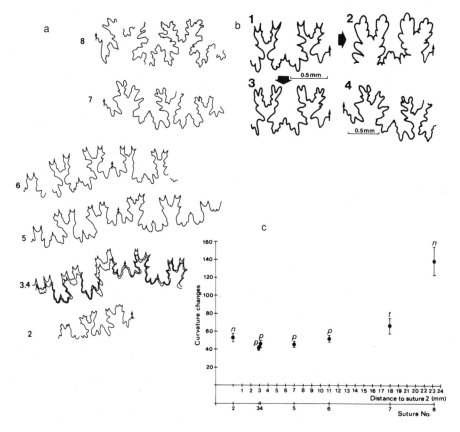

FIGURE 12. (a) Pseudoinverted sutures (nos. 2 to 6) in a specimen of *Glyptoxoceras subcompressum* (Forbes) in Ward and Westermann (1976, Fig. 2); sutures reproduced with the permission of The Lethaia Foundation. (b) Reconstruction of pseudoinverted septal sutures; 1, part of pseudoinverted suture no. 6 in A; 2, Fig. 3B of Ward and Westermann (1976), in which the same suture is reverted back to normal by assuming attachment at definite tie points, same scale as in 1; 3, our reconstruction, in which reversal of pressure differential across the posterior mantle is postulated to have inflated the adapically directed flutes and deflated the adorally directed ones, same scale as in 1; 4, part of normal suture no. 7 in A; note that there is a greater similarity between 3 and 4 than between 2 and 4. (c) Plot of the number of curvature reversals along the suture against suture number (as numbered in Ward and Westermann, 1976) and distance to suture no. 2, measured at the venter, for the same specimen. Because of the uncertainty in recognizing some points of reversal, a range is provided for each value: *n*, normal suture; *p*, pseudoinverted suture; *t*, transitional suture. according to Ward and Westermann (1976).

viscous-fingering pattern, which may revert to either a flat or an invaginated surface. This was not the case in *Glyptoxoceras*, in that its rear mantle was a more or less permanently shaped living tissue; its differentiation proceeded throughout ontogeny and did not restart each time a new septum was to be produced.

4.3. The Pressure System within the Chambers

As mentioned above, pseudosutures reveal that pressure within the form-ing chamber increased late in the translocation stage. Two main possibilities can be explored. (1) Cameral liquid, after remaining at ambient (hydrostatic) pressure during most of the translocation, suddenly rose above this value when the rate of forward movement was decreased by e.g., making the cameral liquid hyperosmotic. (2) Rear body movement slightly anticipated chamber filling by muscular pull, thus creating a certain vacuum in the chamber so as to maintain liquid pressure below ambient value; then, when the motion rate decreased, liquid pressure rose to ambient value (transmitted either via the soft body or the siphuncle). In both cases the viscoelastic tissue forming the rear mantle (see Section 5) was pressed by the cameral liquid, conditions thus being appropriate for the formation of viscous fingers where the interface between the rear body and the cameral liquid made contact with the shell's inner surface.

There is no definitive evidence supporting either possibility, although some suggestions can be made, again from the study of pseudosutures. Lytoceratid and phylloceratid basal pseudosutures consist of a series of apically directed lobulations interrupted by dendrites (which corresponded to invaginated saddles; Section 4.1). Therefore, these pseudosutures are morphological equivalents of the pseudoinverted sutures of *Glyptoxoceras* (Section 4.2), i.e., contracted saddles and saddle-like minor folds and ex-panded lobes and lobe-like folds. The same process of pseudoinversion took place in vascoceratids (see Zaborski, 1986, Text-fig. 2) and in some Permian ammonoids (Fig. 11A,C), although to a much lesser extent. The homology of basal pseudosutures and pseudoinverted sutures suggests that the resultant pressure across the interface between the septal mantle and the cameral liquid was most probably reversed compared to the pressurization stage, such that the cameral liquid pressure was probably lower than hydrostatic pressure during the formation of the basal and preceding pseudosutures of a given chamber. This supports hypothesis 2 above. In this case, a body-supporting function of the cameral liquid (see Lominadze *et al.*, 1993) could have been relevant only late during translocation.

Bayer's (1978a) *Bifericeras* pseudoinversion was associated with a dislo-cation of the siphuncle. If pressure reversion also applied in this case, the siphuncle might be seen as responsible for the pressurization of the septal mantle in the final stages of translocation. Rieber (1979) reported two ex-tremely close final sutures in a specimen of *Brasilia* in which the last suture line was greatly simplified. Simplification of mature interpenetrating sutures is not uncommon according to Westermann (1971, p. 18). In these close septa, the penultimate septal neck might have partly plugged the rear of the sleeve from which the siphuncular cord extruded. If it was the siphuncle that released liquid into the chamber, this partial plugging might have prevented pressurization of the chamber, thus leading to sutural simplification. This hypothesis is weakened by the fact that the siphuncle, as a chamber-filling organ, would have been inefficient, given the high rates of mantle transloca- tion attributed to ammonites (G. E. G. Westermann and R. A. Hewitt, personal communication; Section 6). According to Ward (1987), it has not been possible to decide whether the rear mantle or the siphuncle is the chamber-filling organ in *Nautilus*.

5. Conclusions Concerning the Structure of the Septal Mantle

The septal mantle in *Nautilus* is a thin, translucent epithelium closing the rear body and attached peripherally by the posterior aponeurosis (Ward, 1976). Westermann (1975) proposed that the septal mantle in ammonoids was similarly attached to the shell wall along an aponeurosis-like structure that was better developed and stiffer than that in *Nautilus*. Attachment of the mantle and the septal membrane "was probably still only in the tie points" (Westermann, 1975, p. 49). Seilacher (1975) and Bayer (1977a, 1978a) also compared the rear mantle to a membrane but implied differences as to its degree of attachment (tie points or closed line, respectively) and elasticity. Models based on a membrane-like rear mantle have been well accepted up to the present (Bandel, 1981; Seilacher, 1988; Hewitt *et al.*, 1991), but note the exception of Henderson (1984). In contrast, the viscous-fingering model implies that the septal mantle in ammonoids must have consisted of more

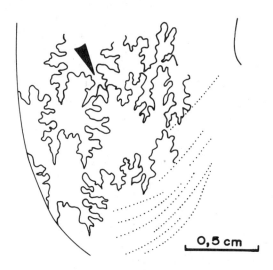

FIGURE 13. Folded lateral branch of the external lobe in *Hoploscaphites nicolleti* (Morton), in Landman and Waage (1986, Fig. 3), caused by a deviation of the last suture from the median plane; oblique ventral view.

than just a thin membrane because, for viscous fingers to develop, two volumes of "fluid" must be in contact.

There is paleontological evidence to suggest that lobes were "filled" with some kind of tissue. Bayer (1977a, Figs. 11–13,15, 1977b, Fig. 16) showed bulges and wrinkles in ammonoid septa. Particularly informative are apically bulging folds on the surface of small septal flutes that bent when crossing prominent transverse ribs on the outer shell. These folds can be explained as a response to bending of the original mantle extensions if they were filled with some tissue and, thus, retained a fixed internal volume. The mode in which sutural elements adapted to ribs was also studied by Arkell (1957), Seilacher (1975, 1988), and Checa (1986), who each gave a different explanation. In all the cases examined by these authors, saddles and lobes displayed different behaviors, the former invariably being much more stretchable. Additionally, during the life of the animal, lobes and lobules on occasion became deviated laterally (as in some interpenetrating sutures, Fig. 13) and did not appreciably deform. In general, all these imperfections could hardly have been formed in a single membrane; they fit better with the fact that lobes were extensions of some rear tissue, whereas saddles were empty spaces between the lobes. Our explanation differs from Bayer's (1978a) suggestion that the lobes must have been filled with visceral mass. Viscerae could neither fit closely into the many small lobes and other lobe-like minor folds nor behave mechanically as a single tissue.

Very thinly striated surfaces on the internal shell wall were reported and illustrated in vascoceratids by Hewitt et al. (1991, Fig. 4). Striations are found some distance adoral of the septal sutures; they run more or less parallel to the direction of growth and, when intruding into the lobes, radiate according to the main elongation of lobules. Because they cross the drag bands associated with some umbilical lobes at acute angles, they seem to be unrelated to movement traces. Similar structures were reported by Hölder (1955), Senior (1971), and Henderson (1984). This last author interpreted them as static lines of attachment in front of the septal suture. Striations are associated with annular color bands developed in front of the sutures in vascoceratids. These bands also represent the crests of low ridges, which roughly reproduce the outline of the septal suture, and were interpreted by Hewitt et al. (1991) as aponeurotic bands similar to those of *Nautilus*. It seems likely that striations and ridges represent unrelated structures, because the former generally extend adorally beyond the aponeurosis. Striations could represent the marginal imprint of some tissue filling the rear body, the rear surface of which replicated the septum.

Our conclusion agrees partly with that of Henderson (1984), who, following different reasoning, suggested that some connective tissue formed the rear mantle. In our interpretation little muscle would have filled the rear mantle except for the aponeurotic band. This tissue could have been to some extent comparable to actinian mesoglea, for example (see Alexander, 1962; Wainwright et al., 1976), in which oriented collagen fibers grow within a gelatinous matrix. The matrix is a long-chain polymer gel with viscoelastic properties; the collagen fibers act as a framework, reinforcing the weak matrix. In the case of ammonoids, these fibers were probably pressure sensitive and grew according to the pressure gradient field developed on the rear surface of the body, i.e., the septal "blueprint" (for a similar idea, see Damiani, 1990).

There are some Paleozoic ammonoids for which the hypothesis of a *Nautilus*-like rear mantle could fit. This is the case with the undifferentiated, apically convex unfrilled septa of some Anarcestina (Anarcestidae) and many Clymeniina, for which a *Nautilus*-like pressure system could be proposed. The Cheiloceratidae and other Anarcestidae displaying adorally convex unfrilled septa could also have borne membranous mantles but inflated by adorally directed ammonoid-like pressures. Finally, the apically pointed lobes and the lobe distribution of many Goniatitina apparently fit neither the viscous-fingering nor the single-membrane model.

6. Revised Chamber Formation Cycle

The margin of the rear body remained attached along the adoral ridge of the folioles (postseptal prismatic zone of the septum of Henderson, 1984) and, at a more adoral position, throughout the posterior aponeurosis. After detachment of the aponeurosis, the liquid began to enter the chamber, and adhesion of the rear mantle to the newly formed septum was lost. In the case of lytoceratids and phylloceratids, this process led to longitudinal contraction and simplification of body lobes, whose extraction from their recently calcified replicas was facilitated by the highly viscoelastic nature of the mantle

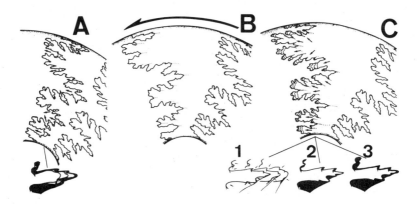

FIGURE 14. Reconstructed chamber formation cycle in ammonites, inferred from pseudosutures and drag bands. (A) Detachment of the soft body from the last-formed septum, with the subsequent relaxation and longitudinal contraction of the body lobes and minor lobulations. (B) Translocation of the soft body. (C) Compression and deformation of the soft body (1), secretion of the preseptal membrane (2), and secretion of the septum (3). Light gray, soft body; medium gray, preseptal membrane; dark gray, septum.

(Fig. 14A). In these forms, contraction continued until the mantle became completely invaginated, so that saddle and foliole walls stuck together and its rear surface formed a more or less flat surface with a wavy margin (see Lominadze *et al.*, 1993; Fig. 11F,H,I). In ammonites in which flutes join the shell wall at narrow angles, body lobules could have slid along the adoral (septal) surface of the opposing folioles during the process of extraction (Fig. 15). This is also likely in Jurassic ammonites with retracted septal sutures (Checa and Sandoval, 1989). In simple-sutured ammonites, elements tend to join the shell wall at wide angles, so that the body lobes must have slid within their aragonitic sheaths, thereafter reorienting following the more advanced position of the body (see Hewitt *et al.*, 1991, Fig. 3).

Extraction was then probably by both intrusion of the cameral liquid and relaxation of the rear body rather than by muscular contraction, as generally assumed (see Westermann, 1975: Fig. 14A). Upon complete relaxation, the dorsal retractor muscle (in front of the dorsal lobe) detached, and the entire posterior mantle crept forward, partly by muscular pull, in close equilibrium with chamber filling (Fig. 14B). At this stage, the pressure differential across the interface between the cameral liquid and the body was most probably below zero, and thus, cameral fluid did not support the rear mantle (Section 4.3). Therefore, the whole of the septal surface must have been more or less apically convex. During translocation, small discontinuities in motion led to peripheral mucus enrichments (which later formed pseudosutures in dehydrating) and, in some instances, to pseudoseptal membranes (Section 3).

The forward motion of the posterior mantle began to slow, while somehow the volume of liquid released into the chamber exceeded that vacated by the rear mantle. This led to compression of the rear body, which yielded and deformed in a viscoelastic manner (Fig. 14C, stage 1; Section 4). In the Lytoceratina, compression apparently took place when the forward motion of the body had ceased completely. Liquid then intruded into the invaginated

FIGURE 15. In two interpenetrating septa of *Aspidoceras longispinum* (Sowerby), some lobules (arrows) of the lateral lobe (L) lean tightly against the external saddle (E_s) of the preceding septum. Upper Kimmeridgian of Alta Coloma, U.AC$_{21}$.5a.107, Departamento de Estratigrafía y Paleontología, University of Granada. Scale bar, 1 cm. Arrow indicates growth direction.

saddles. When motion definitively ceased, the dorsal retractor and the rest of the muscles reattached. When deformation and differentiation into flutes of the posterior body also ceased, the body attached to the inner shell wall along the posterior aponeurosis. According to Bandel and Boletzky (1979) and Bandel (1981), a mineral ramp of prismatic aragonite needles (mural ridge of Blind, 1975; preseptal prismatic zone of Henderson, 1984) was produced at the same time or immediately thereafter, just behind the attachment of the rear body, exactly reproducing the shape of the new suture. This ramp attached one or more organic pellicles, which, further apically, merged with the organic siphuncular tube, forming its outermost layer. Alternatively, Blind (1975, 1980) attributed a function similar to that in *Nautilus* to this mural ridge (attachment of the subepithelial musculature).

Westermann (1975) suggested that, after fixation of the mantle, its margins attached a conchiolin membrane (periostracum) along the suture before mineral deposition. Bayer (1978a) also proposed formation of an organic membrane on which the nacreous septum developed. Bandel (1981) rejected these hypotheses and thought that the first deposits (spherulitic–prismatic) formed in a mucus excreted from the epithelium, "a $CaCO_3$-enriched extra-pallial fluid." Tiny compression and tension wrinkles are sometimes observed on the apical surface of some lobulations (Fig. 16); evidently, these wrinkles could have formed and been preserved only on a preexisting mineralizing membrane (Fig. 14C, stage 2).

Subsequently, the nacreous layer forming most of the septum was secreted, and generally a thin prismatic layer, thickened at the contact with the shell wall, completed the septum (Fig. 14C, stage 3). At this stage, the shape of the rear body was maintained by adhesion both to the septal surface and along the postseptal prismatic zone lining the folioles.

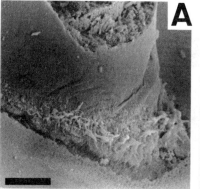

FIGURE 16. (A) Compression wrinkles on the apical septal surface of *Amauroceras* sp., Pliensbachian, Germany. Scale bar, 10 μm. (B) Tension wrinkles on the apical septal surface of *Cardioceras* sp., Callovian, Poland. Scale bar, 20 μm. Unregistered specimens, Departamento de Estratigrafía y Paleontología, University of Granada.

The rear body grew in a differentiated shape, i.e., by fluting in response to the pressure-gradient field (Section 4.1), from the time it began to expand in the last stages of the translocation until the time it detached from the completed septal surface. This could represent a substantial proportion of the time elapsed between detachment from the last septum and the onset of mineralization of the subsequent septum, given the pseudosuture concentration prior to the definitive septal suture (see also Lominadze *et al.*, 1993). When relaxed, the rear body still remained invaginated or deflated, but differentiated, retaining a memory of the "compressed" shape; this accounts for the replication of successive sutures in ammonoid phragmocones.

Cameral liquid emptying through the siphuncle began somewhat prior to or on completion of the septum, giving rise to variously shaped cameral membranes (Section 3).

On the basis that cameral sheets were replicated directly by the mantle, Weitschat and Bandel (1991) supposed a chamber-production period of 1 to 2 days, similar to that in present-day *Sepia*. Hewitt *et al.* (1991), on the contrary, suggested that pseudosepta of shallow-water vascoceratids could have grown at constant intervals of time, possibly following diurnal or semidiurnal growth periodicities (they found between 14 and 28 pseudosutural and sutural growth increments). Pseudosutures also display a more or less constant spacing related to the linear growth rate, as is the case in *Nautilus* septa (Saunders, 1984). The *Sepia*-like chamber production rate for ammonoids was strongly opposed by Westermann (1992). Lominadze *et al.* (1993) assumed the number of pseudosutures per chamber to be a measure of chamber formation periodicity and concluded that phylloceratids and lytoceratids showed the highest values.

7. Conclusions

The viscous-fingering model of septal fabrication provides an attractive theoretical framework for explaining most features of the ammonoid septal suture and surface. For viscous fingers to form, there are two necessary physical conditions: (1) two fluids of different viscosities must be in contact, and (2) the less viscous fluid must push the more viscous one. From the paleontological evidence available, both conditions applied to ammonoids, because the rear soft body acted as a volume of viscoelastic material (Sections 4.1 and 5), and it was pushed by the cameral liquid in the last stages of translocation (Section 4). Therefore, the ammonoid shell acted as an unusually wide Hele–Shaw cell in which a Saffmann–Taylor instability developed, leading to the septal template (Fig. 17). More specifically, the size, number, and position of the septal flutes can be explained on the basis of the investigated topological properties of viscous fingers (Sections 2.3, 2.4, and 2.5; Fig. 17). Contrary to previous models, which assumed that rear mantle fluting occurred on a genetically controlled arrangement of either tie points or

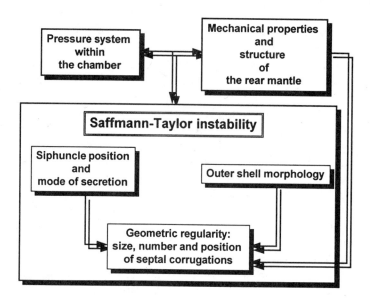

FIGURE 17. Fabricational factors that determined septal morphology. The existence of adorally directed pressures from within the chamber acting on a viscoelastic rear soft body led to a system similar to a Hele–Shaw cell in which a Saffmann–Taylor instability developed. In the ammonoid septal template, certain factors introduced regularity (nonrandomness) in the resultant viscous-fingering pattern. Siphuncle position and secretion mode determined the occurrence of an external lobe and adjacent undulations. Outer shell morphology (cross section and its changes throughout ontogeny) controlled the size, number, and exact position of each undulation within the whorl cross section. Additionally, the viscoelastic properties of the rear mantle influenced undulation size and pattern complexity.

corrugations, our model does not explain septal shape as the product of a genetic coding. Genetics acted through the construction of the biological and biophysical elements and system leading to the Saffmann–Taylor instability, which constituted the septal template.

ACKNOWLEDGMENTS. This research was supported by the EMMI Research Group and Projects PB91-0733.0 (A.C.) and PB92-1137 (J.G.-R.) of the DGICYT. Special thanks are given to Drs. R. A. Hewitt and G. E. G. Westermann (McMaster University, Canada) and the three editors of this volume, who significantly improved the paper through critical reading of the manuscript and constructive comments. Permission for the reproduction of figures was obtained from the editors or directors of *Lethaia* (Dr. C. Franzén, Swedish Museum of Natural History), *Neues Jahrbuch für Geologie und Paläontologie* (Dr. A. Seilacher, University of Tübingen), and *Geobios* (Dr. P. R. Racheboeuf, Université Claude Bernard, Lyon). This chapter is dedicated to Gerd Westermann (McMaster University, Hamilton, Ontario, Canada), who, apart from other fields of ammonoid paleontology, has contributed enormously to an understanding of the "septum problem."

References

Alexander, R. McN., 1962, Viscoelastic properties of the body wall of sea anemones, *J. Exp. Biol.* **39**:373–386.

Arkell, W. J., 1957, Sutures and septa in Jurassic ammonite systematics. *Geol. Mag.* **94**:235–248.

Arkell, A. K., Kummel, B., and Wright, C. W., 1957, Mesozoic Ammonoidea, in: *Treatise on Invertebrate Paleontology*, Part L, *Mollusca 4* (R. C. Moore, ed.), Geological Society of America and University of Kansas Press, Lawrence, KS, pp. 80–437.

Bandel, K., 1981, The structure and formation of the siphuncular tube of *Quenstedtoceras* compared with that of *Nautilus* (Cephalopoda), *N. Jb. Geol. Paläont. Abh.* **161**:153–171.

Bandel, K., 1982, Morphologie und Bildung der frühontogenetischen Gehäuse bei conchiferen Mollusken, *Facies* **7**:1–198.

Bandel, K., 1986, The ammonitella: A model of formation with the aid of the embryonic shell of archaeogastropods, *Lethaia* **19**:171–180.

Bandel, K., and Boletzky, S. von, 1979, A comparative study of the structure, development and morphological relationships of chambered cephalopod shells, *Veliger* **21**:313–354.

Bayer, U., 1975, Organische Tapeten im Ammoniten-Phragmokon und ihr Einfluss auf die Fossilization, *N. Jb. Geol. Paläont. Mh.* **1975**:12–25.

Bayer, U., 1977a, Cephalopoden-Septen Teil 1: Konstruktionsmorphologie des Ammoniten-Septums, *N. Jb. Geol. Paläont. Abh.* **154**:290–366.

Bayer, U., 1977b, Cephalopoden-Septen Teil 2: Regelmechanismen im Gehäuse- und Septenbau der Ammoniten, *N. Jb. Geol. Paläont. Abh.* **155**:162–215.

Bayer, U., 1978a, The impossibility of inverted suture lines in ammonites, *Lethaia* **11**:307–313.

Bayer, U., 1978b, Constructional morphology of ammonite septa, *N. Jb. Geol. Paläont. Abh.* **157**:150–155.

Bayer, U., 1985, *Pattern Recognition Problems in Geology and Paleontology*, Springer-Verlag, Berlin.

Blind, W., 1975, Über die Entstehung und Funktion der Lobenlinie bei Ammonoideen, *Paläontol. Z.* **49**:254–267.

Blind, W., 1980, Über Anlage und Ausformung von Cephalopoden-Septen, *N. Jb. Geol. Paläont. Abh.* **160**:217–240.

Boyajian, G., and Lutz, T., 1992, Evolution of biological complexity and its relation to taxonomic longevity in the Ammonoidea, *Geology* **20**:983–986.

Buckman, S. S., 1919–1921, *Type Ammonites*, Vol. III, Buckman, London.

Checa, A., 1986, Interrelated structural variations in Physodoceratinae (Aspidoceratidae, Ammonitina), *N. Jb. Geol. Paläont. Mh.* **1986**:16–26.

Checa, A., 1991, Sectorial expansion and shell morphogenesis in molluscs, *Lethaia* **24**:97–114.

Checa, A., and Sandoval, J., 1989, Septal retraction in Jurassic Ammonitina, *N. Jb. Geol. Paläont. Mh.* **1989**:193–211.

Damiani, G., 1986, Significato funzionalle dell'evoluzione dei setti e delle linee di sutura dei nautiloidi e degli ammonoidi, in: *Atti I Conv. Int., Fossili, Evoluzione, Ambiente, Pergola, 1984* (G. Pallini, ed.), Tecnoscienza, Roma, pp. 123–130.

Damiani, G., 1990, Computer simulation of some ammonoid suture lines, in: *Atti II Conv. Int., Fossili, Evoluzione, Ambiente, Pergola, 1987* (G. Pallini, F. Cecca, S. Cresta, and M. Santantonio, eds.), Tecnostampa, Ostra Vetere, pp. 221–228.

Denton, E. J., 1974, On buoyancy and lives of modern and fossil cephalopods. *Proc. R. Soc. Lond. B [Biol. Sci.]* **185**:273–299.

Denton, E. J., and Gilpin-Brown, J. B., 1966, On the buoyancy of the pearly *Nautilus*, *J. Mar. Biol. Assoc. U.K.* **46**:365–381.

Denton, E. J., and Gilpin-Brown, J. B., 1973, Floatation mechanisms in modern and fossil cephalopods, *Adv. Mar. Biol.* **11**:197–268.

Doguzhayeva, L., and Mutvei, H., 1986, Retro- and prochoanitic septal necks in ammonoids, and transition between them, *Palaeontogr. Abt. A* **195**(1–3):1–18.

Drushchits, V. V., and Doguzhayeva, L. A., 1974, Some morphogenetic characteristics of phyllo-ceratids and lytoceratids (Ammonoidea), *Paleontol. J.* **8**(1):37–48.

Durand-Delga, M., 1954, À propos de «*Bochianites*» *superstes* Perv.: Remarques sur les ammon-ites droites du Crétacé inférieur, *C. R. Somm. Seances Soc. Geol. Fr.* **7**:134–137.

Erben, H., and Reid, R. E. H., 1971, Ultrastructure of shell, origin of conellae and siphuncular membranes of an ammonite, *Biomineralization* **3**:22–31.

Feder, J., 1988, *Fractals*, Plenum Press, New York.

García-Ruiz, J. M., 1992, "Peacock" viscous fingers, *Nature* **356**:113.

García-Ruiz, J. M., 1993, Natural viscous fingering, in: *Growth Patterns in Physical Sciences and Biology* (J. M. García-Ruiz, E. Louis, P. Meakin, and L. M. Sander, eds.), Plenum Press, New York, pp. 183–189.

García-Ruiz, J. M., and Checa, A., 1993, A model for the morphogenesis of ammonoid septal sutures, *Geobios Mém. Spéc.* **15**:157–162.

García-Ruiz, J. M., and Otálora, F., 1994, Uso de la geometría fractal en las ciencias naturales, *Epsilon* (in press).

García-Ruiz, J. M., Checa, A., and Rivas, P., 1990, On the origin of ammonite sutures, *Paleobiology* **16**:349–354.

Guex, J., 1981, Associations virtuelles et discontinuités dans la distribution des espèces fossiles: Un exemple intéressant. *Bull. Soc. Vaudoise Sci. Nat.* **75**:179–197.

Haniel, C. A., 1915, VI. Die Cephalopoden der Dyas von Timor, in: *Paläontologie von Timor* (J. Wanner, ed.), E. Schweizerbart'sche Verlagsbuchhandlung, Stuttgart, pp. 1–153.

Henderson, R. A., 1984, A muscle attachment proposal for septal function in Mesozoic ammon-ites, *Palaeontology* **27**:461–486.

Hewitt, R. A., 1985, Numerical aspects of sutural ontogeny in the Ammonitina and Lytoceratina, *N. Jb. Geol. Paläont. Abh.* **170**:273–290.

Hewitt, R. A., and Westermann, G. E. G., 1987, Function of complexly fluted septa in ammonoid shells. II. Septal evolution and conclusions, *N. Jb. Geol. Paläont. Abh.* **174**:135–169.

Hewitt, R. A., Checa, A., Westermann, G. E. G., and Zaborski, P. M. P., 1991, Chamber growth in ammonites inferred from colour markings and naturally etched surfaces of Cretaceous vasco-ceratids from Nigeria. *Lethaia* **24**:271–284.

Hölder, H., 1954, Über die Sipho-Anheftung bei Ammoniten, *N. Jb. Geol. Paläont. Mh.* **1954**:372–379.

Hölder, H., 1955, Die Ammoniten-Gattung *Taramelliceras* im Süddeutschen unter- und Mittel-malm, *Palaeontogr. Abt. A* **106**:37–153.

John, R., 1909, *Über die Lebensweise und Organisation des Ammoniten*, Inaug.-Diss., Universität Tübingen, Stuttgart.

Jordan, R., 1968, Zur Anatomie mesozoischer Ammoniten nach den Strukturelementen der Gehäuseinnenwand, *Geol. Jahrb. Beih.* **77**:1–64.

Korvin, G., 1992, *Fractal Models in Earth Sciences*, Elsevier, Amsterdam.

Kulicki, C., 1974, Remarks on the embryogeny and postembryonal development of ammonites. *Acta Palaeontol. Pol.* **19**:201–224.

Kulicki, C., 1979, The ammonite shell: Its structure, development and functional significance, *Palaeontol. Pol.* **39**:97–142.

Kulicki, C., and Mutvei, H., 1982, Ultrastructure of the siphonal tube in *Quenstedtoceras* (Ammonitina), *Stockholm Contrib. Geol.* **37**:129–138.

Landman, N. H., 1988, Heterochrony in ammonites, in: *Heterochrony* (M. L. McKinney, ed.), Plenum Press, New York, pp. 159–182.

Landman, N. H., and Bandel, K., 1985, Internal structures in the early whorls of Mesozoic ammonites, *Amer. Mus. Novit.* **2823**:1–21.

Landman, N. H., and Waage, K. M., 1986, Shell abnormalities in scaphitid ammonites, *Lethaia* **19**:211–224.

Landman, N. H., Tanabe, K., Mapes, R. H., Klofak, S. M., and Whitehill, J., 1993, Pseudosutures in Paleozoic ammonoids, *Lethaia* **26**:99–100.

Lenormand, R., 1985, Différents mechanismes de déplacements visqueuses et capillaires en milieux poreux: Diagramme des phases, *C. R. Acad. Sci. Paris Ser. II* **301**:247–250.

Lominadze, T. A., Sharikadzé, M. Z., and Kvantaliani, I. V., 1993, On mechanism of soft body movement within body chamber in ammonites, *Geobios Mém. Spéc.* **15**:267–273.

Long, C. A., 1985, Intricate sutures as fractal curves, *J. Morphol.* **185**:285–295.

Longley, P. A., and Batty, M., 1989, Fractal measurements and line generalization, *Comput. Geosci.* **15**:167–183.

Mandelbrot, B. B., 1982, *The Fractal Geometry of Nature*, W.H. Freeman, San Francisco.

Miller, A. K., Furnish, W. M., and Schindewolf, O. H., 1957, Paleozoic Ammonoidea, in: *Treatise on Invertebrate Paleontology*, Part L, *Mollusca 4* (R. C. Moore, ed.), Geological Society of America and University of Kansas Press, Lawrence, KS, pp. 11–79.

Mutvei, H., 1975, The mode of life in ammonoids, *Paläontol. Z.* **49**:196–202.

Owen, R., 1832, *Memoir on the Pearly Nautilus*, Royal College of Surgeons, London.

Peitgen, H. O., Jürgens, H., and Saupe, D., 1992, *Fractals for the Classroom. Part I*, Springer-Verlag, New York.

Pfaff, E., 1911, Über Form und Bau der Ammonitensepten und ihre Beziehungen sur Suturlinie. *Niedersachs. Geol. Vereins Hannover* **4**:207–223.

Reif, W.-E., Thomas, R. D. K., and Fischer, M. S., 1985, Constructional morphology: The analysis of constraints in evolution, *Acta Biotheor.* **34**:233–248.

Rieber, H., 1979, Eine abnorme, stark vereinfachte Lobenlinie bei *Brasilia decipiens* (Buckman). *Paläotontol. Z.* **53**:230–236.

Ruzhentshev, V. Y., 1963, Theory of phylogenetic systematics (Part 2 of 4), *Int. Geol. Rev.* **5**:915–944.

Saunders, W. B., 1984, *Nautilus* growth and longevity: Evidence from marked and recaptured animals, *Science* **224**:990–992.

Schindewolf, O. H., 1965, Studien zur Stammesgeschichte der Ammoniten. Lieferung IV. *Akad. Wiss. Lit. Mainz Abh. Math. Natur. Kl.* **1965**:407–508.

Schmidt, M., 1925, Ammonitestudien, *Fortschr. Geol. Paläont.* **10**:275–363.

Seilacher, A., 1973, Fabricational noise in adaptive morphology, *Syst. Zool.* **22**:451–465.

Seilacher, A., 1975, Mechanische Simulation und funktionelle Evolution des Ammoniten-Septums, *Paläontol. Z.* **49**:268–286.

Seilacher, A., 1988, Why are nautiloid and ammonites so different? *N. Jb. Geol. Paläont. Abh.* **177**:41–69.

Senior, J. R., 1971, Wrinkle-layer structures in Jurassic ammonites, *Palaeontology* **14**:107–153.

Solger, F., 1901, Die Lebensweise der Ammoniten, *Naturw. Wochenschr.* **17**:89–94.

Spath, L. F., 1919, Notes on ammonites, *Geol. Mag.* **61**:27–35.

Stanley, E. H., 1987, Role of fluctuations in fluid mechanics and dendritic solidification, in: *The Physics of Structure Formation* (W. Guttinger and H. Dangelmayr, eds.), Springer-Verlag, Berlin, pp. 210–243.

Swinnerton, H. H., and Trueman, A. E., 1918, The morphology and development of the ammonite septum, *Q. J. Geol. Soc. Lond.* **53**:26–58.

Tanabe, K., 1977, Functional evolution of *Otoscaphites puerculus* (Jimbo) and *Scaphites planus* (Yabe), Upper Cretaceous ammonites. *Mem. Fac. Sci. Kyushu Univ. Ser. D* **23**:367–407.

Tanabe, K., Landman, N. H., and Weitschat, W., 1993, Septal necks in Mesozoic Ammonoidea: Structure, ontogenetic development, and evolution, in: *The Ammonoidea: Environment, Ecology, and Evolutionary Change*, Syst. Assoc. Spec. Vol. 47 (M. R. House, ed.), Clarendon Press, Oxford, pp. 57–84.

Tate, R., and Blake, J. F., 1876, *The Yorkshire Lias*, London.

Thompson, D'A. W., 1942, *On Growth and Form*, Cambridge University Press, Cambridge.

Van Damme, H., 1989, Flow and interfacial instabilities in newtonian and colloidal fluids, in: *The Fractal Approach to Heterogeneous Chemistry* (D. Avnir, ed.), John Wiley & Sons, Chichester.

Vicencio, R., 1973, *Models for the Morphology and Morphogenesis of the Ammonoid Shell*, Unpublished Doctoral Thesis, McMaster Univ.

Vogel, K. P., 1959, Zwergwuchs bei Polyptychiten (Ammonoidea), *Geol. Jahrb.* **76**:469–540.

Wainwright, S. A., Biggs, W. D., Currey, J. D., and Gosline, J. M., 1976, *Mechanical Design in Organisms*, Edward Arnold, London.

Ward, P. D., 1987, *The Natural History of Nautilus*, Allen and Unwin, Boston.

Ward, P. D., and Westermann, G. E. G., 1976, Sutural inversion in a heteromorph ammonite and its implication for septal formation, *Lethaia* **9**:357–361.

Weitschat, W., 1986, Phosphatisierte Ammonoideen aus der Mittleren Trias von Central-Spitzbergen, *Mitth. Geol. Paläontol. Inst. Univ. Hamburg* **61**:249–279.

Weitschat, W., and Bandel, K., 1991, Organic components in phragmocones of Boreal Triassic ammonoids: Implications for ammonoid biology, *Paläontol. Z.* **65**:269–303.

Westermann, G. E. G., 1954, Monographie der Otoitidae (Ammonoidea). *Geol. Jahrb. Beih.* **15**:1–364.

Westermann, G. E. G., 1956, Phylogenie der Stephanocerataceae und Perisphinctaceae des Dogger. *N. Jb. Geol. Paläont. Abh.* **103**:233–279.

Westermann, G. E. G., 1958, The significance of septa and sutures in Jurassic ammonite systematics, *Geol. Mag.* **95**:441–455.

Westermann, G. E. G., 1971, Form structure and function of shell and siphuncle in coiled Mesozoic ammonoids, *Life Sci. Contrib. R. Ont. Mus.* **78**:1–39.

Westermann, G. E. G., 1975, Model for origin, function and fabrication of fluted cephalopod septa, *Paläontol. Z.* **49**:235–253.

Westermann, G. E. G., 1982, The connecting rings of *Nautilus* and Mesozoic ammonoids: Implications for ammonite bathymetry, *Lethaia* **15**:373–384.

Westermann, G. E. G., 1992, Formation and function of suspended organic cameral sheets in Triassic ammonoids -discussion, *Paläontol. Z.* **66**:437–441.

Wiedmann, J., and Kullmann, J., 1980, Ammonoid sutures in ontogeny and phylogeny, in: *The Ammonoidea*, Systematics Association Special Volume 18 (M. R. House and J. R. Senior, eds.), Academic Press, London, pp. 215–255.

Zaborski, P. M. P., 1986, Internal mould markings in a Cretaceous ammonite from Nigeria, *Palaeontology* **29**:725–738.

Chapter 10

Architecture and Strength of the Ammonoid Shell

ROGER A. HEWITT

1. Introduction

The present chapter concentrates on those aspects of ammonoid morphology that are directly related to the habitat depth of ammonoids and the strength of the shell. Other chapters in the present volume discuss the role of cameral water and ornamentation in locomotion (Chapters 7 and 16, this volume), the growth of septal sutures (Chapter 9, this volume), and ecology (Chapter 16, this volume). The symbols used in this chapter are defined in Table I.

ROGER A. HEWITT • Department of Geology, McMaster University, Hamilton, Ontario L8S 4M1, Canada.

Ammonoid Paleobiology, Volume 13 of *Topics in Geobiology*, edited by Neil Landman *et al.*, Plenum Press, New York, 1996.

Table I. Symbols Used in Equations 1–5 and Related Mechanical Parameters of
Ammonoid Septa and Siphuncles

---

a Chord ratio; it normally measures septum flatness $= 2r/L = 1/\sin\theta$ if r is constant.

a^* Umbilical flute in which a is just a scaling factor.

b Orthogonal stress ratio with tension around radius r. In the case of typical anticlastic saddles: $b = 2 + (r/R)$ if $\phi R < \theta r$; often b is then about π. In unfluted and adorally concave dorsal septal sutures: $b = 2 + (r/R) = \{T + [T - (1/a)r]\}/[T - (1/a)r]$ and $f = 1$. In unfluted and synclastic, adorally concave septa: $b = 2 - (r/R)$ with local radii $r < R$ measured at d and $f = c = 0$.

c Relative added compressive stress added via shell wall $= r - q/2r = [(r/2d) - (q/2d)]/(r/2d)$. The negative values describing the case $q > r$ are treated as zero stress.

d Central thickness of anticlastic saddle-flute or minimum thickness at the margin of an unfluted septum.

d' Mean undecayed wall thickness of a chitinous cylinder.

E Young's modulus of nacre; 45,800 MPa in *Nautilus*.

E^* Young's modulus of chitinous tubes in megapascals.

E' No elastic instability unless $H > 1000$ m.

f 1 in anticlastic septa failing in shear: $= 0$ elsewhere.

h Maximum sutural amplitude and height H of Batt (1991); $=$ about $L/2 + 2R(1 - \cos\phi) + r(1 - \cos\theta)$.

H Habitat depth limit deduced from septa with equation 1.

H' Depth if working stress of siphuncle was 16.45 MPa. (Data shown in brackets came from a different specimen or an earlier phase of growth than the septum data.)

H_s Depth deduced from septa with buckling equation 4.

H_p Depth deduced from septa with buckling equation 5.

$H^{\#}$ Depth H deduced from another growth phase or species.

H^* Depth deduced from the mean safety factor of 1.86 (the latter was actually 1.857 in the 3017 septa).

i Intersutural span (Jacobs, 1990) or spacing.

i Torus radius (T) of infinity (i.e., compressed whorls).

I Mean estimated implosion depth of septa $= (100 P_2) - 10$.

IST Inverse septal thickness $= L/4d = ar/4d$.

ISC Index of sutural complexity $=$ suture length$/X^*$.

K Approximate transverse area inside whorl, or actual area of an unfluted septum, in square millimeters.

l Length of a chitinous segment of the siphuncle.

L $= 2r/a$ in millimeters $=$ average measured flute length of the series of septa yielding the cited mean IST. In most cases it was either the whorl breadth or the internal distance between the venters in the median section, yielding a series of thickness measurements.

m Weibull modulus ($= 27$ in equation 3), not meters (m).

θ Half the rotation angle of flute radius r forming L. (Table II records the observed minimum angle of θ).

ϕ Half the rotation angle of flute radius R or Q.

P Pressures $(P_2 + P_1)$ in megapascals at the implosion depth I.

q Central axial radius of lobe-flute curvature $= aL/2 - aLc$.

q/d $(2a - 4ac)\, IST\, [(I + 10)/(H + 10)]$.

Q Central radius of curvature normal to q between lobes.

r Axial externally concave radius of curvature in saddles, which becomes R toward the suture, or minimum radius of synclastic curvature, measured in same units as d.

R Local radius of curvature normal to r at d measurement.

S Mean tensile strength of nacre estimated from volume.

S^* Apparent tensile strength of nacre failing by shearing.

Table I. (Continued)

SAI Sutural amplitude index (Batt, 1991) = $5h/X^* = 1.2–5.0$.

SiSI   Siphuncle strength index = $100\ d'/(r - 0.5\ d')$.

τ Nacre ultimate shear strength $\tau = (411\ S)/(411 + S)$.

τ' 9.068 MPa = shear stress in working siphuncle. This is equivalent to a tensile stress of 16.45 MPa (1 σ = 5.70 MPa) deduced from 108 pairs of H and H' estimates.

T Torus radius (mm) defined by central lobe-flutes at d.

V Standard half-volume of 7.5 mm^3 with $S = 136.41$ MPa.

w Width or wavelength of lobe–flute in central area at d.

w/d {sin ϕ $4a$ IST $[(I + 10)/(H + 10)]$}$/(b - 2)$ if $Q = R$.

X^* External whorl circumference around septal sutures. From 0.8–1.2 πL in equidimensional whorls, to 3 πL.

2. The Hydrostatic Loading of Ammonoid Shells

2.1. Background

Much of the previous literature reviewed by Hewitt and Westermann (1987a) and Jacobs (1992) was based on the assumption that no significant external hydrostatic pressure was applied to the phragmocone (Figs. 1–4). It subsequently was found that extant cephalopods with septate shells produce an initial vacuum by simple osmosis in their siphuncular epithelium (Denton, 1974). As in these forms (Denton and Gilpin-Brown, 1966), nitrogen and other gases would have slowly diffused into the vacuum formed by the dehydration of the cameral gel required to support the uncalcified membranes of ammonoid septa (Hewitt and Westermann, 1987a, p. 160). Because there is no significant increase in partial gas pressures with depth in the ocean, it is possible to estimate habitat depth limits from the strength of each successive "last septum" (Figs. 3–5). Extensive reflooding of this chamber via an intact siphuncle produced a minimal reduction in the maximum stress developed in this most recently formed septum. All the septa initially functioned as a "last septum" and were subsequently protected from the water pressure applied via the body chamber. These septa were only loaded via their sutures with the phragmocone wall and may be described as "internal septa." Raup and Takahashi (1966, p. 173) wrote that "stress analysis of the last septum is suggested as a means of evaluating shell strength and hence estimating maximum depths at which fossils lived."

This was first attempted in a correct manner by Westermann (1973). The uniform force of gravity acts on the mass of a building in a constant direction, and it declines to zero on the vertical surfaces forming the edges of a vault or dome. But hydrostatic pressure was applied normal to every point on the surface of a phragmocone and the siphuncular tube. Any corrugation or fluting of this surface increased the total force applied to the shell, but it increased

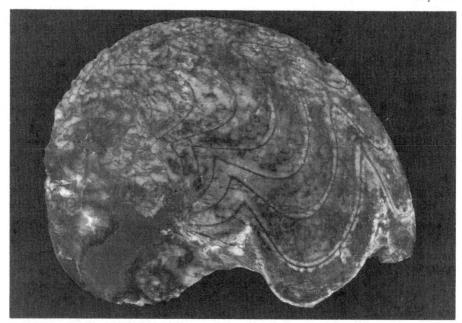

FIGURE 1. Lateral view of the dark septal sutures of a juvenile specimen (diameter = 134 mm) of the oxycone *Carinoceras acutiforme* (Termier), a relative of *Manticoceras* (AMNH 45060). Note the anticlastic curvature of the septal surface within the two V-shaped lobes. Phragmocones of this species are often seen in shops and range up to a diameter of 400 mm without showing any further increase in sutural complexity. They lived in shallow seas during a middle Frasnian marine transgression (lower *Beloceras* Zone, Devonian) and are quarried from a massive lime-stone bed in the Tafilalt of the Anti Atlas Mountains of Morocco (R. T. Becker, personal communication, 1992).

the stiffness of the shell only along the fold axes. The sutures of an externally convex and hemispherical septum, therefore, contracted from the wall like a deflated balloon, while those of morphologically identical domes are thrust outwards. Architectural vaults with a similar shape to an ammonoid septum produce an entirely compressive stress field (Shaaban and Ketchum, 1976). A "vault model" of the ammonoid phragmocone was proposed by Pfaff (1911) and Jacobs (1990). Vaults are loaded by a unidirectional gravitational force rather than hydrostatic pressure. Unidirectional loading could also result from the crushing of a body chamber by a predator or the bending of a stiff and flat plate by water pressure, but it would not have developed during the hydro-static loading of a curved surface.

Ammonoid septa (Tables I and II) were composed of nacreous aragonite with similar microstructure, strength (i.e., maximum stress, Fig. 5), and elastic constants [i.e., stiffness (E) and Poisson's ratio] to those of *Nautilus* nacre (Hewitt and Westermann, 1987b). Engineers and biologists report these mem-brane properties of a material in SI units of pressure. One megapascal (1MPa) is the pressure resulting from a force of 1 N acting uniformly over an area of

FIGURE 2. Initial ventral (right) and final (above) impressions of the virgatotome ribs of *Lithacoceras (Subplanites) rueppellianum* (Quenstedt) on a bedding plane of a centimeter-thick *flinz* layer from the upper Solnhofen Limestone Formation of ?Eichstatt, Germany (ROM 51107, A. Zeiss, personal communication, 1994). The coarse lateral corrugations protected the aperture and subdivided into more numerous ribs over the relatively stiff venter. The septa were strong enough to permit the shell to sink in a vertical orientation before flooding on the sea bed. Middle Hybonotum Zone, lower Tithonian, Jurassic. Scale bar = 1 cm.

1 mm^2. Callomon (1994) confused the small "cross-sectional area" of membrane forces within the nacre with the large "surface area" exposed to the forces produced by hydrostatic loading. The external "surface area" of the shell resisted the combined atmospheric (maximum $P_1 = 0.1$ MPa) and hydrostatic pressures ($P_2 = 1$ MPa increase per 100 m depth) by concentrating the hydrostatic force as two orthogonal membrane forces.

The membrane forces per unit "cross-sectional area" are known as stresses, and they break the shell at certain maximum values known as the tensile and compressive strengths of the nacre. Because these strengths are over two

Table II. Bathymetric Data in Meters (H, I on right) Deduced from the Central Region of the "Last Septa" and the Siphunuclar Tubes of Various Phases in the Ontogeny of Ammonoids[a]

Order ANARCESTIDA	L, mm	IST	a	$\theta > \phi$	T, mm	b	c	S^*	H_p	H_s	H'	H	I
BACTRITINA (unfluted)													
Bactrites quadrilineatus; 51108,51217.	8.6	21.11	1.19	----	0	1.02	0.00	139	---	---	215	324	535
Same species and locality (M1); 51218.	9.2	30.63	1.12	----	0	1.02	0.00	138	---	---	125	128	390
AGONIATITINA													
Agoniatites vaunuxemi; M.R. House Coll.	48.9	73.20	1.84	----	i	1.00	0.00	118	---	---	---	37*	78
ANARCESTINA (unfluted)													
Anarcestes sp., L.Dev. III-D1; Bc 1510.	3.0	19.72	1.05	73>45	4	2.63	0.00	140		dorsal		77	247
Werneroceras testatum from M.Dev. I.	7.2	22.69	1.31	50>27	11	1.41	?+	142		ventral		---	<329
Median regions of a septum; Bc 1511.	7.2	32.54	1.31	50>0	11	2.22	0.00	132		central		---	<129
	7.2	29.52	1.31	50<70	11	2.59	0.00	134		dorsal		77	124
GEPHUROCERATINA (fluted)													
Carinoceras acutiforme; AMNH 45060.	32.4	21.08	1.79	40>28	i	3.12	0.32	108	209	251	---	70	92
Manticoceras cordatum; 51109.	8.03	29.12	1.32	30>22	i	3.09	0.27	128	97	66	---	60	108
	1.42	15.60	?1.32	---	-	?3.51	?0.27	153	---	---	---	126#	219
Mesobeloceras kayseri; Bc 1513.	4.2	22.02	1.15	35>30	i	3.07	0.22	127	76	67	---	73	167
Beloceras tenuistriatum; Bc 1514.	5.0	21.01	1.60	27>40	i	3.74	0.34	135	126	168	---	53	108

Taxon													
CLYMENIINA (unfluted)													
Cymaclymenia striata; Bc 1515.	7.4	27.75	1.06	70>44	i	1.34	0.00	138	central			179*	342
Platyclymenia ruedemanni; Bc 1516.	4.3	22.89	1.56	40>40	i	1.36	0.00	153	central			251	305
	2.9	43.58	1.28	45>30	i	1.22	0.00	158	central			180	220
	1.3	32.00	1.24	40>--	i	?1.22	0.00	176	central			291#	?354
Order GONIATITIDA													
TORNOCERATINA													
Tornoceras uniangulare; 51110.	6.47	22.23	1.21	13<30	i	1.74	0.00	143	---	---	164	155*	296
Maeneceras biferum; Bc 1512.	6.56	18.03	1.06	35>45	5.9	4.73	0.13	130	610	105	---	70*	138
GONIATITINA													
Dimorphoceras poletum; SUI 1755.	1.28	18.12	---	---	-	---	---	-	[(l+10)/(H+10) = 1.85]				
Goniatites kentuckiensis; SUI 1718.	2.96	22.01	1.70	---	-	?2.63	0.26	125	---	41	184	61	131
	1.37	14.00	?1.75	---	-	?2.63	?0.30	139	---	364	---	150	234
Dombarites choctawensis; AMNH 45102a-l.	0.38	8.27	1.75	40>?30	-	2.63	0.31	158	231	689	---	217	460
Ontogeny deduced from 12 specimens from locality M1 of R. H. Mapes.	5.43	17.45	1.51	40>35	7	3.11	0.26	126	84	111	(132)	76	157
	3.33	19.67	0.92	55>30	6	2.73	0.30	130	178	431	(132)	140	295
	1.92	18.83	1.02	----	3	2.73	0.09	139	---	51	132	113	264
	0.60	9.56	1.03	60>?"	1.1	2.58	0.06	151	---	179	235	208	580
	0.33	7.25	?1.03	-----	0.8	?2.58	0.07	160	---	249	385	276	?837
	0.22	3.26	1.01	84>--	0.5	<2.58	0.00	161	---	E'		905	1934
Agathiceras applini; AMNH 45061a-f and SUI 1766.	4.40	12.33	1.13	55>--	5.4	2.47	0.09	125	---	213	185	204	367
	1.57	10.36	1.13	50>--	-	?2.47	0.17	137	---	610	246	303	499

Table II. (Continued)

	L, mm	IST	a	θ > φ	T, mm	b	c	S*	H_p	H_s	H'	H	I
Neoglyphioceras belllineatum; SUI 1713.	1.67	14.29	?1.02	----	-	?2.99	?0.08	146	---	393	---	259	334
	0.52	11.56	?1.02	----	-	?2.99	0.08	162	---	230	219	218	440
	0.41	8.59	1.24	----	-	?2.99	0.05	166	---	428	375	322	520
	0.17	5.36	2.00	----	-	?2.65	0.29	173	---	E'	---	348	671
Glaphyrites sp., Okmulgee; AMNH 45077.	6.3	14.44	1.08	65>35	6.4	3.11	0.10	127	344	148	---	135*	259
G. excelsus; AMNH 45065.	3.03	15.64	1.86	50>26	7.9	3.47	0.23	134	207	73	---	62*	124
	1.85	15.69	1.86	----	5	?3.47	?0.19	142	---	51	82	70*	138
G. welleri; AMNH 45078, SUI 1726.	2.25	17.73	1.75	27>33	4.6	3.76	0.40	141	408	653	116	64	125
	0.69	18.92	2.00	25<26	1.3	3.75	0.19	161	213	42	168	68	110
	0.30	6.64	1.95	----	-	3.10	0.26	170	---	E'	262	289	450
	0.18	5.53	1.72	----	-	3.34	0.34	178	---	E'	313	347	615
G. clinei; SUI 1735.	1.56	18.86	?1.35	----	-	?3.50	?0.09	150	---	51	204	93	163
	0.79	12.17	?1.35	----	-	?3.50	?0.14	160	---	198	197	100	279
	0.47	7.25	?1.36	----	-	3.50	0.21	166	---	483	316	204	502
	0.14	4.51	1.56	?45>30	-	2.84	0.23	179	E'	E'	---	385	965
Gastrioceras adaense; AMNH 45075.	14.2	19.01	1.33	40>--	12	3.58	0.25	118	---	277	---	94	130
Bisatoceras microphalus; AMNH 45063.	8.39	21.50	*0.75	37>35	4.9	3.03	0.24	127	365	344	---	142*	273

Syngastrioceras oblatum; AMNH 45069.	5.99	13.73	0.98	60>50	9.5	3.58	0.03	127	560	155	--	132*	254
S. smithwickensis; AMNH 45070.	5.65	10.36	1.45	40>35	8.8	3.12	0.21	124	536	322	--	142*	273
Eothalassoceras inexpectans;	1.20	11.00	1.64	40>--	-	?2.22	0.04	145	--	76	217	131*	<252
SUI 1759, AMNH 45064.	4.95	9.72	1.17	40>45	8.7	3.13	0.04	126	E'	E'	172	196	348
	3.91	21.39	1.30	38>30	2.7	2.81	0.26	129	54	108	172	80	170
	0.45	10.49	1.20	40>--	-	3.04	0.09	163	--	192	--	206	429
Schistoceras missouriense; AMNH 45067a-b.	13.6	10.94	1.31	---	12.0	2.72	0.11	116	--	146	137	139	302
45080, and L=3.1 from SUI 3119.	3.1	14.09	1.06?	---	-	2.22	0.0?	131	--	288	239	281	?385
Goniolobocerus goniolobum; AMNH 45066.	13.9	27.40	1.13	60>50	i	3.32	0.2?	121	--	40	--	53	114
G. welleri; SUI 1743.	1.02	13.31	1.33	-----	-	?3.32	0.34	152	--	E'	156	170	?278
Pseudogastrioceras simulator; SUI 1740.	0.46	10.67	?1.76	-----	-	?3.14	?0.18	166	--	203	231	170	?287
	0.33	8.04	?1.76	-----	-	?3.14	?0.18	170	--	331	--	217	?396
	0.19	5.27	1.76	-----	-	?3.14	0.27	176	--	E'	325	292	?645
Schumardites cuyleri ; AMNH 45068.	7.3	18.56	1.10	50>35	12	4.00	0.10	129	375	58	--	69	152
; SUI 1791.	3.2	11.89	?1.10	-----	-	?4.00	?0.10	138	--	665	203	206	260
Properrinites boesei; AMNH 45074.	8.50	21.25	1.58	23>22	10	3.02	0.25	122	89	60	--	60*	121
P. bakeri; SUI 1790.	1.26	16.90	1.33	-----	1.6	?3.02	0.21	149	--	218	257	151	225
	0.36	7.39	?1.33	-----	-	?3.02	0.12	164	--	896	--	382	562
Perrinites hilli ; AMNH 40622.	14.3	<24.6	1.54	40>?15	-	>2.22	0.23	114	--	--	--	--	?141
; AMNH 41183.	10.6	14.27	1.03	40>?15	-	?2.80	0.16	119	E'	716	(116)	250	297
	8.8	18.23	?1.03	-----	-	?2.80	?0.16	122	--	272	116	168	?237

Table II. (Continued)

	L, mm	IST	a	θ > φ	T, mm	b	c	S*	H_p	H_s	H'	H	I
; SUI 1785, AMNH 41183.	10.3	14.39	?1.54	----	-	?2.80	?0.23	119	---	53	141	66	?199
	8.0	12.88	?1.54	----	-	?2.80	?0.23	122	---	400	131	164	?229
	3.5	16.79	?1.54	----	-	?2.80	?0.22	134	---	176	175	130	?190
	1.4	16.31	?1.54	----	-	?2.80	?0.19	147	---	127	199	136	?214
Waagenoceras quadalupense; AMNH 28777.	10.3	17.70	0.97	----	22	3.25	0.05	119	---	91	---	106*	206
; SUI 1782.	1.85	11.93	1.0?	----	-	?3.25	0.16	140	---	816	201	266	?370
Subkargalites sp. from s. of Colorado River, Tx.; AMNH 45076.	4.04	10.94	1.12	30>30	4.7	3.62	0.04	131	E'	177	---	149*	286
	3.48	14.52	1.04	35>30	5.0	3.43	0.0?	132	636	89	133	128*	246
Popanoceras annae; SUI 1777.	2.10	15.12	---	----	-	---	---	-		$[(I+10)/(H+10) = 1.49]$			
	0.79	12.24	?1.81	----	-	?2.79	?0.23	152	---	156	210	139	260
	0.19	4.71	1.81	----	-	?2.79	0.23	173	---	E'	---	365	785
Peritrochia sellardsi; 51219.	3.40	17.54	1.69	40>25	4.8	3.23	0.15	135	200	58	---	84	137
Adrianites dunbari; SUI 1764.	2.52	9.31	>1.32	----	-	>2.22	?0.0	140	---	112	238	223	<502
	0.29	7.25	>1.06	----	-	?3.14	?0.0	159	---	401	293	325	?690
Order AMMONITIDA PROLECANITINA													
Pronorites praepermicus; SUI 1686.	1.08	18.29	?1.39	----	-	?2.66	0.23	155	---	103	---	116*	?224
	0.68	14.41	?1.39	----	-	?2.66	0.25	164	---	322	224	188	?306
	0.24	7.89	1.39	----	-	?2.66	0.17	176	---	462	---	297	?590

Taxon													
Medlicottia burckhardti; AMNH 28777.	13.0	11.92	1.38	----	i	2.90	0.17	114	---	196	---	127*	245
AMNH 25934, with siphuncle at L=3.	8.2	16.11	0.98	----	i	4.00	0.00	126	---	90	118	97*	190
OTOCERATINA													
Otoceras boreale; AMNH 45096.	16.0	29.78	0.86	40>40	30	3.12	0.22	118	69	89	---	74	149
ARCESTINA s.l.													
Paranannites aspenensis; AMNH 45053.	4.60	15.82	1.39	55>23	2.4	3.18	0.12	131	308	69	---	105*	185
Juvenites sp.; AMNH 45094.	1.1	10.89	---	----	---	---	---	---			[(l+10)/(H+10) = 1.51]		
J. septentrionalis; AMNH 45057.	4.0	15.0	---	----	---	---	---	---	---	---	154	---	---
Owenites koeneni; AMNH 45055.	4.00	17.68	1.34	35>?	11	>2.22	0.10	128	---	66	---	142	<245
Prosphingites slossi; AMNH 45054.	5.26	<14.5	1.49	40>23	4.3	2.97	0.15	126	231	87	---	112*	198
Umbilical part of same septum.	3.44	18.92	1.53	40>20	4.3	3.03	0.15	126	136	39	---	72*	142
Parapopanoceras tetsa; 51220.	6.87	25.55	0.66	75>18	8.2	2.37	0.07	124	141	141	---	185	310
P. sp.; AMNH 45082, 45097.	4.7	9.57	1.38	20>15	6.3	3.42	0.28	123	E'	798	118	150*	287
Umbilical part of same septum.	4.6	10.54	1.54	23>15	6.3	3.53	0.25	124	E'	370	118	115*	222
Sturia mongolica; BMNH "Nifoekoko".	85.0	8.50	0.71	----	i	2.43	0.0?	96	---	937	---	356*	652
Psilosturia sp.; AH 45091-2.	9.8	23.90	1.07	23>25	2.6	3.19	0.22	125	209	93	177	89*	155
Ptychites sp.; AMNH 45087.	6.6	10.00	1.37	20<57	-	3.54	0.14	130	261	153	85	112	269
Paracladiscites timidus; AMNH 39808.	83.9	22.18	0.93	----	i	2.7	0.18	95	---	119	---	89*	174
Cladiscites beyrichi; AMNH 42440-1.	25.0	>5.54	1.04	----	13	2.29	-0.08	103	---	E'	170	421*	<773
	7.0	9.63	1.05	----	-	2.50	0.00	121	---	274	170	258*	469

Table II. (Continued)

	L, mm	IST	a	θ > φ	T, mm	b	c	S*	H_p	H_s	H'	H	I
C. sp.; AMNH 45095.	6.9	27.95	1.06	60>20	i	2.96	0.17	127	144	52	---	79	144
Hypocladiscites subaratus; AMNH 39580.	69.0	4.74	0.87	----	i	2.55	-0.50	95	---	447	67	482*	<905
H. subaratus; AMNH 42437.	27.4	9.42	1.46	----	i	3.00	0.0?	107	---	129	(67)	140*	251
Proarcestes gabbi; AMNH 26341.	14.1	11.98	0.99	----	12	2.70	0.07	115	---	250	---	198*	359
P. ausseeanus; AMNH 42130.	13.0	4.70	0.92	----	15	2.86	0.00	114	---	E'	---	482*	906
Arcestes andersoni; 51116.	15.7	16.60	0.98	80>30	22	3.31	0.12	115	581	149	---	118*	210
A. sundaicus; BMNH c.21760.	17.6	---	1.14	----	17	2.22	0.25	---	---	---	---	---	---
A. sp. from Saltzkammergut; 51118.	11.2	14.00	1.42	40>?	-	3.3?	0.14	120	---	101	---	92*	?179
	4.6	15.74	1.10	40>?	-	3.3	-0.13	132	---	37	---	124*	?220
A. sp. from Gabbs Fm.; AMNH 45079.	13.4	6.61	1.03	30>15	18	3.27	0.19	113	E'	E'	---	291*	532
Joannites johannisaustriae; 51119.	13.8	8.64	?1.12	70>42	-	?4.51	0.04	117	E'	358	---	147*	>263
	3.8	17.42	1.01	----	-	4.51	0.10	131	---	109	---	90*	158
Megaphyllites sp.; AMNH 45098, W.	2.71	17.06	1.62	20>22	2.8	3.62	0.24	141	255	59	185	59	141
PINACOCERATINA s.l.													
Meekoceras gracilitatis; AMNH 45056a-b.	18.0	29.78	1.34?	?>38	i	3.23	?0.24	119	21	33	--	54*	90
	5.91	20.32	1.34	28<45	i	3.79	0.24	132	145	94	61	64*	127
Dieneroceras knechti; AMNH 45058.	5.04	17.33	1.40	45>30	i	3.86	0.11	136	343	49	289	71*	140
Anasibirites sp.; AMNH 45085.	2.0	17.9	---	----	-	---	---	---	---	--	224	---	---
Buddhaites hagei; 51121.	10.4	12.69	0.81	25>33	14	3.22	0.28	120	E'	E'	101	185	387

Pinacoceras metternichi; BMNH c.547.	80.0	16.6	---	---	i	---	---	96	---	---	---	---	---
;AMNH 27573.	69.0	11.86	1.47	---	i	2.9?	0.25	96	---	311	---	101*	?196
P. parma; BMNH c.21795.	32.8	16.4?	1.83	---	i	2.22	?0.28	102	---	109	---	84*	?165
P. sp. from Bihati; AMNH 41196.	31.7	12.45	1.70	---	i	2.90	0.19	104	---	120	---	87*	171
CERATITINA													
Longobardites nevadanus; 51111, 51221.	6.50	22.67	1.35	52>18	i	2.84	0.23	126	53	60	93	74	147
Paradanubites sp.; AMNH 45093.	9.0	18.82	1.02	27>23	3.1	3.34	0.29	126	E'	E'	191	176	207
Czekanowskites hayesi; W., AMNH 45081.	6.1	16.94	1.09	23>33	2.0	3.67	0.26	132	939	347	(96)	110	199
Hollandites sp.; W. & 51112.	8.4	26.80	1.09	26>32	6.8	3.59	0.34	130	349	312	(196)	72	128
Anagymnotoceras sp.; AMNH 45083.	11.3	24.07	1.22	32>30	4.1	4.05	0.31	127	411	162	152	52*	106
Beyrichites sp., Canada; 51113.	9.0	24.04	1.84	20>25	1.7	3.45	0.41	127	238	464	---	48	85
	3.1	14.24	1.23	20>16	0.8	2.93	0.18	139	E'	244	107	167	280
Gymnotoceras deleeni; 51114.	7.0	24.21	1.01	30>31	1.6	3.61	0.28	131	364	182	139	77*	151
Rimkinites sp.; AMNH 45086.	8.8	22.74	*0.60	25>32	i	2.67	0.25	124	307	559	149	189*	366
Progonoceratites sp.; AMNH 45099-45100.	2.8	24.43	1.31	30>25	i	3.23	0.26	144	268	139	148	94	142
Ceratites nodosus; Tr.31 & 51115.	24.1	23.83	2.27	30>22	i	4.18	0.29	119	117	23	---	21	46
Eutomoceras laubei; 51122.	9.0	24.75	1.59	25>22	3.8	4.11	0.31	129	320	77	---	37*	77
Nevadites whitneyi; 51123.	12.7	34.92	1.05	45>0	2.2	2.22	0.22	118	---	35	---	87*	151
Drepanites sp., Canada; AMNH 45101.	3.4	27.04	---	---	-	---	---	---	---	---	115	---	---
Neotibetites gigantus; AMNH 45089.	17.0	29.76	1.15	35>15	i	2.93	0.16	118	119	25	141	67*	114
Anatibetites sp.; AMNH 45084.	8.1	23.96	---	---	-	---	---	---	---	---	---	---	---

$[(l+10)/(H+10)] = 1.60$

Table II. (Continued)

	L, mm	IST	a	θ > φ	T, mm	b	c	S*	H_p	H_s	H'	H	I
Tropites welleri; AMNH 27630.	12.9	8.45	0.78	----	13	2.63	0.00	117	---	761	---	355*	669
Parajuvavites sp.; AMNH 45088.	5.85	13.41	1.66	32>22	10	3.47	0.23	125	890	186	71	93	163
	0.76	8.40	?1.66	-----	-	?3.47	?0.23	152	---	449	145	171*	?326
PHYLLOCERATINA													
Monophyllites simonyi; BMNH c.34089-90.	26.5	22.1	0.96	---	i	2.63	0.04	113	---	49	---	101*	197
Rhacophyllites neojurensis;	84.7	10.95	0.82	-----	i	2.50	0.3?	98	---	E'	---	255*	484
BMNH c.21794 & AMNH 42419.	51.0	15.58	1.06	-----	i	2.77	0.30	105	---	580	---	129*	248
R. debilis; BMNH c.34092.	38.0	6.3	0.95	-----	i	3.20	0.03	107	---	E'	---	293*	553
R. krumbecki; AMNH 38909.	16.7	<28.7	1.08	-----	i	3.13	0.28	118	---	100	---	>63*	>125
Phylloceras heterophyllum; 51124.	40.7	27.45	0.98	-----	i	3.30	0.25	123	---	93	156	66	139
P. sp. from Alaska; 51106.	119.0	6.71	1.18	55>30	i	3.75	0.31	93	E'	E'	202	184*	332
	65.0	6.89	1.08	45>30	i	3.17	0.33	98	E'	E'	202	240*	454
P. plicatum from Beds C1-6,7,15; 51272-7.	10.1	5.79	1.04	30>23	i	2.82	0.24	118	E'	E'	244	482	754
from Bed C1-3; 51267-51271.	7.5	6.95	1.13	30>17	i	2.85	0.24	123	E'	E'	234	342	591
Holcophylloceras ultramontanum MLP.	15.7	12.76	1.00	30>15	i	2.60	0.03	114	552	125	148	153	337
Haplophylloceras strigile; 51125-6,51222-5.	10.0	9.46	1.61	28>20	5	3.67	0.28	121	E'	858	184	142	224
Ptychophylloceras galoi ; 51259.	15.5	11.68	1.55	30>25	i	3.43	0.30	119	E'	E'	95	123	200
; 51129.	9.33	9.36	?1.55	-----	i	?3.43	?0.30	124	E'	E'	229	187	264
P. plasticum; 51127-8, 51278.	17.7	5.21	1.35	45>33	i	3.62	0.31	113	E'	E'	213	332	476

Calliphylloceras sp. from Alaska; 51132.	68.0	9.35	0.97	-----	i	2.66	-0.09	96	---	238	205	204*	388
C. heterophylloides; 51133.	25.0	8.43	0.88	60>45	i	3.32	0.02	107	E'	457	221	197	428
C. malayanum; 51134, 51261.	18.6	10.09	0.84	40>40	i	3.21	0.24	114	E'	E'	203	216	442
C. sp. from Mendoza; 51135.	13.6	9.26	0.74	35>29	i	2.90	0.21	116	E'	E'	333	325*	614
Hypophylloceras subramosum; 51252, MM 19688.	4.29	8.14	1.19	28>30	i	2.74	0.07	128	305	187	169	186	494
LYTOCERATINA													
Lytoceras sp., Irianjaya; 51143.	60.7	6.40	0.86	30>22	12	3.18	0.15	101	E'	E'	---	315*	595
L. sp., U. Jur. or Cretaceous; 51231.	5.95	7.52	1.05	45>--	2.8	?3.3	0.26	129	---	E'	243	404	548
L. fimbriatum; 51136, 51228-51230.	17.5	13.40	1.85	30>10	2.7	3.17	0.33	118	E'	180	---	60	158
L. jurense, Belmont Bed 6; 51137-8.	54.0	21.32	0.93	-----	30	2.22	0.02	104	---	10	---	68	228
L. cornucopia; 51139 from Bed 3, 21361.	68.7	13.15	1.49	-----	i	3.32	0.36	102	---	860	(193)	85*	166
L. eudesianum, Cerro Puchenque; 51140.	14.6	6.78	1.43	35>15	2.5	3.16	0.28	111	E'	E'	153	164	387
L. eudesianum, Alaska ; 51141.	13.7	10.47	1.56	32>15	2.4	2.88	0.22	119	E'	284	---	137*	264
; 51142.	12.8	9.26	1.47	20>0	3.4	<2.88	0.20	118	---	380	228	163*	312
Lobolytoceras siemensi; 51144.	10.4	10.42	1.21	45>15	3.0	2.81	0.21	121	E'	490	224	189*	361
Pleurolytoceras hircinum; AMNH 27465.	6.85	12.56	1.33	40>50	1.6	3.59	0.20	132	395	232	192	115*	223
Pseudophyllites sp., Vancouver; 51145.	67.2	18.48	0.57	20>20	23	3.20	0.19	102	E'	577	170	155	313
Gaudryceras denseplicatum; MM 19690.	5.78	11.83	1.23	35>13	3.9	3.10	0.25	140	E'	E'	202	249	328
G. tenuiliratum; 51146, 51226-7.	5.34	10.03	1.24	30>20	2.6	3.32	0.20	130	E'	485	206	179	328
AMMONITINA													
Arietites sp. from France; 51147.	64.0	32.24	1.25	45<50	i	3.82	0.30	107	42	64	---	31*	75

Table II. (Continued)

	L, mm	IST	a	$\theta > \phi$	T, mm	b	c	S^*	H_p	H_s	H'	H	I
Promicroceras marstonense; AMNH 45052.	1.20	14.47	1.07	38>20	0.9	2.56	0.19	148	107	116	202	146	392
; AMNH 45051a–c.	0.55	13.09	1.28	50>23	0.4	2.70	0.00	162	189	75	---	181	348
Liparoceras sp., Blockley clays; 51150.	19.0	17.58	1.37	40>20	12	3.40	0.21	116	E'	194	---	94	141
L. sp., Blockley Bed 2;51149,AMNH 45059.	43.6	25.98	1.48	70>25	23	3.41	0.16	108	123	30	---	45	76
L. henleyi from Dornten; 51151.	18.3	18.62	2.01	30>10	9	2.85	0.34	115	968	312	---	84	113
Amaltheus margaritatus; 51153.	5.60	35.0	1.75	35>25	i	3.56	0.32	135	29	25	---	26*	58
Dactylioceras commune;51154,51232-33.	8.08	18.51	1.00	25>25	2	3.62	0.23	129	772	172	91	88	196
D. commune, coarse ribs; 51234,J.18.	7.30	22.18	1.25	45>-	2	2.80	0.18	126	---	19	59	56	164
Phlyseogrammoceras doerntense; 51155.	16.7	21.84	*1.15	45>42	i	4.18	0.16	122	193	48	99	49	110
Hildoceras bifrons, Whitby; R.,51156.	12.2	13.60	1.56	45>30	i	4.80	0.29	124	E'	539	147	82	119
H. bifrons from Hurcot Lane; J.1435.	20.8	18.60	1.26	40>16	i	3.51	0.07	121	624	31	78	61	136
Bouleiceras sp., Namakia River; 51157.	69.0	43.13	*0.35	45>18	17	5.00	?0.25	109	E'	971	69	99#	144
B. nitescens; 51158.	27.7	19.05	0.72	25<45	i	3.00	0.05	113	674	352	(78)	189#	258
; 51159.	29.8	19.51	0.94	41>40	i	3.40	-0.14	114	491	75	78	113	164
Erycitoides howelli; 51160-2.	14.8	15.16	1.17	45>35	i	3.71	0.22	121	E'	356	104	115	185
Costileioceras sp., Achdorf; 51163-4.	13.6	17.49	0.93	16>20	i	3.48	0.25	120	E'	617	93	141	216
Brasilia gigantea; 51166-8.	28.0	16.96	1.11	35>30	i	3.54	0.09	114	955	151	94	107	165
Graphoceras concavum; 51170-1.	18.5	18.08	1.06	50>45	i	3.43	0.18	117	495	261	88	119	178
Euhoploceras ochocoense; 51173,51283-6.	18.7	19.28	1.55	35>20	i	3.27	0.07	117	308	37	78	70	112

Papilliceras espinazitensis; 51235.	27.9	30.18	1.80	45>30	i	3.19	0.28	114	15	25	105	32	62
; W., 51174, 51236-7.	10.3	20.08	1.28	30>25	i	4.61	0.27	128	E'	165	94	55	105
Fissilobiceras zitteli; R., AMNH 45202.	35.1	16.20	1.37	25>20	i	3.44	0.20	109	743	118	86	73	142
Strigoceras truellii; AMNH 1113/1.	17.0	18.0	---	-----	i	>2.22	---	113	---	---	96	---	<240
Oxycerites orbis; 51175.	9.0	14.00	0.70	60>35	i	3.26	0.24	123	E'	E'	---	212*	402
O. tilli, Horsany-Hegy; 51176.	4.5	12.00	1.84	25>28	i	3.25	0.20	134	284	116	100	98*	190
O. sulaensis, Ass. VI; 51238, 51258.	35.2	25.83	0.99	60>45	i	2.94	0.28	107	98	217	119	85	147
O. sulaensis, Ass. X; CPC 33304.	20.7	12.94	1.06	60>35	i	3.37	0.12	113	810	226	152	126*	243
O. (Alcidellus) sp., Katch; 51178.	25.0	16.55	1.08	30>25	i	3.03	0.18	110	425	170	175	116*	206
Uhligites indopicta; 51179.	11.0	10.15	0.91	50>35	i	2.94	0.14	120	E'	891	183	274	453
Aconeceras nisus; 51180.	6.3	27.26	1.11	35>30	i	2.94	0.30	129	1	21	---	35	151
Lissoceras sp. of Oberdorf; 51181.	4.0	35.30	?1.11	-----	i	?2.94	?0.30	136	---	46	93	51	121
Teloceras cf. blagdeni; 51182-3.	12.0	19.37	1.30	20>15	8	3.37	0.25	123	E'	205	---	92	146
Stephanoceras mutabile; 51184-5. AMNH 45203-4.	23.8	14.62	1.33	45>30	53	3.68	0.22	111	699	186	88	79*	155
	12.1	18.28	0.99	40>25	11	3.01	0.17	119	451	174	97	121	221
Pseudotoites argentinus; 51186.	15.5	17.72	1.14	25>10	10	2.75	0.16	113	519	110	65	106*	206
Emileia giebeli; 51250-1, 51256-7.	13.9	11.40	1.10	40>29	10	3.50	0.23	115	E'	766	(111)	159	270
; inner whorls and microconch 51187.	8.22	15.78	1.00	55>45	6	3.57	0.17	123	569	243	111	118	221
Macrocephalites etheridgei; 51264.	9.81	12.82	0.97	62>30	10	3.46	0.16	122	E'	372	131	151	286
M. bifurcatus bifurcatus; 51188, 51265-6.	24.1	22.08	1.58	35>25	9	3.37	0.27	111	135	71	81	47	93
M. bifurcatus intermedius; CPC 33301.	26.2	16.38	1.30	25>20	10	3.36	0.29	109	943	283	---	80*	157
(Ass. II at Wai Kalipu); 51281-2.	10.6	12.36	?1.3	-----	5	?3.4	?0.29	119	---	311	118	92	230

Table II. (Continued)

	L, mm	IST	a	θ > φ	T, mm	b	c	S*	H_p	H_s	H'	H	I
Epinayaites palmarus; 51262-3, CPC 33306.	13.1	19.30	1.06	40>33	7	3.23	0.09	121	194	68	137	90	178
Quenstedtoceras henrici; 74Pl.	19.0	26.12	1.51	25>15	20	2.86	0.22	113	73	28	---	49*	99
Bullatimorphites (Kheraiceras) sp.; J. 1789a.	5.00	17.19	0.88	60>45	3	3.29	-0.12	129	236	58	---	128*	247
B.(K.) bullatus, Espinacito; J.2271.	20.0	17.19	1.10	40>40	22	3.25	0.20	110	259	184	(111)	96#	181
MLP 12642 from E. & Caracoles J. 1149.	7.51	18.60	0.86	35<40	7	3.50	0.13	124	541	177	111	117	220
B. (Bomburites) uhligi, Lechstedt; J.1824.	6.43	22.13	1.08	70>45	13	3.26	0.10	126	47	33	82	69	157
Leptosphinctes s.l. sp.	43.3	8.51	1.62	-----	-	>2.22	0.21	100	---	396	142	---	<353
AMNH unregistered specimen.	37.7	15.86	?1.59	-----	-	>2.22	?0.23	105	---	105	142	---	<198
Choffatia jupiter; J.1147, 51240-3.	22.6	14.54	1.50	25>22	8	3.83	0.28	116	645	138	---	54	139
Kranaosphinctes burui; 51189, 51260.	15.5	15.08	1.14	35>20	10	3.72	0.21	120	E'	262	103	103	188
Paraboliceras sabineanum; 51244.	23.9	15.43	?1.2	-----	i	?2.5	?0.3	110	---	210	---	103	258
; 51244-5.	11.8	18.81	1.20	->22	i	2.49	0.29	119	74	238	77	118	228
; 51244-5.	2.65	10.54	1.14	60>---	-	?2.5	0.19	136	---	371	---	229	479
Choicensphinctes choicensis; 51190.	57.4	19.02	0.87	23>22	14	2.96	0.28	102	411	275	100	89#	220
; 51191, 51239.	10.2	20.17	1.21	50>15	10	3.11	0.17	124	261	40	121	64	162
Kossmatia bifurcata; 51193-4.	39.1	13.23	1.68	30>30	i	4.30	0.33	109	E'	387	100	56*	113
Virgatosphinctes sanchezi;51195-6,J.1528/1	66.4	13.52	1.33	25>13	27	3.32	0.15	102	E'	231	144	109	168
V. aff. communis; 51197.	101.3	18.9	1.50	20>8	30	2.84	0.14	98	606	38	111	58*	117
Aspidoceras haupti; 51198.	82.3	26.94	1.65	15>15	30	3.70	0.34	102	351	86	---	27*	58

Physodoceras circumspinosum; 51199.	25.9	19.37	2.63	18>-	16	2.22	0.21	107	---	9	145	46*	94
Beudanticeras glabrum & tooth mark;51200.	15.2	14.30	0.92	20>20	6	3.21	0.10	118	E'	548	158	215	279
Anapachydiscus sp;; MM 19692.	128.8	9.61	1.07	50>20	-	2.56	0.08	92	500	243	---	152#	349
; 51246 also from locality H6018.	1.89	11.27	1.13	55>-	-	3.35	0.11	147	---	170	77	151	347
Pachydiscus sp. from N. America; 5161.	60.2	14.27	1.43	-----	16	3.43	0.26	104	---	233	---	76*	150
P. cf. soyaensis from Japan; 51247.	23.2	12.88	1.27	40>20	24	3.60	0.16	115	E'	190	95	102	193
Canadoceras newberryanum; 51201.	26.7	12.87	1.00	60>25	15	3.26	0.32	112	E'	E'	159	216	284
Damesites semicostatus; MM 18205.	22.9	10.64	0.79	----	i	2.22	0.14	107	---	815	---	320*	604
D. sugata; MM 19689, 51253.	1.95	13.36	1.19	45>-	2	?2.74	0.17	141	---	245	197	187	336
D. sp., loc. RH1204; 51279-80.	2.06	15.76	1.13	45>15	2	2.74	0.21	141	E'	359	180	199	302
Placenticeras meeki ; W., C.T.	86.0	22.74	2.09	29>25	i	6.1?	0.41	102	E'	664	50	?19#	?28
; W., C.T., 51202-3.	28.2	16.94	1.70	40>35	i	3.96	0.30	112	823	341	50	71	97
; MM 18214 (USGS loc. 16036).	8.5	19.15	1.88	----	i	4.00	?0.3	128	---	212	50	64	86
	29.0	23.09	1.38	----	i	4.00	?0.24	113	---	155	---	62#	84
Vascoceras cauvini; 46121,46124,46135.	20.4	32.00	1.57	30>30	20	3.60	0.22	117	21	9	65	25	55
(H' includes 8 other specimens).	18.3	32.00	1.97	25>23	14	2.90	0.17	116	-2	-2	65	25	57
ANCYLOCERATINA (P. = 51248-9, MM 19687)													
Didymoceras cheyennense; AMNH 62-15.	28.0	--	1.64	-----	-	---	---	---	---	---	174	---	---
Nipponites sp.; MM 18254, 18571.	6.38	<13.9	---	-----	-	---	---	---	---	---	127	---	---
Eubostrychoceras japonicum; MM 18528.	14.4	14.34	---	-----	-	---	---	---	---	---	---	---	---
Bostrychoceras elongatum; 51204.	21.7	15.25	0.88	15>10	0	2.54	-0.06	111	E'	103	139	165*	316

Table II. (Continued)

	L, mm	IST	a	θ > φ	T, mm	b	c	S*	H_p	H_s	H'	H	l
Hyphantoceras orientale; MM 19691.	11.0	12.60	1.44	20>22	0	2.80	0.14	121	283	127	---	124*	240
Polyptychoceras pseudogaltinum; see *P.*	4.86	9.07	1.18	45>20	0	?3.0	0.06	127	E'	283	154	201	393
Baculites grandis; AMNH 9547.	50.7	11.07	1.18	-----	i	3.50	0.29	104	---	907	204	122*	237
Baculites cuneatus; C.T.	33.5	27.03	1.19	50>30	i	3.42	0.21	112	18	10	48	27	98
B. compressus; C.T., 51205-6.	35.2	18.03	1.47	35>20	i	3.23	0.30	108	E'	483	91	98	129
Baculites reesidei; C.T.	14.5	18.07	1.09	29>20	i	2.95	0.20	118	318	121	106	96	209
B. ovatus from Manyberries; C.T.	9.6	26.64	1.57	28>20	i	3.30	0.22	126	74	31	95	48	88
Baculites bailyi, loc.RH1204: 51207.	7.8	14.73	2.00	22>20	i	4.28	0.30	130	E'	143	---	50*	101
Scaphites hippocrepis; 51208.	10.2	18.48	2.42	20>10	6	2.70	0.33	124	62	43	55	42	107
S. preventricosus; AMNH-424.	12.8	11.99	1.21	60>32	10	3.07	0.11	116	882	305	(151)	167	259
;"B228" Landman Coll. AMNH.	7.2	15.96	?1.21	-----	-	?3.07	?0.11	124	---	94	151	114	218
	1.35	18.61	?1.21	-----	-	?3.07	?0.11	149	---	63	151	112	214
Scaphites whitfieldi; AMNH 45049-50.	7.1	15.19	2.19	20<67	0	?3.07	0.25	127	60	213	---	?112	?129
	4.5	24.70	?2.2	-----	0	?3.07	0.25	138	---	9	---	?36	?83
Clioscaphites vermiformis; AMNH D.9931.	8.90	15.30	?1.7	-----	0	?3.07	?0.23	113	---	165	---	?94#	?147
	2.46	25.44	1.70	-----	0	?3.07	0.23	144	---	129	---	?60*	?110
	0.96	19.70	1.51	35>?15	-	?2.61	0.18	153	102	56	102	113#	201

								E'				
Hoploscaphites spp.; 51209 from U.S.A.	15.0	23.86	1.18	20>15	5	3.79	0.19	123	---	199	83	96	111
	3.9	19.91	?1.18	----	-	?3.7	?0.02	141	---	11	83	?53	?149
; R2110 Tanabe Coll. from Obira.	0.46	7.00	?1.18	----	-	?3.79	?0.07	165	---	89	---	?123	?527
	0.27	4.88	?1.18	----	-	?3.79	?0.00	172	---	162	---	?185	?780
Jeletzkytes spp.; AMNH - Landman Coll.	28.5	26.12	?2.67	----	0	?3.52	0.29	112	---	19	---	?20#	?36
	12.3	26.12	1.30	----	0	<4.67	0.13	126	---	37	75	>44#	>71
; 51210 also from U.S.A.	18.7	22.77	1.37	20>15	9	2.42	0.21	113	63	88	---	98#	153
	7.5	17.19	1.29	30>10	4	2.62	0.23	124	701	177	---	133#	236
	2.7	24.75	1.34	30>15	2	2.65	0.20	133	86	50	---	86#	154
Rhaeboceras subglobosus; C.T.	45.7	24.63	1.17	----	35	3.50	0.04	109	---	37	---	63#	100

[a]Individual septa or series of septa are identified by an average size parameter (*L* in mm) corresponding to their mean inverse septal thickness (*IST*) and mean implosion depth (*I*). The lowest of the four maximum habitat depth estimates (i.e., H_P, H_S, H^*, H) would represent the true depth if there were no errors in these estimates. In reality, the buckling-based estimates (H_P, H_S) exaggerate any measurement and diagenetic errors in the septal thickness data and sometimes yield a negative depth, indicative of an implosion pressure of less than 1 atm. They therefore provide a good guide to smaller errors also present in the estimate H. However, the central parts of the deep-water-adapted septa were simply too thick to fail by buckling. The H_P and H_S estimates are, therefore, often replaced by the symbol E. The specimen registration numbers are discussed in the acknowledgments. The symbol "W." refers to additional siphuncle data in publications by G. E. G. Westermann, and the symbol "R." refers to old notes and peels made by the author from other specimens. The septum data of Saunders (1995) appears to refer to their thin margins. Inferior H^* and $H^\#$ estimates (Table I) are indicated by a suffix in column H. A prototype of this column yielded an average safety factor based on 1368 *IST* measurements (Fig. 11) and the similar depths reviewed in Chapter 16 of this volume. The more anomalous depth limit cited there for *Stephanoceras mutabile* was based on the siphuncles ($H' = 64$ m) of additional specimens measured by Geraghty and Westermann (1994).

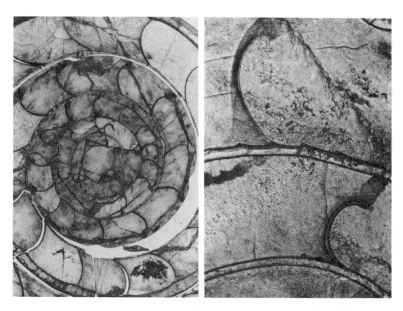

FIGURE 3. Median thin section (left) through the strongest juvenile whorls of *Glaphyrites clinei* (Miller and Owen) showing localized and anticlastic shrinkage of the siphuncle. The 5.5-mm-long enlargement of these whorls (right) shows septum 9 ($L = 0.23$ mm) and the larger but relatively thin septum 20 (numbered from the proseptum). From Carboniferous shales, 0.75 mile south of Collinsville, Oklahoma (SUI 1735; see Miller and Unklesbay, 1943, p. 13).

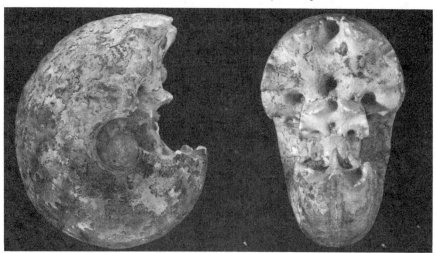

FIGURE 4. Transverse (right) and lateral (left) views of a neoceratite phragmocone (diameter 61 mm) from one of the shallowest ammonoid habitats (ROM 46124). This specimen of *Vascoceras cauvini* Chudeau displays ceratitic sutures of relatively flat and anticlastic septa. It was collected from a layer at the top of the main limestone (unit six) quarried at Ashaka in the Gongila Formation of northeast Nigeria. P. M. P. Zaborski (personal communication, 1987) suggests an "uppermost Cenomanian" (Cretaceous) age.

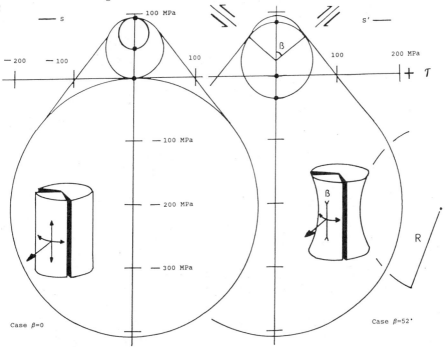

FIGURE 5. Two Mohr diagrams illustrating the reduced apparent tensile strength ($S^* < S$) of the large nacreous flutes (300 cm^3 at maximum stress) resulting from the addition of anticlastic curvature (right) to a cylindrical segment in a state of membrane tension (left). Shear stress (τ) generated by clockwise motion around a shear plane and tensile stress (above) are defined as positive. Both flutes are loaded by hydrostatic pressure applied from the inside of the cylinder, which developed a compressive stress around the circumference of radius R in the anticlastic case. The small circle through the two membrane stresses in the cylinder (95 and 47 MPa) is enlarged, touching the envelope of failure at a dihedral angle (β) of about 52°. This indicates that the anticlastic shell failed by shearing along planes situated some 26° from the tensile stress direction at a reduced and apparent tensile strength $S' = S^*$. The angle will be reduced to zero in a cylinder and will also be somewhat reduced when smaller volumes of nacre are broken at the same tensile-to-compressive stress ratio. Mohr diagrams are reviewed in most textbooks on structural geology and strength of materials.

orders of magnitude greater than the force per unit of external "surface area" applied at a depth of 90 m, it was theoretically possible for a mature and shallow-dwelling ammonoid to have constructed a spherical phragmocone wall with a volume equal to 1% of the total phragmocone volume. In reality, this wall volume would have been increased to 2.5% to prevent the sphere from buckling in compression (i.e., see Ross, 1990, who presents an "empirical equation" that reduces the depths deduced from the "classical equation" reviewed by Hewitt and Westermann, 1988). The density of the phragmocone was further increased by other functions of an ammonoid shell. The combined volumes of the septa and phragmocone wall were actually about 10% of the total phragmocone volume.

Bayer (1977) implied that the body of an ammonoid could resist or modify the pressure applied to the "last septum." A fully muscular mantle produces a tensile stress of 0.26 MPa and a time-averaged tensile stress of 0.13 MPa (Johnsen and Kier, 1993). Each muscle fiber represents a membrane force holding a part of the hydrostatic load in tension. A mixture of muscular tissues and blood vessels might just have been able to hold the pressure of 0.10 MPa acting against the vacuum at sea level. However, when such muscles were aligned parallel to the surface of the mantle, they could have held a maximum pressure of only 0.04 MPa (i.e., 0.13 MPa/π in a cylinder, other morphologies being of equal or reduced strength after muscle realignment). Any increase in the thickness of the muscular layer (i.e., membrane force) also increased their cross-sectional area, and it produced no increase in tensile strength.

2.2. Implosion of the "Last Septum"

The combined pressures acting on the buoyant part of the ammonoid phragmocone were concentrated within the two "cross-sectional areas" (i.e., as defined above, not just the cross-sectional area of the shell) of each point on the phragmocone wall, the "last septum," and the siphuncular tube wall. A small part of this hydrostatic load was also transferred to the central part of all the other septa via their sutures with the phragmocone wall (Westermann, 1975). Major tensile stresses developed in any thin region of the external shell surface and "last septum" displaying externally concave curvature. Compressive stresses developed in the externally convex regions of the phragmocone wall. Both of these kinds of membrane stress varied little within each "cross section" through the local thickness of the shell material (d).

Simple shells with double curvature may be defined by orthogonal radii of curvature (r, R) situated on one side of the shell surface. They are said to display "bowed-out," "elliptic," "synclastic," or "positive Gaussian curvature." The semihemispheric whorls of *Goniatites* and septa of *Bactrites* display this type of curvature at any point on their surface. This curvature generated entirely compressive or entirely tensile membrane stresses, and it produced simple septal sutures potentially able to withstand high hydrostatic pressures. In contrast, the fluted septa (Figs. 4–6) and corrugated ribs (Fig. 2) of many ammonoids illustrate the opposite morphology, variously known as "bowed-in," "hyperbolic," "anticlastic," or "negative Gaussian curvature." In this case, the radii are separated by the plane of the shell surface, and the latter develops opposite displacements and stresses within the orthogonal planes of the radii of curvature. The "last septum" expanded in the orthogonal plane of the externally concave surface and contracted in the orthogonal plane of the externally convex surface. The graphical representation of these stresses by Mohr's circles (Figs. 5–7) shows that the septal nacre would then have broken by shearing at a much reduced, apparent tensile stress (S^* on Tables I and II).

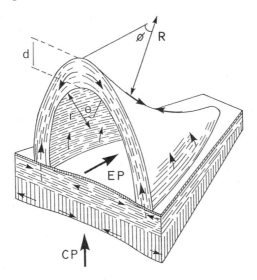

FIGURE 6. Diagram illustrating the anticlastic curvature of the lobule of the last septum of an ammonoid (redrawn from Hewitt and Westermann, 1986, Fig. 7). The maximum and axial radius of curvature of the lobule (R) is rotated around a smaller angle ϕ than the minimum and transverse radius of curvature (r) with a rotation angle θ and a parabolic profile. Note the great local axial thickness (d), the directly applied external hydrostatic pressure (EP) and the circumferentically applied hydrostatic pressure (CP). The orientation of the θ and ϕ angles and the orthogonal radii of curvature are normally reversed in the central part of an ammonoid septum.

    Those ammonoid septa displaying complete anticlastic curvature can be identified by having a central orthogonal stress ratio (b) of more than 2.0. The septa with equal and anticlastic radii of curvature (e.g., b in their central region = 3.0 in Table II) are said to have a mean curvature of zero, and they were stable only at zero pressure before they were calcified (Bayer, 1977, p. 360; Isenberg, 1992, p. 109). This is not a stable morphology, as suggested by Bayer (1985), and yet, it developed on at least three occasions during the Devonian (Korn, 1991). Ammonoids may have been preadapted for this fluting by an increase in the density and stiffness of their "cameral liquid" to form a gel. Such a gel initially increased the capillary attraction of films of water into the posterior chambers of orthocerids containing cameral deposits, and it later became dense enough to balance the pressure applied to fluted cameral membranes by the body of the ammonoid (Hewitt and Westermann, 1987a, p. 160).

    If each "last septum" had generated only bending stresses, it would have wasted most of the weight of the nacre in relatively unstressed middle shell layers and in the sutural regions analogous to the proximal end of a beam. The crude equation for the maximum bending stress in a simply supported flat square (Fig. 8) implies that flat versions of many ammonite septa implode at sea level [e.g., when their inverse septal thickness ($IST = L/4d$) is greater than

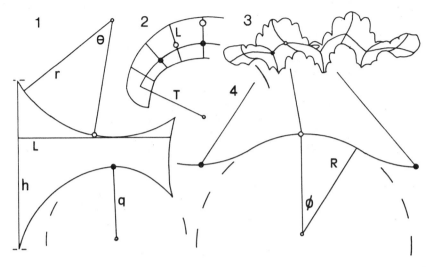

FIGURE 7. (1) Submedian section illustrating the rotation of the axial radius of curvature of the saddle-flute (*r*) through a half-angle θ, and the minimum central radius (*q*) of the more parabolic lobe–flute axis. Note flute length (*L*) and ventral sutural amplitude (*h* as measured by Batt, 1991). (2) Transverse whorl section showing the torus radius (*T*) defined by the central hinges of the lobe flutes and the measured length (*L*) of the largest saddle–flute axis. (3) Oblique view of the same septum showing principal stress directions of the two central saddle and lobe-flutes. (4) Curved transverse section around the circumference of the torus radius and normal to the axes of curvature of the saddle-flute (*R*). The transverse and central radius of curvature of the saddle-flute (*R*) is rotated through a half-angle φ.

19, the maximum span length (*L*) is the length (i) of the flat square, and if *d* is a constant septal thickness]. The central part of real ammonite septa could have resisted hydrostatic pressure without bending. Their minimum radius of curvature (*r*) was over ten times their thickness (*d*).

The central part of a fluted septum consists of alternate concave and convex anticlastic axes described in Tables I and II. Septa situated within relatively weak phragmocone walls, with a large lateral radius of curvature (i.e., "compressed"; Fig. 9), have flat saddle–flute axes (large chord ratio *a*). They also have contrasting radii of curvature between the lobes and saddles, indicated by a large central axial ratio *c*. The folioles and lobules of their sutures (Fig. 9) form three-dimensional bifurcations in which both the externally concave radii of the flutes and the linear distance from the suture are roughly halved. This dendritic pattern is complicated by the parabolic axial profile of the central lobes (Fig. 7-3) and the transverse profile of the marginal lobules (Fig. 6). The projected transverse radii of curvature of the lobules produce V-shaped incisions or "tie points" between the externally convex folioles of the septal suture. It is assumed here that the difference in the central curvature of lobe and saddle-flute axes (e.g., parameter *c*) was proportional to the extra load applied via the phragmocone wall.

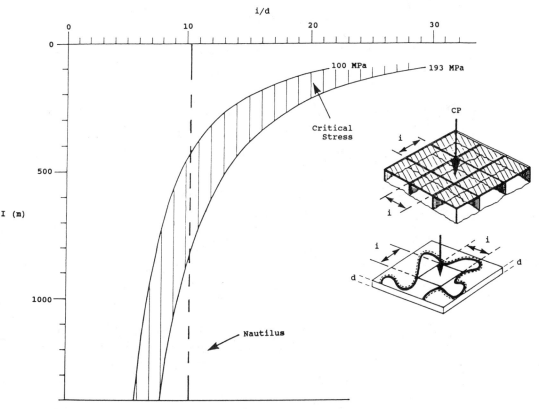

FIGURE 8. Relationship between the implosion depth in meters (I) and the relative support distance of a flat and simply supported whorl or septum area (i is the length of an unsupported square, and d the constant thickness; redrawn from Hewitt and Westermann, 1986, Fig. 6).

    The widespread assumption that thin and nacreous septa were less likely to fail in compression than tension is incorrect. Pfaff (1911, p. 211) determined a tensile strength of 29 MPa and a compressive strength of 176 MPa from *Nautilus* septa. The review by Hewitt and Westermann (1987b) increased these estimates to 131 and 411 MPa, respectively. However, Henderson (1984) rightly suggested that the phragmocone wall and "last septum" were also liable to fail in compression by the local buckling reviewed by Wainwright *et al.* (1976, p. 252). Henderson refuted his own hypothesis by confusing the compressive stress required to buckle the nacre with the hydrostatic pressure (Hewitt and Westermann, 1988). Thin septa broke in tension as soon as they buckled between the lobes. Two alternative and tentative maximum habitat depths (H_p, H_s) are deduced here (Table II) from the morphology of this region of the "last septum" and the stiffness (E) of *Nautilus* nacre.

    Buckling failure promoted the development of the high-amplitude saddles noted by Batt (1991) in deeper facies. The depressed whorls and depressed

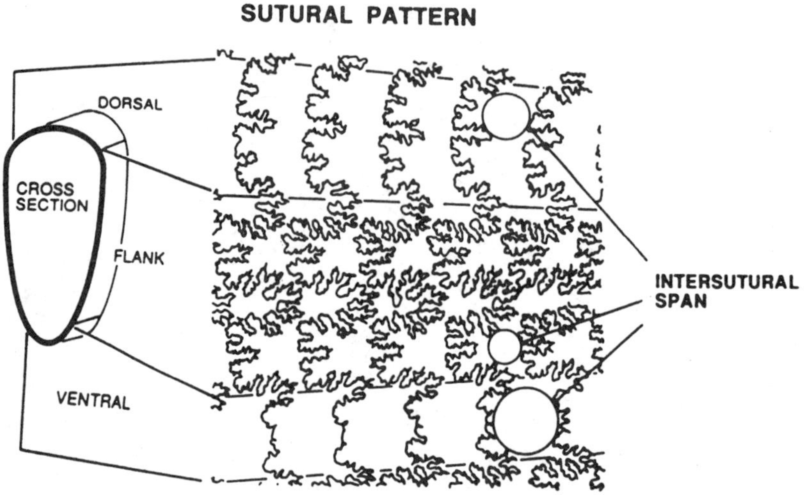

SUTURAL PATTERN

FIGURE 9. Intersutural spans (i) of vault-like surfaces on a phragmocone of *Baculites* are inverse to the local minimum radius of curvature of this straight shell (after Jacobs, 1990, Fig. 3).

septa of tiny juvenile ammonoids (Hewitt and Westermann, 1987a) were particularly liable to buckle. The flutes between their saddles buttressed the potentially toroidal geometry of their septa by introducing bands of tension along the central part of the septum. The radius (T) of this hypothetical torus (Fig. 7-1, Tables I and II), the central lobe–flute radius (q), and the maximum septal thickness (d) were analyzed using Fig. 5 of Sobel and Flugge (1967). These hypothetical unfluted septa would have buckled at depths of 10 to 300 m unless they were supported by the tensile zones between the saddles.

Varices are regularly spaced apertural thickenings associated with constrictions (Fig. 10), producing rings of external shell material of double or triple the normal thickness. Varices reproduced the bending stress concentrations associated with the stiff *Nautilus* sutures studied by Hewitt and Westermann (1987b). This must have destroyed any precise balance of forces produced by complex septal sutures. The inflated curvature of the externally convex, anticlastic elements of the phylloceratid septal suture (i.e., folioles) was also concerned with increasing the strength of thin-walled phragmocones. The small and undifferentiated lobules and folioles of a complex *Lytoceras* suture developed a neutral displacement against the inward motion of the surrounding wall. This type of sutural complexity increased within these strong and rounded whorl sections (Bayer, 1977; Jacobs, 1990), and it should not be confused with that produced by phylloid or ceratitic sutures. The latter pushed outward to support the phragmocone wall, but undifferentiated lobules and folioles acted more like springs. The stiffness of the phragmocone wall and the contrasting flexibility of ammonoid septal sutures reduced the excessive bending moments produced by contrasting radial displacements of

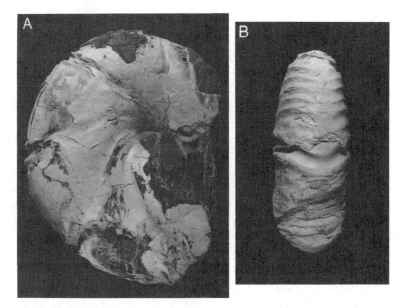

FIGURE 10. Two specimens of *Ptychophylloceras galoi* (Boehm) from the Upper Jurassic of the Sula Islands showing the thick shell material forming varices. (A) ROM 51130, diameter 132 mm. (B) ROM 51131, diameter 122 mm.

planar surfaces joining the septum to the phragmocone wall (Hewitt and Westermann, 1986; Jacobs, 1990). Bayer (1977, p.356) illustrated moderately complex but widely separated septal sutures of ammonites. These sutures reduced the bending moments over the "last septum" rather than within the whole phragmocone.

2.3. Habitat Depth Limits

Callomon (1994) confused the implosion depth (I) of average septa with the reduced implosion depth of thinner septa defining the maximum habitat depth (H). The average implosion depth (I) of the septa was increased to reduce the number of deaths from the implosion of these rare septa. Hewitt and Westermann (1990) and Hewitt (1993) estimated the safety factor against implosion from the ratio $(I + 10)/(H + 10)$ deduced from the measured variance of the septal strength estimates. They found that the measurement error was approximately balanced by the variance related to the variable strength of a standard volume of nacre. A 1-in-20 death rate from implosion was deduced from records of juvenile *Nautilus* living down to a depth of 400 m in Fiji.

Hewitt and Westermann (1990) defined this habitat depth (H) as the maximum depth visited during the growth of every chamber during one or more juvenile phases of ontogeny. Each "phase" of an ammonoid ontogeny may be defined as the time spent growing a series of septa, with a constant mean strength and variance, at one particular maximum depth (*H*). These phases of ontogeny are commonly delineated by sudden changes in septal thickness associated with unusually approximated septal sutures (i.e., reduced cameral angles between septa). Table II presents averages of the inverse septal thickness (*IST*) within three or more phases of an ontogeny for various species. It is also important to separate morphotypes of variable or taxonomically "lumped" species with a significantly different mean *IST*. The variance of small or inappropriately combined samples can yield unrealistically high safety factors (Fig. 11). Ammonoids have been analyzed by assuming that there were 60 septa per juvenile phase and a 1-in-20 risk of an individual ammonoid being killed by implosion during this period of their ontogeny. These assumptions imply that the maximum habitat depth (*H*) was situated at 3.15 standard deviations of *IST* above the mean implosion depth (*I*). All these calculations are based on the $n-1$ or sample definition of one standard deviation. However, an average and population-based safety factor $[(I + 10)/(H + 10)]$ of 1.86 was used to roughly estimate the maximum habitat depth (H^*) of the remaining samples represented by an inadequate number of *IST* measurements (<5). This approach underestimates the habitat depth limits of the initial and mature phases of growth containing a small number of septa (Hewitt and Westermann, 1990) by 20–30%.

2.4. "Internal Septa" and Siphuncle

Reviewers of the function of ammonoid septa contrast the role of flutes as a "buttress" for weak whorls (Raup and Stanley, 1971; Westermann, 1975; Seilacher, 1975; Wainwright *et al.*, 1976, p. 263; Ward, 1980) with other internal functions such as increasing the capillarity (Mutvei, 1967, 1975; Kulicki and Mutvei, 1988; Saunders, 1995; Saunders *et al.*, 1994) or "decoupling" of "cameral liquid" (Ward, 1987, pp. 220–235; Weitschat and Bandel, 1991, 1992).

An earlier review by Rangheard and Theobald (1961) concluded that the strength of an ammonoid shell depended on the septa and ornament, and that "the chief use of the shell is to protect the body." Cowen *et al.* (1973, p. 210) commented on the similar views of Westermann (1971, p. 7) by writing that "we agree that ribbing and tuberculation would have helped to protect the shell during predation. But any ribbing and tuberculation would have achieved this." They incorrectly assumed that the ribs grew obliquely to the aperture (Checa and Westermann, 1989) and seldom faded or bifurcated on the underlying venter of mature macroconchs (Fig. 2). However, it is possible to argue that fluted septa buttressed the phragmocone wall and increased the capillarity of the chambers without presupposing that the morphology or thickness of each "last septum" was modified for internal functions.

Westermann and Weaver (1979) studied two series of compressed synclastic (*a* = 1.07, *b* = 1.67) and fluted septa (*a* = 1.38, *b* = 2.2, *c* = 0.13) formed from the liquid plastic PL-1. Septal models were installed in an elliptical cylinder representing the phragmocone wall and subjected to ambient hydrostatic pressure by production of a partial vacuum within the camera under room atmospheric pressure (e.g., P_1 = 0.09 MPa). The marginal septa and the phragmocone walls both had a thickness (d) of 1.5 mm. The minimum cross-sectional diameter (L) of 107 mm corresponded to the position of the lobes in the fluted model. The orthogonal diameter of 193 mm contained three corrugations producing six lobes and six saddles around the septal suture. The plastic had half the stiffness of nacre and a septal thickness of 1.3 mm at the centrally placed strain gauge between the lobes.

The unpublished notes of G. E. G. Westermann indicate that a pressure of 0.09 MPa produced a tensile strain of +0.000069 between the lobes on the concave surface when the septum was in an "internal" position. This changed to a compressive strain of –0.000133 when the same septum was in the "last septum" position. The synclastic model had a strain of –0.000087, which changed to +0.000005 when it was loaded in the "last septum" position. It is reasonable to assume that only half of the bending stress in the internal and fluted septum was present in the "last septum" adjacent to an unloaded chamber. The relevant equations of Hewitt and Westermann (1987b) suggest that the axial region between the lobes of the model "last septum" developed a compressive stress that was 50 times greater than the hydrostatic pressure and three times greater than the equivalent compressive stress in the "internal septum."

Another region of the fluted septal model was inadvertently broken when loaded by the same pressure in the reverse direction. It showed a single crack running straight over the fluted face of the septum. One end of this crack extended at an angle of about 45° to the axis between the marginal pair of lobes (i.e., the dihedral angle β shown in Fig. 5 is 90°). The other end extended along the flank of the lobes and parallel to their axes. The bend in the crack was situated on the flank and near to the most anticlastic region of the flute (e.g., q = 36 mm, d = 1.5 mm, b = 3). The broken part of the suture consisted of one whole lobe and a shallow saddle, which rested against the tightly curved margin of the wall. The flutes evidently came apart half-way between the lobe and saddle crests and preserved the rest of the septum intact. Bayer (1977, p. 355) illustrated ammonite septa showing similar patterns of breakage.

Sutural complexity (Ward, 1980; Boyajian and Lutz, 1992) correlates with the ratio of the central thickness (d) to the marginal thickness of an anticlastic septum (Westermann, 1975; Hewitt, 1985). The central thickness was enlarged in proportion to the length of the flutes (L) at any particular habitat depth (H). However, Mesozoic ammonoids increased the number of marginal flutes rather than the thickness along the septal margin. These thin and numerous flutes protected the "last septum" like a spring, and they subsequently reduced the risk of lethal damage being inflicted on the "internal septa." A large or ventrally situated fracture could have produced enough cracks and shards to rupture the siphuncle. The explosion of the siphuncle would have started an uncontrolled and lethal cycle of cameral refilling, reverse hydrostatic loading of internal septa, and internal breakage.

The chitinous segments of the siphuncular tube were continuously loaded in tension by the blood and coelomic fluid within the siphuncle. The whole siphuncle can be compared to a garden hose under internal water pressure, and it would have expanded in all directions in a state of tension. It was paradoxically liable to buckle if the axial expansion was opposed to the "last septum." Roark and Young (1982, p. 502) refer to this buckling of a corrugated or cylindrical tube by internal pressure as "squirming instability." A siphuncular tube would have squirmed if the product of the pressure within the tube and the cross-sectional area of the tube became equal to the axial load initiating the Euler buckling of a thin-walled cylinder. These buckling equations are reviewed by Wainwright *et al.* (1976, pp. 249–250) and Roark and Young (1982, p. 503). The relevant siphuncle parameters (Table I) are the mean radius (r) along the tube axis, the wall thickness (d'), the square of the slenderness ratio (l/r), and the unknown stiffness (E^*) of the chitinous wall. The uncalcified segments of an adult *Nautilus* siphuncle are composed of protein (13% by volume), chitin (5% by volume), and water (i.e., 82% as deduced by Lowenstam *et al.*, 1984; Grégoire, 1984, p. 8; and confirmed below). The septal necks evidently had the important functions of interrupting the overall expansion of this curved and flexible tube by subdividing it into straight segments.

Squirming instability could also have ruptured each chitinous segment of the siphuncle. However, these segments were probably able to expand slightly along their axis inside the septal necks (Bayer, 1977, p. 341). The stress analysis by Hewitt *et al.* (1993) originally suggested that the chitinous segments would have inflated in tension rather than expand into the neck and then squirm. The bending moment produced by the inflation of the segment halved the strength of a hypothetical *Michelinoceras* siphuncle with an isotropic stiffness (E^*) of 635 MPa between septal necks (R.A. Hewitt and U. A. Abdelsalam, unpublished data). The highest tensile bending stresses occurred near the necks, as previously suggested by Chamberlain and Moore (1982), and also midway between them. However, Hewitt *et al.* (1993) subsequently found that the *Michelinoceras* siphuncle could have remained straight and in membrane tension if the chitinous material had a more realistic axial stiffness (E^*) of 5900 MPa and a transverse stiffness of 11,800 MPa.

The slender siphuncles of orthocerids had a fine-grained matrix of aragonite, and their stiffness was further increased by a spicular layer of aragonite that is not observed in ammonoids. Ammonoid siphuncles have segments of uncalcified material preserved as a diagenetic phosphate mineral known as francolite (Westermann, 1982; Hewitt and Westermann, 1983). The color, optical properties, and exfoliation textures of these phosphatic tubes suggest that they were originally composed of two layers of approximately the same thickness. They also display more evidence of anticlastic shrinkage and early diagenetic distortion than nautiloid siphuncles (Miller and Unklesbay, 1943;

Joly, 1976; Hasenmueller and Hattin, 1985; Weitschat and Bandel, 1991). The literature cited by Grégoire (1984) has not proved that their chitin fibers formed stiff hoops around the siphuncle axis like the α-chitin fibers of *Nautilus* described by Lowenstam *et al.* (1984). The axial stiffness of α-chitin fibers is known to be about 77 times greater than their tensile strength (Wainwright *et al.*, 1976, p. 109). The transverse and axial tensile strengths of the uncalcified part of an adult *Nautilus* siphuncle are 70 and 35 MPa, respectively (see below), and they imply that the orthotropic stiffnesses (E^*) were 5500 MPa and 2700 MPa, respectively.

The magnitudes of the bending stresses are likely to have increased with the square of the slenderness ratio if the relative thickness (d'/r) and stiffness (E^*) of the tube remained constant. The square of the average slenderness ratio of the chitinous segments increased from about 60 in *Michelinoceras* and 88 in the majority of ammonoids (mean of 29 genera) to 2000 in the larger chambers of a *Lytoceras*. It varied from 25 to 80 in *Nautilus* after removal of the spicular layer. The majority of ammonoid siphuncles probably needed only to have half the stiffness (E^*) of a mature *Nautilus* siphuncle in order to avoid squirming instability within their habitats (e.g., an adult *Nautilus* requires a stiffness (E^*) of more than 460 MPa to prevent squirming at pressures recorded by Chamberlain and Moore, 1982). However, there is a minority group of relatively deep-water-adapted ammonoids with relatively slender siphuncles. Some lytoceratids and phylloceratids probably required a siphuncular tube wall with a greater stiffness (E^*) of 1800 to 4700 MPa. These potentially weak siphuncles were small enough to pass through about one-third of the width of the single anticlastic flute that supported the long septal necks and other nacreous structures (Doguzhaeva, 1988) associated with their complex septal sutures. The nacreous parts of their siphuncles reduced but did not entirely remove the risk of bending or buckling failure.

Segments of an ammonoid siphuncle with a ratio of radius (r) to chitinous wall thickness (d') of more than 10:1 probably also had their strength limited by tensile membrane stresses. The required average radius of each point on the siphuncular tube (r) can be calculated by adding the inner and outer diameters of the tube and dividing the sum by 4. The explosion pressure ($P_1 + P_2$) is then obtained by dividing an assumed tensile strength in MPa units by r/d'. An alternative approach uses the internal shear stress when r/d' is less than 10. This shear stress is equal to the square of the outer radius ($r + 0.5$ d') divided by the difference between the squares of the outer and inner radii (Hewitt and Westermann, 1987b). The average shear stress is multiplied by a constant of 1.814 for direct comparison with the tensile stress deduced from the thin-walled siphuncles. This constant is simply the ratio between the tensile and the shear stress at the boundary between these alternative modes of failure (i.e., at $r/d = 10$). The maximum habitat depth (H') was deduced by dividing this stress into an average working stress of 16.45 MPa deduced from the maximum habitat depth of 108 ammonoid species and growth phases preserving a siphuncle (i.e. H not H^* on Table II).

Denton and Gilpin-Brown (1966, p. 754) studied well-preserved *Nautilus* siphuncles with a wall thickness equal to about 12% of the central radius (i.e., d'/r = probably 0.12). Denton and Gilpin-Brown found that "about three quarters" of the material was water, and the unrolled siphuncle broke at a tensile strength of 35 MPa when loaded along the relatively weak axial direction of the tube. In the experiments of Chamberlain and Moore (1982), cited above, the cylindrical tube apparently split in both directions at a transverse stress of 70 MPa and an axial stress of 35 MPa at an average explosion pressure of 8.33 MPa. Variable or anisotropic water contents may explain why the thin-walled and thick-walled segments of a *Nautilus* siphuncle exploded at a similar pressure in the experiments of Chamberlain and Moore. Both the square of the slenderness ratio (l/r) and the relative wall thickness (d'/r) increased by a factor of two between chambers 19 and 30 of specimen A (e.g., Chamberlain and Moore, 1982, Fig. 8; Hewitt and Westermann, 1987b). Because the nacreous septal necks represented an invariant geometry and stiffness, it can be suggested that any mode of failure of the siphuncle should have produced a significant difference in explosion pressures in these two particular chambers.

The writer removed part of the final siphuncular segment of an adult *Nautilus pompilius* from dilute and neutral-buffered formalin and dried it at 120°C. The wall thickness (d') decreased from 0.18 mm to 0.049 mm, and the sample weight was reduced 7.4 mg to 1.55 mg. A rough estimate of the relative density of the dry tissue confirmed that water initially formed 82% of the volume. The dry absolute (d') and relative thickness (d'/r) were equal to 26.5% of their initial values (e.g., d'/r = 0.0305, reduced from 0.115). This inconsistancy can be explained by postulating that 32% of the original axial length of the tube was also occupied by water. Rehydration by immersion in tap water increased the weight to only 3.4 mg and the thickness (d') to 0.082 mm.

Hydrostatic loading of siphuncular tube walls generates tensile forces that tend to pull apart the hoops of chitin fibers and the axially aligned protein fibers. The cross-sectional area of the chitin fibers is not large enough to hold the whole load applied to longitudinal sections of the tube wall, and the latter is spread into the protein. However, the chitin–protein complex may be too poorly bonded to transfer a significant part of the tension to the aqueous matrix of these fibers. This hypothesis suggests that the tensile strengths of the solid components of a *Nautilus* siphuncle varied from about 140 MPa in the axially aligned protein to about 403 MPa in the hoops of the chitin–protein complex. These strength estimates could be used to deduce the reduced tensile strengths of a *Nautilus* or ammonoid siphuncle with a higher porosity and water content than a mature *Nautilus* siphuncle. However, it is possible that the chitin fibers of an ammonoid siphuncle were realigned or concentrated to increase the axial stiffness of the tube wall. This could explain why the tensile strength of the ammonoid siphuncles was halved in the transverse or hoop-stress direction.

2.5. History of Bathymetric Calculations

Mutvei (1967, p. 165) appears to have made siphuncle strength estimates before the calculation was published by Denton and Gilpin-Brown (1966). He commented that the "conchiolin tubes in *Promicroceras* and its allies are approximately as thick as the inner conchiolin layers" (d') in a *Nautilus* siphuncle. He made the deduction "that the wall of the siphonal tube in the ammonites under discussion could stand as high hydrostatic pressure as that in *Nautilus*." Mutvei (1975, p. 199) translated the r/d' ratio measured by Denton and Gilpin-Brown into the siphuncle strength index (SiSI) of ammonites later defined by Westermann (1971).

This particular "strength index of the siphuncular tube" (Westermann, 1971, p. 26) also usefully measured the surface area of the siphuncular epithelium. It was defined as the wall thickness (d') multiplied by 100 and divided by the inner radius of the tube (e.g., $r - (d'/2)$ not r). It increasingly overestimated the explosion depth of the stronger tubes and took no account of the shear failure of cylinders with an index of more than 10.5.

Mutvei (1975) noted that *Pavlovia* had an SiSI of nine to 12 and that *Promicroceras* and other ammonites had "a similar wall thickness." However, his photographs show that the SiSI was no more than 6.5 in *Pavlovia*, and his diagrams confirm that there was also a twofold error in his *Nautilus* measurements (e.g., d' being halved in ammonites and doubled in *Nautilus*). Mutvei (1975) and Tanabe (1977) also deliberately doubled the index to take account of possible diagenetic shrinkage. Tanabe (1979) later accepted the objection that shrinkage was unlikely to occur during marine diagenesis and corrected his own measurements. However, Mutvei (1975, p. 199) concluded that *Pavlovia* was "able to withstand a hydrostatic pressure at depths of 1000 m or more." A habitat depth limit (H') of 133 m can be deduced from his Fig. 3.

Westermann (1971) deduced that the siphuncle of adult *Ammonitina* "could not generally withstand the pressure at more than 100 m" and that the "thickness of the septum usually varies with the strength estimate of the siphuncular tube." However, Westermann (1971) and Tanabe (1977, 1979) used the axial tensile strength of the *Nautilus* siphuncle to calibrate the SiSI. The valid correction of this calibration to an average tensile strength of 67 MPa (Westermann, 1982) now appears to have overestimated their strength. Westermann (1971, 1982) also assumed that ammonoids maintained a constant safety factor against explosion (I/H) rather than a constant safe distance ($I-H$, see Fig. 11) and rejected the evidence for ontogenetic shallowing provided by his siphuncle measurements.

The relatively large siphuncles seen in the early whorls of ammonoids were actually well adapted to resist pressures equivalent to a habitat depth limit of 300 to 900 m. The relatively slender and thin-walled chitinous segments seen in their outer whorls (Tanabe, 1979) represent various mature habitats with a total range from about 20 m in *Ceratites* to 500 m in a dwarf *Phylloceras*. This ontogenetic shallowing could explain why Hewitt (1985) found that the

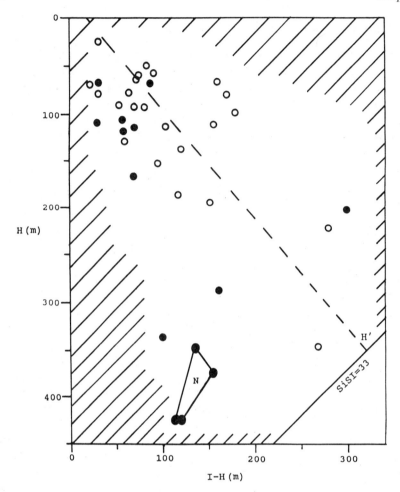

FIGURE 11. Maximum habitat depth (*H*) of various ammonoids and juvenile *Nautilus* (N) deduced from the variance of their septal strength in samples of more than 13 septa (the same samples are revised and enlarged in Table II). The safety factors against siphuncle explosion probably decreased with depth so as to produce a relatively constant safe distance (*I–H*) below the habitat of each ammonoid. The illustrated calibration of the 33 units of the siphuncle strength index (SiSI) of ammonoids assumes that they had a working stress of 15 MPa and a safety factor of 1.86 against siphuncle explosion. Solid circles represent ammonoids with concordant SiSI values in this particular calibration of the septum and siphuncle strength estimates. Table II suggests that ammonoids actually had an average working stress of 16.45 MPa in their siphuncles. The larger solid dots represent juvenile *Nautilus* with a significantly higher working stress in their siphuncles. Their siphuncles have a SiSI of 11 to 16, not 20 to 25.

attachment area of complex septal sutures failed to increase in proportion to the area of the whorl during ontogeny. Most mature ammonoid shells do not show a terminal increase in the thickness of their septa as do *Nautilus*, although there are exceptions to this rule (e.g., *Eothalassoceras, Ptychophyl-*

loceras, Leptosphinctes, Scaphites). However, many mature shells do show evidence of a sudden increase in their septal thickness that was sustained during the growth of their last whorl or half-whorl (e.g., *Dombarites*, macroconchs of *Emileia, Paraboliceras* in Table II). Other genera, such as *Lytoceras*, included both shallow-water- and deep-water-adapted species, and they evidently found it easy to reduce the thickness and spacing of their septa when moving into a shallower habitat.

The maximum habitat depth deduced from siphuncles (H') are compared here (Table II) with estimates from the central region of the septum lying between the largest saddles (H). This was previously attempted by Hewitt and Westermann (1987a), Westermann (1990), Hewitt (1993), Wang and Westermann (1993), and Geraghty and Westermann (1994). But some of their specimens did not yield valid *IST* measurements, and they are just listed here to provide the following estimates of H': *Arcestes* 158 m (ROM 51116 at $L = 2$ mm), *Asteroceras* 91 m (ROM 51148), *Dichotomosphinctes* 101 m (ROM 51192), *Parathisbites* 97 m, *Protrachyceras* 129 m, *Pseudoamaltheus* 36 m (ROM 51152). It is also possible to deduce H' from the middle of the range of some of the SiSI measurements of Ward and Signor (1983). It varies from 62 m in *Submortoniceras*, 70 m in *Acanthohoplites*, and 77 m in *Collignoniceras* to about 262 m in *Lytoceras*.

The most precise depth estimates (H) are based on equations 3.1 to 3.6 of Hewitt (1993). The relative thickness of a septum is now defined as the inverse septal thickness (*IST*). It is obtained by dividing the distance between the saddles on opposite flanks or sides of the whorl (L) by four times the central shell thickness of the septum (d). The morphology of the majority of these septa was measured directly by matching transparent graduated circles to camera lucida images of their flute profiles. The strength of the central saddle–flute is relatively easy to calculate from the tensile strength (S), the projected or actual area of the septum (K), the thickness (d) of the septum and a standard half-volume of nacre (V):

$$H = (I + 10)(1 - 3.15\sigma IST/\overline{IST}) - 10 \tag{1}$$

implosion pressure = (apparent tensile strength)/(tensile stress) =

$$(I + 10) = 100\{S - [f(S - t)/(b - c)]\}/[a(L/4d)(b - c)] \tag{2}$$

$$\text{tensile strength} = S = 136.41\,[V/(Kd/2)]^{1/m} \tag{3}$$

These membrane stress and habitat depth equations differ from those reviewed by Hewitt (1993) in only three significant respects:

1. The central saddle axis is now assumed to fail in membrane shear rather than pure tension at a reduced and apparent tensile strength (S^*)

defined by calculated Mohr envelopes of failure (Fig. 5; Timoshenko, 1983).

2. The *Nautilus*-based Weibull modulus (m) cited in equation 3.4 of Hewitt (1993) is increased from estimates of 13 or 16, based on bending tests (equations reviewed by Timoshenko, 1983), to 27. The new estimate is derived from the variance of implosion test data (Hewitt and Westermann, 1990).

3. A morphological parameter (c) also represents the increase in strength produced by the superimposed compressive stress applied to tensile regions of the septum by hydrostatic loading of the phragmocone wall. The commonly observed difference between the central and axial radius of curvature of the flutes between the lobes (q) and the adjacent axial radius of curvature between the saddles (r) is assumed to maintain equal but opposite lateral displacements of the central septum. These displacements were around the torus radius (T) of depressed septa and along the median plane of compressed septa (Fig. 7). If these lateral displacements were unequal, they would have generated major bending or shear stress concentrations within this central part of the "last septum." The small magnitude of this stress was confirmed by another calculation based on the assumption that all the pressure applied to half the external surface area of the "last chamber" was concentrated in the central cross-sectional area of the underlying "last septum."

The complex failure of septa by local buckling in compression depends on the maintenance of tension by "zero edge displacement under external lateral pressure" (Stein and McElman, 1965). The buckles were parallel to the lobe and saddle-flutes, and Eq. (5) incorrectly assumes that the centrally fluted septa had complete curvature in this direction ($\theta = 360°$). The "edge" is the boundary between the lobe and saddle-flutes crossing the central septum. The tension produces a maximum "buckling coefficient" of 15.5 when $q = Q$ in a fully anticlastic central lobe–flute (e.g., from Fig. 5 of Stein and McElman). This particular maximum buckling pressure yields the habitat depth estimate H_p. The alternative estimate, H_s, assumes the same boundary and loading conditions and is equivalent to the experimentally determined Mungan equation given as equation 3.8 in Hewitt (1993). The significant central width ($\omega = \sin \phi \, 2Q$) is a constant in this estimate ($\phi = 19°$ with $Q/q = 9.25$), but it is permitted to vary in the H_p estimate. Both estimates are based on hypothetical thin septa situated 3.15 standard deviations above the mean IST (e.g., d is now reduced by a factor of 1.86 in an average ammonoid) and assume that the compressive stress was $q/2d$ times $(1 + c)$:

$$H_s + 10 \text{ m} = 1.346 \times 10^6 \; m / [(q/d)^{7/3}(1 + c)] \tag{4}$$

$$H_p + 10 \text{ m} = 6.5 \times 10^7 \; m / [(w/d)^2 (q/d)(1 + c)] \tag{5}$$

3. External Shell Adaptations

The ventral keel of ammonites increased the flexural stiffness of the ventral part of their apertural margin. Smaller spiral edge stiffeners are seen in deeper dwelling goniatites (e.g., *Agathiceras*), large Triassic ammonoids with subrectangular whorl flanks (e.g., *Sturia, Hypocladiscites*), similarly shaped 400-mm-diameter coiled nautiloids (*Cenoceras*) from the Toarcian black shales, and some shallow-dwelling Jurassic ammonites (e.g., *Liparoceras, Amaltheus*, and *Strigoceras* in Table II). The main role of these lirae was probably to attach a thin and melanin-stained periostracum to the shell. However, the majority of Mesozoic ammonoids were either smooth or had a conspicuous ornament of corrugations (Ward, 1981).

Undernourished individuals of the Eocene nautiloid *Cimomia imperialis* (Sowerby) were able to maintain a constant mature diameter by failing to enlarge the breadth of their body chamber and phragmocone. They inflated them later if their food supply improved. Cardioceratid and other Jurassic ammonoids are similarly known to grade from a spherical to an oxyconic, compressed morphology in one locality and stratum (Callomon, 1985). The extensive intraspecific variation of *Quenstedtoceras* and *Cimomia* took place above a depth of 50 to 70 m, where the reduced mechanical strength of the oxyconic whorls had no adverse effects. The same type of intraspecific variation is known from *Macrocephalites bifurcatus* Boehm, and it is much reduced in the spherical species *M. etheridgei* (Spath). The septa of the latter species yielded a habitat depth estimate (H) of 151 m, while the compressed varieties of *M. bifurcatus* lived down to depths of only 50–90 m (Table II).

A unique feature of the intraspecific variation of Mesozoic ammonoids involved the development of large corrugated ribs on the exposed umbilical regions of evolute and rounded whorl sections. This widespread relationship was reviewed as the "first Buckman law of covariation" by Westermann (1971) following studies of mature and compressed genera with a maximum habitat depth of 70 m (i.e., *Euhoploceras* in Table II). It involves thick, radial rib crests and nodes, that bifurcate (Fig. 2) or fade away toward the venter. An empty evolute whorl with a constant radius of curvature and thickness in transverse section developed a constant membrane stress along the spiral direction of growth. This stress was equal to one-quarter of the whorl breadth divided by the wall thickness (Hewitt and Westermann, 1987a). However, this stress changed abruptly from compression to tension at the point of maximum whorl breadth. The thick corrugated ribs and septal sutures of an evolute ammonoid reduced the bending stresses resulting from this separation of the dorsal and ventral parts of a rounded whorl. They also produced a radial increase in stiffness, which helped to stiffen the large umbilicus during radial bending.

Howarth (1978) has noted a similar covariation within the relatively evolute and ventrally corrugated Dactylioceratidae. Various species show a bifurcation of the ribs over the venter and a reduction of the number of corrugations per whorl in the more depressed individuals. The specimens listed in Table II lack any biostratigraphic data, and they are only provisionally identified as *Dactylioceras commune* (Sowerby). Two depressed specimens had 43 and 51 ventral ribs on the last whorl. Two others had a more equidimensional whorl section, finer ventral ribs (i.e., 56 per whorl), and penetrated to a greater depth.

Table II suggests that the coarse-ribbed *Dactylioceras* specimens lived down to depths of only about 60 m. Their corrugations produced a flexible lip around the aperture (Checa and Westermann, 1989), and they turned cracks into orientations parallel to the aperture (Ward, 1981). Mehl (1978) described clusters of shell fragments from other Toarcian ammonites and interpreted them as resting sites of predatory coleoids. Lamont (1982, p. 17) predicted that countershaded ornament became finer with depth and that the corrugations of ammonites reduced the scale of turbulence which might have scoured the sediment. Cowen *et al.* (1973) also implied that the mechanical and the camouflage functions of ventral corrugations were indicative of a *Nautilus*-like habitat. However, an epipelagic habitat is supported by other lines of evidence reviewed by Westermann (Chapter 16, this volume).

ACKNOWLEDGMENTS. The author wishes to thank Prof. G. E. G. Westermann, Dr. D. K. Jacobs, and all of the editors for suggesting improvements to the manuscript. Dr. Westermann also funded much of the work from his N.S.E.R.C. grant and made available specimens then housed at McMaster University (a few of them, prefixed Tr. and J. have remained there). The specimens have been moved to the Royal Ontario Museum (the prefix ROM is omitted in Table II to save space), and the American Museum of Natural History (prefix AMNH). Many of them were collected and donated by Prof. R. H. Mapes (Ohio University), Wang Yi-Gang (Vancouver, BC) and the late E. Noble (San Diego, CA). Additional specimens were loaned from the Australian Geological Survey (CPC) and the La Plata Museum (MLP). C. J. Tsujita loaned specimens from his collection (C.T. in Table II) from the Bearpaw Formation which will be donated to the Tyrell Museum. . The following people provided access to additional specimens: Dr. R. T. Becker (Freie Universität, Berlin, prefix Bc); John Cooper (Natural History Museum, London, prefix BMNH); Prof. B. F. Glenister (University of Iowa, prefix SUI); Dr. N. H. Landman (AMNH); Dr. Y. Shigeta and Prof. K. Tanabe (prefix MM denotes University Museum, Tokyo); and Prof. M. R. House (Southampton University).

References

Batt, R. J., 1991, Sutural amplitude of ammonite shells as a paleoenvironmental indicator, *Lethaia* **24**:219–225.

Bayer, U., 1977, Cephalopoden-Septen Teil 1: Konstruktionsmorphologie der Ammoniten-Septums, *N. Jb. Geol. Paläont. Abh.* **154**:290–366.

Bayer, U., 1985, The biomechanical interpretation of ammonite septa, *Geol. Soc. Am. Abstr. Prog.* **17**(1):4.

Boyajian, G., and Lutz, T., 1992, Evolution of biological complexity and its relation to taxonomic longevity in the Ammonoidea, *Geology* **20**:983–986.

Callomon, J. H., 1985, The evolution of the Jurassic ammonite family Cardioceratidae, *Spec. Pap. Palaeontol.* **33**:49–90.

Callomon, J. H., 1994, The Ammonoidea: Environment, ecology, and evolutionary change, *Hist. Biol.* **7**:342–345.

Chamberlain, J. A., and Moore, W. A., 1982, Rupture strength and flow rate of *Nautilus* siphuncular tube, *Paleobiology* **8**:408–425.

Checa, A., and Westermann, G. E. G., 1989, Segmental growth in planulate ammonites: Inferences on costal function, *Lethaia* **22**:95–100.

Cowen, R., Gertman, R., and Wiggett, G., 1973, Camouflage patterns in *Nautilus*, and their implications for cephalopod paleobiology, *Lethaia* **6**:201–214.

Denton, E. J., 1974, On buoyancy and lives of modern and fossil cephalopods, *Proc. R. Soc. Lond. [Biol.]* **185**:273–299.

Denton, E. J., and Gilpin-Brown, J. B., 1966, On the buoyancy of the pearly *Nautilus*, *J. Mar. Biol. Assoc. U.K.* **46**:723–759.

Doguzhaeva, L., 1988, Siphuncular tube and septal necks in ammonoid evolution, in: *Cephalopods—Present and Past* (J. Wiedmann and J. Kullmann, eds.), Schweizerbart'sche, Verlagsbuchhandlung Stuttgart, pp. 291–301.

Geraghty, M. D., and Westermann, G. E. G., 1994, Origin of Jurassic ammonite concretion assemblages at Alfeld, Germany, a biogenic alternative, *Paläontol. Z.* **68**:473–490.

Grégoire, C., 1984, Remains of organic components in the siphonal tube and in the brown membrane of ammonoids and nautiloids: Hydrothermal simulation of their diagenetic ultrastructural alterations, *Akad. Wiss. Lit. Abh. Math. Naturwiss. Kl. (Mainz)* **5**:1–56.

Hasenmueller, W. A., and Hattin, D. E., 1985, Apatitic connecting rings in moulds of *Baculites* sp. from the middle part of the Smoky Hill Member, Niobrara Chalk (Santonian), of western Kansas, *Cretaceous Res.* **6**:317–330.

Henderson, A., 1984, A muscle attachment proposal for septal function in Mesozoic ammonites, *Palaeontology* **27**:461–486.

Hewitt, R. A., 1985, Numerical aspects of sutural ontogeny in the Ammonitina and Lytoceratina, *N. Jb. Geol. Paläont. Abh.* **170**:273–290.

Hewitt, R. A., 1993, Relation of shell strength to evolution in the Ammonoidea, in: *The Ammonoidea: Environment, Ecology, and Evolutionary Change*, Systematics Association Spec. Vol. 47 (M. R. House, ed.), Clarendon Press, Oxford, pp. 35–56.

Hewitt, R. A., and Westermann, G. E. G., 1983, Mineralogy, structure and homology of ammonoid siphuncles, *N. Jb. Geol. Paläont. Abh.* **165**:378–396.

Hewitt, R. A., and Westermann, G. E. G., 1986, Function of complexly fluted septa in ammonoid shells. I. Mechanical principles and functional models, *N. Jb. Geol. Paläont. Abh.* **172**:47–69.

Hewitt, R. A., and Westermann, G. E. G., 1987a, Function of complexly fluted septa in ammonoid shells. II. Septal evolution and conclusions, *N. Jb. Geol. Paläont. Abh.* **174**:135–169.

Hewitt, R. A., and Westermann, G. E. G., 1987b, *Nautilus* shell architecture, in: *Nautilus: The Biology and Paleobiology of a Living Fossil* (W. B. Saunders and N. H. Landman, eds.), Plenum Press, New York, pp. 435–461.

Hewitt, R. A., and Westermann, G. E. G., 1988, Application of buckling equations to the functional morphology of nautiloid and ammonoid phragmocones, *Hist. Biol.* **1**:225–231.

Hewitt, R. A., and Westermann, G. E. G., 1990, *Nautilus* shell strength variance as an indicator of habitat depth limits, *N. Jb. Geol. Paläont. Abh.* **179**:71–95.

Hewitt, R. A., Abdelsalam, U. A., Dokainish, M. A., and Westermann, G. E. G., 1993, Comparison of the relative strength of siphuncles with prochoanitic and retrochoanitic septal necks by finite element analysis, in: *The Ammonoidea: Environment, Ecology and Evolutionary Change*, Systematics Association Spec. Vol. 47 (M. R. House, ed.), Clarendon Press, Oxford, pp. 85–98.

Howarth, M. K., 1978, The stratigraphy and ammonite fauna of the Upper Lias of Northamptonshire. *Bull. Br. Mus. (Nat. Hist.) Geol.* **29**:235–288.

Isenberg, C., 1992, *The Science of Soap Films and Soap Bubbles*, Dover, New York.

Jacobs, D. K., 1990, Sutural pattern and shell strength in *Baculites* with implications for other cephalopod shell morphologies, *Paleobiology* **16**:336–348.

Jacobs, D. K., 1992, The support of hydrostatic load in cephalopod shells. Adaptive and ontogenetic explanations of shell form and evolution from Hooke 1695 to the present, in: *Evolutionary Biology*, Vol. 26 (M. K. Hecht, B. Wallace, and R. J. MacIntyre, eds.), Plenum Press, New York, pp. 287–349.

Johnsen, S., and Kier, W. M., 1993, Intramuscular crossed connective tissue fibers: Skeletal support in the lateral fins of squid and cuttlefish (Mollusca: Cephalopoda), *J. Zool. (Lond.)* **231**:311–338.

Joly, B., 1976, Les Phylloceratidae malgaches au Jurassique. Generalités sur les Phylloceratidae et quelques Juraphyllitidae, *Doc. Lab. Geol. Fac. Sci. Lyon* **67**:1–471.

Korn, D., 1991, Relationship between shell form, septal construction and suture line in clymeniid cephalopods (Ammonoidea; Upper Devonian), *N. Jb. Geol. Paläont. Abh.* **185**:115–130.

Kulicki, C., and Mutvei, H., 1988, Functional interpretation of ammonoid septa, in: *Cephalopods—Present and Past* (J. Wiedmann and J. Kullmann, eds.), Schweizerbart'sche, Verlagsbuchhandlung Stuttgart, pp. 713–718.

Lamont, A., 1982, Mouth and tooth mimicry, *Scot. J. Sci.* **2**:11–32.

Lowenstam, H. A., Traub, W., and Weiner, S., 1984, *Nautilus* hard parts: A study of the mineral and organic constituents, *Paleobiology* **10**:268–279.

Mehl, J., 1978, Anhaufungen scherbenartiger Fragmente von Ammonitenschalen im suddeutschen Lias und Malm und ihre Deutung als Frassreste, *Ber. Naturforsch. Ges. Freib. Breisgau* **68**:75–93.

Miller, A. K., and Unklesbay, A. G., 1943, The siphuncle of late Paleozoic ammonoids, *J. Paleontol.* **17**:1–25.

Mutvei, H., 1967, On the microscopic shell structure in some Jurassic ammonoids, *N. Jb. Geol. Paläont. Abh.* **129**:157–166.

Mutvei, H., 1975, The mode of life in ammonoids, *Paläontol. Z.* **49**:196–202.

Pfaff, E., 1911, Über Form und Bau der Ammonitenseptum und ihre Beziehungen zur Suturelinie, *Jahresber. Niedersach. Geol. Ver.* **4**:207–223.

Rangheard, Y., and Theobald, N., 1961, Signification biologique de la coquille des ammonites, *Ann. Sci. Univ. Besançon Geol.* **14**:119–133.

Raup, D. M., and Takahashi, T., 1966, Experiments on strength of cephalopod shells, *Geol. Soc. Am. Spec. Pap.* **101**:172–173.

Raup, D. M., and Stanley, S. M., 1971, *Principles of Paleontology*, 1st Ed., Freeman, San Francisco.

Roark, R. J., and Young, W. C., 1982, *Formulas for Stress and Strain*, 5th Ed., McGraw-Hill, New York.

Ross, C. T. F., 1990, *Pressure Vessels under External Pressure: Statics and Dynamics*, Elsevier, Amsterdam.

Saunders, W. B., 1995, The ammonite suture problem: Relationships between shell and septum thickness in Paleozoic ammonoids, *Paleobiology* **21**:343–355.

Saunders, W. B., Ward, P. D., and Daniel, T. L., 1994, Cameral liquid transport: Resolution of the ammonite suture problem?, *Geol. Soc. Am. Abstr. Prog.* **26**:A-375.

Seilacher, A., 1975, Mechanische Simulation und funktionelle Evolution des Ammoniten-Septums, *Paläontol. Z.* **49**:268–286.

Shaaban, A., and Ketcham, M.S., 1976, Design of hipped hypar shells, *J. Struct. Di.* **102**(ST11):2151–2161.

Sobel, L. H., and Flugge, W., 1967, Stability of toroidal shells under uniform external pressure, *AIAAA J.* **5**:425–431.

Stein, M., and McElman, J. A., 1965, Buckling of segments of toroidal shells, *AIAA J.* **3**:1704–1709.

Tanabe, K., 1977, Functional evolution of *Otoscaphites puerculus* (Jimbo) and *Scaphites planus* (Yabe), Upper Cretaceous ammonites, *Mem. Fac. Sci. Kyushu Univ. [D] Geol.* **23**:367–407.

Tanabe, K., 1979, Palaeoecological analysis of ammonoid assemblages in the Turonian *Scaphites* facies of Hokkaido, Japan, *Palaeontology* **22**:609–630.

Timoshenko, C. P., 1983, *The History of Strength of Materials*, Dover, New York.

Wainwright, S. A., Biggs, W. D., Currey, J. D., and Gorsline, J. M., 1976, *Mechanical Design in Organisms*, Edward Arnold, London.

Wang, Y., and Westermann, G. E. G., 1993, Paleoecology of Triassic ammonoids, *Geobios Mém. Spéc.* **15**:373–392.

Ward, P., 1980, Comparative shell shape distributions in Jurassic–Cretaceous ammonites and Jurassic–Tertiary nautilids, *Paleobiology* **6**:32–43.

Ward, P., 1981, Shell sculpture as a defensive adaptation in ammonoids, *Paleobiology* **7**:96–100.

Ward, P. D., 1987, *The Natural History of Nautilus*, Allen and Unwin, London.

Ward, P., and Signor, P., 1983, Evolutionary tempo in Jurassic and Cretaceous ammonites, *Paleobiology* **9**:183–198.

Weitschat, W., and Bandel, K., 1991, Organic components in phragmocones of Boreal Triassic ammonoids: Implications for ammonoid biology, *Paläontol. Z.* **65**:269–303.

Weitschat, W., and Bandel, K., 1992, Formation and function of suspended organic cameral sheets in Triassic ammonoids: Reply, *Paläontol. Z.* **66**(3/4):443–444.

Westermann, G. E. G., 1971, Form, structure and function of shell and siphuncle in coiled Mesozoic ammonoids, *Life Sci. Contrib. R. Ont. Mus.* **78**:1–39.

Westermann, G. E. G., 1973, Strength of concave septa and depth limits of fossil cephalopods, *Lethaia* **6**:373–403.

Westermann, G. E. G., 1975, A model for origin, function and fabrication of fluted cephalopod septa, *Paläontol. Z.* **49**:235–253.

Westermann, G. E. G., 1982, The connecting rings of *Nautilus* and Mesozoic ammonoids: Implications for ammonoid bathymetry, *Lethaia* **15**:373–384.

Westermann, G. E. G., 1990, New developments in ecology of Jurassic–Cretaceous ammonites, *Atti II Convegno Internazionale, Fossili, Evoluzione, Ambiente, Pergola 1987* (G. Pallini, F. Cecca, S. Cresta, and M. Santantonio, eds.), Technostampa, Ostra Vetere, Italy, pp. 459–478.

Westermann, G. E. G., and Weaver, D. S., 1979, Photoelasticity experiments on simplified models of ammonoid shells, in: *The Ammonoidea: Abstracts Systematics Association Symposium* (M. R. House and J. Senior, eds.), York, England, p. 21.

IV

Growth

Chapter 11

Ammonoid Embryonic Development

NEIL H. LANDMAN, KAZUSHIGE TANABE, and YASUNARI SHIGETA

NEIL H. LANDMAN • Department of Invertebrates, American Museum of Natural History, New York, New York 10024. KAZUSHIGE TANABE • Geological Institute, University of Tokyo, Tokyo 113, Japan. YASUNARI SHIGETA • Department of Paleontology, National Science Museum, Tokyo 160, Japan.

Ammonoid Paleobiology, Volume 13 of *Topics in Geobiology*, edited by Neil Landman *et al.*, Plenum Press, New York, 1996.

1. Introduction

Ammonoids retain a record of growth in their shells, and, therefore, material is readily available for studies of early ontogeny. Such studies were performed first in the mid-19th century and have been pursued with vigor ever since. Using optical and scanning electron microscopy, ammonoid workers have described the morphology of the early whorls and have attempted to reconstruct the sequence of early ontogenetic development and to identify the embryonic shell.

Studies of early ontogeny are obviously crucial in understanding the ecology and mode of life of adults. Such factors as population structure and biogeographic distribution grow out of the constraints of early ontogeny. For example, differences in early life history may explain why some ammonoid species are more restricted in their biogeographic distribution than are others. These relationships may bear, in turn, on broader evolutionary issues such as species longevity and extinction.

Studies of early ontogeny are also helpful in trying to reconstruct phylogeny. In the studies of Hyatt (1866, 1883, 1889, 1894), Smith (1898, 1914), and Buckman (1887–1907, 1909, 1918), ontogeny and phylogeny were closely linked together in a theory of recapitulation. According to these authors, the early ontogenetic stages of an individual represented a recapitulation of the adult stages of its ancestors. Although this view no longer is considered valid, there are, nevertheless, numerous characters in early ontogeny that are useful in reconstructing phylogeny.

Much of the information presented in this chapter, especially with respect to the size of the embryonic shell, is new. However, the morphological descriptions and interpretations of ontogenetic development rely heavily on previously published data. Many of these data are based on Mesozoic rather than Paleozoic ammonoids because the former are generally better preserved. Specimens cited in this chapter are reposited in the American Museum of Natural History (AMNH), the University of Iowa (SUI), the University Museum of the University of Tokyo (UMUT), the New York State Museum (NYSM), and the Yale Peabody Museum of Natural History (YPM).

2. Description of the Ammonitella

2.1. Terminology

Figure 1 illustrates the terms used to describe the morphological features of the early whorls. The illustrated specimen represents the early whorls of the Late Cretaceous species *Scaphites whitfieldi* (Ancyloceratina), but the same terms are used for all ammonoids. The ammonitella is defined as the shell up to the end of the primary constriction (Druschits and Khiami, 1970; Druschits *et al.*, 1977a,b; Tanabe *et al.*, 1980; Birkelund, 1981; Landman,

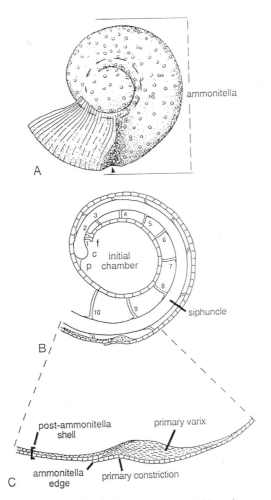

FIGURE 1. (A) Side view of the early whorls of an ammonoid based on a specimen of the Late Cretaceous species *Scaphites whitfieldi*. The ammonitella (about 0.7 mm in diameter) consists of the shell up to the primary constriction (arrow). The dashed line indicates the outline of the initial chamber in median cross section. (B) Median cross section through the same specimen showing the initial chamber (protoconch), flange (f), proseptum (1), primary septum (2), septa 3–10, cecum (c), prosiphon (p), and siphuncle. (C) Close-up of B showing the primary constriction, primary varix, ammonitella edge, and postammonitella shell.

1987). (This term originally was defined as the whole animal up to this point. It commonly is used in this sense as well as in the more restricted sense to mean only the shell of the animal.) The term "initial portion of the shell" ("*Gehäuse-Anfangsteile*," Erben, 1960) refers in a general way to the beginning of the ammonitella. The term "initial chamber" ("*Anfangskammer*," Branco, 1879, 1880; Schindewolf, 1933; Erben, 1960; "protoconch," Owen, 1878;

Hyatt, 1883; "first whorl," Bandel, 1982) refers specifically to the portion of the ammonitella up to the proseptum.

2.2. Shape

The initial chamber ranges in shape from globular to spindle-like and has a circular to lenticular outline in transverse cross section (Fig. 2G–I; Branco, 1879, 1880; Bogoslovsky, 1969, Fig. 2; Erben, 1964, Fig. 1; Erben, 1966, Fig. 3). In median cross section, the initial chamber is U-shaped or, more commonly, forms the beginning of a spiral (Fig. 2D–F). A cicatrix, the scar-like feature on the early portion of the shell of many nautiloids (Arnold *et al.*, 1987), is absent. In ammonoids with a bulbous initial chamber, the succeeding whorls are loosely coiled or even straight. For example, in *Mimagoniatites*, the spherical initial chamber is loosely enveloped by the succeeding whorls, leaving an umbilical perforation (Fig. 2B). In ammonoids with a barrel- to spindle-shaped initial chamber, the succeeding whorls are closely coiled.

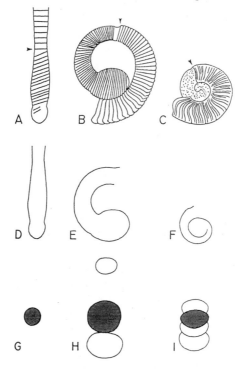

FIGURE 2. (A–C) Side views of the early whorls in *Pseudobactrites*, *Mimagoniatites*, and *Scaphites*. Arrows indicate the end of the ammonitella. (D–F) Median cross sections through the ammonitella in the same three genera. (G–I) Transverse cross-sections through the initial chamber (G) and initial chamber and first whorl (H, I) in the same three genera. The initial chamber is shaded. Scale bar, 1 mm.

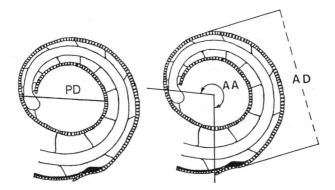

FIGURE 3. Median cross sections through the ammonitella showing measurements of the initial chamber diameter (PD), ammonitella diameter (AD), and ammonitella angle (AA).

They tightly envelop the initial chamber, leaving a shallow to deep dorsal impression in the whorl profile as seen in transverse cross section (Fig. 2I; Erben, 1964, Fig. 3).

The distal end of the initial chamber is marked by an abrupt narrowing of the shell in phylogenetically primitive ammonoids such as bactritids (Fig. 2A; Mapes, 1979, Figs. 10–12), whereas there is only some flattening along the venter at this point in more advanced forms (Fig. 2F; Bandel, 1986). This change in shell shape was called the first growth change ("1. *Wachstums-Änderung*") by Erben (1964), who interpreted it as indicating a major shift in ontogenetic development.

A second change in shell shape occurs at approximately one whorl from the end of the initial chamber in closely coiled ammonitellas; it appears as a groove in the shell wall, which is especially well expressed along the venter (Figs. 1, 2C). In bactritids, this change in shape appears as a gradual narrowing of the shell, followed by a widening (Fig. 2A; Mapes, 1979, Figs. 10–12). This feature has been referred to as the primary constriction (Shul'ga-Nesterenko, 1926; Bogoslovskaya, 1959) or nepionic constriction (Erben *et al.*, 1968; Birkelund and Hansen, 1968), although a variety of other terms also have been used ("*Einschnürung*," Branco, 1879, 1880; "*première varice*," Grandjean, 1910; "*Anfangseinschnürung*," Böhmers, 1936; "2. *Wachstumsänderung*," Erben, 1964; Erben *et al.*, 1969; "primary varix," Druschits and Khiami, 1970). Landman and Waage (1982) emphasized the importance of distinguishing the actual constriction in the shell wall (primary constriction) both from the shell thickening at this point (primary varix) and from the trace of this thickening on the steinkern (varix trace).

2.3. Size

Three measurements were made of the ammonitella in median cross section (Fig. 3). The diameter of the ammonitella (AD) is defined as the distance from the adoral end of the primary constriction through the center of the initial chamber to the opposite side of the ammonitella. (In straight ammonitellas, for example, in bactritids, this dimension is more properly called length.) The diameter of the initial chamber (PD) is measured from the ventral edge of the proseptum through the center of the initial chamber to the opposite side. In closely coiled ammonitellas, the ammonitella angle (AA) is defined as the angle from the ventral edge of the proseptum to the adoral end of the primary constriction.

The diameter of the ammonitella ranges from a minimum of 0.5 mm to a maximum of 2.6 mm in all the suborders studied (Fig. 4; Table I). Most values occur between 0.5 and 1.5 mm (small to medium) except in the Agoniatitina (1.5–2.6 mm), Goniatitina (0.6–2.3 mm), and Lytoceratina (0.8–1.9 mm). In parabactritids, the ammonitella diameter (length) averages 1.4 mm (Hecht, 1991). In the Lytoceratina, ammonitella diameter appears to increase over geological time from the Middle Jurassic to the Late Cretaceous (Fig. 5).

The diameter of the initial chamber covaries with that of the ammonitella and ranges from 0.25 to 1.60 mm, with most values occurring between 0.25 and 0.75 mm (small to medium; Table I; House, 1985, Fig. 3; Lehmann, 1990, Fig. 4.69). The largest initial chambers occur in the Agoniatitina (0.80–1.6 mm). There is a strong positive correlation between initial chamber diameter and ammonitella diameter, both within and among species (Fig. 6; Tanabe *et al.*, 1979; Tanabe and Ohtsuka, 1985; Landman, 1987; Shigeta, 1993). A strong positive correlation also occurs between initial chamber volume and ammonitella volume (Fig. 7). However, the precise nature of this relationship may vary among suborders as, for example, between Goniatitina and Ammonitida, as illustrated in Fig. 7.

The ammonitella angle ranges from as little as 240° in some Ceratitina and Ammonitina to as much as 410° in some Goniatitina (e.g., *Peritrochia* and *Perrinites*; Fig. 8; Table I; Grandjean, 1910; Bogoslovskaya, 1959). The ammonitella angle in the Goniatitina is larger than those in all other suborders (Tanabe *et al.*, 1994). A plot of ammonitella angle versus ammonitella diameter in seven suborders reveals only a weak correlation (Fig. 9).

In closely coiled ammonitellas, the whorl width and radius of the spiral show no significant increase over the first whorl up to the end of the primary constriction. In contrast, after the primary constriction, there is an abrupt increase in both of these dimensions (Fig. 10; Currie, 1942, 1943; Palframan, 1967b; Tanabe, 1975, 1977a; Kulicki, 1974, 1979; Hirano, 1975; Obata *et al.*, 1979; Zell *et al.*, 1979; Landman, 1987, 1988). This change in whorl shape is dramatic in heteromorph ammonoids such as *Baculites*, in which the postammonitella shell becomes orthoconic (Brown, 1891; Bandel *et al.*, 1982, Fig. 1C), and *Eubostrychoceras*, in which the postammonitella shell becomes loosely coiled (Tanabe *et al.*, 1981, Pl. 35, Fig. 1e).

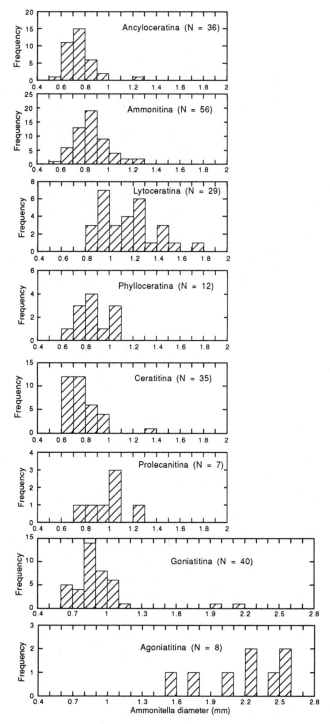

FIGURE 4. Size–frequency histograms of ammonitella diameter in eight ammonoid suborders. For species represented by more than two specimens, the mean was used. *N*, number of species. See Appendices for data sources.

Table I. Comparison of the Ammonitella in 11 Suborders of the Ammonoidea[a]

Suborder	Initial chamber diameter (mm)*	Ammonitella diameter (mm)**	Ammonitella angle (degrees)+	Shape and length (mm) of prosiphon++	Shape of cecum	Initial position of siphuncle	Micro-ornamentation
Agoniatitina	Large to very large (0.80–1.60)	Large to very large (1.50–2.60)	—	—	—	—	Transverse lirae
Anarcestina	—	Medium (1.00)	—	—	—	—	Transverse lirae
Gephuroceratina	Medium to very large (0.65–1.10)	Medium (1.05–1.20)	—	—	—	—	Unknown
Tornoceratina	Medium to large (0.50–1.00)	Medium (1.40–1.50)	—	—	—	—	Transverse lirae
Goniatitina	Small to medium (0.30–0.70) but very large in Perrinites and Gonioloboceras	Small to medium (0.60–1.20) but very large in Perrinites and Gonioloboceras	Medium to large (345–410)	Short and curved (≤ 0.10)	Elliptical in median section	Mostly marginal but central in Bisatoceras and Agathiceras	Longitudinal lirae
Prolecanitina	Small to medium (0.35–0.60)	Small to medium (0.70–1.20)	Medium to large (310–355)	Short and curved (0.05–0.25)	Rectangular in median section	Marginal	Unknown
Ceratitina	Small to medium (0.30–0.65)	Small to medium (0.60–1.30)	Small to large (240–370)	Short and curved (≤ 0.20)	Elliptical in median section	Mostly marginal but central in the Ceratitaceae and Megaphyllitaceae	Tubercles

Phylloceratina	Small to medium (0.40–0.65)	Small to medium (0.65–1.30)	Medium to large (260–380)	Short and curved (≤ 0.15)	Elliptical in median section	Central in the Phylloceratidae and marginal in the Ussuritidae	Tubercles
Lytoceratina	Small to very large (0.30–1.05)	Small to large (0.80–1.90)	Medium to large (270–365)	Short and curved (0.05–0.10)	Hemicircular in median section	Marginal	Tubercles
Ammonitina	Small to medium (0.30–0.70)	Small to medium (0.60–1.25)	Small to large (240–360)	Long and straight but short and curved in the Amaltheidae, Collignoniceratidae, and Placenticeratidae	Elliptical in median section	Central to subcentral	Tubercles
Ancyloceratina	Small to medium (0.25–0.70)	Small to medium (0.50–1.30)	Medium (255–330)	Long and straight (Ancylocerataceae and Parahoplitaceae), short and curved (Scaphitaceae)	Elliptical in median section	Subcentral to marginal	Tubercles

[a]Symbols: *small ($0.25 \leq PD < 0.5$), medium ($0.5 \leq PD < 0.75$), large ($0.75 \leq PD < 1.0$) very large ($PD \geq 1.0$); **small ($0.5 \leq AD < 1.0$), medium ($1.0 \leq AD < 1.5$), large ($1.5 \leq AD < 2.0$), very large ($AD \geq 2.0$); [+] small ($AA < 250$), medium ($250 \leq AA < 350$), large ($AA \geq 350$); [++] short (≤ 0.3), long (> 0.3).

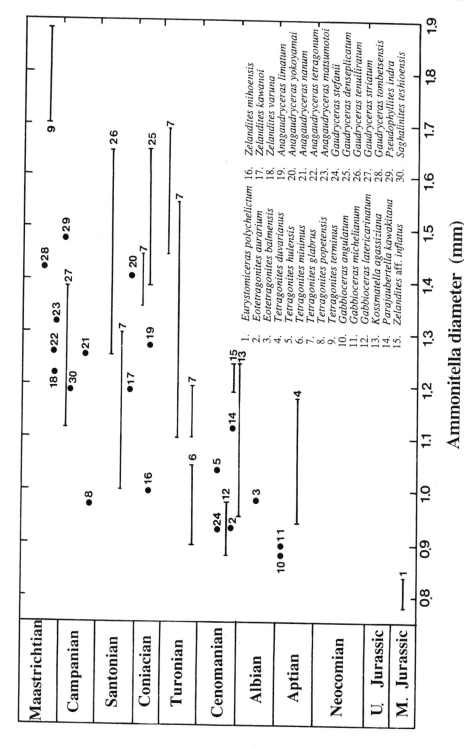

FIGURE 5. Change in ammonitella diameter in the Lytoceratina with respect to geological time. Horizontal bars indicate the range of variation within a single species.

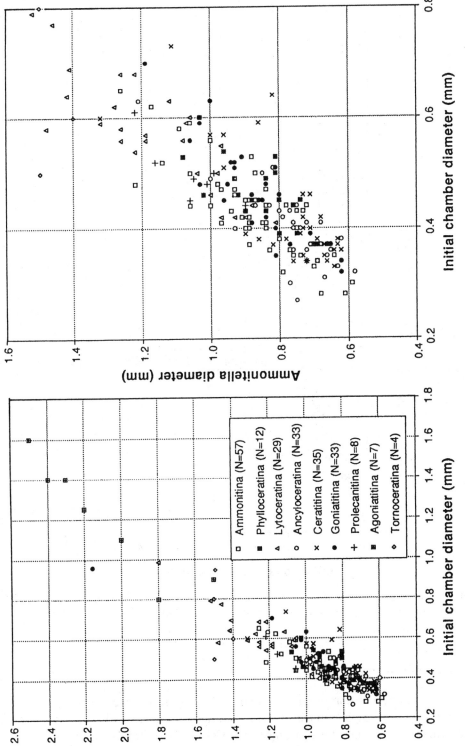

FIGURE 6. Ammonitella diameter versus initial chamber diameter in 223 species of the Ammonoidea. The right plot is a close-up of the lower left portion of the left plot. For species represented by more than two specimens, the mean was plotted. *N*, number of species. See Appendices for data sources.

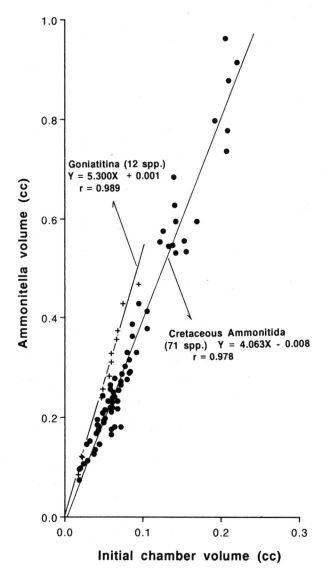

FIGURE 7. Ammonitella volume versus initial chamber volume in 12 species of Carboniferous Goniatitina and 71 species of Cretaceous Ammonitida. (Data from Shigeta, 1993, and Tanabe *et al.*, 1995).

2.4. Ornamentation

The ammonitella is commonly covered with a microornamentation that occurs on the exposed portions of the initial chamber and succeeding whorls and terminates at the end of the primary constriction. Growth lines are absent on the ammonitellas of Mesozoic ammonoids. As pointed out by Bandel

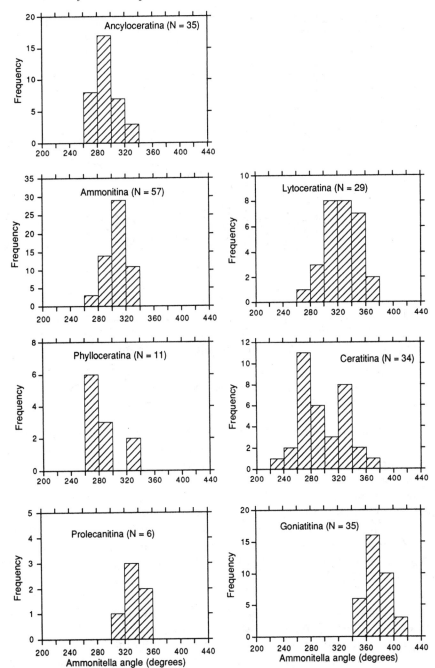

FIGURE 8. Size–frequency histograms of ammonitella angle in seven suborders of the Ammonoidea. For species represented by more than two specimens, the mean was used. *N*, number of species. See Appendices for data sources.

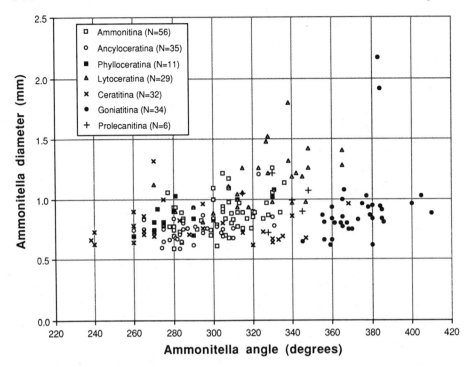

FIGURE 9. Ammonitella diameter versus ammonitella angle in 203 species of the Ammonoidea. For species represented by more than two specimens, the mean was plotted. *N*, number of species. See Appendices for data sources.

(1986), what has sometimes been mistaken for growth lines on steinkerns represents instead the impression of the inside surface of the shell wall on the internal mold. Growth lines have been reported from the ammonitellas of some Paleozoic ammonoids ("*Anwachsstreifen*," Erben *et al.*, 1969), but these features probably are lirae rather than growth lines (see Chapter 12, this volume, for the distinction between growth lines and lirae).

Several kinds of ornamentation have been documented on ammonitellas (Table I). Lirae are present on the ammonitellas of many Paleozoic forms. In the Agoniatitina, the ammonitella is covered with fine transverse lirae parallel to the aperture (Fig. 11C; Babin, 1989, Pl. 1, Fig. 2; Wissner and Norris, 1991, Pl. 3.1, Fig. 1; Erben, 1964, Pl. 7, Figs. 6, 7, Pl. 8, Figs. 1, 2, 6, 7, Pl. 9, Fig. 1; Göddertz, 1989, Pl. 2, Fig. 2). In *Mimagoniatites*, these lirae develop a slight backward projection along the venter at the end of the initial chamber (Erben, 1964, Pl. 8, Figs. 3–5). In the Tornoceratina, the ammonitella also is covered with fine transverse lirae (Fig. 11E,F; Beecher, 1890; House, 1965, Fig. 2). Transverse lirae also have been reported in the Anarcestina (see Miller, 1938, Fig. 8). In the Goniatitina, in contrast, the ammonitella is ornamented with evenly spaced, longitudinal lirae; these disappear just before the end of the primary constriction (Fig. 11D; Tanabe *et al.*, 1993).

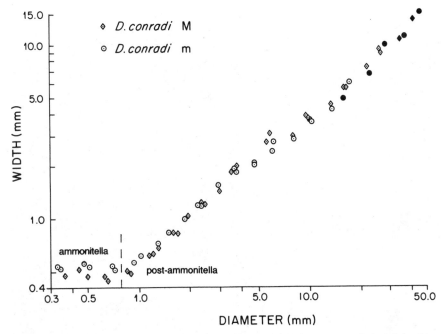

FIGURE 10. Whorl width versus shell diameter through the ontogeny of six adults [three macroconchs (M) and three microconchs (m)] of *Discoscaphites conradi* showing the abrupt change in whorl width at the end of the ammonitella. Black symbols indicate measurements near the base of or in the mature body chamber. (After Landman and Waage, 1993, Fig. 162)

In the Ceratitina, Lytoceratina, Phylloceratina, Ammonitina, and Ancylo-ceratina, the ammonitella is covered with a tuberculate microornamentation rather than with lirae (Figs. 12 and 13; Kulicki, 1974, 1979; Bandel, 1982; Bandel *et al.*, 1982; Landman, 1985, 1987; Landman and Waage, 1993; Tanabe, 1989; for data on the Ceratitina, W. Weitschat, personal communication, 1993). The tubercles range in diameter from 2 to 10 μm and, in general, are irregularly distributed over the exposed surface of the ammonitella. They die out at the end of the primary constriction. In some ammonoids, the tubercles coalesce into a single layer covering part of the initial chamber (Fig. 13E; Tanabe, 1989; Chapter 4, this volume).

In other ammonoids, the surface of the ammonitella appears smooth (Fig. 11A,B). For example, Miller (1938) described smooth ammonitellas in some Gephuroceratina, although this smoothness may simply reflect poor preservation (see also Clausen, 1969). In the Bactritina, the shaft after the initial chamber is ornamented with transverse lirae, but it is unclear whether these are also present on the initial chamber (Erben, 1964; Mapes, 1979).

There is an abrupt change in ornamentation at the end of the primary constriction (Figs. 11A, 12C–F, and 13F). For example, at this point in *Tornoceras*, the lirae abruptly become biconvex, with a forward projection

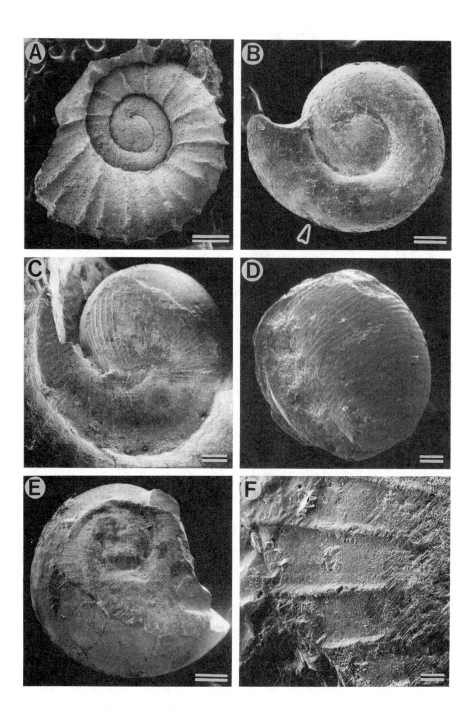

FIGURE 11. Early whorls of Paleozoic ammonoids. (A) *Manticoceras sinuosum* (Gephuroceratina), Upper Devonian, New York State, NYSM 3755 (12306/7). No ornamentation is visible on the ammonitella, possibly because of poor preservation, but prominent subcostae appear immediately afterward. Scale bar, 500 μm. (B) *Probeloceras lutheri* (Gephuroceratina), Upper Devonian, New York State, NYSM 12726. The specimen is a steinkern and shows the varix trace (arrow). Scale bar, 200 μm. (C) *Agoniatites vanuxemi* (Agoniatitina), Middle Devonian, New York State, NYSM 3545 (12000/6). The ammonitella is covered with fine transverse lirae. Scale bar, 200 μm. (D) *Vidrioceras* sp. (Goniatitina), Upper Pennsylvanian, Kansas, UMUT PM 19014. The ammonitella is covered with evenly spaced longitudinal lirae. Scale bar, 100 μm. (E,F) *Tornoceras (Tornoceras) uniangulare aldenense* (Tornoceratina), Middle Devonian, New York State, NYSM 12553. (E) The ammonitella is covered with fine transverse lirae. Scale bar, 200 μm. (F) Close-up of lirae. Scale bar, 20 μm.

along the ventrolateral margin and a backward projection along the venter (Beecher, 1890; House, 1965, Fig. 2). In *Scaphites*, the shell just adoral of the primary constriction is covered with fine ribs and growth lines (Fig. 12C–F), but in *Gaudryceras*, this part of the shell is covered with evenly spaced subcostae (Fig. 13F; Tanabe, 1989).

2.5. Microstructure of the Shell Wall

The microstructure of the shell wall of the ammonitella has been documented in the Ammonitina, Phylloceratina, Lytoceratina, and Ancyloceratina (Erben *et al.*, 1969; Kulicki, 1974, 1979; Birkelund and Hansen, 1974; Birkelund, 1981; Druschits and Khiami, 1970; Druschits and Doguzhaeva, 1974, 1981; Druschits *et al.*, 1977a,b; Chapter 4, this volume).

The shell wall of the ammonitella is thin. For example, in the Late Cretaceous heteromorph *Baculites*, it is approximately 2 μm thick at the proximal end of the initial chamber and increases to a thickness of approximately 4 μm at the distal end of the initial chamber. It reaches a thickness of approximately 8 μm just adapical of the primary varix.

The shell wall is constructed of several prismatic layers, but the number and the position of these layers are subject to debate (Fig. 14). Erben *et al.* (1968, Fig. 1; 1969, Fig. 5) reported five layers (Fig. 14A, p_1–p_5) in the wall of the initial chamber, all but one of which (Fig. 14A, p_4) wedge out on the outer side before or at the distal end of the initial chamber (Fig. 14A, arrow). According to these authors, a new layer (Fig. 14A, p_6) appears on the inner side at this point and eventually forms most of the wall of the first whorl. Birkelund and Hansen (1968; 1974, Fig. 2) reported only two layers in the wall of the initial chamber, both of which wedge out on the outer side at the distal end of the initial chamber (Fig. 14B, arrow; see also Druschits *et al.*, 1977a, Fig. 6; Tanabe *et al.*, 1980, Fig. 4, for slight variations). According to these authors, two new layers appear on the inner side at this point and form the wall of the first whorl. Kulicki (1979, Figs. 6, 7) confirmed that there are two principal layers in the wall of the initial chamber, but he identified the outer

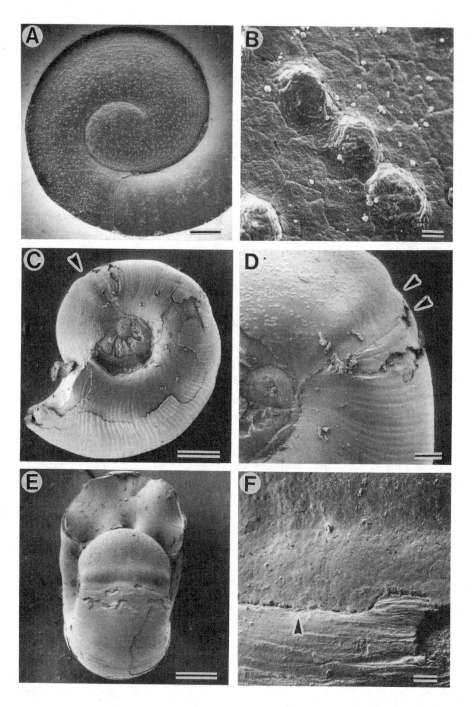

FIGURE 12. Early whorls of Mesozoic ammonoids. (A,B) Species of *Hoploscaphites* or *Jeletzkytes* (Ancyloceratina), Upper Cretaceous, South Dakota, YPM 34113. (A) View of the ammonitella showing the tuberculate ornamentation. Scale bar, 100 μm. (B) Close-up of tubercles. Scale bar, 2 μm. (C–F) *Scaphites whitfieldi* (Ancyloceratina), Upper Cretaceous, South Dakota, AMNH 44833. (C) View of part of the ammonitella and first whorl showing the primary constriction (arrow). The region of the initial chamber is poorly preserved. Scale bar, 200 μm. (D) Close-up of the primary constriction (upper arrow) and ammonitella edge (lower arrow). Scale bar, 50 μm. (E) Ventral view of the primary constriction. Scale bar, 200 μm. (F) Close-up of the ammonitella edge (arrow) and postammonitella shell covered with fine ribs and growth lines. Scale bar, 10 μm.

---

one as the dorsal wall of the first whorl (Fig. 14C, dp) and the inner one as the actual wall of the initial chamber (Fig. 14C, pi; he also recognized two other layers of more limited extent, ip and ml). According to him, the actual wall of the initial chamber does not wedge out but forms the external layer of the wall of the first whorl (Fig. 14C, op). This wall also includes two additional layers (Fig. 14C, ip, mp), which first appear on the inner side at the distal end of the initial chamber. In contrast, Bandel (1982, Figs. 41, 43, 46–48) and Tanabe (1989, Fig. 7) argued that the wall of the initial chamber wedges out, but on the inner side, and that the external layer of the wall of the first whorl (Fig. 14D, op) first appears on the outer side near the distal end of the initial chamber (Fig. 14D, arrow).

The most marked change in microstructure in all ammonoids whose microstructure has been studied occurs at the primary constriction (Erben *et al.*, 1968, 1969; Birkelund and Hansen, 1968, 1974; Birkelund, 1981; Kulicki, 1974, 1979; Druschits *et al.*, 1977a). The prismatic layer of the first whorl decreases in thickness, and a large pad of nacre develops on the inside of the shell. This pad of nacre is known as the primary varix (Druschits and Khiami, 1970; Druschits and Doguzhaeva, 1974; Landman and Waage, 1982; also called "*première varice*," Grandjean, 1910; Dauphin, 1975, 1977; "nepionic ridge," Druschits *et al.*, 1977a,b, 1980; and "nepionic swelling," Kulicki, 1979; we include the primary varix as part of the ammonitella, although this feature was excluded in the original definition of this term by Druschits and Khiami, 1970, p. 30). It parallels the primary constriction and lies close to its adapical end (first illustrated in Hyatt, 1872, Pl. 4, Fig. 11). In some specimens the outer prismatic layer doubles back along the inside edge of the primary varix (Kulicki, 1974, 1979). The postammonitella shell emerges from below the primary varix and consists of both an outer prismatic and an inner nacreous layer (Figs. 1, 12C–F and 14).

2.6. Septa

Schindewolf (1928, 1929, 1951, 1954) called the first septum the proseptum to emphasize its uniqueness relative to all other septa. The proseptum develops at the distal end of the initial chamber (Fig. 15A,C; Erben *et al.*, 1969).

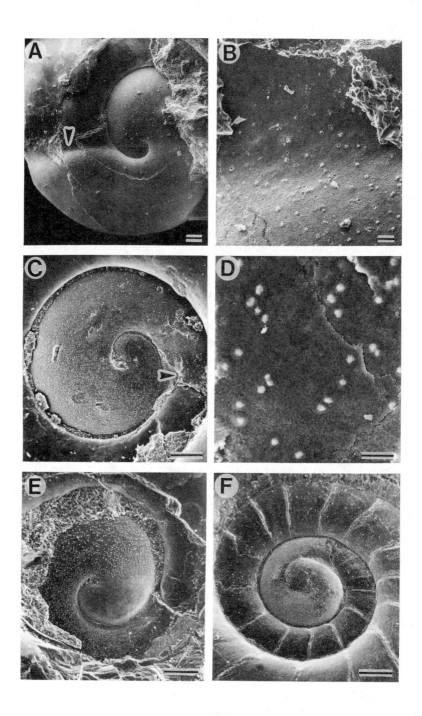

FIGURE 13. Early whorls of Mesozoic ammonoids. (A,B) *Sphenodiscus lenticularis* (Ammonitina), Upper Cretaceous, South Dakota, YPM 34985. (A) View of the ammonitella and primary constriction (arrow). Scale bar, 100 μm. (B) Close-up of tubercles on the ammonitella. Scale bar, 10 μm. (C,D) *Metaplacenticeras subtilistriatum* (Ammonitina), Upper Cretaceous, Hokkaido, UMUT MM 18328. (C) View of the ammonitella and primary constriction (arrow). Scale bar, 130 μm. (D) Close-up of tubercles on the ammonitella. Scale bar, 13 μm. (E) *Anapachydiscus* sp. (Ammonitina), Upper Cretaceous, Hokkaido, UMUT MM 18327. The tubercles have coalesced into a single layer covering part of the initial chamber. Scale bar, 90 μm. (F) *Gaudryceras denseplicatum* (Lytoceratina), Upper Cretaceous, Hokkaido, UMUT MM 18322. The postammonitella shell is ornamented with prominent subcostae. Scale bar, 330 μm.

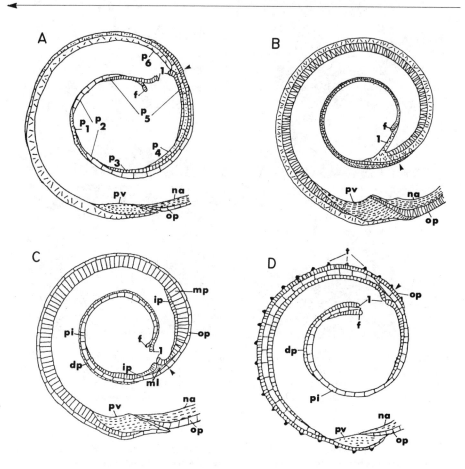

FIGURE 14. Microstructure of the shell wall of the ammonitella as reported by (A) Erben *et al.* (1968, 1969), (B) Birkelund and Hansen (1968, 1974), (C) Kulicki (1979), and (D) Bandel (1982). Abbreviations: 1, proseptum; dp, dorsal prismatic layer of the first whorl; f, flange; ip, inner prismatic layer of the initial chamber or of the first whorl; ml, middle prismatic layer of the initial chamber; mp, middle prismatic layer of the first whorl; na, nacreous layer of the postammonitella shell; op, outer prismatic layer of the first whorl or of the postammonitella shell; p_1–p_6, prismatic layers of the initial chamber and first whorl; pi, prismatic layer of the initial chamber; pv, primary varix; t, tubercles. Arrows indicate the distal end of the initial chamber. See text for explanation.

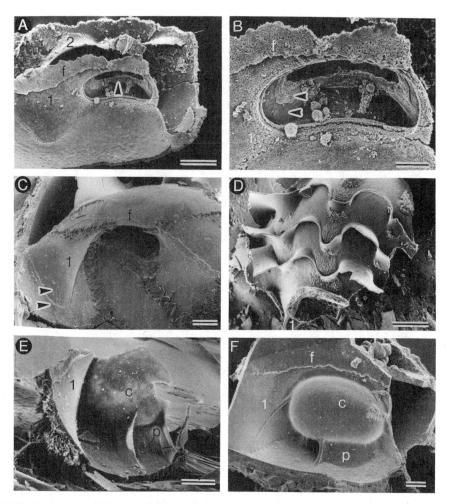

FIGURE 15. Internal features of the ammonitella observed in specimens free of matrix. (A,B) *Scaphites* sp. cf. *S. whitfieldi* (Ancyloceratina), Upper Cretaceous, South Dakota, AMNH 42900. (A) Interior of the initial chamber and first whorl showing the proseptum (1), its neck-like attachment (arrow), flange (f), and second septum (2). Scale bar, 50 μm. (B) Close-up of the proseptum (1), its neck-like attachment (lower arrow), and flange (f). Prismatic attachment deposits (upper arrow) of the siphuncle occur on the neck-like attachment of the proseptum. Scale bar, 20 μm. (C) *Baculites* sp. (Ancyloceratina), Upper Cretaceous, Wyoming, AMNH 42905. Interior of the initial chamber showing the proseptum (1) and flange (f). A prismatic ridge (lower arrow) occurs at the base of the proseptum, and the surface of the proseptum is marked by wrinkles (upper arrow). Scale bar, 40 μm. (D) *Euhoplites* sp. (Ammonitina), Lower Cretaceous, England, AMNH 27261a. Muscle scars are visible on the inside surface of the dorsal wall adoral of the proseptum (1) and the lobes of the next few septa. The first scar actually consists of two separate but connecting scars. Scale bar, 100 μm. (E) *Quenstedtoceras* sp. (Ammonitina), Middle Jurassic, Poland, AMNH 42911. Close-up of the proseptum (1), cecum (c), and prosiphon (p). Note the wrinkles in the prosiphon. Scale bar, 40 μm. (F) *Hypacanthoplites* sp. (Ammonitina), Lower Cretaceous, Germany, AMNH 20952a. Close-up of the proseptum (1), flange (f), cecum (c), and prosiphon (p). Scale bar, 40 μm.

It closes off the initial chamber and appears to form a continuation of the flange, i.e., the inner lip of the initial chamber (Bandel, 1982). There is an opening in the middle of the proseptum that approximately equals the whorl height.

The shape of the proseptum is different from that of all subsequent septa (Schindewolf, 1954; Erben *et al.*, 1969). Variation in the shape of the proseptum and its corresponding suture, the prosuture, was documented first by Branco (1879, 1880) and later by Schindewolf (1928, 1929). Branco described three character states of the prosuture: asellate, latisellate, and angustisellate, depending on the size of the dorsal and ventral saddles. However, this categorization probably needs to be expanded because it does not accommodate the full range of variation observed within the Ammonoidea (see, e.g., House, 1965; Bensaïd, 1974).

The proseptum is prismatic in microstructure (Erben *et al.*, 1969; Birkelund and Hansen, 1974; Druschits *et al.*, 1977a,b). In median cross section, it shows a complex relationship with the shell wall (Erben *et al.*, 1969; Birkelund and Hansen, 1974; Kulicki, 1979). For example, in median cross sections of *Quenstedtoceras*, the ventral portion of the proseptum forms a continuation of the middle prismatic layer of the initial chamber (Fig. 14C, ml; Kulicki, 1979, Figs. 7, 10). In well-preserved specimens free of interior matrix, a prismatic ridge appears at the base of the proseptum, and the surface of the proseptum is marked by wrinkles along the lateral lobes (Fig. 15A,C; Landman and Bandel, 1985).

In some ammonoids, an adorally directed neck-like attachment develops around the proseptal opening (Fig. 15A,B; Landman, 1985, 1987; Landman and Bandel, 1985; Bandel, 1986). This neck-like attachment forms a suture where it joins the shell wall and can easily be mistaken for a second proseptum. Two prosepta have been reported in the Prolecanitina and Goniatitina by Böhmers (1936). This author noted that the first two septa in these forms differ from subsequent septa in having short amphichoanitic necks (necks directed both adapically and adorally). Based on this evidence, he called both septa prosepta, a terminology later adopted by Miller and Unklesbay (1943), Miller *et al.* (1957), and Arkell (1957, p. L101). However, other studies have suggested that these two septa represent the proseptum and second septum with an amphichoanitic neck and a retrochoanitic neck, respectively (Schindewolf, 1954; Tanabe *et al.*, 1994).

The second septum, sometimes called the primary septum, has a shape completely different from that of the proseptum (Fig. 15A; Schindewolf, 1928, 1929, 1951, 1954; Erben *et al.*, 1969). It is characterized by ventral and dorsal lobes and as many as three lateral and umbilical lobes, depending on the suborder (Schindewolf, 1954; Wiedmann and Kullmann, 1981). The second septum is the developmental basis in ontogeny for all subsequent septa.

The distance between the proseptum and second septum varies markedly among suborders. The second septum may be separated from the proseptum and form its own suture. In other ammonoids, such as *Quenstedtoceras*, the second septum rides dorsally on the proseptum, although the two septa are distinct ventrally (Druschits and Khiami, 1970; Kulicki, 1979; Bandel, 1982; Landman and Bandel, 1985). As a result, the second septum in this genus forms an incomplete internal suture (Bandel, 1986).

The microstructure of the second septum, like that of all subsequent septa, differs from that of the proseptum. In all ammonoids in which septal microstructure has been studied (Ammonitina, Lytoceratina, Phylloceratina, and Ancyloceratina), the second and all later septa are composed mainly of nacre (Birkelund and Hansen, 1974; Kulicki, 1979; Bandel, 1982; Landman and Bandel, 1985). The observation of a prismatic second septum by Erben *et al.* (1969) has not been substantiated (Bandel, 1986).

2.7. Siphuncle

The bulb-like beginning of the siphuncle, called the cecum, is located in the initial chamber (Fig. 15E,F). Like the rest of the siphuncle, presumably the cecum was originally organic (Bandel, 1982; Tanabe and Ohtsuka, 1985; Ohtsuka, 1986) and, in well-preserved specimens, retains traces of fine wrinkles (Kulicki, 1979).

The shape of the cecum in median cross section is elliptical, hemicircular, or rectangular (Table I). It is elliptical in the Bactritina, Goniatitina (Fig. 16A), Ceratitina (Fig. 16B), Phylloceratina (Fig. 17A), Ancyloceratina (Fig. 17B), and Ammonitina (Fig. 17C,D); this shape probably is the phylogenetically primitive condition. In contrast, the cecum is rectangular in the Prolecanitina (Fig. 16C,D) and hemicircular in the Lytoceratina (Fig. 16E,F).

The cecum is attached to the inside surface of the initial chamber by means of the prosiphon, which consists of one or more bands (Figs. 15E,F, 16, and 17; Munier-Chalmas, 1873; Crickmay, 1925; Zakharov, 1972; Druschits and Doguzhaeva, 1981; this feature was called the "fixator" by Druschits *et al.*, 1977b, 1980). The prosiphon was originally organic, and wrinkles are commonly present along its length (Fig. 15E; Kulicki, 1979). Although there is variation in the morphology of the prosiphon within a single species (Kulicki, 1979; Bandel, 1982, 1986; Landman and Bandel, 1985), it is possible to distinguish two main types in the Ammonoidea as a whole (Grandjean, 1910; Zakharov, 1972, 1974, 1989; Druschits and Doguzhaeva, 1974, 1981; Vavilov and Alekseyev, 1979; Tanabe *et al.*, 1979, 1980; Birkelund, 1981; Landman, 1987; Blind, 1988; Table I). In most Ammonitina, excluding the Amaltheidae, Collignoniceratidae (Fig. 17F), and Placenticeratidae, the prosiphon is long and nearly straight (Fig. 17C–E). In contrast, it is short and curved in the Bactritina, Goniatitina (Fig. 16A), Prolecanitina (Fig. 16C,D), Lytoceratina (Fig. 16E,F), Phylloceratina (Fig. 17A), and some Ancyloceratina (Fig. 17B). In the Ceratitina, the shape and size of the prosiphon are highly variable (Fig. 16B; Weitschat and Bandel, 1991).

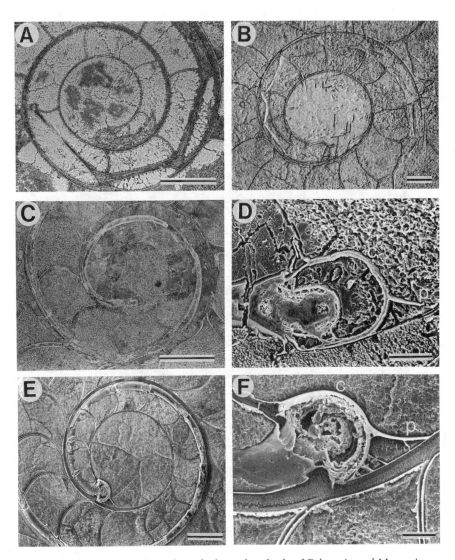

FIGURE 16. Median cross sections through the early whorls of Paleozoic and Mesozoic ammonoids showing the shape of the cecum and prosiphon. (A) *Glaphyrites warei* (Goniatitina), Middle Pennsylvanian, Oklahoma, UMUT PM 19026-1. Scale bar, 250 μm. (B) *Indigirites tozeri* (Ceratitina), Middle Triassic, Spitsbergen, AMNH 44353. Scale bar, 100 μm. (C,D) *Artinskia electraensis* (Prolecanitina), Middle Permian, Nevada, UMUT PM 19040-2. (C) Overall view. Scale bar, 500 μm. (D) Close-up of cecum (c) and prosiphon (p). Scale bar, 50 μm. (E,F) *Gaudryceras striatum* (Lytoceratina), Upper Cretaceous, Hokkaido, UMUT MM (= EES 11). (E) Overall view. Scale bar, 230 μm. (F) Close-up of cecum (c) and prosiphon (p). Scale bar, 42 μm.

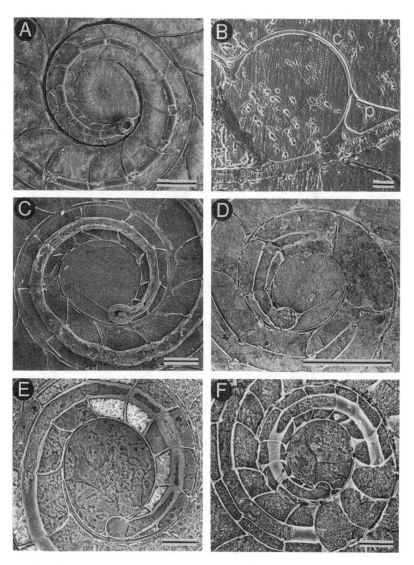

FIGURE 17. Median cross sections through the early whorls of Paleozoic and Mesozoic ammonoids showing the shape of the cecum and prosiphon. (A) *Hypophylloceras subramosum* (Phylloceratina), Upper Cretaceous, Hokkaido, UMUT MM 19683 (= EES 19). Scale bar, 300 μm. (B) *Scaphites preventricosus* (Ancyloceratina), Upper Cretaceous, Montana, AMNH 43035. Close-up of the cecum (c) and prosiphon (p). Scale bar, 10 μm. (C) *Damesites sugata* (Ammonitina), Upper Cretaceous, Hokkaido, UMUT MM 18326 (= EES 37). Scale bar, 160 μm. (D) *Eleganticeras elegantulum* (Ammonitina), Lower Jurassic, England, UMUT MM 19066-1. Scale bar, 500 μm. (E) *Promicroceras* sp. (Ammonitina), Lower Jurassic, England, UMUT MM 19069-1. Scale bar, 124 μm. (F) *Subprionocyclus neptuni* (Ammonitina), Upper Cretaceous, Hokkaido, UMUT MM 19075. Scale bar, 160 μm.

The cecum and siphuncle are attached to the septa by means of prismatic attachment deposits; these have been referred to as "auxiliary deposits" (Fig. 15B; Kulicki, 1979; Bandel, 1982; Landman and Bandel, 1985; Chapters 4 and 6, this volume). The initial position of the siphuncle ranges from marginal to central, depending on the suborder (Figs. 16 and 17; Table 1; Druschits and Doguzhaeva, 1974, 1981; Tanabe and Ohtsuka, 1985).

2.8. Muscle Scars

Muscle scars are rarely preserved in the ammonitella, although they have been detected in a few genera of Ceratitina and Ammonitina (see Chapter 3, this volume). Bandel (1982) documented the ontogenetic progression of muscle scars in *Quenstedtoceras*. He identified a muscle scar on the inside surface of the flange, a pair of muscle scars on the adoral face of the proseptum on either side of the proseptal opening, and another pair of muscle scars on the inside surface of the dorsal wall adoral of the second septum. He noted that, adoral of the third septum, these two muscle scars united into a single muscle field. A similar sequence has been reported in *Euhoplites* (Fig. 15D; Landman and Bandel, 1985) and in several genera of Triassic Ceratitina (Weitschat and Bandel, 1991).

3. Sequence of Embryonic Development

Information about the embryonic development of ammonoids comes from two sources: examination of specimens actually preserved at early ontogenetic stages and study of the morphology and microstructure of the early whorls of larger specimens. In order to reconstruct early ontogenetic stages using the second method, it usually is necessary to break down specimens to expose the inner whorls. These two approaches are complementary and provide the best evidence available for determining the sequence of embryonic development.

3.1. Reconstructions Based on the Early Whorls of Larger Specimens

Reconstructions based on the morphology of the early whorls of larger specimens have been suggested by numerous workers (Branco, 1879, 1880; Hyatt, 1894; Smith, 1901; Grandjean, 1910; Shul'ga-Nesterenko, 1926; Schindewolf, 1929; Spath, 1933; Böhmers, 1936; Trueman, 1941; Currie, 1944; Shimansky, 1954; Arkell, 1957; Erben, 1962, 1964, 1966; Erben *et al.*, 1969; Palframan, 1967a; Druschits and Khiami, 1970; Druschits *et al.*, 1977a; Druschits and Doguzhaeva, 1981; Makowski, 1971; Zakharov, 1972; Birkelund and Hansen, 1974; Birkelund, 1981; Kulicki, 1974, 1979; Tanabe *et al.*, 1980;

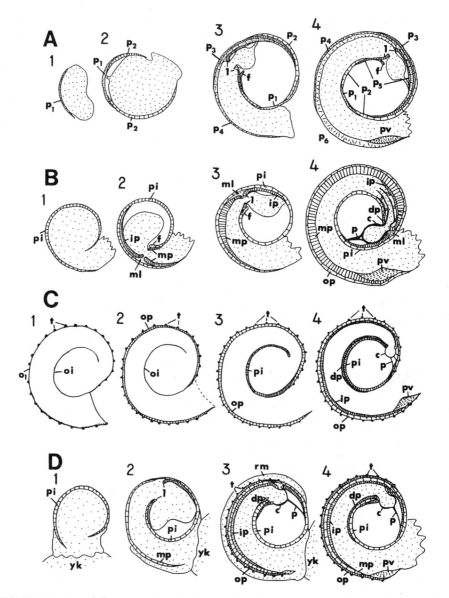

FIGURE 18. Four different models (A–D) depicting the sequence of ammonoid embryonic development (1–4). Animals are represented in median cross section with soft tissues shaded. (A) Erben *et al.* (1969, Fig. 5) described three stages in early ontogeny: an embryonic stage (1,2), a larval stage (3), which was followed by metamorphosis (4), and a postlarval stage (not shown). Six prismatic layers (p_1–p_6) comprise the wall of the initial chamber and first whorl and were secreted sequentially. (B) Kulicki (1979, Fig. 7) emphasized a *Nautilus*-like mode of embryonic shell development. (C) Bandel (1982, Figs. 40, 46, 47) argued that the ammonitella originally consisted of an organic, unmineralized shell. (D) Tanabe (1989, Fig. 7) proposed that the ammonitella was temporarily enveloped by the outer mantle late in embryonic development.

Tanabe, 1989; Lehmann, 1981; Bandel, 1982, 1986; Landman, 1982, 1987; Blind, 1988). We review several of these reconstructions in the next few pages.

The first detailed reconstruction based on SEM data was that of Erben (1962, 1964, 1966) and Erben *et al.* (1968, 1969), who recognized three phases in early ontogeny (Fig. 18A). According to them, in the first phase, the initial chamber was secreted inside the egg capsule. The animal then hatched as a veliger larva with a ciliated velum. During this larval stage, the first whorl, proseptum, flange, cecum, and prosiphon formed. The primary constriction and accompanying varix were thought to have developed during metamorphosis. Following metamorphosis, additional septa were secreted, and a nacreous layer was added to the shell wall.

This reconstruction was based primarily on four lines of evidence:

1. Microstructure of the shell wall. Erben *et al.* (1968, 1969) observed that several of the layers comprising the wall of the initial chamber wedged out on the outer side at the distal end of the initial chamber and were replaced by a new layer that formed most of the wall of the first whorl. They related this change to hatching. However, Kulicki (1979) has suggested that the wall of the initial chamber does not, in fact, wedge out but forms the external layer of the wall of the first whorl (but see also Bandel, 1982). In addition, there is no evidence of a discontinuity on the shell surface at the transition from the initial chamber to the first whorl; if there had been a break in secretion at this point, a discontinuity would be present (Kulicki, 1979, p. 128).

2. Ornamentation. Erben (1962, 1964, 1966) noted a change in the pattern of growth lines ("*Anwachsstreifen*") in phylogenetically primitive ammonoids at the distal end of the initial chamber. For example, he reported that the growth lines in *Mimagoniatites* developed a slight backward projection along the venter at this point; this projection became more pronounced over the course of the first whorl (Erben, 1964, Fig. 4, Pl. 8, Figs. 3–5). Erben (1966, p. 651) interpreted this change as indicating the development of a locomotor organ such as a velum. Bogoslovsky (1969, p. 66) argued that this change reflected instead the development of a funnel in the embryonic stage. Close

---

FIGURE 18. (Continued) According to his model, the prismatic layer of the initial chamber (pi) and the middle (mp) and inner (ip) prismatic layers of the first whorl were formed by the interior epithelium, whereas the outer prismatic layer of the first whorl (op) with tubercles (t) was secreted by the exterior epithelium of the reflected mantle (rm). Abbreviations: 1, proseptum; c, cecum; dp, dorsal prismatic layer of the first whorl; f, flange; ip, inner prismatic layer of the initial chamber or of the first whorl; ml, middle prismatic layer of the initial chamber; mp, middle prismatic layer of the first whorl; oi, organic wall of the initial chamber; o1, organic wall of the first whorl; op, outer prismatic layer of the first whorl; p, prosiphon; pi, prismatic layer of the initial chamber; pv, primary varix; rm, reflected mantle; t, tubercles; yk, yolk mass. See text for more details.

inspection of the shell surface reveals that the features in question are not, in fact, growth lines but lirae (see Chapter 12, this volume, for the distinction between growth lines and lirae). The changes in the lirae are gradual, and such gradual changes in ornamentation have been documented in the embryonic development of other molluscs, for example, modern *Nautilus* (see Arnold *et al.*, 1987). In any event, the most marked change in ornamentation in all ammonoids occurs at the primary constriction, not before.

3. Primary varix. Erben *et al.* (1969) interpreted the appearance of nacre late in shell development, in the form of the primary varix, as an indication of metamorphosis because, in contrast, nacre appears early on in the ontogenetic development of *Nautilus*, where metamorphosis is absent. But even in the embryonic development of *Nautilus*, the initial shell material at the cicatrix is not nacreous but prismatic (Arnold *et al.*, 1987). Nacre appears only later, lining the interior of the cap-shaped initial shell.

4. The shape of the proseptum. According to Erben *et al.* (1969; see also Schindewolf, 1954, pp. 230–231), the change in shape from the proseptum to the second septum implies a complete metamorphosis of the ammonoid soft body. However, Bandel (1982, p. 68) has argued that, because the proseptum formed before the formation of the siphuncle, the proseptum has a shape different from that of all later septa. Hewitt (1985) has also pointed out that the shape of the proseptum in Mesozoic ammonoids is an adaptation to "resist circumferential stresses imposed by subsequent whorls."

The alternative model of early ontogeny is that of direct development in which there are only two phases, embryonic and postembryonic. This model has been suggested by many workers and is widely accepted today (Grandjean, 1910; Böhmers, 1936; Druschits and Khiami, 1970; Druschits *et al.*, 1977a; Druschits and Doguzhaeva, 1981; Zakharov, 1972; Kulicki, 1974, 1979; Birkelund and Hansen, 1974; Tanabe *et al.*, 1980; Tanabe, 1989; Bandel, 1982, 1986; Landman, 1982, 1987).

In the model of direct development, the ammonitella is the embryonic shell. In many species, therefore, the newly hatched ammonoid more or less resembles a miniature adult. The most compelling pieces of evidence for this model are (1) the uniform surface of the ammonitella, without any indication of a discontinuity in secretion, and (2) the abrupt changes in ornamentation, shell shape, and microstructure at the end of the primary constriction. Similarly abrupt changes coincide with hatching in many other molluscs (Bandel, 1975, 1982; Jablonski and Lutz, 1980).

This model is also consistent with the fact that development is direct, without a larval phase, in all living cephalopods whose early development has been studied (Arnold and Williams-Arnold, 1977; Arnold *et al.*, 1987; Wells and Wells, 1977; Bandel and Boletzky, 1979; Boletzky, 1988). Although the term "larva" sometimes is used in cephalopods to refer to individuals immediately after hatching, these individuals do not undergo any metamorphosis (Ruzhentsev and Shimansky, 1954; Boletzky, 1974, 1993; Wells and Wells, 1977). Hence, such an animal is not a "larva" in the strict sense of the term.

Within the framework of direct development, it is difficult to reconstruct the exact sequence in which the ammonitella formed. We review three models that cover most of the possible variations.

Based on his study of *Quenstedtoceras*, Kulicki (1979) suggested that the ammonitella formed by accretionary growth (Fig. 18B). According to him, the initial chamber was secreted by a cap-like secretory zone. Subsequently, this zone differentiated into two subzones, an anterior one, which formed the outer layer of the wall of the first whorl, and a posterior one, which formed the inner layers of the walls of the initial chamber and first whorl. Kulicki inferred that, during formation of the proseptum, the soft body withdrew from the initial chamber, after which the cecum and prosiphon developed. The primary varix was thought to have formed right before hatching during a temporary withdrawal of the mantle margin, at which time secretion of nacreous material occurred.

Bandel (1982, 1986, 1989, 1991) introduced a new concept in his model of embryonic development (Fig. 18C). He argued that in the early stages of embryonic development, the ammonitella consisted of an organic, unmineralized shell. He based this argument on studies of well preserved specimens of *Quenstedtoceras* and the observation that in living cephalopods with small embryonic shells (<2 mm in size), e.g., *Spirula*, the embryonic shell is initially entirely organic. According to Bandel, the organic ammonitella was secreted in uninterrupted contact with the gland cells of the mantle. The surface of this organic shell was devoid of growth lines and, in Mesozoic ammonoids, was covered with a tuberculate microornamentation. This shell was thought to have been mineralized rapidly by prismatic needles from the inside; this formed an outer layer of uniform thickness, preserving the original ornamentation of the organic shell. A similar process of rapid mineralization has been reported in modern archaeogastropods (Bandel, 1986). It is important to note that this outer layer was inferred to have formed only on the exposed portions of the ammonitella. In closely coiled ammonitellas, the portion of the initial chamber covered by the first whorl still would have been unmineralized at this stage. Subsequently, several prismatic layers supposedly were secreted from the inside, starting backward from the aperture; this served to thicken the original, outer layer and complete the rest of the wall of the initial chamber.

Bandel (1982) also reconstructed the developmental sequence of the internal features. According to his model, a portion of the visceral mass first differentiated to form the cells of the siphuncle. Subsequently, the rest of the visceral mass withdrew from the initial chamber, remaining attached to it only by retractor muscles, thought to have been located on the inside surface of the flange, and by siphuncular tissue, thought to have been located on the inside surface of the initial chamber. The visceral mass then formed the organic precursor of the proseptum, which assumed the shape of the apical end of the visceral sac. Thus, according to Bandel, the proseptum formed before the formation of the actual siphuncle, explaining the unique shape of the proseptum relative to all other septa. After mineralization of the proseptum, the retractor muscles reattached in two bundles onto the adoral face of the proseptum. Finally, the cecum and prosiphon developed, which would have permitted the removal of cameral liquid from the initial chamber.

Tanabe (1989) presented an alternative model to explain the presence of tubercles and absence of growth lines on the ammonitellas of Mesozoic ammonoids (Fig. 18D). He proposed that the ammonitella was enveloped temporarily by the outer mantle late in embryonic development, a process similar to that which occurs in modern *Spirula*. According to this model, the outer mantle secreted a thin prismatic layer with tuberculate ornamentation on the exposed portions of the shell. Tanabe based this hypothesis on his observation that tubercles commonly cover several contiguous prisms on the outer layer, suggesting that the tubercles developed after the completion of the underlying prisms. Following the secretion of this outer layer, the mantle was thought to have migrated back toward the aperture, resuming its earlier position.

3.2. Evidence from Specimens Preserved at Early Ontogenetic Stages

Several fossils interpreted as ammonoid eggs have been reported from the Mesozoic (Dreyfuss, 1933; Wetzel, 1959; Lehmann, 1966, 1981; Müller, 1969). These structures appear as hollow spheres approximately 0.5 mm in diameter. They are filled with calcite or the same material as the surrounding matrix and show no evidence of embryonic shells inside. They occur either scattered in the rock associated with ammonitellas and small juveniles (Dreyfuss, 1933; Wetzel, 1959) or clustered as a mass within the body chambers of adults (Lehmann, 1966, 1981; Müller, 1969). With their small size and lack of embryonic shells inside, these small spheres may represent eggs at an early stage of development (Kulicki, 1979). This interpretation is consistent with the fact that the eggs of many modern cephalopods grow in size during embryogenesis (Zuev and Nesis, 1971).

These ammonoid eggs, if truly that, would shed some light on mode of development (e.g., possible brooding within the body chamber in some ammonoids) but provide no information about the embryonic development of the shell. The best source of such information comes from accumulations of embryonic shells preserved at different developmental stages. These accumulations may represent egg masses in which the individual embryos developed at different rates (asynchronous development). These eggs may have been deposited on the sea floor (Mapes *et al.*, in prep.) or, alternatively, may have been laid originally as gelatinous masses in midwater, settling to the bottom only afterward, a phenomenon similar to that observed in the midwater squid *Illex illecebrosus* (see O'Dor, 1983; Mangold, 1987; Hewitt, 1993; Chapter 16, this volume).

Kulicki (1989) and Kulicki and Doguzhaeva (1994) documented the development of the embryonic shell in the Early Cretaceous genus *Aconeceras* based on actual specimens from the Symbirsk area, Russia (see Chapter 4, this

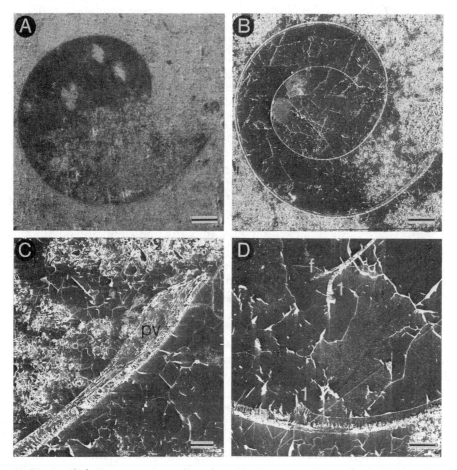

FIGURE 19. Median cross sections through embryonic shells of *Baculites* (Ancyloceratina) showing two developmental stages. (A) Embryonic shell of *Baculites* sp. cf. *B. asper* or *B. codyensis*, Upper Cretaceous, Montana, AMNH 44834. This specimen is at an early stage of development; the portion of the initial chamber covered by the first whorl is still unmineralized. Scale bar, 100 μm. (B–D) Embryonic shell of *Baculites* sp. cf. *B. mariasensis* or *B. sweetgrassensis*, Upper Cretaceous, Montana, AMNH 43203. This specimen is at a much later stage of development; both the initial chamber and first whorl are now completely mineralized. (B) Overall view showing the proseptum and primary varix. Scale bar, 100 μm. (C) Close-up of the primary varix (pv). Scale bar, 20 μm. (D) Close-up of the proseptum (1) and flange (f). Scale bar, 20 μm.

volume; see also Ruzhentsev, 1974, Pl. 1, Fig. 1; Druschits and Khiami, 1970). They described three successive stages in embryonic development: (1) mineralization of the exposed portions of the initial chamber and first whorl ending at the primary constriction, (2) mineralization of the wall of the initial chamber near the site of the future proseptum, and (3) mineralization of the rest of the wall of the initial chamber and formation of the primary varix and proseptum.

Similar developmental stages have been observed in specimens of Late Cretaceous *Baculites* from North America (Fig. 19; see Landman, 1982). In specimens corresponding in their degree of development to the first stage described by Kulicki and Doguzhaeva (1994), the portion of the initial chamber covered by the first whorl is not preserved and is presumed to have been still organic (Fig. 19A). Alternatively, it is possible that this part of the initial chamber was mineralized but simply broke off. However, these specimens are so similar to those described by Kulicki (1989) and Kulicki and Doguzhaeva (1994, Fig. 6A) that a more likely hypothesis is that all of these specimens represent the same stage in the embryonic development of the shell.

Both of these studies strongly support Bandel's (1982, 1986) model according to which the ammonitella of Mesozoic ammonoids initially consisted only of an organic shell. As Bandel hypothesized, the exposed portions of the ammonitella mineralized first. Thereafter, mineralization proceeded backward from the aperture, thickening the wall of the first whorl and completing the rest of the wall of the initial chamber. However, it is unclear from these studies whether the tuberculate ornamentation of the ammonitella was originally present on the surface of the organic shell as suggested by Bandel (1982) or whether it developed later as proposed by Tanabe (1989).

In contrast to these data supporting the existence of an originally organic ammonitella in Mesozoic ammonoids, Tanabe *et al.* (1993) presented evidence suggesting an accretionary mode of growth in Carboniferous Goniatitina (Fig. 20). Based on actual specimens of *Aristoceras* and *Vidrioceras*, these authors identified three successive stages in the formation of the embryonic shell: (1) mineralization of the initial chamber, (2) mineralization of part of the first whorl, and (3) mineralization of the rest of the first whorl and formation of the primary varix and proseptum.

In addition to the accumulations of embryonic shells referred to in the previous paragraphs, there are numerous reports of intact ammonitellas that provide further information about shell development [as listed in geological order: Mississippian Goniatitidae from Alberta (Schindewolf, 1959); Carboniferous Goniatitidae from Britain and Ireland (Ramsbottom, 1981; Tanabe *et al.*, 1995); Pennsylvanian Bactritoidea from Texas and Kansas (Hecht and Mapes, 1990); Early Permian ammonoids from the Aktyubinsk area, Russia (Ruzhentsev, 1974, Pl. 1, Fig. 2); Middle Triassic Ceratitidae from Nevada (H. Bucher, personal communication, 1993); Late Triassic ammonoids from Austria (Wiedmann, 1973); Early Jurassic ammonoids from France (Dreyfuss,

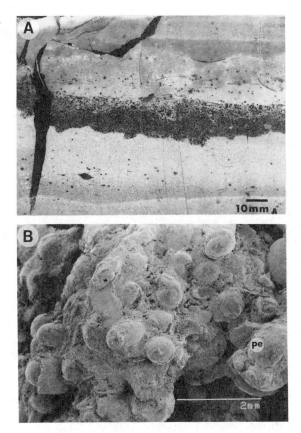

FIGURE 20. (A) Cross section through a mass of preserved ammonitellas of *Vidrioceras* sp. and *Aristoceras* sp. (both Goniatitina) in a carbonate concretion from the Upper Pennsylvanian of Kansas. (B) Close-up of a weathered portion of the concretion showing densely packed ammonitellas and small postembryonic shells (pe). (After Tanabe *et al.*, 1993, Fig 1.)

1933); Early Jurassic *Arnioceras* from England (Trueman, 1941); Early Jurassic *Harpoceras elegans* from Germany (Wetzel, 1959); Middle Jurassic *Quenstedtoceras* from Poland (Blind, 1979; Chapter 4, this volume); Late Jurassic Perisphinctidae from Cuba (Kulicki and Wierzbowski, 1983); Late Cretaceous *Baculites* from North America (Smith, 1901) and Jordan (Bandel, 1982); Late Cretaceous *Scaphites* from North America (Landman, 1985); and Late Cretaceous *Baculites* or *Hoploscaphites* from Denmark (Birkelund, 1979, 1981)]. These ammonitellas all terminate at the primary constriction and accompanying varix, suggesting that hatching occurred after formation of these features. The primary varix probably developed just prior to hatching during a temporary withdrawal of the mantle margin (Kulicki, 1979).

Internal features are sometimes present in these ammonitellas and consist of the cecum, prosiphon, and at least one septum. However, the cecum and prosiphon are rarely preserved, probably because these structures were originally organic (see Wetzel, 1959; Landman, 1982). [Alternatively, the absence of a prosiphon may indicate that this feature had not formed yet. According to R.A. Hewitt (personal communication, 1993), the prosiphon formed by shrinkage of cameral membranes as the fluid (or gel) was pumped out of the initial chamber.] The proseptum is always present and probably formed near the end of embryonic development, implying that nacreous septa developed only postembryonically (Smith, 1901; Druschits and Khiami, 1970; Landman, 1982; Tanabe *et al.*, 1993). However, ammonitellas with more than one septum also occur, indicating that in some species additional septa may have formed before hatching (Blind, 1979; Bandel, 1982; Kulicki and Wierzbowski, 1983). Such species-specific variation in the number of embryonic septa has, in fact, been reported in modern sepioids (Bandel and Boletzky, 1979). However, Landman (1985) has cautioned that some supposedly embryonic shells with more than one septum may actually represent fragments of larger specimens that have broken at the embryonic–postembryonic shell boundary.

4. Posthatching Mode of Life

Like the adults of most ammonoids, ammonitellas were probably neutrally buoyant at or soon after hatching and, consequently, could have lived in the water column rather than on the bottom (Zakharov, 1972; Kulicki, 1974, 1979; Druschits *et al.*, 1977a; Tanabe *et al.*, 1980; Tanabe and Ohtsuka, 1985; Bandel, 1982, 1986; Landman, 1985; Ward and Bandel, 1987; Westermann, 1990; Weitschat and Bandel, 1991; Kakabadzé and Sharikadzé, 1993; Chapter 16, this volume; but compare Wetzel, 1959). The initial chamber represents a relatively large float that originally was filled with liquid (or gel; R.A. Hewitt, personal communication, 1993); this liquid (or gel) subsequently was removed via the cecum and prosiphon (Trueman, 1941; Zakharov, 1972, 1989; Tanabe *et al.*, 1980; Landman, 1987; Hewitt, 1988). This event probably occurred just prior to or immediately after hatching; in the latter case, it would have resulted in a slight delay before entry into the water column.

The size relationships among the component parts of the ammonitella provide additional support for the hypothesis of neutral buoyancy. As noted previously, there is a strong positive correlation between the volume of the initial chamber (phragmocone) and the volume of the ammonitella (phragmocone plus body chamber) both within and among species (Fig. 7; Shigeta, 1993). In contrast, there is a negative correlation between ammonitella angle and the whorl expansion rate of the ammonitella so that larger ammonitella angles are associated with more closely coiled ammonitellas (Tanabe and Ohtsuka, 1985, Fig. 6). Both of these correlations probably reflect volumetric relationships necessary to maintain neutral buoyancy.

Density calculations performed on actual ammonitellas are also consistent with the hypothesis of neutral buoyancy. Shigeta (1993) reported measurements on ammonitellas of several Cretaceous ammonoids, assuming that animals hatched with a single septum and without any cameral liquid in the initial chamber. According to his measurements, the density of these ammonitellas was less than that of sea water. However, such measurements are subject to error because of the difficulty of estimating the volume of shell material and cameral liquid and the weight of the soft body (Westermann, 1993; Chapter 16, this volume). Allowing for the limitations of this methodology, the values calculated fall within the range of neutral buoyancy.

With the possible exception of the Agoniatitina and Bactritina, most ammonoids probably followed a planktic mode of life at hatching (Zakharov, 1972; Kulicki, 1974; "pseudolarval stage" of Kulicki, 1979; Druschits et al., 1977a; Landman, 1982, 1985; Tanabe and Ohtsuka, 1985; Morton, 1988; Westermann, 1990; Shigeta, 1993; see Chapter 16, this volume). This hypothesis is based on two functional arguments: the small size of the ammonitella and its nearly spherical shape, both of which are presumably adaptations to life in the plankton (Kulicki, 1979). A planktic mode of life is common among many modern coleoids including nektic squids and sepioids and benthic octopods with relatively small eggs (Boletzky, 1974, 1977, 1987b; Vecchione, 1987). Among octopods, this mode of life is considered the phylogenetically primitive condition, whereas a benthic mode of life at hatching is considered derived (Boletzky, 1987b, 1992).

In the plankton newly hatched ammonoids may have been active swimmers or more probably, passive vertical migrators, drifting with surface currents (Birklund and Hansen, 1974; Kulicki, 1974; Ward and Bandel, 1987; Westermann, 1990; Weitschat and Bandel, 1991; Chapter 16, this volume; see Sturani, 1971, p. 46, for a description of another mode of life in algal meadows, especially for Lytoceratina and Phylloceratina). This planktic stage may have lasted several weeks or months, depending on the size of the ammonitella (Kulicki, 1979), its rate of growth (Westermann, 1990; Shigeta, 1993), and the mode of life of the adult (see Boletzky, 1974, 1977, 1987b). In addition, Shigeta (1993) suggested that the duration of the planktic stage was dependent on the rate of increase in the density of the newly hatched ammonoid (but see Westermann, 1993). In the plankton, ammonoids may have secreted as many as two whorls, reaching shell diameters of 3–5 mm (Westermann, 1954, 1990; "neanoconch" of Westermann, Chapter 16, this volume; Kulicki, 1974; Landman, 1987).

A planktic mode of life at hatching is consistent with a number of observations on the mode of occurrence of specimens preserved at this stage. There are several occurrences of ammonitellas and very small juveniles with older juveniles and adults and mostly nektic and planktic organisms in environments in which the bottom was anaerobic with oxygenated water above (Upper Jurassic Jagua Formation in Cuba, Kulicki and Wierzbowski, 1983; Middle Triassic Fossil Hill Member of the Favret Formation in Nevada, H.

Bucher, personal communication, 1993; Silberling and Nichols, 1982; Upper Cretaceous Sharon Springs Member of the Pierre Shale in Wyoming, Landman, 1988). These assemblages are "quasiautochthonous" and strongly suggest that the newly hatched ammonoids were planktic or at least nektic. In addition, there are numerous examples of mixed assemblages of juvenile and adult ammonoids from presumably well-oxygenated environments in which very small juveniles (<3–4 mm shell diameter) are rare or absent (Middle Jurassic Bearreraig Sandstone Formation in northwest Scotland, Morton, 1988; Upper Cretaceous Yezo Group in Hokkaido, Japan, Shigeta, 1993; Upper Cretaceous Fox Hills Formation in South Dakota, Landman and Waage, 1993; Landman and Klofak, in prep.). These data suggest that newly hatched ammonoids may have lived in a different environment from that of older juveniles and adults, although their absence may also result from taphonomic processes. A planktic mode of life at hatching is also consistent with the fact that some Mesozoic ammonoids such as the Late Cretaceous heteromorph *Turrilites costatus* have broad biogeographic distributions despite the fact that the adults of these species are presumed to have been poor swimmers (Ward and Bandel, 1987; but see Chapter 16, this volume, for an alternative explanation).

5. Reproductive Strategy

The embryos of cephalopods are generally larger than those of other molluscs (Naef, 1922; Berthold and Engeser, 1987; Engeser, 1990). Two size classes can be distinguished within the cephalopods as a whole: embryos less than about 2 mm in size versus embryos greater than about 2 mm in size (Bandel and Boletzky, 1979; Bandel, 1991; Engeser, 1990). The embryos of ammonoids, to judge from their embryonic shells, fall into the first category. In marked contrast, the quintessential example of the second category is the embryo of *Nautilus*, which measures approximately 30 mm in size (Arnold *et al.*, 1987; Landman, 1988).

The reproductive strategy of ammonoids was similar to that of many coleoids (Engeser, 1990; Tanabe *et al.*, 1993). The common occurrence of ammonitellas in dense concentrations suggests that ammonoids produced a large number of offspring, probably thousands of embryos per female (Ward and Bandel, 1987; Landman, 1988; see Mangold, 1987, pp. 172–178, for a comparison of the number of offspring in coleoids). The ammonoid embryonic shell is small relative to that of the adult, implying little parental investment per egg. As a corollary, many ammonoids, like the majority of coleoids, were probably semelparous, reproducing once and then dying. The length of embryonic development probably also was similar to that in many coleoids, which lasts several tens of days depending on temperature conditions (Boletzky, 1974, 1977, 1987a; Hewitt, 1988; Weitschat and Bandel, 1991). The occurrence of large numbers of preserved ammonitellas further suggests that many species experienced a high degree of mortality at hatching (Kulicki, 1979; Kulicki and Wierzbowski, 1983; Bandel, 1982; Landman, 1987). Juveniles also are abundant in some ammonoid assemblages (Ziegler, 1962; Callomon, 1963; Lehmann, 1966; Kennedy and Cobban, 1976; Kulicki and Wierzbowski, 1983; Morton, 1988; Shigeta, 1993; Landman and Waage, 1993), implying that this age group comprised a large portion of the population.

Several authors have characterized the reproductive strategy of ammonoids relative to that of present-day *Nautilus* as r-selected versus K-selected (House, 1985; Landman, 1988; Hewitt, 1988; see also Vermeij, 1978). *Nautilus* is iteroparous, with females laying a few large eggs over several breeding seasons (Landman, 1988). The eggs are extremely yolk-rich, and the embryonic shell gradually forms by accretionary growth (Arnold *et al.*, 1987). The eggs develop very slowly, with hatching in aquaria taking more than 1 year after egg laying (Carlson, 1991). In addition, in *Nautilus*, as in other K-selected species, juveniles comprise a small portion of the population (Saunders and Ward, 1987).

Although ammonoids are clearly r-selected relative to *Nautilus*, within the ammonoids themselves, as in modern coleoids (Boletzky, 1977; Mangold, 1987), there is a wide range of variation. For example, the Agoniatitina, with ammonitellas nearly 3 mm in diameter, contrast with most other ammonoids with ammonitellas approximately 0.5–1.5 mm in diameter. Even among these more typical ammonoids, variation in ammonitella size may correlate with differences in number of offspring and length of embryonic development (e.g., deep-water Lytoceratina versus shallow-water Ammonitina). Moreover, differences in the embryonic size of species may correlate with differences in biogeographic distribution and population structure. For example, in Late Cretaceous ammonoids from Japan, Tanabe and Ohtsuka (1985) and Shigeta (1993) reported that species with smaller embryonic shells (e.g., collignoniceratids, whose embryonic shells are ≤700 µm) were more restricted in their facies distribution and displayed more juveniles in their preserved assemblages than species with larger embryonic shells (e.g., Lytoceratina, Phylloceratina, and Desmoceratidae, whose embryonic shells are ≥1000 µm). These correlations are probably part of a still larger picture that includes differences in environmental tolerance, ecological specialization, and adult mode of life.

6. Future Research

One of the most conspicuous gaps in any review of ammonoid embryonic development is the lack of studies on Paleozoic ammonoids. There are few or no recent data on the microstructure and internal features of such groups as the Anarcestina and Bactritina. There also is little information on the ornament of the embryonic shells in these groups. However, the hypothesis that there is more than one pattern of embryonic development within the Ammonoidea (Tanabe *et al.*, 1993) requires further testing using additional data from as many ammonoid groups as possible.

In addition, the reproductive strategy of ammonoids clearly links them more closely to coleoids than to nautiloids (Engeser, 1990; Tanabe *et al.*, 1993; Jacobs and Landman, 1993). This relationship between ammonoids and coleoids needs to be more fully explored by explicit studies comparing embryonic development and posthatching mode of life in both these groups. Within the Ammonoidea, the diversity, albeit limited, in the size and shape of embryonic shells also suggests that more attention must be given to studies linking early ontogeny with other species-specific traits such as biogeographic distribution, population structure, and evolutionary longevity.

ACKNOWLEDGMENTS. We thank Royal Mapes (Ohio University), W. Bruce Saunders (Bryn Mawr College), and Wolfgang Weitschat (Geologisch-Paläontologisches Insititut der Universität, Hamburg) for providing us with many excellent specimens for study; Tim White (Yale Peabody Museum), Copeland MacClintock (YPM), Julia Golden (University of Iowa), and Ed Landing (New York State Museum) for kindly arranging loans of museum material; Roger Hewitt (McMaster University), Gerd Westermann (McMaster University), Klaus Bandel (Geologisch-Paläontologisches Institut der Universität, Hamburg), and Wolfgang Weitschat for reviewing an earlier draft of this manuscript and making many helpful suggestions; and Susan Klofak, Kathleen Sarg, Andrew Modell, and Stephanie Crooms (all AMNH) for help in research, manuscript preparation, or both. This project was supported in part by NSF grant No. EAR 9104888 to N. Landman.

Appendix I. Dimensions of the Ammonitella in 11 Suborders of the Ammonoidea

Suborder	Superfamily	Family	Species	PD (mm)	AD (mm)	AA (deg)	PL (mm)	PL/PD	Remarks
Anarcestina	Anarcestaceae	Anarcestidae	*Archoceras (A.) paeckelmanii* Schindewolf	—	1.0*				Bensaïd (1974)
			Archoceras (A.) tataense Bensaïd	—	1.0*				Bensaïd (1974)
Gephuroceratina	Gephurocerataceae	Gephuroceratidae	*Probeloceras lutheri* (Clarke)	—	1.06				NYSM 12726
			Manticoceras (M.) bullatum Wedekind	0.68–0.85	1.2*				Clausen (1969)
			Manticoceras (M.) affine (Stein.)	0.75–1.10	—				Clausen (1969)
			Manticoceras (M.) orbiculum (Beyr.)	0.68–0.88	—				Clausen (1969)
			Manticoceras (M.) adorfense Wedekind	0.65–0.83	—				Clausen (1969)
			Manticoceras (M.) serratum (Stein.)	0.63–0.70	—				Clausen (1969)
			Manticoceras (M.) intumescens (Beyr.)	0.65–0.80	—				Clausen (1969)
			Manticoceras (M.) cordatum (Sdgbr.)	0.68–0.83	—				Clausen (1969)
			Manticoceras (M.) crassum Wedekind	0.68–0.83	—				Clausen (1969)
			Manticoceras (M.) drevermanni Wedekind	0.68–0.78	—				Clausen (1969)
			Manticoceras (M.) galeatum Wedekind	0.70–0.90	—				Clausen (1969)
Agoniatitina	Agoniatitaceae	Agoniatitidae	*Agoniatites obliquus* (Whidborne)	1.4*	2.4*				Wissner and Norris (1991)
			Agoniatites holzapfeli Wedekind	1.4*	2.3*				Erben (1964)
			Agoniatites fulguralis (Whidborne)	1.6*	2.5*				Erben (1964)
			Agoniatites sp.	—	2.6*				Erben (1964)
		Mimagoniatitidae	*Mimagoniatites (M.)* cf. *zorgensis* (Roemer)	1.2–1.3*	2.2*				Erben (1964)

Suborder	Superfamily	Family	Species	PD (mm)	AD (mm)	AA (deg)	PL (mm)	PL/PD	Remarks
			Mimagoniatites (M.) fecundus (Barrande)	1.1*	1.9–2.1*				Erben (1964)
	Mimo-cerataceae		*Convoluticeras lardeuxi* Erben	0.8*	1.8*				Erben (1964)
Tornoce-ratina			*Gyroceratites gracilis* (Meyer)	0.9*	1.5*				Erben (1964)
	Tornoce-rataceae		*Tornoceras (T.) arkonense* House	0.8	1.5*				House (1965)
			Tornoceras (T.) uniangulare widderi House	—	1.5				House (1965)
			Tornoceras (T.) uniangulare aldenense House	0.89–1.0	1.49				House (1965)
			Tornoceras (T.) uniangulare uniangulare (Conrad)	0.8	—				House (1965)
			Tornoceras (T.) uniangulare obesum Clarke	0.98	—				House (1965)
			Tornoceras (T.) concentricum House	0.6	1.4				House (1965)
			Tornoceras (T.) arcuatum House	0.71	—				House (1965)
			Aulatornoceras bicostatum (Hall)	0.5	1.4–1.5*				House (1965)
Goniatitina	Dimorpho-cerataceae	Dimorpho-ceratidae	*Dimorphoceras politum* (Shumard)	—	0.90	—	—	—	SUI 1755
		Girtyocera-tidae	*Girtyoceras meslerianum* (Girty)	0.45	0.96	400	—	—	UMUT PM 19023-1
			Eumorphoceras plummeri (Miller and Youngquist)	—	1.04	—	—	—	UMUT PM 19030
	Goniatitaceae	Goniatitidae	*Gatherites morrowensis* (Miller and Moore)	0.42	0.84	385	—	—	UMUT PM 19032
			Goniatites sp. aff. *G. crenistria* Phillips	0.48	0.95	382	0.11	0.23	UMUT PM 19019-1
				0.60	1.10	372	0.08	0.13	UMUT PM 19019-2
			Goniatites choctawensis Shumard	0.56	1.06	345	—	—	UMUT PM 19020-1
				0.55	1.09	386	0.06	0.07	UMUT PM 19020-2
			Goniatites multiliratus Gordon	0.53	0.94	384	—	—	UMUT PM 19033
		Agathicera-tidae	*Agathiceras applini* Plummer and Scott	0.48	1.03	—	0.05	0.10	SUI 1766

Superfamily	Family	Species						Reference
Neoglyphio-cerataceae	Neoglyphio-ceratidae	*Neoglyphioceras abramovi* Popow	0.49–0.52	0.92–0.94	360	0.05	0.10	Zakharov (1974)
	Cravenocera-tidae	*Cravenoceras richardsonianum* (Girty)	0.46	0.80	367	—	—	UMUT PM 19021
		Cravenoceras incisum (Hyatt)	0.53	0.95	380	—	—	UMUT PM 19022-1
Gastrio-cerataceae	Gastriocera-tidae	*Pseudogastrioceras simulator* (Girty)	0.40	0.80	364	0.04	0.17	SUI 1740
		Pseudogastrioceras fedorowi (Karpinsky)	0.38–0.40	0.74–0.76	370	—	—	Bogoslovskaya (1959)
		Paragastrioceras sp.	0.48	0.84	360	—	—	Bogoslovskaya (1959)
		Owenoceras bellilineatum (Miller and Owen)	—	0.87	355	—	—	SUI 1713
	Reticulo-ceratidae	*Arkanites relictus* (Quinn, McCaleb and Webb)	0.51	0.81	386	—	—	UMUT PM 19029
	Glaphyritidae	*Glaphyrites hyattianus* (Girty)	0.59	1.03	405	0.05	0.08	UMUT PM 19025-2
		Glaphyrites warei (Miller and Owen)	0.46	0.88	385	0.06	0.13	UMUT PM 19026-1
			0.45	0.86	372	0.05	0.12	UMUT PM 19026-2
		Glaphyrites jonesi (Miller and Owen)	0.54	0.96	375	—	—	UMUT PM 19027
		Glaphyrites clinei (Miller and Owen)	0.39	0.71	382	—	—	UMUT PM 19028
		Glaphyrites welleri (Smith)	0.52	0.94	364	—	—	SUI 1735
			0.35	0.81	356	0.03	0.09	SUI 1726
Goniolobo-cerataceae	Homoceratidae	*Homoceras subglobosum* (Bisat)	0.53	0.91	385	0.10	0.19	UMUT PM 19024-2
	Goniolo-boceratidae	*Gonioloboceras welleri* Smith	—	1.92	384	—	—	SUI 1743
Shumardi-taceae	Perrinitidae	*Properrinites bakeri* (Plummer and Scott)	0.70	1.19	—	—	—	SUI 1790
		Perrinites sp.	1.03	2.31	383	—	—	AMNH 41183a
			0.89	2.00	382	0.16	0.18	AMNH 41183b
Adriani-taceae	Adrianitidae	*Texoceras* sp.	—	0.97	—	—	—	UMUT PM 19037-1
		Adrianites dunbari Miller and Furnish	0.63	1.00	365	0.12	0.19	SUI 1764

Suborder	Superfamily	Family	Species	PD (mm)	AD (mm)	AA (deg)	PL (mm)	PL/PD	Remarks
			Crimites elkoensis Miller, Furnish and Clarke	0.37	0.65	345	—	—	UMUT PM 19038
			Crimites krotowi Karpinsky	0.34–0.39	0.73–0.81	365	—	—	Bogoslovskaya (1959)
	Popano-cerataceae	Popano-ceratidae	*Popanoceras annae* Ruzhencev	—	0.66	—	—	—	SUI 1777
			Peritrochia erebus Girty	0.45	0.88	410	—	—	UMUT PM 19039-1
			Peritrochia typicus Ruzhencev	0.40–0.42	0.88	340	—	—	Bogoslovskaya (1959)
			Stacheoceras subinterruptum (Krot.)	0.39–0.46	0.84	380	—	—	Bogoslovskaya (1959)
	Neoicocera-taceae	Metalegocera-tidae	*Metalegoceras baylorense* White	0.45	0.85	365	0.02	0.05	UMUT PM 19035
	Cyclolo-baceae	Cyclolobidae	*Mexicoceras guadalupense* (Girty)	0.52	0.93	378	—	—	SUI 1782
		Vidrioceratidae	*Vidrioceras* sp.	0.44	0.80	361	—	—	UMUT PM 19329
	Thalasso-cerataceae	Thalasso-ceratidae	*Eothalassoceras inexpectans* (Miller and Owen)	0.37	0.66	356	0.03	0.08	UMUT PM 19036-1
				0.37	0.66	360	—	—	UMUT PM 19036-2
			Thalassoceras gemmellaroi Karpinsky	0.30–0.34	0.58–0.66	380	—	—	Bogoslovskaya (1959)
			Aristoceras sp.	0.36	0.75	368	—	—	UMUT PM 19010
		Bisatoceratidae	*Bisatoceras* n. sp.	0.34	0.62	354	0.11	0.31	UMUT PM 19033-1
Prole-canitina	Medlicot-tiaceae	Pronoritidae	*Pronorites praepermicus* Karpinsky	0.61	1.22	330	0.26	0.43	SUI 1686
			Neopronorites vulgaris (Karpinsky)	0.44–0.54	1.00–1.10	310–320	—	—	Bogoslovskaya (1959)
			Neopronorites permicus (Tschernow)	0.44–0.56	0.96–1.10	340	—	—	Bogoslovskaya (1959)
		Medlicottiidae	*Artinskia electraensis* (Plummer and Scott)	0.48	1.01	340	—	—	UMUT PM 19040-1
				0.52	1.16	355	0.05	0.10	UMUT PM 19040-2
			Artinskia artiensis (Grünewaldt)	0.43–0.44	0.90	345	—	—	Bogoslovskaya (1959)

Prolecani-taceae		Medlicottia orbignyana (Verneuil)	0.34	0.72	326–330	—	—	Bogoslovskaya (1959)
	Daraelitidae	Daraelites elegans Tschernow	0.45	1.06	—	—	—	Bogoslovskaya (1959)
Ceratitina	Noritaceae							
	Olenikitidae	Olenikites spiniplicatus (Mojsisovics)	0.35	0.73	325	0.05	0.14	AMNH 44347
		Subolenekites altus (Mojsisovics)	0.40	0.72	315	0.11	0.28	UMUT MM 19041
		Svalbardiceras spitzbergensis Frebold	0.37	0.70	290	—	—	AMNH 44349
		Svalbardiceras sibiricum (Mojsisovics)	0.39	0.89–0.91	260	—	—	Zakharov (1971)
	Ophiceratidae	Nordophiceras schmidti (Mojsisovics)	0.38	0.78	260	0.04	0.10	Zakharov (1971)
		Ophiceras sp.	0.39	0.74	—	—	—	Zakharov (1974)
	Meekocera-tidae	Arctoceras septentrionale ((Diener)	0.51	0.68	—	—	—	Zakharov (1974)
		Kingites sp.	0.42	0.68	347	0.28	0.67	Zakharov (1974)
		Boreomeekoceras keyserlingi (Mojsisovics)	0.35	0.73	325	—	—	UMUT MM 19045-1
		Arctomeekoceras rotundatum (Mojsisovics)	0.37	0.82	280–285	0.07	0.19	Zakharov (1971)
		Wyomingites spathi (Kummel)	0.29	—	—	—	—	Zakharov (1974)
		Wyomingites chaoi (Kiparisova)	0.29	—	—	—	—	Zakharov (1974)
	Paranamitidae	Paranannites aspenensis Hyatt and Smith	0.37	0.66	238	0.05	0.14	Zakharov (1974)
		Paranannites spathi (Frebold)	0.35	0.67	330	—	—	UMUT MM 19343
		Prosphingites grambergi Popow	0.38–0.41	0.68–0.79	265	0.03–0.09	0.22	Zakharov (1971)
Megaphyl-litaceae	Parapopano-ceratidae	Parapopanoceras paniculatum Popow	0.34	0.66	333	0.06	0.18	AMNH 44352
		Stenopopanoceras mirabile Popow	0.38	0.72	270	—	—	UMUT MM 19054
		Amphipopanoceras asseretoi Dagys and Konstantinov	0.38	0.69	335	—	—	UMUT MM 19055
	Megaphyl-litidae	Megaphyllites prometheus Shevyrev	0.53–0.60	0.94–1.05	270–280	—	—	Druschits and Doguzhaeva (1981)

Suborder	Superfamily	Family	Species	PD (mm)	AD (mm)	AA (deg)	PL (mm)	PL/PD	Remarks
	Nathorstitaceae	Nathorstitidae	Indigirites tozeri Weitschat and Lehmann	0.32–0.33	0.59–0.68	330	0.15	0.47	AMNH 44353, UMUT MM 19056
			Stolleyites tenuis (Stolley)	0.36	0.62	320	0.06	0.17	AMNH 44354
	Arcestaceae	Cladiscitidae	Phyllocladiscites busarginensis Zakharov	0.59	0.76–0.96	265	0.10	0.17	Zakharov (1974)
		Arcestidae	Arcestes sp.	0.37	0.63	–	–	–	Zakharov (1974)
	Ceratitaceae	Ceratitidae	Frechites laqueatus (Lindstroem)	0.46	0.71	288	0.05	0.11	AMNH 44357
			Frechites humboldtensis (Hyatt and Smith)	0.59	–	–	–	–	Arkad'yev and Vavilov (1984)
			Frechites sp.	0.60	1.32	270	–	–	Arkad'yev and Vavilov (1984)
			Gymnotoceras falciforme (Smith)	0.38	0.69	270	–	–	Arkad'yev and Vavilov (1984)
			Gymnotoceras meeki (Mojsisovics)	0.39–0.45	0.72–0.73	270	0.22	0.49	Arkad'yev and Vavilov (1984)
			Gymnotoceras rotelliforme (Meek)	0.45	–	–	–	–	Arkad'yev and Vavilov (1984)
			Anagymnotoceras varium (McLearn)	0.38	0.86	340	–	–	UMUT MM 19344
		Sibiritidae	Sibirites eichwaldi Mojsisovics	0.35–0.37	0.60–0.67	260	0.04	0.11	Zakharov (1971)
			Parasibirites grambergi Popow	0.40	0.71	265	0.02	0.05	Zakharov (1971)
		Keyserlingitidae	Keyserlingites sp.	0.64	0.81–0.82	300	0.26	0.40	Zakharov (1971)
	Dinaritaceae	Columbitidae	Columbites sp.	0.46	0.73	240	0.02	0.04	Zakharov (1974)
			Subcolumbites multiformis Kiparisova	0.34–0.41	0.63	240	0.05	0.13	Zakharov (1971)
	Danubitaceae	Danubitidae	Czekanowskites rieberi Dagys and Weitschat	0.39	0.80	305	–	–	AMNH 44358
		Longobarditidae	Grambergia taimyrensis Popow	0.34	0.76	295	–	–	Zakharov (1971)
	Sagecerataceae	Sageceratidae	Pseudosageceras sp.	0.57	0.96	295	0.07	0.07	Zakharov (1971)

Suborder	Superfamily	Family	Species						Reference
Phylloceratina	Pinacocerataceae	Hedenstroemiidae	Hedenstroemia hedenstroemi (Keyserling)	0.63	—	287–296	0.13	0.21	Zakharov (1974)
			Hedenstroemia mojsisovicsi Diener	0.45–0.53	—	—	—	—	Zakharov (1974)
		Ptychitidae	Aristoptychites kolymensis Kiparisova	0.54	0.96	368	0.12	0.22	AMNH 44360
		Gymnitidae	Placites polydactylus oldhami Mojsisovics	0.39	—	—	0.04	0.09	Zakharov (1974)
	Phyllocerataceae	Ussuritidae	Eophyllites sp.	0.042–0.43	—	—	—	—	Zakharov (1971)
			Indigirophyllites spitsbergensis (Oeberg)	0.46	1.02	330	0.07	0.15	AMNH 44362
			Calliphylloceras velledae (Michelin)	0.43–0.56	0.80–0.88	280–300	0.08	0.19	Druschits and Doguzhaeva (1981)
			Calliphylloceras subalpinum (Anthula)	0.45	0.80	275	—	—	Druschits and Khiami (1970)
			Ptychophylloceras ptychoicum (Quenstedt)	0.38–0.41	0.69	260	—	—	Druschits and Doguzhaeva (1981)
			Holcophylloceras sp.	0.46–0.59	0.70–0.91	270	0.07–0.10	0.15–0.17	Druschits and Doguzhaeva (1981)
			Holcophylloceras guettardi (Rasp.)	0.45–0.55	0.70–0.91	270–280	0.04–0.07	0.18	Druschits and Doguzhaeva (1981)
		Phylloceratidae	Partschiceras sp.	0.41–0.49	0.63–0.85	260–280	0.04–0.08	0.14	Druschits and Doguzhaeva (1981)
			Partschiceras japonicum (Matsumoto)	0.46	0.92	272	—	—	Shigeta (1993)
			Hypophylloceras subramosum (Shimizu)	0.53–0.66	0.90–1.15	270–292	0.06–0.12	0.10–0.12	Tanabe et al. (1979)
			Hypophylloceras hetonaiensis (Matsumoto)	0.43	0.90	280	—	—	Shigeta (1993)
			Phyllopachyceras ezoense (Yokoyama)	0.47–0.58	0.85–1.30	284–377	0.02–0.13	0.04–0.27	Tanabe et al. (1979)

Suborder	Superfamily	Family	Species	PD (mm)	AD (mm)	AA (deg)	PL (mm)	PL/PD	Remarks
Lytoceratina			Phyllopachyceras sp.	0.44	0.76	—	—	—	Druschits and Khiami (1970)
	Lytocerataceae	Lytoceratidae	Eurystomiceras polychelictum Bockh	0.39–0.42	0.78–0.84	290–300	0.03	0.08	Druschits and Doguzhaeva (1981)
			Biasaloceras subsequens (Karakasch)	0.32–0.34	—	—	—	—	Druschits and Khiami (1970)
	Tetragonitacae	Protetragonitidae	Protetragonites tauricus Kul.-Voron.	0.65	—	—	—	—	Druschits and Khiami (1970)
		Tetragonitidae	Tetragonites duvalianus d'Orbigny	0.50–0.70	0.94–1.18	300–330	0.03–0.07	0.06–0.11	Druschits and Doguzhaeva (1981)
			Tetragonites hulensis Murphy	0.50	1.04	312	—	—	Shigeta (1993)
			Tetragonites glabrus (Jimbo)	0.56	1.08	331	—	—	Shigeta (1993)
			Tetragonites popetensis Yabe	0.42	0.97	340	—	—	Shigeta (1993)
			Tetragonites minimus Shigeta	0.50–0.60	0.90–1.05	320–340	0.06	0.10	Shigeta (1989)
			Tetragonites terminus Shigeta	0.93–1.05	1.7–1.90	330–345	0.12	0.13	Shigeta (1989)
			Gabbioceras latericarinatum Anthula	0.49	0.88–0.98	280–300	0.04	0.10	Druschits and Doguzhaeva (1981)
			Gabbioceras angulatum Anderson	0.44	0.88	300	0.09	0.20	AMNH 44370
			Gabbioceras michelianum (d'Orbigny)	0.42	0.90	281	—	—	Shigeta (1993)
			Pseudophyllites indra (Forbes)	0.58	1.48	327	—	—	Shigeta (1993)
			Saghalinites teshioensis Matsumoto	0.56	1.19	340	—	—	Shigeta (1993)
		Gaudryceratidae	Eotetragonites aureum (Anderson)	0.29	0.93	312	—	—	Druschits and Doguzhaeva (1981)
			Eotetragonites balmensis (Breistroffer)	0.50	0.98	347	0.09	0.18	AMNH 44368

Species						Reference
Gaudryceras stefaninii Venzo	0.40	0.93	318	—	—	Shigeta (1993)
Gaudryceras denseplicatum (Jimbo)	0.71–0.86	1.39–1.65	320–335	0.11	0.16	Tanabe et al. (1979)
Gaudryceras striatum (Jimbo)	0.66–0.69	1.12–1.39	342–353	0.09	0.15	Ohtsuka (1986)
Gaudryceras tombetsense Matsumoto	0.64	1.42	348	—	—	Shigeta (1993)
Anagaudryceras limatum (Yabe)	0.62	1.28	365	0.07	0.11	AMNH 44373
Anagaudryceras yokoyamai (Yabe)	0.69	1.41	365	—	—	Ohtsuka (1986)
Anagaudryceras nanum Matsumoto	0.56	1.26	315	—	—	Shigeta (1993)
Anagaudryceras tetragonum Matsumoto and Kanie	0.58	1.26	323	—	—	Shigeta (1993)
Anagaudryceras matsumotoi Morozumi	0.59	1.32	338	—	—	Shigeta (1993)
Kossmatella agassiziana Pictet	0.52–0.67	0.94–1.27	300–315	0.04–0.07	0.08–0.10	Druschits and Doguzhaeva (1981)
Parajaubertella kawakitana Matsumoto	0.60–0.66	1.12	270	0.04–0.07	0.12	Druschits and Doguzhaeva (1981)
Zelandites sp. aff. Z. inflatus Matsumoto	0.67–0.68	1.19–1.24	320–336	0.06	0.09	Tanabe et al. (1979)
Zelandites mihoensis Matsumoto	0.46	1.00	312	—	—	Shigeta (1993)
Zelandites kawanoii (Jimbo)	0.57	1.19	345	—	—	Shigeta (1993)
Zelandites varuna (Forbes)	0.54	1.22	342	—	—	Shigeta (1993)
Karsteniceras obatai Matsukawa	0.27	0.75	305	—	—	Shigeta (1993)
Ptychoceras renngarteni Egonin	0.50	0.85	330	0.33	0.66	Druschits and Doguzhaeva (1981)
Luppovia sp.	0.37	0.70	290	—	—	Doguzhaeva and Mikhailova (1982)
Diadochoceras nodosocostatiforme (Shimizu)	0.30	0.77	291	—	—	Shigeta (1993)

Ancyloceratina — Ancylocerataceae — Ancyloceratidae, Ptychoceratidae

Douvilleiceratina — Douvilleicerataceae — Douvilleiceratidae

Suborder	Superfamily	Family	Species	PD (mm)	AD (mm)	AA (deg)	PL (mm)	PL/PD	Remarks
	Deshayesitaceae	Deshayesitidae	*Diadochoceras* sp.	0.38–0.45	0.70–0.80	260–270	0.08	0.16	Druschits and Doghuzhaeva (1981)
			Deshayesites deshayesi (d'Orbigny)	0.42–0.56	0.91–1.09	300	0.10–0.21	0.24–0.38	Druschits and Doghuzhaeva (1981)
		Parahoplitidae	*Acanthohoplites* sp.	0.33–045	0.62–0.88	270–280	0.07–0.13	0.21–0.29	Druschits and Doguzhaeva (1981)
			Colombiceras sp.	0.35–0.41	0.60–0.63	280–300	0.07	0.30	Druschits and Doguzhaeva (1981)
			Nolaniceras sp.	0.38–0.45	0.73–0.85	270–280	0.08–0.14	0.21–0.35	Druschits and Doguzhaeva (1981)
			Hypacanthoplites subcornuenianus (Shimizu)	0.40	0.93	290	—	—	Shigeta (1993)
			Hypacanthoplites sp.	0.38–0.49	0.73–0.92	260–270	0.18	0.33	Druschits and Doguzhaeva (1981)
			Nodosohoplites sinuosocostatus Egoian	0.38–0.41	0.71–0.76	250–270	0.10–0.11	0.27	Druschits and Doguzhaeva (1981)
			Parahoplites melchioris Anthula	0.55–0.70	1.13–1.29	315–330	0.25–0.26	0.50–0.59	Druschits and Doguzhaeva (1981)
	Scaphitaceae	Scaphitidae	*Scaphites planus* (Yabe)	0.40–0.55	0.73–0.95	280	0.08	0.15	Tanabe (1977a), Tanabe et al. (1979)
			Scaphites yonekurai Yabe	0.44	0.87	295	—	—	Shigeta (1993)
			Scaphites pseudoaequalis Yabe	0.35–0.46	0.64–0.80	295	0.12	0.27	Tanabe (1977b)
			Scaphites larvaeformis Meek and Hayden	0.28–0.36	0.52–0.65	266–300	—	—	Landman (1987)

Species						Reference
Scaphites carlilensis Moreman	—	0.59–0.60	274	—	—	Landman (1987)
Scaphites warreni Meek and Hayden	0.27–0.38	0.60–0.72	266–290	—	—	Landman (1987)
Scaphites whitfieldi Cobban	0.25–0.40	0.55–0.74	260–308	0.04	0.12	Landman (1987)
Scaphites nigricollensis Cobban	0.34–0.40	0.60–0.76	253–308	—	—	Landman (1987)
Scaphites corvensis Cobban	0.31	0.67	282	—	—	Landman (1987)
Scaphites preventricosus Cobban	0.29–0.42	0.58–0.71	257–292	0.03	0.09	Landman (1987)
Scaphites depressus Reeside	0.32–0.40	0.68–0.83	282–292	—	—	Landman (1987)
Clioscaphites vermiformis (Meek and Hayden)	0.29–0.44	0.60–0.82	259–306	—	—	Landman (1987)
Hoploscaphites nicolletii (Morton)	0.39–0.46	0.72–0.81	288–316	—	—	Landman and Waage (1993)
Hoploscaphites comprimus (Owen)	—	0.65–0.67	—	—	—	Landman and Waage (1993)
Jeletzkytes spedeni Landman and Waage	0.40–0.42	0.72–0.80	292–302	—	—	Landman and Waage (1993)
Jelezkytes nebrascensis (Owen)	0.41	0.68	310	—	—	Landman and Waage (1993)
Discoscaphites conradi (Morton)	0.37–0.43	0.72–0.80	299–320	—	—	Landman and Waage (1993)
Discoscaphites gulosus (Morton)	0.34–0.38	0.67–0.77	294–313	—	—	Landman and Waage (1993)
Discoscaphites rossi Landman and Waage	0.31–0.37	0.65–0.72	296–314	—	—	Landman and Waage (1993)
Otoscaphites puerculus (Jimbo)	0.43–0.58	0.71–0.92	285	—	—	Tanabe (1977a), Tanabe et al. (1979)
Otocaphites klamathensis (Anderson)	0.37–0.48	0.67–0.83	285	—	—	Tanabe (1977b)
Otoscaphites matsumotoi Tanabe	0.43	0.80	285	—	—	Shigeta (1993)

Suborder	Superfamily	Family	Species	PD (mm)	AD (mm)	AA (deg)	PL (mm)	PL/PD	Remarks
Ammonitina	Turrilitaceae	Baculitidae	Baculites sp.	—	0.78	331	—	—	Landman (1982)
	Hildocerataceae	Hildoceratidae	Eleganticeras elegantulum (Young and Bird)	0.40–0.42	0.80–0.88	277–292	0.28	0.71	Tanabe and Ohtsuka (1985)
		Graphoceratidae	Graphoceras opalinum (Rein)	0.36	—	—	0.21	0.58	Grandjean (1910)
	Psilocerataceae	Arietitidae	Arietites sp.	0.35	0.64	284	—	—	Tanabe and Ohtsuka (1985)
			Arietites kridion Hehl.	0.37	—	—	0.05	0.14	Grandjean (1910)
			Coroniceras reynsei (Spath)	0.30	0.59	280	0.04	0.13	UMUT MM 19684
	Eoderocerataceae	Amaltheidae	Amaltheus margaritatus d'Orbigny	0.42	—	310	0.06	0.14	Grandjean (1910)
			Amauroceras ferrugineum (Simpson)	0.56	1.00	300	0.05	0.09	Unregistered Hamburg Univ. specimen
			Pleuroceras sp.	0.50	1.06	277	—	—	Tanabe and Ohtsuka (1985)
		Eoderoceratidae	Promicroceras sp.	0.42–0.43	0.71–0.75	280–286	0.25	0.58	UMUT MM 19069-1-2
	Dactyliocerataceae	Dactylioceratidae	Peronoceras fibulatum (Sowerby)	0.45	0.89	312	0.22	0.49	Ohtsuka (1986)
	Stephanicerataceae	Sphaeroceratidae	Sphaeroceras brongniarti (Sowerby)	0.34	—	—	0.25	0.73	Grandjean (1910)
	Spirocerataceae	Spiroceratidae	Spiroceras calloviense Morris	0.47	0.84	325	0.25	0.53	YPM 01854
	Haplocerataceae	Oppeliidae	Aconeceras trautscholdi Sinzov	0.30–0.35	0.63	295	—	—	Druschits and Khiami (1970)
				0.34–0.38	0.64–0.90	275–360	0.08–0.17	0.24–0.50	Druschits and Doguzhaeva (1981)
			Sanmartinoceras sp.	0.42–0.45	0.84	300	0.14	0.31	Druschits and Doguzhaeva (1981)

Superfamily	Family	Species						Reference
Perisphinc-taceae	Craspiditidae	*Simbirskites coronatiformis* Pavlow	0.55–0.60	1.05–1.12	300–315	0.11–0.40	0.20–0.67	Druschits and Doguzhaeva (1981)
		Simbirskites elatus Pavlow	0.57–0.60	1.05–1.06	—	—	—	Druschits and Doguzhaeva (1981)
		Simbirskites sp.	0.48	0.98	308	0.23	0.48	Druschits and Doguzhaeva (1981)
		Speetoniceras versicolor (Trautschold)	0.53–0.63	1.02–1.15	300	0.21–0.38	0.39–0.60	Druschits and Doguzhaeva (1981)
		Craspedodiscus discofalcatus Lahusen	0.55–0.70	1.13–1.20	300–315	0.15–0.56	0.27–0.72	Druschits and Doguzhaeva (1981)
		Craspedodiscus sp.	0.65	1.26	330	0.40	0.62	Druschits and Doguzhaeva (1981)
Desmocera-taceae	Desmocera-tidae	*Beudanticeras laevigatum* Sowerby	0.51–0.55	0.99	310	—	—	Druschits and Khiami (1970)
			0.49–0.55	0.77–0.99	330	0.17	0.31–0.35	Druschits and Doguzhaeva (1981)
		Beudanticeras beudanti (Brongniart)	0.44	1.06	330	0.38	0.86	Dauphin (1975)
		Zurcherella falcistriata (Anthula)	0.38–0.42	0.76–0.80	282–290	—	—	Druschits and Khiami (1970)
			0.38–0.42	0.66–0.84	270–290	0.21–0.27	0.57–0.64	Druschits and Doguzhaeva (1981)
		Desmoceras kossmati Matsumoto	0.40	0.90	305	—	—	Shigeta (1993)
		Desmoceras japonicum (Yabe)	0.45–0.48	0.95–0.98	317–337	—	—	Ohtsuka (1986), Shigeta (1993)
		Desmoceras ezoanum Matsumoto	0.48	1.22	305	—	—	Shigeta (1993)
		Damesites latidorsatus (Michelin)	0.41	0.85	320	0.29	0.71	Dauphin (1975)

Suborder	Superfamily	Family	Species	PD (mm)	AD (mm)	AA (deg)	PL (mm)	PL/PD	Remarks
			Damesites ainuanus Matsumoto	0.35–0.38	0.67–0.73	290–308	0.15	0.40	Tanabe *et al.* (1979), Ohtsuka (1986)
			Damesites damesi (Jimbo)	0.36–0.46	0.83–0.95	307–360	0.15–0.26	0.32–0.57	Tanabe *et al.* (1979), Shigeta (1993)
			Damesites semicostatus Matsumoto	0.34–0.47	0.77–0.91	313–320	0.26	0.66	Tanabe *et al.* (1979), Shigeta (1993)
			Tragodesmoceroides subcostatus Matsumoto	0.36–0.47	0.83–0.92	304–314	—	—	Tanabe *et al.* (1979), Tanabe and Ohtsuka (1985)
			Desmophyllites diphylloides (Forbes)	0.44–0.47	0.83–0.89	298–320	—	—	Ohtsuka (1986)
			Desmophyllites sp.	0.43	0.84	317	0.27	0.62	Tanabe and Ohtsuka (1985), Ohtsuka (1986)
			Microdesmoceras tetragonum Matsumoto and Muramoto	0.43	0.94	305	—	—	Shigeta (1993)
			Melchiorites sp.	0.32–0.36	0.67–0.70	270–290	0.17–0.25	0.53–0.71	Druschits and Doguzhaeva (1981)
			Valdedorsella akuschaensis (Anthula)	0.28	0.68	308	—	—	Shigeta (1993)
			Pseudohaploceras nipponicus (Shimizu)	0.32	0.79	302	—	—	Shigeta (1993)
			Puzosia orientale Matsumoto	0.36	0.83	310	—	—	Shigeta (1993)
			Mesopuzosia pacifica Matsumoto	0.37–0.43	0.83–0.84	302–310	0.28	0.66	Tanabe *et al.* (1979), Shigeta (1993)
			Mesopuzosia yubarensis (Jimbo)	0.28	0.61	302	—	—	Shigeta (1993)
			Bhimaites takahashii Matsumoto	0.37	0.89	306	—	—	Shigeta (1993)
			Hauericeras angustum Yabe	0.33	0.70	312	—	—	Shigeta (1993)
			Hauericeras gardeni (Baily)	0.38–0.50	0.70–0.73	315	0.15	0.30	Tanabe *et al.* (1979), Tanabe and Ohtsuka (1985)

Superfamily	Family	Species						Source
	Pachydiscidae	*Anapachydiscus yezoensis* Matsumoto	0.35	0.76	315	—	—	Shigeta (1993)
		Eupachydiscus haradai (Jimbo)	0.37–0.53	0.73–1.07	306–333	0.25	0.48	Tanabe et al. (1979), Tanabe and Ohtsuka (1985)
		Menuites pusillus Matsumoto	0.50	0.87	328	0.15	0.30	Tanabe et al. (1979)
		Canadoceras kossmati Matsumoto	0.41	0.89	315	—	—	Shigeta (1993)
		Canadoceras mystricum Matsumoto	0.44	1.00	312	—	—	Shigeta (1993)
		Teshioites sp.	0.46	0.92	305	—	—	Shigeta (1993)
	Kossmati-ceratidae	*Eogunnarites unicus* (Yabe)	0.35	0.76	319	—	—	Shigeta (1993)
		Marshallites compressus Matsumoto	0.41	0.97	313	—	—	Shigeta (1993)
		Yokoyamaoceras ishikawai (Jimbo)	0.39–0.61	0.80–0.97	297–344	0.28	0.53	Tanabe et al. (1979), Shigeta (1993)
Acantho-cerataceae	Acantho-ceratidae	*Mantelliceras japonicum* Mat., Muramoto and Takahashi	0.40	0.89	284	—	—	Shigeta (1993)
		Calycoceras orientale Matsumoto, Saito and Fukada	0.46	0.93	281	—	—	Shigeta (1993)
	Collignoni-ceratidae	*Collignoniceras woollgari* (Mantell)	0.45	0.82	294	—	—	UMUT MM 19074
		Subprionocyclus bakeri (Anderson)	0.36–0.42	0.74–0.75	270–313	—	—	Unregistered UMUT specimens
		Subprionocyclus neptuni (Geinitz)	0.37–0.50	0.70–0.85	242–307	0.03–0.09	0.08–0.21	Unregistered UMUT specimens
		Subprionocyclus minimum (Hayasaka and Fukada)	0.35–0.50	0.59–0.89	250–348	0.03–0.15	0.09–0.39	Unregistered UMUT specimens
		Protexanites minimus Matsumoto	0.35	0.74	280	—	—	Shigeta (1993)
		Texanites kawasakii (Kawada)	0.47	0.93	280	—	—	Shigeta (1993)
Hoplitaceae	Placenti-ceratidae	*Metaplacenticeras subtilistriatum* (Jimbo)	0.49–0.54	1.09–1.19	315–356	0.12	0.22	Tanabe et al. (1979), unregistered UMUT sp.

*Estimate

Appendix II. Age and Locality Data of Species Cited in the Text and Not Listed in Appendix I. See Also Appendix, Chapter 6, this volume.

Suborder	Species	Horizon	Locality	Sample
Goniatitina	*Gonioloboceras welleri* Smith	Pennsylvanian	Jacksboro, Texas	N=1 (SU1 1743)
	Vidrioceras sp.	U. Pennsylvanian	Pomona, Kansas	N=1 (UMUT PM 19329)
	Aristoceras sp.	U. Pennsylvanian	Pomona, Kansas	N=1 (UMUT PM 19010)
Prolecanitina	*Artinskia electraensis* (Plummer and Scott)	M Permian	Buck Mt., Nevada	N=2 (UMUT PM 19040-1,2)
Ceratitina	*Paranannites spathi* (Frebold)	Smithian	Spitsbergen	N=1 (UMUT MM 19343)
	Anagymnotoceras varium (McLean)	M. Anisian	Spitsbergen	N=1 (UMUT MM 19344)
Ammonitina	*Collignoniceras woolgari* (Mantell)	M. Turonian	Black Hills, S. Dakota	N=1 (UMUT MM 19074)
	Subprionocyclus bakeri (Anderson)	M. Turonian	Obira, Hokkaido	N=11
	Subprionocyclus neptuni (Geinitz)	U. Turonian	Manji, Hokkaido	N=66
	Subprionocyclus minimus (Hayasaka and Fukada)	U. Turonian	Manji, Hokkaido	N=44

References

Arkad'yev, V. V., and Vavilov, M. N., 1984, The internal structure and ontogeny of the Late Anisian Beyrichitidae (Ammonoidea) of Central Siberia, *Paleontol. J.* **4**:61–72.

Arkell, W. J., 1957, Introduction to Mesozoic Ammonoidea, in: *Treatise on Invertebrate Paleontology*, Part L, *Mollusca 4* (R. C. Moore, ed.), Geological Society of America and University of Kansas Press, Lawrence, KS, pp. 81–129.

Arnold, J. M., and Williams-Arnold, L. D., 1977, Cephalopods: Decapoda, in: *Reproduction of Marine Invertebrates*, Vol. 4 (A. C. Giese and J. S. Pearse, eds.), Academic Press, New York, pp. 243–284.

Arnold, J. M., Landman, N. H., and Mutvei, H., 1987, Development of the embryonic shell of *Nautilus*, in: *Nautilus, The Biology and Paleobiology of a Living Fossil* (W. B. Saunders and N. H. Landman, eds.), Plenum Press, New York, pp. 373–400.

Babin, C., 1989, Les goniatites du Dévonien du Synclinorium Médian Armoricain et leur signification paléobiogéographique, *Palaeontogr. Abt. A* **206**:25–48.

Bandel, K., 1975, Embryonalgehäuse karibischer Meso- und Neogastropoden (Mollusca), *Akad. Wiss. Lit. Mainz Abh. Math. Naturwiss. Kl.* **1975**(1):1–133.

Bandel, K., 1982, Morphologie und Bildung der frühontogenetischen Gehäuse bei conchiferen Mollusken, *Facies* **7**:1–198.

Bandel, K., 1986, The ammonitella: A model of formation with the aid of the embryonic shell of archaeogastropods, *Lethaia* **19**:171–180.

Bandel, K., 1989, Cephalopod shell structure and general mechanisms of shell formation, in: *Skeletal Biomineralization: Patterns, Processes, and Evolutionary Trends, Short Course in Geology*, Vol. 5, Pt. II (J. G. Carter, ed.), American Geophysical Union, Washington, DC, pp. 97–115.

Bandel, K., 1991, Ontogenetic changes reflected in the morphology of the molluscan shell, in: *Constructional Morphology and Evolution* (N. Schmidt-Kittler and K. Vogel, eds.), Springer-Verlag, Berlin, pp. 211–230.

Bandel, K., and Boletzky, S. v., 1979, A comparative study of the structure, development, and morphological relationships of chambered cephalopod shells, *Veliger* **21**:313–354.

Bandel, K., Landman, N. H., and Waage, K. M., 1982, Microornament on early whorls of Mesozoic ammonites: Implications for early ontogeny, *J. Paleontol.* **56**(2):386–391.

Beecher, C. E., 1890, On the development of the shell in the genus *Tornoceras* Hyatt, *Am. J. Sci.* **60**:71–75.

Bensaïd, M., 1974, Étude sur des goniatites à la limite du Dévonien moyen et supérieur du Sud Marocain, *Notes Serv. Géol. Maroc* **36**(264):81–140.

Berthold, T., and Engeser, T., 1987, Phylogenetic analysis and systematization of the Cephalopoda (Mollusca), *Verh. Naturwiss. Ver. Hamb. N.F.* **29**:187–220.

Birkelund, T., 1979, The last Maastrichtian ammonites, in: *Cretaceous–Tertiary boundary events* (T. Birkelund and R. G. Bromley, eds.), University of Copenhagen, Copenhagen, pp. 51–57.

Birkelund, T., 1981, Ammonoid shell structure, in: *The Ammonoidea*, Systematics Association Spec. Vol. 18 (M. R. House and J. R. Senior, eds.), Academic Press, London, pp. 177–214.

Birkelund, T., and Hansen, H. J., 1968, Early shell growth and structures of the septa and the siphuncular tube in some Maastrichtian ammonites, *Medd. Dan. Geol. Foren.* **18**:71–78.

Birkelund, T., and Hansen, H. J., 1974, Shell ultrastructures of some Maastrichtian Ammonoidea and Coleoidea and their taxonomic implications, *K. Dan. Vidensk. Selsk. Biol. Skr.* **20**(6):1–34.

Blind, W., 1979, The early ontogenetic development of ammonoids by investigation of shell-structures, in: *Symposium on Ammonoidea, Systematics Association York, England, Abstracts*, p. 32.

Blind, W., 1988, Über die primäre Anlage des Siphos bei ectocochleaten Cephalopoden, *Palaeontogr. Abt. A* **204**:67–93.

Bogoslovskaya, M. F., 1959, The internal structure of certain Artinskian ammonoid shells, *Paleontol. Zh.* **1**:49–59 (in Russian).

Bogoslovsky, B. I., 1969, Devonskie Ammonoidei. I. Agoniatity, *Trans. Paleont. Inst. Akad. Nauk SSSR* **124**:1–341 (in Russian).

Böhmers, J. C. A., 1936, *Bau und Struktur von Schale und Sipho bei permischen Ammonoidea*, Dissertation, Drukkerij University, Amsterdam.

Boletzky, S. v., 1974, The "larvae" of Cephalopoda: A review, *Thalassia Jugosl.* **10**:45–76.

Boletzky, S. v., 1977, Post-hatching behaviour and mode of life in cephalopods, in: *The Biology of Cephalopods, Symposia of the Zoological Society of London,* No. 38 (M. Nixon and J. B. Messenger, eds.), Academic Press, London, pp. 557–567.

Boletzky, S. v., 1987a, Embryonic phase, in: *Cephalopod Life Cycles,* Vol. II (P. R. Boyle, ed.), Academic Press, London, pp. 5–31.

Boletzky, S. v., 1987b, Juvenile behavior, in: *Cephalopod Life Cycles,* Vol. II (P. R. Boyle, ed.), Academic Press, London, pp. 45–60.

Boletzky, S. v., 1988, Characteristics of cephalopod embryogenesis, in: *Cephalopods—Present and Past* (J. Wiedmann and J. Kullmann, eds.), Schweizerbart'sche Verlagsbuchhandlung, Stuttgart, pp. 167–179.

Boletzky, S. v., 1992, Evolutionary aspects of development, life style, and reproductive mode in incirrate octopods (Mollusca, Cephalopoda), *Rev. Suisse Zool.* **99**:755–770.

Boletzky, S. v., 1993, Development and reproduction in the evolutionary biology of Cephalopoda, *Geobios Mem. Spec.* **15**:33–38.

Branco, W., 1879, Beiträge zur Entwicklungsgeschichte der fossilen Cephalopoden, *Palaeontographica* **26**:15–50.

Branco, W., 1880, Beiträge zur Entwicklungsgeschichte der fossilen Cephalopoden, *Palaeontographica* **27**:17–81.

Brown, A. P., 1891, On the young of *Baculites compressus* Say, *Nautilus* **5**(2):19–21.

Buckman, S. S., 1887–1907, A monograph of the ammonites of the Inferior Oolite Series, *Palaeontogr. Soc.* **40-61**:1–456.

Buckman, S. S., 1909, *Yorkshire Type Ammonites,* Vol. 1, No. 1, Wesley, London, pp. 1–12.

Buckman, S. S., 1918, Jurassic chronology: I-Lias, *Q. J. Geol. Soc. Lond.* **73**:257–327.

Callomon, J. H., 1963, Sexual dimorphism in Jurassic ammonites, *Trans. Leicester Lit. Phil. Soc.* **57**:21–56.

Carlson, B. A., 1991, *Nautilus* hatches at Waikiki Aquarium, *Chambered Nautilus Newsl.* **63**:2–3.

Clausen, C. D., 1969, Oberdevonische Cephalopoden aus dem Rheinischen Schiefergebirge, II Gephuroceratidae, Beloceratidae, *Palaeontogr. Abt. A* **132**:95–178.

Crickmay, C. H., 1925, The discovery of the prosiphon in Cretaceous ammonites from California with remarks upon the function of the organ, *Am. J. Sci.* **9**:229–232.

Currie, E. D., 1942, Growth changes in the ammonite *Promicroceras marstonense* Spath, *Proc. R. Soc. Edinb. [B]* **61**:344–367.

Currie, E. D., 1943, Growth stages in some species of *Promicroceras, Geol. Mag.* **80**:15–22.

Currie, E. D., 1944, Growth stages in some Jurassic ammonites, *Trans. R. Soc. Edinb.* **61**(6):171–198.

Dauphin, Y., 1975, Anatomie de la protoconque et des tours initiaux de *Beudanticeras beudanti* (Brongniart) et *Desmoceras latidorsatum* (Michelin) (Desmoceratidae, Ammonitina)—Albien de Gourdon (Alpes-Maritimes), *Ann. Paleontol. Invertebr.* **61**(1):3–16.

Dauphin, Y., 1977, Anatomie de la protoconque et des tours initiaux de *Uhligella walleranti* Jacob (Desmoceratidae, Ammonitina)—Albien de Gourdon (Alpes-Martimes), *Ann. Paleontol. Invertebr.* **63**(2):77–83.

Doguzhaeva, L., and Mikhailova, I., 1982, The genus *Luppovia* and the phylogeny of Cretaceous heteromorphic ammonites, *Lethaia* **15**:55–65.

Dreyfuss, M., 1933, Découverte de nodules phosphatés à jeunes ammonites dans le Toarcien de Créveney (Haute-Saône), *C. R. Somm. Seances Soc. Geol. Fr.* **14**:224–226.

Druschits, V. V., and Doguzhaeva, L. A., 1974, Some morphogenetic characteristics of phylloceratids and lytoceratids (Ammonoidea), *Paleontol. J.* **8**(1):37–48.

Druschits, V. V., and Doguzhaeva, L. A., 1981, *Ammonites Under the Electron Microscope,* Moscow University Press, Moscow (in Russian).

Druschits, V. V., and Khiami, N., 1970, Structure of the septa, protoconch walls and initial whorls in Early Cretaceous ammonites, *Paleontol. J.* **4**(1):26–38.

Druschits, V. V., Doguzhaeva, L. A., and Mikhailova, I. A., 1977a, The structure of the ammonitella and the direct development of ammonites, *Paleontol. J.* **11**(2):188–199.

Druschits, V. V., Doguzhaeva, L. A., and Lominadze, T. A., 1977b, Internal structural features of the shell of Middle Callovian ammonites, *Paleontol. J.* **11**(3):271–284.

Druschits. V. V., Mikhailova, I. A., Kabanov, G. K., and Knorina, M. V., 1980, Morphogenesis of the *Simbirskites* group, *Paleontol. J.* **14**(1):42–57.

Engeser, T., 1990, Major events in cephalopod evolution, in: *Major Evolutionary Radiations*, Systematics Association Spec. Vol. 42 (P. D. Taylor and G. P. Larwood, eds.), Clarendon Press, Oxford, pp. 119–138.

Erben, H. K., 1960, Primitive Ammonoidea aus dem Unterdevon Frankreichs und Deutschlands, *N. Jb. Geol. Paläont. Abh.* **110**(1):1–128.

Erben, H. K., 1962, Über den Prosipho, die Prosutur und die Ontogenie der Ammonoidea, *Paläontol. Z.* **36**:99–108.

Erben, H. K., 1964, Die Evolution der ältesten Ammonoidea, *N. Jb. Geol. Paläont. Abh.* **120**(2):107–212.

Erben, H. K., 1966, Über den Ursprung der Ammonoidea, *Biol. Rev.* **41**:641–658.

Erben, H. K., Flajs, G., and Siehl, A., 1968, Ammonoids: Early ontogeny of ultra-microscopical shell structure, *Nature* **219**:396–398.

Erben, H. K., Flajs, G., and Siehl, A., 1969, Die frühontogenetische Entwicklung der Schallenstruktur ectocochleaten Cephalopoden, *Palaeontogr. Abt. A* **132**:1–54.

Göddertz, B., 1989, Unterdevonische hercynische Goniatiten aus Deutschland, Frankreich und der Türkei, *Palaeontogr. Abt. A* **208**:61–89.

Grandjean, F., 1910, Le siphon des ammonites et des belémnites, *Soc. Geol. Fr. Bull. Ser. 4* **10**:496–519.

Hecht, G., 1991, *Paleoecology and paleobiology of bactritoid cephalopods from the Pennsylvanian (Missourian) Eudora Shale (Kansas) and Wolf Mountain Shale*, MS Dissertation, Ohio University, Athens, OH.

Hecht, G. D., and Mapes, R. H., 1990, Paleobiology of bactritoid cephalopods from the Pennsylvanian (Missourian) of Texas and Kansas, *Geol. Soc. Am. Abstr. Programs* **22**(7):A221.

Hewitt, R. A., 1985, Numerical aspects of sutural ontogeny in the Ammonitina and Lytoceratina. *N. Jb. Geol. Paläont. Abh.* **170**:273–290.

Hewitt, R. A., 1988, Significance of early septal ontogeny in ammonoids and other ectocochliates, in: *Cephalopods—Present and Past* (J. Wiedmann and J. Kullmann, eds.), Schweizerbart'sche Verlagsbuchhandlung, Stuttgart, pp. 207–214.

Hewitt, R. A., 1993, Relation of shell strength to evolution in the Ammonoidea, in: *The Ammonoidea: Environment, Ecology and Evolutionary Change*, Systematics Association Special Volume 47 (M.R. House, ed.), Clarendon Press, Oxford, pp. 35–56.

Hirano, H., 1975, Ontogenetic study of Late Cretaceous *Gaudryceras tenuiliratum, Mem. Fac. Sci. Kyushu Univ. Ser. D. Geol.* **22**:165–192.

House, M. R., 1965, A study in the Tornoceratidae: The succession of *Tornoceras* and related genera in the North American Devonian, *Phil. Trans. R. Soc. Lond. [B]* **250**(763):79–130.

House, M. R., 1985, The ammonoid time-scale and ammonoid evolution, in: *The Chronology of the Geological Record, The Geological Society, Memoir* 10 (N. J. Snelling, ed.), Blackwell Scientific Publications, Oxford, pp. 273–283.

Hyatt, A., 1866, On the agreement between the different periods in the life of the individual shell and the collective life of the tetrabranchiate cephalopods, *Proc. Bost. Soc. Nat. Hist.* **10**:302–303.

Hyatt, A., 1872, Fossil cephalopods of the Museum of Comparative Zoology: Embryology, *Bull. Mus. Comp. Zool.* **3**:59–111.

Hyatt, A., 1883, Fossil cephalopods in the Museum of Comparative Zoology, *Am. Assoc. Adv. Sci. Pr.* **32**:323–361.

Hyatt, A., 1889, Genesis of the Arietidae, *Smithson. Contrib. Knowledge* **26**(673):1–238.

Hyatt, A., 1894, Phylogeny of an acquired characteristic, *Proc. Am. Philos. Soc.* **32**(143):349–647.

Ivanov, A. N., 1971, On the problem of periodicity of the formation of septa in ammonoid shells and in that of other cephalopods, *Uch. Zap. Yarslv. Pedagog. Inst. Geol. Paleontol.* **87**:127–130 (in Russian).

Jablonski, D., and Lutz, R. A., 1980, Larval shell morphology: Ecology and paleoecological applications, in: *Skeletal Growth of Aquatic Organisms* (D. C. Rhoads and R. A. Lutz, eds.), Plenum Press, New York, pp. 323–377.

Jacobs, D. K., and Landman, N. H., 1993, *Nautilus*—a poor model for the function and behavior of ammonoids? *Lethaia* **26**:101–111.

Kakabadzé, M. V., and Sharikadzé, M. Z., 1993, On the mode of life of heteromorph ammonites (heterocone, ancylocone, ptychocone), *Geobios Mém. Spec.* **15**:209–215.

Kennedy, W. J., and Cobban, W. A., 1976, Aspects of ammonite biology, biogeography, and biostratigraphy, *Spec. Pap. Palaeontol.* **17**.

Kulicki, C., 1974, Remarks on the embryogeny and postembryonal development of ammonites, *Acta Palaeontol. Pol.* **19**:201–224.

Kulicki, C., 1979, The ammonite shell: Its structure, development and biological significance, *Palaeontol. Pol.* **39**:97–142.

Kulicki, C., 1989, Archaeogastropod model of mineralization of ammonitella shell, in: *Skeletal Biomineralization: Patterns, Processes, and Evolutionary Trends, Short Course in Geology*, Vol. 5, Pt. II (J. G. Carter, ed.), American Geophysical Union, Washington, DC, p. 324.

Kulicki, C., and Doguzhaeva, L. A., 1994, Development and calcification of the ammonitella shell, *Acta Palaeontol. Pol.* **39**:17–44.

Kulicki, C., and Wierzbowski, A., 1983, The Jurassic juvenile ammonites of the Jagua Formation, Cuba, *Acta Palaeontol. Pol.* **28**(3,4):369–384.

Landman, N. H., 1982, Embryonic shells of *Baculites, J. Paleontol.* **56**(5):1235–1241.

Landman, N. H., 1985, Preserved ammonitellas of *Scaphites* (Ammonoidea, Ancyloceratina), *Am. Mus. Novit.* **2815**:1–10.

Landman, N. H., 1987, Ontogeny of Upper Cretaceous (Turonian–Santonian) scaphitid ammonites from the Western Interior of North America: Systematics, developmental patterns, and life history, *Bull. Am. Mus. Nat. Hist.* **185**(2):118–241.

Landman, N. H., 1988, Early ontogeny of Mesozoic ammonites and nautilids, in: *Cephalopods— Present and Past* (J. Wiedmann and J. Kullmann, eds.), Schweizerbart'sche Verlagsbuchhandlung, Stuttgart, pp. 215–228.

Landman, N. H., and Bandel, K., 1985, Internal structures in the early whorls of Mesozoic ammonites, *Am. Mus. Novit.* **2823**:1–21.

Landman, N. H., and Klofak, S. M., in prep., Size frequency studies in Late Cretaceous ammonoids: Evidence for rate of growth.

Landman, N. H., and Waage, K. M., 1982, Terminology of structures in embryonic shells of Mesozoic ammonites, *J. Paleontol.* **56**(5):1293–1295.

Landman, N. H., and Waage, K. M., 1993, Scaphitid ammonites of the Upper Cretaceous (Maastrichtian) Fox Hills Formation in South Dakota and Wyoming, *Bull. Am. Mus. Nat. Hist.* **215**:1–257.

Lehmann, U., 1966, Dimorphismus bei Ammoniten der Ahrensburger Lias-Geschiebe, *Paläontol. Z.* **40**:26–55.

Lehmann, U., 1981, *The Ammonites: Their Life and Their World*, Cambridge University Press, Cambridge.

Lehmann, U., 1990, *Ammonoideen—Leben zwischen Skylla und Charybdis*, 2nd ed., Ferdinand Enke Verlag, Stuttgart.

Makowski, H., 1971, Some remarks on the ontogenetic development and sexual dimorphism in the Ammonoidea, *Acta Geol. Pol.* **21**(3):321–340.

Mangold, K., 1987, Reproduction, in: *Cephalopod Life Cycles*, Vol. II (P. R. Boyle, ed.), Academic Press, New York, pp. 157–200.

Mapes, R. H., 1979, Carboniferous and Permian Bactritoidea (Cephalopoda) in North America, *Univ. Kans. Paleontol. Contrib. Artic.* **64**:1–75.

Mapes, R. H., Tanabe, K., Landman, N. H., and Faulkner, C. J., In prep., Ammonoid cephalopod egg clusters from the Carboniferous of Kansas.

Miller, A. K., 1938, Devonian ammonoids of America, *Geol. Soc. Am. Spec. Pap.* **14**:1–262.

Miller, A. K., and Unklesbay, A. G., 1943, The siphuncle of Late Paleozoic ammonoids, *J. Paleontol.* **17**:1–25.

Miller, A. K., Furnish, W. M., and Schindewolf, O. H., 1957, Paleozoic Ammonoidea, in: *Treatise on Invertebrate Paleontology*, Part L, *Mollusca 4* (R. C. Moore, ed.), Geological Society of America and University of Kansas Press, Lawrence, KS, pp. 11–79.

Morton, N., 1988, Segregation and migration patterns in some *Graphoceras* populations (Middle Jurassic), in: *Cephalopods—Present and Past* (J. Wiedmann and J. Kullman, eds.), Schweizerbart'sche Verlagsbuchhandlung, Stuttgart, pp. 377–385.

Müller, A. H., 1969, Ammoniten mit "Eierbeutel" und die Frage nach dem Sexualdimorphismus der Ceratiten (Cephalopoda), *Monatsber. Dtsch. Akad. Wiss. Berl.* **11**(5/6):411–420.

Munier-Chalmas, E., 1873, Sur le développement du phragmostracum des Céphalopodes et sur les rapports zoologiques des Ammonites avec les Spirules, *C. R. Acad. Sci.* **77**(1).

Naef, A., 1922, *Die fossilen Tintenfische—eine Paläozoologische Monographie*, G. Fischer, Jena.

Obata, I., Tanabe, K., and Futakami, H., 1979, Ontogeny and variation in *Subprionocyclus neptuni*, an Upper Cretaceous collignoniceratid ammonite, *Bull. Natl. Sci. Mus. Ser. C (Geol.)* **5**(2):51–88.

O'Dor, R. K., 1983, *Illex illecebrosus*, in: *Cephalopod Life Cycles*, Vol. I (P. R. Boyle, ed.), Academic Press, New York, pp. 175–199.

Ohtsuka, Y., 1986, Early internal shell microstructure of some Mesozoic Ammonoidea: Implications for higher taxonomy, *Trans. Proc. Palaeontol. Soc. Jpn. N.S.* **141**:275–288.

Owen, C. B., 1878, On the relative positions to their constructors of the chambered shells of cephalopods, *Proc. Zool. Soc. Lond.* **1878**:955–975.

Palframan, D. F. B., 1967a, Modes of early shell growth in the ammonite *Promicroceras marstonense* Spath, *Nature* **216**:1128–1130.

Palframan, D. F. B., 1967b, Variation and ontogeny of some Oxford Clay ammonites: *Distichoceras bicostatum* (Stahl) and *Horioceras bougieri* (D'Orbigny), from England, *Palaeontology (Lond.)* **10**(1):60–94.

Ramsbottom, W. H. C., 1981, Eustatic control in Carboniferous ammonoid biostratigraphy, in: *The Ammonoidea*, Systematics Association Spec. Vol. 18 (M. R. House and J. R. Senior, eds.), Academic Press, London, pp. 369–388.

Ruzhentsev, V. E., 1974, Superorder Ammonoidea, General Section, in: *Fundamentals of Paleontology*, Vol. 5, *Mollusca–Cephalopoda* (V. E. Ruzhentsev, ed.), Keter Press, Jerusalem, pp. 371–511.

Ruzhentsev, V. E., and Shimansky, V. N., 1954, Nizhnepermskie svernutye i sognutie nautiloidei yuzhnogo Urala [Coiled and curved Lower Permian nautiloids of the southern Urals], *Trans. Paleontol. Inst. Akad. Nauk SSSR* **50**:1–152.

Saunders, W. B., and Ward, P. D., 1987, Ecology, distribution, and population characteristics of *Nautilus*, in: *Nautilus—The Biology and Paleobiology of a Living Fossil* (W. B. Saunders and N. H. Landman, eds.), Plenum Press, New York, pp. 137–162.

Schindewolf, O. H., 1928, Zur Terminologie der Lobenlinie, *Paläontol. Z.* **9**:181–186.

Schindewolf, O. H., 1929, Vergleichende Studien zur Phylogenie, Morphogenie und Terminologie der Ammoneenlobenlinie, *Abh. Preuss. Geol. Landesanst. N.F.* **115**:1–102.

Schindewolf, O. H., 1933, Vergleichende Morphologie und Phylogenie der Anfangskammern tetrabranchiater Cephalopoden, *Abh. Preuss. Geol. Landesanst. N.F.* **148**:1–115.

Schindewolf, O. H., 1951, Zur Morphogenie und Terminologie der Ammoneen-Lobenlinie, *Paläontol. Z.* **25**:11–34.

Schindewolf, O. H., 1954, On development, evolution and terminology of ammonoid suture line, *Bull. Mus. Comp. Zool.* **112**(3):217–237.

Schindewolf, O. H., 1959, Adolescent cephalopods from the Exshaw Formation of Alberta, *J. Paleontol.* **33**(6):971–976.

Shigeta, Y., 1989, Systematics of the ammonite genus *Tetragonites* from the Upper Cretaceous of Hokkaido, *Trans. Proc. Paleontol. Soc. Jpn. N.S.* **156**:319–342.

Shigeta, Y., 1993, Post-hatching early life history of Cretaceous Ammonoidea, *Lethaia* **26**:133–145.

Shimansky, V. N., 1954, Pryamye nautiloidei i baktritoidei sakmarskogo i artinskogo yarusov Yuzhnogo Urala [Straight nautiloids and bactritoids from the Sakmarian and Artinskian stages of the southern Urals], *Trans. Paleontol. Inst. Akad. Nauk SSSR* **44**:1–156.

Shul'ga-Nesterenko, M., 1926, Nouvelles données sur l'organisation intérieure des conques des ammonites de l'étage d'Artinsk, *Bull. Soc. Nat. Moscou Sec. Géol.* **2**(34):81–99.

Silberling, N. J., and Nichols, K. M., 1982, Middle Triassic molluscan fossils of biostratigraphic significance from the Humboldt Range, northwestern Nevada, *U.S. Geol. Surv. Prof. Pap.* **1207**:1–77.

Smith, J. P., 1898, The development of *Lytoceras* and *Phylloceras*, *Proc. Calif. Acad. Sci. (Geol.)* **1**(4):129–160.

Smith, J. P., 1901, The larval coil of *Baculites*, *Am. Nat.* **35**(409):39–49.

Smith, J. P., 1914, Acceleration of development in fossil Cephalopoda, *Stanford Univ. Publ. Univ. Ser.* **1914**:1–30.

Spath, C. F., 1933, The evolution of the Cephalopoda, *Biol. Rev.* **8**:418–462.

Sturani, C., 1971, Ammonites and stratigraphy of the "Poseidonia Alpina" beds of the Venetian Alps, *Mem. Ist. Geol. Mineral. Univ. Padova* **28**:1–190.

Tanabe, K., 1975, Functional morphology of *Otoscaphites puerculus* (Jimbo), an Upper Cretaceous ammonite, *Trans. Proc. Palaeontol. Soc. Jpn. N.S.* **99**:109–132.

Tanabe, K., 1977a, Functional evolution of *Otoscaphites puerculus* (Jimbo) and *Scaphites planus* (Yabe), Upper Cretaceous ammonites, *Mem. Fac. Sci. Kyushu Univ. Ser. D. Geol.* **23**:367–407.

Tanabe, K., 1977b, Mid-Cretaceous scaphitid ammonites from Hokkaido, *Palaeontol. Soc. Jpn. Spec. Pap.* **21**:11–22.

Tanabe, K., 1989, Endocochliate embryo model in the Mesozoic Ammonitida, *Hist. Biol.* **2**:183–196.

Tanabe, K., and Ohtsuka, Y., 1985, Ammonoid early internal shell structure: Its bearing on early life history, *Paleobiology* **11**(3):310–322.

Tanabe, K., Obata, I., Fukuda, Y., Futakami, M., 1979, Early shell growth in some Upper Cretaceous ammonites and its implications to major taxonomy, *Bull. Natl. Sci. Mus. Ser. C (Geol.)* **5**(4):155–176.

Tanabe, K., Fukuda, Y., and Obata, I., 1980, Ontogenetic development and functional morphology in the early growth stages of three Cretaceous ammonites, *Bull. Natl. Sci. Mus. Ser. C (Geol.)* **6**:9–26.

Tanabe, K., Obata, I., and Futakami, M., 1981, Early shell morphology in some Upper Cretaceous heteromorph ammonites, *Trans. Proc. Palaeontol. Soc. Jpn. N.S.* **124**:215–234.

Tanabe, K., Landman, N. H., Mapes, R. H., and Faulkner, C. J., 1993, Analysis of a Carboniferous embryonic ammonoid assemblage from Kansas, U.S.A.—Implications for ammonoid embryology, *Lethaia* **26**:215–224.

Tanabe, K., Landman N. H., and Mapes, R. H., 1994, Early shell features of some Late Paleozoic ammonoids and their systematic implications, *Trans. Proc. Palaeontol. Soc. Jpn. N.S.* **173**:383–400.

Tanabe, K., Shigeta, Y., and Mapes, R. H., 1995, Early life history of Carboniferous ammonoids inferred from analysis of fossil assemblages and shell hydrostatics, *Palaios* **10**:80–86.

Trueman, A. E., 1941, The ammonite body chamber with special reference to the buoyancy and mode of life of the living ammonite, *Q. J. Geol. Soc. Lond.* **96**:339–383.

Vavilov, M. N., and Alekseyev, S. N., 1979, Ontogenetic development and internal structure of the Middle Triassic genus *Aristoptychites*, *Paleontol. J.* **13**(3):312–318.

Vecchione, M., 1987, Juvenile ecology, in: *Cephalopod Life Cycles*, Vol. II (P. R. Boyle, ed.), Academic Press, London, pp. 61–84.

Vermeij, G. J., 1978, *Biogeography and Adaptation*, Harvard University Press, Cambridge, MA.

Ward, P. D., and Bandel, K., 1987, Life history strategies in fossil cephalopods, in: *Cephalopod Life Cycles*, Vol. II (P. R. Boyle, ed.), Academic Press, London, pp. 329–350.

Weitschat, W., and Bandel, K., 1991, Organic components in phragmocones of Boreal Triassic ammonoids: Implications for ammonoid biology, *Paläontol. Z.* **65**:269–303.

Wells, M. J., and Wells, J., 1977, Cephalopoda: Octopoda, in: *Reproduction of Marine Invertebrates*, Vol. 4 (A. C. Geise and J. S. Pearse, eds.), Academic Press, New York, pp. 291–330.

Westermann, G. E. G., 1954, Monographie der Otoitidae (Ammonoidea), *Geol. Jahrb. Beih.* **15**:1–364.

Westermann, G. E. G., 1990, New developments in ecology of Jurassic–Cretaceous ammonoids, in: *Atti del secondo convegno internazionale, Fossili, Evoluzione, Ambiente, Pergola, 1987* (G. Pallini, F. Cecca, S. Cresta, and M. Santantonio, eds.), Tecnostampa, Osta Vetere, Italy, pp. 459–478.

Westermann, G. E. G., 1993, On alleged negative buoyancy of ammonoids, *Lethaia* **26**:246.

Wetzel, W., 1959, Über Ammoniten-Larven, *N. Jb. Geol. Paläont. Abh.* **107**(2):240–252.

Wiedmann, J., 1973, Ammoniten–Nuklei aus Schlaemmproben der nordalpinen Obertrias—ihre stammesgeschichtliche und stratigraphische Bedeutung, *Mitt. Ges. Geol. Bergbaustud.* **21**:521–616.

Wiedmann, J., and Kullmann, J., 1981, Ammonoid sutures in ontogeny and phylogeny, in: *The Ammonoidea*, Systematics Association Spec. Vol. 18 (M. R. House and J. R. Senior, eds.), Academic Press, New York, pp. 215–256.

Wissner, U. F. G., and Norris, A. W., 1991, Middle Devonian goniatites from the Dunedin and Besa River formations of northeastern British Columbia, *Geol. Surv. Can. Bull.* **412**:45–79.

Zakharov, Y. D., 1971, Some features of the development of the hydrostatic apparatus in early Mesozoic ammonoids, *Paleontol. J.* **5**(1):24–33.

Zakharov, Y. D., 1972, Formation of the caecum and prosiphon in ammonoids, *Paleontol. J.* **6**(2):201–206.

Zakharov, Y. D., 1974, New data on internal shell structure in Carboniferous, Triassic and Cretaceous ammonoids, *Paleontol. J.* **8**(1):25–36.

Zakharov, Y. D., 1989, New data on biomineralization of the Ammonoidea, in: *Skeletal Biomineralization Patterns, Processes, and Evolutionary Trends, Short Course in Geology*, Vol. 5, Pt. II (J. G. Carter, ed.), American Geophysical Union, Washington, DC, p. 325.

Zell, H., Zell, I., and Winter, S., 1979, Das Gehäusewachstum der Ammonitengattung *Amaltheus* De Montfort während der frühontogentischen Entwicklung, *N. Jb. Geol. Paläont. Mh.* **10**:631–640.

Ziegler, B., 1962, Die Ammoniten-Gattung *Aulacostephanus* im Oberjura (Taxionomie, Stratigraphie, Biology), *Palaeontogr. Abt. A* **119**:1–172.

Zuev, G. V., and Nesis, K. N. (eds.), 1971, *Squids (Biology and Fishery)*, Pishchevaya Promyshlennost, Moscow (in Russian).

Chapter 12

Mode and Rate of Growth in Ammonoids

HUGO BUCHER, NEIL H. LANDMAN, SUSAN M. KLOFAK, and JEAN GUEX

HUGO BUCHER • URA CNRS 157, Centre des Sciences de la Terre, Université de Bourgogne, 21000 Dijon, France. Present address: Centre des Sciences de la Terre, Université Claude-Bernard, Lyon I, 69622 Villeurbanne Cedex, France.   NEIL H. LANDMAN and SUSAN M. KLOFAK • Department of Invertebrates, American Museum of Natural History, New York, New York 10024 and Department of Biology, CUNY, New York, New York 10036. JEAN GUEX •   Institut de Géologie, Université de Lausanne, BFSH-2, CH-1015, Lausanne, Switzerland.

Ammonoid Paleobiology, Volume 13 of *Topics in Geobiology*, edited by Neil Landman *et al.*, Plenum Press, New York, 1996.

1. Introduction

In this chapter we discuss the mode and rate of growth in ammonoids, focusing primarily on postembryonic growth. We first discuss the general mode of growth and then describe the ontogenetic sequence of growth stages. These stages are recognized on the basis of changes in morphology. For example, a graph of the increase in size of whorl width versus shell diameter in an individual reveals changes through ontogeny that pinpoint the end of one growth stage and the beginning of another. We next discuss the overall rate of growth through ontogeny and establish a generalized growth curve. In this discussion, we refer to other cephalopods whose rate of growth is known. Fluctuations in the rate of growth that are superimposed on this growth curve are indicated in ammonoids by the presence of such shell features as varices and constrictions.

The absolute rate of growth in ammonoids depended on a variety of factors, including temperature, food availability, and injuries to the individual animal. In addition, ambient pressure and the permeability of the siphuncle governed the rate at which cameral liquid was removed and, hence, the growth rate. Most methods to determine the actual rate of growth in ammonoids assume that particular morphological features were secreted at a known rate or periodicity. Other methods attempt to identify an environmental signal that was captured in the morphology or chemistry of the shell. It is also possible to study epizoans that grew on the shell of an ammonoid while the ammonoid was alive. In this method, the epizoans are used as chronometers to measure the rate of ammonoid growth. We compile the data from these various methods to arrive at some general estimates of the age at maturity, especially for shallow-water taxa.

Specimens illustrated in this chapter are reposited at the University of Lausanne (here indicated by the prefix HB), the American Museum of Natural History (AMNH), the Yale Peabody Museum (YPM), the United States Geological Survey (USGS), and the British Museum (Natural History) [BM(NH)].

2. Mode of Growth

In this section, we describe the overall mode of growth of ammonoids with reference to *Nautilus*, the only externally shelled cephalopod that is still extant. Ammonoids are, in fact, phylogenetically more closely related to coleoids than they are to *Nautilus* (Engeser, 1990; Jacobs and Landman, 1993; Chapter 1, this volume). However, the retention of an external shell in ammonoids implies that these extinct forms shared with *Nautilus* basic similarities in their processes of growth, although not necessarily a similarity in their rate of growth or age at maturity.

Therefore we begin our discussion by briefly reviewing the mode of growth of *Nautilus*. We are primarily interested in three aspects: (1) secretion of shell material at the apertural margin as the soft body increases in size, (2) secretion of additional shell layers on the inside surface of the body chamber, and (3) secretion of septa at the rear of the body, thereby forming chambers.

The cycle of chamber formation in *Nautilus* consists of several steps: secretion of a septum, formation of a siphuncular segment, removal of cameral liquid, and forward movement of the body to the position of the next septum (translocation) (Ward, 1987; Ward et al., 1981; Ward and Chamberlain, 1983). Most of the time involved in chamber formation (approximately 90%) is devoted to septal secretion with the rear of the soft body fixed against the face of the septum being secreted. Removal of cameral liquid begins when the septum reaches approximately 60% of its final thickness. When the septum is completed, the soft body moves forward to the position of the next septum. Throughout the chamber formation cycle, growth of the shell at the apertural margin is continuous (see references cited above).

The rate of weight increase of the shell and soft body is coordinated with the rate of chamber formation to achieve near-neutral buoyancy as the animal grows (Denton and Gilpin-Brown, 1966; Ward, 1987). In other words, the rate at which chambers form and at which cameral liquid is removed balances the rate of weight increase of the shell and soft body. This system of coordinated balances also entails maintaining a balance between the volumetric increase of the body chamber and soft body and the volumetric increase of new buoyancy chambers.

This system is flexible enough to withstand a certain amount of perturbation. The system responds to perturbations by modifying the rate of removal, and sometimes the direction of flow, of cameral liquid and the volume of buoyancy chambers. For example, in the case of shell breakage at the apertural margin, the animal responds to the increase in positive buoyancy by reducing the rate of cameral liquid removal, even possibly refilling a previously emptied chamber, reducing the volume of the succeeding chamber (septal approximation), and repairing the injury at the apertural margin (Ward, 1986). Changes in cameral volume and the rate of cameral liquid removal also occur at maturity in association with an overall decrease in the rate of growth, a change in shell shape, and the development of the reproductive organs (Collins et al., 1980).

This overall mode of growth was probably also characteristic of ammonoids. In ammonoids, growth occurred at the apertural margin and commonly involved the formation of ribs and other ornamental features. Additional shell layers were secreted on the inside surface of the body chamber (Chapter 4, this volume). The cycle of chamber formation probably consisted of septal and siphuncular secretion, removal of cameral liquid or possibly gel (Hewitt and Westermann, 1987), and movement of the soft body to the position of a new septum (Chapters 6 and 9, this volume). On the basis of the location and shape of muscle scars, we can infer that the rear of the soft body of the ammonoid was fixed against the septum during most of the chamber formation cycle (Tanabe and Landman, in prep.). It is likely that translocation occurred in incremental steps to judge from observations of closely spaced pseudosutures between actual sutures (Zaborski, 1986; Hewitt *et al.*, 1991; Weitschat and Bandel, 1991, 1992; Westermann, 1992; Landman *et al.*, 1993; Lominadzé *et al.*, 1993). During the chamber formation cycle, growth of the soft body and of the shell margin was probably continuous.

The time of chamber formation and the volumetric increase in chamber volume within an individual ammonoid were, no doubt, also coordinated with growth at the aperture to insure near-neutral buoyancy as the animal grew. As in *Nautilus*, we suspect that this system did not function as a means to regulate the depth of the animal in the water column but, rather simply, to maintain near-neutral buoyancy during growth (Chapter 7, this volume). Variation in the spacing between septa (chamber volume) and the time of chamber formation within an individual ammonoid may have represented a response to variation in the overall rate of growth (e.g., more rapid growth because of favorable conditions), changes in the shape of the shell (which were generally more marked than those in *Nautilus*), and modifications in the volume and density of the hard and soft tissues (Westermann, 1971, 1975).

3. Growth Stages

The growth program of ammonoids consisted of several stages. We recognize these stages on the basis of more or less abrupt changes in the shape, position, and size of individual morphological features. These morphological features include, for example, the umbilicus, the pattern of ornamentation, and the suture. In some specimens, the growth of these and other features was gradual, and, therefore, it is difficult to identify when and where changes occurred. As a result, the transition from one stage to another is unclear. In other instances, however, changes in morphology were abrupt (called critical points; "*Knickpunkte*" of Kullmann and Scheuch, 1972) and occurred in nearly all specimens within a species, indicating that growth was polyphasic (Gould, 1966). For example, in the shells of some planispirally coiled ammonoids, there are several changes in the pattern of coiling, each of which occurs at a consistent number of whorls; these are recognized on the basis of measurements of the spiral radius versus the angle of rotation (whorl number) (Currie, 1942, 1943, 1944; Obata 1959, 1960; Kant and Kullmann, 1973; Landman, 1987). Commonly, changes in a number of morphological features occurred at the same time in ontogenetic development and, thereby, delimit the boundaries of a growth stage.

The first stage in all ammonoids was the embryonic stage (Fig. 1A; Chap. 11, this volume). The embryonic shell is called the ammonitella. (This term is also used to describe the embryonic animal as a whole.) The ammonitella extends adaperturally to the primary constriction and accompanying varix. The ornamentation of the ammonitella, if present, ends at this point, and a new ornamentation begins. Commonly, there is also a change in shell shape

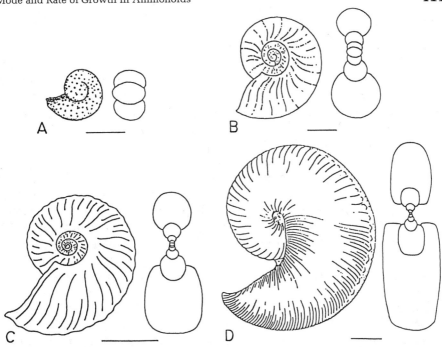

FIGURE 1. Schematic drawings of four growth stages in the ontogeny of *Hoploscaphites nicolletii*, macroconch (Late Cretaceous), in lateral and transverse cross-sectional views. (A) Ammonitella. Scale bar, 500 µm. (B) Neanoconch. Scale bar, 1 mm. (C) Juvenile. Scale bar, 5 mm. (D) Adult. Scale bar, 1 cm.

at this point (Currie, 1942, 1943; Lehmann, 1966, Fig. 1; Kulicki, 1974, 1979; Landman, 1987, 1988; Maeda, 1993); this change is particularly dramatic in heteromorph ammonoids (Bandel *et al.*, 1982, Fig. 1C). On the basis of discoveries of preserved ammonitellas, we know that as few as one septum, i.e., the proseptum, formed during embryogenesis (Chap. 11, this volume).

The rest of growth was, by definition, postembryonic. The first postembryonic stage is called the neanic stage (Westermann, 1958), and the animal or shell that formed during this stage is called the neanoconch (Fig. 1B; Chap. 16, this volume). This stage is equivalent to the late nepionic plus early neanic stages of Hyatt (1894) and Smith (1898), the larval stage of Zell *et al.* (1979), "the pseudolarval stage" of Kulicki (1979), the "goniatitic growth phase" of Hewitt (1985), the *phase juvénile* of Mignot (1993), and the stage up to the "neanic–ephibic developmental rupture" of Checa and Sandoval (1989, abbreviated NEDR). ("Ephibic" is spelled "ephebic" by Hyatt, 1894, and elsewhere.)

The neanoconch is commonly planorbiconic or cadiconic in planispiral ammonoids and displays little or no ornamentation. It extends adaperturally up to approximately two whorls beyond the end of the ammonitella, corresponding to a shell diameter of 3–5 mm. In general, it shows an interval of widely spaced septa. In many taxa, there is an abrupt increase in septal spacing at the beginning of the neanoconch, leading to a maximum midway through, followed by an equally abrupt decrease at the end (e.g., in Middle Jurassic *Quenstedtoceras*, Kulicki, 1974, Fig. 8; Middle Triassic *Gymnotoceras*, Fig. 2A; and Late Cretaceous *Clioscaphites*, Fig. 2C). In other taxa, this pattern is not as well developed (e.g., in Middle Triassic *Parafrechites*, Fig. 2B). In still other taxa, as noted by Mignot (1993), an interval of wide septal spacing is absent altogether (e.g., in Middle Jurassic *Sphaeroceras*).

The end of the neanic stage was marked by changes in several other features in addition to septal spacing, including whorl shape, degree of whorl overlap, ornamentation, and sutural complexity (Currie, 1942, 1943, 1944; Westermann, 1954; House, 1965, Figs. 17–19; Obata, 1965; Palframan, 1966, 1967; Kullmann and Scheuch, 1970, 1972; Kant, 1973a–c; Kulicki, 1974; Tanabe, 1975, 1977; Zell *et al.*, 1979; Landman, 1987; Checa and Sandoval, 1989; Maeda, 1993). In heteromorph ammonoids in which the early whorls are uncoiled, the end of the neanic stage may have corresponded to the first appearance of ornamentation (Tanabe *et al.*, 1981, Pl. 37, Fig. 3, Pl. 38, Fig. 2b; Matsukawa, 1987, Figs. 2, 3). These morphological changes imply further changes in the mechanical strength of the shell (Hewitt, 1988) and its hydrodynamic efficiency (Landman, 1987; Jacobs, 1992; Chapter 7, this volume). During the neanic stage, the ammonoid may have lived in the plankton, and, at the end of this stage, it may have assumed another mode of life in a different habitat (Westermann, 1958; Zell *et al.*, 1979; Landman, 1987, 1988; Checa and Sandoval, 1989; Shigeta, 1993; Chapter 16, this volume).

The next growth stage is referred to as the juvenile stage and the animal or shell as the juvenile (Fig. 1C; Chapter 16, this volume). This stage included all of the remaining whorls up to the start of the mature body chamber and is equivalent to the late neanic plus early ephebic stages of Hyatt (1894), the later part of the juvenile stage of Westermann (1954) and Landman and Waage (1993), and the early part of the *phase submature* of Mignot (1993). Some authors also distinguish among early, middle, and late juvenile stages, mainly on the basis of shell size (e.g., Morton, 1988, described small, medium, and

FIGURE 2. (A) Plot of septal angle (in degrees) versus septum number in a juvenile specimen of *Gymnotoceras rotelliformis* (Middle Triassic, Nevada, HB 45582). The end of the neanoconch is marked by a minimum in septal spacing around septum 25 (arrow). The first few septa in this specimen were not preserved, and so the septum number is an approximation. The septal angle is defined as the angle between septa n and $n-1$. (B) Plot of septal angle (in degrees) versus whorl number in a juvenile specimen of *Parafrechites meeki* (Middle Triassic, Nevada, HB 23745). The arrow indicates a change in the pattern of septal spacing, which correlates with the end of the neanoconch. (C) Plot of septal angle (in degrees) versus whorl number in a mature specimen of *Clioscaphites vermiformis* (Upper Cretaceous, Montana, USGS 21426). The end of the neanoconch is marked by a minimum in septal spacing (arrow). The septal approximation at the end of ontogeny indicates the attainment of maturity (modified from Landman, 1987, Fig. 59).

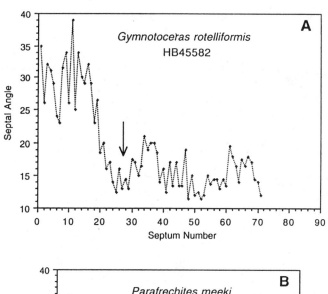

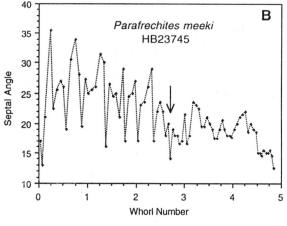

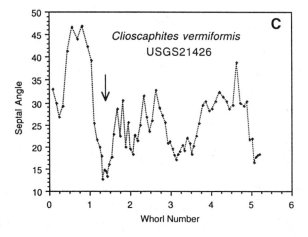

large juveniles of Middle Jurassic *Graphoceras*). Another term in the literature, adolescent, also refers to a late juvenile stage (Hyatt, 1894; see Collins
and Ward, 1987, for the use of this term as applied to *Nautilus*).

In general, the juvenile shell is more compressed and involute than the
neanoconch, and ornamentation is almost always present (Westermann, 1954;
Landman, 1987). In the juvenile shells of many taxa, there is a phase of more
or less uniform septal spacing, punctuated by occasional fluctuations (Fig. 2;
corresponds to the interval between depressions II and III in Kulicki, 1974,
Fig. 8). There is some variation in the number of septa that formed during the
juvenile stage among individuals within the same species (Lehmann, 1966;
Kulicki, 1974; Mignot, 1993).

Maturity or adulthood is the next stage, and the animal or shell is called
the adult (Fig. 1D; Callomon, 1963; Makowski, 1962). This stage is demarcated
by a number of morphological changes, most of which affected the final body
chamber and last few septa. It is also common to describe a subadult or
submature stage in which the sequence of morphological changes culminating
in maturity has only just begun or is in progress (e.g., Mignot, 1993). In
contrast, an immature stage lacks any sign of maturity (Howarth, 1992) and is
equivalent to the neanic and juvenile stages.

The morphological changes associated with maturity are described briefly
below (but see Chapter 13, this volume, for a more complete discussion). The
pattern of septal spacing generally changed in late ontogeny and showed an
initial increase followed by a final decrease (Fig. 2C; Oechsle, 1958; Rieber,
1963; Lehmann, 1966; Clausen, 1968; Bayer, 1972a,b, 1977; Kulicki, 1974;
Druschits *et al.*, 1977; Blind and Jordan, 1979; Checa, 1987; Dommergues,
1988; Bucher and Guex, 1990; Mignot, 1993). The closer spacing or approximation of the last few septa is widely interpreted as indicating the onset and
attainment of maturity (Callomon, 1963; Crick, 1978). Microconchs showed
septal approximation in the last one or over the last two chambers, whereas
macroconchs showed progressively reduced septal spacing over many more
chambers (Chapter 13, this volume; Makowski, 1962; Crick, 1978; Lehmann,
1981; Landman and Waage, 1993). Commonly, but not invariably, there were
also concurrent changes in the shape and length of the body chamber including a reduction in angular length, a change in the pattern of coiling, an increase
in whorl width, either an appearance or strengthening of ornamentation or,
conversely, an attenuation or even disappearance of ornamentation, and
modifications in the shape and thickness of the apertural lip.

As indicated, these changes are interpreted as reflecting sexual maturation
in analogy with present-day *Nautilus* (Makowski, 1962; Callomon, 1963).
However, it is important to note that the exact timing of these changes relative
to the growth of the reproductive organs is unclear. In *Nautilus*, for example,
the secretion of the last septum begins before the complete development of
the reproductive organs (Collins and Ward, 1987). The final thickening of the
apertural lip, however, coincides with the attainment of full sexual maturity.

All of these morphological changes in the ontogeny of ammonoids indicate that growth was determinate and that maturity occurred at a more or less prescribed size for each species (Makowski, 1962; Callomon, 1963). As mentioned, many of these changes were initiated before the final size was reached in order to achieve the desired adult shape, an inevitable consequence of accretionary growth (Callomon, 1963). Thus, it is reasonable to interpret these changes in terms of the mode of life of the adult (Chapter 16, this volume; Seilacher and Gunji, 1993). For example, development of a hook-like body chamber in some Cretaceous heteromorph ammonoids culminated in an upturned aperture, which has been interpreted as an adult adaptation facilitating vertical locomotion (Chapter 16, this volume).

In many modern cephalopods, the release of a hormone from the optic gland system induces sexual maturation (Wells and Wells, 1959; Richard, 1970; Boyle and Thorpe, 1984; Mangold, 1987). When the gonads are large enough to respond to this hormone, the maturation program takes over, entraining a sequence of morphological modifications leading to the attainment of the final mature size, and, ultimately, the completion of the life cycle. Such a sequence of morphogenetic events has been referred to as a "countdown" terminating in the cessation of growth (Seilacher and Gunji, 1993). In some large ammonoids, late ontogenetic modifications are absent, suggesting that growth may have been more or less indeterminate in these forms (Westermann, 1971; Stevens, 1988).

In the 19th century, the morphological features we now recognize as indicating maturity were considered to have formed during a stage following maturity and indicative of old age. This stage is referred to as the senescent, senile, or gerontic stage (Hyatt, 1894). This stage of life in many modern cephalopods is characterized by a deterioration in locomotor performance, an inability to heal injuries, a decrease in weight, and a reduction in metabolic activity (Van Heukelem, 1978). These phenomena may also have characterized the postreproductive life of ammonoids. However, such a stage is unlikely to have been expressed in shell secretory products, so it is probably never directly observable in fossil material (see also discussion in Miller *et al.*, 1957, p. L14, and Matyja, 1986, p. 42).

4. Growth Curve

In this section, we discuss the overall pattern of growth through ontogeny, identifying those periods when the rate of growth accelerated, decelerated, or remained the same. These changes were controlled by the organism's internal clock, but also may have been affected by external factors (see Section 6). At the end of this section, we present a generalized growth curve for all ammonoids, although, given their diversity, this curve will not necessarily apply to every taxon.

The rate of growth of an organism is defined as the change in its overall size through time. It is, therefore, important to choose an appropriate measure of size. In studies of modern coleoids, the variables "weight" and "mantle length" are usually chosen (Mangold, 1983; Forsythe and Van Heukelem, 1987). Because soft parts are not preserved in ammonoids, this metric is unavailable to us. Instead, the diameter of the shell, the number of whorls, or the number of septa is generally used as a measure of size (Matyja, 1986). These same parameters are also employed in studies of the rate of growth of present-day *Nautilus* (Saunders, 1983; Ward, 1985; Landman and Cochran, 1987). The volume of the body chamber or its cube root is, perhaps, a more accurate measure of size in ammonoids, especially if there were significant modifications in the shape of the shell during ontogeny, as there were in heteromorph forms (R.A. Hewitt, personal communication, 1994). However, the volume of the body chamber is seldom used in the literature, and in order to facilitate comparisons with other studies, the standard parameters are used in this chapter.

In practice, determination of the ontogenetic pattern of growth in ammonoids depends on reference to *Nautilus*, whose rate of growth is relatively well known. As explained earlier (Section 2), the fact that both *Nautilus* and ammonoids share an external shell implies basic similarities in their processes and patterns of growth. Therefore, we will begin with a short review of what is known about the ontogenetic pattern of growth in *Nautilus*.

In *Nautilus*, the length of time required for chamber formation (referred to simply as the time of chamber formation in the rest of the chapter) generally increases throughout ontogeny. This is because cameral volume increases faster than the surface area of the siphuncle during ontogeny, requiring increasingly longer times to pump out larger chambers (Chamberlain, 1978; Ward, 1982, Fig. 2). In addition, septa become thicker throughout ontogeny, and, assuming a constant rate of carbonate secretion, it takes increasingly longer times to secrete thicker septa (Westermann, 1990).

The increase in the time of chamber formation appears to be exponential throughout most of ontogeny, judging from growth studies of *Nautilus pompilius* and *N. macromphalus* in aquaria (Ward, 1985; Landman and Cochran, 1987). This increase in time corresponds with a more or less exponential increase in chamber volume, expressed by septa spaced at equal angular intervals. Departures from this pattern of uniform septal spacing, for example, at maturity (see below), indicate variation in the time of chamber formation. Such variation, which also occurs in *Sepia*, as described by Wiedmann and Boletzky (1982), reflects changes in the overall rate of growth and in the shape of the shell and soft body.

Consistent with the exponential increase in the time of chamber formation in *Nautilus*, the rate of growth of the venter at the apertural margin is more or less constant during most of ontogeny (Saunders, 1983; Ward, 1985; Carlson *et al.*, 1992). As a result, the diameter of the shell generally increases as a linear function of time, whereas the number of whorls increases as a negatively logarithmic function of time.

These general patterns also probably characterized the rate of growth of ammonoids because similar physiological processes operated in these animals as well. The same ontogenetic relationships between cameral volume and the surface area of the siphuncle and between septal thickness and shell diameter also existed in most ammonoids (Chamberlain, 1978; Ward, 1982; Westermann, 1990). In addition, in many ammonoids, the diameter of the siphuncle grew with a strong negative allometry (Westermann, 1971. 1990; Tanabe, 1977; Landman, 1987). All of this evidence suggests that the time of chamber formation in ammonoids increased exponentially throughout most of ontogeny. Between any two adjacent chambers, the time of formation may have been only slightly different, but between chambers at the beginning and at the end of ontogeny, the difference in time of formation was undoubtedly significant. The exact rate of increase in the time of chamber formation probably varied among individuals within the same species and especially among individuals of different species.

An exponential increase in the time of chamber formation in ammonoids correlated with a nearly exponential increase in chamber volume, which is expressed by the fact that septa are spaced at approximately equal angular intervals (Fig. 3). However, as we have seen, septal spacing in ammonoids is very variable and much less uniform than it is in *Nautilus*. As in *Nautilus*, it

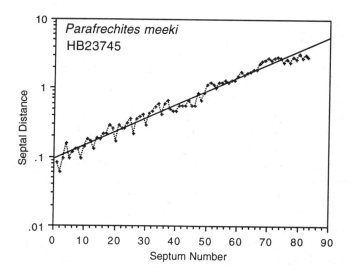

FIGURE 3. Semilog plot of the distance between septa (mm) versus septum number in a juvenile specimen of *Parafrechites meeki* (Middle Triassic, Nevada, HB 23745). The septal distance is defined as the distance between septa n and $n - 1$. For consistency, this distance is measured at one-third the whorl height from the venter on a median cross section. Note that the distance between septa increases exponentially over ontogeny (same specimen as in Fig. 2B.)

is likely that fluctuations in septal spacing were related to variations in the overall rate of growth and in the shape of the shell and soft body.

An exponential increase in the time of chamber formation in ammonoids implies that the rate of growth of the venter at the apertural margin was probably constant during most of ontogeny, with shell diameter increasing as a linear function of time. Increase in angular length (whorl number), in contrast, was probably a negatively logarithmic function of time, with equal angular increments taking increasingly longer to form.

Departures from this general pattern of growth occurred during the neanic and mature stages. The neanic stage, as noted, was characterized by an interval of wide septal spacing. In addition, in many ammonoids, the beginning of this stage was marked by a pronounced increase in whorl width. These observations both suggest that there may have been an acceleration in the rate of growth during this time interval.

In contrast, the rate of growth probably decelerated at the approach of maturity (Callomon, 1963). Observations of this stage in *Nautilus* indicate a rapid decrease in the rate of growth of the shell at the apertural margin (Ward, 1985). This decrease coincides with the formation of a number of morphological features, including thickening of the apertural margin and approximation of the last few septa (Davis, 1972; Collins and Ward, 1987).

The explanation of septal approximation at maturity is not entirely clear. In *Nautilus*, this feature is associated with, among other changes, a decrease in the rate of growth at the apertural margin, a reduction in the forward movement of the soft body, an enlargement of the reproductive organs, and a change in the shape of the shell. It is possible that the reduced spacing and extra thickness of the last few septa serve to provide additional weight necessary to compensate for a decrease in the density of the soft body as a result of the growth of the reproductive organs, which are presumably composed of fatty tissue. Ward (1987) has also suggested that septal approximation in *Nautilus* allows for a final trimming of buoyancy.

The presence of many of the same morphological features in ammonoids as in *Nautilus* also indicates that there was probably a deceleration in the rate of growth of ammonoids at the approach of maturity. In many ammonoids the final apertural lip is thickened and bears ventral and lateral lappets. The adult body chamber commonly shows a change in shape and a reduction in angular length. In addition, ribs and other ornamental features tend to become more closely spaced toward the aperture (Matyja, 1986). All of this evidence suggests a deceleration in the rate of growth and, ultimately, the cessation of growth at a more or less prescribed size. However, the time interval over which the rate of growth decelerated may have varied depending on the taxon and dimorph (possibly indicated by differences in the patterns of septal approximation).

Additional information about the overall pattern of growth in ammonoids comes from a study of growth increments on the lower jaws of Mesozoic ammonoids. In this study, Hewitt *et al.* (1993) investigated several specimens

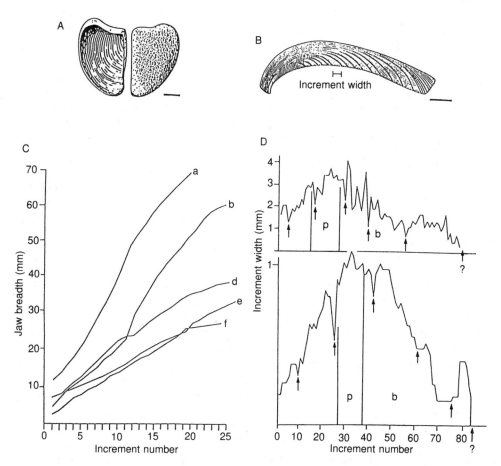

FIGURE 4. (A) View of the concave inner and convex outer surfaces of the two valves composing a specimen of *Laevapytchus*, here interpreted as a lower jaw. The concave surfaces of these valves are marked with concentric ridges called growth increments. Scale bar, 1 cm. (After Lehmann, 1981, Fig. 73.) (B) Longitudinal cross section through one of these valves showing the internal lamellae that correspond to the growth increments. Note the measurement of increment width. Scale bar, 5 mm. (After Schindewolf, 1958, Fig. 2.) (C) Plot of the ontogenetic increase in jaw breadth with respect to increment number in five specimens of *Lamellaptychus*, also interpreted as lower jaws. The three smallest specimens (d,e,f) show a nearly constant increase in jaw breadth with a slight decrease near the end of ontogeny; the two largest specimens (a,b) show a more sigmoidal pattern. Jaw breadth is measured across the jaw (perpendicular to the straight edges of the valves) when the two valves are folded into a U-shape, approximating their position in life. (After Hewitt *et al.*, 1993, Fig. 1.) (D) Plot of increment width versus increment number through ontogeny in two specimens of *Laevaptychus*. There is an increase in increment width followed by a decrease starting at the point of the jaw corresponding to the formation of the middle or the end of the penultimate whorl (p) and continuing through the part of the jaw corresponding to the formation of the mature body chamber (b). The arrows indicate possible winter events characterized by reductions in increment width. According to this interpretation, the animals in which these jaws once occurred attained maturity in about 6 years. However, the decrease in increment width at the end of ontogeny may simply be related to the attainment of maturity. (After Hewitt *et al.*, 1993, Fig. 2.)

of *Lamellaptychus* and *Laevaptychus*, here interpreted as lower jaws, belonging to Late Jurassic oppeliids and Late Jurassic and Early Cretaceous aspidoceratids, respectively. The growth of these jaws was related to the growth of the entire animal because they were located within the body chamber. For example, measurements of specimens of *Lamellaptychus* indicate that the length of these jaws approximately equals the whorl height at the aperture, so, if the whorl height at the aperture doubled in, for example, 5 years, then half the length of the jaw would have formed in that same time interval (Hewitt *et al.*, 1993).

The concave, chitinous surfaces of these lower jaws are marked with concentric ridges (growth increments) that correspond to internal lamellae (Fig. 4A,B). Hewitt *et al.* (1993) plotted jaw breadth versus the number of growth increments on the surface of the jaw in several specimens of *Lamellaptychus* (Fig. 4C). Smaller specimens (labeled d, e, and f on Fig. 4C, possibly belonging to microconchs) showed a nearly constant increase in jaw breadth versus increment number, with a slight decrease in the slope of the curve near the end of ontogeny. Larger specimens (a and b on Fig. 4C, possibly belonging to macroconchs) showed a more sigmoidal pattern consisting of two phases, an early phase of increasing slope and a later phase of decreasing slope. Hewitt *et al.* (1993) also plotted increment width versus increment number in several specimens of *Laevaptychus* (Fig. 4D). They detected an increase in increment width in early ontogeny followed by a decrease in later ontogeny. According to their calculations, the point at which this change occurred corresponded to the end of the penultimate whorl and the beginning of the adult body chamber in the ammonoids in which these jaws grew.

Assuming that these growth increments were secreted at equal time intervals, plots of increment width versus increment number and jaw breadth versus increment number indicate that the rate of growth in these ammonoids was constant or accelerated up to the end of the penultimate whorl and then decelerated at the start of the mature body chamber. This suggests that a major portion of the life span of these ammonoids involved a protracted period during which the rate of growth decreased, beginning with the onset of maturity.

In summary, we present a generalized growth curve for all ammonoids (Fig. 5). We postulate that the rate of growth accelerated during the neanic stage, was more or less constant during the juvenile stage, and decelerated at the approach of maturity. However, given the wide diversity of ammonoids, this curve represents only an approximation. Differences in the slope of this curve probably resulted from variation in those features affecting the rate of cameral liquid removal and shell secretion, namely, the diameter of the siphuncle, the length of the septal necks, and the thickness of the shell wall and septa as well as from variation in environment and mode of life (Westermann, 1990). In some ammonoids there may also have been variation in the duration of the juvenile and adult stages, with some forms showing indeterminate growth (Ivanov, 1975).

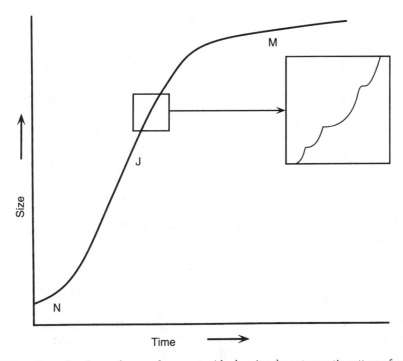

FIGURE 5. Generalized growth curve for ammonoids showing the ontogenetic pattern of growth through ontogeny. Size is usually expressed as shell diameter or whorl number. The cube root of the volume of the body chamber is a more accurate measure of size, especially if there were pronounced changes in the shape of the shell during ontogeny (R.A. Hewitt, 1994, personal communication). The enlarged view on the right reveals numerous fine-scale fluctuations in the rate of growth. Abbreviations: N, neanic stage; J, juvenile stage; M, mature stage.

5. Intrinsic Fluctuations in the Rate of Growth

Superimposed on the ontogenetic growth curve were short-term fluctuations in the rate of growth, which are indicated by the presence of certain shell features. Such fluctuations were probably an intrinsic part of the growth program, although there also may have been an environmental component. These fluctuations indicate that, on a finer scale, the growth curve of ammonoids was never as smooth as it is shown in Fig. 5.

5.1. Growth Lines, Lirae, and Ribs

Growth lines represent discontinuities in secretion and indicate that growth was intermittent at some time scale (Vermeij, 1993). Growth lines appear in the periostracum and on the shell surface (Checa, 1994). Okamoto (1989, Fig. 1) illustrated growth lines spaced at intervals of approximately 100 μm in a specimen of Late Cretaceous nostoceratids. We have observed still finer growth lines on a number of other species (Fig. 6). These lines appear on the surface of the outer prismatic layer but are not visible in cross section (compare with growth lines in *Nautilus*, Fig. 7). The smallest such growth lines measure 1 μm in width, suggesting that they reflect intermittent growth at the cellular level (G.R. Clark, II, personal communication, 1993).

The distinction between growth lines and lirae is somewhat arbitrary, and these two terms have been used interchangeably by many authors (Doguzhaeva, 1982; Korn and Price, 1987). However, the general sense is that lirae are larger and more widely spaced than growth lines—they characteristically occur at intervals of 0.1 to 1.0 mm, although their precise spacing depends on where they occur on the shell (Fig. 8). Lirae may uniformly cover the shell surface or show variations in spacing throughout ontogeny (Doguzhaeva, 1982).

On the next coarser scale are ribs. In most ammonoids, ribs and lirae tend to parallel growth lines, indicating that both of these features mark former positions of the apertural margin (Fig. 6D,E; see Chapter 8, this volume, for a discussion of how the orientation of the living ammonoid may have influenced rib direction; Checa and Westermann, 1989; Okamoto, 1989; Checa, 1994). However, in some ammonoids, ribs intersect growth lines at low angles (Arkell *et al.*, 1957; Cowen *et al.*, 1973). In general, ribs represent plications (corrugations) of the shell wall and, therefore, are expressed on the steinkern.

The formation of ribs was probably controlled mostly by the growth program rather than by the environment. Rib formation may have involved a system of activation and inhibition at the mantle edge, similar to that which produces color patterns in living gastropods and bivalves (Meinhardt and Klinger, 1987). Checa (1994) proposed that, in many ammonoids, an entire rib may have formed during a single secretory event (see Section 5.6). This would have produced episodic advances in the growth of the apertural margin. There is some evidence suggesting that such a process may have been reflected in septal spacing. Checa (1987, Fig. 2) and Dommergues (1988, Figs. 2–8)

FIGURE 6. Ammonoid ribs, lirae, and growth lines. (A–C) Ventral view of part of the mature body chamber of a specimen of *Scaphites carlilensis*, Upper Cretaceous, Kansas, USGS loc. D5140. (A) The body chamber is marked by strong ribs. The adoral direction is toward the lower left-hand corner of the photo. Scale bar, 500 μm. (B) Close-up of a well-preserved portion of this shell reveals the existence of growth lines. Scale bar, 20 μm. (C) These growth lines occur at intervals of approximately 1 μm. Scale bar, 2 μm. (D–F) Lateral view of part of the ammonitella and first whorl of the neanoconch of a specimen of *Scaphites whitfieldi*, Upper Cretaceous, South Dakota, AMNH 44833. (D) The arrow indicates the end of the ammonitella. Scale bar, 50 μm. (E) Close-up of ribs and growth lines on the neanoconch shows that ribs parallel growth lines. Scale bar, 10 μm. (F) A further enlargement reveals that the growth lines occur at intervals of approximately 1 μm. Scale bar, 5 μm. (G–I) Lateral view of part of the mature body chamber of a specimen of *Parafrechites meeki*, Middle Triassic, Nevada, AMNH 44994. (G) Ribs, lirae, and megastriae (arrows) are visible. The adoral direction is toward the bottom of the photo. Scale bar, 200 μm. (H) Close-up showing lirae (arrows) and growth lines between them. Scale bar, 20 μm. (I) The growth lines appear to be spaced at intervals of about 1 μm. Scale bar, 5 μm.

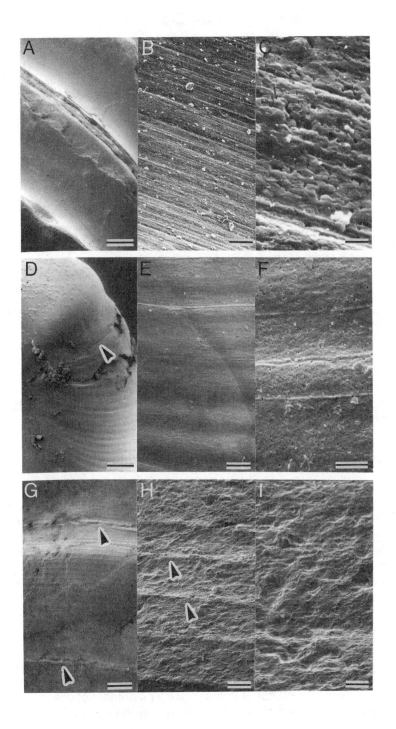

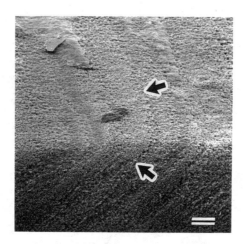

FIGURE 7. Growth lines along the venter of the mature body chamber of a specimen of *Nautilus belauensis*, Palau, AMNH 45073. A piece of the shell was broken off near the apertural margin and sectioned perpendicular to the growth lines. It was photographed at an angle so that the surface of the shell (lighter area on top) as well as the cross section through the spherulitic prismatic layer (darker area on bottom) are visible. The growth lines (arrows) appear both on the surface and in cross section. The visibility of the growth lines was enhanced by etching with acid (EDTA). The adoral direction is toward the right. Scale bar, 20 μm.

observed a positive correlation between septal spacing and rib spacing in many Jurassic ammonoids. This correlation suggests that septal spacing may have responded to the growth spurts and concomitant changes in shell shape associated with rib formation.

5.2. Constrictions, Varices, and Stretch Pathology

Constrictions are defined as grooves on the surface of the shell and may be rectiradiate, prorsiradiate, sigmoid, or angular (Fig. 9; Arkell *et al.*, 1957). They generally represent flexures in the shell wall and hence form corresponding grooves on the steinkern. They are sometimes accompanied by a varix, that is, a thickening of the shell wall, caused primarily by an increase in the thickness of the nacreous layer (Birkelund, 1981; see Arkell *et al.*, 1957, for a description of the different kinds of varices, for example, labial ridges). Varices may also occur without an associated constriction; such varices have been called internal ridges or pseudo-constrictions (Westermann, 1990) and appear as grooves on the steinkern.

Constrictions have generally been interpreted as reflecting discontinuities in growth (Simoulin, 1945; Arkell *et al.*, 1957; Kulicki, 1974; Kennedy and Cobban, 1976; Obata *et al.*, 1978; Bogoslovsky, 1982; Westermann, 1990; see Section 5.6). This interpretation is supported by a number of observations: (1) the association of constrictions with varices, (2) the change in direction of

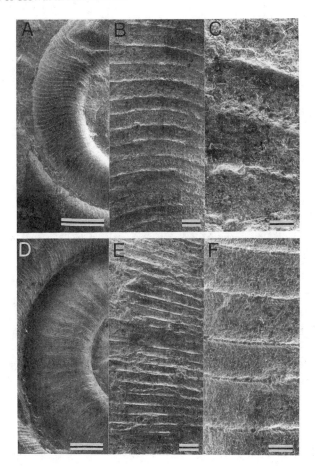

FIGURE 8. Ammonoid lirae. (A–C) Lateral view of part of the juvenile whorls of a specimen of *Ussurites* sp., Triassic, Greece, AMNH 45072. (A) Lirae are visible on the whorl flanks. Coiling is counterclockwise. Scale bar, 2 mm. (B) Enlarged view of A. Scale bar, 200 μm. (C) Enlarged view of B. Scale bar, 50 μm. (D–F) Lateral view of part of the juvenile whorls of a specimen of *Clymenia undulata*, Devonian, Germany, YPM 2513. (D) Lirae are visible on the whorl flanks. Coiling is clockwise. Scale bar, 2 mm. (E) Enlarged view of D. Scale bar, 200 μm. (F) Enlarged view of E. Scale bar, 100 μm.

ribbing following a constriction (Fig. 9; Simoulin, 1945; Arkell *et al.*, 1957; Checa and Westermann, 1989; Maeda, 1993), (3) the pronounced increase in whorl height following a constriction, as shown in many perisphinctids (Arkell *et al.*, 1957), and (4) the disappearance of repaired shell injuries at a constriction (Simoulin, 1945).

In contrast to varices and constrictions, a stretch pathology refers to a portion of a shell characterized by a decrease in whorl width and height, an attenuation of ornamentation, and a thinning of the shell wall (Landman and Waage, 1986). Stretch pathologies are common in macroconchs of some Late

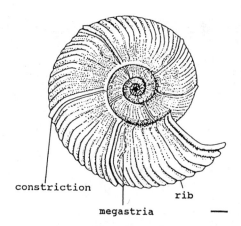

constriction   rib

megastria   ▬

FIGURE 9. Sketch of a macroconch of *Yokoyamaoceras ishikawai*, Late Cretaceous, Japan, showing ribs, constrictions, and megastriae (after Maeda, 1993, Fig. 2). Scale bar, 10 mm.

Cretaceous scaphitid ammonoids and occur at the point where the shell uncoils from the phragmocone into the shaft of the adult body chamber. This phenomenon may indicate a period of rapid growth just prior to the onset of maturity or, alternatively, may record a diversion of resources toward the development of the reproductive organs. A period of rapid growth would explain the extreme rarity of submature specimens of these ammonoids in which the hook-like body chamber is only half-formed (Landman and Waage, 1993).

5.3. Spines

The presence of spines may indicate that there were variations in the rate of forward growth of the apertural margin. For example, on the basis of the pattern of growth lines, Checa and Martin-Ramos (1989) reconstructed the formation of long, hollow spines in Late Jurassic *Aspidoceras*. They concluded that the mantle must have remained in a more or less stationary position until each spine was completed, implying occasional pauses in forward growth of the apertural margin. In other instances, however, the process of spine formation may have occurred without any interruptions in forward growth. For example, in Late Jurassic *Orthaspidoceras*, the formation of mammiform spines probably occurred as the mantle continuously moved forward (Checa and Martin-Ramos, 1989).

It is possible that some spines, depending on their location on the shell, may have been resorbed during ontogeny. This is because in closely coiled ammonoids, the dorsum of one whorl is secreted against the venter of the preceding whorl, and, therefore, the ventral surface of the preceding whorl had to have been relatively smooth. In Early Triassic *Monacanthites*, spines are present on the venter of the last whorl of mature specimens but are absent on the venter of earlier whorls where parabolic nodes occur instead (Tozer, 1965). If these nodes were once the site of spines, the spines would have had to have been resorbed before secretion of the next whorl, in a process similar to that which occurs in some gastropods (Carriker, 1972; Vermeij, 1993). An alternative explanation is that spines developed only on the last whorl of mature specimens (Tozer, 1965).

In other ammonoids, spines may have been retained throughout ontogeny even though they evidently interfered with the secretion of succeeding whorls. In Late Cretaceous *Euomphaloceras septemseriatum*, ventrolateral tubercles form the septate bases of long spines (Kennedy, 1988, Pl. 8, Figs. 4, 6, 9; Pl. 9, Fig. 11). These spines are accommodated in radial grooves along the flanks of the succeeding whorls. In some instances, the spines are almost entirely concealed within these grooves. In other instances, the spines are missing altogether and may have broken off during life, leaving only the ventrolateral tubercles behind.

Seilacher and Gunji (1993) described another example in which ornamental features may have been discarded during ontogeny. In Early Jurassic *Lytoceras lythense*, the apertural margin displays a crenulated flare. These apertural flares have been found as isolated fragments in the Posidonia Shale of Germany. One explanation is that the animal purposely shed these flares during ontogeny. Alternatively, these features may have simply broken off soon after they were formed.

5.4. Megastriae

Megastriae are defined as distinctive thick lines, different from growth lines and lirae, that extend continuously around the flanks and venter (Figs. 6G, 9; Bucher and Guex, 1990). This term is used to refer to all such features that have previously been referred to by different terms: *alte Mundränder* (Pompeckj, 1884; Teisseyre, 1889; Mojsisovics, 1886; Wähner, 1894; Diener, 1895), demarcation lines (Matsumoto *et al.*, 1972; Obata *et al.*, 1978), parabolic lines (Arkell *et al.*, 1957; Matsumoto, 1991; Maeda, 1993), and transitional mouth borders (Tozer, 1991). Megastriae highlight the discontinuous nature of shell secretion in that they represent intrinsic pauses in growth superimposed on the overall growth curve. There are many different kinds of megastriae, and they occur in a wide variety of taxa (Fig. 10).

Microstructural studies of megastriae reveal that these features represent actual breaks in secretion. In median cross section, a megastria is seen to result from the overlap of two consecutive shell segments (Fig. 11). The old segment may either wedge out gradually or end abruptly as a thickening or an upturned flare visible on the surface of the shell (Fig. 6G). The new segment appears beneath the previous one and thickens progressively until it outgrows the former apertural margin. Secretion of this new segment involved both the

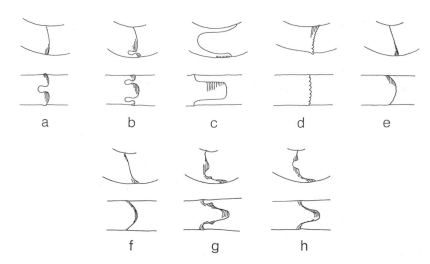

FIGURE 10. Types of megastriae shown in lateral (above) and ventral view (below). The adoral direction is toward the right. Some representative taxa are listed for each type of megastriae: (a) parabolic, *Keyserlingites, Monacanthites* (both Early Triassic); (b) parabolic, *Nordophiceras* (Early Triassic), *Ussurites* (Middle Triassic), *Perisphinctes* (Late Jurassic); (c) concave, *Pleuroacanthites* (Early Jurassic); (d) crinkled and flared, *Lytoceras* (Early Jurassic); (e) straight, *Pachylytoceras* (Middle Jurassic); (f) sigmoidal, *Eotetragonites* (Cretaceous); (g, h) subparabolic, *Anagymnotoceras, Eogymnotoceras, Gymnotoceras, Parafrechites* (all Middle Triassic).

FIGURE 11. Median cross section through a specimen of *Anagaudryceras limatum*, Upper Cretaceous, Hokkaido, AMNH 44374, showing the arrangement of shell layers in a megastria. The old shell segment stops abruptly, and the new shell segment appears beneath it. The arrow indicates the adoral direction. The exterior is toward the top of the photo. When the relief created by a megastria is strong enough, the inner prismatic layer of the new shell segment may not perfectly line the inside surface of the shell wall. Abbreviations: OP, outer prismatic layer; NA, nacreous layer; IP, inner prismatic layer. Scale bar, 10 μm.

outer prismatic and nacreous layers, implying a retreat of the secreting edge of the mantle (see Kulicki, 1979, Pl. 47, Fig. 7; compare to Bucher and Guex, 1990, Fig. 4). In contrast, the inner prismatic layer (= preseptal layer of Guex, 1970) is perfectly continuous, indicating that it was laid down subsequently. Thus, the structural relationships in megastriae appear to be analogous to those of apertural shell repairs (for example, in pelecypods, Tevesz and Carter, 1980, Figs. 6, 7).

Megastriae are associated with particular morphological features and occur more commonly on certain portions of a shell than others. In general, there are more megastriae on juvenile than on mature whorls. In Middle Triassic *Parafrechites meeki*, the juvenile whorls display numerous, variably spaced megastriae (Fig. 12). However, these features are generally absent on the adult body chamber, implying that there were more growth pauses in early than in later ontogeny. In only a few ammonoid groups, e.g., some lytoceratids, did megastriae persist to maturity.

Megastriae are commonly associated with ribs, suggesting a relationship between pauses in secretion and rib formation (see also Checa, 1994). For example, on the juvenile whorls of *Parafrechites meeki*, megastriae occur on the adapical sides of ribs (Fig. 12A). On later whorls, where there are bifurcating ribs, a megastria always occurs on the adapical side of the more adoral rib of the bifurcating pair (Fig. 12C). This megastria cuts across the whorl at the point of bifurcation. The presence of closely spaced megastriae results in composite ribs consisting of a number of juxtaposed elements (Fig. 12D).

In many ammonoids, open spines also are associated with megastriae. For example, in Early Jurassic *Analytoceras articulatum*, megastriae occur in association with hollow ventrolateral spines on the adult body chamber

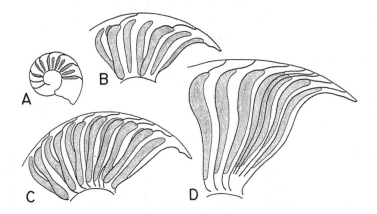

FIGURE 12. The relationship between megastriae and ribs shown diagrammatically in lateral view at successively later ontogenetic stages (A–D) of *Parafrechites meeki*, Middle Triassic, Nevada. The shaded areas represent ribs, the dark lines, megastriae. Coiling is clockwise. (Modified from Bucher and Guex, 1990, Fig. 3.)

(Wähner, 1894). This suggests that secretion of this kind of spine coincided with a pause in forward growth of the apertural margin (see Section 5.3).

In addition to growth pauses, megastriae also imply that some resorption of the shell may have occurred because these features demarcate angular discontinuities that affect growth lines and ribs. In some lytoceratids, megastriae cut sharply and obliquely across roughly parallel ribs at the umbilical margin. In areas of the shell not covered by subsequent whorls, the only possible opportunity for creating this type of discontinuity was during the growth break itself. Similarly, in Middle Triassic *Parafrechites* and *Gymnotoceras*, subparabolic megastriae cut sharply across growth lines, implying that partial resorption occurred, especially along the ventrolateral shoulder and, to a smaller extent, on the flanks (Fig. 10g; Bucher and Guex, 1990, Fig. 2A). There is also evidence of shell resorption in some Jurassic perisphinctids. In *Grossouvria*, for example, parabolic megastriae swing backward along the ventrolateral shoulder cutting across preexisting growth lines to form notches on each side of the venter (Fig. 10b; Arkell *et al.*, 1957, Fig. 140, 1a, b). However, in other perisphinctids such as *Dichotomoceras?*, there is a gradual change in the shape of growth lines that culminates in the formation of a parabolic margin at megastriae (R. Enay, personal communication, 1995). This suggests that the shape of the aperture at these points was the result of normal growth processes rather than secondary resorption.

FIGURE 13. Plots of septal spacing (crosses) and megastriae spacing (open circles) versus whorl number in three ammonoid species. Curves of septal spacing have been translated to the right with respect to those of megastriae spacing to compensate for the known length of the body chamber so that the megastriae and septa that formed at approximately the same time appear near one another. (A) *Gymnotoceras rotelliformis*, Middle Triassic, Nevada, HB 45670. The patterns of septal spacing and megastriae spacing are broadly similar. The ratio of the number of megastriae to that of septa averages approximately 1.5 between whorls 1.5 and 3.25. The distance between septa (septal distance) and the distance between megastriae (megastriae distance) were measured at one-third the whorl height from the venter. The diameter used to calculate each of the ratios on the y-axes was measured at a point midway between successive septa or megastriae, depending on the particular ratio. The early whorls of this specimen were not preserved, and hence, the whorl number is an approximation. The septal curve has been translated 0.5 whorls to the right. (B) *Parafrechites meeki*, Middle Triassic, Nevada, HB 23718. Both curves show a peak in spacing between whorls 2.5 and 3.0. The ratio of the number of megastriae to that of septa averages approximately 3.0 for the episode of megastriae crowding at 3.7 whorls. The diameter, septal distance, and megastriae distance are defined as in A. The early whorls of this specimen were also not preserved, and hence, the whorl number is an approximation. The septal curve has been translated 0.6 whorls to the right. (C) *Eotetragonites* sp., Upper Cretaceous, Madagascar, HB 45989. The pattern of septal spacing covaries with that of megastriae spacing. The ratio of the number of megastriae to that of septa equals 0.2 between whorls 1.5 and 3.5. Septal spacing and megastriae spacing are measured in degrees as indicated on the y-axes. The number of whorls is counted starting at the primary constriction. The amount of translation of the curve of septal spacing, 230°, was determined graphically so that the patterns of megastriae spacing and septal spacing matched. However, this value is a good approximation of the angular length of the body chamber in this species (see Okamoto, Chapter 8, this volume, Fig. 3 for an estimate of the angular length of the body chamber in a species of the closely related genus *Tetragonites*).

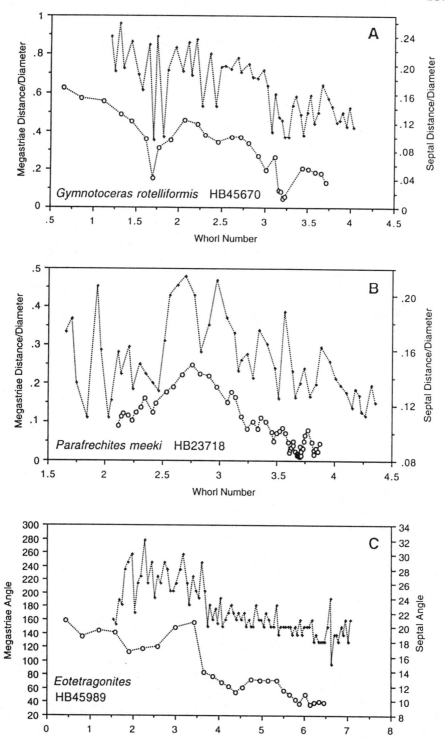

The pauses in growth associated with megastriae appear to be reflected in septal spacing. We investigated the relationship between the spacing of septa and the spacing of megastriae in specimens of Middle Triassic *Gymnotoceras* and *Parafrechites* and Late Cretaceous *Eotetragonites* (Fig. 13). After the angular length of the body chamber had been taken into account, a marked similarity in the patterns of septal and megastriae spacing within each specimen was observed. Bucher and Guex (1990) also observed a correlation

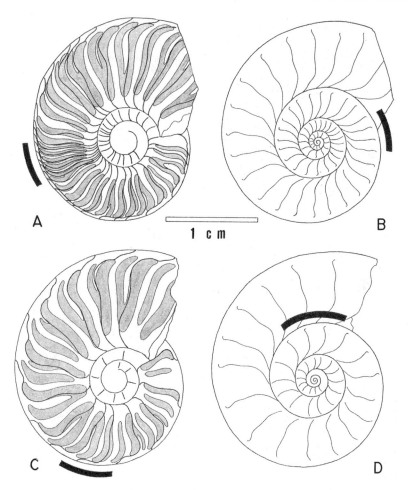

FIGURE 14. Diagrammatic lateral views (A,C) and median cross sections (B,D) of two specimens of *Gymnotoceras rotelliformis*, Middle Triassic, Nevada. (A,B) Compressed specimen, HB 45582. (C,D) Depressed specimen, HB 45663. In lateral view, the shaded areas represent ribs, the dark lines, megastriae. In median cross section, the siphuncle is omitted for simplicity, and only the septa are shown. Megastriae, ribs, and septa are all more closely spaced in the more compressed specimen. The black arcs indicate intervals of closely spaced megastriae (A,C) that correspond in time of formation with intervals of closely spaced septa (B,D) within each specimen. (From Bucher and Guex, 1990, Fig. 6.)

between the spacing of septa and that of megastriae, on the one hand, and the degree of shell compression, on the other hand, among individuals within a single species of *Gymnotoceras*. According to them, septa and megastriae are more closely spaced in more compressed specimens, suggesting that pauses in growth were more numerous, with respect to angular length, in these specimens than in more depressed specimens (Fig. 14).

It has been suggested that, during the pause in growth represented by a megastria, the animal moved forward preparatory to forming a new septum (Tozer, 1991). However, the number of megastriae that corresponded in time of formation with that of a single septum usually does not equal one. For example, the ratio of the number of megastriae to the number of septa that formed at approximately the same time averages 1.5 in *Gymnotoceras rotelliformis* (Fig. 13A), 3.0 in *Parafrechites meeki* (Fig. 13B), and 0.2 in *Eotetragonites* sp. (Fig. 13C). Thus, although the spacing of septa appears to mimic the spacing of megastriae within a specimen, there is clearly no one-to-one correspondence between the number of megastriae and the number of septa. This suggests that translocation to the site of a new septum probably did not coincide with the pause in growth associated with a single megastria.

There are several lines of evidence suggesting that the formation of megastriae was intrinsic in origin and was not the result of injuries or other external factors: (1) the association of megastriae with open spines and ribs, which were themselves presumably intrinsic in origin, (2) the presence of megastriae in some taxa (e.g., berychitids, lytoceratids, and perisphinctids) but not others (e.g., arcestids, paraceratids, and baculitids), (3) consistent changes in the spacing of megastriae through ontogeny, and (4) the correlation between megastriae spacing and the degree of shell compression in certain species. Additional support for an intrinsic origin of megastriae comes from study of a sample of Middle Triassic *Parafrechites meeki* thought to represent a census population (Bucher and Guex, 1990). If megastriae were caused by the environment, and if all the specimens in the sample experienced this same environment, one might expect to observe a similarity in the pattern of megastriae spacing among individuals within any given size class. However, no such similarity appears.

5.5. Pseudosutures

Suture-like lines called pseudosutures occur between sutures and follow the general outline of the sutures (Fig. 15). Pseudosutures have been observed in many ammonoids and have been interpreted as imprints of the adapical end of the soft body (Zaborski, 1986; Hewitt *et al.*, 1991; Weitschat and Bandel, 1991, 1992; Westermann, 1992; Landman *et al.*, 1993; Lominadzé *et al.*, 1993; Chapter 6, this volume). The spacing of pseudosutures provides some information on variations in the rate at which the body moved forward, provided that these features formed at equal intervals of time (see Section 7.1 for actual

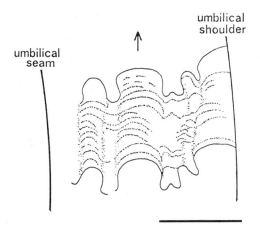

FIGURE 15. Specimen of *Paravascoceras cauvini*, Upper Cretaceous, Nigeria, BM(NH) C.90409, showing approximately 15 pseudosutures between two successive sutures. The arrow indicates the adoral direction. (After Zaborski, 1986, Text-Fig. 4.)

estimates of the periodicity of pseudosutures). For example, Hewitt *et al.* (1991) observed that pseudosutures in Late Cretaceous vascoceratids are evenly spaced along the venter, suggesting a constant rate of translocation. In contrast, Lominadzé *et al.* (1993) noted that pseudosutures in phylloceratids and lytoceratids are more closely spaced near the actual septum, indicating a decrease in the rate of translocation at that point. However, both of these inferences rely on the untested assumption that the formation of pseudosutures was periodic.

5.6. Episodic Growth

Many of the fluctuations in the rate of growth described above are related to secretion of portions of a whorl as distinct units. We refer to this type of growth as episodic. Arkell *et al.* (1957) used the term "segmental growth" to describe this same phenomenon. (Note, however, that this term is also used in ontogenetic studies of arthropods, but in a completely different sense.) It is important to bear in mind that such whorl "segments" did not necessarily form with any particular temporal regularity. Secretion of any one portion of a whorl may have been rapid and may have been followed by a long interval of no apertural growth except perhaps for a thickening of the apertural lip.

One example in which episodic growth may have occurred was in rib formation. On the basis of a study of growth lines and ribs in Jurassic and Cretaceous ammonoids, Checa and Westermann (1989, Fig. 1) postulated that, in the formation of branching ribs, secretion proceeded by adding alternately wedge-shaped and parallel-sided whorl segments. Checa (1994) also analyzed the microscopic sculptural features on the shell surfaces of these ammonoids and, on the basis of these new data, argued that the formation of an entire rib may have occurred during a single secretory event. According to him, the free edge of the mantle initially extended completely beyond the apertural margin to the position of a new rib while, at the same time, secreting a periostracum. Subsequently, the mantle retracted, producing a rib-like fold in the periostracal layer (which was presumably attached to the mantle). Calcification of this fold occurred just before or at the time of initiation of a new episode of mantle extension. This reconstruction is supported by the fact that ribs are sometimes bound, at least on one side, by megastriae, indicating pauses in secretion.

Constrictions, along with their accompanying varices, also mark off distinct portions of a whorl, implying "segmental" or episodic growth (Arkell *et al.*, 1957). The discordant nature of constrictions, noted earlier, as well as the fact that megastriae occur on the adoral sides of some constrictions (Maeda, 1993; Checa, 1994) are consistent with this notion. Seilacher and Gunji (1993) interpreted the process of shell secretion leading to the formation of constrictions as representing "iterative countdown cycles" superimposed on the overall growth pattern. In addition, Checa (1994) proposed that the whorl "segments" bounded by constrictions may have formed as single units, implying an episodic pattern of shell formation comparable to that in many gastropods (Mackenzie, 1960; Linsley and Javidpour, 1980; Vermeij, 1993). However, the relative time of formation of the whorl "segment" between constrictions versus that of the constriction itself is unknown. This relationship may have been similar to that in many gastropods, in which the time interval between the formation of varices is much shorter than the time of formation of the varix itself (Seilacher and Gunji, 1993; Vermeij, 1993).

This process of "segmental growth" may have necessitated occasional readjustments in the shape of the shell. These readjustments seem to have occurred at the sites of constrictions and megastriae. As we noted earlier, these features demarcate angular discontinuities that affect growth lines and ribs, thereby implying changes in the trajectory of shell secretion. Such readjustments served to maintain the shape of the shell and the orientation of the apertural margin throughout ontogeny (Seilacher and Gunji, 1993, Fig. 2; Checa, 1994). In some taxa, these readjustments in shell shape were pronounced, resulting in a polygonal outline (e.g., Early Jurassic *Lytoceras cornucopiae* and Late Devonian *Wocklumeria sphaeroides*). The magnitude of the readjustment appears to have covaried with the angular distance between constrictions (Checa, 1994) or megastriae.

The pattern of growth in heteromorph ammonoids was similar, in some respects, to "segmental growth." In these forms, there were marked changes in the shape of the shell during ontogeny. For example, Okamoto (1993; Chapter 8, this volume) has described changeovers from one mode of coiling to another in Late Cretaceous *Nipponites* and related genera. Seilacher and Gunji (1993) speculated that these changeovers in mode of coiling occurred rapidly and were followed by long intervals of slower growth. They described this pattern as consisting of a series of morphogenetic countdowns leading to iterative changes in shell shape.

6. Environmental Control on the Rate of Growth

Many environmental factors affect the rate of growth of marine organisms, including temperature, food availability, light intensity, pressure, dissolved oxygen, day length, and the abundance and kind of predators. As an introduction to this section, we present several examples of the effects of the environment on the rate of growth of modern cephalopods.

Temperature is one of the main factors affecting the rate of growth of cephalopods. For example, specimens of *S. officinalis* from the English Channel experience a lower temperature during the winter months when they migrate to an offshore habitat (Boletzky, 1983). This decrease in temperature is reflected in a deceleration in the rate of growth, a reduction in septal spacing, and an increase in the time of chamber formation (Hewitt and Stait, 1988). Food is another important factor affecting growth. Wiedmann and Boletzky (1982) demonstrated that lack of food during the life cycle of *S. officinalis* results in lower rates of growth and closer septal spacing. Indeed, temperature and food are interrelated because feeding rate is usually a function of temperature (Boletzky, 1983; Mangold, 1983; Hewitt and Stait, 1988).

The developmental program in many cephalopods is extremely plastic and can be modified easily by the environment, leading to variation in the age and size at maturity within a single species (Mangold, 1987). For example, certain environmental conditions (e.g., undernutrition, short day length) tend to promote early maturation at a small size whereas others (e.g., long day length, high light intensity) tend to promote late maturation at a large size (Mangold, 1987, table, p. 189). The outcome also depends on when during the life cycle the particular conditions prevail, that is, before or after the gonads attain the size at which they are receptive to the growth hormone (Richard, 1970; Mesnil, 1977). As a result, individuals within the same population that experience different environmental conditions (perhaps during migrations) at different points in their ontogeny mature at different ages and sizes (Wells and Wells, 1977; Zuev, 1975). One example of this phenomenon in ectocochliate cephalopods is shown by a study of *Nautilus* in an aquarium. In this study, several specimens achieved maturity at one-half their normal size in approximately 3 years as compared to an estimated age at maturity of 10 years in nature (Ward, 1985). This case of precocious maturation probably resulted from a favorable combination of low water pressure (because of the shallowness of the aquarium) and abundant food (provided by the aquarist).

Many of these same environmental factors (temperature, nutrition) may also have influenced the rate of growth of ammonoids. Ammonoids, in comparison with *Nautilus*, show much more variation in septal spacing. Some of this variation has been attributed to seasonality, with episodes of closer septal spacing interpreted as slowdowns in growth associated with colder temperatures, etc. during the winter (Westermann, 1971; Zakharov, 1977; Kulicki, 1974). Other variations in septal spacing may have been caused by changes in habitat during ontogeny. Hewitt (Chapter 10, this volume) and Westermann (Chapter 16, this volume) have argued that animals at different growth stages may have lived in different environments. In addition, some species may have undergone seasonal onshore–offshore migrations (Morton, 1988) as well as daily vertical migrations (Westermann, 1990). These changes in habitat may also have affected the rate of growth and been reflected in septal spacing.

Injuries to the individual animal clearly reduced the rate of growth. Injuries in ammonoids are identified by scars on the shell surface and, sometimes, in addition, by a reduction in septal spacing (septal approximation) at an angular distance of approximately one body chamber length adapical of the scar (Chapter 15, this volume; Lehmann, 1966; Bayer, 1977; Landman and Waage, 1986). Experimental observations on *Nautilus* help explain why septal approximation is associated with an injury. During the repair of an injury in *Nautilus*, the rate of cameral liquid removal from the most recently formed chamber decreases, or this process stops altogether, effectively allowing liquid to flow back into the chamber (Ward and Greenwald, 1982). As a result, there is a delay in the formation of the next septum, which is eventually secreted at a reduced distance from the preceding septum (Ward, 1985). The resultant septal approximation appears to be a result of (1) the cessation of forward movement of the soft body coincident with the injury, (2) the interruption in the normal cycle of chamber formation, and (3) the need for additional weight to counteract the effect of positive buoyancy resulting from the breakage of shell material at the apertural margin.

Hydrostatic pressure, which covaries with water depth, must also have affected ammonoid growth rates. In *Nautilus*, there is a strong negative correlation between the rate of cameral liquid removal, thus the rate of growth, and hydrostatic pressure (Ward, 1982). We can therefore infer that ammonoids that lived in deep water grew more slowly than those that lived in shallow water. In addition, many deep-water ammonoids (some lytoceratids, phylloceratids, and desmoceratids) show thicker septa and narrower but thicker-walled siphuncles than do many shallow-water forms, implying additional differences in rates of cameral liquid removal and septal secretion and, therefore, rate of growth (Doguzhaeva, 1988; Westermann, 1990; Tanabe *et al.*, 1993). Moreover, in some deep-water ammonoids, the development of long septal necks further decreased the ratio of siphuncular surface area to chamber volume through ontogeny, thus depressing even more the rate of cameral liquid removal (Tanabe *et al.*, 1993).

Several studies have also suggested that environmental factors may have controlled the size at maturity within an ammonoid species. Matyja (1986) reported that in some Jurassic ammonoids there are three adult morphs, which are distinguished primarily on the basis of their size. He argued that the size at which maturity occurred depended on the particular environmental conditions. In a study of Early Jurassic liparoceratids from England, Hewitt and Hurst (1977) documented changes in the size of adults over geological time. They attributed these changes to fluctuations in climatic conditions, which, according to them, also affected feeding rates. Elmi and Benshili (1987) reported differences in adult size among populations of a single species of Early Jurassic *Hildoceras* from Northwest Europe and North Africa. They related these differences in size to differences in environmental conditions, namely water depth, oxygen content, and abundance of nutrients. Mignot *et al.* (1993) further suggested that the small adult size of this species in some areas was an adaptive response (paedomorphism) to a less than optimal environment. Mancini (1978) suggested this same interpretation to explain a fauna of dwarfed ammonoids from the Lower Cretaceous of Texas. In contrast, Stevens (1988), in analogy with what we know about giant squids, speculated that gigantism in ammonoids was related to life in cold deep-water environments. Such animals may have grown slowly but may have lived for a long period of time. Along these same lines, Kemper and Wiedenroth (1987) observed that in the Lower Cretaceous of Northwest Germany, ammonoids that lived in the Boreal Realm tended to be larger than those that lived in the warmer water of the Tethyan Realm. (See also Landman and Waage, 1993, p. 230, for an additional example of variation in adult size within a single species, possibly related to differences in the environment.)

Environmental factors may also have affected the shape of the ammonoid shell, notably how robust it was (Chapter 10, this volume). Such variation in shape as a result of environmental conditions is common in gastropods (Vermeij, 1980, 1993). In ammonoids, this may have produced a wide range of intraspecific variation (Kennedy and Cobban, 1976; Callomon, 1985; Landman and Waage, 1993). However, in some instances at least, the degree of whorl compression appears to have been controlled by selection for hydrodynamic efficiency (Jacobs *et al.*, 1994; Chapter 7, this volume).

If ammonoids showed as much developmental plasticity in response to environmental conditions as do modern cephalopods, this would explain the high incidence of heterochrony in ammonoid evolution (Landman, 1988). Heterochrony depends on the dissociablility of three processes: size increase, morphological differentiation, and maturation (Gould, 1977). If the environment affected this dissociability in ammonoids, it could have produced an acceleration or retardation in the program of growth. Although the literature on heterochrony in ammonoids is enormous, we present one example that bears on the question of the time of maturation. There are a number of scaphitid species in the Upper Cretaceous of North America that develop all of the morphological features characteristic of maturity at a diameter of only approximately 10 mm (Cobban, 1951). Landman (1989) interpreted these forms as progenetic offshoots of larger cooccurring species. It is unclear if the environment was responsible for this acceleration in maturation. However, what is interesting about this study is that the formation of the mature body chamber in these progenetic species began exactly at the end of the neanic stage. This may indicate the earliest point in ontogeny when maturation could have occurred with all of the ensuing implications about the time of initiation of gonad development.

7. Determination of the Actual Rate of Growth

Now that we have described the overall growth curve and the fluctuations in it caused by internal and external factors, we are ready to introduce actual numbers into our discussion. There are several ways to determine the actual rate of ammonoid growth, each of which is described below. Following the description of these methods, we synthesize the data and provide an estimate of the age at maturity of various taxa (Table I).

7.1. Assumptions about the Periodicity of Shell Secretion

It has commonly been assumed that ammonoid lirae, ribs, and septa formed with a constant periodicity (see general discussions in Checa, 1987, and Dommergues, 1988). Specific estimates of the time of septal formation range from 1 day to 1 month. For example, Ivanov (1971) suggested that ammonoid septa formed every lunar month. (See Kahn and Pompea, 1978, for a similar assumption about the time of septal formation in nautilids.) Doguzhaeva (1982) hypothesized that ammonoid septa formed with a fortnightly periodicity, yielding an estimated age at maturity of 2.5 to 4.5 years, depending on the number of septa present (Table I). Weitschat and Bandel (1991) suggested that septa formed every 1 to 2 days, similar to that in *Sepia*, and estimated an age at maturity in ammonoids of 1 to 2 years (Table I).

The time of formation of other morphological features has also generally been assumed to have been constant. Doguzhaeva (1982) suggested that in some ammonoids the formation of lirae occurred daily. Hirano (1981) hypothesized that constrictions in Late Cretaceous *Gaudryceras denseplicatum* formed every year and calculated an age at maturity of 20 years for this species (Table I). Hewitt *et al.* (1991) assumed that pseudosutures in Late Cretaceous *Vascoceras* formed diurnally or semidiurnally and estimated an age at maturity of 6 years for species within this genus (Table I; see also Seilacher, 1988). Although they did not postulate a specific time of pseudosuture formation, Lominadzé *et al.* (1993) nevertheless argued that, on the basis of the number of pseudosutures per chamber, the time of chamber formation was longer in phylloceratids and lytoceratids than in other ammonoids.

All of these assumptions about periodicity of shell secretion are unconfirmed. As discussed earlier (Section 4), the time of septal formation in ammonoids probably increased exponentially through most of ontogeny. Therefore, it is not possible to multiply the number of septa by some constant

Table I. Estimate of Age at Maturity of Various Ammonoids Based on the Methods Described in the Text

Method	Source	Taxon	Adult size/number of whorls[a]	Age at maturity (years)
Assumption of periodicity (septa)	Doguzhaeva (1982)	Ammonoidea	6 whorls	2.5–3.6
			7 whorls	3.4–3.7
			8 whorls	3.8–4.5
Assumption of periodicity (septa)	Weitschat and Bandel (1991)	Ammonoidea	—	1–2
Assumption of periodicity (constrictions)	Hirano (1981)	*Gaudryceras denseplicatum*	140 mm/7 whorls	20
Assumption of periodicity (pseudosutures)	Hewitt *et al.* (1991)	*Vascoceras* spp.	100 mm	6
Seasonality (shell volume)	Trueman (1941)	*Dactylioceras commune*	50 mm	4
Seasonality (rate of whorl expansion)	Westermann (1971)	*Hammatoceras insigne*	70 mm/4.5 whorls[b]	4–6
		Paracravenoceras ozarkense	25 mm/8.5 whorls[b]	4–6
Seasonality (septal spacing)	Westermann (1971)	*Leioceras* spp.	50 mm	4–6
		Ludwigia spp.	70 mm	4–6
		Sonninia spp.	100 mm	4–6
Seasonality (septal spacing)	Zakharov (1977)	*Pinacoceras* aff. *regiforme*	40 mm[b]	7
Seasonality (septal spacing)	Kulicki (1974)	*Quenstedtoceras* spp.	50 mm/5.5 whorls (♂)	2
			130 mm/7.5 whorls (♀)	3
Seasonality (jaw increments)	Hewitt *et al.* (1993)	Aspidoceratids	300 mm	5–6
		Oppeliids (♂,♀)	50 mm	1
Seasonality (oxygen isotopes)	Jordan and Stahl (1970)	*Staufenia staufensis*	100 mm[b]	4[c]
		Quenstedtoceras sp.	80 mm[b]	5[c]
Size classes	Trueman (1941)	*Promicroceras marstonense*	25 mm	5
Size classes	Landman and Klofak (in prep.)	*Hoploscaphites nicolletii* (♂)	50 mm	6
Size classes	This chapter	*Schreyerites* n. sp.	55 mm/6 whorls	6
Epizoans (*Serpula*)	Schindewolf (1934)	*Arietites* cf. *kridion*	40 mm[b]	3.5[c]
Epizoans (*Nanogyra nana*)	Hirano (1981)	*Leioceras opalinum*	60 mm	1–7.5[c]
Epizoans (*Liostrea*)	Merkt (1966)	*Euagassiceras* sp.	30 mm[b]	3–4.5[c]
Epizoans (*Placunopsis ostracina*)	Meischner (1968)	*Ceratites semipartitus*	370 mm[b]	7[c]

[a]These values represent gross estimates based on many specimens unless the author provided size data for a single specimen.
[b]Data based on a single specimen.
[c]The age at maturity is calculated on the basis of the time of formation of a certain number of whorls at the end of ontogeny, as cited in this chapter, and the observed rate of whorl expansion (Raup's *W*), using equation 10 in the Appendix. This method assumes a constant rate of growth of the venter at the apertural margin. Because the rate of growth is not constant throughout ontogeny and, notably, decreases at maturity, this calculation yields only an estimate. Values are approximated to the nearest half-year.

value to obtain an absolute age at maturity (Landman, 1983, 1986). Similarly, there is no evidence to suggest that the time interval between the formation of successive lirae, ribs, or constrictions was constant during ontogeny, especially if, as supposed in some ammonoids, growth at the apertural margin was episodic. The spacing of such features on the shell was probably controlled by the growth program, although it may have been subject to some environmental influence.

On the basis of a study of the spacing of lirae and septa in various ammonoids, Doguzhaeva (1982) documented that in some ammonoids a nearly constant number of lirae on the outer shell corresponded in time of formation with that of a chamber at the back of the body. However, this does not imply a constant period of chamber formation. If anything, the time interval between the formation of successive lirae probably increased during ontogeny, implying a concomitant increase in the time of chamber formation.

7.2. Detection of Seasonal Signals in Morphology

There have been several attempts to identify seasonal signals in the morphology of ammonoid shells and jaws. These studies presume that the environment was seasonal, that the rate of ammonoid growth was fast enough to record seasonal changes, and that one can differentiate seasonal signals from intrinsic fluctuations in the rate of growth.

Trueman (1941, Fig. 5) noted ontogenetic changes in shell volume in several specimens of Early Jurassic *Dactylioceras commune* at shell diameters of 12–15 mm, 25 mm, and 40–50 mm. He speculated that these changes may have been seasonal in origin, suggesting an age at maturity of approximately 4 years for this species (Table I). Westermann (1971) observed ontogenetic fluctuations in the rate of whorl expansion in a specimen of Early Jurassic *Hammatoceras insigne* and in a specimen of Pennsylvanian *Paracravenoceras ozarkense* (illustrated in Raup and Chamberlain, 1967, Fig. 2, and Raup, 1967, Fig. 10, respectively). He also interpreted these fluctuations as seasonal, suggesting an age at maturity of 4 to 6 years in these ammonoids (Table I).

Studies of septal spacing have also provided evidence of seasonality. Westermann (1971) identified four to six possible annual cycles in graphs of septal spacing in several species of Middle Jurassic *Leioceras*, *Ludwigia*, and *Sonninia* (Table I) (see Rieber, 1963, Figs. 9, 10, and Oechsle, 1958, Fig. 7, for illustrations of the original graphs). Zakharov (1977, Fig. 3) recognized six episodes of reduced septal spacing in a specimen of Late Triassic *Pinacoceras* aff. *regiforme* with about 80 septa. He therefore estimated an age at maturity of 7 years in this species (Table I). Kulicki (1974, Fig. 8) investigated the pattern of septal spacing in several specimens of Middle Jurassic *Quenstedtoceras*. He (p. 218) argued that "the termination of the last sexual cycle," as indicated by a decrease in septal spacing at the end of ontogeny, occurred in the autumn or winter. On the basis of this assumption, he concluded that microconchs (males?) reached maturity in 2 years, and macroconchs (females?) in 3 years (Table I).

In a study of growth increments on two specimens of *Laevaptychus*, here interpreted as lower jaws, Hewitt *et al.* (1993) hypothesized that fluctuations in increment width reflected seasonal changes. As we have already observed, a plot of increment width versus increment number in these specimens shows an increase in width in early ontogeny followed by a decrease in later ontogeny (Fig. 13D). However, superimposed on this overall ontogenetic pattern are several episodes of reduced increment width, which Hewitt *et al.* interpreted as marking winter events (indicated by arrows on Fig. 13D). There is some uncertainty as to the exact number of such events, although the best estimate seems to be five or six. (One complication is that the episode of reduced increment width at the end of ontogeny may simply be related to a deceleration in the rate of growth associated with maturity). Hewitt *et al.* (1993) concluded that the ammonoids in which these jaws once occurred, probably Late Jurassic and Early Cretaceous aspidoceratids, reached maturity in 5 to 6 years.

On the basis of this figure, Hewitt *et al.* (1993) also calculated the rate at which these growth increments formed. Using this value, they determined the age of several specimens of *Lamellaptychus*, here also interpreted as lower jaws, for which they had already counted the number of growth increments (Fig. 13C). They estimated that the ammonoids in which these jaws once occurred, most probably Late Jurassic oppeliids, attained maturity in approximately 1 year.

7.3. Detection of Seasonal Signals in Isotopic Data

A number of studies have investigated the record of oxygen isotopes in ammonoid shells to learn about the rate of growth. This method assumes that the growing shell recorded fluctuations in water temperature that were related to seasonal changes in the environment. However, the isotopic composition of biogenic $CaCO_3$ depends on a number of factors other than the temperature during secretion, including the isotopic composition of the ambient water (Grossman and Ku, 1986; Geary *et al.*, 1992), species-specific effects related to the growth of the particular organism (McConnaughey, 1989a,b), and the state of preservation of the sampled shell material (Stahl and Jordan, 1969; Jordan and Stahl, 1970; Buchart and Weiner, 1981). These factors influence the extent to which the isotopic data faithfully reproduce a seasonal temperature signal.

Oxygen isotopic analyses have been performed on several well-preserved specimens of *Baculites* from the Upper Cretaceous of North America (Tourtelot and Rye, 1969; Forester *et al.*, 1977; Rye and Sommer, 1980; Whittaker *et al.*, 1987). Samples taken along the shell length revealed cyclic fluctuations

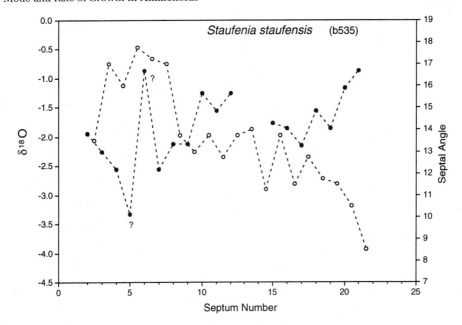

FIGURE 16. Plots of $\delta^{18}O$ (in parts per thousand) of septa (filled circles) and of the angular spacing between septa (in degrees; open circles) versus septum number in a mature specimen of Middle Jurassic *Staufenia staufensis* from Germany. (Values of $\delta^{18}O$ are from Jordan and Stahl, 1970, p. 48; values of the septal angle were measured from their Fig. 5A; the specimen is referred to there as b535.) These authors questioned the $\delta^{18}O$ values of septa 5 and 6 because of the small size of the samples; they did not analyze septa 1, 13, and 14 for the same reason. The angle between two successive septa is plotted at the midpoint between their respective septum numbers. Septa are numbered starting from the last, most recently formed septum (1).

in $\delta^{18}O$, but there are too few data points to permit an interpretation in terms of rate of growth.

The most complete isotopic data set for determination of the rate of growth of ammonoids appears in a study by Jordan and Stahl (1970). These authors measured the oxygen isotopic composition of the seven most recently formed septa of an adult specimen of Middle Jurassic *Quenstedtoceras* sp. and the 21 most recently formed septa of an adult specimen of Middle Jurassic *Staufenia staufensis*. These authors interpreted the isotopic values in both specimens as reflecting seasonal temperature signals. Based on this interpretation, they inferred that the specimen of *Quenstedtoceras* sp. formed five septa (0.3 whorls) in 1 year, that the specimen of *S. staufensis* formed 17 septa (0.6 whorls) in 1.5 years, and that septa formed at a constant, albeit different, periodicity in each specimen (Table I).

If one accepts that the cyclic pattern in the *S. staufensis* data is seasonal in origin, it is interesting to investigate the relationship between septal spacing and the values of $\delta^{18}O$. The spacing between septa shows a slight decrease at the end of ontogeny (Jordan and Stahl, 1970, Fig. 5A), which is associated with an increase in $\delta^{18}O$ (Fig. 16). If septal spacing provides a measure of variation in rate of growth, this negative correlation between septal spacing and $\delta^{18}O$ suggests that a decrease in the rate of growth corresponded to a decrease in temperature, which is consistent with the temperature–growth rate relationship observed in *Sepia* (see Boletzky, 1983). However, the decline in septal spacing in *S. staufensis* corresponds to the attainment of maturity and may not be related to the environment.

There is, however, some uncertainty whether the isotopic data record a seasonal signal. Theoretically, as long as the time of septal formation is rapid, that is, considerably shorter than a year, the isotopic composition of a septum will accurately reflect the temperature at a discrete point in time. As the time of septal formation becomes longer, however, the isotopic composition of a septum will reflect a temperature value that is averaged over a longer and longer time interval (see Section 7.6). As a result, the isotopic pattern will represent a distorted version of the original signal (Fig. 17; Landman, 1987). This may apply to the isotopic pattern in *S. staufensis* because the data were derived from septa in the submature to mature stages of ontogeny when the

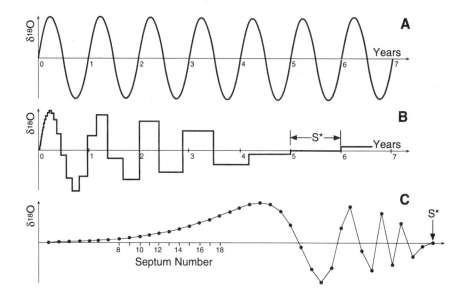

FIGURE 17. Pattern produced by nonperiodic sampling of a seasonal signal of $\delta^{18}O$. (A) Idealized seasonal signal of $\delta^{18}O$ shown over time (years). (B) Distortion of the original signal resulting from increasingly longer intervals of septal formation; same axes as in A. Each horizontal step represents the average value of the signal during the time of formation of a single septum. In this simulation, we assumed 41 septa and an exponential increase in the time of septal formation. Only a single septum (S*) is formed during the sixth year. (C) Plot of $\delta^{18}O$ versus septum number. The signal is dilated for rapidly formed septa (left side) and contracted for slowly formed septa (right side). The value of $\delta^{18}O$ for any particular septum is the same as its value plotted in B.

time of septal formation was relatively long. If this is the case, the isotopic pattern in this specimen does not faithfully record a seasonal signal and is much more difficult to interpret in terms of time.

Another explanation for the pattern of $\delta^{18}O$ in *S. staufensis* is that it reflects some sort of migrational behavior (Kennedy and Cobban, 1976). It is well established that the $\delta^{18}O$ values of *Nautilus* septa record a change in temperature after hatching, reflecting a migration to colder, deeper water at this point (Oba *et al.*, 1992; Landman et al., 1994). If the isotopic pattern in *S. staufensis* results from migration, there is little hope of deriving any information from it about rate of growth, unless the migrations themselves corresponded to seasons.

7.4. Size Classes

The use of size classes to determine the rate of growth of ammonoids is possible if the particular ammonoid species bred (and spawned) at a definite time of the year and if this periodicity resulted in distinct age classes or cohorts (Rounsefell and Everhart, 1953). In a census population, as represented in the fossil record by a single mass-mortality event (Deevey, 1947; Fagerstrom, 1964; Hutchinson, 1978; Brett and Seilacher, 1991; Kidwell and Bosence, 1991), such behavior ideally produces a polymodal size–frequency histogram in which the succession of modes corresponds to the succession of year classes (see, for example, Boucot, 1953; Sheldon, 1965; Speden, 1970; Surlyk, 1972; Richards and Bambach, 1975).

Trueman (1941, p. 367) identified size classes in a sample of Early Jurassic *Promicroceras marstonense* from England. These size classes ranged in diameter from 1.5 mm (ammonitellas) to 25 mm (adults). If these size classes represented the young of five successive annual breeding seasons, then this species reached maturity in 5 years (Table I). Trueman (1941, p. 367) also identified size classes in a sample of Early Jurassic *Arnioceras* sp. from England, but the largest individuals were only 7 mm in shell diameter, which precluded an estimate of the age at maturity.

We have obtained preliminary results from analysis of samples of *Hoploscaphites nicolletii* from the Upper Cretaceous of South Dakota (Landman and Klofak, in prep.). Juveniles and adults of this species are sometimes preserved together in concretions, which have been interpreted as resulting from mass mortality events followed by rapid burial (Waage, 1968; Landman and Waage, 1993). We measured the shell diameters of all the specimens in each of three concretions, totaling more than 1000 specimens, and identified what appear to be the same six size classes in each concretion, ranging in shell diameter from 5 mm (neanoconchs) to 50 mm (adults). If it had an annual breeding cycle, this species may therefore have reached maturity in 6 years (Table I). A comparable number of size classes also have been observed in a study of Middle Triassic *Schreyerites* n. sp. from the Fossil Hill Formation of Nevada

(Table I). This sample consisted of several hundred individuals and probably represented a census population (H. Bucher, unpublished data).

These size-class studies clearly rely on a number of assumptions. Breeding (and spawning) is assumed to have been periodic, usually annual, and the duration of the breeding (and spawning) season is assumed to have been relatively short. In present-day cephalopods, however, the period of spawning is variable, ranging from seasonal to year-round, depending on the species and its environment (Mangold, 1987, Table 9.7). In addition, these studies assume that all ontogenetic stages were present in the population sample, although it is possible that animals at different ontogenetic stages lived in different environments (Chapter 16, this volume). Given these and other assumptions, it is probably best to regard age estimates from these studies as approximations.

7.5. Epizoans

The rate of ammonoid growth can also be estimated by studying epizoans that grew on the shell of an ammonoid while the ammonoid was still alive. This method depends on working out the relationship between the rate of growth of the epizoan and that of the ammonoid. In general, we assume that the rate of growth of the epizoan species was identical to that of the same species today, or of a closely related living form if the epizoan species is extinct. If performed properly, this method can provide reliable estimates of the rate of ammonoid growth.

Several examples have been described of Early Jurassic ammonoids encrusted by serpulids (Lange, 1932; Schindewolf, 1934; Merkt, 1966; Seilacher, 1982). The serpulids grew along the venter of the ammonoid, supposedly in synchronism with the ammonoid. To the extent that this was true, one can calculate the rate of growth of the ammonoid on the basis of the growth rate of the serpulid. Schindewolf (1934) used this approach to estimate the time of formation of the last whorl of a small adult specimen of *Arietites* encrusted by a serpulid. On the basis of the rate of growth of several species of modern serpulids, he calculated a range in the time of formation of this whorl from 0.5 to 3.0 years. We favor the midway figure of 1.5 years because this is based on the rate of growth of the modern species of *Serpula* that is most closely related to the fossil form (Table I).

Ammonoids encrusted by bivalves provide additional evidence of the rate of ammonoid growth. Hirano (1981) described a specimen of Middle Jurassic *Leioceras opalinum* encrusted by the oyster *Nanogyra nana*. On the basis of the age of the oyster and where it settled relative to the apertural margin of the ammonoid, he estimated that the last whorl of this ammonoid formed in 0.5 to 4.0 years (Table I). Merkt (1966) described an adult specimen of Early Jurassic *Euagassiceras* sp. encrusted by the oyster *Liostrea* sp. He argued that in the last 2 to 3 years of the oyster's life it attained a size large enough to affect the formation of the last 1.5 whorls of the ammonoid, as indicated by a marked whorl asymmetry. This suggests that these 1.5 whorls formed in 2 to 3 years (Table I).

Meischner (1968) described a complex case of encrustation of an adult specimen of Middle Triassic *Ceratites semipartitus* by the bivalve *Placunopsis ostracina*. The bivalves occur on approximately the last two-thirds of the final whorl. Meischner hypothesized that the bivalves represented six annual spatfalls on the basis of their orientation, size, and distribution on the ammonoid shell and the probable orientation of the ammonoid during life. According to this author, four of the spatfalls settled on the ammonoid while it was still alive, implying that the last two-thirds of the final whorl formed in about 3 years (Table I). The other two spatfalls settled after the ammonoid died, suggesting that this individual did not live more than 1 year after reaching maturity.

Seilacher (1960) also presented an estimate of the number of years between the attainment of maturity and death in an adult specimen of Late Cretaceous *Buchiceras bilobatum* encrusted by *Ostrea* sp. The orientation and distribution of the oysters suggest that they may have grown after the ammonoid was mature but before its death (but see Heptonstall, 1970, and Westermann, 1971, for another interpretation). On the basis of the age of the oysters, this specimen may have lived 1/2 to 3 years after reaching maturity.

7.6. Estimates of the Age at Maturity

Estimates of the age at maturity of various ammonoids, derived from the methods discussed above, are presented in Table I (see Chapter 11, this volume, for the duration of the embryonic stage). Most of the species listed are Mesozoic except for *Paracravenoceras ozarkense*, which is Pennsylvanian. In addition, most of the species described lived in shallow water, with the possible exception of the aspidoceratids and *Gaudryceras denseplicatum*. The estimates of the age at maturity vary in their reliability, depending on the method used. The most reliable estimates are probably those derived from studies of epizoans. The least reliable estimates are probably those derived from studies assuming periodic shell secretion because this method lacks an independent means of determining time.

Many of these studies provide an estimate of the time of formation of only one or two whorls at the end of ontogeny. In order to calculate the age at maturity in these cases, a constant rate of growth of the venter at the apertural margin throughout ontogeny was assumed. We used the value of the time of formation of the portion of the whorl given in a particular study and, on the basis of the whorl expansion rate of the shell (Raup's W, measured from photos or actual specimens), calculated the age at maturity according to equation 10 in the Appendix. Because the rate of growth was not constant throughout ontogeny and, notably, decelerated at the onset of maturity, this calculation only yields an estimate. To obtain a more precise result, one would have to take into account variation in the rate of growth throughout ontogeny, but, inasmuch as the data are so sparse, only a general estimate is practical now.

The age at maturity of the ammonoids listed in Table I averages about 7 years or less. For example, studies of epizoans yield estimates ranging from 1 to 7.5 years. Studies based on identifying a seasonal signal in patterns of septal spacing yield estimates ranging from 2 to 7 years. The highest estimate of the age at maturity (20 years) is given for *Gaudryceras denseplicatum*. This species lived in an offshore environment (Tanabe, 1979) in which the rate of growth may well have been low. However, this estimate is based on the unproven assumption that constrictions formed with an annual periodicity and is probably unreliable. There are two estimates of the age at maturity of *Quenstedtoceras* spp. based on two different methods, septal spacing and oxygen isotopes. The two estimates are comparable, ranging from 2 to 5 years.

A few other estimates of the age at maturity of ammonoids have been published in the literature and are similar to those in Table I. Westermann (1990, 1992; Chapter 16, this volume) suggested that small shallow-water ammonoids reached maturity in 1 to 2 years, whereas midshelf forms of average size reached maturity in 3 to 5 years. In addition, Ward (1982, 1992) argued that the rate of growth of shallow-water ammonoids was higher than that of *Nautilus* on the basis of the value of the ratio of siphuncular area to chamber volume through ontogeny. (Note, however, that the rate of growth of some Paleozoic nautiloids may have been more rapid than that of *Nautilus*: Hewitt and Watkins, 1980; Hewitt and Hurst, 1983). There is also a general consensus that deep-water ammonoids reached maturity later than shallow-water forms. Westermann (1990, 1992) estimated that the age at maturity of large specimens of *Lytoceras* may have equaled as much as 50 years.

If 5 years is accepted as a reasonable estimate of the age at maturity of a shallow-water ammonoid, the duration of each growth stage in ontogeny can be calculated using equation 10 in the Appendix, assuming that the rate of growth was constant throughout ontogeny. For example, in an ammonoid with W of 2, the neanic stage, represented by the first two postembryonic whorls, would have lasted about 0.5 years, the juvenile stage, represented by the next two whorls, about 2 years, and the submature to mature stage, represented by the last whorl, about 2.5 years. The time of chamber formation in this ammonoid can also be calculated using equation 12 in the Appendix, assuming that the time of chamber formation increased exponentially through ontogeny. For example, if there were 90 chambers in this ammonoid, and the time of formation of the first postembryonic chamber is assumed to have been 1 day based on estimates of the time of chamber formation in *Sepia*, then the time of formation of the last chamber (90) would have been about 95 days. If there were only 70 chambers in this ammonoid, then the time of formation of the last chamber would have been about 135 days. However, it is important to remember that all of these estimates assume a constant rate of growth.

Table I indicates differences in the age at maturity of dimorphs within the same species ("antidimorphs" of Davis, 1972). In ammonoids, microconchs are generally smaller and have fewer whorls than do macroconchs (Makowski, 1971). Kulicki (1974) suggested that microconchs of Middle Jurassic *Quenstedtoceras* reached maturity earlier than did macroconchs (see also Guex, 1970; Makowski, 1962, 1971; Westermann, 1990). This is consistent with observations in present-day cephalopods in which males tend to mature earlier and usually at a smaller size than do females (Mangold, 1987). In contrast, Hewitt *et al.* (1993) suggested that macroconchs of oppeliids reached maturity at the same age as the smaller microconchs because they grew more rapidly. Matyja (1986) has also pointed out that, in some species of Jurassic ammonoids, there are three adult morphs, each of a different size and, presumably, age.

There is little information about the length of time that ammonoids lived after reaching maturity. This may have varied widely. Studies of ammonoids encrusted by epizoans suggest that some species of Ceratitina and Ammonitina did not live longer than 1 to 2 years after reaching maturity (Meischner, 1968; Seilacher, 1960, respectively). This is consistent with speculations that the formation of the mature body chamber was designed for a single egg-laying episode followed by death. In contrast, Kulicki (1974) hypothesized that some macroconchs of Middle Jurassic *Quenstedtoceras* experienced two "sexual cycles" separated by a year. Westermann (1971) suggested that the pronounced modifications of the adult body chamber in many ammonoids indicate that these animals lived a long time after reaching maturity. Callomon (1963) made a similar suggestion on the basis of his observation that specimens of adult shells are far more common in the fossil record than those of immature shells, implying that the long duration of the adult stage increased the chances of fossilization. However, these deposits may equally reflect age segregation, postspawning mortality, migrations, or hydrologic accumulations (Chapter 16, this volume).

8. Summary

Changes in shell shape, ornamentation, septal spacing, and other morphological features in the ontogeny of ammonoids indicate the presence of three stages in postembryonic growth: a neanic stage immediately after hatching, a juvenile stage, and a mature stage. Growth was determinate in most ammonoids, and maturity occurred at a more or less prescribed size for each species.

The generalized growth curve shows that the rate of growth accelerated during the neanic stage, was nearly constant during the juvenile stage, and decelerated at the onset of maturity (Fig. 5). As in present-day *Nautilus*, the time of chamber formation probably increased more or less exponentially over most of ontogeny. Microconchs, with generally fewer whorls than macro-conchs, may have reached maturity earlier.

Observations of several morphological features support this generalized growth curve. The neanic stage was characterized by an interval of wide septal spacing. It also coincided, as shown in some ammonoids, with the formation of widely spaced increments on the lower jaw. These features both suggest that there was an acceleration in the rate of growth during this stage. In contrast, at the approach of maturity, the rate of growth decelerated, as shown by a thickening of the apertural margin and the approximation of the last few septa.

Superimposed on this generalized growth curve were fluctuations in the relative rate of growth, many of which were caused by intrinsic processes. Such fluctuations are indicated by the presence of ribs, constrictions, varices, and megastriae. Although many of these features are regularly spaced with respect to their angular distance on the shell, there are no data to suggest that they were periodic with respect to their time of formation. In some am-monoids, the growth of the shell was episodic, meaning that successive portions of a whorl were secreted as separate units.

In present-day cephalopods, the environment affects how quickly animals grow and how soon they reach maturity (Mangold, 1987). Environmental factors must also have affected the rate of growth of ammonoids and their age at maturity. These factors included light, temperature, pressure, dissolved oxygen, and food availability. Pressure (depth) may have played an especially important role because it controlled the rate at which cameral liquid was removed and, hence, the rate of growth.

As a result of intrinsic and extrinsic factors, the absolute rate of growth varied within and among species. Growth was rapid in most shallow-water ammonoids; they probably attained maturity in about 5 years. This is consis-tent with the fact that in many of these forms, the wall of the siphuncle was thin, permitting rapid removal of cameral liquid (Westermann, 1990). The growth rate of these ammonoids may have approached that of coleoids, most of which grow rapidly and reach maturity in 1 to 2 years (Mangold, 1983; Wells, 1983). In contrast, the rate of growth of deep-water ammonoids may have been more similar to that of *Nautilus*, which grows slowly and reaches maturity in 10–15 years (Saunders, 1983; Landman and Cochran, 1987).

The length of time an ammonoid lived after reaching maturity probably varied among taxa in accordance with whether reproduction was semelparous or iteroparous. The presence of distinctive monospecific assemblages consist-ing of nearly all adults in shallow-water deposits of the Mesozoic (Callomon, 1981; Landman and Waage, 1993) suggests that some species may have been semelparous and died soon after spawning. (However, there are many other explanations for these occurrences; see Chapter 16, this volume.) Deep-water forms, however, may have been iteroparous and may have lived for several years after reaching maturity.

ACKNOWLEDGMENTS. This research was supported by a Fellowship Grant from the Swiss National Science Foundation (No. 8220-030667) and a travel grant from the American Museum of Natural History to H. Bucher. J. Guex also acknowledges the support of the Swiss National Science Foundation (No. 2000-037327-93-1). Some of the material used in this study was generously made available to us by Tim White and Dr. Copeland MacClintock (both Yale Peabody Museum) and by Drs. Jean-Louis Dommergues (Université de Bourgogne) and Serge Elmi (Université Claude-Bernard). We thank Dr. M. Queyranne (UBC, Vancouver) for his help in simulating the effect of nonperiodic sampling of a seasonal signal. We also thank Drs. Roger Hewitt (McMaster University), Gerd Westermann (McMaster University), Antonio Checa (Universidad de Granada), David Jacobs (UCLA), and John Chamberlain (Brooklyn College), who read earlier drafts of this manuscript and made many, many helpful suggestions. Kathy Sarg prepared most of the figures in this manuscript, Peling Fong-Melville assisted in scanning electron microscopy, Andrew Modell prepared the photographs, and Stephanie Crooms (all AMNH) word-processed the manuscript.

Appendix

A.1. Calculation of the Age of an Ammonoid When the Time of Formation of the Last Whorl is Given

If the rate of growth along the venter of an ammonoid was constant through ontogeny, then

$$A = cL \tag{1}$$

where A is the age of the ammonoid, c is a constant expressed in units of time per unit length along the venter, and L is the total length along the venter. The ratio of the age of the whole ammonoid to the time of formation of a portion of the shell at the end of ontogeny (A_p) is expressed as

$$A / A_p = L / L_p \tag{2}$$

where L_p is the length along the venter of the portion of the shell at the end of ontogeny. Therefore,

$$A = \left(\frac{L}{L_p}\right) A_p \tag{3}$$

In a planispirally coiled ammonoid in which shell shape did not change during ontogeny, the length along the venter (L) is defined as

$$L = r\sqrt{(1/k^2) + 1} \tag{4}$$

where r is the radius of the spiral and k is the rate of spiral expansion (Thompson, 1917, p. 810; Hewitt, 1986, p. 338; Landman, 1987, p. 212). The ratio of the length along the venter of the whole ammonoid to that of a portion of the shell at the end of ontogeny is equal to

$$L/L_p = \frac{r_\theta}{r_\theta - r_{\theta-n\pi}} \tag{5}$$

where r_θ is the radius at angle θ, and $r_{\theta-n\pi}$ is the radius at angle $\theta - n\pi$, and n is some multiple of π (e.g., 2π radians = 1 whorl). Substituting

$$r = ae^{k\theta} \tag{6}$$

where a is the value of r when $\theta = 0$

$$\frac{L}{L_p} = \frac{ae^{k\theta}}{ae^{k\theta} - ae^{k(\theta-n\pi)}} = \frac{ae^{k\theta}}{ae^{k\theta}(1 - e^{-nk\pi})} = \frac{1}{1 - \dfrac{1}{e^{nk\pi}}} = \frac{e^{nk\pi}}{e^{nk\pi} - 1} \tag{7}$$

Substituting

$$W = e^{2k\pi} \tag{8}$$

where W is the rate of whorl expansion (Raup, 1967),

$$L/L_p = \frac{W^{n/2}}{W^{n/2} - 1} \tag{9}$$

Therefore,

$$A = \left(\frac{W^{n/2}}{W^{n/2} - 1}\right) A_p \tag{10}$$

If A_p is the time of formation of the last whorl ($n = 2$), and if $W = 2$, this expression becomes

$$A = \left(\frac{W}{W - 1}\right) A_p = 2A_p \tag{11}$$

In this case, the age of the whole ammonoid is equal to twice the time of formation of the last whorl. Because the rate of growth was probably not constant throughout ontogeny, and, notably, decreased at maturity, this calculation yields only an estimate.

A.2. Calculation of the Time of Chamber Formation When the Age of an Ammonoid Is Given

If the time of chamber formation increased exponentially through ontogeny, then

$$y = fe^{b(x-1)} \tag{12}$$

where y is the time of chamber formation, x is the chamber number, f is the time of formation of the first postembryonic chamber ($x = 1$), and b is a constant that is inversely proportional to the total number of chambers in the ammonoid.

The integral of this equation is

$$A = \frac{f}{b}\left[e^{b(x-1)} - 1 \right] \tag{13}$$

where A is the age of the ammonoid at any value of $x \geq 2$ (if $x =$ the chamber number of the last chamber = the total number of chambers in the ammonoid, then A is the age of the whole ammonoid). If A, f, and the total number of chambers are given, this equation can be solved for b. For example, if $A = 1865$ days (5 years), $f = 1$ day, and the total number of chambers = 90, then $b = 0.051$. Using this value of b, the time of chamber formation (y) can be calculated for any value of $1 \leq x \leq 90$. Note that this method yields only an estimate because, in fact, the rate of growth was not constant during ontogeny.

References

Arkell, W. J., Kummel, B., and Wright, C. W., 1957, Mesozoic Ammonoidea, in: *Treatise on Invertebrate Paleontology*, Part L, *Mollusca 4* (R. C. Moore, ed.), Geological Society of America and University of Kansas Press, Lawrence, KS, pp. 80–465.

Bandel, K., Landman, N. H., and Waage, K. M., 1982, Micro-ornament on early whorls of Mesozoic ammonites: Implications for early ontogeny, *J. Paleont.* 56(2):386–391.

Bayer, U., 1972a, Zur Ontogenie und Variabilität des jurassischen Ammoniten *Leioceras opalinum, N. Jb. Geol. Paläont. Abh.* 140:306–327.

Bayer, U., 1972b, Ontogenie der liassischen Ammonitengattung *Bifericeras, Paläontol. Z.* 46:225–241.

Bayer, U., 1977, Cephalopoden-Septen. Teil 2. Regelmechanismen im Gehäuse- und Septenbau der Ammoniten, *N. Jb. Geol. Paläont. Abh.* 155:162–215.

Birkelund, T., 1981, Ammonoid shell structure, in: *The Ammonoidea*, Systematics Association Spec. Vol. 18 (M. R. House and J. R. Senior, eds.), Academic Press, London, pp. 177–214.

Blind, W., and Jordan, R., 1979, "Septen-Gabelung" an einer *Dorsetensia romani* (Oppel) aus dem nordwestdeutschen Dogger, *Paläontol. Z.* 53:137–141.

Bogoslovsky, B. I., 1982, An interesting form of apertural formation in the shell of clymeniids, *Dokl. Akad. Nauk SSSR* 264(6):1483–1486 (in Russian).

Boletzky, S. v., 1983, *Sepia officinalis*, in: *Cephalopod Life Cycles*, Vol. I (P. R. Boyle, ed.), Academic Press, New York, pp. 31–52.

Boucot, A. J., 1953, Life and death assemblages among fossils, *Am. J. Sci.* **251**:25–40.

Boyle, P. R., and Thorpe, R. S., 1984, Optic gland enlargement and female gonad maturation in a population of the octopus *Eledone cirrhosa*: A multivariate analysis, *Mar. Biol.* **79**:127–132.

Brett, C. E., and Seilacher, A., 1991, Fossil Lagerstätten: A taphonomic consequence of event sedimentation, in: *Cycles and Events in Stratigraphy* (G. Einsele, W. Ricken, and A. Seilacher, eds.), Springer-Verlag, New York, pp. 283–297.

Buchardt, B., and Weiner, S., 1981, Diagenesis of aragonite from Upper Cretaceous ammonites: A geochemical case-study, *Sedimentology* **28**:423–438.

Bucher, H., and Guex, J., 1990, Rythmes de croissance chez les ammonites triasiques, *Bull. Soc. Vaudoise Sci. Nat.* **80**(2):191–209.

Callomon, J. H., 1963, Sexual dimorphism in Jurassic ammonites, *Trans. Leicester Lit. Phil. Soc.* **57**:21–56.

Callomon, J. H., 1981, Dimorphism in ammonoids, in: *The Ammonoidea*, Systematics Association Spec. Vol. 18 (M. R. House and J. R. Senior, eds.), Academic Press, London, pp. 257–273.

Callomon, J. H., 1985, The evolution of the Jurassic ammonite family Cardioceratidae, *Spec. Pap. Palaeont.* **33**:49–90.

Carlson, B., Awai, M., and Arnold, J., 1992, Waikiki Aquarium's Chambered *Nautilus* reach their first "Hatch-day" anniversary, *Hawaiian Shell News* **40**(1):1,3–4.

Carriker, M. R., 1972, Observations on removal of spines by muricid gastropods during shell growth, *Veliger* **15**:69–74.

Chamberlain, J. A., Jr., 1978, Permeability of the siphuncular tube of *Nautilus*: Its ecologic and paleoecologic implications, *N. Jb. Geol. Paläont. Mh.* **3**:129–142.

Checa, A., 1987, Morphogenesis in ammonites—differences linked to growth pattern, *Lethaia* **20**:141–148.

Checa, A., 1994, A model for the morphogenesis of ribs in ammonites inferred from associated microsculptures, *Palaeontology (Lond.)* **37**(4):863–888.

Checa, A., and Martin-Ramos, D., 1989, Growth and function of spines in the Jurassic ammonite *Aspidoceras, Palaeontology (Lond.)* **32**:645–655.

Checa, A., and Sandoval, J., 1989, Septal retraction in Jurassic Ammonitina, *N. Jb. Geol. Paläont. Mh.* **4**:193–211.

Checa, A., and Westermann, G. E. G., 1989, Segmental growth in planulate ammonites: Inferences on costal function, *Lethaia* **22**:95–100.

Clausen, C.-D., 1968, Oberdevonische Cephalopoden aus dem Rheinischen Schiefergebirge. I. Orthocerida, Bactritida, *Palaeontogr. Abt. A* **128**:1–86.

Cobban, W. A., 1951, Scaphitoid cephalopods of the Colorado group, *U.S. Geol. Surv. Prof. Pap.* **239**:1–42.

Collins, D., and Ward, P. D., 1987, Adolescent growth and maturity, in: *Nautilus—The Biology and Paleobiology of a Living Fossil* (W. B. Saunders and N. H. Landman, eds.), Plenum Press, New York, pp. 421–432.

Collins, D., Ward, P. D., and Westermann, G. E. G., 1980, Function of cameral water in *Nautilus, Paleobiology* **6**:168–172.

Cowen, R., Gertman, R., and Wright, G., 1973, Camouflage patterns in *Nautilus* and their implications for cephalopod paleobiology, *Lethaia* **6**:201–213.

Crick, R. E., 1978, Morphological variations in the ammonite *Scaphites* of the Blue Hill Member, Carlile Shale, Upper Cretaceous, Kansas, *Univ. Kans. Paleontol. Contrib. Pap.* **88**:1–28.

Currie, E. D., 1942, Growth changes in the ammonite *Promicroceras marstonense* Spath, *Proc. R. Soc. Edinb. Sect. B* **61**:344–367.

Currie, E. D., 1943, Growth stages in some species of *Promicroceras, Geol. Mag.* **80**:15–22.

Currie, E. D., 1944, Growth stages in some Jurassic ammonites, *Trans. R. Soc. Edinb.* **61**:171–198.

Davis, R. A., 1972, Mature modification and dimorphism in selected Late Paleozoic ammonoids, *Bull. Am. Paleont.* **62**(272):27–130.

Deevey, E. S., 1947, Life tables for natural populations of animals, *Q. Rev. Biol.* **22**:283–314.

Denton, E., and Gilpin-Brown, J., 1966, On the buoyancy of the pearly *Nautilus, J. Mar. Biol. Assoc. U.K.* **46**:723–759.

Diener, C., 1895, Himalayan fossils—the Cephalopoda of the Muschelkalk, *Palaeont. Indica Ser.* 15, II **2**:1–118.

Doguzhaeva, L., 1982, Rhythms of ammonoid shell secretion, *Lethaia* **15**:385–394.

Doguzhaeva, L. A., 1988, Siphuncular tube and septal necks in ammonite evolution, in: *Cephalopods—Present and Past* (J. Wiedmann and J. Kullmann, eds.), Schweizerbart'sche Verlagsbuchhandlung, Stuttgart, pp. 291–302.

Dommergues, J.-L., 1988, Can ribs and septa provide an alternate standard for age in ammonite ontogenetic studies?, *Lethaia* **21**:243–256.

Druschits, V. V., Doguzhaeva, L. A., and Mikhailova, I. A., 1977, The structure of the ammonitella and the direct development of ammonites, *Paleontol. J.* **11**(2):188–199.

Elmi, S., and Benshili, K., 1987, Relation entre la structuration tectonique, la composition des peuplements et l'évolution; exemple du Toarcien du Moyen-Atlas méridional (Maroc), *Boll. Soc. Paleontol. Ital.* **26**:47–62.

Engeser, T. S., 1990, Major events in cephalopod evolution, in: *Major Evolutionary Radiations*, Systematics Association Spec. Vol. 42 (P. D. Taylor and G. P. Larwood, eds.), Clarendon Press, Oxford, pp. 119–138.

Fagerstrom, J. A., 1964, Fossil communities in paleoecology: Their recognition and significance, *Geol. Soc. Am. Bull.* **75**:1197–1216.

Forester, R. W., Caldwell, W. G. E., and Oro, F. H., 1977, Oxygen and carbon isotopic study of ammonites from the Late Cretaceous Bearpaw Formation in southwestern Saskatchewan, *Can. J. Earth Sci.* **14**:2086–2100.

Forsythe, J. W., and Van Heukelem, W. F., 1987, Growth, in: *Cephalopod Life Cycles*, Vol. II (P. R. Boyle, ed.), Academic Press, New York, pp. 135–156.

Geary, D. H., Breiske, T. A., and Bemis, B. E., 1992, The influence and interaction of temperature, salinity, and upwelling on the stable isotopic profiles of strombid gastropod shells, *Palaios* **7**:77–85.

Gould, S. J., 1966, Allometry and size in ontogeny and phylogeny, *Biol. Rev.* **41**:587–640.

Gould, S. J., 1977, *Ontogeny and Phylogeny*, Belknap Press, Harvard University, Cambridge, MA.

Grossman, E. L., and Ku, T. L., 1986, Oxygen and carbon isotope fractionation in biogenic aragonite: Temperature effects, *Chem. Geol.* **59**:59–72.

Guex, J., 1970, Sur les moules internes des Dactyliocératides, *Bull. Lab. Geol. Mineral. Geophys. Mus. Geol. Univ. Lausanne* **70**(182):1–7.

Heptonstall, W., 1970, Buoyancy control in ammonoids, *Lethaia* **3**:317–328.

Hewitt, R. A., 1985, Numerical aspects of sutural ontogeny in the Ammonitina and Lytoceratina, *N. Jb. Geol. Paläont. Abh.* **170**(3):273–290.

Hewitt, R. A., 1986, Terminology of ammonoid coiling equations, *Lethaia* **19**:338.

Hewitt, R. A., 1988, Significance of early septal ontogeny in ammonoids and other ectocochliates, in: *Cephalopods—Present and Past* (J. Wiedmann and J. Kullmann, eds.), Schweizerbart'sche Verlagsbuchhandlung, Stuttgart, pp. 207–214.

Hewitt, R. A., and Hurst, J. M., 1977, Size changes in Jurassic liparoceratid ammonites and their stratigraphical and ecological significance, *Lethaia* **10**:287–301.

Hewitt, R. A., and Hurst, J. M., 1983, Aspects of the ecology of actinocerid cephalopods, *N. Jb. Geol. Paläont. Abh.* **165**(3):362–377.

Hewitt, R. A., and Stait, B., 1988, Seasonal variation in septal spacing of *Sepia officinalis* and some Ordovician actinocerid nautiloids, *Lethaia* **21**:383–394.

Hewitt, R. A., and Watkins, R., 1980, Cephalopod ecology across a late Silurian shelf tract, *N. Jb. Geol. Paläont. Abh.* **160**(1):96–117.

Hewitt, R. A., and Westermann, G. E. G., 1987, Function of complexly fluted septa in ammonoid shells, II, Septal evolution and conclusions, *N. Jb. Geol. Paläont. Abh.* **174**(2):135–169.

Hewitt, R. A., Checa, A., Westermann, G. E. G., and Zaborski, P. M., 1991, Chamber growth in ammonites inferred from color markings and naturally etched surfaces of Cretaceous vascoceratids from Nigeria, *Lethaia* **24**:271–287.

Hewitt, R. A., Westermann, G. E. G., and Checa, A., 1993, Growth rates of ammonites estimated from aptychi, *Geobios Mem. Spec.* **15**:203–208.

Hirano, H., 1981, Growth rates in *Nautilus macromphalus* and ammonoids: Its implications, in: *International Symposium on Conceptions and Methods in Paleontology, Barcelona* (J. Martinell, ed.), University of Barcelona, Barcelona, pp. 141–146.

House, M. R., 1965, A study in the Tornoceratidae: The succession of *Tornoceras* and related genera in the North American Devonian, *Phil. Trans. R. Soc. Lond. B* **250**(763):79–130.

Howarth, M. K., 1992, The ammonite family Hildoceratidae in the Lower Jurassic of Britian, Part 1, *Palaeontologr. Soc. Monogr. (Lond.)* **145**:1–106.

Hutchinson, G. E., 1978, *An Introduction to Population Ecology*, Yale University Press, New Haven.

Hyatt, A., 1894, Phylogeny of an acquired characteristic, *Proc. Am. Philos. Soc.* **32**(143):349–647.

Ivanov, A. N., 1971, On the problem of periodicity of the formation of septa in ammonoid shells and in that of other cephalopods, *Uch. Zap. Yarvsl. Pedagog. Inst. Geol. Paleontol.* **87**:127–130 (in Russian).

Ivanov, A. N., 1975, Late ontogeny in ammonites and its characteristics in micro-, macro- and megaconchs, *Yarosl. Pedagog. Inst. Sb. Nauchn. Tr.* **142**:5–57 (in Russian).

Jacobs, D. K., 1992, Shape, drag, and power in ammonoid swimming, *Paleobiology* **18**(2):203–220.

Jacobs, D. K., and Landman, N. H., 1993, *Nautilus*—a poor model for the function and behavior of ammonoids? *Lethaia* **26**:101–111.

Jacobs, D. K., Landman, N. H., and Chamberlain, J. A., Jr., 1994, Ammonite shell shape covaries with facies and hydrodynamics: Iterative evolution as a response to changes in basinal environment, *Geology* **22**:905–908.

Jordan, R., and Stahl, W., 1970, Isotopische Paläotemperatur-Bestimmungen an jurassischen Ammoniten und grundsätzliche Voraussetzungen für diese Methode, *Geol. Jb.* **89**:33–62.

Kahn, P. G. K., and Pompea, S. M., 1978, Nautiloid growth rhythms and dynamical evolution of the Earth–Moon system, *Nature* **275**:606–611.

Kant, R., 1973a, "Knickpunkte" im allometrischen Wachstum von Cephalopoden-Gehäusen, *N. Jb. Geol. Paläont. Abh.* **142**(1):97–114.

Kant, R., 1973b, Allometrisches Wachstum paläozoischer Ammonoideen: Variabilität und Korrelation einiger Merkmale, *N. Jb. Geol. Paläont. Abh.* **143**(2):153–192.

Kant, R., 1973c, Untersuchungen des allometrischen Gehäusewachstums paläozoischer Ammonoideen unter besonder Berücksichtigung einzelner "Populationen," *N. Jb. Geol. Paläont. Abh.* **144**(2):206–251.

Kant, R., and Kullmann, J., 1973, "Knickpunkte" im allometrischen Wachstum von Cephalopoden-Gehäuse, *N. Jb. Geol. Paläont. Abh.* **142**:7–114.

Kemper, E., and Wiedenroth, K., 1987, Klima und Tier-Migrationen am Beispiel der frühkretazischen Ammoniten Nordwestdeutschlands, *Geol. Jahr. A* **96**:315–363.

Kennedy, W. J., 1988, Late Cenomanian and Turonian ammonite faunas from north-east and central Texas, *Spec. Pap. Palaeontol.* **39**:1–131.

Kennedy, W. J., and Cobban, W. A., 1976, Aspects of ammonite biology, biogeography, and biostratigraphy, *Spec. Pap. Palaeontol.* **17**:1–94.

Kidwell, S. M., and Bosence, D. W. J., 1991, Taphonomy and time-averaging of marine shelly faunas, in: *Taphonomy—Releasing the Data Locked in the Fossil Record* (P. A. Allison and D. E. G. Briggs, eds.), Plenum Press, New York, pp. 115–209.

Korn, D., and Price, J. D., 1987, Taxonomy and phylogeny of the Kosmoclymeniinae subfam. nov. (Cephalopoda, Ammonoidea, Clymeniida), *Cour. Forschungsinst. Senckenb.* **92**:5–75.

Kulicki, C., 1974, Remarks on the embryogeny and postembryonal development of ammonites, *Acta Palaeontol. Pol.* **19**:201–224.

Kulicki, C., 1979, The ammonite shell: Its structure, development and biological significance, *Palaeontol. Pol.* **39**:97–142.

Kullmann, J., and Scheuch, J., 1970, Wachstums-Änderungen in der Ontogenese paläozoischer Ammonoideen, *Lethaia* **3**:397–412.

Kullmann, J., and Scheuch. J., 1972, Absolutes und relatives Wachstum bei Ammonoideen, *Lethaia* 5:129–146.

Landman, N. H., 1983, Ammonoid growth rhythms, *Lethaia* 16:248.

Landman, N. H., 1986, Developmental criteria for comparing ammonite ontogenies, *Geol. Soc. Am. Abst. Prog.* 18(6):665.

Landman, N. H., 1987, Ontogeny of Upper Cretaceous (Turonian–Santonian) scaphitid ammonites from the Western Interior of North America: Systematics, developmental patterns, and life history, *Bull. Am. Mus. Nat. Hist.* 185(2):117–241.

Landman, N. H., 1988, Early ontogeny of Mesozoic ammonites and nautilids, in: *Cephalopods— Present and Past* (J. Weidmann and J. Kullmann, eds.), Schweitzerbart'sche Verlagsbuchhandlung, Stuttgart, pp. 215–228.

Landman, N. H., 1989, Iterative progenesis in Upper Cretaceous ammonites, *Paleobiology* 15(2):95–117.

Landman, N. H., and Cochran, J. K., 1987, Growth and longevity of *Nautilus*, in: *Nautilus—The Biology and Paleobiology of a Living Fossil* (W. B. Saunders and N. H. Landman, eds.), Plenum Press, New York, pp. 401–420.

Landman, N. H., and Klofak, S. M., Size frequency studies in Late Cretaceous ammonoids: Evidence for rate of growth, in prep.

Landman, N. H., and Waage, K. M., 1986, Shell abnormalities in scaphitid ammonites, *Lethaia* 19:211–224.

Landman, N. H., and Waage, K. M., 1993, Scaphitid ammonites of the Upper Cretaceous (Maastrichtian) Fox Hills Formation in South Dakota and Wyoming, *Bull. Am. Mus. Nat. Hist.* 215:1–257.

Landman, N. H., Tanabe, K., Mapes, R. H., Klofak, S. M., and Whitehill, J., 1993, Pseudosutures in Paleozoic ammonoids, *Lethaia* 26:99–100.

Landman, N. H., Cochran, J. K., Rye, D. M., Tanabe, K., and Arnold, J. M., 1994, Early life history of *Nautilus*: Evidence from isotopic analysis of aquarium-reared specimens, *Paleobiology* 20(1):40–51.

Lange, W., 1932, Über Symbiosen von *Serpula* mit Ammoniten im unteren Lias Norddeutschlands, *Z. Dtsch. Geol. Ges.* 84:229–234.

Lehmann, U., 1966, Dimorphismus bei Ammoniten der Ahrensburger Lias-Geschiebe, *Paläontol. Z.* 40:26–55.

Lehmann, U., 1981, *The Ammonites: Their Life and Their World*, Cambridge University Press, Cambridge.

Linsley, R. M., and Javidpour, M., 1980, Episodic growth in Gastropoda, *Malacologia* 20:153–160.

Lominadzé, T. A., Sharikadzé, M. Z., and Kvantaliani, I. V., 1993, On mechanism of soft body movement within body chamber in ammonites, *Geobios Mem. Spec.* 15:267–273.

Mackenzie, C. L., Jr., 1960, Interpretation of varices and growth ridges on shells of *Eupleura caudata*, *Ecology* 41(4):783–784.

Maeda, H., 1993, Dimorphism of Late Cretaceous false-puzosiine ammonites, *Yokoyamaoceras* Wright and Matsumoto, 1954 and *Neopuzosia* Matsumoto, 1954, *Trans. Proc. Palaeont. Soc. Jpn. N.S.* 169:97–128.

Makowski, H., 1962, Problem of sexual dimorphism in ammonites, *Palaeontol. Pol.* 12:1–92.

Makowski, H., 1971, Some remarks on the ontogenetic development and sexual dimorphism in the Ammonoidea, *Acta Geol. Pol.* 21(3):321–340.

Mancini, E. A., 1978, Origin of micromorph faunas in the geologic record, *J. Paleontol.* 52(2):311–322.

Mangold, K., 1983, Food, feeding and growth in cephalopods, *Mem. Natl. Mus. Victoria* 44:81–93.

Mangold, K., 1987, Reproduction, in: *Cephalopod Life Cycles*, Vol. II (P. R. Boyle, ed.), Academic Press, London, pp. 157–200.

Matsukawa, M., 1987, Early shell morphology of *Karsteniceras* (ancyloceratid) from the Lower Cretaceous Choshi Group, Japan and its significance to the phylogeny of Cretaceous heteromorph ammonites, *Trans. Proc. Palaeont. Soc. Jpn. New Ser.* 148:346–359.

Matsumoto, T., 1991, The Mid-Cretaceous ammonites of the family Kossmaticeratidae from Japan, *Palaeont. Soc. Jpn. Spec. Pap.* **33**:1–143.

Matsumoto, T., Muramoto, T., and Inoma, A., 1972, Two small desmoceratid ammonites from Hokkaido, *Trans. Proc. Palaeont. Soc. Jpn. New Ser.* **87**:377–394.

Matyja, B. A., 1986, Developmental polymorphism in Oxfordian ammonites, *Acta Geol. Pol.* **36**(1–3):37–68.

McConnaughey, T., 1989a, ^{13}C and ^{18}O isotopic disequilibrium in biological carbonates, *I.* Patterns, *Geochim. Cosmochim. Acta* **53**:151–162.

McConnaughey, T., 1989b, ^{13}C and ^{18}O isotopic disequilibrium in biological carbonates. II. *In vitro* simulation of kinetic isotope effects, *Geochim. Cosmochim. Acta* **53**:163–171.

Meinhardt, H., and Klinger, M., 1987, A model for pattern formation on the shells of molluscs, *J. Theor. Biol.* **126**:63–89.

Meischner, D., 1968, Perniciöse Epökie von *Placunopsis* auf *Ceratites*, *Lethaia* **1**:156–174.

Merkt, J., 1966, Über Austern und Serpeln als Epöken auf Ammonitengehäusen, *N. Jb. Geol. Paläont. Abh.* **125**:467–479.

Mesnil, B., 1977, Growth and life cycle of squid, *Loligo pealei* and *Illex illecebrosus*, from the Northwest Atlantic, *ICNAF Sel. Papers* **2**:55–69.

Mignot, Y., 1993, Un problème de paléobiologie chez les ammonoïdes (Cephalopoda): Croissance et miniaturisation en liaison avec les environnements, *Docum. Lab. Geol. Lyon* **124**:1–113.

Mignot, Y., Elmi, S., and Dommergues, J.-L., 1993, Croissance et miniaturisation de quelques *Hildoceras* (Cephalopoda) en liaison avec des environnements contraignant de la Téthys toarcienne, *Geobios Mem. Spec.* **15**:305–312.

Miller, A. K., Furnish, W. M., and Schindewolf, O. H., 1957, Paleozoic Ammonoidea, in: *Treatise on Invertebrate Paleontology*, Part L, *Mollusca 4* (R. C. Moore, ed.), Geological Society of America and University of Kansas Press, Lawrence, KS, pp. 11–80.

Mojsisovics, E. v., 1886, Arktische Triasfaunen, *Mem. Acad. Imp. Sci. St. Petersbourg* **7**:33.

Morton, N., 1988, Segregation and migration patterns in some *Graphoceras* populations (Middle Jurassic), in: *Cephalopods—Present and Past* (J. Wiedmann and J. Kullmann, eds.), Schweizerbart'sche Verlagsbuchhandlung, Stuttgart, pp. 377–385.

Oba, T., Kai, M., and Tanabe, K., 1992, Early life history and habitat of *Nautilus pompilius*, *Kagoshima Univ. Res. Center S. Pac. Occas. Pap.* **1**:26–29.

Obata, I., 1959, Croissance relative sur quelques espèces des Desmoceratidae, *Mem. Fac. Sci. Kyushu Univ., Ser. D Geol.* **9**(1):33–45.

Obata, I., 1960, Spirale de quelques ammonites, *Mem. Fac. Sci., Kyushu Univ., Ser. D Geol.* **9**(3):151–163.

Obata, I., 1965, Allometry of *Reesidites minimus*, a Cretaceous ammonite species, *Trans. Proc. Palaeont. Soc. Japan, New Ser.* **58**:39–63.

Obata, I., Futakami, M., Kawashita, Y., and Takahashi, T., 1978, Apertural features in some Cretaceous ammonites from Hokkaido, *Bull. Natl. Sci. Mus. Ser. C (Geol.)* **4**(3):139–155.

Oechsle, E., 1958, Stratigraphie und Ammonitenfauna der Sonninien-Schichten des Filsgebiets unter besonderer Berücksichtigung der Sowerbyi-Zone (Mittlerer Dogger, Württemberg), *Palaeontogr. Abt. A* **111**:47–129.

Okamoto, T., 1989, Changes in life orientation during the ontogeny of some heteromorph ammonoids, *Palaeontology (Lond.)* **31**(2):281–294.

Okamoto, T., 1993, Theoretical modelling of ammonite morphogenesis, *N. Jb. Geol. Paläont. Abh.* **190**(2/3):183–190.

Palframan, D. F. B., 1966, Variation and ontogeny of some Oxfordian ammonites: *Taramelliceras richei* (de Loriol) and *Creniceras renggeri* (Oppel), from Woodham, Buckinghamshire, *Palaeontology (Lond.)* **9**(2):290–311.

Palframan, D. F. B., 1967, Mode of early shell growth in the ammonite *Promicroceras marstonense* Spath, *Nature (Lond.)* **216**:1128–1130.

Pompeckj, J. F., 1884, Über Ammonoideen mit anomaler Wohnkammer, *J. Ver. Vaterl. Naturk. Wurtt.* **49**:220–290.

Raup, D., 1967, Geometric analysis of shell coiling: Coiling in ammonoids, *J. Paleontol.* **41**:43–65.

Raup, D., and Chamberlain, J. A., Jr., 1967, Equations for volume and center of gravity in ammonoid shells, *J. Paleontol.* **41**:566–574.

Richard, A., 1970, Analyse du cycle sexual chez les céphalopodes mise en évidence expérimentale d'un rythme conditionné par les variations des facteurs externes et internes, *Bull. Soc. Zool. Fr.* **95**:461–469.

Richards, R. P., and Bambach, R. K., 1975, Population dynamics of some Paleozoic brachiopods and their paleoecological significance, *J. Paleontol.* **49**(5):775–798.

Rieber, H., 1963, Ammoniten und Stratigraphie des Braunjura β der Schwaebischen Alb., *Palaeontogr. Abt. A* **122**:1–89.

Rounsefell, G. A., and Everhart, W. H., 1953, *Fishery Science—Its Methods and Applications*, John Wiley & Sons, New York.

Rye, D. M., and Sommer, M. A., 1980, Reconstructing paleotemperature and paleosalinity regimes with oxygen isotopes, in: *Skeletal Growth of Aquatic Organisms* (D. C. Rhoads and R. A. Lutz, eds.), Plenum Press, New York, pp. 169–202.

Saunders, W. B., 1983, Natural rates of growth and longevity of *Nautilus belauensis*, *Paleobiology* **9**(3):280–288.

Schindewolf, O. H., 1934, Über Epöken auf Cephalopoden-Gehäusen, *Palaeontol. Z.* **16**:15–31.

Schindewolf, O. H., 1958, Über Aptychen (Ammonoidea), *Palaeontogr. Abt. A* **111**:1–46.

Seilacher, A., 1960, Epizoans as a key to ammonoid ecology, *J. Paleontol.* **34**:189–193.

Seilacher, A., 1982, Ammonite shells as habitats in the Poseidonia Shales of Holzmaden—floats or benthic islands? *N. Jb. Geol. Paläont. Abh.* **159**:98–114.

Seilacher, A., 1988, Why are nautiloid and ammonite sutures so different? *N. Jb. Geol. Paläont. Abh.* **177**:41–67.

Seilacher, A., and Gunji, P. Y., 1993, Morphogenetic countdowns in heteromorph shells, *N. Jb. Geol. Paläont. Abh.* **190**(2/3):237–265.

Sheldon, R. W., 1965, Fossil communities with multi-modal size–frequency distributions, *Nature* **206**(4991):1336–1338.

Shigeta, Y., 1993, Post-hatching early life history of Cretaceous Ammonoidea, *Lethaia* **26**(2):133–145.

Simoulin, E., 1945, Observations sur la croissance de la coquille chez quelques Stéphanocératides, *Ann. Soc. Géol. Nord* **65**:9–19.

Smith, J. P., 1898, The development of *Lytoceras* and *Phylloceras*, *Proc. Calif. Acad. Sci. (Geol.)* **1**(4):129–161.

Speden, I. G., 1970, The type Fox Hills Formation, Cretaceous (Maestrichtian), South Dakota, Part 2, Systematics of the Bivalvia, *Peabody Mus. Nat. Hist. Yale Univ. Bull.* **33**:1–222.

Stahl, W., and Jordan, R., 1969, General considerations on isotopic paleotemperature determinations and analyses on Jurassic ammonites, *Earth Planet. Sci. Lett.* **6**:173–178.

Stevens, G. R., 1988, Giant ammonites: A review, in: *Cephalopods—Present and Past* (J. Wiedmann and J. Kullmann, eds.), Schweizerbart'sche Verlagsbuchhandlung, Stuttgart, pp. 141–166.

Surlyk, F., 1972, Morphological adaptations and population structures of the Danish Chalk brachiopods (Maastrichtian, Upper Cretaceous), *K. Dan. Vidensk. Selsk. Biol. Skr.* **19**(2):1–57.

Tanabe, K., 1975, Functional morphology of *Otoscaphites puerculus* (Jimbo), an Upper Cretaceous ammonite, *Trans. Proc. Palaeontol. Soc. Jpn, New Ser.* **99**:109–132.

Tanabe, K., 1977, Functional evolution of *Otoscaphites puerculus* (Jimbo) and *Scaphites planus* (Yabe), Upper Cretaceous ammonites, *Mem. Fac. Sci. Kyushu Univ. Ser. D (Geol.)* **23**:367–407.

Tanabe, K., 1979, Palaeoecological analysis of ammonoid assemblages in the Turonian *Scaphites* facies of Hokkaido, Japan, *Palaeontology (Lond.)* **22**(3):609–630.

Tanabe, K., and Landman, N. H., Translocation of the soft body in Mesozoic ammonoids, in prep.

Tanabe, T., Obata, I., and Futakami, H., 1981, Early shell morphology in some Upper Cretaceous heteromorph ammonites, *Trans. Proc. Palaeontol. Soc. Jpn. New Ser.* **124**:215–234.

Tanabe, K., Landman, N. H., and Weitschat, W., 1993, Septal necks in Mesozoic Ammonoidea: Structure, ontogenetic development, and evolution, in: *The Ammonoidea: Environment,*

Ecology, and Evolutionary Change, Systematics Association Spec. Vol. 47 (M. R. House, ed.), Clarendon Press, Oxford, pp. 57–84.

Teisseyre, L., 1889, Über die systematische Bedeutung der sog. Parabeln der Perisphincten, *N. Jb. Miner. Geol. Paläont.* **6**(1):570–643.

Tevesz, M. J. S., and Carter, J. G., 1980, Environmental relationships of shell form and structure of unionacean bivalves, in: *Skeletal Growth of Aquatic Organisms* (D. C. Rhoads and R. A. Lutz, eds.), Plenum Press, New York, pp. 295–322.

Thompson, D. W., 1917, *On Growth and Form*, Cambridge University Press, London.

Tourtelot, H. A., and Rye, R. O., 1969, Distribution of oxygen and carbon isotopes in fossils of Late Cretaceous age, Western Interior region of North America, *Geol. Soc. Am. Bull.* **80**:1903–1922.

Tozer, E. T., 1965, Latest Lower Triassic ammonoids from Ellesmere Island and northeastern British Columbia, *Geol. Surv. Can. Bull.* **123**.

Tozer, E. T., 1991. Relationship between spines, parabolic nodes, rhythmic shell secretion and formation of septa in some Triassic ammonoids, in: *The Ammonoidea: Evolution and Environmental Change*, Systematics Association Symp. London Prog. Abstr., pp. 23–24.

Trueman, A. E., 1941, The ammonite body-chamber, with special reference to the buoyancy and mode of life of the living ammonite, *Q. J. Geol. Soc. Lond.* **96**:339–383.

Van Heukelem, W. F., 1978, Aging in lower animals, in: *Biology of Aging* (J. A. Behnke, C. E. Finch, and B. C. Moment, eds.), Plenum Press, New York, pp. 115–130.

Vermeij, G. J., 1980, Gastropod shell growth rate, allometry, and adult size—environmental implications, in: *Skeletal Growth of Aquatic Organisms* (D. C. Rhoads and R. A. Lutz, eds.), Plenum Press, New York, pp. 379–394.

Vermeij, G. J., 1993, *A Natural History of Shells*, Princeton University Press, Princeton.

Waage, K. M., 1968, The type Fox Hills Formation, Cretaceous (Maestrichtian), South Dakota, Part 1, stratigraphy and paleoenvironments, *Peabody Mus. Nat. Hist. Yale Univ. Bull.* **27**:1–175.

Wähner, F., 1894, Beiträge zur Kenntniss der tieferen Zonen des unteren Lias in der nordöstlische Alpen, *Beitr. Paläontol. Österr. Ungarns. Orients* **9**(I–II):1–54.

Ward, P. D., 1982, The relationship of siphuncle size to emptying rates in chambered cephalopods: Implications for cephalopod paleobiology, *Paleobiology* **8**:426–433.

Ward, P. D., 1985, Periodicity of chamber formation in chambered cephalopods: Evidence from *Nautilus macromphalus* and *Nautilus pompilius, Paleobiology* **11**:438–450.

Ward, P. D., 1986, Rates and processes of compensatory buoyancy change in *Nautilus macromphalus, Veliger* **28**:356–368.

Ward, P. D., 1987, *The Natural History of Nautilus*, Allen and Unwin, Boston.

Ward, P. D., 1992, *On Methuselah's Trail*, W. H. Freeman, New York.

Ward, P. D., and Chamberlain, J. A., Jr., 1983, Radiographic observation of chamber formation in *Nautilus pompilius, Nature (Lond.)* **304**:57–59.

Ward, P. D., and Greenwald, L., 1982, Chamber refilling in *Nautilus, J. Mar. Biol. Assoc. U.K.* **62**:469–475.

Ward, P. D., Greenwald, L., and Magnier, Y., 1981, The chamber formation cycle in *Nautilus macromphalus, Paleobiology* **7**(4):481–493.

Weitschat, W., and Bandel, K., 1991, Organic components in phragmocones of Boreal Triassic ammonoids: Implications for ammonoid biology, *Paläontol. Z.* **65**:269–303.

Weitschat, W., and Bandel, K., 1992, Formation and function of suspended organic cameral sheets in Triassic ammonoids: Reply, *Paläontol. Z.* **66**:443–444.

Wells, M. J., 1983, Cephalopods do it differently, *New Sci.* **100**:332–338.

Wells, M. J., and Wells, J., 1959, Hormonal control of sexual maturity in *Octopus, J. Exp. Biol.* **36**:1–33.

Wells, M. J., and Wells, J., 1977, Cephalopoda: Octopoda, in: *Reproduction of Marine Invertebrates*, Vol. IV (A. C. Giese and J. S. Pearse, eds.), Academic Press, New York, pp. 291–336.

Westermann, G. E. G., 1954, Monographie der Otoitidae (Ammonoidea), *Geol. Jahrb. Beih.* **15**:1–364.

Westermann, G. E. G., 1958, The significance of septa and sutures in Jurassic ammonite systematics, *Geol. Mag.* **95**(6):441–455.

Westermann, G. E. G., 1971, Form, structure, and function of shell and siphuncle in coiled Mesozoic ammonoids, *Life Sci. Contr. R. Ont. Mus.* **78**:1–39.

Westermann, G. E. G., 1975, Architecture and buoyancy of simple cephalopod phragmocones and remarks on ammonites, *Paläontol. Z.* **49**:221–234.

Westermann, G. E. G., 1990, New developments in ecology of Jurassic–Cretaceous ammonoids, in: *Atti del secondo convegno inernazionale, Fossili, Evoluzione, Ambiente, Pergola, 1987* (G. Pallini, F. Cecca, S. Cresta, and M. Santantonio, eds.), Tectnostampa, Ostra Vetere, Italy, pp. 459–478.

Westermann, G. E. G., 1992, Formation and function of suspended organic cameral sheets in Triassic ammonoids—discussion, *Paläontol. Z.* **66**(3/4):437–441.

Whittaker, S. G., Kyser, T. K., and Caldwell, W. G. E., 1987, Paleoenvironmental geochemistry of the Clagett marine cyclothem in south-central Saskatchewan, *Can. J. Earth Sci.* **24**:967–984.

Wiedmann, J., and Boletzky, S. v., 1982, Wachstum und Differenzierung des Schlups von *Sepia officinalis* unterkünstlichen Aufzuchtbedingungen—Grenzen der Anwendung im palökologischen Modell, *N. Jb. Geol. Paläont. Abh.* **164**(1/2):118–133.

Zaborski, P. M. P., 1986, Internal mould markings in a Cretaceous ammonite from Nigeria, *Palaeontology* **29**:725–738.

Zakharov, Y. D., 1977, Ontogeny of ceratites of the genus *Pinacoceras* and developmental features of the suborder Pinacoceratina, *Paleontol. J.* **4**:445–451.

Zell, H., Zell, I., and Winter, S., 1979, Das Gehäusewachstum der Ammonitengattung *Amaltheus* De Montfort während der frühontogenetischen Entwicklung, *N. Jb. Geol. Paläont. Mh.* **10**:631–640.

Zuev, G. V., 1975, Physiological variation in female squids *Symplectoteuthis pteropus* (Steenstrup), *Biol. Moyra* **38**:55–62.

Chapter 13

Mature Modifications and Dimorphism in Ammonoid Cephalopods

RICHARD ARNOLD DAVIS, NEIL H. LANDMAN,
JEAN-LOUIS DOMMERGUES, DIDIER MARCHAND, and
HUGO BUCHER

RICHARD ARNOLD DAVIS • Department of Chemistry and Physical Sciences, College of Mount St. Joseph, Cincinnati, Ohio, 45233-1670.   NEIL H. LANDMAN • Department of Invertebrates, American Museum of Natural History, New York, New York, 10024.   JEAN-LOUIS DOMMERGUES and DIDIER MARCHAND • URA CNRS 157, Centre des Sciences de la Terre, Université de Bourgogne, F-21100 Dijon, France.   HUGO BUCHER • Centre des Sciences de la Terre, Université Claude-Bernard, Lyon I, 69622 Villeurbanne Cedex, France.

Ammonoid Paleobiology, Volume 13 of *Topics in Geobiology*, edited by Neil Landman *et al.*, Plenum Press, New York, 1996.

1. Introduction

The shell of an ammonoid is a kind of autobiography of the animal that once occupied it. Different parts of the shell tell different parts of the life history. The growth lines and the tiny intervals in between, along with the shape of the shell itself, record what was happening at the anterior end of the body. The septa and their sutures relate the tale of the other extremity.

But the saga is not complete. Even in those specimens to which taphonomy has been least unkind, many aspects of the life of the individual have not been recorded in the shell. (Or, perhaps, we do not yet have the tools and knowledge to decipher them.) Nonetheless, at least the outline of the autobiography is there.

Many ammonoid shells indicate that, toward the end of the recorded life span, there were changes in the animal. Growth did not carry on unchanged until brought to death's end. On the contrary. The last-formed portion of a shell differs from those portions formed earlier in life. Perhaps there is evidence of a modification in the rate of coiling, or the apertural margin (the peristome) took on a configuration different from that which it had earlier in ontogeny. At the other end of the body, it may be that the septa became more closely spaced. Just which of these changes, and others, occurred varied from taxon to taxon. In some groups, any modifications present were negligible; in others, profound.

Moreover, in some species, these late-ontogenetic shell modifications took more than one path. For example, two conchs that were identical in early growth stages may have become strikingly different from one another later in life, revealing dimorphism within the taxon.

The purpose of this chapter is to discuss these late-ontogenetic shell modifications in ammonoids and any dimorphism or related phenomena that resulted. However, the chapter definitely is not intended as an exhaustive reiteration of all that has been written about these subjects. As Callomon (1981) wrote: "To review yet again the topic of dimorphism as a whole would be unnecessarily repetitive. . . ." If one wished to undertake such a task, the works of Callomon, Makowski, and Westermann listed in the bibliography would be admirable places to start, with Lehmann (1981) thrown in for a more recent perspective.

At the outset, a few logistical remarks are appropriate. This chapter is decidedly a committee effort. Different parts of the geological column are dealt with by different people: R. A. Davis compiled the Paleozoic part; H. Bucher, the Triassic; J.-L. Dommergues and D. Marchand, the Jurassic; and N. H. Landman, the Cretaceous. Everyone contributed ideas for the general, intro-ductory, and closing matter, but R. A. Davis bears the primary responsibility (? blame) for putting it into its present form. An exception to this is Section 3.6, which is the work of J.-L. Dommergues and D. Marchand.

Over the years, the convention has built up that fossil cephalopods should be illustrated with their adapertural ends uppermost. This convention, of course, flies in the face of the probable orientation of the animals in life, at least for those in which the final body chamber is preserved. We prefer an orientation that coincides with that in life, and most of the illustrations in this chapter reflect this. Some of the illustrations that were published previously, however, are oriented with the adapertural end uppermost, to be faithful to the original publications.

The specimens illustrated in this chapter are reposited in a number of institutions, as indicated in the acknowledgments.

2. Late-Ontogenetic Shell Modifications

2.1. Kinds of Late-Ontogenetic Shell Modifications

Long before the beginning of the present century, it was recognized that the apertural regions of many ammonoids are different from the previously formed parts of the same conchs. A whole series of these late-ontogenetic shell modifications have been documented:

1. Changes in coiling.
2. Changes in whorl cross section.
3. Changes in ornament.
4. Development of an apertural constriction.
5. Development of an apertural shell thickening.
6. Formation of a characteristic ultimate peristome.
7. Changes in nature of shell deposition.
8. Septal approximation (i.e., progressive reduction of the interseptal distance).
9. Progressive simplification of the ultimate few sutures.
10. Thickening of the last septum.
11. Development of muscle scars.

The spectacular changes in coiling exhibited by some of the heteromorphs are well known. In some of the scaphitids, for example, the rate of coiling diminishes to none at all and then goes back up, resulting in the distinctive "shaft" and "hook" (Fig. 9C–J). Almost as dramatic are the geniculations that result in the egg-like shape of the Permian species *Hyattoceras abichi* Gemmellaro, 1887 (Fig. 2C–E). More subtle is the "flattening" of the venter present in individuals of many other Paleozoic taxa. In these cases, there is a slight increase in the rate of coiling, followed immediately by a slightly greater decrease; this is located some two-thirds to three-quarters of the way through the last whorl (i.e., one-quarter to one-third of a whorl adapical of the aperture (Figs. 1B,C, 2I, 3C,G, for example)).

Not uncommonly, these modifications in coiling are related to changes in whorl cross section. In the Permian species *Hyattoceras geinitzi* Gemmellaro, 1887, for example, the rounded venter (bottom of Fig. 2G) narrows at the change in coiling (bottom of Fig. 2H) and then broadens again toward the aperture (top of Fig. 2G).

Such a broadening of the body chamber (for example, Fig. 2B) is reminiscent of the gentle lateral swelling in the shell of present-day *Nautilus*. There may be a concomitant reduction in whorl height, and the dorsum of the body chamber may appear to pull away from the previous whorl at the umbilicus, in a phenomenon Lehmann (1981) termed "retraction" (Fig. 1F).

Many ammonoid taxa exhibit changes in ornament in the most recently formed body chamber. In some instances, ribbing, etc. diminish; in others, they increase; in even others, first the one and then the other. In the Permian species *Pseudagathiceras spinosum*, for example, spines project from the dorsum of the last whorl but are absent from the venter (Fig. 3F,G). On the other hand, in the Permian species *Agathiceras suessi*, there is a pit developed in each side of the body chamber, a feature not found either adapical or adapertural of that point (Fig. 3H).

It is possible that individuals of some taxa exhibited a change in color pattern late in ontogeny. Mapes and Sneck (1987), in their discussion of *Owenites* Hyatt & Smith, 1905, from the Triassic of Nevada, reported approximation of transverse color bands in the adapertural parts of shells that show late-ontogenetic shell modifications.

FIGURE 1. Permian ammonoids. (A) *Cyclolobus walkeri* Diener, 1903. Ankitohazo, Madagascar; maximum diameter 93 mm, ×3/4; MNHN B 7520 (specimen shown: Vaillant-Couturier-Treat, 1933, Pl. 1, Fig. 4; Davis *et al.*, 1969; Davis, 1972). (B,E) *Adrianites* sp. Maoen Mollo, Timor, Indonesia; ×$2\frac{1}{4}$; GIUA Drawer 55, T328 (specimen shown: Davis *et al.*, 1969; Davis, 1972). (C,D) *Adrianites* sp. cf. *A. insignis* Gemmellaro, 1887. Sosio Limestone; Province of Palermo, Sicily, Italy; maximum diameter 21$\frac{1}{2}$ mm, ×$2\frac{1}{4}$; BMNH C37654 (specimen shown: Davis *et al.*, 1969; Davis, 1972). (F,G) *Mexicoceras guadalupense guadalupense* (Girty, 1908) Ruzhencev, 1955. South Wells Limestone; about 2 miles SE of D Ranch, South Wells, Culberson County Texas (USGS loc. 7649); maximum diameter 48 mm, ×$1\frac{1}{2}$; USNM 144423 (specimen shown: Miller and Furnish, 1940; Davis *et al.*, 1969; Davis, 1972).

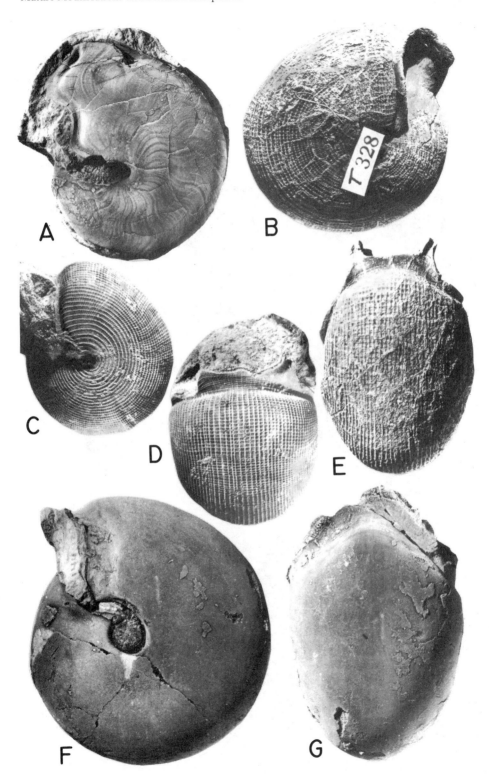

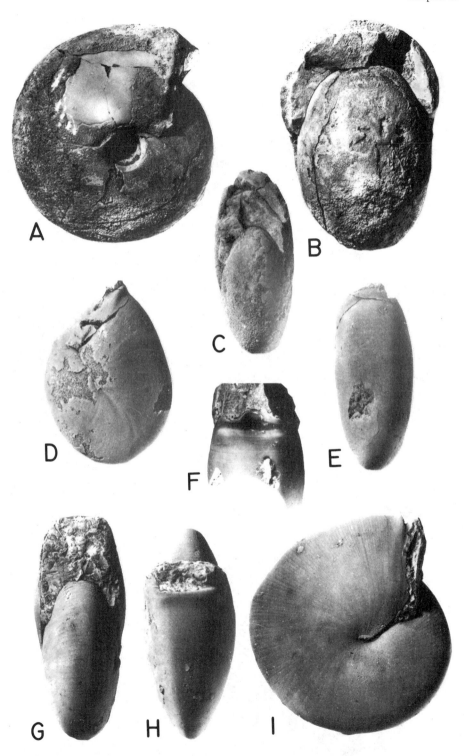

FIGURE 2. Permian ammonoids. (A,B) *Waagenoceras mojsisovicsi* Gemmellaro, 1887. Sosio Limestone; Passo di Burgio, Province of Palermo, Sicily, Italy; maximum diameter 160 mm, ×$\frac{1}{2}$; IGUP #32 (syntype; shown: Gemmellaro, 1887; Davis *et al.*, 1969; Davis, 1972). (C,D,E) *Hyatto-ceras abichi* Gemmellaro, 1887. Sosio Limestone; Passo di Burgio, Province of Palermo, Sicily, Italy; maximum diameter 22 mm, ×$2\frac{1}{4}$; IGUP #55 (syntype; figured: Gemmellaro, 1887; Davis *et al.*, 1969; Davis, 1972). (F–I) *Hyattoceras geinitzi* Gemmellaro, 1887. (F) Unlabeled but almost certainly from the Sosio region; ×$1\frac{1}{2}$; GPIT unnumbered (specimen shown: Davis *et al.*, 1969; Davis, 1972). (G–I) Sosio Limestone; Rocca di San Benedetto, Province of Palermo, Sicily, Italy; maximum diameter 40 mm, ×$1\frac{1}{2}$; IGUP #52 (syntype; shown: Gemmellaro, 1887; Davis *et al.*, 1969; Davis, 1972).

A relatively common late-ontogenetic shell modification is development of an apertural constriction, a thickening of the shell adjacent to the aperture, or both (Figs. 1–3). Of course, in specimens preserved only as steinkerns, it may not be possible to determine whether a depression on the internal mold is the result of just a shell thickening or marks an actual constriction. In present-day *Nautilus*, there is a slight but distinct thickening of the shell adjacent to the peristome of the fully grown shell.

One of the late-ontogenetic shell modifications earliest recognized was the development of a characteristic peristome. For example, in many Jurassic forms, there is a ventral prolongation, called a rostrum; alternatively, or in addition, there may be a pair of lateral apophyses, one on each side, called lappets or sometimes "ears." Even in some Paleozoic taxa, however, there are apertural modifications suggestive of the rostra and lappets of later forms (see, for example, Fig. 1). The apertures of some ammonoids can be as apparently bizarre as are some of those of the discosorid nautiloids of the Silurian (compare Fig. 131 in Arkell *et al.*, 1957, to Fig. 11 in Teichert *et al.*, 1964b).

In some instances, there seems to have been an actual change in the nature of shell deposition late in ontogeny. For example, Davis (1972) reported spherical bodies within the shell material near the aperture of individuals of the Permian species *Adrianites isomorphus* Gemmellaro, 1888; these left minute pits in the internal mold. Such pits near the apertures occur in the steinkerns of specimens of other Permian taxa too. Nettleship and Mapes (1993) also reported a change in the *Runzelschicht* near the aperture of a specimen of *Glaphyrites* Ruzhencev, 1936 from the Upper Carboniferous of Oklahoma. These occurrences bring to mind the fact that "black matter" rings the apertures of fully grown shells of present-day *Nautilus* but does not do so in ontogenetically younger individuals.

In general, the distance between adjacent septa in an ammonoid phragmo-cone increases in an adapertural direction. However, commonly, the ultimate two septa of an ammonoid shell are more closely spaced than are those adapical of them, or the last several septa are progressively more closely spaced. Septal approximation, of course, is a well-known phenomenon in fully grown shells of present-day *Nautilus* (see, for example, Willey, 1902).

FIGURE 3. Permian ammonoids. (A–D) *Agathiceras uralicum* (Karpinsky, 1874). Road Canyon Formation; saddle and slope W of Dugout Mountain, 0.3 miles NNW of Hill 4861, Brewster County, Texas (USNM loc. 732z). (A,C) Maximum diameter 32 mm, $\times 1\frac{3}{4}$; SUI 32460 BE (specimen shown: Davis *et al.*, 1969; Davis, 1972). (B,D) Maximum diameter 19 mm, $\times 1\frac{3}{4}$; SUI 32460 BN (specimen shown: Davis *et al.*, 1969; Davis, 1972). (E) *Texoceras texanum* (Girty, 1908) Miller and Furnish, 1940. Bone Spring Limestone; road-cut on the west side of US Highway 62-180, 1.8 miles NNE of the junction with US Highway 54, Culberson County, Texas; $\times 5\frac{1}{2}$; SUI 32461 AJ (specimen shown: Davis, 1972). (F,G) *Pseudagathiceras spinosum* Miller, 1944. Zone of *Waagenoceras* (beds 5, 6, and/or 7 of the Malascachas Section); 1200 m S 17°W of Noria de Malascachas, Valle de Las Delicias, Coahuila, Mexico; maximum diameter 21 mm, $\times 2\frac{1}{4}$; YPM 16310 (holotype; shown: Miller, 1944; Davis, 1972). (H–J) *Agathiceras suessi* Gemmellaro, 1887. Passo di Burgio, Province of Palermo, Sicily, Italy; maximum diameter 35 mm, $\times 1\frac{3}{4}$; IGUP #119 (syntype; shown: Gemmellaro, 1887; Davis *et al.*, 1969; Davis, 1972).

Concomitant with septal approximation may be a simplification of the suture (Fig. 3E). According to Furnish (1966), in ammonoids in which the overall trend of the suture is parabolic, there is a tendency for this trace to straighten in the ultimate few sutures.

Not uncommonly, the ultimate septum is distinctly thicker than are those adapical of it. Again, this phenomenon is common in *Nautilus*. On occasion, the last two septa in an ammonoid may be so close as to give the appearance of being a single, very thick, bifurcate septum (Blind and Jordan, 1979).

Finally, in the last-formed body chamber, muscle scars may be apparent for the first time or may be decidedly more distinct than in the phragmocone.

Not all of the above-enumerated late-ontogenetic shell modifications are likely to be present in a given conch. In fact, some taxa are not known to have developed any such modifications; this may reflect an actual lack (Ruzhencev, 1962) or result from either unavailability of suitably preserved material or inadequate study.

In general, an individual ammonoid may have at least several aspects of its conch affected. However, which late-ontogenetic shell modifications are present, if any, and to what extent, varies from taxon to taxon. Note, too, that different modifications occurred at different times in the ontogeny of the animal. For example, changes in coiling a quarter-whorl back from the aperture took place well before the formation of an apertural constriction in the same conch.

2.2. Causes and Meanings

In years past, many authors (for example, Hyatt, 1874) attributed late-ontogenetic shell modifications to senility—the gerontic individual just could not maintain discipline in its shell-forming faculties, so chaos resulted. Alternatively, some authors have considered some of these modifications to be otherwise pathological (for example, Hyatt, 1874; Coëmme, 1917).

Of course, the question immediately arises as to whether these shell modifications really are late ontogenetic. Some of the modifications are similar to varices in the shells of some gastropods. Moreover, some resemble features of parabolic nodes, etc., present in certain Mesozoic ammonoids and of constrictions and pseudoconstrictions that occur in the phragmocones of many Mesozoic and Paleozoic ammonoids (see Chapter 12, this volume). Apparently because of these similarities, some authors have suggested that the modifications are not late-ontogenetic features at all but were produced early or periodically and then eroded or resorbed to allow further growth (for example: Nikitin, 1884; Siemiradzki, 1925; Davitashvili and Khimshiashvili, 1954; Sazonov, 1957; *fide* Makowski, 1962).

There is a great overall similarity of these modifications within a given taxon or subdivision of that taxon. Hence, it is unlikely that they are the result of senility, disease, or genetic problems.

Resorption and erosion should have left some individuals with partially resorbed or partially eroded shell modifications; according to Makowski (1962), none are known. Moreover, some modifications (changes of coiling, for example) are located so far adapical of the aperture that more than one full whorl would have had to have been resorbed or eroded to remove them (see the medial section of the Permian species *Hyattoceras abichi* in Arkell *et al.*, 1957, Fig. 2c). These facts strongly indicate that the late-ontogenetic shell modifications discussed here truly are late ontogenetic, in fact, that they were terminal with respect to shell growth. Moreover, in animals that have them, late-ontogenetic shell modifications indicate that growth was determinate.

Comparable shell modifications in present-day *Nautilus* are associated with the onset of sexual maturity. Summing up, it must be concluded that the late-ontogenetic shell modifications in ammonoids are, in fact, mature modifications. All of this is no startlingly new revelation, of course (Pompeckj, 1894).

However, the distinction between maturity and gerontism may be moot. Many present-day coleoid cephalopods reach maturity, reproduce, and die, all in short order. If ammonoids did the same, then, in effect, "mature" and "gerontic" are the same.

2.3. Function of Mature Modifications

As so well stated by Lehmann (1981, p. 85), "It is difficult to resist the temptation to look for a purpose for these structures . . . ." Such searches tend to be either what the late Kenneth E. Caster used to call "high in so-whatness" or even absurd. This is especially so in light of the absence of detailed stratigraphic and paleoecological analyses. Until more such investigations appropriate to ammonoid mature modifications have been done, the discussion presented by Davis (1972) probably is sufficient. After all, Kerr (1931) was so impressed with the apparent uselessness of most mature modifications that he concluded that ammonoid shells must have been internal.

Two points seem worth emphasizing, however. Given the fact that the "mature" shell differs morphologically from the immature, it may well be that the adult faced different physiological and, hence, ecological needs than did the juvenile (for a thorough-going discussion of the ecology and way of life of ammonoids, see Chapter 16, this volume.). Alternatively, if the animal, on reaching adulthood, quickly bred and died, mature modifications may have served mainly a sexual or reproductive function.

3. Dimorphism

Dimorphism may be defined as the existence in a given species of individuals of two different adult morphologies. The two opposing members of a dimorphic pair have been termed antidimorphs (Davis, 1972).

The history of the concept of dimorphism with respect to ammonoids extends back more than a century and a half (de Blainville, 1840; d'Orbigny, 1847). From the very beginning, discussions of ammonoid dimorphism have been linked to sex. (The history of the study of dimorphism in ammonoids has been recounted excellently and in detail in the not-too-far-distant past and, so, does not need to be retold here; the historically inclined reader may want to peruse Makowski, 1962, and Callomon, 1963, and, for a later perspective, Lehmann, 1981.)

What might be thought of as the "modern age" of ammonoid dimorphism began with the almost simultaneous publication of works by Makowski (1962) and Callomon (1963). Both authors worked mainly with Jurassic ammonoids from two different areas and, independently, came up with nearly the same conclusions regarding dimorphism and its biological meaning.

3.1. Kinds of Dimorphism

Not all dimorphism in ammonoids is the same. A number of different schemes have been proposed.

Makowski (1962) recognized two sorts of dimorphism. In doing so, he emphasized primarily the number of whorls. In type A, one antidimorph has five to six whorls but may have as few as four. The other antidimorph has seven to nine volutions but may have as few as six. In type B, one antidimorph has seven to nine whorls but may have as few as six. The other antidimorph has at least one more whorl than its opposite.

Guex (1968) proposed the addition of a third kind of dimorphism in which the smaller of the pair has only three or four volutions. This kind was called type O.

Like Makowski, Westermann (1964) recognized two kinds of dimorphism. In the less common of the two, antidimorphs differ only in size but not in other morphological characteristics. In the more abundant kind of dimorphism, one antidimorph not only is smaller than its opposite but has a more highly modified peristome. The two antidimorphs also may differ in other mature modifications. In addition to the two main categories, there was said to be dimorphism, intermediate in nature, in which the antidimorphs differ both in size and in minor morphological traits, for example, the nature of the ribbing on the mature shell.

Houša (1965) also discussed two varieties of dimorphism. In type I, the antidimorphs differ in size only; both have simple apertures. In type II, they differ both in size and in the nature of the peristome.

Zeiss (1969) recognized three categories. In group I, the antidimorphs differ in size and ribbing. In group II, there are differences in size and in the nature of the apertural region, but the antidimorphs are ornamented similarly. In group III, dimorphism is not clearly recognizable.

Obviously the several schemes do not correspond. Each of Makowski's categories, for example, includes representatives of both of Westermann's. Nevertheless, some generalities emerge.

There are some ammonoids in which dimorphism has not been recognized. In those taxa in which dimorphism has been documented, generally one antidimorph is larger than the other. For these, Callomon (1955) proposed the convenient and most apt terms "macroconch" and "microconch" for the larger and smaller morphs, respectively. Commonly, but not always, the two antidimorphs of a pair differ not only in size but also in mature modifications, such as the shape of the peristome, ornament, septal approximation, and so on. However, in the portions of the conchs adapical of the onset of mature modifications, shells of the two antidimorphs are closely similar and, in fact, may not be able to be differentiated.

It well may be that dimorphism was expressed in parts of the ammonoid less commonly preserved than the shell. For example Till (1909, 1910, *fide* Teichert *et al.*, 1964b) discussed putative dimorphism in Jurassic and Cretaceous cephalopod jaws (*Hadrocheilus* Till, 1907).

In addition to morphological characteristics such as shell size and shape, it is possible that antidimorphs may have differed in color. Mapes and Sneck (1987) reported two different color patterns in specimens of the Triassic genus *Owenites* Hyatt & Smith, 1905 from the same locality; they suggested that the different color patterns marked different sexes.

3.2. Recognition of Dimorphism

Criteria for the identification of dimorphism in ammonoids are well known (see, for example: Makowski, 1962; Callomon, 1963; Westermann, 1964; Davis, 1972):

1. There should be distinct differences between antidimorphs.
2. Antidimorphs should have identical early ontogenies.
3. Antidimorphs should have the same stratigraphic ranges.

4. The numerical ratio of one antidimorph to the other should be consistent across the total stratigraphic and geographic range of the taxon.
5. Antidimorphs should have identical phylogenies.

The first of the above is patently obvious. If one can not tell antidimorphs apart, one certainly can not claim that dimorphism exists. Or, if one were to find more than two morphs of mature individuals in a species, then, by definition, one would be dealing with something other than simple dimorphism.

It is generally recognized that the inner whorls of antidimorphs are identical or closely similar. As maturity was approached, the microconchs and macroconchs diverged from one another morphologically, even as antidimorphs of *Homo sapiens* L. do as puberty is reached. (Note, however, that McCaleb, 1968, concluded that what he considered to be antidimorphs of the Late Carboniferous species *Syngastrioceras oblatum* are more different from one another as juveniles than as ontogenetically older individuals—see the discussion in Section 4.2.1b.)

The very fact that antidimorphs differ from one another morphologically suggests that they differed ecologically as well and, hence, may have occupied different geographic areas. Thus, they need not occur together in a given lithostratigraphic unit consistently, nor in constant ratios. After all, the opposite sexes of certain present-day cephalopods spend most of their lives apart, coming together only for procreation.

It would seem a given that all individuals of a particular species are the result of the same evolutionary history. However, if the traits that serve to differentiate antidimorphs had a purely sexual function in life, then those of the males could have evolved in a different way than did those of the females. This, in turn, could have resulted in a lack of a one-to-one correlation of the stratigraphic ranges of the antidimorphs. Indeed, as reiterated by Lehmann (1981), macroconchs in a given stratigraphic sequence not uncommonly show greater morphological variation than do microconchs in the same sequence; this has resulted in more taxa having been named among macroconchs than amongst microconchs. Thus, if one attempts to "pair up" macroconchs and microconchs, taxon to taxon, the would-be matchmaker is doomed to frustration. (See also Section 6.2.)

It is critical that the phylogeny and, thus, the taxonomy of the animals under consideration be clearly understood. After all, if one notes the existence of two distinct morphologies within a "species," before positing the existence of dimorphism, one needs to eliminate that possibility that the "species" is, in fact, two species.

3.3. Sex and Which Is Which

Present-day cephalopods are dioecious. Hence, it is hardly surprising that dimorphism in cephalopods almost universally has been considered to be sexual dimorphism. Particularly when dealing with a very diverse community of ammonoids, one might have difficulty in deciding which microconch is the antidimorph of which macroconch, but there is virtually no question that there were male ammonoids and female ammonoids.

In taxa in which dimorphism is readily apparent, the question immediately arises as to which antidimorph is the female and which is the male. A number of lines of evidence have been used to solve the problem (see, for example, summaries in Davis, 1972, and Lehmann, 1981.).

Most workers have considered that the macroconch and microconch are female and male, respectively. It is, of course, true that the female of the octopod *Argonauta* L. is many times the size of the male. However, among present-day cephalopods, which sex, if any, is the larger, and by how much, differs widely from taxon to taxon (see, for example: Makowski, 1962; Wells, 1962, 1966; Mangold-Wirz, 1963; Westermann, 1969a; Mangold-Wirz *et al.*, 1969). The fact that the male in *Nautilus* is somewhat larger and broader than the female has been known for many years (Willey, 1895, 1902; Saunders and Ward, 1987; Hayasaka *et al.*, 1987). In short, size alone does not provide an unequivocal answer to the general question as to which ammonoid antidimorph is female and which is male.

Mature modifications commonly are better developed in the so-called microconchs and, therefore, commonly are interpreted, without further discussion, as secondary sexual features. This may well be a valid interpretation in many cases, but it has yet to be proved as a general rule. Contrary to what frequently has been asserted in the literature, the presence of highly differentiated mature modifications in a small ammonoid is not enough to demonstrate that it is a microconch, let alone a male, and to claim a case of sexual dimorphism even in the absence of any possible macroconch in the assemblage. It is possible that what may look like a microconch is only an individual of a species in which males and females have similar, very small adult sizes.

In the mid-1800s, aptychi were regarded by some workers as indicators of gender. Keferstein (1866) was of the opinion that aptychi were for the protection of the nidamental glands and, hence, would be found only in females. On the other hand, Siebold (1848), apparently impressed by the size difference between male and female *Argonauta* and by the occurrence of the males within the mantle-cavities of the females of that genus, concluded that aptychi were the shells of dwarf male ammonoids. Thus, according to those ideas, an ammonoid shell containing aptychi would, most likely, have been a female.

The ratio of the number of individuals of one antidimorph to that of the other has been used as evidence. According to Coëmme (1917, *fide* Makowski, 1962), the microconch of a pair is less abundant than the macroconch, and, in present-day cephalopods, the male is less common; therefore, microconchs are males. Unfortunately, neither of the premises in this argument is invariably correct. Geraghty and Westermann (1994) reported ratios of antidimorphs in two species of *Stephanoceras* Waagen, 1869 in the Jurassic of Germany as 100 : 1 in one and 1 : 100 in the other (see also Sections 4.2.3 and 7.2.3 of this chapter). Reported sex ratios in present-day coleoids vary from at least 3 : 1 in favor of males to at least 7 : 1 against (see, for example: Pelseneer, 1926; Mangold-Wirz, 1963; Westermann, 1969a; Mangold-Wirz et al., 1969). In live-trapped *Nautilus*, females are virtually always outnumbered by males—by ratios of 0.43 : 1 to 0.05 : 1 (Willey, 1902; Saunders and Ward, 1987; Hayasaka et al., 1987), although Willey did report 16 females to 10 males (1.6 : 1) in *Nautilus macromphalus* Sowerby, 1849. In other words, the numerical ratio of one antidimorph to its opposite, taken alone, is not very helpful in determining which sex is which.

Lehmann (1981) argued that maturation of eggs in cephalopods generally takes longer than does that of spermatophores and that it took macroconchs longer to reach their maximum size than it did the microconchs, which are smaller. Therefore, according to Lehmann, it is reasonable to conclude that the macroconchs are female and the microconchs are male.

At least three instances are known of putative ammonoid egg capsules within body chambers of ammonoids (Lehmann, 1966; Müller, 1969; Zakharow, quoted in Lehmann, 1981, q.v.). In all three, the ammonoid is a macroconch. If these finds have been interpreted correctly, this would add weight to the "macroconch is female" argument. (Maeda's 1991 report of small ammonoids within the body chamber of a larger ammonoid was considered by him to be of a postmortem accumulation.)

3.4. Monomorphism and Polymorphism?

A number of reports have appeared that indicate that simple sexual dimorphism may not be the whole story. Brochwicz-Lewiński and Rózak (1976) have pointed to the possibility of transitions from microconchs to macroconchs in the Jurassic perisphinctids. Meléndez and Fontana (1993) have described both dimorphism and nonsexual polymorphism in the same family.

Ivanov (1971, 1975, 1983) has argued that, among the Jurassic ammonoids that he has studied, instead of two morphs per taxon, there are three. In addition to the expected microconchs and macroconchs, there is a significantly larger form that Ivanov called a megaconch. All three of these morphs have identical inner whorls. Unlike the micro- and macroconchs, however, megaconchs do not have septal approximation or other traits that indicate determinate growth. In other words, it would appear as though megaconchs continued to grow, essentially unaffected, right past the sizes at which first microconchs and then macroconchs matured and stopped growing.

Matyja (1986, 1994) also reported that, in some of the taxa of Jurassic ammonoids that he studied, there are three morphs. Again, there are the expected macroconchs and microconchs, but, in addition, there is a group of even smaller individuals that Matyja designated miniconchs.

In addition to trimorphism, Matyja indicated that, in some Jurassic taxa (for example, the middle Oxfordian genus *Gregoryceras* Spath, 1924), there is essentially but a single morph. This form resembles the microconch of other taxa.

This reported occurrence of one, two, or three morphs within ammonoid taxa led Matyja to postulate the existence in the Ammonoidea of what he has called developmental polymorphism. According to this hypothesis, for each species of ammonoid there is a particular path of ontogeny/shell-growth. Depending on circumstances, maturation and, thus, departure from the particular path might have been triggered at different times. For example, under certain circumstances, all individuals of a population might have matured and stopped growing at the same time, resulting in what might be called monomorphism. The morphology (including size) of the mature individuals would depend on where on the ontogeny/shell-growth path they were when maturation was triggered.

Under other circumstances, the males and females, being at least somewhat different from one another physiologically, might have matured at two different points on the path and, hence, be morphologically different (including size). This would have resulted in dimorphism in a population of fossils. Thus, "sexual" dimorphism is not caused by gender, *per se*, but by the fact that the two sexes matured and diverged from the ontogeny/shell-growth path at two different points.

Under even other circumstances, maturation/divergence might have occurred at three different times. This would have resulted in trimorphism in the population of fossils. The morphologies (including sizes) of the three morphs would have depended on when each of the three matured with respect to the ontogeny/shell-growth path of their species. According to this theory, miniconchs are from animals that matured very early, and megaconchs from those that matured very late, or not at all. The occasional animal might have reached a very large size before death intervened, which could be the explanation for anomalously large individuals occasionally reported (for example: Mapes, 1976).

Just what are the circumstances on which the time or times of maturation depended? Matyja (1986) related the nature of the mono-, di-, or trimorphism exhibited in a given population to geography and lithofacies and, hence, to environment.

One environmental factor that has been related to the timing of maturation is the presence of parasites. In some present-day molluscs, infestation of an animal by parasites can lead to the destruction of the gonads. This, in turn, can result in a delay or even a lack of both maturation and cessation of growth. Such pathological gigantism might manifest itself in the occurrence of megaconchs. (Gigantism is one of the pathological conditions discussed in Chapter 15, this volume; see also Chapter 12, this volume, on growth.)

Polymorphism related to variation in the age of maturation is known in present-day coleoid cephalopods. In many of these animals, sexual maturation has been demonstrated to be controlled by a hormone released by the optic glands (Mangold, 1987; Chapter 12, this volume). The optic glands, in turn, are connected to the brain by way of nerves from the subpedunculate lobes. Although the hormone affects the gonads they must be in a particular developmental state for the hormone to exert any effect. If the gonads are in the appropriate state to respond to the hormone, then, when it is released, the maturation process is initiated and leads to growth to the final size of the animal and, in due course, to the completion of the life cycle and death.

Thus, the initiation of the maturation process is controlled by both the developmental state of the gonads and secretion of the optic-gland hormone. Any factors that affect the development of the gonads, or the production and release of the hormone or both would affect the start of the maturation process and, hence, the rest of the life of the animal, including the size to which it might grow.

Although the details are far from completely known, a number of environmental factors may be involved. Both light intensity and photoperiod have effects on maturation. High temperatures seem to result in a gonad's more rapidly attaining the stage at which the optic-gland hormone will affect it. Too little food early in ontogeny means that the gonad takes longer to become receptive to the hormone. Too little food later in life results in spawning at a smaller body size than in well-fed animals. Note that most of the results like these have been obtained from captive animals in the laboratory, where they would not have experienced the effects of seasonality and migrations these creatures might have encountered in the wild.

Mangold (1987) has reviewed a mass of information about maturation and spawning of present-day cephalopods. Among the examples pertinent to the dimorphism/polymorphism question is the ommastrephid squid *Dosidicus gigas* from the Pacific Ocean; there are three populations: small sized and early maturing, intermediate sized and late maturing, and large sized and very late maturing. It has been suggested that temperature differences among the areas these populations inhabit is the significant factor, although animals undergo complicated patterns of migration, both geographic and in depth.

Hirano (1978, 1979) described possible trimorphism in the Cretaceous genus *Gaudryceras* Grossouvre, 1894 and proposed what he called "transient polymorphism." This is discussed in the Cretaceous part of this chapter (Section 7.2.2).

Callomon (1988), in his review of Matyja (1986), took exception to at least some of the interpretations and conclusions expressed therein. As pointed out by Callomon, in order for developmental polymorphism to be convincing as a general rule for ammonoids, it needs to be supported by extensive studies of numerically large populations of precisely known geologic ages derived from a wide variety of lithofacies, chronostratigraphic units, geographic areas, and taxa. Matyja's more recent work (1994) obviously is part of an attempt to fill that need. Callomon's advice applies just as well to other intriguing ideas about polymorphism that have been proposed for the ammonoids.

3.5. Taxonomic Implications and Perspectives

Once one recognizes a pair of antidimorphs as such, then one is faced with the question of what to call them taxonomically. Many different ways of distinguishing antidimorphs taxonomically have been used. Only some of these have been biologically straightforward and meaningful.

If one is convinced that two morphs are opposite sexes of a pair, the conclusion is that they could have interbred. Thus, by definition, the two members of a sexually dimorphic pair are members of the same species. Hence, both must bear the same binomen, according to the *International Code of Zoological Nomenclature*.

If, prior to the recognition of sexual dimorphism, both antidimorphs had been recognized as belonging in the same species, the application of the single binomen is easy. Various methods could be used to designate which is which. Demanet (1943), for example, designated antidimorphs thus: *Gastrioceras listeri* forme *listeri* and *Gastrioceras listeri* forme *subcrenatum*; the former was considered to be the female. On the other hand, Furnish and Knapp (1966) chose to use designations that do not look so much like the names of subspecies: *Gastrioceras occidentale* Form α and *Gastrioceras occidentale* Form β. Alternatively, an M (for macroconch) or an m (for microconch) could be placed after the species name. If one is convinced that one knows the actual sexes, then a ♂ or ♀ could serve the purpose.

If, however, the two groups now recognized as antidimorphs originally had been assigned to different genera or other supraspecific taxonomic levels, the name situation is decidedly more complicated. The potential nomenclatorial morass has been admirably plumbed in the discussions by Callomon (1969) and Lehmann (1981). There does not seem to be an easy way to drain the swamp, either (but see Westermann (1969b) for one proposal).

Nor does the problem evaporate once the two antidimorphs have their proper names. The systematic implications are manifold. For example, the demonstration of dimorphism in one species not uncommonly leaves in doubt the status of closely related species in which sexual dimorphism has not yet been discovered.

On the other hand, the type of dimorphism, etc., that a group of ammonoids exhibits might be helpful in elucidating relationships and, thus, be useful in reconstructing phylogenetic history. Houša (1965), for example, used type of dimorphism as a taxonomic indicator.

3.6. Sexual Dimorphism and Ontogeny

The consideration of sexual dimorphism in the context of ontogenetic development is a fairly unbeaten track that currently is littered with more questions than answers.

3.6.1. Heterochrony and Dimorphism

Are microconchs progenetic relative to macroconchs because of accelerated maturation, and/or are they hypomorphic (*sensu* Shea, 1986) because of retardation in the rate of size increase (Gould, 1977)? If, as is usual in paleontological approaches, size and age are not dissociated, microconchs must be thought of as progenetic. Conversely, if these two yardsticks are dissociated, progenesis is no longer the only possibility, and hypomorphosis also must be considered. Various rhythmic criteria such as septa, growth lines, megastriae, and rib densities can be used to pinpoint age (Doguzhaeva, 1982; Dommergues, 1988; Bucher and Guex, 1990; Mignot, 1993). Attempting to apply such criteria to the problem of dimorphism, Neige (1992) concluded that, among some Jurassic taxa, the small adult size of microconchs resulted either from straightforward progenesis or from a combination of progenesis and hypomorphosis (examples given were the Morphoceratidae and the Cardioceratidae, respectively). Moreover, the occurrence of mature modifications in many taxa complicates the identification of progenesis by inducing a combination of juvenile and mature features (Landman *et al.*, 1991).

As discussed in Section 3.3, the presence of highly differentiated mature modifications in a small ammonoid commonly has been accepted as a demonstration that sexual dimorphism exists in the taxon and that the bearer of the highly differentiated mature modifications is a microconch. Unfortunately, there are virtually no studies dealing with the ontogenetic expression of mature modifications versus size, and the question as to whether mature modifications are more a matter of sex or of small size remains unanswered.

3.6.2. The Relationship of Juvenile and Adult Morphologies

From a heterochronic point of view, is the morphogenesis of mature modifications predictable from the ontogenetic patterns of juveniles? In numerous cases, it commonly is obvious that the mature modifications are an outgrowth of growth-line patterns in earlier ontogeny. In some instances, it seems that the construction of mature modifications results from a nonhomogeneous preadult reduction in growth rate. For example, during the last phase of ontogeny, the growth rate may decrease faster or even stop earlier in the concave parts of the growth lines than in their convex parts (resulting in rostra or lappets). In these cases, the adult peristome of the microconch is an extension of the juvenile peristome (e.g., the Hildoceratidae and Hecticoceratinae). Sometimes, growth lines and ornament are disconnected, and only growth lines can be used to predict adult apertural morphologies (e.g., the Perisphinctidae and Otoitidae). However, in extreme cases, the morphology of the mature modifications is so derived that its relationship to the shape of previous growth lines is unclear (e.g., the Oecoptychiidae and Leptosphinctinae).

Do both microconchs and macroconchs diverge from juvenile patterns of growth as they attain maturity, or do only microconchs do so? Models for comparison of ontogenies of sexual antidimorphs have been proposed by Brooks (1991), who identified six possible mature male and female patterns from a single theoretical common juvenile pattern. Unfortunately, this approach has not been applied to dimorphism in ammonoids.

4. Mature Modifications and Dimorphism in Paleozoic Ammonoids

As related in the excellent historical surveys of Makowski (1962) and Callomon (1963), most students of ammonoid mature modifications and dimorphism have concentrated on the Mesozoic, and especially on the Jurassic. Most descriptions, mentions, and illustrations of these phenomena in the Paleozoic are widely scattered in the literature, mostly in taxonomic works. According to Makowski, sexual dimorphism in goniatites was discussed for the first time by Foord and Crick (1897), although Haug, in the same year, suggested a case of potential dimorphism in ammonoids from the Permian of Italy.

4.1. Examples of Mature Modifications in Paleozoic Ammonoids

4.1.1. Mature Modifications in Devonian Ammonoids

In his discussions of dimorphism in Devonian goniatites, Makowski (1962, 1991) described several mature modifications. In *Tornoceras* Hyatt, 1884, for example, there are septal approximation and sutural simplification; in macroconchs, there are, in addition, a reduction of the diameter of the umbilicus and an increase in whorl height. In *Manticoceras* Hyatt, 1884, there also is an increase in whorl height in the apparent macroconch. In addition, in *Cheiloceras subpartitum* (Münster, 1839), the type species of the genus, constrictions on the internal mold are more crowded on the last body chamber of the apparent macroconch than they are on the phragmocone.

Korn (1992) reported mature modifications in individuals of other Devonian taxa. Mature specimens of *Balvia* Lange, 1929 and *Mimimitoceras* Korn, 1988 each have an apertural shell thickening. Of interest, Korn also indicated that, in the latter genus, juveniles differ from one another more than do adults of the same species.

On the other hand, juveniles and adults of *Prionoceras* Hyatt, 1884 do not differ significantly in their morphology. In fact, according to Korn (1992), there is a trend in the Prionoceratidae toward morphological differences between preadults and adults.

Ruzhencev (1962) illustrated mature specimens of two anarcestids. In *Prolobites* Karpinsky, 1885, the mature shell is egg-shaped with a closed umbilicus; there is a deep apertural constriction so that the animal would have been almost completely enclosed inside the shell. In *Timanites keyserlingi* Miller, 1938, the peristome has lappets and a ventral sinus.

4.1.2. Mature Modifications in Carboniferous Ammonoids

Trewin (1970), in his treatment of apparent dimorphism in *Eumorphoceras yatesae* Trewin, 1970 from the Namurian of England, dealt with specimens, all of which are crushed. Nonetheless, he was able to discern that there is a diminution in ornament in adapertural regions of larger shells.

Ruzhencev (1962, p. 255, 257) reported that the body chamber of *Glaphyrites* Ruzhencev, 1936 widens considerably near the aperture and becomes flatter ventrally; there is an apertural constriction, but this is confined to the umbilical region, although the edge of the aperture turns in on the venter. A number of mature modifications in specimens of the same genus from the Upper Carboniferous Wewoka Formation of Oklahoma were found by Nettleship and Mapes (1993). These include septal approximation, septal simplification, a thickened last septum, muscle scars, changes in coiling, and a change in the *Runzelschicht.*

According to Ruzhencev (1962) individuals of at least some species of the Early Carboniferous (Namurian) goniatites *Dombarites* Librovitch, 1947 and *Homoceras* Hyatt, 1884 developed a large ventral keel, as did the figured specimen identified as *Praedaraelites aktubensis* Ruzhencev, a prolecanitid, also from the Namurian.

Frest et al. (1981) reported mature-modifications in *Maximites cherokeensis* (Miller and Owen, 1939) Miller and Furnish, 1957 from the Upper Pennsylvanian of Oklahoma and Missouri. In mature individuals of this species there is a narrow apertural constriction with a broad hyponomic sinus and conspicuous ventrolateral salients; in addition, septal approximation is present. Mature-modifications in *M. oklahomensis* Frest, Glenister, and Furnish, 1981, from the Upper Pennsylvanian of Oklahoma, are similar, although there is no apertural constriction; the venter is slightly flattened in the last one-third whorl, and a slight divergence in coiling is present (Frest *et al.*, 1981).

4.1.3. Mature Modifications in Permian Ammonoids

Davis (1972) discussed at some length mature modifications in Permian ammonoids from a number of taxa. Examples of most of the categories of late-ontogenetic shell modifications itemized in the introductory part of Section 2 of this chapter were found to occur in one or more of these taxa (Figs. 1–3). Table I is, in effect, a checklist of which kind of mature modification is present in which taxon.

In addition to those examples, there is a report by Nassichuk (1970) of apparent mature modifications in *Spirolegoceras fischeri* Miller, Furnish, and Clark, 1957 from the Lower Permian Phosphoria Formation of Idaho. Transverse constrictions are present on the internal molds of the phragmocones but are reduced in intensity or absent in the body chambers of the largest specimens. In a specimen of *Popanoceras* sp. cf. *P. sobolewskyanum* (de Verneuil, 1845) from the Permian of the Canadian Arctic Archipelago, Nas-

Table I. Mature Modifications in Selected Permian ammonoids[a]

Taxon	Coiling change	Aper. constr.	Ornament change	"Lappets"	Whorl-sec. ch.	"Runzel-schicht"	Ventral salient	Thick septum	Approx-imation	Suture simpl.
Adrianitidae										
Adrianites	X	X		X	X	X				
Crimites	X	X			X					
Epadrianites	X	X								
Hoffmannia	X	X	X		X					
Neocrimites	X	X	X	X						
Palermites	X	X		X	X					
Pseud-agathiceras	X	X	X							
Sizilites	X	X	X		X					
Texoceras	X	X				X			X	X
Cyclolobidae										
Cyclolobus	X	X		X	X					
Mexicoceras	X	X			X		X		X	
Waagenoceras	X	X			X	X		X	X	
Hyattoceratidae										
Hyattoceras	X	X	X		X					
Marathonitidae										
Marathonites	X	X			X					
Pseudovidrioceras		X	X		X				X	
Vidrioceratidae										
Peritrochia		X		X						
Stacheoceras	X	X		X	X					
Agathiceratidae										
Agathiceras	X	X	X	X	X					

[a]Summarized from Davis (1972). Note that *Waagenina*, as used in that work, is included in *Stacheoceras* here; note also that familial assignments have been updated, too.

sichuk found a mature modification in the form of a deep, constricted ventral sinus; Ruzhencev (1962) also reported a very deep ventral sinus in this species. On the other hand, Ruzhencev identified a specimen of *Paragastrioceras* Tschernow, 1907 in which there is a pronounced ventral projection in the peristome with a small sinus right in its tip.

Schiappa *et al.* (1995) reported mature-modifications in *Nevadoceras steelei* Schiappa, Spinosa, and Snyder, 1995. In this adrianitid from the Permian of Nevada, the conch is sublenticular at small diameters, but is pachyconic at maturity, there is a change in coiling, and there is an apertural constriction; the mature body-chamber is 330° long. In the same paper, specimens of *Veruzhites* Leonova, 1988, from the Permian of the Pamir Mountains, Tajikistan, were illustrated for purposes of comparison; some of these appear to show changes in coiling and an apertural constriction.

In *Zhonglupuceras celestre* (Yabe, 1928) Zhou Zuren, 1985, a pseudo-haloritid from the Artinskian of Hunan, there are transverse tubercles near the umbilici of mature individuals as well as a longitudinal tubercle on each side just adapical of the aperture. The aperture is first constricted and then expanded, and the apertural rim is thickened (Zhou Zuren, 1985). In that same paper, similar mature modifications were reported in *Zhonglupuceras longilobum* Zhou Zuren, 1985.

Mature-modifications in the pseudohaloritid subfamily Shouchangoceratinae are variable (Frest *et al.*, 1981). In *Neoaganides*, the presumed ancestor of the lineage, mature-modifications are not conspicuous; in the type-species, *N. grahamensis* Plummer and Scott, 1937, from the Upper Pennsylvanian of the United States, the venter is flattened slightly at maturity, and there are a distinct ventral sinus and weak ventrolateral salients in the peristome. In other genera of the subfamily there is a range of mature ornaments, peristomal configurations, changes in whorl cross-section, and changes in coiling. For example, Frest *et al.* (1981) figured spectacular mature-modifications in some Upper Permian forms from South China. In *Elephantoceras nodosum* Zhao and Zheng, 1977 and *Shangraoceras falcoplicatum* Zhao and Zheng, 1977, there are long ventrolateral lappets that curve toward the dorsum. In *Sangzhites aberrans* Zhao and Zheng, 1977, there is a pair of similar ventrolateral lappets, along with a shorter lateral or dorsolateral lappet on each side. On the other hand, mature specimens of *Shouchangoceras* Zhao and Zheng, 1977 do not have conspicuous lappets. In a number of genera of the subfamily, there are changes in the radius of curvature of the coiling, so that the venter of the ultimate one-quarter to one-third of a whorl appears flat in lateral view, and the conch has a weakly geniculate profile. However, in mature specimens of *Neoaganides* Plummer and Scott, 1937, which originated in the Late Pennsylvanian, and *Erinoceras* Zhao and Zheng, 1977, from the Lower Permian, the ultimate whorl shows a tendency to diverge from the penultimate volution (Frest *et al.*, 1981).

In *Pseudohalorites*, mature individuals each have conspicuous apertural constriction; weak ventrolateral salients may be present in the peristome, and there is a weak, broad ventral sinus (Frest *et al.*, 1981). In *P. arabicus* Miller and Furnish, 1957, from the Sakmarian of Oman, for example, there is a double apertural constriction, but no lappets.

4.2. Dimorphism in Paleozoic Ammonoids

As mentioned previously, the first discussion of sexual dimorphism in goniatites seems to have been by Foord and Crick (1897). For example, in their discussion of the Carboniferous species *Gastrioceras carbonarium* (von Buch, 1832), they recognized antidimorphs in which one is more inflated and more widely umbilicate than the other.

4.2.1. Examples of Dimorphism in Paleozoic Ammonoids

4.2.1a. Dimorphism in Devonian Ammonoids. This is best known through the works of Makowski (1962, 1991). He described examples of his type-A dimorphism in a number of taxa of goniatites: *Manticoceras, Tornoceras,* and *Cheiloceras* Frech, 1897. In all of these, of course, antidimorphs differ in the number of whorls and in adult diameter. The inner whorls of individuals of each pair are identical; the outermost whorl of each macroconch of at least some of these taxa has a somewhat greater whorl height than does the corresponding microconch. In *Tornoceras frechi parvum* Makowski, 1991, the venter of the adult macroconch is slightly flattened as compared to the microconch. In *Tornoceras subacutum* Makowski, 1991, the venter of the macroconch is flattened close to the aperture, whereas that of the microconch is rounded. The species of *Cheiloceras* in which Makowski found type-A dimorphism was the type-species *C. subpartitum*; however, in other species that have been referred to the genus, he found type-B dimorphism (see Section 3.1 above).

Korn (1992) speculated that specimens referred to *Mimimitoceras* might be antidimorphs of others referred to *Balvia*. He rejected this interpretation because the relative abundances of the two genera reverse through the course of the *Wocklumeria* Stufe in Germany.

According to Makowski (1962), dimorphism in the clymeniid *Clymenia involuta* Wedekind, 1908 was discovered by Perna (1914). Walliser (1963) reported dimorphism in anarcestids in which antidimorphs differ in ornament, but he gave no details or examples.

4.2.1b. Dimorphism in Carboniferous Ammonoids. Trewin (1970) identified dimorphism in the Namurian species *Eumorphoceras yatesae* Trewin, 1970. Unfortunately, all the specimens are crushed, so he was unable to determine actual diameters, numbers of whorls, or whorl cross sections. In one of what he considered to be antidimorphs, the smaller of the pair, he discovered that transverse ribs first reduce to nodes and then die out adaperturally. In the other antidimorph, the ribs weaken but do not disappear or reduce to nodes; moreover, they extend to a larger diameter. Both putative antidimorphs occur in the same strata and, apparently, only there; as far as could be determined from the material, the earlier ontogenetic parts of both are identical. At two other stratigraphic levels in the same part of England, there are two other pairs of ammonoids that may be sets of antidimorphs: *Eumorphoceras bisulcatum ferrimontanum* Yates, 1962 and *E. bisulcatum erinense* Yates, 1962; and *E. bisulcatum grassingtonense* Dunham and Stubblefield, 1945 and a smaller, undescribed form.

McCaleb *et al.* (1964) reported that specimens of the population of *Arkanites relictus* (Quinn, McCaleb, and Webb, 1962) they studied could be divided into three morphological categories on the basis of shell shape but that all differences in shape are intergradational. Note that the authors did not report any specific mature modifications; thus, decisions about ontogenetic stage seem to have been based on the size of the specimen. In contrast, Manger and Rice (1975) reported dimorphism in specimens of the same species from the Morrowan of Arkansas and Oklahoma. According to them, antidimorphs differ in shell proportions and ornament.

McCaleb *et al.* (1964) also reported that *Tumulites varians* McCaleb, Quinn, and Furnish, 1964 exhibits variation similar to that they described for *Arkanites relictus.*

Furnish and Knapp (1966) suggested that dimorphism is common in the families Gastrioceratidae and Schistoceratidae of the Pennsylvanian. In the former, there have been a number of accounts of dimorphism in *Gastrioceras* Hyatt, 1884 in addition to Foord and Crick's (1897) discussion mentioned in the introduction to this section. In *Gastrioceras listeri* (Martin, 1809), Demanet (1943) recognized "*G. listeri* forme *listeri*" and "*G. listeri* forme *subcrenatum*" as antidimorphs. The two occur in intimate association in diverse marine rocks in both Europe and North America. "Forme *listeri*" was considered to be the female; specimens are swollen, with flattened whorls and large umbilici. Individuals of "forme *subcrenatum*," on the other hand, are less inflated, have "more-elevated" whorls, and have smaller umbilici. Unfortunately, Demanet did not give measurements of diameter. However, the author did cite other species that were considered either to be members of a dimorphic pair or to include antidimorphs: *G. rurae* Schmidt, 1925 and *G. martini* Schmidt, 1925; *G. cumbriense* Bisat, 1924; *G. crenatum weristerense* Demanet, 1943; and *Homoceras beyrichianum biplex.*

Sexual dimorphism in *Gastrioceras occidentale* (S. A. Miller and Faber, 1892) was discussed by Furnish and Knapp (1966). They designated the following antidimorphs: *G. occidentale* form α and *G. occidentale* form β, of which the latter grew to a larger size. Strangely, morphological differences were found to be greater in the immature whorls than in the adult part of the shell. Other citations of dimorphism in *Gastrioceras* include Ramsbottom and Calver (1962) and McCaleb (1968).

Sexual dimorphism was suggested to occur in *Axinolobus* Gordon, 1960 by McCaleb and Furnish (1964) and mentioned by McCaleb (1968). According to the former publication, none of the specimens available for the study retains a complete body chamber or aperture; hence, large specimens were assumed to be mature. McCaleb and Furnish speculated that *A. quinni* Miller and Furnish, 1964 and *A. modulus* Gordon, 1960 from the Lower Pennsylvanian (Morrowan) of Arkansas and Oklahoma are antidimorphs. The venter of mature individuals of the former is concave, whereas that of the latter is rounded; the umbilicus of *A. quinni* is more open, its whorl cross section is thinner, and it has more prominent transverse ribbing. Moreover, *A. quinni* apparently reached a larger diameter.

McCaleb and Furnish also suggested that *Branneroceras branneri* Smith, 1896 and *B. halense* Miller and Moore, 1938 are antidimorphs. Both occur together, and one has a relatively broad umbilicus and strong ornament, whereas the other has a narrower umbilicus and weaker ornament. McCaleb (1968) indicated that the antidimorphs differ in shell proportions but that there is some intergradation between the two forms.

McCaleb (1968) discussed potential dimorphism in *Syngastrioceras oblatum* (Miller and Moore, 1938) from the Early Pennsylvanian. The two putative antidimorphs formerly were called *Glaphyrites oblatus* Miller and Moore, 1938 and *Cravenoceras? morrowense* Miller and Moore, 1938. There are fewer volutions in the latter, and it has a narrower umbilicus; the differences in whorl cross section result in differences in the sutures of the two forms. There is overlap in the range of variation of the cross sections, and, as reported for some other Carboniferous taxa, the morphological differences are greater between juveniles than among the apparent adults.

In addition to the above-enumerated examples, McCaleb (1968) referred to dimorphism in *Pseudogastrioceras* Miller, 1934 and *Retites* McCaleb, 1964, both from the Early Pennsylvanian. On the other hand, although they found a number of mature modifications in a population of *Glaphyrites* from the Upper Carboniferous of Oklahoma, Nettleship and Mapes (1993) found a wide variation in size of individuals at maturity and no apparent macroconchs and microconchs. On the other hand, the study by Kant (1973) would seem to point to dimorphism with respect to certain conch dimensions in *Glaphyrites* and other Carboniferous taxa and even polymorphism in some.

The dimorphism reported by Frest *et al.* (1981) in *Maximites oklahomensis*, from the Upper Pennsylvanian of Oklahoma, is unusual. What they interpreted to be antidimorphs ("form *a*" and "form *b*") differ only in details of the mature suture.

4.2.1c. Dimorphism in Permian Ammonoids. Both Haug (1897) and McCaleb and Furnish (1964) suggested that *Hyattoceras (Hyattoceras) geinitzi* (Fig. 2F–I) and *H. (Abichia) abichi* (Fig. 2C–E), from the Permian of Italy, might be antidimorphs; they differ both in size and in the nature of the mature modifications. In addition to this example, Davis (1972) suggested at least the possibility of dimorphism in *Marathonites* Böse, 1919, *Waagenina* Krotow, 1888 (now considered a junior synonym of *Stacheoceras* Gemmellaro, 1887), and *Agathiceras* Gemmellaro, 1887.

There do seem to be two morphologies of mature individuals in *Marathonites*; these differ in shell shape but not in size. Unfortunately, there are so few specimens of each known, and they were collected at different localities, that it simply is not clear whether the two forms are antidimorphs or belong to separate taxa. In *Stacheoceras darae* Gemmellaro, 1887, possible antidimorphs differ both in size and in the configuration of the apertural constriction, but there are too few mature specimens known to allow one to do other than just suggest the possibility of dimorphism.

On the other hand, dimorphism in a population of *Agathiceras uralicum* from the Permian of Texas seems more strongly indicated (Davis *et al.*, 1969; Davis, 1972). The shell proportions, ornament, and mature modifications of the antidimorphs are the same (Fig. 3A–D); however, they differ in size—the smaller averages 20.7 mm in diameter, and the larger, 30.0 mm. Even this case, however, is established on the basis of a single collection from one stratigraphic level at one locality, and the early whorls are not well enough preserved to allow the ontogenies to be deciphered.

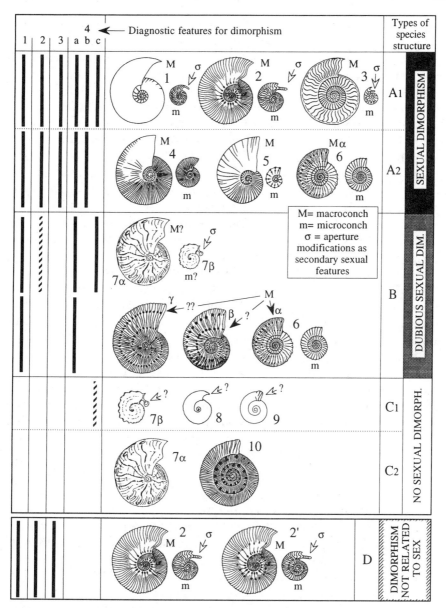

FIGURE 4. Significance of the main types of species structure in Jurassic ammonoids related to dimorphism, with an indication of their diagnostic features (striped bars indicate a level of uncertainty) and illustrations of some selected characteristic examples: 1 M and m, *Cardioceras* Neumayr and Uhlig, 1881; 2 M and m, *Kosmoceras* Waagen, 1869; 3 M and m, *Hildaites* Buckman, 1921; 4 M and m, *Macrocephalites* Zittel, 1884; 5 M and m, *Phricodoceras* Hyatt, 1900; 6 Mα and 6 m, *Aegoceras* Waagen, 1869; 6 Mβ, *Liparoceras* Hyatt, 1867; 6 Mγ, *Becheiceras* Trueman, 1918; 7α, *Taramelliceras* Del Campana, 1904; 7β, *Creniceras* Munier-Chalmas, 1892; 8, *Cymbites* Neumayr, 1878; 9, *Gemmellaroceras* Hyatt, 1900; 10, *Prodactylioceras* Spath, 1923. The types of species structure (A–D) and the diagnostic features for dimorphism (1–4) are discussed in the text.

Ruzhencev (1962) showed specimens of two species of *Popanoceras* Hyatt, 1884: *P. sobolewskyanum* (de Verneuil, 1845) and *P. annae* Ruzhencev; the ventral sinus in the peristome of the former is much more pronounced than is that of the latter. It is not clear whether this difference is taxonomic or dimorphic.

Frest, *et al.* (1981) suggested the possibility that *Pseudohalorites celestris* Yabe, 1928 and *P. celestris* var. *densistriatus* Chao, 1954, from the Artinskian of China, are antidimorphs. Zhou Zuren (1985), on the basis of more material, came to the same conclusion; both were assigned to *Zhonglupuceras celestre* (Yabe, 1928) Zhou Zuren, 1985. Mature microconchs (originally assigned to *P. celestris* var. *densistriatus*) have diameters of 17.4 to 21.1 mm, compared to 29.4 to 32.4 mm in macroconchs (Zhou Zuren, 1985). In the smaller antidimorph, lappets are developed, and the aperture apparently is thickened; the aperture of the larger form is less ornate. Zhou Zuren (1985) also reported dimorphism in *Z. longilobum*, a species he named that year.

In their discussion of the pseudohaloritid subfamily Shouchangoceratinae, Frest *et al.* (1981) described dimorphism that involves no difference in the size of the putative antidimorphs. In *Sosioceras pygmaeum* (Gemmellaro, 1888) Miller and Furnish, 1957, from the Upper Permian of Sicily, specimens with two different kinds of mature modifications occur in about equal proportions. Both forms have a double constriction near the peristome, which is strongest ventrally, but fades midway up the flanks; at that point, there is a gentle salient in the peristome on each side. In what they called "form *a*," the shell extends only slightly adapertural of the constrictions, and the ventral sinus in the peristome is slight. In "form *b*," on the other hand, the shell extends farther beyond the constrictions, and the hyponomic sinus is deeper.

4.2.2. Possible Polymorphism in Paleozoic Ammonoids

In his study of *Tornoceras frechi* Wedekind, 1918 from the Upper Devonian (Famennian) of Poland, Makowski (1991) found the expected macroconchs and microconchs. In addition, however, there are specimens of the species that are distinctly larger than the macroconchs. Makowski likened these to what Ivanov has called megaconchs (see Section 3.4 above).

In their study of *Arkanites relictus* (Quinn, McCaleb, and Webb, 1962) from the Morrowan of Arkansas, McCaleb *et al.* (1964) found three "morphs" that differ in whorl cross section and ornament. However, the differences in shape are intergradational, and, hence, the "morphs" are not discrete from one another. The same authors reported the same sort of variation of shell shape in *Tumulites* McCaleb, Quinn, and Furnish, 1964.

In a study on allometric growth in Paleozoic ammonoids, Kant (1973) presented data that might be taken as indicating the existence of polymorphism with respect to certain conch dimensions in certain Carboniferous ammonoids.

4.2.3. Proportion of Antidimorphs within a Species

Makowski (1991) reported the numerical ratios of macroconchs (M) to microconchs (m) in a number populations of Devonian goniatites: *Tornoceras frechi parvum*, 28M to 67m, equal to 0.42 : 1; *T. subacutum*, 31M : 34m (0.91 : 1); and *T. sublentiforme* (Sobolew, 1914), 53M : 80m in one bed and 28M : 34m in another (0.66 : 1 and 0.82 : 1, respectively).

In his discussion of putative dimorphism in the Early Pennsylvanian *Syngastrioceras oblatum*, McCaleb (1968) indicated that one antidimorph, not identified as to gender or macroconch/microconch, comprised 75% of the population.

In *Maximites oklahomensis*, from the Upper Pennsylvanian of Oklahoma, what Frest *et al.* (1981) called "form *a*" consititues about 40% of a population, as contrasted with "form *b*". The microconchs of *Zhonglupuceras celestre*, from the Artinskian of China, are much less abundant that are the macroconchs (Zhou Zuren, 1985).

In a population of *Agathiceras uralicum* from a single level in the Permian of Texas, Davis *et al.* (1969) and Davis (1972) found a ratio of 82M : 28m (2.9:1).

4.3. Patterns through Time in the Paleozoic

Relative to the ammonoids of the Jurassic and Cretaceous, few examples of mature modifications and very few of dimorphism have been documented in the Paleozoic, too few really to outline any patterns in time.

It may be that mature modifications either were lacking or were very subtle in many Devonian and Carboniferous taxa. With the tremendous number of studies on Paleozoic ammonoids that have been done in the last century and more, it seems unlikely that they would have been almost universally overlooked. One trend was noted by Korn (1992), who indicated that morphological differences between preadults and adults in the Devonian family Prionoceratidae increased over time.

It is possible that antidimorphs in the Devonian differed mainly in size and number of whorls and that dimorphism involving other mature modifications came only later. Unfortunately, there do not seem to be enough data points, at least in the Paleozoic, really to test McCaleb's contention that "dimorphism is a predominant feature at the inception of an evolutionary lineage and decreases through phylogeny" (1968, p. 29).

5. Mature Modifications and Dimorphism in Triassic Ammonoids

5.1. Mature Modifications in Triassic Ammonoids

Mature modifications and evolutionary patterns of Triassic ammonoids are poorly documented in comparison with those of the rest of the Mesozoic.

Although a systematic review is desirable, many families or even suborders would remain blank in any such compilation. As an alternative, the main types of mature modifications are presented here in relationship to the overall shell geometry of the phragmocone. Despite the absence of exhaustive documentation on the mature morphology of animals of many Triassic ammonoid groups, our qualitative approach seeks to highlight any morphological constraints or correspondence between broadly defined immature and mature morphologies. In a few well-documented cases, the main types of mature transformation can be related to heterochronies and evolutionary trends.

5.1.1. Ornament and Shape of the Adult Body Chamber

5.1.1a. Oxycones. As exemplified by the Middle Triassic Longobarditidae, oxycones with declining sculpture or without any sculpture when submature display only minor changes in their mature shape. The mature body chamber is generally thicker, with a less-acute venter. Some narrowly umbilicated oxycones tend to acquire an occluded umbilicus at maturity. There is little modification in the shape of the mature aperture, which follows either sinuous or biconcave growth lines. As an additional mature character of some longobarditids such as *Intornites*, achievement of maturity also is manifested by swollen upper flanks.

5.1.1b. Discocones. Members of the Middle Triassic Beyrichitinae provide a good illustration of mature modifications in discoidal forms. In the vast majority of beyrichitids, the whorl section generally becomes thicker at maturity, with simultaneous fading of sculpture, as shown by *Eogymnotoceras* (Fig. 5A,B). However, decreasing strength or even disappearance of sculpture is not a general rule, as there are cases, such as *Favreticeras* (Fig. 5H–J), where attainment of maturity is accompanied by the occurrence of prominent ornament that is absent in earlier growth stages. Of interest, the proterogenetic origin of the combination of a heavily sculptured mature stage with a smooth immature stage is well documented in the *Favreticeras* lineage (Bucher, 1992). Egression of the mature body chamber generally is conspicuous.

5.1.1c. Platycones. In the vast majority of platyconic forms, sculpture tends to fade at maturity. However, there are a few exceptions, such as *Platycuccoceras*, a member of the Middle Triassic family Balatonitidae. Persistence of hollow lateral spines on the mature body chamber of individuals of some species of *Platycuccoceras*, a genus derived from the entirely spinose *Augustaceras*, is interpreted as resulting from the proterogenetic introduction of the weakly sculptured platycone shape found in all immature stages of representatives of *Platycuccoceras*.

5.1.1d. Serpenticones. Smooth serpenticones do not display any striking changes at maturity. Sculptured serpenticones generally show decreasing strength and approximation of ornament at maturity, as well exemplified by the variocostate *Kashmirites* (Fig. 5F,G) of the Early Triassic.

The Early Triassic *Palaeophyllites* (Fig. 6I,J), the oldest known phylloceratid, already shows a ventral thickening of the ribs at maturity, a common trait

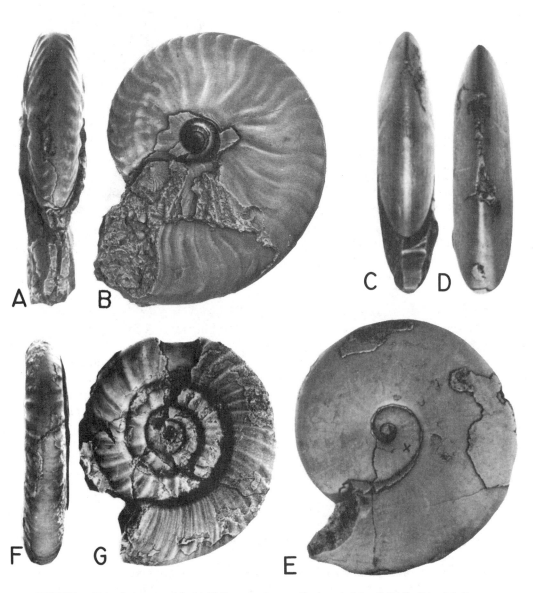

FIGURE 5. Triassic ammonoids. (A,B) *Eogymnotoceras* Bucher. Anisian (Middle Triassic). From Tozer, 1994, Pl. 60, Fig. 9; ×1. (C–E) *Amphipopanoceras* Voinova. Anisian (Middle Triassic). From Tozer, 1994, Pl. 61, Fig. 8; ×1. (F,G) *Kashmirites* Diener, 1913. Smithian (Lower Triassic). From Tozer, 1994, Pl. 36, Fig. 14; ×1. (H–J) *Favreticeras* Bucher. Anisian (Middle Triassic). From Bucher, 1992, Pl. 5, Figs. 4–6; ×1.

H I J

FIGURE 5. (Continued)

of Early Jurassic juraphyllitids. This peculiar character is so far not known from Middle and Late Triassic phylloceratids.

5.1.1e. Sphaerocones. The Late Triassic haloritids provide some of the most elaborate mature modifications among Triassic sphaerocones. Accompanying the loss of ribbing and the concomitant development of marginal nodes, the mature transition of *Halorites* (Fig. 6A–C) shows a dramatic change of the whorl section, with introduction of a low, subtabulate ventral shape before a reversal to the rounded shape near the final aperture. Drastic mature modifications occur in *Homerites* (see Mojsisovics, 1893, Pl. 89, Fig. 4), another haloritid. Coiling becomes abruptly elliptical, with a sudden appearance of ribs and ventrolateral spines. The first pair of spines is remarkably strong, mimicking a pair of "horns" above the animal's head.

Dramatic changes in the ventral outline of smooth sphaerocones at maturity are also well documented in some arcestids of the Late Triassic; the outline changes from subcircular to a lanceolate shape, followed by the final apertural reversion to a rounded shape. Although less pronounced, a comparable mature modification occurs in the Middle Triassic *Amphipopanoceras* Voinova, 1947 (Fig. 5C–E). A purely morphofunctional interpretation of the ventral shape of the smooth-shelled individuals of these taxa would see it as acting as a stern or at least being a peculiar feature restricted to smooth or weakly sculptured involute and more or less globose forms.

5.1.1f. Cadicones. Egression and contraction of the whorl shape are the chief mature modifications encountered among cadicones, such as the Late Triassic *Tropites* Mojsisovics, 1875 (see Mojsisovics, 1893, Pl. 108, Figs. 1, 3).

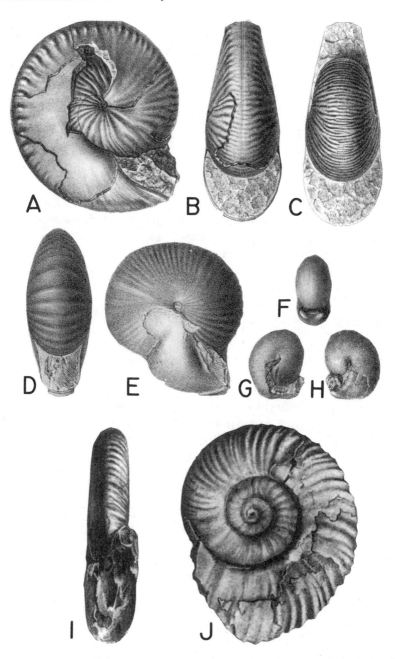

FIGURE 6. Triassic ammonoids. (A–C) *Halorites* Mojsisovics, 1879. Middle Norian (Upper Triassic). From Mojsisovics, 1896, Pl. 3, Fig. 3; ×1. (D,E) *Coroceras* Hyatt, 1877. Lower Carnian (Upper Triassic). From Mojsisovics, 1875, Pl. 70, Fig. 17; ×1. (F–H) *Arcestes* Suess, 1865. Carnian–upper Norian (Upper Triassic). From Mojsisovics, 1896, Pl. 20, Fig. 10; ×1. (I,J) *Palaeophyllites* Welter, 1922. Spathian (uppermost Lower Triassic). From Welter, 1922, Pl. 8, Figs. 5,6; ×1.

5.1.2. Apertural Modifications

Lappets comparable to those of some Jurassic species as well as hyponomic sinuses like those of some Paleozoic taxa are not known among Triassic ammonoids. The shape of the mature peristome generally reflects more-or-less closely the trajectory of the preceding growth lines, although the attitude of the final peristome generally is more prorsiradiate than is that of the growth lines adapical of it. Discoidal shells with a ventral projection of growth lines generally end with a peristome of similar shape that cannot properly be referred to as having a rostrum. A significant percentage of smooth Triassic ammonoids, ranging from serpenticones to sphaerocones, tend to signal completion of growth with an apertural collar. A remarkable trait of Triassic ammonoids is that the most elaborate apertural modifications are encountered among sphaeroconic forms. For instance: individuals of some species of *Arcestes* Suess, 1865 (Fig. 6F–H) develop a pair of gutter-shaped ventrolateral "lappets"; individuals of *Isculitoides* Spath, 1930 each have a rostrum that has a curved and rounded spatula-shape; and individuals of *Coroceras* Hyatt, 1877 (Fig. 6D,E) each have a hood-shaped aperture terminating an elliptical body chamber.

5.2. Dimorphism in Triassic Ammonoids

Although sexual dimorphism must have existed among Triassic ceratites, the only evidence of it is the report of a purported egg mass in a specimen of *Ceratites* de Haan, 1825 (see Müller, 1969). Lack of recognition of antidimorphs in this group probably reflects both few studies concentrating on dimorphism in ceratites and the fact that dimorphism evidently is not expressed in the Ceratitina in the same way as it is in the Ammonitina, for example, by the presence of lappets in microconchs. As was formerly the case in the Ammonitina, it is possible that antidimorphs presently are assigned to separate species or, on the other hand, are lumped together as part of a continuous pattern of variation within a single species.

Dzik (1990) recently described dimorphism in the middle Anisian Acrochordiceratidae. He interpreted the relatively smaller and compressed species of *Epacrochordiceras* Spath, 1934 as microconchs and the larger, depressed species of *Acrochordiceras* Hyatt, 1877 as macroconchs. However, extensive collections from Nevada, collected with detailed stratigraphic data, indicate that there is a chronological succession of highly variable species. In each of these, the taxa assigned to *Epacrochordiceras* and *Acrochordiceras* are simply end members of the one species that existed during that particular time interval. Within each time interval, the intraspecific variation, with respect to both size and shape, corresponds to a normal distribution. This implies that it is difficult to discriminate antidimorphs on this basis.

6. Mature Modifications and Dimorphism in Jurassic Ammonoids

6.1. Introduction

In 1962, Makowski reviewed the old assumption of sexual dimorphism in ammonites on the basis of convincing and well-documented examples selected mainly from Jurassic taxa. At about the same time, Callomon (1963), Tintant (1963), Westermann (1964), and Elmi (1967) provided extensive illustrations of sexual dimorphism among Jurassic ammonoids that commonly display unambiguous secondary sexual features.

Because extensive compilations of dimorphism in Jurassic ammonoids are readily available (see Callomon, 1981, and references cited therein), we will not repeat that information here. Instead, we propose to apply some of the material discussed earlier in this chapter in a consideration of the topic in a more critical and analytical framework and to explore a number of paleobiological aspects.

6.2. A Protocol for Identifying Dimorphism in Jurassic Ammonoids

A minimum number of the diagnostic features or conditions discussed in Section 3.2 of this chapter have to be recognized before ammonoids can be paired as sexual antidimorphs. The first three conditions reiterated below pertain to the recognition of dimorphism, whereas the fourth condition relates to the sexual nature of that dimorphism.

1. There must be some objectively recognizable difference between suspected antidimorphs. This means that, among fossil assemblages, members of a suspected dimorphic pair must be distinguishable by at least one trait (adult size, shell spiral parameter, ornament, suture, etc.). Nevertheless, such differences may be cryptic and have a biometric basis only. If it is not possible to identify the two morphological entities objectively, the search for dimorphism should be called off.

2. The next step in analyzing dimorphism is to demonstrate, in cladistic terms, that the two morphological entities considered are sister groups. This condition must be observed throughout the entire sequence of assemblages where the suspected dimorphic lineage occurs. As a consequence, at each level, the two entities must share at least one apomorphy. Moreover, with gradual apomorphies (trends), as with unexpected apomorphies (events), morphological changes must follow the same chronological pattern in both entities. If such apomorphies exist, the search for dimorphism can continue.

3. The third condition to be established is one of similar paleobiogeographic and paleoecological history. This means that any significant disjunctions in space and time are serious obstacles to a hypothesis of

dimorphism. On the contrary, such disjunctions generally point to iterative branching of species within a continuous lineage.

4. When considering the sexual nature of a previously established case of dimorphism, the paleontologist can use only secondary sexual features preserved in the shell. An actualistic approach can help in deciphering this category of features.

    a. A difference in size between the two adult sexual antidimorphs is a common pattern in present-day cephalopods. By analogy, a difference in size between two ammonoid antidimorphs reinforces the conclusion about the sexual nature of the phenomenon in a given case.

    b. Likewise, we generally find that juvenile ontogenies of sexual antidimorphs in living cephalopods are similar. Consequently, similar ontogenetic patterns in juvenile stages of suspected ammonoid antidimorphs strongly support the hypothesis of sexual dimorphism. In fact, it generally is impossible to attribute a juvenile ammonoid to either of the antidimorphs.

    c. Ammonoids, especially Jurassic forms, are unusual among cephalopods in that they have highly differentiated and complex final ontogenetic stages. Apertures bearing variously developed rostra and/or lappets are the most characteristic examples of these mature modifications. However, many other features may be observed, such as constricted and/or collared apertures, contracted and/or very loosely coiled body chambers, various alterations of ornament, or modification (usually simplification) of sutures. As discussed in Section 3.3, the presence of highly differentiated mature-modifications in a small ammonoid is not enough by itself to demonstrate the existence of sexual dimorphism in a given taxon.

6.3. Examples of Dimorphism in Jurassic Ammonoids

In view of the uncertainties in identifying dimorphism, we propose four main categories (A to D) to encompass actual cases observed in Jurassic ammonoids. The first three categories concern the sexual nature of dimorphism, whereas the last one brings together examples of dimorphism not related to sex. These are diagramatically represented in Fig. 4.

6.3.1. Sexual Dimorphism

For all the taxa placed in this category (Fig. 4, category A), it is possible to identify the first three diagnostic features (1–3 in Section 6.2 above), namely, (1) objective difference between suspected antidimorphs, (2) antidimorphs as sister groups, and (3) similar paleobiogeographic and paleoecological history; this proves at least that dimorphism exists. Moreover, it also is possible in these taxa to recognize 4a (difference in adult size) and 4b (similar juvenile ontogenies) under paragraph 4, which suggest a sexual basis for that dimorphism. In the most characteristic cases (Fig. 4, A1; taxa 1—*Cardioceras*, 2—*Kosmoceras*, and 3—*Hildaites*), the microconchs bear mature modifications (e.g., rostrum, lappets), unknown or indistinct in the less characteristic cases (Fig. 4, A2; taxa 4—*Macrocephalites*, 5—*Phricodoceras*, and 6—*Aegoceras*). This reinforces the hypothesis of a sexual basis for the dimorphism.

6.3.2. Dubious Sexual Dimorphism

In this category (Fig. 4, category B), we group what look like sexually dimorphic taxa. Nevertheless, it is impossible to demonstrate diagnostic features 2, 3, and 4b for these ammonoids. In such conditions, it is not even certain that members of each proposed couple are antidimorphs. In some instances, the presence of what seem to be microconchs bearing mature modifications is noted, but, unfortunately, these features do not improve phyletic understanding of the group. Two examples are described here.

The first concerns the Late Jurassic *Taramelliceras* and *Creniceras*, which are suspected of being a dimorphic couple (Fig. 4, taxa 7α and 7β). These two taxa, which sometimes are collected together, display great differences in adult size but have similar inner whorls. The suspected microconch (*Creniceras*) bears apertural mature modifications (well-developed lappets). Unfortunately, the juvenile stages are smooth and provide little information for a cladistic analysis. Moreover, can we accept as homologous characters (and implicitly as synapomorphies) the single ventral row of tubercles of *Creniceras* and the set of three rows of tubercles, one ventral and two lateral, that characterize *Taramelliceras*? If not, the dimorphic assumption (Palframan, 1966) must be abandoned. But, even if we accept this apomorphy, we are forced to recognize that *Taramelliceras* and *Creniceras* commonly are separated in space and time and that, consequently, the third diagnostic feature is missing. *Taramelliceras* seems to be a eurytopic taxon (present in almost all environments), whereas *Creniceras*, appears to be a stenotopic taxon (restricted to distal platform environments). Moreover, *Creniceras* disappears from the geological column before *Taramelliceras*.

The second example involves the Liparoceratidae (Fig. 4, taxon 6), a widely discussed taxon with regard to sexual dimorphism in ammonoids (Callomon, 1963, 1981; Dommergues, 1987). This Early Jurassic family includes, in the middle and late Carixian, a succession of usually abundant, small or medium-sized, rather evolute, and simply ribbed ammonoids; this strongly suggests a succession of microconchs (Fig. 4, taxon 6 m). Three genus names (*Beaniceras* Buckman, 1913, *Aegoceras* Waagen, 1869, and *Oistoceras* Buckman, 1911) can be used to designate the three main stages of these "capricorn" ammonoids [after the species *Aegoceras capricornis* (Schlotheim, 1813)].

In almost all assemblages, larger liparoceratids also can be collected. Compared with the suspected microconchs, they all display, at the end of growth, a rather thick, involute shell and complex, bituberculate ornament; these shells are here called "liparoceras" (after the genus *Liparoceras* Hyatt, 1867). The inner whorls of the "liparoceras" forms are identical to those of the coexisting, simply ribbed "capricorn" forms (Dommergues, 1987). Such ammonoids, with two very distinct ontogenetic stages, "capricorn" in the juvenile whorls and "liparoceras" later, are termed "androgyns" (after the genus name *Androgynoceras* Hyatt, 1867, commonly used in the literature to designate them). Despite the usual inconsistency of mature-modifications in liparoceratids, their "capricorn" juvenile morphologies have an apomorphic significance and lead us to suspect that "capricorn" and "androgyn" entities may be sexual antidimorphs (Callomon, 1963, 1981).

In the middle Carixian, close to the appearance of the "capricorn" forms in the fossil record, the variability of the large liparoceratids is so great and complex that it is tempting to expand the macroconch interpretation to include certain specimens in which the "capricorn" juvenile stage is either extremely reduced or even completely missing (Fig. 4, taxon 6 Mβ, = *Liparoceras* part., in litt.). To consider the "capricorns" (Fig. 4, taxon 6 m) as microconchs and the "androgyns" (Fig. 4, taxon 6 Mα) and certain evolute "*Liparoceras*" (Fig. 4, taxon 6 Mβ) as macroconchs is one of the possible assumptions for the middle Carixian assemblages of northwestern Europe (Callomon, 1963, 1981). However, this interpretation cannot be expanded farther in space and time without encountering strongly conflicting data. In fact, the classic dimorphic liparoceratids are only one individual lineage that is clearly restricted in space and time. The resounding success of these Euroboreal forms during the middle and late Carixian and of their direct issue (the Amaltheidae) during the Domerian, must not conceal the fact that from the latest Sinemurian to the middle Domerian, the liparoceratids are known over a large part of the world through a genus of large sphaerocones (Fig. 4, taxon 6 Mγ) (*Vicininodiceras* Trueman, 1918, *Becheiceras* Trueman, 1918, etc.) for which dimorphism is highly improbable.

6.3.3. No Sexual Dimorphism

If there is too much doubt about the nature of the sexual dimorphism, or if it is simply impossible to propose any pairing, we must admit a monomorphic species structure (Fig. 4, category C). Two cases are possible. In the first (Fig. 4, C1), the ammonoids are small and generally bear mature modifications. These features may vary in shape and development, e.g., well-developed lappets in the Middle and Late Jurassic genus *Creniceras* (Fig. 4, taxon 7β), rostrum and uncoiled body chamber in the Early Jurassic *Cymbites* (Fig. 4, taxon 8), and collared aperture in the Early Jurassic *Gemmellaroceras* (Fig. 4, taxon 9). Such probably monomorphic ammonoids, which look like microconchs, support the idea that apertural mature modifications are not necessarily sexual features.

In the second case (Fig. 4, C2), the ammonoids are medium-sized or large, without apertural mature modifications, and so look like macroconchs. Two examples are described here. *Taramelliceras* is now interpreted as *not* being paired with *Creniceras*. Thus considered, *Taramelliceras* displays no dimorphism either in adult size or in ornamentation. The case is even clearer for

Prodactylioceras, because no potential microconch has ever been found in any assemblage where specimens of this genus occur. Moreover, all biometric analyses of adult size or ornament indicate a monomorphic taxon.

6.3.4. Nonsexual Dimorphism and Polymorphism

Numerous populations of Jurassic ammonoids collected from the same stratigraphic level have been studied in paleobiological terms (see: Marchand, 1976; Charpy and Thierry, 1976; Tintant, 1976; Thierry, 1978; Contini *et al.*, 1984). A lot of them show a polymorphic structure independent of any sexual dimorphism (Fig. 4, category D). *Kosmoceras* Waagen, 1869 is the example described in this chapter. As seen above (Fig. 4, taxon 2), it possesses a clear sexual dimorphism with the microconch bearing mature modifications. Nevertheless, there is in this taxon another category of polymorphism that, without preference, affects the ornament of both sexual antidimorphs. Traditionally, those morphs that display one or two rows of lateral tubercles are designated as two different genera or subgenera: *Zugokosmoceras* Buckman, 1923 (Fig. 4, taxon 2 M) and *Kosmoceras* (Fig. 4, taxon 2' M and m), respectively. Nevertheless, as early as 1963, Tintant suggested that these two ornamental entities might be no more than expressions of intraspecific polymorphism. This assumption has been confirmed both by studies of numerous populations and by the discovery of a macroconch specimen displaying a *Zugokosmoceras* pattern on one side and a *Kosmoceras* pattern on the opposite side (Tintant, 1976).

6.4. Patterns of Dimorphism through Time in the Jurassic

In the first synthesis of sexual dimorphism in Jurassic ammonoids, Callomon (1963) pointed out that dimorphism becomes clearly recognizable from the late Toarcian (Early Jurassic) onwards, although it probably was present before that. Earlier observations can be improved by attempting a quantification using our five categories A1 through C2 (Fig. 7). This analysis highlights five interesting points:

1. Even though sexual dimorphism is not the dominant pattern during the Hettangian, it is soon clearly expressed but without microconchs bearing mature modifications. We notice that dubious cases (B) and nondimorphic micromorphs (C1) do not yet exist and appear only with the Sinemurian.
2. Obvious sexual dimorphism, with microconchs bearing mature modifications, starts during the Toarcian among the Hildoceratinae. As this subfamily is the ancestor of all Middle and Late Jurassic Ammonitina, the apertural mature modifications must be considered as an autapomorphy of the Hildoceratinae and of their issue. This is true even if these features subsequently disappear by reversion.

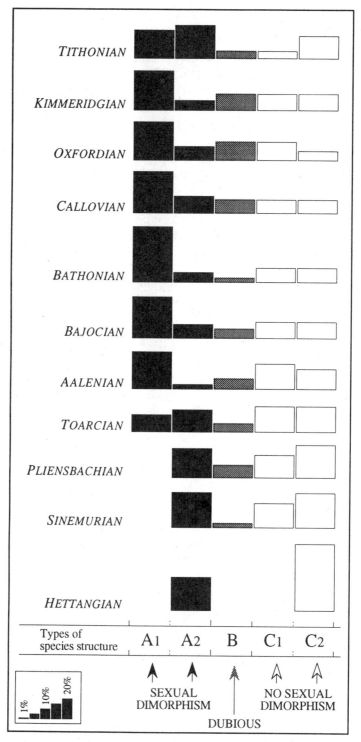

FIGURE 7. History of dimorphism in Jurassic ammonoids, taking into account the types of species structure (see Fig. 4). All of the bargraphs are given in percentages.

3. The maximum number of dimorphic taxa with microconchs bearing mature modifications occurs during the Middle Jurassic (Aalenian to Callovian). At the same time, dimorphic taxa without microconchs bearing mature modifications are especially rare.

4. From the Callovian to the Kimmeridgian, we note a significant increase in the percentage of doubtful cases of sexual dimorphism. This phenomenon is partly associated with the surge in the numbers of micromorph ammonoids that are very difficult to pair with any larger ammonoids, if we take into account the ranges and paleobiogeographic patterns.

5. In the Tithonian, we note a marked increase in dimorphic taxa without mature modifications. This trend presages the pattern that was to prevail during most of the Cretaceous.

7. Mature Modifications and Dimorphism in Cretaceous Ammonoids

7.1. Mature Modifications in Cretaceous Ammonoids

Late-ontogenetic shell modifications in Cretaceous ammonoids are generally similar to those in other ammonoids and involve changes in the shape and ornament of the final body chamber, modifications of the peristome, and septal approximation. In this section, we describe mature modifications in all of the suborders in the Cretaceous with the exception of the Phylloceratina. No mature modifications have been reported in Cretaceous representatives of this suborder, possibly because large enough (mature) specimens are scarce (W. J. Kennedy, personal communication, 1995).

We describe below examples of mature modifications in a variety of species, arranged by suborder and superfamily. Additional examples appear in the section on sexual dimorphism in Cretaceous ammonoids (Section 7.2).

7.1.1. Ammonitina

7.1.1a. Haplocerataceae. Mature modifications of the peristome have been reported in some members of this superfamily. In Early Cretaceous (late Barremian–Aptian) *Sanmartinoceras groenlandicum* Rosenkrantz, 1934, the apertural margin is modified into a pair of lateral lappets and a ventral rostrum (Kennedy and Klinger, 1979, Fig. 1A,B). The lappets are associated with a spiral groove that extends along the midflank of the shell.

7.1.1b. Desmocerataceae. One interesting example of a mature modification in this superfamily involves the loss and reappearance of ornament on the final body chamber (Fig. 8). In microconchs of Late Cretaceous (Campanian) *Menuites oralensis* Cobban and Kennedy, 1993 from the western interior of North America, ribs and umbilical bullae are present on the early whorls of the shell; these disappear on the adapical half of the final body

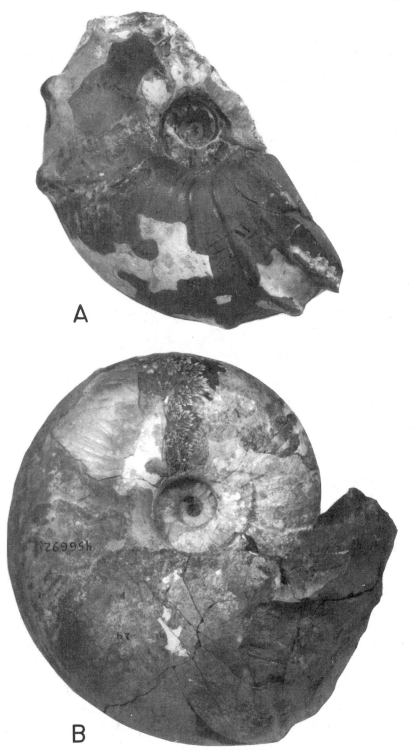

A

B

FIGURE 8. Macroconch (B, USNM 456692) and microconch (A, USGS D715), respectively, of Late Cretaceous (Campanian) *Menuites oralensis* Cobban and Kennedy, 1993 from the western interior of North America, ×1. Note that in the microconch, the ornament disappears midway on the final body chamber but reappears near the aperture. In the macroconch, which is larger, the final body chamber is smooth.

chamber (Cobban and Kennedy, 1993). Ventrolateral tubercles, which appear at the end of the phragmocone, also disappear midway on the final body chamber. As a result, the adoral half of the final body chamber is smooth except for some lirae and growth lines. However, ornament suddenly reappears near the aperture, where several strong ribs occur.

Obata *et al.* (1978) described mature modifications of the peristome in the Desmoceratidae and Kossmaticeratidae. For example, in microconchs of Late Cretaceous (Santonian–Campanian) *Hauericeras (Gardeniceras) angustum* Yabe, 1904 from Hokkaido, Japan, the peristome displays a ventral rostrum and a pair of lateral lappets. The rostrum is subdivided by a shallow lobe at its termination (Obata *et al.*, 1978, Fig. 4, Pl. 3). The lappets are slightly shorter than the rostrum and are located at one-third of the whorl height from the umbilical seam.

7.1.1c. Hoplitaceae. A slight uncoiling of the final body chamber occurs in some members of this superfamily. For example, in Late Cretaceous (Turonian–Coniacian) *Placenticeras kaffarium* Etheridge, 1904, there is a slight egression of the umbilical seam at maturity, leading to a scaphitoid-like uncoiling of the final body chamber (Klinger and Kennedy, 1989, Fig. 79). There are associated changes in the ornament and in the shape of the shell. The venter becomes rounded, and the umbilical tubercles migrate away from the umbilical seam and weaken or even disappear.

7.1.1d. Acanthocerataceae. Mature modifications are common in this superfamily and involve changes in the shape and ornament of the final body chamber. In Late Cretaceous (late Albian) *Mortoniceras (Durnovarites) collignoni* Cooper and Kennedy, 1979 from Angola, tubercles abruptly disappear near the adapical end of the final body chamber (Cooper and Kennedy, 1979, Fig. 66). At the same time, ribs develop a strong, adorally convex curvature, and the whorl section changes from subrectangular to lanceolate. A recurved rostrum develops on the ventral part of the apertural margin.

Additional examples of mature modifications occur in the Vascoceratidae. In Late Cretaceous (late Cenomanian–middle Turonian) *Fagesia catinus* (Mantell, 1822), there is a change in the shape of the shell at maturity (Kennedy *et al.*, 1987, Pl. 8, Figs. 3, 4). The umbilical wall becomes strongly inclined outward, and the whorl section becomes very depressed and coronate. In Late Cretaceous (Turonian) *Neoptychites cephalotus* (Courtiller, 1860), there is also a change in the shape of the shell at maturity. In some specimens, the adapical end of the body chamber becomes very inflated along the midflank (Kennedy and Wright, 1979, Pl. 84, Fig. 3). This gives a fusiform profile to the shell when viewed from the venter.

7.1.2. Ancyloceratina

7.1.2a. Scaphitaceae. The best-known examples of uncoiling at maturity are found in this superfamily (referred to as scaphites in the rest of the chapter). For example, in Late Cretaceous (Turonian) *Scaphites whitfieldi* Cobban, 1951, the final body chamber uncoils to form an elongate shaft and recurved hook (Cobban, 1951; Landman, 1987). In geologically younger species such as Late Cretaceous (Maastrichtian) *Hoploscaphites nicolletii* (Morton, 1842), the final body chamber also uncoils, but to a smaller extent, remaining in close contact with the phragmocone (Fig. 9I,J). Along with uncoiling, there are changes in ornament. For example, in *H. nicolletii*, the ornament weakens on the adapical part of the final body chamber. In addition, ribbing becomes finer and more closely spaced, and ventrolateral tubercles diminish in size or disappear altogether toward the aperture (Fig. 9I,J).

Lappets are present in some scaphite species. For example, microconchs of the Late Cretaceous (Cenomanian) *Worthoceras vermiculus* (Shumard, 1860) bear long, lateral lappets characterized by strong, convex growth lines (Fig. 9D; Kennedy, 1988, Text-Fig. 39). Lappets also appear in microconchs of Late Cretaceous (Turonian–Santonian) *Yezoites puerculus* (Jimbo, 1894) from Hokkaido, Japan (Fig. 9H; Tanabe, 1977). These lappets are relatively broader than those in desmoceratids and kossmaticeratids (see Section 7.2.1a; Obata *et al.*, 1978, Fig. 11).

Modifications of the apertural margin also occur in some species of *Scaphites (Pteroscaphites)* Wright, 1953 (Fig. 9E,F). For example, in Late Cretaceous (Santonian) *Scaphites (Pteroscaphites) coloradensis* (Cobban, 1951), the apertural margin is modified into a pair of lateral projections that

---

FIGURE 9. Dimorphism in the Ancyloceratina. (A,B) Macroconch (USNM 411537) and microconch (USNM 411539), respectively, of Late Cretaceous (Cenomanian) *Sciponoceras gracile* (Shumard, 1860) from Texas, ×1. The two antidimorphs differ mainly in size. (C,D) Macroconch (USNM 411552) and microconch (USMN 411553), respectively, of Late Cretaceous (Cenomanian) *Worthoceras vermiculus* (Shumard, 1860) from Texas, ×2. The two antidimorphs differ in size and in the form of the apertural margin. (E,F) Macroconch (USNM 106714) and microconch (USNM 106715), respectively, of Late Cretaceous (Santonian) *Pteroscaphites (Scaphites) coloradensis* (Cobban, 1951) from Montana, ×2. Both antidimorphs show the same mature modifications at the apertural margin. (G,H) Macroconch (AMNH 45280) and microconch (AMNH 45281), respectively, of Late Cretaceous (Turonian) *Yezoites puerculus* (Jimbo, 1894) from Hokkaido, Japan, ×2. The apertural margin in the microconch is modified into a pair of lateral lappets. (I,J) Macroconch (YPM 27245) and microconch (YPM 27235), respectively, of Late Cretaceous (Maastrichtian) *Hoploscaphites nicolletii* (Morton, 1842) from South Dakota, ×1. In the macroconch, the umbilical shoulder of the body chamber is straight in side view, whereas it is curved in the microconch. (K) Mature apertural margin of a microconch (?) of Late Cretaceous (Santonian) *Baculites thomi* Reeside, 1927 (USGS 21419) from Montana showing a short dorsal and an unusually long (pathologic?), ventral rostrum, ×1.

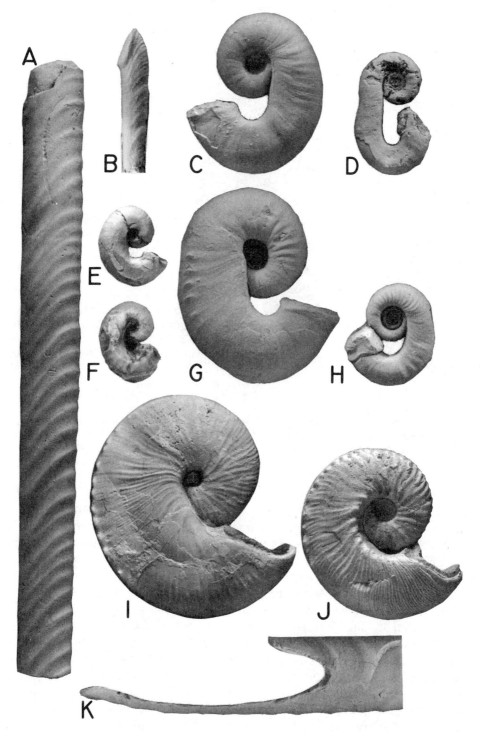

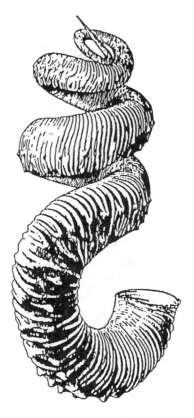

FIGURE 10. Late Cretaceous (Campanian) *Didymoceras nebrascense* (Meek and Hayden, 1856) from the western interior of North America, ×0.5. The final body chamber curves downward and then upward, terminating almost against the base of the spire. From Scott and Cobban, 1965.

are slightly enrolled, suggesting hollow spines (Kennedy, 1988; Landman, 1989). These projections differ in morphology from the lappets in *Worthoceras* Adkins, 1928 and *Yezoites* Yabe, 1910 and are present in both antidimorphs.

7.1.2b. Turrilitaceae. The straight-shelled and more irregularly coiled members of the Ancyloceratina also show modifications at maturity. In microconchs of Late Cretaceous (Cenomanian) *Sciponoceras gracile* (Shumard, 1860) from Texas, maturity is marked by a strengthening of ribs along the venter and the development of a flared apertural margin (Fig. 9B; Kennedy, 1988, p. 108). In Late Cretaceous (late Santonian–early Campanian) *Baculites thomi* Reeside, 1927 from the western interior of North America, the apertural margin at maturity exhibits a short dorsal and a long ventral rostrum (Fig. 9K; Cobban and Kennedy, 1991c). The ventral rostrum tapers gradually to a rounded tip. In Late Cretaceous (Cenomanian) *Allocrioceras annulatum* (Shumard, 1860), maturity is indicated by a crowding and weakening of ribs, together with a loss of tubercles (Kennedy, 1988, Fig. 36). In Late Cretaceous

(Campanian) *Nostoceras (N.) hyatti* Stephenson, 1941, the adult body chamber uncoils into a twisted U, so that the aperture faces the base of the spire (Cobban and Kennedy, 1994a, Pl. 1). Similarly, in Late Cretaceous (Campanian) *Didymoceras nebrascense* (Meek and Hayden, 1856), the adult body chamber forms a retroversal hook (Fig. 10; Cobban *et al.*, 1996). The body chamber curves downward and then upward, terminating almost against the base of the spire. There is a concomitant change in ornament from closely spaced ribbing on the adapical part of the body chamber to more widely spaced ribbing on the adoral part.

7.1.3. Lytoceratina

Among Tetragonitaceae, Cooper and Kennedy (1979, Fig. 7) reported that in the Cretaceous (Albian) *Eogaudryceras (E.) italicum* Wiedmann and Dieni, 1968, the whorls become flat-sided and compressed at maturity, and fine, thread-like grooves appear (see Section 7.2.2 for a discussion of mature modifications in the related genus *Gaudryceras* Grossouvre, 1894). In Late Cretaceous (Albian–Cenomanian) *Tetragonites subtimotheanus subtimotheanus* Wiedmann, 1973, maturity is indicated by a rounding of the whorl section and a slight uncoiling of the body chamber (Wiedmann, 1973, Pl. 3).

7.2. Dimorphism in Cretaceous Ammonoids

Dimorphism has been reported in many Cretaceous ammonoids, notably in the Desmocerataceae and Acanthocerataceae (Ammonitina), in the Scaphitaceae and Turrilitaceae (Ancyloceratina), and in the Tetragonitaceae (Lytoceratina) (see the Appendix to this chapter for a list, albeit incomplete, of reports of sexual dimorphism in Cretaceous ammonoids; Kennedy and Wright, 1985a, Fig. 5). In this section we describe several examples within each of these three suborders. Dimorphism has not been reported in the Phylloceratina, possibly because of the lack of adequate samples.

The criteria used to recognize dimorphism are not the same in all Cretaceous ammonoids (Kennedy, 1989). The full suite of characters includes differences in adult size, ornament, shell shape, and the form of the apertural margin. Sometimes all these aspects are involved, whereas at other times, only a single criterion, such as a difference in size, serves to distinguish antidimorphs.

7.2.1. Examples of Dimorphism in Cretaceous Ammonoids

7.2.1a. Ammonitina.

Desmocerataceae. In the Cretaceous (Campanian) *Menuites oralensis* from the western interior of North America, antidimorphs differ in adult size and ornament (Fig. 8). Cobban and Kennedy (1993) measured the size of specimens in one collection from South Dakota. They noted that the diameter at the base of the body chamber in microconchs ranges between 32 and 64 mm.

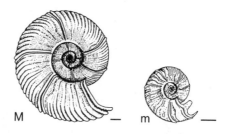

FIGURE 11. Macroconch (M) and microconch (m) of Late Cretaceous (Santonian–lower Campanian) *Yokoyamoceras ishikawai* (Jimbo, 1894) from Hokkaido, Japan and Sakhalin, Russia. The apertural margin in the microconch is modified into a ventral rostrum and a pair of lateral lappets. Scale bar, 10 mm. (After Maeda, 1993, Fig. 2.)

Macroconchs in the same collection attain a diameter of as much as 224 mm, including part of the body chamber (equivalent to a diameter at the base of the body chamber of approximately 140 mm; see Cobban and Kennedy, 1993, Pl. 14). The differences in ornament are also very clear-cut. In microconchs, ribs and umbilical and ventrolateral tubercles are present on the adapical half of the body chamber, and strong, narrow ribs occur near the aperture. In contrast, the adult body chamber of macroconchs is mostly smooth except for growth lines and low, obscure ribs.

Maeda (1993) described sexual dimorphism in a species of the Kossmaticeratidae (Fig. 11). In Late Cretaceous (Santonian–early Campanian) *Yokoyamaoceras ishikawai* (Jimbo, 1894) from Hokkaido, Japan, and Sakhalin, Russia, macroconchs develop eight to nine whorls and range in diameter from 90 to 140 mm. Microconchs, on the other hand, exhibit six to seven whorls and range in diameter from 25 to 40 mm, indicating no overlap in either number of whorls or size between antidimorphs. Ventrolateral tubercles appear on the adapical half of the adult body chamber in some microconchs but are absent in all macroconchs. Finally, in macroconchs, the apertural margin is simple, whereas, in microconchs, it is modified into a ventral rostrum and a pair of lateral lappets, each of which is 2–5 mm long.

Acanthocerataceae. Dimorphism has been reported in Late Cretaceous (Cenomanian) *Metoicoceras mosbyense* Cobban, 1953 from the western interior of North America (Fig. 12). Cobban and Kennedy (1991b) documented a difference in size and ornament between antidimorphs. Microconchs are as small as 125 mm in diameter, whereas macroconchs can reach a diameter of as much as 300 mm (Cobban, 1953). In addition, microconchs are more robust and more strongly ornamented than are macroconchs. For example, ventro-

FIGURE 12. Macroconch (B, USNM 108323b) and microconch (A, USNM 108318a), respectively, of Late Cretaceous (Cenomanian) *Metoicoceras mosbyense* Cobban, 1953 from Montana, ×1. The microconch is smaller, more robust, and more strongly ornamented than the macroconch.

A

B

lateral tubercles persist to the base of the body chamber in microconchs but die out well before this point in macroconchs.

Futakami (1990) recognized dimorphism in Late Cretaceous (Turonian) *Subprionocyclus minimus* (Hayasaka and Fukuda, 1951) on the basis of differences in diameter, ornament, and whorl width. Macroconchs attain a shell diameter of about 120 mm, approximately twice that of microconchs. The ornament on the adult body chamber of macroconchs is generally coarser than that on the adult body chamber of microconchs, and the macroconch whorl section also tends to be more robust. This is just the opposite of the situation in *Metoicoceras mosbyense* described above.

7.2.1b. Ancyloceratina.

Scaphitaceae. Some of the best-documented examples of dimorphism are in this superfamily. Dimorphism involves differences in size, shape, ornament, and sometimes even the form of the apertural margin. Because the scaphites are better known than some of the other Cretaceous ammonoid groups, they are covered in somewhat more detail.

Antidimorphs in this group are easily distinguished at maturity by differences in the shape of the body chamber. For example, in macroconchs of Late Cretaceous (Maastrichtian) *Hoploscaphites nicolletii* from South Dakota, the umbilical shoulder of the adult body chamber is straight in side view and partially occludes the umbilicus of the phragmocone (Fig. 9I,J). A dorsal swelling sometimes develops along the umbilical shoulder. In contrast, in microconchs, the umbilical shoulder of the adult body chamber parallels the curve of the venter, forming a flat, broad dorsal shelf that slopes gently outward at an obtuse angle to the flanks. As a result, the umbilicus of the phragmocone is well exposed.

The ornament on the adult body chamber is similar in both antidimorphs, but there are minor differences. For example, in *Hoploscaphites nicolletii*, the distinctive pattern of fine ribbing on the body chambers of macroconchs is echoed only weakly on the body chambers of microconchs. In addition, umbilical bullae are rare or absent in macroconchs, whereas they are common in microconchs. Furthermore, because of the larger shell size and expanded whorl height of macroconchs, the number of ribs, especially secondary ribs, is generally higher in macroconchs than it is in microconchs.

Scaphite antidimorphs differ in adult size, although there is overlap. In species of *Jeletzkytes* Riccardi, 1983, the average adult size of macroconchs is 30% to 50% larger than that of microconchs (Landman and Waage, 1993). Bar graphs of the sizes of many species, such as Late Cretaceous (Maastrichtian) *Jeletzkytes spedeni* Landman and Waage, 1993 from South Dakota, show a bimodal distribution, with peaks representing the two antidimorphs (Fig. 13). However, the size ranges of the two antidimorphs commonly overlap. For example, the extent of overlap in *J. spedeni* is approximately 30% of the total combined size range of both antidimorphs together. In *Hoploscaphites nicolletii*, the extent of size overlap is approximately 40% (Landman and Waage, 1993, Fig. 57). Comparable amounts of size overlap between antidimorphs

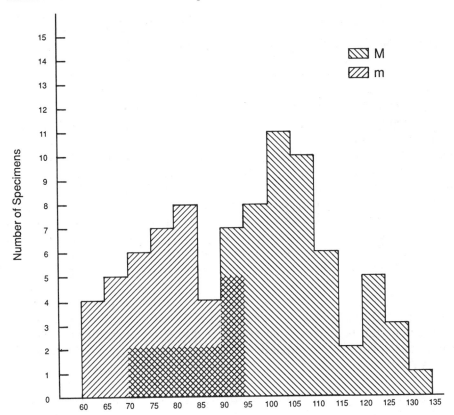

FIGURE 13. Bar graph of adult size in a collection of Late Cretaceous (Maastrichtian) *Jeletzkytes spedeni* Landman and Waage, 1993 from South Dakota. In general, macroconchs are larger than microconchs, but the size ranges of the two antidimorphs overlap. X-axis is maximum length (shell diameter) in mm. From Landman and Waage, 1993, Fig. 112.

have been observed in other scaphite species [≈30% in *Discoscaphites conradi* (Morton, 1834) from South Dakota, Landman and Waage, 1993; ≈20% in what Cobban (1969) called *Scaphites leei* III from the western interior of North America, Cobban, 1969; ≈50% in *Scaphites whitfieldi* Cobban, 1951 from Wyoming and South Dakota, Landman, 1987, Fig. 77; and ≈50% in *Trachyscaphites spiniger spiniger* (Schlüter, 1872) from Texas, Cobban and Kennedy, 1992.]

In contrast, no size overlap has been reported between antidimorphs in Late Cretaceous (Maastrichtian) *Hoploscaphites constrictus* (J. Sowerby, 1817) from either Poland or France (Makowski, 1962; Kennedy, 1986b) or in Late Cretaceous (Campanian) *Trachyscaphites pulcherrimus* (Roemer, 1841) from Austria (Kennedy and Summesberger, 1984). [Editorial note: Cobban, 1969, designated subspecies of *Scaphites leei* Reeside, 1927 and *Scaphites hippocrepis* (DeKay, 1827) using the Roman numerals I, II, and III. Although this practice does not conform to that allowed in the *International Code of*

Zoological Nomenclature, for the purposes of this chapter and as a service to the reader, these "subspecies" are listed as originally published.]

In some scaphite species, the pattern of septal approximation differs between antidimorphs (Landman and Waage, 1993). For example, in Late Cretaceous (Maastrichtian) *Hoploscaphites nicolletii* from South Dakota, septal approximation occurs over more chambers and is more marked in macroconchs than in microconchs (Fig. 14). In macroconchs, septal approximation commonly occurs over the last three chambers, whereas, in microconchs, it commonly occurs in only the last chamber. In contrast, there is no difference in the pattern of septal approximation between antidimorphs in *Jeletzkytes spedeni*. Septal approximation tends to occur over the last four or five chambers in both antidimorphs.

In scaphites, as in other ammonoids, the ontogenetic development of the shell is similar in both antidimorphs until the onset of maturity. For example, this is true in *Jeletzkytes spedeni*, in which the pattern of growth of the umbilical diameter with respect to that of the shell diameter is approximately the same in both antidimorphs until the onset of maturity (Landman and Waage, 1993, Fig. 117). However, there are exceptions. In Late Cretaceous (Maastrichtian) *Hoploscaphites comprimus* (Owen, 1852) from South Dakota, microconchs, throughout ontogeny, tend to exhibit a larger umbilical diameter than do macroconchs at comparable shell diameters (Fig. 15). As a result, it is usually possible to distinguish antidimorphs of this species even at preadult stages.

In another genus of scaphites, *Worthoceras*, dimorphism is expressed by differences not only in adult shape, size, and ornament but also in the form of the apertural margin (Fig. 9C,D). In Late Cretaceous (Cenomanian) *Worthoceras vermiculus* from Texas, macroconchs range in size from 15.5 to 21.4 mm, and microconchs from 12.5 to 17.5 mm, indicating a size overlap between antidimorphs of approximately 20% of the total combined size range of the two antidimorphs together (Kennedy, 1988). Macroconchs are characterized by a stout body chamber, which partially conceals the umbilicus of the phragmocone, and a simple apertural margin. In contrast, microconchs have a much more slender body chamber, and, as a consequence, the umbilicus of the phragmocone is completely exposed. In addition, the microconch body

---

FIGURE 14. Variation in the pattern of septal approximation between antidimorphs in four species of Late Cretaceous (Maastrichtian) scaphites from South Dakota. The number of chambers over which septal approximation occurs is given for 40 specimens of *Hoploscaphites nicolletii* (Morton, 1842), 13 specimens of *Discoscaphites gulosus* (Morton, 1834), 15 specimens of *Discoscaphites conradi* (Morton, 1834), and 23 specimens of *Jeletzkytes spedeni* Landman and Waage, 1993. Chambers are numbered starting with the last, most recently formed chamber of the phragmocone (which is designated 1). In each of the three species on top, septal approximation affects more chambers in macroconchs than it does in microconchs. In contrast, there is no difference in the pattern of septal approximation between antidimorphs in *J. spedeni*. From Landman and Waage, 1993, Fig. 36.

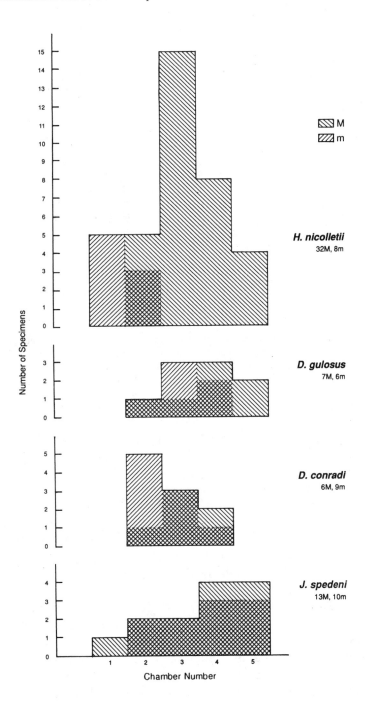

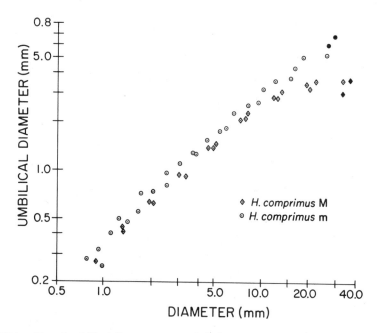

FIGURE 15. Plot of umbilical diameter versus shell diameter through the ontogeny of six adults (three macroconchs and three microconchs) of Late Cretaceous (Maastrichtian) *Hoploscaphites comprimus* (Owen, 1852) from South Dakota. Microconchs tend to exhibit a larger umbilical diameter than do macroconchs at comparable shell diameters. Black symbols indicate measurements near the base of or in the mature body chamber. From Landman and Waage, 1993, Fig. 80.

chamber is longer in proportion to the phragmocone than is the macroconch body-chamber. The apertural margin in microconchs also displays a pair of prominent lateral lappets.

In *Yezoites*, dimorphism is recognized on the basis of the same criteria. In Late Cretaceous (Turonian) *Yezoites puerculus* from Hokkaido, Japan, macroconchs are approximately 1.5 to 2.5 times the size of microconchs (Fig. 9G,H; Tanabe, 1977). The body chamber in macroconchs is more robust than that in microconchs and conceals the umbilicus of the phragmocone. The apertural margin in macroconchs is simple, whereas in microconchs, it is modified into a pair of lateral lappets.

Turrilitaceae. In many other ancyloceratids such as *Didymoceras* Hyatt, 1894, *Bostrychoceras* Hyatt, 1900, and *Oxybeloceras* Hyatt, 1900, antidimorphs are distinguished mainly by differences in adult size (Cobban and Kennedy, 1994a; Cobban *et al.*, 1996). For example, in one collection of the orthocone *Sciponoceras gracile* from the Britton Formation of north-central Texas, specimens that range between 7 and 12 mm in whorl height and exhibit a mature apertural margin are interpreted as microconchs, whereas specimens that reach whorl heights of up to 42 mm are interpreted as macroconchs (although none of these shows evidence of a mature apertural margin) (Fig. 9A,B; Kennedy, 1988).

    7.2.1c. Lytoceratina. Dimorphism has been reported in *Gaudryceras* Grossouvre, 1894, *Costidiscus* Uhlig, 1882, and *Macroscaphites* Meek, 1876 (see Section 7.2.2 below and Appendix). Marked differences in adult size also occur among some species of *Tetragonites* Kossmat, 1895 (see, for example, Wiedmann, 1973, Pl. 1–8), which may indicate the existence of dimorphism (W. J. Kennedy, 1995, personal communication), although this has not yet been proved.

7.2.2. Possible Polymorphism in Cretaceous Ammonoids

    Hirano (1978, 1979) described a possible example of trimorphism in *Gaudryceras* from Hokkaido, Japan, and Sakhalin, Russia. He described three varieties of *Gaudryceras*: *Gaudryceras denseplicatum* var. *denseplicatum* Hirano, 1978 and *Gaudryceras denseplicatum* var. *intermedium* Hirano, 1978, both of which appeared in the Turonian and persisted into the early Campanian, and *Gaudryceras denseplicatum* var. *tenuiliratum* Hirano, 1978, which appeared in the Coniacian in the same geographic area as the other two varieties and persisted into the early or middle Campanian. Starting at the beginning of the Santonian, the abundance of *G. d.* var. *tenuiliratum* increased with respect to that of the other two varieties, and, eventually, it replaced them.

    Hirano interpreted this situation as a case of sexual dimorphism coupled with what he called "transient polymorphism." He regarded *Gaudryceras denseplicatum* var. *denseplicatum* and *G. d.* var. *intermedium* as sexual antidimorphs because they cooccur stratigraphically and geographically and because the early ontogenetic development of the shell is similar in both forms. They are distinguishable at maturity only on the basis of differences in shell size and rib shape. He considered *G. d.* var. *intermedium*, which attains a diameter of more than 300 mm, as the macroconch, and *G. d.* var. *denseplicatum*, which attains a diameter of at most 200 mm, as the microconch. The third form, *G. d.* var. *tenuiliratum*, attains the same size as *G. d.* var. *denseplicatum* but differs from it in some aspects of shell shape and ornament. Hirano interpreted *G. d.* var. *tenuiliratum* as a "transient polymorph" that arose in the Santonian and eventually replaced the other two forms.

    This is an intriguing hypothesis, but one that is difficult to prove. A possible alternative, if we accept that *G. d.* var. *denseplicatum* and *G. d.* var. *intermedium* constitute a dimorphic pair, is that *G. d.* var. *tenuiliratum* is a separate species altogether. In this case, one might argue that sexual dimorphism has not yet been detected in *G. d.* var. *tenuiliratum*.

    The possibility of trimorphism also was raised in scaphites with respect to the relationship between *Scaphites* Parkinson, 1811 and *Clioscaphites* Cobban, 1951, on the one hand, and *Scaphites (Pteroscaphites)* Wright, 1953, on the other. The latter taxon consists of about eight rare species, each of which cooccurs with a larger species of *Scaphites* or *Clioscaphites* at a distinctive horizon in the Late Cretaceous (Turonian–Santonian) of the western interior of North America (Fig. 9E,F). Species of *Scaphites (Pteroscaphites)* are characterized by a small adult size, an uncoiled, hook-like body chamber, and the presence, in some members, of a modified apertural margin.

On the basis of their adult morphology, Wiedmann (1965) suggested that these forms may represent microconchs of the cooccurring larger species of *Scaphites* and *Clioscaphites*. Indeed, the ontogenetic development of the shell prior to maturity in each species of *Scaphites (Pteroscaphites)* is identical to that in the cooccurring species of *Scaphites* or *Clioscaphites*. However, dimorphism is already apparent in the larger scaphites (Cobban, 1951; Landman, 1987). It is based on differences in the degree of robustness, the extent to which the final body-chamber uncoils, the outline of the umbilical shoulder of the body chamber in side view, and adult size. In light of this, it would appear that the species of *Scaphites (Pteroscaphites)* represent yet a third morph.

However, using the same dimorphic criteria as those used in the larger species of *Scaphites* and *Clioscaphites*, Landman (1989) recognized dimorphism in each of the species of *Scaphites (Pteroscaphites)*. Macroconchs in these species tend to be larger and more robust than are the corresponding microconchs. The umbilical wall of the shaft of the body chamber is straight in side view in macroconchs, whereas it is curved in microconchs. In some macroconchs, a dorsal swelling develops along the umbilical shoulder. Both antidimorphs show the same mature modifications at the apertural margin.

The fact that the ontogenetic development of the shell prior to maturity in each species of *Scaphites (Pteroscaphites)* is identical to that in the cooccurring species of *Scaphites* or *Clioscaphites* may reflect progenesis (Landman, 1989). Species of *Scaphites (Pteroscaphites)* initiate maturity at a smaller size than do species of the larger scaphites. The mature modifications at the apertural margin, which are present in most species of this subgenus, probably represent outgrowths of the morphological patterns of growth in effect at the reduced size at which these small forms became mature. Thus, according to this explanation, each species of *Scaphites (Pteroscaphites)* and the cooccurring species of *Scaphites* or *Clioscaphites* constitute a pair of sister species, each of which is dimorphic. (There are, of course, taxonomic ramifications, but these are not discussed here.)

Some species of Cretaceous ammonoids show a considerable variation in shell shape, size, and ornament that may have sexual as well as other causes. Such a high degree of variation is common in the Vascoceratidae. Reyment (1988) described a number of polymorphs in Late Cretaceous (late Cenomanian–early Turonian) vascoceratid species from North Africa. He interpreted these polymorphs as reflecting selection across a variety of habitats and speculated that sexual dimorphism also may be present but that it is difficult to document. Kennedy and Wright (1979) reported wide variation in adult diameter among species of the Late Cretaceous (Turonian) vascoceratid *Neoptychites* Kossmat, 1895 across a broad geographic range. They suggested that this variation could result from geographic or sexual causes but that the data were inadequate to resolve the issue. (See Kassab and Hamama, 1991, for another example of polymorphism, this one, from Late Cretaceous sphenodiscids.)

7.2.3. Proportion of Macroconchs to Microconchs within a Species

The proportion of macroconchs to microconchs within a species varies widely among Cretaceous ammonoids, but in most collections, macroconchs are more common than microconchs. For example, macroconchs outnumber microconchs in collections of *Trachyscaphites pulcherrimus* (Roemer, 1841) from Austria (Kennedy and Summesberger, 1984), *Yokoyamaoceras ishikawai* (Jimbo, 1894) from Hokkaido, Japan, and Sakhalin, Russia (Maeda, 1993), *Yezoites delicatulus* (Warren, 1930) from Texas (Kennedy, 1988), and *Worthoceras vermiculus* (Shumard, 1860) from Texas (Kennedy *et al.*, 1989).

In several studies, the actual ratio of macroconchs (M) to microconchs (m) has been calculated, or the number of specimens of each antidimorph in a collection has been given, allowing us to calculate this ratio. In these studies, the value of this ratio varies from as much as 20M : 1m to as little as 0.5M : 1m. For example, this ratio equals 1.9M : 1m in *Hoploscaphites constrictus* from France (Kennedy, 1986b), 2.2M : 1m in a collection of this species from Poland (Makowski, 1962), 20M : 1m in *Hoploscaphites nicolletii* from South Dakota (Landman and Waage, 1993), 1.5M : 1m in *Hoploscaphites comprimus*, also from South Dakota (Landman and Waage, 1993), 2M : 1m in *Menuites oralensis* from Colorado (Cobban and Kennedy, 1993), 3.2M : 1m in *Menuites portlocki complexus* (Hall and Meek, 1856) from Wyoming (Cobban and Kennedy, 1993), 0.5M : 1m in each of two collections of *Scaphites hippocrepis* I Cobban, 1969 from Wyoming (Cobban, 1969), 0.8M : 1m in *Scaphites hippocrepis* III Cobban, 1969 from Montana (Cobban, 1969), and 0.7M : 1m in *Scaphites leei* III from New Mexico (Cobban, 1969).

Variation in the proportion of macroconchs to microconchs within collections of the same species from different localities may provide clues about the geographic distribution of antidimorphs during life. For example, in Late Cretaceous (Cenomanian) *Metoicoceras geslinianum* (d'Orbigny, 1850), Kennedy (1988) reported that macroconchs occur almost to the exclusion of microconchs in the Tropic Shale of Utah, whereas both antidimorphs commonly occur together in the Mancos Shale, of equivalent age, in New Mexico. In Late Cretaceous (Cenomanian) *Metoicoceras mosbyense*, macroconchs and microconchs are equally abundant in the Mosby Sandstone Member of the Belle Fourche Shale of central Montana, but farther south, in Wyoming, macroconchs are scarce (Cobban *et al.*, 1989). In Late Cretaceous (Maastrichtian) *Hoploscaphites nicolletii* from the Fox Hills Formation of South Dakota, macroconchs are much more abundant than are microconchs in one assemblage zone; the ratio of macroconchs to microconchs in this zone equals 20 : 1 (Landman and Waage, 1993). However, in another assemblage zone, which represents a more nearshore environment, the reverse is true: microconchs are more abundant than macroconchs. The above data suggest the possibility that antidimorphs may have lived in different regions or environments, at least at one point during their lifetimes.

7.2.4. Patterns of Dimorphism through Time in the Cretaceous

Klinger and Kennedy (1989, p. 253) suggested that the morphological expression of dimorphism varied with time in Late Cretaceous (late Albian–Maastrichtian) placenticeratids. In older forms (from the Albian to Cenomanian), dimorphism is mainly expressed in terms of differences in adult size. In geologically younger forms (from the Santonian), dimorphism is expressed, in addition, by differences in the strength of ornament. These differences arose in the Turonian and Coniacian, when new species appeared with lateral tubercles. For example, in Santonian *Placenticeras polyopsis* (Dujardin, 1837), microconchs are smaller and, in addition, are more strongly ornamented than are macroconchs.

Scaphites also show variation in the expression of dimorphism over time. More primitive scaphites tend to show a less-pronounced dimorphism. For example, in a large sample of Late Cretaceous (Turonian) *Scaphites whitfieldi* from Wyoming and South Dakota, one can easily distinguish antidimorphs that occur at each end of the size range of the sample (Landman, 1987, Fig. 77). Small, slender, evolute forms are interpreted as microconchs, and large, robust, relatively involute forms are interpreted as macroconchs. However, there are many intermediate-sized specimens for which it is difficult to identify the correct antidimorph confidently. These specimens are intergradational in shape, and the morphological criteria used to discriminate antidimorphs in the rest of this sample are not effective for these specimens.

In contrast, geologically younger, more derived species, such as Late Cretaceous (Maastrichtian) *Hoploscaphites nicolletii* and *Discoscaphites conradi* display a strong, invariant dimorphism (Fig. 9I,J; Landman and Waage, 1993). As indicated previously, the criteria used to distinguish antidimorphs involve differences in a number of morphological features. Size is not an overriding consideration, although macroconchs generally are significantly larger than are microconchs. However, even among those specimens of intermediate size, the criteria used to discriminate antidimorphs are almost infallible.

8. Future Studies

Much work remains to be done in the areas of mature modifications and dimorphism in ammonoids. In the last few decades there has been tremendous success in linking up antidimorphs, and more of that work needs to be done. Of course, to go beyond simple "matchmaking," there need to be many more extensive studies of numerically large populations of precisely known geological ages derived from a wide variety of lithofacies, chronostratigraphic units, geographic areas, well-understood taxa, and taphonomic conditions.

A real boon to studies of mature modifications and dimorphism would be for all students of fossil cephalopods to observe and record data more consistently, and with these subjects in mind. For example, all illustrations of shells of ammonoids that have partial or complete body chambers preserved should be marked to show the position of the last septum, and shell measurements at that point should be given on a routine basis, including the number of whorls in the phragmocone. Workers need to be precise in the use of the word "mature"; if mature modifications are present, they should be described meticulously and illustrated. If none is present, then the criteria used to justify the label "mature" need to be spelled out in detail.

Question marks loom especially large in the Paleozoic and Triassic. Are there really so many taxa there that lack mature modifications? that lack dimorphism?

Finger-wagging aside, there are some really exciting research opportunities. Is dimorphism just one aspect of a larger polymorphic phenomenon, as Matyja and some others would have us believe? The existence of more than one adult morph within a single species offers an exceptional opportunity to study morphogenetic processes. How do those different adult morphologies arise from similar juveniles ontogenetically? In discussing two putative antidimorphs in the Carboniferous genus *Axinolobus*, McCaleb and Furnish (1964, p. 254) opined that ". . . *A. quinni* is a slightly more evolved form than is *A. modulus*." How do the different adult morphologies of antidimorphs arise phylogenetically? What is the significance of reported cases of antidimorphs in which the juveniles are more different than are the adults?

What is the ecological significance of mature modifications and dimorphism? What is the biogeographic significance? Are dimorphic species more eurytopic than are monomorphic ones? Are they more abundant? Do they encompass more individuals per species? Were they more "successful"?

It appears that nearly all Cretaceous ammonoids that show dimorphism are shallow-water to midshelf forms. Moreover, there does not seem to be any correlation between mode of life and the expression of dimorphism in those ammonoids. Is this really true? And, if so, is it general across the entire geological record of ammonoid dimorphism?

In fact, what is the picture through geological time? What is the actual taxonomic distribution (and, hence, phylogenetic history) of mature modifications and dimorphism? Were dimorphic species more subject to extinction events than were monomorphic species, or less so?

There is no doubt that there are exciting times ahead in the study of mature modifications and dimorphism in ammonoid cephalopods!

Appendix: Reports of Sexual Dimorphism in Cretaceous Ammonoids

Suborder	Superfamily	Family	Genus	Source
Ammonitina	Haplocerataceae	Aconeceratidae	*Gyaloceras*	Kennedy and Klinger, 1979
Ammonitina	Haplocerataceae	Aconeceratidae	*Koloceras*	Riccardi et al., 1987; Klinger and Kennedy, 1990
Ammonitina	Haplocerataceae	Aconeceratidae	*Protaconeceras*	Riccardi et al., 1987
Ammonitina	Haplocerataceae	Aconeceratidae	*Sanmartinoceras*	Kennedy and Klinger, 1979; Riccardi et al., 1987; Parent, 1991
Ammonitina	Haplocerataceae	Aconeceratidae	*Sinzovia*	Riccardi et al., 1987
Ammonitina	Haplocerataceae	Binneyitidae	*Borrisiakoceras*	Kennedy and Cobban, 1976
Ammonitina	Haplocerataceae	Haploceratidae	*Haploceras*	Bujtor, 1993
Ammonitina	Perisphinctaceae	Neocomitidae	*Berriasella*	Howarth, 1992
Ammonitina	Perisphinctaceae	Neocomitidae	*Neocomites*	Fatmi, 1969
Ammonitina	Perisphinctaceae	Neocomitidae	*Thurmanniceras*	Bujtor, 1993
Ammonitina	Perisphinctaceae	Olcostephanidae	*Groebericeras*	Howarth, 1992[*]
Ammonitina	Perisphinctaceae	Olcostephanidae	*Olcostephanus*	Fatmi, 1969; Bujtor, 1993
Ammonitina	Perisphinctaceae	Olcostephanidae	*Saynoceras*	Bulot et al., 1990
Ammonitina	Desmocerataceae	Desmoceratidae	*Anapuzosia*	Matsumoto, 1988
Ammonitina	Desmocerataceae	Desmoceratidae	*Achilleoceras*	Matsumoto, 1988[*]
Ammonitina	Desmocerataceae	Desmoceratidae	*Austiniceras*	Matsumoto, 1988[*]
Ammonitina	Desmocerataceae	Desmoceratidae	*Bhimaites*	Matsumoto, 1988
Ammonitina	Desmocerataceae	Desmoceratidae	*Epipuzosia*	Matsumoto, 1988
Ammonitina	Desmocerataceae	Desmoceratidae	*Grandidiericeras*	Matsumoto and Saito, 1987; Matsumoto, 1988
Ammonitina	Desmocerataceae	Desmoceratidae	*Hauericeras*	Matsumoto et al., 1990b
Ammonitina	Desmocerataceae	Desmoceratidae	*Jimboiceras*	Matsumoto, 1988
Ammonitina	Desmocerataceae	Desmoceratidae	*Kitchinites*	Matsumoto, 1988; Matsumoto et al., 1990b
Ammonitina	Desmocerataceae	Desmoceratidae	*Matsumotoceras*	Matsumoto, 1988; Matsumoto et al., 1990a[*]
Ammonitina	Desmocerataceae	Desmoceratidae	*Mesopuzosia*	Matsumoto, 1988; Matsumoto et al., 1990a,b
Ammonitina	Desmocerataceae	Desmoceratidae	*Neopuzosia*	Matsumoto, 1988; Matsumoto et al., 1990b
Ammonitina	Desmocerataceae	Desmoceratidae	*Pachydesmoceras*	Matsumoto, 1987b, 1988
Ammonitina	Desmocerataceae	Desmoceratidae	*Parapuzosia*	Matsumoto, 1988[*]

Suborder	Superfamily	Family	Genus	Reference
Ammonitina	Desmocerataceae	Desmoceratidae	Puzosia	Marcinowski, 1980; Wright and Kennedy, 1984; Cooper and Kennedy, 1987; Matsumoto, 1988; Matsumoto et al., 1990a,b; Matsumoto and Skwarko, 1993
Ammonitina	Desmocerataceae	Kossmaticeratidae	Grossouvrites	Olivero and Medina, 1989
Ammonitina	Desmocerataceae	Kossmaticeratidae	Gunnarites	Kennedy and Klinger, 1985[*]
Ammonitina	Desmocerataceae	Kossmaticeratidae	Kossmaticeras	Matsumoto, 1991b
Ammonitina	Desmocerataceae	Kossmaticeratidae	Maorites	Macellari, 1986
Ammonitina	Desmocerataceae	Kossmaticeratidae	Yokoyamaoceras	Matsumoto, 1991b; Maeda, 1993
Ammonitina	Desmocerataceae	Pachydiscidae	Anapachydiscus	Kennedy, 1986b,d, 1989; Kennedy et al., 1986; Kennedy and Henderson, 1992; Kennedy and Klinger, 1993
Ammonitina	Desmocerataceae	Pachydiscidae	Canadoceras	Kennedy, 1986d
Ammonitina	Desmocerataceae	Pachydiscidae	Eupachydiscus	Kennedy, 1986d
Ammonitina	Desmocerataceae	Pachydiscidae	Lewesiceras	Wright and Kennedy, 1984; Kennedy, 1986d
Ammonitina	Desmocerataceae	Pachydiscidae	Menuites	Kennedy, 1986d, 1989; Kennedy and Henderson, 1992; Kennedy and Klinger, 1993; Cobban and Kennedy, 1993
Ammonitina	Desmocerataceae	Pachydiscidae	Pachydiscoides	Kennedy, 1984[*]
Ammonitina	Desmocerataceae	Pachydiscidae	Pachydiscus	Kennedy and Summesberger, 1984; Kennedy, 1986d; Jagt, 1989
Ammonitina	Desmocerataceae	Pachydiscidae	Teshioites	Kennedy, 1986d
Ammonitina	Desmocerataceae	Silesitidae	Silesites	Cecca and Landra, 1994
Ammonitina	Hoplitaceae	Hoplitidae	Anahoplites	Marcinowski and Wiedmann, 1990
Ammonitina	Hoplitaceae	Hoplitidae	Callihoplites	Marcinowski and Wiedmann, 1990
Ammonitina	Hoplitaceae	Hoplitidae	Euhoplites	Amedro, 1992
Ammonitina	Hoplitaceae	Hoplitidae	Hyphoplites	Wright and Kennedy, 1984
Ammonitina	Hoplitaceae	Placenticeratidae	Hoplitoplacenticeras	Kennedy, 1986d
Ammonitina	Hoplitaceae	Placenticeratidae	Hypengonoceras	Klinger and Kennedy, 1989
Ammonitina	Hoplitaceae	Placenticeratidae	Karamaites	Kennedy and Wright, 1982, 1983; Klinger and Kennedy, 1989

Suborder	Superfamily	Family	Genus	Source
Ammonitina	Hoplitaceae	Placenticeratidae	Placenticeras	Kennedy and Wright, 1981, 1983; Kennedy, 1984, 1986d, 1988; Kennedy and Cobban, 1991a; Cobban et al., 1989; Klinger and Kennedy, 1989; Matsumoto and Skwarko, 1991; Ganguly and Bardhan, 1993
Ammonitina	Acanthocerataceae	Acanthoceratidae	Acanthoceras	Moreau et al., 1983; Wright and Kennedy, 1987; Kennedy and Cobban, 1990; Matsumoto and Skwarko, 1991[*]
Ammonitina	Acanthocerataceae	Acanthoceratidae	Acompsoceras	Wright and Kennedy, 1987
Ammonitina	Acanthocerataceae	Acanthoceratidae	Benueites	Reyment, 1971, 1988
Ammonitina	Acanthocerataceae	Acanthoceratidae	Calycoceras	Wright and Kennedy, 1987, 1990; Cobban et al., 1989; Cobban and Kennedy, 1990; Matsumoto and Skwarko, 1991; Kennedy and Juignet, 1994
Ammonitina	Acanthocerataceae	Acanthoceratidae	Conlinoceras	Kennedy and Cobban, 1990
Ammonitina	Acanthocerataceae	Acanthoceratidae	Cunningtoniceras	Wright and Kennedy, 1987; Matsumoto et al., 1989[*]; Kennedy and Cobban, 1990
Ammonitina	Acanthocerataceae	Acanthoceratidae	Eucalycoceras	Cobban, 1988a; Wright and Kennedy 1990
Ammonitina	Acanthocerataceae	Acanthoceratidae	Euomphaloceras	Kennedy, 1988; Wright and Kennedy, 1990
Ammonitina	Acanthocerataceae	Acanthoceratidae	Mantelliceras	Wright and Kennedy, 1984; Kennedy, 1989; Matsumoto and Takahashi, 1992[*]
Ammonitina	Acanthocerataceae	Acanthoceratidae	Metoicoceras	Kennedy et al., 1981a, 1989; Kennedy, 1988, 1989; Förster et al., 1983; Cobban and Kennedy, 1991b
Ammonitina	Acanthocerataceae	Acanthoceratidae	Mrhiliceras	Kennedy and Wright, 1985b; Wright and Kennedy, 1987
Ammonitina	Acanthocerataceae	Acanthoceratidae	Nannometoioceras	Kennedy, 1988, 1989; Cobban et al., 1989
Ammonitina	Acanthocerataceae	Acanthoceratidae	Neocardioceras	Cobban, 1988a
Ammonitina	Acanthocerataceae	Acanthoceratidae	Paraburoceras	Cobban et al., 1989
Ammonitina	Acanthocerataceae	Acanthoceratidae	Paraconlinoceras	Kennedy and Cobban, 1990
Ammonitina	Acanthocerataceae	Acanthoceratidae	Plesiacanthoceras	Kennedy and Cobban, 1990
Ammonitina	Acanthocerataceae	Acanthoceratidae	Plesiacanthoceratoides	Kennedy and Cobban, 1990

Suborder	Superfamily	Family	Genus	References
Ammonitina	Acanthocerataceae	Acanthoceratidae	Protacanthoceras	Wright and Kennedy, 1980, 1987; Kennedy and Wright, 1985a
Ammonitina	Acanthocerataceae	Acanthoceratidae	Pseudaspidoceras	Matsumoto, 1991a[*]
Ammonitina	Acanthocerataceae	Acanthoceratidae	Pseudocalycoceras	Kennedy, 1988
Ammonitina	Acanthocerataceae	Acanthoceratidae	Spathites	Kennedy and Cobban, 1988b
Ammonitina	Acanthocerataceae	Acanthoceratidae	Sumitomoceras	Cobban, 1988a
Ammonitina	Acanthocerataceae	Acanthoceratidae	Tarrantoceras	Kennedy, 1988; Kennedy and Cobban, 1990; Cobban, 1988a
Ammonitina	Acanthocerataceae	Acanthoceratidae	Thomelites	Wright and Kennedy, 1990
Ammonitina	Acanthocerataceae	Acanthoceratidae	Watinoceras	Zaborski, 1987[*]; Cobban, 1988b
Ammonitina	Acanthocerataceae	Brancoceratidae	Euhystrichoceras	Kennedy and Wright, 1981; Wright and Kennedy, 1984
Ammonitina	Acanthocerataceae	Brancoceratidae	Hysteroceras	Amedro, 1992[*]
Ammonitina	Acanthocerataceae	Brancoceratidae	Mortoniceras	Marcinowski and Wiedmann, 1990; Amedro, 1992[*]
Ammonitina	Acanthocerataceae	Coilopoceratidae	Coilopoceras	Kennedy and Wright, 1984b; Kennedy, 1988; Reyment, 1988; Luger and Gröschke, 1989; Meister et al., 1992
Ammonitina	Acanthocerataceae	Coilopoceratidae	Hoplitoides	Kennedy and Cobban, 1988b; Reyment, 1988
Ammonitina	Acanthocerataceae	Collignoniceratidae	Barroisiceras	Immel, 1987[*]
Ammonitina	Acanthocerataceae	Collignoniceratidae	Collignoniceras	Kennedy et al., 1980, 1989
Ammonitina	Acanthocerataceae	Collignoniceratidae	Forresteria	Kennedy et al., 1983; Kennedy, 1984; Kennedy and Cobban, 1991a
Ammonitina	Acanthocerataceae	Collignoniceratidae	Gauthiericeras	Kennedy, 1984[*]
Ammonitina	Acanthocerataceae	Collignoniceratidae	Paratexanites	Kennedy, 1984
Ammonitina	Acanthocerataceae	Collignoniceratidae	Peroniceras	Kennedy, 1984[*]
Ammonitina	Acanthocerataceae	Collignoniceratidae	Prionocyclus	Kennedy, 1988
Ammonitina	Acanthocerataceae	Collignoniceratidae	Protexanites	Kennedy, 1984; Kennedy and Cobban, 1991a
Ammonitina	Acanthocerataceae	Collignoniceratidae	Submortoniceras	Kennedy et al., 1981b[*]
Ammonitina	Acanthocerataceae	Collignoniceratidae	Subprionocyclus	Reyment, 1982[*]; Futakami, 1990
Ammonitina	Acanthocerataceae	Collignoniceratidae	Yabeiceras	Kennedy et al., 1983[*]
Ammonitina	Acanthocerataceae	Flickiidae	Litophragmatoceras	Kennedy and Cobban, 1988a

Suborder	Superfamily	Family	Genus	Source
Ammonitina	Acanthocerataceae	Flickiidae	Noskytes	Kennedy and Wright, 1984a[*]
Ammonitina	Acanthocerataceae	Flickiidae	Salaziceras	Kennedy and Wright, 1984a[*]
Ammonitina	Acanthocerataceae	Forbesiceratidae	Forbesiceras	Kennedy and Juignet, 1984; Kennedy and Cobban, 1990; Wright and Kennedy, 1984; Matsumoto. 1987a
Ammonitina	Acanthocerataceae	Lyelliceratidae	Neophlycticeras	Kennedy and Delamette, 1994; Wright and Kennedy, 1994[*]
Ammonitina	Acanthocerataceae	Lyelliceratidae	Stoliczkaia	Wright and Kennedy, 1984
Ammonitina	Acanthocerataceae	Sphenodiscidae	Libycoceras	Kassab and Hamama, 1991
Ammonitina	Acanthocarataceae	Sphenodiscidae	Manambolites	Luger and Gröschke, 1989
Ammonitina	Acanthocerataceae	Sphenodiscidae	Sphenodiscus	Kennedy, 1986c,d[*]
Ammonitina	Acanthocerataceae	Tissotiidae	Metatissotia	Kennedy, 1984
Ammonitina	Acanthocerataceae	Vascoceratidae	Fagesia	Kennedy and Wright, 1985a; Kennedy et al., 1987
Ammonitina	Acanthocerataceae	Vascoceratidae	Hourcquia	Matsumoto and Toshimitsu, 1984[*]
Ammonitina	Acanthocerataceae	Vascoceratidae	Microdiphasoceras	Cobban et al., 1989
Ammonitina	Acanthocerataceae	Vascoceratidae	Neoptychites	Kennedy and Wright, 1979[*]; Kennedy and Cobban, 1988b; Cobban and Hook, 1983[*]; Zaborski, 1987[*]
Ammonitina	Acanthocerataceae	Vascoceratidae	Pseudobarroisiceras	Matsumoto and Toshimitsu, 1984[*]
Ammonitina	Acanthocerataceae	Vascoceratidae	Rubroceras	Cobban et al., 1989
Ammonitina	Acanthocerataceae	Vascoceratidae	Thomasites	Zaborski, 1987[*]
Ammonitina	Acanthocerataceae	Vascoceratidae	Vascoceras	Cobban et al., 1989; Meister et al., 1992
Ancyloceratina	Ancylocerataceae	Ancyloceratidae	Ancyloceras	Klinger and Kennedy, 1977; Förster and Weier, 1983[*]
Ancyloceratina	Ancylocerataceae	Ancyloceratidae	Acrioceras	Klinger and Kennedy, 1992[*]
Ancyloceratina	Ancylocerataceae	Ancyloceratidae	Crioceratites	Klinger and Kennedy, 1992[*]
Ancyloceratina	Ancylocerataceae	Heteroceratidae	Colchidites	Aguirre-Urreta and Klinger, 1986
Ancyloceratina	Ancylocerataceae	Heteroceratidae	Heteroceras	Aguirre-Urreta and Klinger, 1986
Ancyloceratina	Douvilleicerataceae	Astiericeratidae	Astiericeras	Kennedy, 1986a
Ancyloceratina	Douvilleicerataceae	Douvilleiceratidae	Douvilleiceras	Amedro, 1992
Ancyloceratina	Douvilleicerataceae	Douvilleiceratidae	Paraspiticeras	Aguirre-Urreta and Rawson, 1993

Suborder	Superfamily	Family	Genus	References
Ancyloceratina	Deshayesitaceae	Parahoplitidae	*Hypacanthoplites*	Kemper, 1982
Ancyloceratina	Scaphitaceae	Scaphitidae	*Acanthoscaphites*	Birkelund, 1982; Kennedy, 1986c; Kennedy and Summesberger, 1987; Jagt and Kennedy, 1989; Jagt et al., 1992
Ancyloceratina	Scaphitaceae	Scaphitidae	*Clioscaphites*	Landman, 1987
Ancyloceratina	Scaphitaceae	Scaphitidae	*Discoscaphites*	Jeletzky and Waage, 1978; Kennedy and Cobban, 1993c; Landman and Waage, 1993
Ancyloceratina	Scaphitaceae	Scaphitidae	*Hoploscaphites*	Makowski, 1962; Birkelund, 1982; Kennedy, 1986b,c, 1989, 1993; Kennedy et al., 1986; Kennedy and Summesberger, 1987; Kennedy and Cobban, 1993a; Landman and Waage, 1993
Ancyloceratina	Scaphitaceae	Scaphitidae	*Jeletzkytes*	Landman and Waage, 1993; Kennedy and Cobban, 1993a; Cobban and Kennedy, 1994a; Jagt and Kennedy, 1994; Kennedy and Cobban, 1993a
Ancyloceratina	Scaphitaceae	Scaphitidae	*Rhaeboceras*	Cobban, 1987
Ancyloceratina	Scaphitaceae	Scaphitidae	*Scaphites*	Cobban, 1969, 1984; Cobben and Kennedy, 1991a; Kennedy, 1984, 1986d, 1988, 1989; Kennedy et al., 1989, 1992; Kennedy and Christensen, 1991; Kennedy and Cobban, 1991a,b; Marcinowski, 1980, 1983; Immel, 1987; Kaplan et al., 1987; Landman, 1987, 1989; Jagt, 1989
Ancyloceratina	Scaphitaceae	Scaphitidae	*Trachyscaphites*	Kennedy and Summesberger, 1984; Kennedy, 1986d; Cobban and Kennedy, 1992, 1994b
Ancyloceratina	Scaphitaceae	Scaphitidae	*Worthoceras*	Förster et al., 1983; Kennedy, 1988; Kennedy and Cobban, 1988a; Kennedy et al., 1989; Cobban et al., 1989; Bujtor, 1991
Ancyloceratina	Scaphitaceae	Scaphitidae	*Yezoites*	Tanabe, 1977; Kennedy, 1984, 1988; Kennedy and Christensen, 1991; Kaplan et al., 1987
Ancyloceratina	Turrilitaceae	Anisoceratidae	*Allocrioceras*	Kennedy, 1988
Ancyloceratina	Turrilitaceae	Baculitidae	*Baculites*	Kennedy, 1984,* 1986b,*c; Jagt, 1989; Cobban and Kennedy, 1991b
Ancyloceratina	Turrilitaceae	Baculitidae	*Boehmoceras*	Kennedy and Cobban, 1991b
Ancyloceratina	Turrilitaceae	Baculitidae	*Eubaculites*	Klinger and Kennedy, 1993

Suborder	Superfamily	Family	Genus	Source
Ancyloceratina	Turrilitaceae	Baculitidae	*Lechites*	Cooper and Kennedy, 1977; Kennedy and Wright, 1985a
Ancyloceratina	Turrilitaceae	Baculitidae	*Sciponoceras*	Marcinowski, 1980; Kennedy and Juignet, 1983; Kennedy, 1988
Ancyloceratina	Turrilitaceae	Diplomoceratidae	*Oxybeloceras*	Cobban et al., 1996
Ancyloceratina	Turrilitaceae	Diplomoceratidae	*Solenoceras*	Cobban and Kennedy, 1994a; Cobban et al., 1996
Ancyloceratina	Turrilitaceae	Hamitidae	*Metaptychoceras*	Cobban et al., 1989
Ancyloceratina	Turrilitaceae	Labeceratidae	*Labeceras*	Klinger, 1989
Ancyloceratina	Turrilitaceae	Labeceratidae	*Myloceras*	Klinger, 1989
Ancyloceratina	Turrilitaceae	Nostoceratidae	*Anaklinoceras*	Cobban et al., 1996
Ancyloceratina	Turrilitaceae	Nostoceratidae	*Axonoceras*	Cobban et al., 1996
Ancyloceratina	Turrilitaceae	Nostoceratidae	*Bostrychoceras*	Kennedy, 1986d; Cobban et al., 1996
Ancyloceratina	Turrilitaceae	Nostoceratidae	*Didymoceras*	Cobban and Kennedy, 1994a; Cobban et al., 1996
Ancyloceratina	Turrilitaceae	Nostoceratidae	*Eubostrychoceras*	Kennedy, 1986d
Ancyloceratina	Turrilitaceae	Nostoceratidae	*Exiteloceras*	Cobban et al., 1996
Ancyloceratina	Turrilitaceae	Nostoceratidae	*Hyphantoceras*	Kennedy and Wright, 1985a
Ancyloceratina	Turrilitaceae	Nostoceratidae	*Nostoceras*	Luger and Gröschke, 1989[*]; Kennedy and Cobban, 1993a,b; Cobban and Kennedy, 1994a; Cobban et al., 1996
Ancyloceratina	Turrilitaceae	Ptychoceratidae	*Lytocrioceras*	Delanoy and Poupon, 1992[*]
Ancyloceratina	Turrilitaceae	Turrilitidae	*Mariella*	Kennedy and Wright, 1985a
Lytoceratina	Lytocerataceae	Macroscaphitidae	*Costidiscus*	Cecca and Landra, 1994
Lytoceratina	Lytocerataceae	Macroscaphitidae	*Macroscaphites*	Cecca and Landra, 1994
Lytoceratina	Tetragonitaceae	Gaudryceratidae	*Gaudryceras*	Hirano, 1978, 1979

[+]This table undoubtedly represents an incomplete listing but it is presented as a baseline for further research. In some studies, antidimorphs have been assigned to two different genera, both of which are here listed separately. This table is organized more-or-less according to the classification of Wright (1981), except that families are listed alphabetically within superfamilies and genera are listed alphabetically within families.

[*]Dimorphism is suggested but can not be proved on the basis of available data.

ACKNOWLEDGMENTS. A number of institutions made specimens available for this chapter or for projects reported here; we are grateful for their aid: American Museum of Natural History (AMNH); British Museum of Natural History (BMNH); Geologisch Instituut der Universiteit van Amsterdam (GIUA); Institut und Museum für Geologie und Paläontologie der Universität Tübingen (GPIT); Department of Geology, University of Iowa (SUI); Istituto di Geologia dell'Università de Palermo (IGUP); Muséum National d'Histoire Naturelle, Paris (MNHN); United States Geological Survey, Denver (USGS); United States National Museum (USNM); and Yale Peabody Museum (YPM).

A number of the photographs, although they have appeared in other publications, originally were made and made available by W. M. Furnish and Brian F. Glenister of the University of Iowa. The latter and Royal H. Mapes of Ohio University provided taxonomic and nomenclatorial information. The help af all three is gratefully acknowledged.

We also thank Susan M. Klofak, Andrew S. Modell, Kathleen Sarg, and Rebecca Chamberlain, all from AMNH, for help in the preparation of the figures and for aid in compiling information for the section on Cretaceous ammonoids, and Dr. W. J. Kennedy, who reviewed that section.

References

Aguirre-Urreta, M. B., and Klinger, H. C., 1986, Upper Barremian Heteroceratinae (Cephalopoda, Ammonoidea) from Patagonia and Zululand, with comments on the systematics of the subfamily, *Ann. S. Afr. Mus.* **96**(8):315–358.

Aguirre-Urreta, M. B., and Rawson, P. F., 1993, The Lower Cretaceous ammonite *Paraspiticeras* from the Neuquen Basin, west-central Argentina, *N. Jb. Geol. Palaont. Abh.* **188**(1):51–69.

Amedro, F., 1992, L'Albien du bassin anglo-parisien: Ammonites, zonation phylétique, séquences, *Bull. Centres Rech. Explor. Prod. Elf-Aquitaine* **16**(1):187–233.

Arkell, W. J., Furnish, W. M., Kummel, B., Miller, A. K., Moore, R. C., Schindewolf, O. H., Sylvester-Bradley, P. C., and Wright, C. W., 1957, *Treatise on Invertebrate Paleontology*. Part L. *Mollusca 4*, Geological Society of America and University of Kansas Press, Lawrence, KS.

Birkelund, T., 1982, Maastrichtian ammonites from Hemmoor, Niederelbe (NW-Germany), *Geol. J.* **A61**:13–33.

Blainville, M. H. D. de, 1840, Prodrome d'une monographie des ammonites, in: *Supplement du Dictionnaire des Sciences Naturelles*, Bertrand, Paris, pp. 1–31.

Blind, W., and Jordan, R., 1979, "Septen-Gabelung" an einer *Dorsetensia romani* (Oppel) aus dem nordwestdeutschen Dogger, *Palaont. Z.* **53**(3/4):137–141.

Brochwicz-Lewiński, W., and Rózak, Z., 1976, Some difficulties in recognition of sexual dimorphism in Jurassic perisphinctids (Ammonoidea), *Acta Palaeontol. Pol.* **21**(1):115–124.

Brooks, M. J., 1991, The ontogeny of sexual dimorphism: Quantitative models and a case study in Labrisomid Blennies (Teleostei: Paraclinus), *Syst. Zool.* **40**(3):71–283.

Bucher, H., 1992, Ammonoids of the Shoshonensis Zone (Middle Anisian, Middle Triassic) from NW Nevada, *Jahrb. Geol. Bundesanst.* **135**(2):423–466.

Bucher, H., and Guex, J., 1990, Rythmes de croissance chez les ammonites triasiques, *Bull. Geol. Lausanne* **308**:191–209.

Bujtor, L., 1991, A new *Worthoceras* (Ammonoidea, Cretaceous) from Hungary, and remarks on the distribution of *Worthoceras* species, *Geol. Mag.* **128**:537–542.

Bujtor, L., 1993, Valanginian ammonite fauna from the Kisújbánya Basin (Mecsek Mts., South Hungary) and its paleobiogeographical significance, *N. Jb. Geol. Palaont. Abh.* **188**(1):103–131.

Bulot, L., Company, M., and Thieuloy, J. P., 1990, Origine, évolution et systématique du genre Valanginien *Saynoceras* (Ammonitina, Olcostephaninae), *Geobios* **23**(4):399–413.

Callomon, J. H., 1955, The ammonite succession in the Lower Oxford Clay and Kellaway beds at Kidlington, Oxfordshire, and the zones of the Callovian Stage, *Phil. Trans. R. Soc. Lond. [Biol.]* **239**:215–264.

Callomon, J. H., 1963, Sexual dimorphism in Jurassic ammonites, *Trans. Leicester Lit. Phil. Soc.* **57**:21–56.

Callomon, J. H., 1969, Dimorphism in Jurassic ammonoidea. Some reflections, in: *Sexual dimorphism in fossil metazoa and taxonomic implications* (G. E. G. Westermann, ed.), International Union of Geological Sciences, series A, number 1, E. Schweizerbart'sche, Stuttgart, pp. 111–125.

Callomon, J. H., 1981, Dimorphism in ammonoids, in: *The Ammonoidea*, Syst. Assoc. Spec. Vol. 18 (M. R. House and J. R. Senior, eds.), Academic Press, London, pp. 257–273.

Callomon, J. H., 1988, [Review of] Matyja, B. A., 1986. Developmental polymorphism in Oxfordian ammonites, *Acta Geol. Pol.* **36**:37–68, *Cephalopod Newsl.* **9**:14–16.

Cecca, F., and Landra, G., 1994, Late Barremian–Early Aptian ammonites from the Maiolica Formation near Cesana Brianza (Lombardy Basin, northern Italy), *Riv. It. Paleont. Strat.* **100**(3):395–422.

Charpy, N., and Thierry, J., 1976, Dimorphisme et polymorphisme chez *Pachyceras* Bayle (Ammonitina, Stephanocerataceae) du Callovien supérieur (Jurassique moyen), *Haliotis* **6**:185–218.

Cobban, W. A., 1951, Scaphitoid cephalopods of the Colorado Group, *U.S. Geol. Surv. Prof. Pap.* **239**:1–42.

Cobban, W. A., 1953, Cenomanian ammonite fauna from the Mosby Sandstone of central Montana, *U.S. Geol. Surv. Prof. Pap.* **243D**:45–55.

Cobban, W. A., 1969, The late Cretaceous ammonites *Scaphites leei* Reeside and *Scaphites hippocrepis* (DeKay) in the western interior of the United States, *U.S. Geol. Surv. Prof. Pap.* **619**:1–29.

Cobban, W. A., 1984, Molluscan record from a mid-Cretaceous borehole in Weston County, Wyoming, *U.S. Geol. Surv. Prof. Pap.* **1271**:1–24.

Cobban, W. A., 1987, The Upper Cretaceous ammonite *Rhaeboceras* Meek in the western interior of the United States, *U.S. Geol. Surv. Prof. Pap.* **1477**:1–15.

Cobban, W. A., 1988a, *Tarrantoceras* Stephenson and related ammonoid genera from Cenomanian (Upper Cretaceous) rocks in Texas and the western interior of the United States, *U.S. Geol. Surv. Prof. Pap.* **1473**:1–30.

Cobban, W. A., 1988b, The Upper Cretaceous ammonite *Watinoceras* Warren in the western interior of the United States, *U.S. Geol. Surv. Bull.* **1788**:1–15.

Cobban, W. A., and Hook, S. C., 1983, Mid-Cretaceous (Turonian) ammonite fauna from Fence Lake area of west-central New Mexico, *N.M. Bur. Mines Miner. Resour. Mem.* **41**:1–50.

Cobban, W. A., and Kennedy, W. J., 1990, Variation and ontogeny of *Calycoceras (Proeucalycoceras) canitaurinum* (Haas, 1949) from the Upper Cretaceous (Cenomanian) of the western interior of the United States, *U.S. Geol. Surv. Bull.* **1881**:B1–B7.

Cobban, W. A., and Kennedy, W. J., 1991a, A giant scaphite from the Turonian (Upper Cretaceous) of the western interior of the United States, *U.S. Geol. Surv. Bull.* **1934**:A1–A2.

Cobban, W. A., and Kennedy, W. J., 1991b, Evolution and biogeography of the Cenomanian (Upper Cretaceous) ammonite *Metoicoceras* Hyatt, 1903, with a revision of *Metoicoceras praecox* Haas, 1949, *U.S. Geol. Surv. Bull.* **1934**:B1–B11.

Cobban, W. A., and Kennedy, W. J., 1991c, *Baculites thomi* Reeside 1927, an Upper Cretaceous ammonite in the Western Interior of the United States, *U.S. Geol. Surv. Bull.* **1934**:C.

Cobban, W. A., and Kennedy, W. J., 1992, Campanian *Trachyscaphites spiniger* ammonite fauna in north-east Texas, *Palaeontology (Lond.)* **35**(1):63–93.

Cobban, W. A., and Kennedy, W. J., 1993, The Upper Cretaceous dimorphic pachydiscid ammonite *Menuites* in the western interior of the United States, *U.S. Geol. Surv. Prof. Pap.* **1533**:1–14.

Cobban, W. A., and Kennedy, W. J., 1994a, Upper Cretaceous ammonites from the Coon Creek Tongue of the Ripley Formation at its type locality in McNairy County, Tennessee, *U.S. Geol. Surv. Bull.* **2073**:B1–B12.

Cobban, W. A., and Kennedy, W. J., 1994b, Middle Campanian (Upper Cretaceous) ammonites from the Pecan Gap Chalk of central and northeastern Texas, *U.S. Geol. Surv. Bull.* **2073**:D1–D9.

Cobban, W. A., Hook, S. C., and Kennedy, W. J., 1989, Upper Cretaceous rocks and ammonite faunas of southwestern New Mexico, *N.M. Bur. Mines Miner. Resour. Mem.* **45**:1–137.

Cobban, W. A., Kennedy, W. J., and Scott, G. R., 1996, Campanian and Maastrichtian ammonoid genera of the families Nostoceratidae and Diplomoceratidae from the Western Interior, Gulf and Atlantic Coasts, *J. Paleontol.* (in press).

Coëmme, S., 1917, Note critique sur le genre *Cadmoceras, Bull. Soc. Geol. France Ser.* 4 **17**:44–54.

Contini, D., Marchand, D., and Thierry, J., 1984, Réflexion sur la notion de genre et de sous-genre chez les ammonites: Exemples pris essentiellement dans le Jurassique moyen, *Bull. Soc. Geol. Fr.* **26**(4):653–661.

Cooper, M. R., and Kennedy, W. J., 1977, A revision of the Baculitidae of the Cambridge Greensand, *N. Jb. Geol. Palaont. Mh.* **11**:641–658.

Cooper, M. R., and Kennedy, W. J., 1979, Uppermost Albian (*Stoliczkaia dispar* zone) ammonites from the Angolan littoral, *Ann. S. Afr. Mus.* **77**(10):175–308.

Cooper, M. R., and Kennedy, W. J., 1987, A revision of the Puzosiinae (Cretaceous ammonites) of the Cambridge Greensand, *N. Jb. Geol. Palaont. Abh.* **174**(1):105–121.

Davis, R. A., 1972, Mature modification and dimorphism in selected late Paleozoic ammonoids, *Bull. Am. Paleontol.* **62**(272):23–130.

Davis, R. A., Furnish, W. M., and Glenister, B. F., 1969, Mature modification and dimorphism in late Paleozoic ammonoids, in: *Sexual Dimorphism in Fossil Metazoa and Taxonomic Implications*, International Union of Geological Sciences, Ser. A, No. 1 (G. E. G. Westermann, ed.), E. Schweizerbart'sche, Stuttgart, pp. 101–110.

Davitashvili, L. Sh., and Khimshiashvili, N. G., 1954, On the question of the biological significance of the apertural formation of ammonites, *Works Paleobio. Sect. Acad. Sci. Georgian Soviet Socialist Rep.* **2**:44–76 (in Russian).

Delanoy, G., and Poupon, A., 1992, Sur le genre *Lytocrioceras* Spath, 1924 (Ammonoidea, Ancyloceratina), *Geobios* **25**(3):367–382.

Demanet, F., 1943, Les horizons marins du Westphalien de la Belgique et leurs faunes, *Mem. Mus. R. Hist. Nat. Belg.* **101**:1–166.

Doguzhaeva, L., 1982, Rhythms of ammonoid shell secretion, *Lethaia* **15**:385–394.

Dommergues, J.-L., 1987, L'évolution chez les Ammonitina du Lias moyen (Carixien, Domérien basal) en Europe occidentale, *Docum. Lab. Geol. Lyon* **98**:1–297.

Dommergues, J.-L., 1988, Can ribs and septa provide an alternative standard for age in ammonite ontogenetic studies? *Lethaia* **21**(3):243–256.

Dommergues, J.-L., 1993, The Jurassic ammonite *Coeloceras*: An atypical example of dimorphic progenesis elucidated by cladistics, *Lethaia* **27**(2):143–152.

Dzik, I., 1990, The ammonite *Acrochordiceras* in the Triassic of Silesia, *Acta Palaeontol. Pol.* **35**(1,2):49–65.

Elmi, S., 1967, Le Lias supérieur et le Jurassique moyen de l'Ardèche, *Docum. Lab. Geol. Lyon* **19**(1–3):1–845.

Fatmi, A. N., 1969, Dimorphism in some Jurassic and Lower Cretaceous ammonites from West Pakistan, *Geonews (Geol. Surv. Pakistan)* **1**(2):6–13.

Foord, A. H., and Crick, G. C., 1897, *Catalogue of the Fossil Cephalopoda in the British Museum (Natural History). Part III. Containing the Bactritidae and Part of the Suborder Ammonoidea*, British Museum (Natural History), London.

Förster, R., and Weier, H., 1983, Ammoniten und Alter der Niongala-Schichten (Unterapt, Süd-Tanzania), *Mitt. Bayer. Staatsslg. Palaont. Hist. Geol.* **23**:51–76.

Förster, R., Meyer, R., and Risch, H., 1983, Ammoniten und planktonische Foraminiferen aus den Eibrunner Mergeln (Regensburger Kreide, Nordostbayern), *Zitteliana* **10**:123–141.

Frest, T. J., Glenister, B. F., and Furnish, W. M., 1981, Pennsylvanian-Permian Cheiloceratacean Ammonoid families Maximitidae and Pseudohaloritidae. *Paleontol. Soc. Mem.* **11** (*J. Paleontol.* suppl. to **55**(3)).

Furnish, W. M., 1966, Ammonoids of the Upper Permian *Cyclolobus*-zone, *N. Jb. Paläont. Abh.* **125**:265–296.

Furnish, W. M., and Knapp, W. D., 1966, Lower Pennsylvanian fauna from eastern Kentucky; Part 1, Ammonoids, *J. Paleontol.* **40**(2):296–308.

Futakami, M., 1990, Turonian collignoniceratid ammonites from Hokkaido, Japan. Stratigraphy and paleontology of the Cretaceous in the Ishikari province, central Hokkaido. Part 3, *J. Kawamura Gakuen Women's Univ.* **1**:235–260.

Ganguly, T., and Bardhan, S., 1993, Dimorphism in *Placenticeras mintoi* from the Upper Cretaceous Bagh Beds, central India, *Cretaceous Res.* **14**:747–756.

Gemmellaro, G. G., 1887, La fauna dei calcari con *Fusulina* della valle del Fiume Sosio nella Provincia di Palermo. Fascio I—Cephalopoda, Ammonoidea, *G. Sci. Nat. Econ.* **19**:1–106.

Geraghty, M. D., and Westermann, G. E. G., 1994, Origin of Jurassic ammonite concretions assemblages at Alfeld, Germany: A biogenic alternative, *Palaeont. Z.* **68**(3/4):473–490.

Gould, S. J., 1977, *Ontogeny and Phylogeny*, Harvard University Press, Cambridge, MA, pp. 1–501.

Guex, J., 1968, Note préliminaire sur le dimorphisme sexuel des Hildocerataceae du Toarcien moyen et supérieur de l'Aveyron (France), *Soc. Vaudoise de Sciences Naturelles, Lausanne, Bull.* **70**(327):57–84.

Haug, E., 1897, Observations à la suite d'une note de Ph. Glangeaud sur las forme de l'ouverture de quelques Ammonites, *Bull. Soc. Geol. France Ser. 3* **25**:107.

Hayasaka, S., Ōki, K., Tanabe, K., Saisho, T., and Shinomiya, A., 1987, On the habitat of *Nautilus pompilius* in Tañon Strait (Philippines) and the Fiji Islands, in: *Nautilus. The Biology and Paleobiology of a Living Fossil* (W. Bruce Saunders and Neil H. Landman, eds.), Plenum, Press, New York, pp. 179–200.

Hirano, H., 1978, Phenotypic substitution of *Gaudryceras* (a Cretaceous ammonite), *Trans. Proc. Palaeont. Soc. Jpn. N.S.* **109**:235–258.

Hirano, H., 1979, Importance of transient polymorphism in systematics of Ammonoidea, *The Gakujutsu Kenkyu, Sch. Educ., Waseda Univ. Ser. Biol. Geol.* **28**:35–43.

Houša, V., 1965, Sexual dimorphism and the system of Jurassic and Cretaceous Ammonoidea (Preliminary note), *Casas. Nar. Muz.* **134**:33–35.

Howarth, M. K., 1992, Tithonian and Berriasian ammonites from the Chia Gara Formation in northern Iraq, *Palaeontology (Lond.)* **35**(3):597–655.

Hyatt, A., 1874, Abstract of a memoir on the "Biological relations of the Jurassic ammonites," *Proc. Bost. Soc. Nat. Hist.* **17**:236–241.

Immel, H., 1987, Die Kreideammoniten der nördlichen Kalkalpen, *Zitteliana* **15**:3–163.

Ivanov, A. N., 1971, About some growth alterations in ammonite shells, *Bull. Moscow Soc. Nat. Hist. Geol. Sect.* **46**:155 (in Russian).

Ivanov, A. N., 1975, Late ontogeny of ammonites and, in particular, of the micro-, macro-, and megaconchs, *Coll. Stud. Notes Sci. Works Yarosl. St. Ped. Inst.* **142**:5–57 (in Russian).

Ivanov, A. N., 1983, Were micro- and macroconchs of ammonites sexual dimorphs? in: *Taxonomy and Ecology of Cephalopoda*. Scientific Papers, Academy of Sciences of the USSR, Zoological Institute, Scientific Council on the Problem of "Biological Bases of Utilization, Remaking, and Protection of the Animal World," Malacological Committee, Leningrad, pp. 32–34 (in Russian).

Jagt, J. W. M., 1989, Ammonites from the early Campanian Vaals Formation at the CPL Quarry (Haccourt, Liège, Belgium) and their stratigraphic implications, *Meded. Rijks Geol. Dienst* **43**(1):1–18.

Jagt, J. W. M., and Kennedy, W. J., 1989, *Acanthoscaphites varians* (Lopuski, 1911) (Ammonoidea) from the Upper Maastrichtian of Haccourt, NE Belgium, *Geol. Mijnbouw* **68**:237–240.

Jagt, J. W. M., and Kennedy, W. J., 1994, *Jeletzkytes dorfi* Landman and Waage 1993, a North American ammonoid marker from the lower Upper Maastrichtian of Belgium, and the numerical age of the Lower/Upper Maastrichtian boundary, *N. Jb. Geol. Palaont. Mh.* **4**:239–245.

Jagt, J. W. M., Kennedy, W. J., and Burnett, J., 1992, *Acanthoscaphites tridens* (Kner, 1848) (Ammonoidea) from the Vijlen Member (Lower Maastrichtian) of Gulpen, Limburg, The Netherlands, *Geol. Mijnbouw* **71**:15–21.

Jeletzky, J. A., and Waage, K. M., 1978, Revision of *Ammonites conradi* Morton 1834, and the concept of *Discoscaphites* Meek 1870, *J. Paleontol.* **52**(5):1119–1132.

Kant, R., 1973, Konstruktionsmorphologie, Nr. 20: Untersuchungen des allometrischen Gehäuse-wachstums paläozoischer Ammonoideen unter besonderer Berücksichtigung einzelner "Populationen," *N. Jb. Geol. Palaont. Abh.* **144**(2):206–251.

Kaplan, U., Kennedy, W. J., and Wright, C. W., 1987, Turonian and Coniacian Scaphitidae from England and North-Western Germany, *Geol. J.* **103**:5–39.

Kassab, A. S., and Hamama, H. H., 1991, Polymorphism in the Upper Cretaceous ammonite *Libycoceras ismaeli* (Zittel), *J. Afr. Earth Sci.* **12**(3):437–448.

Keferstein, W., 1866, Cephalopoden, in: *Die Klassen und Ordnungen des Thierreichs wissen-schaftlich dargestellt in Wort und Bild von Hans Georg Bronn, Fortgesetzt von Wilhelm Keferstein*, Vol. 3, Part 2, Verlag Winter, Leipzig, Heidelberg, pp. 1337–1406.

Kemper, E., 1982, Die Ammoniten des späten Apt und frühen Alb Nordwestdeutschlands, *Geol. Jahrb. A* **65**:553–557.

Kennedy, W. J., 1984, Systematic paleontology and stratigraphic distribution of the ammonite faunas of the French Coniacian, *Spec. Pap. Palaeontol.* **31**:1–160.

Kennedy, W. J., 1986a, Observations on *Astiericeras astierianum* (d'Orbigny, 1842) (Cretaceous Ammonoidea), *Geol. Mag.* **123**(5):507–513.

Kennedy, W. J., 1986b, The ammonite fauna of the Calcaire à *Baculites* (Upper Maastrichtian) of the Cotentin Peninsula (Manche, France), *Palaeontology (Lond.)* **29**(1):25–83.

Kennedy, W. J., 1986c, The ammonite fauna of the type Maastrichtian with a revision of *Ammonites colligatus* Binkhorst, 1861, *Bull. Inst. R. Sci. Nat. Belg. Sci. Terre* **56**:151–267.

Kennedy, W. J., 1986d, Campanian and Maastrichtian ammonites from northern Aquitaine, France, *Spec. Pap. Palaeontol.* **36**:1–145.

Kennedy, W. J., 1988, Late Cenomanian and Turonian ammonite faunas from north-east and central Texas, *Spec. Pap. Palaeontol.* **39**:1–131.

Kennedy, W. J., 1989, Thoughts on the evolution and extinction of Cretaceous ammonites, *Proc. Geol. Assoc.* **100**(3):251–279.

Kennedy, W. J., 1993, Campanian and Maastrichtian ammonites from the Mons Basin and adjacent areas (Belgium), *Bull. Inst. R. Sci. Nat. Belg. Sci. Terre* **63**:99–131.

Kennedy, W. J., and Christensen, W. K., 1991, Coniacian and Santonian ammonites from Bornholm, Denmark, *Bull. Geol. Soc. Den.* **38**:203–226.

Kennedy, W. J., and Cobban, W. A., 1976, Aspects of ammonite biology, biogeography, and biostratigraphy, *Spec. Pap. Palaeontol.* **17**:1–93.

Kennedy, W. J., and Cobban, W. A., 1988a, *Litophragmatoceras incomptum* gen. et sp. nov. (Cretaceous Ammonoidea), a cryptic micromorph from the Upper Cenomanian of Arizona, *Geol. Mag.* **125**(5):535–539.

Kennedy, W. J., and Cobban, W. A., 1988b, Mid-Turonian ammonite faunas from northern Mexico, *Geol. Mag.* **125**:593–612.

Kennedy, W. J., and Cobban W. A., 1990, Cenomanian ammonite faunas from the Woodbine Formation and lower part of the Eagle Ford Group, Texas, *Palaeontology (Lond.)* **33**(1):75–154.

Kennedy, W. J., and Cobban, W. A., 1991a, Coniacian ammonite faunas from the United States western interior, *Spec. Pap. Palaeontol.* **45**:1–96.

Kennedy, W. J., and Cobban, W. A., 1991b, Upper Cretaceous (upper Santonian) *Boehmoceras* fauna from the Gulf Coast region of the United States, *Geol. Mag.* **128**(2):167–189.

Kennedy, W. J., and Cobban, W. A., 1993a, Ammonites from the Saratoga Chalk (Upper Creta-ceous), Arkansas, *J. Paleontol.* **67**(3):404–434.

Kennedy, W. J., and Cobban, W. A., 1993b, Campanian ammonites from the Annona Chalk near Yancy, Arkansas, *J. Paleontol.* **67**(1):83–97.

Kennedy, W. J., and Cobban, W. A., 1993c, Maastrichtian ammonites from the Corsicana Formation in northeast Texas, *Geol. Mag.* **130**(1):57–67.

Kennedy, W. J., and Delamette, M., 1994, *Neophlycticeras* Spath, 1922 (Ammonoidea) from the Upper Albian of Ain, France, *N. Jb. Geol. Palaont. Abh.* **191**(1):1–24.

Kennedy, W. J., and Henderson, R. A., 1992, Non-heteromorph ammonites from the Upper Maastrichtian of Pondicherry, South India, *Palaeontology (Lond.)* **35**(2):381–442.

Kennedy, W. J., and Juignet, P., 1983, A revision of the ammonite faunas of the type Cenomanian. I. Introduction, Ancyloceratina, *Cretaceous Res.* **4**:3–83.

Kennedy, W. J., and Juignet, P., 1984, A revision of the ammonite faunas of the type Cenomanian. 2. The families Binneyitidae, Desmoceratidae, Engonoceratidae, Placenticeratidae, Hopliti-dae, Schloenbachiidae, Lyelliceratidae and Forbesiceratidae, *Cretaceous Res.* **5**:93–161.

Kennedy, W. J., and Juignet, P., 1994, A revision of the ammonite faunas of the type Cenomanian, 5. Acanthoceratinae *Calycoceras (Calycoceras), C. (Gentoniceras)* and *C. (Newboldiceras), Cretaceous Res.* **15**:17–57.

Kennedy, W. J., and Klinger, H. C., 1979, Cretaceous faunas from Zululand and Natal, South Africa. The ammonite superfamily Haplocerat.aceae Zittel, 1884, *Ann. S. Afr. Mus.* **77**(6):85–121.

Kennedy, W. J., and Klinger, H. C., 1985, Cretaceous faunas from Zululand and Natal, South Africa. The ammonite family Kossmaticeratidae Spath, 1922, *Ann. S. Afr. Mus.* **95**(5):165–231.

Kennedy, W. J., and Klinger, H. C., 1993, On the affinities of *Cobbanoscaphites* Collignon, 1969 (Cretaceous Ammonoidea), *Ann. S. Afr. Mus.* **102**(7):265–271.

Kennedy, W. J., and Summesberger, H., 1984, Upper Campanian ammonites from the Gschliefgra-ben (Ultrahelvetic, Upper Austria), *Beitr. Palaont. Osterr.* **11**:149–206.

Kennedy, W. J., and Summesberger, H., 1987, Lower Maastrichtian ammonites from Nagoryany (Ukrainian SSR), *Beitr. Palaont. Osterr.* **13**:25–78.

Kennedy, W. J., and Wright, C. W., 1979, Vascoceratid ammonites from the type Turonian, *Palaeontology (Lond.)* **22**(3):665–683.

Kennedy, W. J., and Wright, C. W., 1981, *Euhystrichoceras* and *Algericeras*, the last mortonicera-tine ammonites, *Palaeontology (Lond.)* **24**(2):417–435.

Kennedy, W. J., and Wright, C. W., 1983, *Ammonites polyopsis* Dujardin, 1837, and the Cretaceous ammonite family Placenticeratidae Hyatt, 1900, *Palaeontology (Lond.)* **26**(4):855–873.

Kennedy, W. J., and Wright, C. W., 1984a, The Cretaceous ammonite *Ammonites requienianus* d'Orbigny, 1841, *Palaeontology (Lond.)* **27**(2):281–293.

Kennedy, W. J., and Wright, C. W., 1984b, The affinities of the Cretaceous ammonite *Neosayno-ceras* Breistroffer, 1947, *Palaeontology (Lond.)* **27**(1):159–167.

Kennedy, W. J., and Wright, C. W., 1985a, Evolutionary patterns in Late Cretaceous ammonites, *Spec. Pap. Palaeontol.* **33**:131–143.

Kennedy, W. J., and Wright, C. W., 1985b, *Mrhiliceras* n.g. (Cretaceous Ammonoidea), a new Cenomanian mantelliceratine, *N. Jb. Geol. Palaont. Mh.* **9**:513–526.

Kennedy, W. J., Wright, C. W., and Hancock, J. M., 1980, Collignoniceratid ammonites from the mid-Turonian of England and northern France, *Palaeontology (Lond.)* **23**(3):557–603.

Kennedy, W. J., Juignet, P., and Hancock, J. M., 1981a, Upper Cenomanian ammonites from Anjou and the Vendée, western France, *Palaeontology (Lond.)* **24**(1):25–84.

Kennedy, W. J., Klinger, H. C., and Summesberger, H., 1981b, Cretaceous faunas from Zululand and Natal, South Africa. Additional observations on the ammonite subfamily Texanitinae Collignon, 1948, *Ann. S. Afr. Mus.* **86**(4):115–155.

Kennedy, W. J., Wright, C. W., and Klinger, H. C., 1983, Cretaceous faunas from Zululand and Natal, South Africa. The ammonite subfamily Barroisiceratinae Basse, 1947, *Ann. S. Afr. Mus.* **90**(6):241–324.

Kennedy, W. J., Bilotte, M., Lepicard, B., and Segura, F., 1986, Upper Campanian and Maas-trichtian ammonites from the Petites-Pyrénées, southern France, *Eclogae Geol. Helv.* **79**:1001–1037.

Kennedy, W. J., Wright, C. W., and Hancock, J. M., 1987, Basal Turonian ammonites from west Texas, *Palaeontology (Lond.)* **30**(1):27–74.

Kennedy, W. J., Cobban, W. A., Hancock, J. M., and Hook, S. C., 1989, Biostratigraphy of the Chispa Summit Formation at its type locality: A Cenomanian through Turonian reference section for trans-Pecos Texas, *Bull. Geol. Inst. Univ. Uppsala, N.S.* **15**:39–119.

Kennedy, W. J., Hansotte, M., Bilotte, M., and Burnett, J., 1992, Ammonites and nannofossils from the Campanian of Nalzen (Ariège, France), *Geobios* **25**(2):263–278.

Kerr, J. G., 1931, Notes upon the Dana specimens of *Spirula* and upon certain problems of cephalopod morphology, *Oceanographical Reports Edited by the "Dana" Committee* **8**:1–34.

Klinger, H. C., 1989, The ammonite subfamily Labeceratinae Spath, 1925. Systematics, phylogeny, dimorphism and distribution (with a description of a new species), *Ann. S. Afr. Mus.* **98**(7):189–219.

Klinger, H. C., and Kennedy, W. J., 1977, Cretaceous faunas from Zululand, South Africa and southern Mozambique. The Aptian Ancyloceratidae (Ammonoidea), *Ann. S. Afr. Mus.* **73**(9):215–359.

Klinger, H. C., and Kennedy, W. J., 1989, Cretaceous faunas from Zululand and Natal, South Africa. The ammonite family Placenticeratidae Hyatt, 1900, with comments on the systematic position of the genus *Hypengonoceras* Spath, 1924, *Ann. S. Afr. Mus.* **98**(9):241–408.

Klinger, H. C., and Kennedy, W. J., 1990, Cretaceous faunas from Zululand and Natal, South Africa. A *Koloceras* (Cephalopoda, Ammonoidea) from the Mzinene Formation (Albian), *Ann. S. Afr. Mus.* **99**(2):15–21.

Klinger, H. C., and Kennedy, W. J., 1992, Cretaceous faunas from Zululand and Natal, South Africa. Barremian representatives of the ammonite family Ancyloceratidae Gill, 1871, *Ann. S. Afr. Mus.* **101**(5):71–138.

Klinger, H. C., and Kennedy, W. J., 1993, Cretaceous faunas from Zululand and Natal, South Africa. The heteromorph ammonite genus *Eubaculites* Spath, 1926, *Ann. S. Afr. Mus.* **102**(6):185–264.

Korn, D., 1992, Heterochrony in the evolution of Late Devonian ammonoids, *Acta Palaeontol. Pol.* **37**(1):21–36.

Landman, N. H., 1987, Ontogeny of Upper Cretaceous (Turonian-Santonian) scaphitid ammonites from the Western Interior of North America: Systematics, developmental patterns, and life history, *Bull. Am. Mus. Nat. Hist.* **185**(2):117–241.

Landman, N. H., 1989, Iterative progenesis in Upper Cretaceous ammonites, *Paleobiology* **15**(2):95–117.

Landman, N. H., and Waage, K. M., 1993, Scaphitid ammonites of the Upper Cretaceous (Maastrichtian) Fox Hills Formation in South Dakota and Wyoming, *Bull. Am. Mus. Nat. Hist.* **215**:1–257.

Landman, N. H., Dommergues, J.-L., and Marchand, D., 1991, The complex nature of progenetic species—examples from Mesozoic ammonites, *Lethaia* **24**:409–421.

Lehmann, U., 1966, Dimorphismus bei Ammoniten der Ahrensburger Lias-Geschiebe, *Palaeont. Z.* **40**(1–2):26–55.

Lehmann, U., 1981, *The Ammonites. Their Life and Their World* (translation by J. Lettau of *Ammoniten. Ihr Leben und ihre Umwelt*), Cambridge University Press, Cambridge.

Luger, P., and Gröschke, M., 1989, Late Cretaceous ammonites from the Wadi Qena area in the Egyptian Eastern Desert, *Palaeontology (Lond.)* **32**(2):355–407.

Macellari, C. E., 1986, Late Campanian–Maastrichtian ammonite fauna from Seymour Island (Antarctic Peninsula), *Paleont. Soc. Mem.* **18**:1–55.

Maeda, H., 1991, Sheltered preservation: A peculiar mode of ammonite occurrence in the Cretaceous Yezo Group, Hokkaido, north Japan, *Lethaia* **24**(1):69–82.

Maeda, H., 1993, Dimorphism of Late Cretaceous false-puzosiine ammonites, *Yokoyamaoceras* Wright and Matsumoto, 1954 and *Neopuzosia* Matsumoto, 1954, *Trans. Proc. Palaeont. Soc. Jpn. N.S.* **169**:97–128.

Makowski, H., 1962, Problem of sexual dimorphism in ammonites, *Palaeont. Pol.* **12**:1–92.

Makowski, H., 1971, Some remarks on the ontogenetic development and sexual dimorphism in the Ammonoidea, *Acta Geol. Pol.* **21**(3):321–340.

Makowski, H., 1991, Dimorphism and evolution of the goniatite *Tornoceras* in the Famennian of the Holy Cross Mountains, *Acta Palaeontol. Pol.* **36**(3):241–254.

Manger, W. L., and Rice, W. R., 1975, Intraspecific variation in *Arkanites relictus* (Cephalopoda, Ammonoidea) from Morrowan strata, Arkansas and Oklahoma (abstr.), *Geol. Soc. Am. Abstr. Prog.* **7**(2):212–213.

Mangold, K., 1987, Reproduction, in: *Cephalopod Life Cycles.* Vol. II. *Comparative Reviews* (P. R. Boyle, ed.), Academic Press, London, pp. 157–200.

Mangold-Wirz, K., 1963, Biologie des Céphalopodes benthiques et nectoniques de la Mer Catalane, *Vie Milieu [Suppl]* **13**:1–285.

Mangold-Wirz, K., Lu, C. C., and Aldrich, F. A., 1969, A reconsideration of forms of squid of the genus *Illex* (Illicinae, Ommastrephidae). II. Sexual dimorphism, *Can. J. Zool.* **47**(6):1153–1156.

Mapes, R. H., 1976, An unusually large Pennsylvanian ammonoid from Oklahoma, *Oklahoma Geol. Notes* **36**(2):47–51.

Marchand, D., 1976, Quelques précisions sur le polymorphisme dans la famille des Cardiocera-tidae Douvillé (Ammonoides), *Haliotis* **6**:119–140.

Marcinowski, R., 1980, Cenomanian ammonites from German Democratic Republic, Poland, and the Soviet Union, *Acta Geol. Pol.* **30**(3):215–325.

Marcinowski, R., 1983, Upper Albian and Cenomanian ammonites from some sections of the Mangyshlak and Tuarkyr regions, Transcaspia, Soviet Union, *N. Jb. Geol. Palaont. Mh.* **3**:156–180.

Marcinowski, R., and Wiedmann, J., 1990, The Albian ammonites of Poland, *Palaeont. Pol.* **50**:1–94.

Matsumoto, T., 1987a, Notes on *Forbesiceras* (Ammonoidea) from Hokkaido (Studies of Creta-ceous ammonites from Hokkaido-LX), *Trans. Proc. Palaeont. Soc. Jpn. N.S.* **145**:16–31.

Matsumoto, T. 1987b, Notes on *Pachydesmoceras*, a Cretaceous ammonite genus. *Proc. Jpn. Acad.* **63B**:5–8.

Matsumoto, T., 1988, A monograph of the Puzosiidae (Ammonoidea) from the Cretaceous of Hokkaido, *Palaeont. Soc. Jpn. Spec. Pap.* **30**:1–131.

Matsumoto, T., 1991a, On some acanthoceratid ammonites from the Turonian of Hokkaido (Studies of the Cretaceous ammonites from Hokkaido—LXIX), *Trans. Proc. Palaeont. Soc. Jpn. N.S.* **164**:910–927.

Matsumoto, T. (compiler), 1991b, The mid-Cretaceous ammonites of the family Kossmaticeratidae from Japan, *Palaeont. Soc. Jpn. Spec. Pap.* **33**:1–143.

Matsumoto, T., and Saito, R., 1987, Little known ammonite *Grandidiericeras* from Hokkaido (Studies of Cretaceous ammonites from Hokkaido—LVIII), *Trans. Proc. Palaeont. Soc. Jpn. N.S.* **145**:1–9.

Matsumoto, T., and Skwarko, S. K., 1991, Ammonites of the Cretaceous Ieru Formation, western Papua New Guinea, *BMR J. Aust. Geol. Geophys.* **12**(3):245–262.

Matsumoto, T., and Skwarko, S. K., 1993, Cretaceous ammonites from south-central Papua New Guinea, *AGSO J. Aust. Geol. Geophys.* **14**(4):411–433.

Matsumoto, T., and Takahashi, T., 1992, Ammonites of the genus *Acompsoceras* and some other acanthoceratid species from the Ikushunbetsu Valley, central Hokkaido, *Trans. Proc. Palaeont. Soc. Jpn. N.S.* **166**:1144–1156.

Matsumoto, T., and Toshimitsu, S., 1984, On the systematic positions of the two ammonite genera *Hourcquia* Collignon, 1965 and *Pseudobarroisiceras* Shimizu, 1932, *Mem. Fac. Sci. Kyushu Univ. Ser. D, Geol.* **25**(2):229–246.

Matsumoto, T., Suekane, T., and Kawashita, Y., 1989, Some acanthoceratid ammonites from the Yubari Mountains, Hokkaido—Part 2, *Sci. Rep. Yokosuka City Mus.* **37**:29–44.

Matsumoto, T., Nemoto, M., and Suzuki, C., 1990a, Gigantic ammonites from the Cretaceous Futaba Group of Fukushima Prefecture, *Trans. Proc. Palaeont. Soc. Jpn. N.S.* **157**:366–381.

Matsumoto, T., Toshimitsu, S., and Kawashita, Y., 1990b, On *Hauericeras* de Grossouvre, 1894, a Cretaceous ammonite genus, *Trans. Proc. Palaeont. Soc. Jpn. N.S.* **158**:439–458.

Matyja, B. A., 1986, Developmental polymorphism in Oxfordian ammonites, *Acta Geol. Pol.* **36**(1–3):37–68.

Matyja, B. A., 1994, Developmental polymorphism in the Oxfordian ammonite subfamily Peltoceratinae, in: *Palaeopelagos Special Publication 1, Proceedings of the 3rd Pergola International Symposium*, Rome, pp. 277–286.

McCaleb, J. A., 1968, Lower Pennsylvanian ammonoids from the Bloyd Formation of Arkansas and Oklahoma, *Geol. Soc. Am. Spec. Pap.* **96**:1–123.

McCaleb, J. A., and Furnish, W. M., 1964, The Lower Pennsylvanian ammonoid genus *Axinolobus* in the southern midcontinent, *J. Paleontol.* **38**(2):249–255.

McCaleb, J. A., Quinn, J. H., and Furnish, W. M., 1964, Girtyoceratidae in the southern midcontinent, *Oklahoma Geol. Surv. Circ.* **67**:1–41.

Meister, C., Alzouma, K., Lang, J., and Mathey, B., 1992, Les ammonites du Niger (Afrique occidentale) et la transgression transsaharienne au cours du Cénomanien-Turonien, *Geobios* **25**(1):55–100.

Meléndez, G., and Fontana, B., 1993, Intraspecific variability, sexual dimorphism, and non-sexual polymorphism in the ammonite genus *Larcheria* Tintant (Perisphinctidae) from the Middle Oxfordian of western Europe, in: *The Ammonoidea: Environment, Ecology, and Evolutionary Change*, Systematics Association. Spec. Vol. 47 (M. R. House, ed.), Clarendon Press, Oxford, pp. 165–186.

Mignot, Y., 1993, Un problème de paléobiologie chez les Ammonoïdes (Cephalopoda): Croissance et miniaturisation en liaison avec les environnements, *Docum. Lab. Geol. Lyon* **124**:1–113.

Miller, A. K., 1944, Permian cephalopods, in: Geology and paleontology of the Permian area northwest of Las Delicias, southwestern Coahuila, Mexico (by R. E. King, C. O. Dunbar, P. E. Cloud, Jr., and A. K. Miller), *Geol. Soc. Am. Spec. Pap.* **52**:71–127.

Miller, A. K., and Furnish, W. M., 1940, Permian ammonoids of the Guadalupe Mountain region and adjacent areas, *Geol. Soc. Am. Spec. Pap.* **26**:1–242.

Mojsisovics von Mojsvár, E., 1875, Das Gebirge um Hallstatt, Theil I, Die Mollusken-Faunen der Zlambach- und Hallstätter-Schichten, *K.-K. Geol. Reichanst. Wien Abh.* **6**(1).

Mojsisovics von Mojsvár, E., 1893, Das Gebirge um Hallstatt, Theil I, Die Cephalopoden der Hallstätter Kalke, *K.-K. Geol. Reichanst. Wien Abh.* **6**(2).

Mojsisovics von Mojsvár, E., 1896, Beiträge zur Kenntniss der obertriadischen Cephalopoden-Faunen des Himalaya, *Denks. Akad. Wiss. Wien* **63**:575–701.

Moreau, P., Francis, I. H., and Kennedy, W. J., 1983, Cenomanian ammonites from northern Aquitaine, *Cretaceous Res.* **4**:317–339.

Müller, A. H., 1969, Ammoniten mit "Eierbeutel" und die Frage nach dem Sexualdimorphismus der Ceratiten (Cephalopoda), *Monatsber. Deutsch. Akad. Wiss. Berl.* **11**(5,6):411–420.

Nassichuk, W. W., 1970, Permian ammonoids from Devon and Melville Islands, Canadian Arctic Archipelago, *J. Paleontol.* **44**(1):77–97.

Neige, P., 1992, Mise en place du dimorphisme (sexuel) chez les Ammonoides. Approche ontogénétique et interprétation hétérochronique, *D.E.A. Univ. Bourgogne*, pp. 1–50.

Nettleship, M. T., and Mapes, R. H., 1993, Morphologic variation, maturity, and sexual dimorphism in an Upper Carboniferous ammonoid from the Midcontinent, *Geol. Soc. Am. Abstr. Prog.* **25**(2):67.

Nikitin, S., 1884, General geological map of Russia, Sheet 56, *Works Geol. Comm.* **1**(2):1–153 (in Russian, German summary).

Obata, I., Futakami, M., Kawashita, Y., and Takahashi, T., 1978, Apertural features in some Cretaceous ammonites from Hokkaido, *Bull. Natl. Sci. Mus. (Tokyo) Ser. C (Geol.)* **4**(3):139–155.

Olivero, E. B., and Medina, F. A., 1989, Dimorfismo en *Grossouvrites gemmatus* (Huppe) (Ammonoidea) del Cretácico superior de Antártida, in: *Act. IV Congr. Argent. Paleont. Biostratigr Mendoza*, pp. 65–74.

Orbigny, A. d', 1847, *Paléontologie Française. Terraines jurassiques. Part I: Céphalopodes*, Masson, Paris, pp. 43–46, 433–464.

Palframan, D. F., 1966, Variation and ontogeny of some Oxfordian ammonites. *Taramelliceras richei* (de Loriol) and *Creniceras renggeri* (Oppel) from Woodham Buckinghamshire, *Palaeontology (Lond.)* **9**(2):290–311.

Parent, H., 1991, Ammonites cretácicos de la Formación Rio Mayer (Patagonia austral) *Hatchericeras patagonense* Stanton (Barremiano) y *Sanmartinoceras patagonicum* Bonarelli (Albiano), *Inst. Fisiogr. Univ. Nac. Rosario Notas* **A15**:1–8.

Pelseneer, P., 1926, La proportion relative des sexes chez les animaux et particulièment chez les Mollusques, *Acad. R. Belg. Cl. Sci. Mem. Zième Ser.* **8**(11):1–258.

Perna, E. Ya., 1914, Die Ammoneen des oberen Neodevon von Ostabhang des Südurals, *Mem. Comm. Geol. St. Petersburg, N.S.* **99**:1–87.

Pompeckj, J. F., 1894, Über Ammonoideen mit "anormaler wohnkammer," *J. Ver. Vaterl. Naturk. Württ.* **49**:220–290.

Ramsbottom, W. H. C., and Calver, M. A., 1962, Some marine horizons containing *Gastrioceras* in north west Europe, *C. R. 4 Cong. Strat. Geol. Carb. (Heerlen 1958)* **3**:571–576.

Reyment, R. A., 1971, Vermuteter Dimorphismus bei der Ammonitengattung *Benueites*, *Bull. Geol. Inst. Univ. Uppsala N.S.* **3**(1):1–18.

Reyment, R. A., 1982, Size and shape variation in some Japanese upper Turonian (Cretaceous) ammonites, *Stock. Contrib. Geol.* **37**(16):201–214.

Reyment, R. A., 1988, Does sexual dimorphism occur in Upper Cretaceous ammonites? *Senckenb. lethaea* **69**(1/2):109–119.

Riccardi, A. C., Aguirre Urreta, M. B., and Medina, F. A., 1987, Aconeceratidae (Ammonitina) from the Hauterivian–Albian of southern Patagonia, *Palaeontogr. A* **196**:105–185.

Ruzhencev, V. E., 1962, Superorder Ammonoidea. The ammonoids—general part, in: *Molluscs-Cephalopods. I.* (V. E. Ruzhencev, ed.), *Fundamentals of Paleontology* (Yu. A. Orlov, overall ed.), Publishing House of the Academy of Science of the USSR, Moscow, pp. 243–334 (in Russian).

Saunders, W. B., and Ward, P. D., 1987, Ecology, Distribution, and Population Characteristics of *Nautilus*, in: *Nautilus. The Biology and Paleobiology of a Living Fossil* (W. B. Saunders and N. H. Landman, eds.), Plenum Press, New York, pp. 137–162.

Sazonov, N. T., 1957, *Jurassic deposits of the central region of the Russian Platform*, Leningrad (in Russian).

Schiappa, T. A., Spinosa, C., and Snyder, W. S., 1995, *Nevadoceras*, a new Early Permian adrianitid (Ammonoidea) from Nevada. *J. Paleontol.* **69**(6):1073–1079.

Scott, G. R., and Cobban, W. A., 1965, Geologic and biostratigraphic map of the Pierre Shale between Jarre Creek and Loveland, Colorado, *U.S. Geol. Surv. Misc. Geol. Inv. Map I-439*, scale 1:48,000, separate text.

Shea, B. T., 1986, Ontogenetic approaches to sexual dimorphism in anthropoids, *Hum. Evol.* **1**(2):97–110.

Siebold, C. T., 1848, *Lehrbuch der vergleichenden Anatomie*, Vol. 1, Veit, Berlin.

Siemiradzki, J., 1925, *Podrecznik paleontologii.* Cz1: Paleozoologia, Warsaw.

Tanabe, K., 1977, Functional evolution of *Otoscaphites puerculus* (Jimbo) and *Scaphites planus* (Yabe), Upper Cretaceous ammonites, *Mem. Fac. Sci. Kyushu Univ. Ser. D. Geol.* **23**(3):367–407.

Teichert, C., Kummel, B., Sweet, W. C., Stenzel, H. B., Furnish, W. M., Glenister, B. F., Erben, H. K., Moore, R. C., and Nodine Zeller, D. E., 1964a, *Treatise on Invertebrate Paleontology.* Part K. *Mollusca 3*, Geological Society of America and University of Kansas Press, Lawrence, KS.

Teichert, C., Moore, R. C., and Nodine Zeller, D. E., 1964b, Rhyncholites, in: *Treatise on Invertebrate Paleontology.* Part K. *Mollusca 3* (R. C. Moore, ed.), Geological Society of America and University of Kansas Press, Lawrence, KS, pp. K467–K484.

Thierry, J., 1978, Le genre *Macrocephalites* au Callovien inférieur (Ammonites, Jurassique moyen), *Mem. Geol. Univ. Dijon* **4**:1–490.

Till, A., 1909, Die fossilen Cephalopodengebisse, *K.-K. Geol. Reichanst. Jahrb.* **58**(4):573–608.

Till, A., 1910, Die fossilen Cephalopodengebisse, Folge 3. *K.-K. Geol. Reichanst. Jahrb.* **59**:407–426.

Tintant, H., 1963, *Les Kosmoceratidés du Callovien inférieur et moyen d'Europe occidentale*, University of Dijon, France.

Tintant, H., 1976, Le polymorphisme intraspécifique en paléontologie. Exemple pris chez les ammonites, *Haliotis* **6**:49–69.

Tozer, E. T., 1994, Canadian Triassic ammonoid faunas, *Geol. Surv. Canada Bull.* **467**:1–663.

Trewin, N. H., 1970, A dimorphic goniatite from the Namurian of Cheshire, *Palaeontology (Lond.)* **13**(1):40–46.

Vaillant-Couturier-Treat, I., 1933, Paléontologie de Madagascar. XIX—Le Permo-Trias Marin, *Ann. Paleontol.* **22**(2):37–96.

Walliser, O. H., 1963, Dimorphismus bei Goniatiten, *Palaontol. Z.* **37**(1–2):21.

Wells, M. J., 1962, *Brain and Behavior in Cephalopods*, Stanford University Press, Stanford, CA.

Wells, M. J., 1966, The brain and behavior of cephalopods, in: *Physiology of Mollusca* (K. M. Wilbur and C. M. Younge, eds.), Academic Press, New York, Vol. 2, pp. 547–590.

Welter, O. A., 1922, Die Ammoniten der unteren Trias von Timor, in: *Paläontologie von Timor* (J. Wanner, ed.), E. Schweizerbart'sche, Stuttgart, pp. 81–160.

Westermann, G. E. G., 1964, Sexual-Dimorphismus bei Ammonoideen und seine Bedeutung für Taxionomie der Otoitidae (Einschliesslich Sphaeroceratinae; Ammonitina, M. Jura), *Palaeontogr. Abt. A* **124**(1–3):33–73.

Westermann, G. E. G., 1969a, Supplement: Sexual dimorphism, migration, and segregation in living cephalopods, in: *Sexual Dimorphism in Fossil Metazoa and Taxonomic Implications, International Union of Geological Sciences*, Ser. A, No. 1 (G. E. G. Westermann, ed.), E. Schweizerbart'sche, Stuttgart, pp. 18–20.

Westermann, G. E. G., 1969b, Proposal: Classification and nomenclature of dimorphs at the genus-group level [with discussion], in: *Sexual Dimorphism in Fossil Metazoa and Taxonomic Implications, International Union of Geological Sciences*, Ser. A, No. 1 (G. E. G. Westermann, ed.), E. Schweizerbart'sche, Stuttgart, pp. 234–238.

Wiedmann, J., 1965, Origins, limits, and systematic position of *Scaphites, Palaeontology (Lond.)* **8**(3):397–453.

Wiedmann, J., 1973, The Albian and Cenomanian Tetragonitidae (Cretaceous Ammonoidea), with special reference to the Circum-Indic species, *Eclogae Geol. Helv.* **66**(3):585–616.

Willey, A., 1895, In the home of the Nautilus, *Nat. Sci. (Lond.)* **6**(40):405–414.

Willey, A., 1902, Contribution to the natural history of the pearly nautilus, in: *Zoological Results Based on Material from New Britain, New Guinea, Loyalty Islands, and Elsewhere, Collected During the Years 1895, 1896, and 1897*, Cambridge University Press, Cambridge, pp. 691–830.

Wright, C. W., 1981, Cretaceous Ammonoidea, in: *The Ammonoidea*, Systematics Association Spec. Vol. 18 (M. R. House and J. R. Senior, eds.) Academic Press, London, pp. 157–174.

Wright, C. W., and Kennedy, W. J., 1980, Origin, evolution and systematics of the dwarf Acanthoceratid *Protacanthoceras* Spath, 1923 (Cretaceous Ammonoidea), *Bull. Br. Mus. (Nat. Hist.) Geol.* **34**(2):65–107.

Wright, C. W., and Kennedy, W. J., 1984, The Ammonoidea of the Lower Chalk. Part I. *Monogr. Palaeontogr. Soc. Lond.* pp. 1–126. (Publ. 567, part of Vol 137 for 1983).

Wright, C. W., and Kennedy, W. J., 1987, The Ammonoidea of the Lower Chalk. Part 2. *Monogr. Palaeontogr. Soc. Lond.* pp. 127–218. (Publ. 573, part of Vol. 139 for 1985).

Wright, C. W., and Kennedy, W. J., 1990, The Ammonoidea of the Lower Chalk. Part 3, *Monogr. Palaeontogr. Soc. Lond.* pp. 219–294. (Publ. 585, part of Vol. 144 for 1990).

Wright, C. W., and Kennedy, W. J., 1994, Evolutionary relationships among Stoliczkaiinae (Cretaceous ammonites) with an account of some species from the English *Stoliczkaia dispar* Zone. *Cretaceous Res.* **15**:547–582.

Zaborski, P. M. P., 1987, Lower Turonian (Cretaceous) ammonites from south-east Nigeria, *Bull. Br. Mus. (Nat. Hist.) Geol.* **41**(2):31–66.

Zeiss, A., 1969, Dimorphismus bei Ammoniten des Unter-Tithon. Mit einigen allgemeinen Bemerkungen zum Dimorphismus-Problem, in: *Sexual Dimorphism in Fossil Metazoa and Taxonomic Implications*. International Union of Geological Sciences, ser. A, no. 1 (G. E. G. Westermann, ed.), E. Schweizerbart'sche, Stuttgart, pp. 155–164.

Zhou Zuren, 1985, Several problems on the Early Permian ammonoids from south China, *Palaeontol. Cathayana* **2**:179–210.

V
Taphonomy

Chapter 14

Ammonoid Taphonomy

HARUYOSHI MAEDA and ADOLF SEILACHER

HARUYOSHI MAEDA • Department of Geology and Mineralogy, Faculty of Science, Kyoto University, Kyoto 606-01, Japan. ADOLF SEILACHER • Kline Geology Laboratory, Yale University, New Haven, Connecticut 06511-8161.

Ammonoid Paleobiology, Volume 13 of *Topics in Geobiology*, edited by Neil Landman *et al.*, Plenum Press, New York, 1996.

1. Why Should Taphonomy Be Part of Paleobiology?

Taphonomy deals with the differential decay of organismic shapes, tissues, and skeletons under the influence of biological and physical agents and their fixation in the fossil record. This helps us to understand cases of exceptional preservation, sometimes including soft parts (*Konservat-Lagerstätten*). On the other hand, it provides a measure of the amount of distortion and time averaging that fossil assemblages and shell beds (*Konzentrat-Lagerstätten*) have undergone compared to the original biocoenosis. In this sense, taphonomy is a geological discipline and an important tool in facies analysis.

As pelagic organisms, ammonoids never could be preserved *in situ* (as "fossil snapshots"). Nevertheless, their mode of life and the shape and complex structure of their shells made them behave very specifically after death. The story starts when the dead shell touches bottom in a vertical attitude rather than in the horizontal one that experiments with present-day *Nautilus* shells had predicted (Raup and Chamberlain, 1967; Raup, 1973). The tale continues with the bottom transport that allows waterlogged or sediment-filled shells to roll like wheels and still be deposited in a preferred orientation to the current. The filling process follows, in which the phragmocone chambers act as a chain of narrowing but isomorphic sedimentary basins interconnected by the siphuncular openings. The material filling the camerae may become selectively cemented, so that it survives shell dissolution, compaction, and subsequent reworking. Alternatively, the phragmocone may remain empty and provide the space for bacterial pyrite coating. Without such prefossilization, empty shells are preserved in three dimensions only if the surrounding sediment is cemented, so that calcareous druses can form in the individual chambers or (if the septa have been already dissolved) in the phragmocone as a whole. Without reinforcement, the shell becomes compactionally flattened, in which case, fracture patterns tell us about the timing of the two-stage collapse (body chamber before phragmocone) relative to shell dissolution and sediment stiffening. Even in late diagenesis, double suture lines provide the best clue to the process of pressure solution.

Such considerations may appear out of place in a book on ammonoid paleobiology. However, there exists no ammonoid that does not carry a taphonomic overprint that must be intellectually removed before paleobiological analysis can begin. Taphonomic phenomena also provide us with clues to ammonoid biology "through the back door." In particular, they point to basic differences in the function of buoyancy from that of present-day *Nautilus*, which often is used too literally as a model for ammonoid paleobiology. Thus, ammonoid workers should care about taphonomic processes as much as they do about the stratigraphic sequence.

2. Death

The cause of ammonoid death remains unknown in most cases. Only a few examples show the immediate signs of lethal events, e.g., Late Cretaceous *Placenticeras* bitten by mosasaurs (Kauffman and Kesling, 1960; Kauffman, 1990; Hewitt and Westermann, 1990) and Early Jurassic *Eleganticeras* predated by the decapod crustacean *Eryon* (Lehmann, 1976). Early Carboniferous ammonoid mandibles stacked in a phosphatic nodule might have resulted from the behavior of a cephalopod-eating predator that concentrated the indigestible parts in its feces (Mapes, 1987; Boston and Mapes, 1991).

On the other hand, fossil assemblages ("Konzentrat-Lagerstätten"; Seilacher, 1970, 1990; Seilacher *et al.*, 1985) of ammonoids are widespread in both time and space. Some of these may correspond to mass mortalities. Kauffman (1978), for instance, noted intense fluctuations of ammonoid occurrence within one sedimentary cycle of the Jurassic Posidonienschiefer (Germany) and attributed the ammonoid-rich strata to mass mortality from oxygen reversal. On the other hand, shell-scattered horizons may simply result from a physical concentration of fresh or prefossilized shells.

As in the Posidonienschiefer, fluctuations of ammonoid occurrence are very common in Mesozoic strata, e.g., the Jurassic Toyora Group of Japan (Tanabe *et al.*, 1982, 1984) and the Cretaceous Mowry Shale of the United States (Reeside and Cobban, 1960). Such shell concentrations generally contain a high percentage of juvenile shells. Thus, mass mortality somehow might have enhanced the fossilization potential of juvenile ammonoids.

Worldwide oceanic anoxia developed at various times during the Mesozoic Era (Schlanger and Jenkyns, 1976; Arthur, 1979; Jenkyns, 1985, 1988). Therefore, repeated mass suffocation caused by periodic upwelling of anoxic bottom water or toxic clouds stirred up from the deep anaerobic zone by storm-generated turbidity currents might have caused mass mortality in Mesozoic ammonoids.

3. Soft-Part Preservation

Perfect preservation of soft tissue remains is well known among Mesozoic squids from, e.g., the Lower Jurassic Posidonienschiefer of Germany (Reitner and Urlichs, 1983), the Middle Jurassic Oxford Clay of England (Allison, 1988, 1990a), and the Upper Jurassic Solnhofen Plattenkalk of Germany (Barthel *et al.*, 1990). Muscle fibers of the soft tissues commonly are "pseudomorphed" by a microspherical apatite that might have been precipitated in a high-pH microenvironment around decomposing proteinaceous squid carcasses near the sediment–water interface (Allison, 1988, 1990a). In these "*Konservat-Lagerstätten*," early bacterial phosphatization and/or pyritization might have prevented complete decomposition of the soft tissues (see Allison, 1990a–c).

In contrast, soft-part preservation is very rare in contemporaneous ammonoids. X-ray pictures suggest that some Devonian goniatites from the Wissenbacher Schiefer (Germany) show partly preserved soft tissue, including tentacles and buccal masses within the shells (Stürmer, 1969; Zeiss, 1969; Bandel, 1988). However, several reports of such "softpart" preservation of invertebrates are now questioned: e.g., images of "tentacles" and "mantle sacks" of goniatites and bactritids may be the fractured wall of the body chamber (Otto, 1994). In general, ammonoid musculature was probably so weak that it had a much lower fossilization potential than that in strong swimmers such as squids.

Several authors have reported the remains of a digestive system consisting of mouth parts (jaws and radula) and stomach (Closs, 1967; Lehmann, 1966, 1971, 1972, 1975, 1976, 1981; Lehmann and Weitschat, 1973; Bandel, 1988; see also other chapters in this volume). Moreover, the jaw apparatus (or its aptychus modification) is commonly preserved in the body chamber of "*Lagerstätten*" ammonoids of various ages (Lehmann, 1981; Tanabe, 1983b). This strongly suggests that these carcasses quickly sank to the sea floor and became buried rapidly (discussed later).

4. Necrolysis

Early decay processes have left their records in some well-preserved ammonoid fossils. "Pseudomorphs" of encrusting microbes are commonly preserved in siphuncular tubes and on septal and mural surfaces. Figure 1 shows an example of fungal propagation in a cameral lining. After death, fungal filaments invaded the chamber and attacked the thin organic membrane that covered the septal surface. Subsequent rapid phosphatization might have stopped the fungal activity suddenly and fixed the shapes of the microorganisms. Such cases provide a "snapshot" of early fossilization processes.

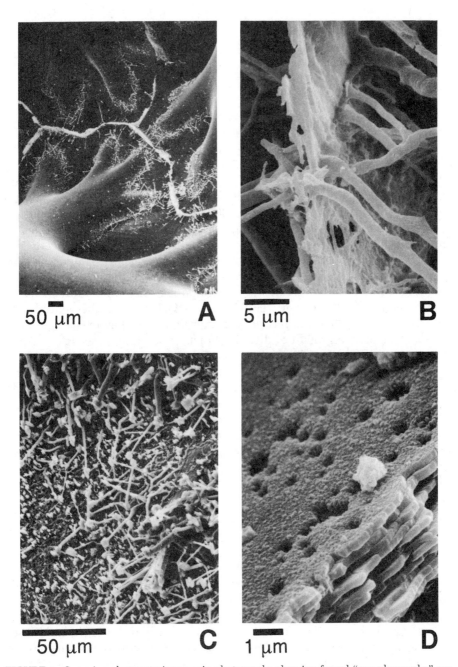

FIGURE 1. Scanning electron microscopic photographs showing fungal "pseudomorphs" preserved in camerae of *Damesites sugata* (Forbes) (Ammonitina, Desmocerataceae), UMUT (Univ. Mus. Univ. Tokyo), specimen without catalog number, from the Coniacian of the Yezo Group in Haboro, Hokkaido, Japan. Photographs courtesy of K. Tanabe (Univ. Tokyo). (A) "Pseudomorphosed" fungi on a cameral membrane. (B) Close-up of "pseudomorphosed" fungi showing their bioerosive effect. (C) "Pseudomorphosed" fungi attached to the organic lining (pellicle) of a septal surface. (D) Septal surface bioeroded by fungal "attack."

5. Encrustation

5.1. Before Death

Epizoans (serpulids, bryozoans, barnacles, foraminifers, and scyphozoans) occur on more than half the live *Nautilus* that inhabit steep forereef environments (Landman *et al.*, 1987). As a measure of ammonoid shell growth, worm tubes and other epibionts overgrown by later whorls have been used (Schindewolf, 1934). This bioassociation also provides a valuable record of life-associated epibiont communities, especially of nonmineralized types (e.g., Taylor, 1979; Kidwell and Jablonski, 1983). Observations of the distribution of encrusters may also suggest living positions or postmortem attitudes of ammonoid shells. For example, in an adult shell of Cretaceous *Buchiceras*, all oysters attached to the two flanks and the venter show similar growth directions. This suggests (1) a *Nautilus*-like position of the host *Buchiceras*, and (2) attachment only after the host had attained maximum adult size, and growth had ceased (Seilacher, 1960). Oyster attachment on the ventral side might also indicate that *Buchiceras* was pelagic rather than crawling on the sea floor. Dense overgrowth of discinid brachiopods on a Jurassic *Lytoceras* suggests a similar history (Seilacher, 1982).

In addition, subcircular pits up to 2 cm in diameter that can be attributed to limpet home depressions are common on both flanks of large pachydiscids and puzosiines (more than 30 cm in shell diameter) in the Cretaceous Yezo Group of Japan and Russia (Kase *et al.*, 1994). The presence of pits that are healed from inside the shell suggests that the limpets were dwelling on live ammonoids. Taking the depth limit of algal growth into consideration (the limpets presumably fed on algae), Kase *et al.* (1994) inferred that these mature ammonoids had periodically visited the upper layer of the euphotic zone, which was probably less than about 20 m in depth.

5.2. After Death

Encrustation by attached organisms on dead ammonoid shells also yields prolific information. Gill and Cobban (1966) estimated the degree of ammonoid shell dissolution in the Cretaceous Pierre Shale (United States) from bryozoan attachment scars on the inner surface of a body chamber of *Baculites*. Perforations in the ammonoid shell-wall were induced by boring of endolithic organisms (clionid sponges, thallophytes, polychaete worms, etc.) in chalk of the Cretaceous Miria Formation (Australia; Henderson and McNamara, 1985).

In the Jurassic Posidonienschiefer (Germany), oysters and other bivalves are attached to both flanks of *Lytoceras* shells in equal densities (Seilacher, 1982). This is in conflict with the "benthic island" model (Kauffman, 1978) and suggests attachment before the shells had sunk to the bottom (discussed later).

Lepas (Thoracia) attached only to the lateral side of a present-day *Spirula* shell (Coleoidea) reminds us of another scenario for postmortem attitudes of ammonoids. Donovan (1989) attributed oyster growth on only one side of a Kimmeridgian *Pectinatites* shell to the floating of the dead ammonoid shell in a horizontal position at the water surface, so that only the side of the shell low down in the water was accessible to oysters.

6. Did Ammonoid Carcasses Surface, Float, or Sink?

To be preserved as fossils, ammonoid carcasses must finally have sunk to the sea bottom. How and why did they sink? This basic question is the starting point of ammonoid biostratinomy. It revolves around the fact that the air chambers of ammonoids were filled with gas during life to provide buoyancy but that these chambers, after death, were not necessarily sealed hermetically.

6.1. *Nautilus* Model?

In life, ammonoid animals probably managed to maintain neutral buoyancy in the water column as does the living *Nautilus*. After death, present-day *Nautilus* shells are believed to surface because their soft tissue, having acted as ballast, drops out within a few hours or days (Chamberlain *et al.*, 1981). Subsequent long drifting in ocean currents and wide dispersal of empty shells have been demonstrated (Reyment, 1973; Toriyama *et al.*, 1966; Hamada, 1977; Matsushima, 1990; etc.). However, does this "surfacing-and-drifting" model apply also to ammonoids, or did ammonoid shells behave differently after death?

6.2. Pressure of the Cameral Gas

In present-day shelled cephalopods, the cameral gas, which counteracts the weight of the shell and soft tissue, diffuses from the cameral liquid to fill the camerae during evacuation of the liquid. Osmosis by way of Na^+,K^+-ATPase (sodium transport enzyme) is the major emptying mechanism in present-day *Nautilus* (Denton and Gilpin-Brown, 1966; Ward and Westermann, 1985; Ward, 1987; Greenwald and Ward, 1987; Jacobs, 1992).

Because gas diffusion is a passive process, the cameral gas pressure is inevitably low (0.6–0.8 atm in *Nautilus*: Denton and Gilpin-Brown, 1966; 0.8–0.9 atm: Ward, 1987). Present-day shelled cephalopods (*Sepia*, *Spirula*, and *Nautilus*) are incapable of adjusting gas pressure to ambient water pressures (Ward, 1987). It is, therefore, assumed that extinct ammonoids also lacked a functional equivalent of the teleost fish gas gland (but see Seilacher and LaBarbera, 1995).

After death, sea water was able to penetrate into the underpressurized camerae by way of the natural permeability of the siphuncular sheath

(Lehmann, 1976). Thus, the ammonoid phragmocone would eventually become waterlogged, as illustrated by Solnhofen ammonoids lying horizontally next to the landing marks left by their first hitting the sea floor in a vertical orientation.

6.3. Waterlogging

Water depth (ambient hydrostatic pressure) and the permeability of the siphuncle to sea water (porosity of the connecting rings) control the mode of waterlogging. If the permeability was sufficient, high hydrostatic pressure might strongly enhance the process (Fig. 2B). As sea water entered, the cameral gas became much compressed, and its volume might have decreased until gas pressure within the camerae equaled the ambient hydrostatic pressure. If the original pressure of the cameral gas was 1 atm, its volume became reduced to less than a tenth at 100 m depth (about 11 atm). This volume reduction of the cameral gas caused the dead ammonoid to lose buoyancy. Therefore, an ammonoid dying at depth probably would not have surfaced but would have sunk directly to even greater depth.

6.4. Implosion

Some orthoconic nautiloid fossils from the Paleozoic have intact shell walls but broken septa. Such selective destruction might be attributable to

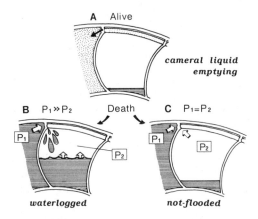

FIGURE 2. Models for variable cameral conditions of an ammonoid shell (A–C). As *Nautilus* do today, the ammonoid animal pumped out cameral liquid by osmosis after the formation of a new camera. Low-pressure gas diffused into the camera and filled the evacuated space while the animal was alive (A). After death, the mode of waterlogging depended on the ambient hydrostatic pressure. When the ambient pressure (P_1) exceeded the cameral gas pressure (P_2), sea water flowed back into the empty camera via the porous region of the shell until P_1 equaled P_2 (B). In cases in which P_1 equaled P_2 (C), the camera did not become waterlogged. Model B might be applicable to the deep-water realm; model C, to shallow surface waters. The soft tissue is dotted.

septal implosion by hydrostatic pressure during postmortem sinking (Wester-mann, 1977, 1985). Implosion during sinking would have been serious if air chambers were hermetically sealed. However, it conflicts with the hypothesis that an ammonoid carcass with hermetically sealed air chambers would not have lost buoyancy and would not have sunk, particularly after loss of the soft tissue. Undamaged phragmocones needed to have been weighted with some-thing (e.g., sea water) to sink after death and to be buried.

In the process of waterlogging, sea water floods the shell until the cameral gas pressure (P_2) equals ambient water pressure (P_1, Fig. 2B). If so, the phragmocone does not implode, even at great water depths (Boston and Mapes, 1991). The intact preservation of many ammonoid phragmocones in deep-water environments of various geological ages without any sign of implosion supports this view. Thus, implosion by hydrostatic pressure may be a subordinate factor in ammonoid taphonomy.

6.5. Buoyancy of Empty Ammonoid Shells

The buoyancy and orientation of ammonoid animals have been discussed by many authors (see Trueman, 1941; Raup and Chamberlain, 1967; Mutvei and Reyment, 1973; Ward and Westermann, 1977; Okamoto, 1988; and also Chapters 7 and 8, this volume). What about dead ammonoids? To roughly model postmortem behavior, we calculated the buoyancy of a late Cretaceous tetragonitid *Tetragonites glabrus* (Jimbo). This species has a moderately evo-lute planispiral shell with a smooth, rounded venter. Figure 3 shows the results of our buoyancy calculation using detailed measurements. We used formulas and equations proposed by Shigeta (1993) for the buoyancy calcula-tion.

An average-size shell of *Tetragonites glabrus* (Jimbo) attains a diameter of 10 cm and displaces 200 cm^3 (including the body chamber). This volume (200 cm^3 = 100%) consists of 136.2 cm^3 (68.10%) of soft tissue (1.067 g/cm^3: Denton and Gilpin-Brown, 1966), 22.70 cm^3 (11.35%) of shell material (2.62 g/cm^3: Reyment, 1958), and 41.14 cm^3 (20.57%) of empty camerae. The shell must weigh 205.2 g in total to maintain neutral buoyancy when sea water density is 1.026 g/cm^3.

If a *Tetragonites* shell did not become immediately waterlogged, it might have surfaced following the removal of soft tissues by scavengers (*Nautilus* model; Fig. 3B). However, our calculations show that only 5.84 cm^3 of sea water (14% waterlogging of the empty camerae) would have counteracted the buoyancy of the empty *Tetragonites* shell (Fig. 3C). Waterlogging in excess of 14% of the cameral volume not only would have prevented the dead *Tetrago-nites* shell from rising to the surface but would have made it sink to the bottom (Fig. 3D). If fully waterlogged, the density of the model *Tetragonites* shell increases to 1.212 g/cm^3 (Fig. 3D; discussed later).

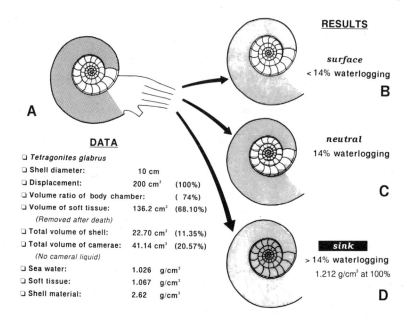

RESULTS

surface
< 14% waterlogging

B

A

DATA

❏ *Tetragonites glabrus*
❏ Shell diameter: 10 cm
❏ Displacement: 200 cm³   (100%)
❏ Volume ratio of body chamber: (74%)
❏ Volume of soft tissue: 136.2 cm³ (68.10%)
   (Removed after death)
❏ Total volume of shell: 22.70 cm³ (11.35%)
❏ Total volume of camerae: 41.14 cm³ (20.57%)
   (No cameral liquid)
❏ Sea water: 1.026 g/cm³
❏ Soft tissue: 1.067 g/cm³
❏ Shell material: 2.62   g/cm³

neutral
14% waterlogging

C

sink
> 14% waterlogging
1.212 g/cm³ at 100%

D

FIGURE 3. Schematic buoyancy calculation of a dead ammonoid shell. As an example, *Tetrago-nites glabrus* (Jimbo), a Late Cretaceous, moderately sized (10 cm in shell diameter), planispiral ammonoid possessing a smooth, moderately evolute, rounded shell (see Fig. 6) is shown here. We use Shigeta's (1993) formula and equations. Measurements of this species made by Shigeta (1993) and ourselves are as follows: body chamber length 1.6π radians; shell thickness (ST) 2% of whorl height (WH); volume of whorl tube (excluding septa) 8% of total displacement; volume of septa 14% of cameral volume; volume of phragmocone 26% of total displacement; density of soft tissue from Denton and Gilpin-Brown (1966). The volume of the camerae does not include septal volume. Only 5.84 cm³ of sea water (14% waterlogging of total cameral volume) was sufficient to cancel out the buoyancy of a dead *Tetragonites* shell of 10 cm diameter and 200 cm³ total displacement. If cameral liquid was originally present, buoyancy could have been counter-acted by a much smaller amount of sea water. At 100% waterlogging, the density of the dead shell would have reached 1.212 g/cm³. This value is close to that of foundered wood particles and of pieces of pumice.

Buoyancy calculations depend on estimates for several factors for which we do not have clear measurements (e.g., the amount of cameral liquid and the volume and density of the soft parts). In any case, we should expect that only slight waterlogging would have been necessary to counteract the buoy-ancy of a dead ammonoid shell, particularly at the high hydrostatic pressures present at greater depths.

6.6. Dispersal Patterns of Dead Ammonoid Shells

Based on actual modes of occurrence and preservation (Fig. 4), we can schematically summarize the possible preburial case histories of ammonoid carcasses (Fig. 5). Many ammonoids probably actively descended and as-

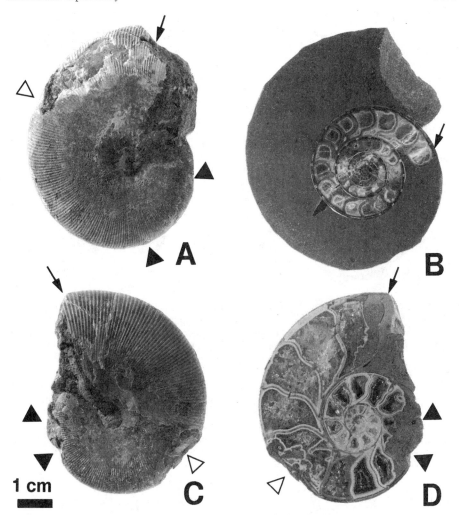

FIGURE 4. Preservational states of *Hypophylloceras* (A, B, and D: Phylloceratina) and *Gaudryceras* (C: Lytoceratina) recording different postmortem histories. Arrow, position of the last septum; solid triangle, preburial puncture; open triangle, post-burial puncture by compaction. All specimens courtesy of Y. Shigeta (Natl. Sci. Mus. Tokyo) and deposited in DGMKU (Dept. Geol. and Mineral., Kyoto Univ.). (A) Punctured phragmocone of *Hypophylloceras* sp. from very shallow-water marine sandstones, 30 m above a beach rock (Hanai and Oji, 1981); Miyako Group (upper Aptian), Japan. (B) Medial longitudinal section of *Gaudryceras tenuiliratum* Yabe from an offshore, deep-water mudstone facies of the Yezo Group (lower Campanian), Japan. (C) Punctured phragmocone of *Hypophylloceras subramosum* (Shimizu) from an offshore, deep-water mudstone facies of the Yezo Group (upper Santonian), Japan. (D) Medial longitudinal section of *H. subramosum* (Shimizu) (same specimen as C). Both leiostracan genera are common in the Cretaceous offshore mudstone facies in Japan. *Hypophylloceras* (A) also occurs in the coastal facies. Note that the body chamber is completely lost in *Hypophylloceras*. Some camerae, filled with sediment, had been punctured before burial; the others are well preserved, each with a siphuncular tube. In contrast, *Gaudryceras tenuiliratum* Yabe retains an almost complete body chamber. Note the intact phragmocone filled with sparry calcite (B).

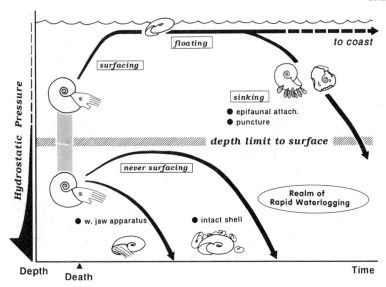

FIGURE 5. Schematic diagram showing dispersal patterns of dead ammonoids in relation to ambient pressure. Probably, many ammonoids migrated up and down in the water column, so that their postmortem histories varied according to depth. Ammonoids that died in deep water might never have surfaced because the empty camerae were quickly waterlogged by high hydrostatic pressure. Ammonoid shells housing a jaw apparatus *in situ* might represent this "never-surfacing" case. Likewise, intact shells without signs of puncture or other damage might belong to this same taphonomic category. At shallower depths, ammonoid carcasses still could have maintained buoyancy. Modes of occurrence and preservation of some ammonoid fossils suggest surfacing, drifting, and resinking, even in deep, offshore mudstone facies. Sinking could have been triggered by the weight of epifaunal overgrowth or by cameral puncture. In *Hypophylloceras* species, the mode of occurrence suggests the latter. The depth limit for surfacing determined the postmortem fates of dead bodies in both ammonoids and marine tetrapods.

cended in the water column as vertical migrants during life. However, a certain critical depth might divide their postmortem histories into two taphonomic pathways, the never-surfacing and the surfacing ones.

6.6.1. Never Surfacing

For a damaged submarine, it is not easy to raise buoyancy sufficiently to surface from great depths. Likewise, ammonoids that died in deep waters may never have surfaced. As soon as the animal died, the high ambient pressure would have caused the camerae to become waterlogged quickly. Because cameral gas reduces in volume, the shell would have lost its buoyancy immediately and sunk to the bottom.

Ammonoids preserving a mouth apparatus (jaws and radula) or its modification (aptychus) *in situ* probably represent the "never-surfacing" pathway followed by rapid burial (Fig. 5). In addition, the abundance of ammonoids without signs of damage or puncture in offshore, deep-water sediments suggests that many either did not float at all after death or became waterlogged rapidly (Kennedy and Cobban, 1976).

Buoyancy calculations also show that a conch of present-day *Nautilus* that dies in water deeper than 200–300 m will not reach the surface (Chamberlain *et al.*, 1981). A diving survey of a CALSUB cruise off New Caledonia has recently found numerous empty undamaged *Nautilus* shells on the sea floor at depths between 300 and 900 m, which are very close to their habitat (Roux, 1990; Roux *et al.*, 1991). This observation suggests that surfacing and subsequent extensive shell drift after death have not occurred in these specimens. Thus, it well may be that not all *Nautilus* shells follow the "*Nautilus*" model.

6.6.2. Surfacing

Some ammonoids undoubtedly underwent a long postmortem drift by currents, as was described in *Borissiakoceras* sp. by Kennedy and Cobban (1976). In shallow water, the waterlogging process might have been slow because ambient pressures were not much higher than the cameral gas pressure. At this depth, dead ammonoid shells may have maintained their buoyancy until the soft parts were lost, and the shell surfaced and drifted. Ammonoid fossils occurring in coastal facies reflect this process (Reyment, 1958).

Even in deep, offshore, mudstone facies, the features of some ammonoid shells suggest that they had surfaced, floated, and sunk again. How did they reach the bottom? Once a dead ammonoid shell had surfaced and floated, it could not have sunk again until buoyancy became insufficient or something weighted the shell. The same question naturally arises about several ammonoids presumed to have been "planktic" or "pelagic" in shallow seas.

Observations in the Posidonienschiefer (Germany) reveal that settlement of attached organisms, in particular, the byssate bivalve *Gervillia* (Bivalvia), occurred when the host ammonoid, *Lytoceras*, was still afloat, in a vertical position (Seilacher, 1982). Tanabe (1983a, 1991) reported a similar overgrowth of *Pseudomytiloides lunaris* (Hayami) (Bivalvia) on ammonoids from the Jurassic Toyora Group (Japan). Heavy overgrowth of attached organisms easily weighted the host ammonoid shell and eventually may have made it sink (Fig. 5; Donovan, 1989, p. 701, Fig. 3).

On the other hand, flooding by cameral puncture also may have happened in a floating ammonoid shell and caused resinking. Specimens of Cretaceous *Hypophylloceras*, a direct descendant of *Phylloceras* (Phylloceratina), uniformly exhibit signs of damage and puncture even though the shells are otherwise well preserved. *Hypophylloceras* is distributed across facies boundaries, suggesting different environmental settings. It occurs commonly in various lithologies, e.g., the offshore mudstone facies of the Yezo Group (Japan and Russia), which was deposited in a forearc basin (Matsumoto, 1942; Matsumoto and Okada, 1973; Tanabe, 1979; Okada, 1983), and the nearshore deposits in the Miyako Group of Japan (Shimizu, 1931; Hanai *et al.*, 1968; Hanai and Oji, 1981). In both cases, the body chambers have been removed completely (Fig. 4). In addition, one or two camerae in each phragmocone commonly are found to have been punctured before burial, probably by predation or during transport (Fig. 4A,C,D). Almost all of the 150 specimens examined show these taphonomic characteristics.

These characteristics contrast strikingly with those of other offshore leiostracans and suggest the "never-surfacing" process. For example, *Gaudryceras tenuiliratum* Yabe (Lytoceratina) cooccurs with *Hypophylloceras subramosum* (Shimizu) in offshore, deep-water facies. Although intact body chambers are rare in *H. subramosum*, perfect body chambers are frequently preserved in *G. tenuiliratum*. The phragmocones of *G. tenuiliratum* usually show intact preservation without any sign of puncture (Fig. 4B). Phylloceratids are believed to have occupied deep-water habitats during Jurassic and Cretaceous times (Scott, 1940; Ziegler, 1967; Hallam, 1969; Cecca, 1992). However, their occasional occurrence in clearly shallow-water deposits (Hallam, 1969; Kennedy and Cobban, 1976) and signs of shell damage are not necessarily in conflict with the inference of a deep-water habitat, because these features may suggest the peculiar postmortem history of phylloceratids: surfacing and floating first, followed by resinking by cameral puncture (Fig. 5).

Phylloceratids have long amphichoanitic septal necks, which differ from the prochoanitic necks of the Lytoceratina and the Ammonitina (Tanabe *et al.*, 1993; Chapter 6, this volume). Possibly, their extended septal necks could have slowed the process of waterlogging and, thus, might have allowed shells to remain buoyant for a longer time after death. If so, it seems possible that dead shells of phylloceratids would have had a greater postmortem dispersal potential than did other deep-water leiostracans.

6.7. Depth Limit on Postmortem Surfacing

In some respects, the postmortem behavior of ammonoid shells resembled that of marine vertebrates. In both cases, gas produced buoyancy, and depth controlled the fates of dead carcasses. Allison *et al.* (1991) found on the sea bottom at a depth of 1240 m a large cetacean carcass that never had surfaced but still preserved good skeletal articulation. The high hydrostatic pressure at that depth prevented the whale carcass from floating up because the decay gas was not able to produce sufficient volume for buoyancy.

Similarly, a critical depth for taphonomic flotation (Allison *et al.*, 1991) must also have existed for ammonoids. We do not numerically fix this depth limit here, because it probably varied depending on conditions. For example, removal of soft tissues or damage of the body chamber by predation or both could have produced buoyancy in the newly emptied shells even in deep water. Roughly speaking, however, the distance above or below this threshold depth may have determined the taphonomic fate of empty ammonoid shells (Fig. 5).

7. Transport

7.1. Landing and Roll Marks

Rothpletz (1909) was the first to illustrate outstanding examples of landing marks from the Solnhofen Plattenkalk of Germany (lower Portlandian): an imprint of an ammonoid shell that first stood in a vertical position on the sediment and then fell over onto its side. Seilacher (1963) showed that many puzzling "traces" in the Solnhofen Plattenkalk can be interpreted as roll-and-drag marks of waterlogged ammonoid shells as they were transported by bottom currents. Barthel *et al.* (1990, Figs. 2.11, 2.12) experimentally reproduced identical markings on wet mud.

On the other hand, peculiar roll marks that were made by corroded shells prove that some of the rolling shells previously had been deposited and became reworked (Seilacher, 1963, 1971). Such processes should not be ignored in evaluating the postmortem behavior of ammonoid shells and in determining taphofacies (Brett and Baird, 1986).

7.2. Ammonoids, Plants, and Pieces of Pumice

Ammonoid shells were sometimes concentrated into extensive shell beds, particularly in shallow, storm-dominated environments. On the other hand, shells frequently accumulated in discontinuous patches, together with many plant remains and pieces of pumice in offshore mudstone facies. Sometimes, calcareous concretions have formed around such accumulations. Figure 6 shows such a concretion from an offshore, relatively deep-water mudstone of the Cretaceous Yezo Group (Japan and Russia). Ammonoid shells, inoceramid bivalve shell fragments, and a large amount of plant remains are concentrated together in a mushroom-shaped, calcareous concretion. Figure 7 is another example showing an accumulation of ammonoids, wood particles, and many pieces of pumice. Some body chambers have suffered slight damage during transport and/or by predation. Although these assemblages seem mismatched, they are not peculiar to the Yezo Group but are widespread in offshore mudstone facies (Maeda, 1987). Thus, we are probably dealing with accumulations of particles that behaved similarly during near-bottom transport.

7.3. Accumulations of Lightweight Materials

The conceptual framework for fossil concentrations as proposed by Kidwell (1986), Kidwell *et al.* (1986), and Fürsich (1990) is useful in classifying the actual modes of occurrence of bivalves, gastropods, and brachiopods. Waterlogged ammonoid shells, however, belong to a somewhat different taphonomic guild, as do plant remains and pieces of pumice. Because of their

Up.

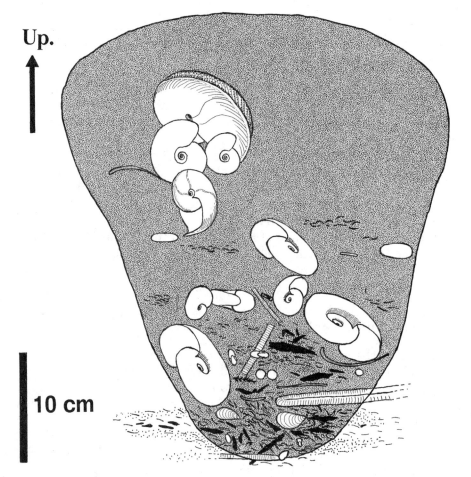

10 cm

FIGURE 6. Ammonoids in a mushroom-shaped, calcareous concretion from an intensely bioturbated mudstone of the Yezo Group (lower Santonian), Hokkaido, north Japan. (Modified from Maeda, 1987, with permission from the Palaeontological Society of Japan.) Keeled involute ammonoids, *Damesites damesi* (Jimbo); smooth-shelled, moderately evolute ammonoids, *Tetragonites glabrus* (Jimbo); heteromorph, *Polyptychoceras pseudogaultinum* (Yokoyama); inoceramid, *Sphenoceramus naumanni* (Yokoyama). Many shells, shell fragments, and numerous plant remains (black) are concentrated together, particularly in the lower part of the concretion. As a result of bioturbation, the shells do not show preferred orientations.

extremely thin walls, the waterlogged ammonoid shells had a density much lower than that of thick-shelled bivalves and gastropods.

On the other hand, underwater photographs show that foundered plant remains accumulate as patches on present-day offshore mud bottoms (Ohta, 1983; S. Ohta, personal communication). Density of the dredged wood particles from a mud bottom of Suruga Bay (250 m deep) is about 1.11 to 1.23 g/cm^3 (Maeda, 1987), i.e., almost identical to that of fully waterlogged ammonoid shells (calculated as 1.212 g/cm^3; Fig. 3D) and of pumice.

FIGURE 7. Accumulation of ammonoid shells together with wood particles and many pieces of pumice (light colored) in a calcareous concretion. Specimen deposited in DGMKU, from Loc. NB-3030, Krasnoyarka Formation (upper Campanian), Yezo Group, Sakhalin, Russia. The coin (lower left) is 2 cm in diameter. Ammonoids: *Desmophyllites diphylloides* (Forbes) and *Tetragonites popetensis* Yabe. The common association of ammonoid shells, plant remains, and/or pieces of pumice suggests that they had similar transport properties in very weak bottom currents, so that they eventually became deposited at the same place on the sea floor.

If so, it seems natural that waterlogged ammonoid shells, plant remains, and pieces of pumice behaved as hydrodynamically similar lightweight materials. Their cooccurrence, therefore, suggests that they were similarly transported by weak bottom currents and finally deposited and buried together on the sea floor (Maeda, 1987, 1990).

7.4. Sheltered Preservation

Waterlogged ammonoid shells and other lightweight particles are sometimes accumulated around large ammonoid shells as "dust heaps." Also, the body chamber may "shelter" many small "refugees" ("sheltered preservation"; Maeda, 1991). For example, a Cenomanian *Calycoceras* shell of 40-cm diameter houses more than 200 well-preserved juveniles of *Desmoceras*, together with a large amount of plant remains, within its body chamber, the punctured phragmocone, and the lower umbilical void. In this hydrodynamic shelter, immature shells greatly outnumber adults. Their phragmocones have

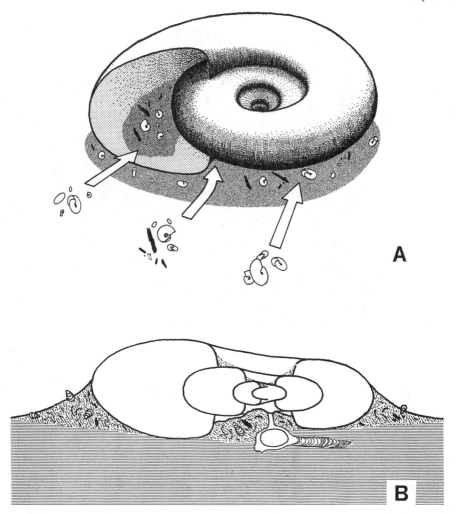

FIGURE 8. Possible biostratinomic effects of a large ammonoid shell that is associated with numerous small ammonoid shells in its body chamber, punctured camerae, and with or without a void below the umbilicus. (A) A large, empty shell acting as a "concentration trap" for small ammonoid shells and plant remains during postmortem transport on a soft, muddy sea bottom. This process may lead to "sheltered preservation" (From Maeda, 1991, p. 80, with permission of Scandinavian University Press). (B) Spatangoids commonly are found *in situ* in the umbilical void below large ammonoid shells, where these deposit feeders might have assembled in search of eutrophic sediments. (From Maeda, 1987, with permission of the Palaeontological Society of Japan.)

retained their original geometries (as well as their shell walls). The proto-conchs of these shells are also intact, but the shells lack the jaw apparatus and show signs of preburial damage caused by transport. Such "sheltered preservation" is widespread in the Cretaceous Yezo Group of Japan and Russia (Maeda, 1991).

Acting as concentration traps, large empty ammonoid shells probably collected small ammonoid shells during postmortem transport along the sea floor (Fig. 8A; Maeda, 1991). On the other hand, the "shelter" also might have enhanced the preservation of adequate size groups by early diagenetic cementation of the fill sediments ("body chamber concretion") in horizons where ammonoids are otherwise poorly preserved.

This type of accumulation and diagenetic protection also applies to spatangoid echinoids, which are regularly found in the lower umbilical voids of flat-lying, large ammonoid shells (Maeda, 1987, 1990). The echinoid tests are almost intact. Also, spatangoid traces (*Scolicia*) are sometimes found around large ammonoid shells. These burrowing echinoids might have been attracted by the organic matter that accumulated below the large ammonoid shell and then became preserved *in situ* (Fig. 8B). Alternatively, they simply may have belonged to the same lightweight category of particles, pumice, and small ammonoids.

7.5. Reworking of Ammonoid Shells

Ammonoid shells are commonly accumulated into condensed sequences. Wendt (1971, 1973), who studied ammonoid concentrations in Triassic red, nodular limestones (Hallstatt facies) of Yugoslavia and Greece, suggested that imbricated ammonoid shells accumulated in long depressions perpendicular to the current. Shell damage and discordant geopetal voids indicate that the shells suffered repeated reworking. Minimal sedimentation rates and episodic water currents were responsible for reworking of cephalopod shells and their accumulation in depressions on a hardground surface (Wendt, 1973).

Abraded internal molds of ammonoid shells, showing truncational facets, are good indexes of reworking in the Jurassic condensed section of the Iberian Range of Spain (Fernández-López, 1984, 1991; Fernández-López and Meléndez, 1994; Gómez and Fernández-López, 1994). Being sediment-filled, these shells probably behaved somewhat differently than waterlogged ones. Similar cephalopod shell accumulations have been observed in the Upper Devonian of Morocco (Wendt and Aigner, 1982) and the Upper Jurassic "Ammonitico Rosso Superior" of north Italy (Hollman, 1962).

On the other hand, ceratite shells show a variety of postmortem histories (Fig. 9; Seilacher, 1963, 1966, 1971). The sedimentary infills and associated geopetal voids show that many of the shells had been buried initially in a vertical position. Such vertical landing of ceratite shells may have been comparable to the results of experiments with the dead shells of present-day *Nautilus*, which retained a vertical position on shallow sea floors less than 8.5 m deep (Raup, 1973; Chamberlain and Weaver, 1978). After primary burial, shells became reexposed, truncated, corroded, reworked, and then redeposited, presumably during storms (Seilacher, 1971).

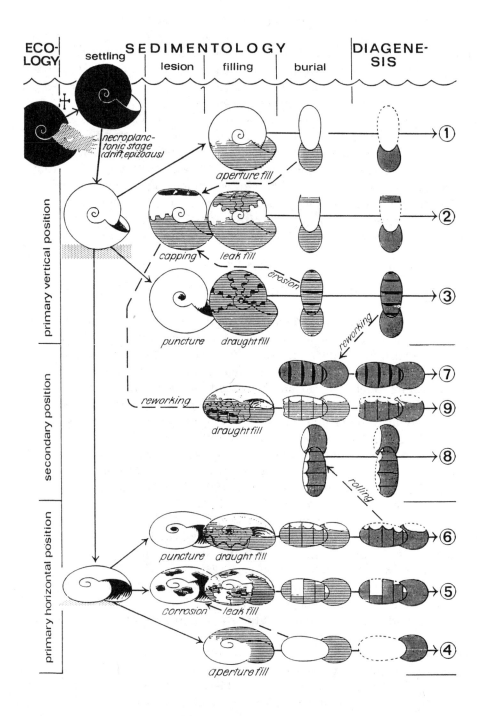

FIGURE 9. Case histories of ceratite preservation in the German Upper Muschelkalk. Here, decay of organic components (periostracum, siphonal tube) and early aragonite dissolution excluded fragments and hollow shells of ceratites from fossilization. What we find are only sedimentary infills converted into pressure-shadow concretions (Seilacher, 1971; Seilacher *et al.*, 1976). These steinkerns reveal a variety of preservational histories. (1) Because of rapid burial of the upright shell, the phragmocone remained unfilled and disappeared during shell dissolution. The surviving body chamber steinkern is easily overlooked. (2) Upright shells became reexposed and "capped" by erosion so that sediment could fill some of the chambers; collectional bias as in case 1. (3) A puncture allowed the phragmocone to draft-fill before it became completely buried, retaining the upright position and with lobe voids recording the filling mechanism. (4) As in case 1, but after the shell had been tipped into a horizontal position. (5) Like case 2, but leaks were situated on the flank. (6) Like case 3, but without lobe voids; instead, draft filling commonly left a sinusoidal fill channel above the plane of symmetry of the phragmocone. (7) Reworking brought specimens of case 3 into a horizontal orientation, but, because of prefossilization, they preserve the lobe voids from the primary upright position. (8) Reworking of prefossilized case 6 from horizontal into upright position, probably by "wheeling" into sticky mud. (9) Case 2 reworked before hardening of the sedimentary infill. It can be distinguished from case 6 by capping inconsistent with the present position and new infill. (Light hatching, soft fill sediment; dark hatching, fill diagenetically hardened; from Seilacher, 1971.)

---

8. Sedimentary Infill

8.1. Sediments in Isolated Camerae

Sedimentary infilling combined with shell dissolution and prefossilization may have played a decisive role in the preservation of ammonoid fossils (Figs. 9 and 10). Several eccentric preservations, e.g., half-ammonoids (discussed later), probably resulted from partial sedimentary infill and subsequent selective dissolution.

The process by which the open body chamber was filled with sediment is self-evident (Lehmann, 1976), but no sediment can intrude an undamaged phragmocone. Intact camerae with siphuncular tubes are either filled with drusy calcite (Fig. 4B) or still open. If the upper side of the shell became damaged, sediment was able to enter the camerae directly. On the other hand, sediment-filled ammonoids with undamaged camerae, but without siphuncles, are also common. How could the sediment have entered the camerae?

8.2. Sinusoidal Lines in Ceratites and Draft Filling

Seilacher (1968) demonstrated by experiment that chambered shells can become completely sediment-filled once the siphuncular tube is lost and there is a puncture somewhere in an inner whorl. The draft current produced by outside turbulence can pile up sediment to levels well above the siphonal connections between the camerae. In horizontally embedded shells, only a sinusoidal fill channel may eventually remain open in the otherwise sediment-filled phragmocone.

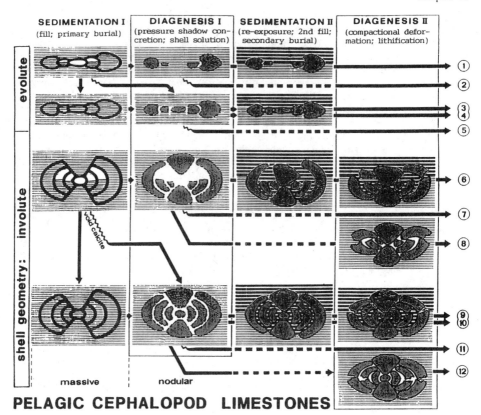

| SEDIMENTATION I (fill; primary burial) | DIAGENESIS I (pressure shadow concretion; shell solution) | SEDIMENTATION II (re–exposure; 2nd fill; secondary burial) | DIAGENESIS II (compactional deformation; lithification) |

PELAGIC CEPHALOPOD LIMESTONES

FIGURE 10. Case histories of ammonoid preservation in a Paleozoic example of cephalopod limestones of the "Ammonitico Rosso" type. Ammonoid shells became draft-filled as they did in the Muschelkalk. However, because of low rates of background sedimentation and lower levels of water turbulence, prefossilized steinkerns were able to become partially exposed during later sedimentational events without lateral dislocation. (1–4) Depending on the degree of infilling, whole steinkerns or only the outer whorls are preserved in evolute forms. (6–12) In bullate shells, additional concretionary plugs formed in the deep umbilical funnels. If reexposed, the steinkern ensemble may have collapsed by gravity (6) or secondary compaction (8–12), while voids left from shell dissolution were filled either by drusy calcite or by sediment of the second generation. Earlier stages in this flow diagram may also have been preserved by early cementation of the matrix or by drusy calcite filling of the voids. (After Neumann *et al.*, 1976, Fig. 24.2.)

The sinusoidal fill channel is seen perfectly in many ceratite steinkerns from the Triassic Muschelkalk of Germany (Fig. 9, part 6; Seilacher, 1966, 1971; Hagdorn and Mundlos, 1983). As in experiments, the top surfaces of incomplete sedimentary infills have various levels, which depend on the septation of the shell. Both experiments and actual specimens underline the importance of the draft current for ceratite preservation.

The draft-current process also may affect the body chamber of flat-lying shells. Here, the sediment accumulates in meniscoid laminae that are appar-

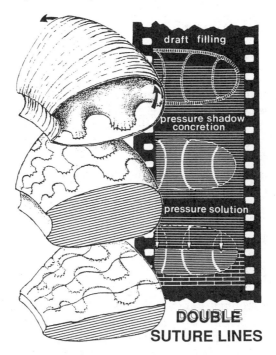

draft filling

pressure shadow concretion

pressure solution

DOUBLE SUTURE LINES

FIGURE 11. Double sutures, common in ceratites of the Middle Triassic Muschelkalk of Germany, resulted from the following processes: (1) Early decomposition of the organic siphuncular tube and high levels of turbulence allowed camerae to have been filled with fine mud by draft through the siphuncular openings. (2) Selective cementation of the fill sediment (pressure-shadow concretion; Seilacher, 1971; Seilacher *et al.*, 1976) produced a rigid steinkern that survived the dissolution of the outer shell. Thus, the surrounding plastic mud was able to replicate the suture lines on the internal mold. (3) During later pressure solution of the steinkern, the mud replica acted as a stencil that projected the "relief" of the original suture lines into deeper levels, where they now intersect with less undulatory cross sections of the respective septa. (From Seilacher, 1988.)

ent as sickle-shaped markings on the surface of incomplete fills after pressure dissolution (Mundlos, 1970; Hagdorn and Mundlos, 1983).

In the Muschelkalk, the sedimentary infills of ceratite shells are particularly clear because fill structures have become enhanced by necrolytic and diagenetic processes. These include (1) bacterial disintegration of the periostracum and siphuncular tubes, (2) concretionary lithification of sedimentary infills ("pressure-shadow concretions"; Seilacher, 1971; Seilacher *et al.*, 1976), (3) aragonite dissolution, followed, in some cases, by (4) lithification of the micritic matrix, and (5) pressure solution (Figs. 9, 11; Seilacher, 1971). Except in some grainstone beds, one never finds shell fragments, apertural rims, growth lines, or embryonic shell parts. According to the degree of sedimentary infilling, Muschelkalk ceratites are preserved only as partial or complete internal molds.

9. Preservation in Nodules

9.1. Carbonate Concretions

Ammonoids are generally better preserved in carbonate (calcite or siderite) nodules than in the host sediment (e.g., Lehmann, 1976, Fig. 57; Kennedy and Cobban, 1976, Pl. 8; Dagys and Weitschat, 1993, Fig. 3; Reeside and Cobban, 1960, Pl. 27). In this situation, phragmocones are usually not collapsed, although partial collapse of body chambers may occur. Camerae are secondarily filled with sparry calcite, but the shell wall, septa, and siphuncular tube are still present (Fig. 4B) because of the early cementation of the carbonate nodules (Allison, 1990b). In some cases, however, the aragonitic shell and septa (but not the siphuncular tubes) subsequently dissolved, so that drusy calcite lines the phragmocone space without subdivisions.

9.2. Consolidation at an Early Stage

Concretionary carbonate rarely replaces sediment grains, but instead, it precipitates in sediment pore spaces. Therefore, the original sediment porosity can be estimated from the volume of the cement (Raiswell, 1976). In calcareous concretions containing well-preserved ammonoids, the calcite cement of the matrix consistently attains 80–90% of volume (our observation; see also Raiswell, 1976). Although there may be different generations of concretions in a given stratigraphic unit, the fossiliferous ones tend to be formed in very early diagenetic stages preceding compaction.

In the Cretaceous Yezo Group (Japan and Russia), fossiliferous calcareous concretions also contain permineralized leaves of angiosperms and gymnosperms with preserved cuticles (Ohana and Kimura, 1991). They generally are embedded obliquely to the bedding plane (30–60°) and "float" in the matrix. This supports the view that the concretion grew close to the sediment surface prior to compaction and horizontal alignment of particles. In such concretions, ammonoid shells are also randomly oriented (Fig. 6). In light of the cooccurrence of benthic body fossils and burrows, this may be largely attributable to intense bioturbation by deposit feeders (Maeda, 1987).

9.3. Pyritization

In dysaerobic muds, framboidal pyrite linings are commonly found in camerae of ammonoid shells that were not sediment filled (Seilacher *et al.*, 1976). This confirms that bacterial sulfate reduction participated in the formation of carbonate concretions (Raiswell, 1976; Allison, 1990b,c). Bacterial activity requires dissolved organic matter from outside (Raiswell, 1976), for which shell accumulations of ammonoids and plant debris were a possible source.

In other cases, phosphatic micronodules have formed in the body chambers, in the sediment-filled innermost whorls, and in the umbilical depression
(Seilacher *et al.*, 1976). The organic parts of the siphuncular tubes also tend
to be phosphatized. Readers are referred to Berner (1968), Raiswell (1976),
Allison (1990a–c), and Prévôt and Lucas (1990) for detailed discussions of
concretionary processes.

10. Compaction

10.1. Compactional Sequence

Ammonoid shells embedded in argillaceous sediments commonly are
flattened, even though the aragonitic shell materials remained unaltered
(Jurassic *Psiloceras* and *Kosmoceras* in England or *Leioceras* in Germany to
Cretaceous *Hypophylloceras* in deep sea cores: Renz, 1979). The particular
modes of compactional flattening depend on shell morphology as well as
depositional conditions.

Oxyconic shells (e.g., *Harpoceras* in the Posidonienschiefer and *Oppelia*
in the Solnhofen Plattenkalk of Germany) clearly show a two-phase collapse
(Fig. 12; Seilacher *et al.*, 1976). Oxyconic shells were almost invariably buried
in a horizontal position. Their septate phragmocones, originally "designed"
to withstand hydrostatic pressure, were also strong against compactional
stress. The body chamber, in contrast, was not "designed" as a pressure vessel
and therefore collapsed first, with fracture patterns characteristic of rigid
shells, e.g., telescopic, keel, and longitudinal fractures (Seilacher *et al.*, 1976,
Figs. 2–7).

On the other hand, serpenticonic shells (e.g., *Dactylioceras* and *Perisphinctes*) collapsed as a whole after shell dissolution because the body chamber
was stronger than in oxycones as a result of its almost circular whorl section
and hollow radial ribs (Fig. 12).

10.2. Soft Deformation

Although phragmocones were kept open by reinforcing septa, the camerae
remained empty in low-energy settings. Compactional collapse may have
been retarded but ended in more complete flattening of the phragmocone than
of the body chamber. Collapsed phragmocones of *Harpoceras* in the Posidonienschiefer (Germany) lack fractures but appear dull from minute wrinkles
and irregularities that mask the original shell sculpture. Evidence suggests
that the shell aragonite had been dissolved during the time period between
body chamber and phragmocone collapse, so that only the periostracal film
remained (Fig. 12; Seilacher *et al.*, 1976, Fig.5; discussed below).

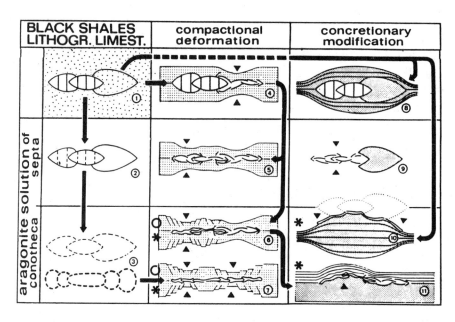

FIGURE 12. Case histories of ammonoid preservation in the Posidonienschiefer and the Solnhofen Plattenkalk (Jurassic of Germany). In anoxic facies, such as the bituminous Posidonienschiefer of Holzmaden (asterisks) and the Solnhofen Plattenkalk, organic shell components (including the siphuncular tube) did not decay. Accordingly, the phragmocones could not become sediment filled. In contrast to the Muschelkalk example, however, a periostracal foil did survive after aragonite dissolution. Different preservational histories resulted from the heterochronous interplay of shell strength, compactional flattening, aragonite dissolution, mud stiffening, and concretionary cementation. What is shown here in various cross sections correlates to distinctive features in bedding-plane views (see Seilacher *et al.*, 1976). Undeformed shell geometries (1) could be preserved only by early cementation of the matrix (8, "nucleus concretion") or of the sedimentary infill (9, body chamber concretion). On concretions formed below the lower umbilicus (10), only the central part of the shell left an undeformed impression, while the marginal periostracum foil became compactionally plastered onto the geode after aragonite dissolution. In the absence of early cementation, oxyconic shells suffered a two-phase collapse: first the body chamber (4), followed (after dissolution of the septa, 2) by the phragmocone (5). Because aragonite dissolution occurred between the two collapse events, oxyconic shells from Holzmaden show rigid fracture patterns on the body chamber and soft compaction on the phragmocone (6). The same is true for Solnhofen oxycones, except that the surrounding mud became stiffened during aragonite dissolution. Therefore, "collapse calderas" on adjacent bedding planes show soft contours on the body chamber and microfaulting over and below the phragmocone (6). Associated serpenticonic genera suffered only a one-phase collapse (3) because the circular cross section and strong ribs made the body chamber strong enough to withstand sediment pressure. Thus, the outer shell (conotheca) collapsed as a whole after aragonite dissolution. In the "domed" preservation of the Posidonienschiefer ammonites (11), already flattened shells became pushed up by a cementation front ascending from an underlying limestone bed. The body chamber, however, did not become "domed" because it had become fractured during brittle collapse (6) and therefore did not act as a coherent periostracal shield. Consistent with this model, larger shells became "domed" more readily than smaller ones, which remained unaffected at comparable distances above the limestone bed.

11. Dissolution

11.1. Differential Durability

The durability of different parts of an ammonoid differs. The decomposition order possibly depended on the sedimentary conditions. Normally, taphonomic decomposition seems to occur in the following order:

1. Soft tissue.
2. Organic periostracum and siphuncular tubes (under aerobic conditions).
3. Aragonitic inner whorls and septa.
4. Aragonitic upper shell region.
5. Whole aragonitic shell (or periostracal "phantom" under anoxic conditions).
6. Calcitic aptychus.

11.2. Dissolution of Inner Whorls

Selective dissolution of inner whorls may have occurred before burial. In the Cretaceous Yezo Group (Japan and Russia), large ammonoids exceeding 50 cm in diameter, e.g., puzosiines, pachydiscids, and acanthoceratids, regularly lack the umbilical center (Fig. 13B,B'). The Santonian genera *Anapachydiscus* and *Eupachydiscus* show dissolution of the inner whorls in shells less than 15 cm in diameter (Maeda, 1987). Siphuncular decomposition and dissolution of inner whorls on the sea floor may have allowed isolated camerae of the outer whorl to be draft-filled (see above). However, it is still uncertain whether the dissolution of these inner whorls began while the animal was still alive or not.

In contrast, associated *Gaudryceras* species have preserved inner whorls in specimens up to 30 cm in diameter because in this genus, the shell was coated by a thick periostracum.

11.3. Dissolution of Septa

If not completely sealed by internal sedimentation, septa remained exposed to pore water on both sides after the shell had been buried. This caused septa to dissolve earlier than the outer shell.

Some *Dactylioceras* shells from the Jurassic "Laibstein-Bank" (Germany) had their septa dissolved even though the phragmocones did not collapse ("hollow phragmocone"; Seilacher *et al.*, 1976). In this case, drusy calcite fills the whole phragmocone space without any trace of septal partitions, whereas the phosphatic siphuncle is preserved in its life position (Seilacher *et al.*, 1976, Fig. 18). This suggests very early cementation of the matrix followed by

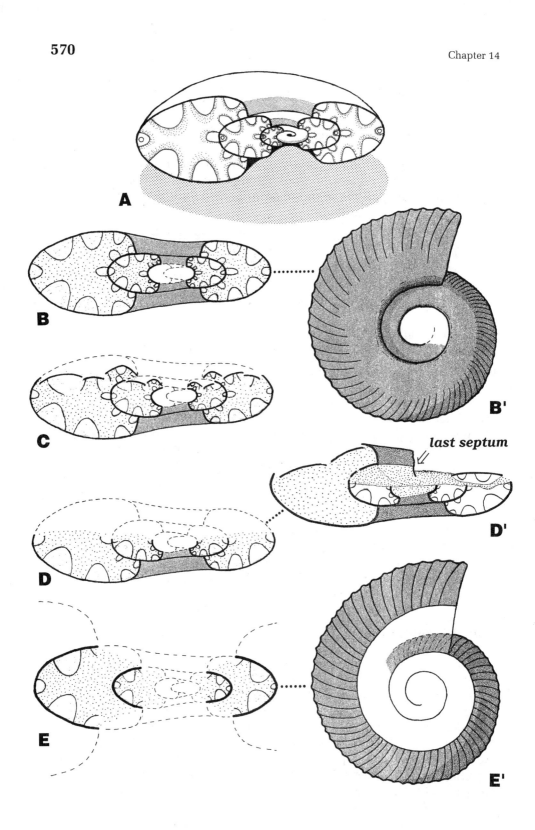

last septum

FIGURE 13. Characteristic mode of preservation of large puzosiine shells in the Upper Cretaceous Yezo Group, Hokkaido, Japan. (Modified from Maeda, 1987, with permission from the Palaeontological Society of Japan.) Lateral (B',E') and cross-sectional (A–E,D') views. (A) Original morphology of the phragmocone. (B,B') Selective dissolution of inner whorls. (C) Crushing of the upper shell region. (D) Half-ammonoid preservation. (D') Intermediate preservation between upper surface crushing (body chamber) and half-ammonoid (phragmocone). (E,E') "Ventral-tire" preservation. Half-ammonoids (D,D') may have been formed through partial sedimentary infilling of phragmocone chambers by weak draft currents and subsequent shell dissolution. The "ventral tires" originally might have been covered with an outer shell whorl. Thus, only the ventral part, which was not directly exposed to the external environment, might have escaped shell dissolution during diagenesis.

◄——

dissolution of septal aragonite. Without earlier cementation of the surrounding matrix, septal dissolution would have triggered shell collapse.

11.4. Half-Ammonoids

Half-ammonoid preservation (= half-ammonite; Fig. 13D) occurs in various Paleozoic and Mesozoic strata, and its origin has been discussed by many scholars (Minato, 1953; Onuki and Bando, 1959; Hollman, 1962; Shikama and Hirano, 1970; Lehmann, 1976; Seilacher et al., 1976; Tanabe et al., 1984). This phenomenon has puzzled us for a long time. Our observations suggest that partial sedimentary infilling of the phragmocone is the cause.

In a horizontally lying, large ammonoid shell, draft currents would have deposited sediment predominantly in the lower halves of the camerae. At low levels of turbulence, deposition could not have proceeded above the level of the septal necks without having blocked the passageway of the currents. The upper parts of the camerae and septa remained exposed to pore water after burial of the shell. They, therefore, could have been dissolved more readily than the parts of the shell sealed by sediments. Septal dissolution was probably followed by collapse of the upper part of the shell (Fig. 13C) and by gradual transition into the the half-ammonoid state (Fig. 13D,D'). It was, therefore, not "external" but "internal" half-burial of the shell, as determined by septal neck positions, that caused the uniform half-ammonoid preservation in various taxa (Maeda, 1987).

11.5. Ventral Tires

Puzosiines and pachydiscids in the Cretaceous Yezo Group (Japan and Russia) exhibit eccentric shell preservation in which only the ventral parts of the outer one or two whorls are preserved ("ventral tires"; Fig. 13E,E'; Maeda, 1987). The "ventral tires" originally might have been covered by an outer whorl. Eventually, only the ventral parts of the partly sediment-filled portions of the phragmocone that were not directly exposed to sea water escaped shell dissolution. "Ventral tires" are widespread, e.g., in the Cretaceous Gault in England.

11.6. Dissolution of the Entire Aragonitic Shell

Compared to calcitic shells, the aragonitic shells of ammonoids readily dissolved in the sediment. For example, *Chondrites* burrows perforated a flattened *Harpoceras* in a sediment-feeder horizon of the Jurassic Posidonien-schiefer of Germany (Seilacher *et al.*, 1976; Brenner and Seilacher, 1978). This suggests that dissolution occurred at a very shallow depth below the sediment–water interface. On the other hand, a thin periostracal film could have exceptionally survived under oxygen-depleted conditions without being decomposed by microbes (Fig. 12).

Aragonite dissolution took place even in calcareous sediments. In the Triassic Muschelkalk (Germany), for instance, the aragonitic shells of ceratites became completely dissolved, so that calcitic or phosphatic shells of attached organisms now appear directly superimposed on ceratite steinkerns (Seilacher, 1971). Wright and Kennedy (1981, 1984) reported similar differential survival of calcitic shells of attached organisms as opposed to aragonitic ammonoid shells from the Cretaceous Chalk in England.

Differential survival of aragonitic shells versus calcitic aptychi is also common. Ammonoids with their aptychi still in the body chambers are common in the Jurassic Solnhofen Plattenkalk of Germany (e.g., Barthel *et al.*, 1990). But in contrast to the perfectly preserved aptychi, the shells occur merely as flattened, periostracal impressions on the bedding planes (Seilacher *et al.*, 1976). The selective dissolution of shells and later reworking could account for the formation of aptychus coquinas in deeper-water environments.

12. Concluding Remarks

Intact ammonoids enclosed in carbonate nodules provide a snapshot of the original thanatocoenosis in uncompacted and highly porous mud. In contrast, poorly preserved and damaged specimens may be compared to a multiexposed film, in which subsequent taphonomic processes have overprinted previous images. Some characteristic damages of ammonoid shells can be attributed to the peculiar modes of life of the animals or to their shell design. Thus, taphonomic damage and preservational bias become a key to the interpretation of ammonoid ecology, if emphasis is switched from biogeological questions to biologically meaningful information contained in time-averaged and ecologically mixed assemblages (Behrensmeyer and Kidwell, 1985).

ACKNOWLEDGMENTS. We dedicate this chapter to I. Hayami in honor of his graceful retirement from the University of Tokyo. We thank R. A. Davis (College of Mount St. Joseph, Cincinnati, Ohio), S. M. Kidwell (University of

Chicago), N. H. Landman (American Museum of Natural History), K. Tanabe (University of Tokyo), and an anonymous reviewer for valuable discussion about this work and for improvement of an earlier draft of this chapter. Criticism and advice by G. E. G. Westermann (McMaster University) were very useful and contributed to improving this chapter. We are also much indebted to Y. Shigeta (Natural Science Museum, Tokyo), and M. Tashiro (Kochi University), who kindly provided important specimens for this study.

References

Allison, P. A., 1988, Phosphatized soft-bodied squids from the Jurassic Oxford Clay, *Lethaia* 21:403–410.

Allison, P. A., 1990a, Decay processes, in: *Palaeobiology—A Synthesis* (D. E. G. Briggs and P. R. Crowther, eds.), Blackwell Science Publications, Oxford, pp. 213–216.

Allison, P. A., 1990b, Carbonate nodules and plattenkalks, in: *Palaeobiology—A Synthesis* (D. E. G. Briggs and P. R. Crowther, eds.), Blackwell Science Publications, Oxford, pp. 250–253.

Allison, P. A., 1990c, Pyrite, in: *Palaeobiology—A Synthesis* (D. E. G. Briggs and P. R. Crowther, eds.), Blackwell Science Publications, Oxford, pp. 253–255.

Allison, P. A., Smith, C. R., Kukert, H., Deming, J. W., and Bennett, B. A., 1991, Deep-water taphonomy of vertebrate carcasses: A whale skeleton in the bathyal Santa Catalina Basin, *Paleobiology* 17:78–89.

Arthur, M. A., 1979, North Atlantic Cretaceous black shales: The record at Site 398 and a brief comparison with other occurrences, *Init. Rep. DSDP* 47:719–738.

Bandel, K., 1988, Operculum and buccal mass of ammonites, in: *Cephalopods—Present and Past* (J. Wiedmann and J. Kullmann, eds.), Schweizerbart'sche Verlagsbuchhandlung, Stuttgart, pp. 653–678.

Barthel, K. W., Swinburne, N. H. M., and Conway Morris, S., 1990, *Solnhofen: A Study in Mesozoic Palaeontology*, Cambridge University Press, Cambridge.

Behrensmeyer, A. K., and Kidwell, S. M., 1985, Taphonomy's contributions to paleobiology, *Paleobiology* 11:105–119.

Berner, R. A., 1968, Calcium carbonate concretions formed by the decomposition of organic matter, *Science* 159:195–197.

Boston, W. B., and Mapes, R. H., 1991, Ectocochleate cephalopod taphonomy, in: *The Process of Fossilization* (S. K. Donovan, ed.), Belhaven Press, London, pp. 220–240.

Brenner, K., and Seilacher, A., 1978, New aspects about the origin of the Toarcian Posidonia Shales, *N. Jb., Geol. Paläont. Abh.* 157:11–18.

Brett, C. E., and Baird, G. C., 1986, Comparative taphonomy: A key to paleoenvironmental interpretation based on fossil preservation, *Palaios* 1:207–227.

Cecca, F., 1992, Ammonite habitats in the Early Tithonian of Western Tethys, *Lethaia* 25:257–267.

Chamberlain, J. A., Jr., and Weaver, J. S., 1978, Equations of motion for post-mortem sinking of cephalopod shells, *Math. Geol.* 10:673–689.

Chamberlain, J. A., Jr., Ward, P. D., and Weaver, J. S., 1981, Postmortem ascent of *Nautilus* shells: Implications for cephalopod paleo-biogeography, *Paleobiology* 7:494–509.

Closs, D., 1967, Goniatiten mit Radula und Kiefferapparat in der Itararé-Formation von Uruguay, *Paläontol. Z.* 41:19–37.

Dagys, A. S., and Weitschat, W., 1993, Extensive intraspecific variation in a Triassic ammonoid from Siberia, *Lethaia* 26:113–121.

Denton, E. J., and Gilpin-Brown, J. B., 1966, On the buoyancy of the pearly nautilus, *J. Mar. Biol. Assoc. U.K.* 46:723–759.

Donovan, S. K., 1989, Taphonomic significance of the encrustation of the dead shell of Recent *Spirula spirula* (Linné) (Cephalopoda: Coleoidea) by *Lepas anatifera* Linné (Cirripedia: Thoracia), *J. Paleontol.* **63**:698–702.

Fernández-López, S., 1984, Criterios elementales de reelaboración taphonómica en ammonites de la Cordillera Ibérica, *Acta Geol. Hisp.* **19**:105–116.

Fernández-López, S., 1991, Taphonomic concepts for a theoretical biochronology, *Rev. Esp. Paleont.* **6**:37–49.

Fernández-López, S., and Meléndez, G., 1994, Abrasion surfaces on internal moulds of ammonites as palaeobathymetric indicators, *Palaeogeogr. Palaeoclimatol. Palaeoecol.* **110**:29–42.

Fürsich, F. T., 1990, Fossil concentrations and life and death assemblages, in: *Palaeobiology—A Synthesis* (D. E. G. Briggs and P. R. Crowther, eds.), Blackwell Scientific Publications, Oxford, pp. 235–239.

Gill, J. R., and Cobban, W. A., 1966, The Red Bird section of the Upper Cretaceous Pierre Shale in Wyoming, *U. S. Geol. Surv. Prof. Pap.* **393-A**:1–75.

Gómez, J. J., and Fernández-López, S., 1994, Condensation process in shallow platforms, *Sediment. Geol.* **92**:147–159.

Greenwald, L., and Ward, P. D., 1987, Buoyancy in *Nautilus*, in: *Nautilus—The Biology and Paleobiology of a Living Fossil* (W. B. Saunders and N. H. Landman, eds.), Plenum Press, New York, pp. 547–560.

Hagdorn, H., and Mundlos, R., 1983, Aspekte der Taphonomie von Muschelkalk-Cephalopoden. Teil 1. Siphozerfall und Füllmechanismus, *N. Jb. Geol. Paläont. Abh.* **166**:369–403.

Hallam, A., 1969, Faunal realms and facies in the Jurassic, *Palaeontology* **12**:1–18.

Hamada, T., 1977, Distribution and some ecological barriers of modern *Nautilus* species, *Sci. Pap. Coll. Gen. Educ. Univ. Tokyo* **27**:89–102.

Hanai, T., and Oji, T., 1981, Early Cretaceous beachrock from the Miyako Group, northeast Japan, *Proc. Jpn. Acad.* **57**(B):362–367.

Hanai, T., Obata, I., and Hayami, I., 1968, Notes on the Cretaceous Miyako Group, *Mem. Nat. Sci. Mus. (Tokyo)* **1**:20–28 (in Japanese with English abstract).

Hewitt, R. A., and Westermann, G. E. G., 1990, Mosasaur tooth marks on the ammonite *Placenticeras* from the Upper Cretaceous Bearpaw Formation of Alberta, *Can. J. Earth Sci.* **27**:469–472.

Henderson, R. A., and McNamara, K. J., 1985, Taphonomy and ichnology of cephalopod shells in a Maastrichtian chalk from Western Australia, *Lethaia* **18**:305–322.

Hollman, R., 1962, Über die Subsolution und die "Knollenkalke" des Calcare Ammonitico Rosso Superiore im Monte Baldo (Malm, Norditalien), *N. Jb. Geol. Paläont. Mh.* **1962**:1963–1972.

Jacobs, D. K., 1992, The support of hydrostatic load in cephalopod shells: Adaptive and ontogenetic explanations of shell form and evolution from Hooke 1695 to the present, in: *Evolutionary Biology*, Vol. 26 (M. K. Hecht, B. Wallace, and R. J. MacIntyre, eds.), Plenum Press, New York, pp. 287–349.

Jenkyns, S. O., 1985, The early Toarcian and Cenomanian–Turonian anoxic events in Europe: Comparisons and contrasts, *Geol. Rundsch.* **74**:505–518.

Jenkyns, S. O., 1988, The early Toarcian (Jurassic) anoxic event: Stratigraphic, sedimentary, and geochemical evidence, *Am. J. Sci.* **288**:101–151.

Kase, T., Shigeta, Y., and Futakami, M., 1994, Limpet home depressions in Cretaceous ammonites, *Lethaia* **27**:49–58.

Kauffman, E. G., 1978, Benthic environments and paleoecology of the Posidonienschiefer (Toarcian), *N. Jb. Geol. Paläont. Abh.* **157**:18–36.

Kauffman, E. G., 1990, Mosasaur predation on ammonites during the Cretaceous: An evolutionary history, in: *Evolutionary Paleobiology of Behavior and Coevolution* (A. J. Boucot, ed.), Elsevier, New York, pp. 184–189.

Kauffman, E. G., and Kesling, R., 1960, An Upper Cretaceous ammonite bitten by a mosasaur (South Dakota), *Univ. Mich. Mus. Pal. Contrib.* **15**:193–248.

Kennedy, W. J., and Cobban, W. A., 1976, Aspects of ammonite biology, biogeography, and biostratigraphy, *Spec. Pap. Palaeontol.* **17**:1–94.

Kidwell, S. M., 1986, Models for fossil concentrations: Paleobiologic implications, *Paleobiology* **12**:6–24.

Kidwell, S. M., Fürsich, F. T., and Aigner, T., 1986, Conceptual framework for the analysis and classification of fossil concentrations, *Palaios* **1**:228–238.

Kidwell, S. M., and Jablonski, D., 1983, Taphonomic feedback—Ecological consequences of shell accumulation, in: *Biotic Interactions in Recent and Fossil Benthic Communities* (M. J. S. Tevesz and P. L. McCall, eds.), Plenum Press, New York, pp. 195–248.

Landman, N. H., Saunders, W. B., Winston, J. E., and Harries, P. J., 1987, Incidence and kinds of epizoans on the shells of live *Nautilus*, in: *Nautilus—The Biology and Paleobiology of a Living Fossil* (W. B. Saunders and N. H. Landman, eds.), Plenum Press, New York, pp. 163–177.

Lehmann, U., 1966, Ammoniten mit Kieferapparat und Radula aus Lias-Gehäusen, *Paläontol. Z.* **44**:25–31.

Lehmann, U., 1971, Jaws, radula, and crop of *Arnioceras* (Ammonoidea), *Palaeontology* **14**:338–341.

Lehmann, U., 1972, Aptychen als Kieferelemente der Ammoniten, *Paläontol. Z.* **46**:34–48.

Lehmann, U., 1975, Über Nahrung und Ernährungsweise von Ammoniten, *Paläontol. Z.* **49**:187–195.

Lehmann, U., 1976, *Ammoniten: Ihr Leben und ihre Umwelt*, Ferdinand Enke Verlag, Stuttgart.

Lehmann, U., 1981, Ammonite jaw apparatus and soft parts, in: *The Ammonoidea*, Systematics Association Spec. Vol. 18 (M. R. House and J. R. Senior, eds.), Academic Press, London, pp. 275–287.

Lehmann, U., and Weitschat, W., 1973, Zur Anatomie und Ökologie von Ammoniten: Funde von Kropf und Kiemen, *Paläontol. Z.* **47**:69–76.

Maeda, H., 1987, Taphonomy of ammonites from the Cretaceous Yezo Group in the Tappu area, northwestern Hokkaido, Japan, *Trans. Proc. Palaeont. Soc. Jpn. N.S.* **148**:285–305.

Maeda, H., 1990, Mechanism of fossilization: Introduction to taphonomy, *Kagaku* **60**:159–163 (in Japanese).

Maeda, H., 1991, Sheltered preservation: A peculiar mode of ammonite occurrence in the Cretaceous Yezo Group, Hokkaido, north Japan, *Lethaia* **24**:69–82.

Mapes, R. H., 1987, Upper Paleozoic cephalopod mandibles: Frequency of occurrence, modes of preservation, and paleoecological implications, *J. Paleontol.* **61**:521–538.

Matsumoto, T., 1942, Fundamentals on the Cretaceous stratigraphy of Japan. Part I. *Mem. Fac. Sci. Kyushu Imp. Univ. (D)* **1**:130–280.

Matsumoto, T., and Okada, H., 1973, Saku Formation of the Yezo geosyncline, *Sci. Rep. Dep. Kyushu Univ. (Geol.)* **11**:275–309 (in Japanese with English abstract).

Matsushima, Y., 1990, *Nautilus pompilius* drifts on the northeast coast of Sumba Island, Indonesia, *Bull. Kanagawa Prefect. Mus. Nat. Sci.* **19**:33–43.

Minato, M., 1953, *Sedimentary Geology*, Iwanami Book Co., Tokyo (in Japanese).

Mundlos, R., 1970, Wohnkammerfüllung bei Ceratitengehäusen, *N. Jb. Geol. Paläont. Mh.* **1970**:18–27.

Mutvei, H., and Reyment, R. A., 1973, Buoyancy control and siphuncle function in ammonoids, *Palaeontology* **16**:623–636.

Neumann, N., Schumann, D., and Wendt, J., 1976, Geosynklinale Knollenkalke, *Zentralbl. Geol. Paläontol.* **2**:358–360.

Ohana, T., and Kimura, T., 1991, Permineralized *Otozamites* leaves (Bennettitales) from the Upper Cretaceous of Hokkaido, Japan, *Trans. Proc. Palaeont. Soc. Jpn. N.S.* **164**:944–963.

Ohta, S., 1983, Photographic census of large-sized benthic organisms in the bathyal zone of Suruga Bay, central Japan, *Bull. Ocean Res. Inst. Univ. Tokyo* **15**:1–155.

Okada, H., 1983, Collision orogenesis and sedimentation in Hokkaido, Japan, in: *Accretion Tectonics in the Circum-Pacific Regions* (M. Hashimoto and S. Uyeda, eds.), Terrapub, Tokyo, pp. 91–105.

Okamoto, T., 1988, Changes in life orientation during the ontogeny of some heteromorph ammonoids, *Palaeontology* **31**:281–294.

Onuki, Y., and Bando, Y., 1959, On the Inai Group of the Lower and Middle Triassic System (stratigraphical and paleontological studies of the Triassic System in the Kitakami Massif, northeast Japan:-3), *Contrib. Inst. Geol. Paleontol. Tohoku Univ.* **50**:1–69 (in Japanese).

Otto, M., 1994, Zur Frage der "Weichteilerhaltung" im Hunsrückschiefer, *Geol. Palaeontol.* **28**:45–63.

Prévôt, L., and Lucas, J., 1990, Phosphate, in: *Palaeobiology—A Synthesis* (D. E. G. Briggs and P. R. Crowther, eds.), Blackwell Science Publications, Oxford, pp. 256–257.

Raiswell, R., 1976, The microbiological formation of carbonate concretions in the Upper Lias of N. E. England, *Chem. Geol.* **18**:227–244.

Raup, D. M., 1973, Depth inferences from vertically embedded cephalopods, *Lethaia* **6**:217–226.

Raup, D. M., and Chamberlain, J. A., Jr., 1967, Equations for volume and center of gravity in ammonoid shells, *J. Paleontol.* **41**:566–574.

Reeside, J. B., and Cobban, W. A., 1960, Studies of the Mowry Shale (Cretaceous) and contemporary formations in the United States and Canada, *U.S. Geol. Surv. Prof. Pap.* **355**:1–126.

Reitner, J., and Urlichs, M., 1983, Echte Weichteilbelemniten aus dem Untertoarcium (Posidonienschiefer) Südwestdeutschlands, *N. Jb. Geol. Paläont. Abh.* **165**:450–465.

Renz, O., 1979, Lower Cretaceous Ammonoidea from the northern Atlantic, Leg 47B, Hole 398D, D.S.D.P., *Init. Rep. DSDP* **47**:361–365.

Reyment, R. A., 1958, Some factors in the distribution of fossil cephalopods, *Stock. Contrib. Geol.* **1**:97–184.

Reyment, R. A., 1973, Factors in the distribution of fossil cephalopods. Part 3. Experiments with exact models of certain shell types, *Bull. Geol. Inst. Univ. Upps. N.S.* **4**:7–41.

Rothpletz, A., 1909, Über die Einbettung der Ammoniten in den Solnhofener Schichten, *Koningl. Bayer. Akad. Wiss. Abh.* **24**(2):311–337.

Roux, M., 1990, Underwater observations of *Nautilus macromphalus* off New Caledonia, *Chambered Nautilus Newsl.* **60**.

Roux, M., Bouchet, P., Bourseau, J. P., Gaillard, C., Grandperrin, R., Guille, A., Laurin B., Monniot, C., Richer de Forges, B., Rio, M., Segonzac, M., Vacelet, J., and Zibrowius, H., 1991, L'environnement bathyal au large de la Nouvelle-Calédonie: Résultes préliminaires de la campagne CALSUB et conséquences paléoécologiques, *Bull. Soc. Geol. Fr.* **162**:675–685.

Schindewolf, O. H., 1934, Über Epöken auf Cephalopoden-Gehäusen, *Paläont. Z.* **16**:15–31.

Schlanger, S. O., and Jenkyns, H. C., 1976, Cretaceous oceanic anoxic events: Causes and consequences, *Geol. Mijnbouw* **55**:179–185.

Scott, G., 1940, Paleontological factors controlling the distribution and mode of life of Cretaceous ammonoids in the Texas area, *J. Paleontol.* **14**:299–323.

Seilacher, A., 1960, Epizoans as a key to ammonoid ecology, *J. Paleontol.* **34**:189–193.

Seilacher, A., 1963, Umlagerung und Rolltransport von Cephalopoden-Gehäusen, *N. Jb. Geol. Paläont. Mh.* **1963**:593–615.

Seilacher, A., 1966, Lobenlibellen und Füllstruktur bei Ceratiten, *N. Jb. Geol. Paläont. Abh.* **125**:480–488.

Seilacher, A., 1968, Sedimentationprozesse im Ammoniten gehäusen, *Akad. Wiss. Lit. Abh. Math. Naturwiss. Kl.* **1967**(9):191–203.

Seilacher, A., 1970, Begriff und Bedeutung der Fossil-Lagerstätten, *N. Jb. Geol. Paläont. Mh.* **1970**:34–39.

Seilacher, A., 1971, Preservational history of ceratite shells, *Palaeontology* **14**:16–21.

Seilacher, A., 1982, Ammonite shells as habitats in the *Posidonia* Shale—floats or benthic islands? *N. Jb. Geol. Paläont. Mh.* **1982**:98–114.

Seilacher, A., 1988, Schlangensterne (*Aspidura*) als Schlüssel zur Entstehungsgeschichte des Muschelkalks, in: *Neue Forschungen zur Erdgeschichte von Crailsheim*, Vol. 1 (H. Hagdorn, ed.), Goldschneck Verlag, Stuttgart, pp. 85–98.

Seilacher, A., Andalib, F., Dietl, G., and Gocht, H., 1976, Preservational history of compressed Jurassic ammonites from Southern Germany, *N. Jb. Geol. Paläont. Abh.* **152**:303–356.

Seilacher, A., and LaBarbera, M., 1995, Ammonites as cartesian divers, *Palaios* **10**:493–506.

Seilacher, A., Reif, W. E., and Westphal, F., 1985, Sedimentological, ecological, and temporal patterns of fossil Lagerstätten, *Phil. Trans. R. Soc. Lond.* **B311**:5–23.

Shigeta, Y., 1993, Post-hatching early life history of Cretaceous Ammonoidea, *Lethaia* **26**:133–145.

Shikama, T., and Hirano, H., 1970, On the mode of occurrence of ammonites in the Nishinakayama Formation, Toyora Group, *Sci. Rep. Yokohama Nat. Univ. Second Sect.* **16**:61–71.

Shimizu, S., 1931, The marine Lower Cretaceous deposits of Japan, with special reference to the ammonites-bearing zones, *Sci. Rep. Tohoku Imp. Univ. Second Ser.* **15**:1–40.

Stürmer, W. 1969, Pyrit-Erhaltung von Weichteilen bei devonischen Cephalopoden, *Paläontol. Z.* **43**:10–12.

Tanabe, K., 1979, Palaeoecological analysis of ammonoid assemblages in the Turonian *Scaphites* facies of Hokkaido, *Palaeontology* **22**:609–630.

Tanabe, K., 1983a, Mode of life of an inoceramid bivalve from the Lower Jurassic of west Japan, *N. Jb. Geol. Paläont. Mh.* **1983**:419–428.

Tanabe, K., 1983b, The jaw apparatuses of Cretaceous desmoceratid ammonites, *Palaeontology* **26**:677–686.

Tanabe, K., 1991, Early Jurassic macrofauna of the oxygen-depleted epicontinental marine basin in the Toyora area, west Japan, *Saito Ho-on Kai Spec. Pub.* **3**:147–161.

Tanabe, K., Inazumi, A., Ohtsuka, Y., Katsuta, T., and Tamahama, K., 1982, Litho- and biofacies and chemical composition of the Lower Jurassic Nishinakayama Formation (Toyora Group) in west Japan, *Mem. Ehime Univ. Sci. Ser. D* **9**:47–62 (in Japanese with English abstract).

Tanabe, K., Inazumi, A., Tamahama, K., and Katsuta, T., 1984, Taphonomy of half and compressed ammonites from the Lower Jurassic black shales of the Toyora area, west Japan, *Palaeogeogr. Palaeoclimatol. Palaeoecol.* **47**:329–346.

Tanabe, K., Landman, N. H., and Weitschat, W., 1993, Septal necks in Mesozoic Ammonoidea: Structure, ontogenetic development, and evolution, in: *The Ammonoidea: Environment, Ecology, and Evolutionary Change*, Systematics Association Spec. Vol. 47 (M. R. House, ed.), Clarendon Press, Oxford, pp. 57–84.

Taylor, P. D., 1979, Paleoecology of the encrusting epifauna of some British Jurassic bivalves, *Palaeogeogr. Palaeoclimatol. Palaeoecol.* **28**:241–262.

Toriyama, R., Sato, T., and Hamada, T., 1966, *Nautilus pompilius* drifts on the coast of Thailand, *Jpn. J. Geol. Geogr.* **36**:149–161.

Trueman, A. E., 1941, The ammonite body chamber, with special reference to the buoyancy and mode of life of the living ammonite, *Geol. Soc. Lond. Q. J.* **96**:339–383.

Ward, P. D., 1987, *The Natural History of Nautilus*, Allen and Unwin, Boston.

Ward, P. D., and Westermann, G. E. G., 1977, First occurrence, systematics and functional morphology of *Nipponites* from the Americas, *J. Paleontol.* **51**:367–372.

Ward, P. D., and Westermann, G. E. G., 1985, Cephalopod paleoecology, in: *Mollusks, Notes for a Short Course, University of Tennessee Studies in Geology*, Vol. 13 (D. J. Bottjer, C. S. Hickman, and P. D. Ward, eds.), University of Tennessee, Knoxville Publication, Knoxville, pp. 215–229,

Wendt, J., 1971, Genese und Fauna submariner sedimentärer Spaltenfüllungen im mediterranen Jura, *Paläontogr. A* **136**:121–192.

Wendt, J., 1973, Cephalopod accumulations in the Middle Triassic Hallstatt-Limestone of Yugoslavia and Greece, *N. Jb. Geol. Paläont. Mh.* **1973**:624–640.

Wendt, J., and Aigner, T., 1982, Condensed Griotte facies and cephalopod accumulation in the Upper Devonian of the eastern Anti-Atlas, Morocco, in: *Cyclic and Event Stratification* (G. Einsle and A. Seilacher, eds.), Springer-Verlag, Berlin, pp. 326–332.

Westermann, G. E. G., 1977, Form and function of orthoconic cephalopod shells with concave septa, *Paleobiology* **3**:300–321.

Westermann, G. E. G., 1985, Post-mortem descent with septal implosion in Silurian nautiloids, *Paläontol. Z.* **59**:79–97.

Wright, C. W., and Kennedy, W. J., 1981, The Ammonoidea of the Plenus Marls and the Middle Chalk, *Palaeontogr. Soc. Monogr.*, No. 560:1–148.

Wright, C. W., and Kennedy, W. J., 1984, The Ammonoidea of the Lower Chalk. I. *Palaeontogr. Soc. Monogr.*, No. 567:1–126.

Zeiss, A., 1969, Weichteile ectocochleater paläozoischer Cephalopoden in Röntgenaufnahmen und ihre paläontologische Bedeutung, *Paläontol. Z.* **43**:13–27.

Ziegler, B., 1967, Ammoniten-Ökologie am Beispiel des Oberjura, *Geol. Rundsch.* **56**:439–464.

VI
Ecology

Chapter 15

Ammonoid Pathology

RAINER HENGSBACH

1. Introduction

Paleopathology is that branch of paleontology that concerns itself with pathological deviations of organisms, the so-called "anomalies" and "abnormalities." Paleopathology of ammonoids deals, therefore, with the pathological phenomena within this group. In this chapter, I use the term "pathology" instead of "paleopathology" in reference to ammonoids because the Ammonoidea became extinct some time near the end of the Cretaceous Period and the beginning of the Tertiary, making the prefix "paleo-" superfluous for this group.

Paleopathology has its roots in the medical sciences, i.e., in pathology. That is why some of the most prominent people who studied paleopathology were physicians (e.g., Moodie and Breuer) or had medical connections (for example, Tilly Edinger, who, as the daughter of the famous neurophysiologist Ludwig Edinger, was strongly influenced by medical science). Largely because of this, Cenozoic vertebrates have been the preferred subjects for paleopathological research. They allow a more or less direct comparison with related present-day animals.

RAINER HENGSBACH • Perwang 95, A-5163 Mattsee, Austria.

Ammonoid Paleobiology, Volume 13 of *Topics in Geobiology,* edited by Neil Landman *et al.,* Plenum Press, New York, 1996.

In contrast to that, the field of ammonoid pathology is a different situation in that the so-called anomalies and abnormalities can be interpreted only indirectly and understood through paleobiological analysis. Of course, this is often possible only within a realm of various degrees of probability. For this reason, a thorough comparison with similar conditions in present-day organisms, especially molluscs, is important.

2. Remarks on Terminology

Pathological phenomena in ammonoids have been known for a long time. Although these initially were called "monstrosities" or "aberrations," the terms "anomaly" and "abnormality" gradually became prevalent. In medical science, anomalies generally are considered to be more or less slight irregularities or malformations. However, an ammonoid that is deformed to the extent that it is almost impossible to identify to species can hardly be considered as having a slight deviation from the norm. On the other hand, because paleopathology is rooted in the medical sciences and cannot be studied meaningfully without constant reference to present-day medical and other biological findings, I recommend a standardization of a technical terminology that approximates the usage in the medical or biological sciences.

In order to avoid ambiguities, overlappings, and confusions of one term with another, I have recently suggested the term "paleopathy" for any abnormal condition regardless of its severity or etiology (Hengsbach, 1990, 1991a,b). Thus, the concept of paleopathies includes fractures, repaired fractures, deformities, hypertrophies, etc., in other words, anomalies and abnormalities as previously defined.

In this context, the term "teratology" should be discussed. For the purposes of paleontology, Lehmann defined "teratologic" as referring to "deformity caused by injury or disease" (1977, p. 380). However, in biology and medicine, teratology is that discipline that deals with developmental malformations and deviations, i.e., with prenatal diseases, whether they are hereditary or ecological (metabolic/physiological) in origin. Strictly speaking, teratology does not deal with postnatal injuries or diseases. However, in fossils, it can be extremely difficult to distinguish abnormalities that are prenatal in origin from those that are postnatal, particularly in the absence of obvious exogenous injuries. For example, prenatal diseases commonly are indistinguishable from postnatal endogenous diseases. Thus, although the term "teratologic" should be restricted to its narrow biological and medical meaning, it may well be impossible to do so with respect to a given paleopathy in ammonoids.

A look at human history illustrates the significance of parasitism (plagues and epidemics) in the world of organisms. In the same sense that teratologic paleopathies are subjects of teratology, parasitism and presumed parasitoses are subjects of their own discipline: paleoparasitology. I have proposed a concept for this field with respect to invertebrates. This paleoparasitologic concept deals with parasitism in fossils and with its relevance for related fields of research (Hengsbach, 1990). Thus, paleoparasitology takes its place with other special branches of paleobiological research, such as paleoecology and paleoneurology (see Hengsbach, 1990, 1991a).

3. Historical Review

The number of references related to paleopathies in ammonoids is truly huge. Numerous examples have been described or shown in stratigraphic and taxonomic monographs. Even works that deal specifically with pathological phenomena in ammonoids have appeared in such numbers that only a few important ones can be mentioned here. The choice must be, of necessity, subjective.

Numerous references to paleopathies in ammonoids appeared as far back as the previous century by, among others, von Stahl (1824), Zieten (1831–1833), A. d'Orbigny (1842–1849), Quenstedt (1858, 1883–1888), and Reynes (1879). A summary and review, especially of Quenstedt's works, was produced by Engel (1894). Most of the paleopathies that are known today were described in this work. Some of the attempts at clarification that were presented therein appear quite modern in light of present-day knowledge. For example, Engel discussed the possibility, even the likelihood, of parasites being disease-causing agents in ammonoids. At the same time, Pompecki (1893, 1894) published two oft-quoted works on paleopathies.

Of those published after the turn of the century, works on asymmetric and other paleopathic ammonoids by Vadasz (1908, 1909), von Staff (1909), and von Bülow (1917) are of special note. Nicolesco (1921) investigated the phenomenon of asymmetry in the body chamber and phragmocone of the Jurassic ammonite *Bigotites* Nicolesco, 1918.

In 1935, Roll published a much-acclaimed paper on "feeding marks" in ammonoid shells (*Fraßspuren*, in German); in this paper, brachyurid decapod crustaceans were proposed as the cause of the shell fragmentation. Maubeuge (1949, 1957) put forth the possibility of mutations as the cause of certain paleopathies that seem teratologic in nature.

In 1956, about a half-century after Engel's main work (1894) appeared, Hölder published his important paper on anomalies in Jurassic ammonites; this is another, summarizing review and has become a basic work on the pathology of ammonoids. In it, Hölder introduced a number of groupings (called "forma-types") for the description and categorization of specific paleopathies. He also discussed ecological variations and the possibility that bacterial infestations may have caused some paleopathies. Theobald (1958) included further observations.

Kauffman and Kesling (1960) presented a much-acclaimed, exemplary analysis of bite marks on an ammonite. The authors concluded from the tooth marks that the bites were those of a mosasaur. Additional conclusions on the paleoecology of the locality (in Cretaceous rocks of South Dakota) were made. In the same year, Jessen (1960) described shell injuries in Upper Carboniferous goniatites, and House (1960) discussed growth paleopathies in Devonian goniatites with a probable indication of pearl formation. Guex (1967) studied injuries to Jurassic ammonoids from southern France and concluded that the enemies of such ammonoids appear to have been cephalopods (belemnites and cannibalistic ammonoids). He attempted a classification of paleopathies based on their location on the shell. One year later, he supplemented his research with observations on conellae (Guex, 1968).

Bayer (1970) investigated paleopathies of Jurassic ammonoids and tried to form conclusions on the life styles of these animals on the basis of the relative frequency of their paleopathies. In the same year, Hölder compared paleopathies in ammonoids to those in *Nautilus* and in the shells of other molluscs. A short time later, the same author presented a paper on the anatomy and shell construction of fossil cephalopods; in this paper, Hölder (1973) documented scars in Devonian ammonoids. Lehmann (1975) presented an analysis of a fracture in a specimen of *Dactylioceras* Hyatt, 1867. Keupp (1976) found some new paleopathic forms, one of which he attributed to parasites in the vicinity of the peristome. Earlier, Rieber (1963) had described asymmetry of the keel and phragmocone in a specimen of *Cardioceras* Neumayr and Uhlig, 1881; he considered this asymmetry to have been induced by parasites. In 1977, Keupp's important contribution on paleopathology in amaltheids of the Fran-conian Lias appeared and recognized additional new forma-types. Hengsbach (1979a) investigated asymmetry in ammonoid suture lines and concluded that some sutural asymmetry seems to have been caused by parasitism. After further research, he was able to corroborate the probability of parasitism (Hengsbach, 1986a,b).

Morton (1983) attributed a deformity of ammonoid shells resulting in deviations from the plane of bilateral symmetry to parasitism (a "sym-metropathy," as designated by Hengsbach, 1991a). Keupp (1986) reported on pearl formation in Jurassic ammonoids. Also in 1986, an investigation by Landman and Waage featured a statistical analysis of paleopathies in scaphites. Bond and Saunders (1989) presented a corresponding study on shell injuries in Lower Carboniferous ammonoids in Arkansas. Recently, Keupp (1992) described paleopathies that he attributed to epizoa, and Keupp and Ilg (1992) gave a statistical analysis of the paleopathies in ammonoids from southern France.

4. Methods and Procedures

The methods of paleopathology are basically those of paleobiological inference and reconstruction. Comparisons with present-day organisms are essential. With the help of these, improbable explanations as to the causes of paleopathies may be able to be recognized and excluded. The specific form and occurrence of a paleopathy may correlate with a certain cause or group of causes, with greater or less probability, depending, of course, on the degree of reliability of the evidence.

Great consideration must be given to instances in which a large number of individuals deviate from the norm, especially when the extent of the deviation is great. Indeed, this is as much the case within an entire taxon as within a local population. In some cases, a paleopathy that at first appears etiologically unclear can be attributed to a specific cause or group of causes when one considers the variation of the paleopathy in one taxon as compared to that in others and when present-day diseases are taken into account.

Known paleopathies may be divided into groups on the basis of their accessibility to analysis. A majority exhibit more or less distinct indications of exogenous injury; these may reveal the causal agent. For example, from bite-mark evidence in a specimen of *Placenticeras*, Kauffman and Kesling (1960) were able to reconstruct the fact that the attacker of the ammonoid had delivered at least 16 bites before it met with success. In their analysis, the authors demonstrated that the ammonoid must have been seized at an angle from above. In order to bring the victim under control, the attacker apparently then rapidly and repeatedly snapped at the ammonoid, as some present-day lizards attack their prey. The body chamber was broken only with the final two bites. To identify the predator, Kauffman and Kesling compared the bite marks with the skulls of known animals of the same geological age as the ammonoid. On the basis of the shape and arrangement of the tooth marks, they came to the conclusion that a mosasaur must have been the predator and that these animals hunted ammonoids on a regular basis. From the number of bites, they presumed that the size of the ammonoid caused problems for the attacking mosasaur. This may be the reason why comparable cases of attacked ammonoids are uncommon; however, Keupp (1985, 1991) described two similar examples.

Paleopathies in which direct comparisons, such as being able to match mosasaur teeth and bite marks, are not possible are much more difficult to interpret. For example, Hölder (1973) described a form of fracture repair in which the fractured edge is substructed (that is, new shell material grew out from under the broken edge). The fractured edge recognizable in the fossil lends significant credence to Hölder's explanation that the paleopathy was the result of exogenous injury. Later, other authors (e.g., Keupp, 1977; Hengsbach, 1979b) found similar paleopathies, which, however, do not exhibit fractured edges. These latter cases fit Hölder's description in that they have a comparable discordance of ribbing, a local depression in the steinkern, or both. The former case, described by Hölder, can be explained as caused by the withdrawal of the edge of the mantle back to the base of the injured area, followed by secretion of new shell matter; the lack of a precise match-up of the epithelium with the sculpture in the original part of the shell resulted in a discordance of the ribbing in the newly secreted shell matter (Keupp, 1977, 1985; Landman and Waage, 1986). The latter cases (those of Keupp and

Hengsbach) can be attributed to more-intensive secretion of sculptureless secondary shell at the site of damage by epithelium so injured as to have been incapable of forming the characteristic ornament. Similar cases are known from present-day molluscs.

Paleopathies that do not exhibit evidence of obvious exogenous physical injury to the shell or related "repair marks" are much more difficult to interpret. In these cases, normally only a careful comparison with similar phenomena in present-day organisms (preferably closely related forms) may provide clues. It goes without saying that the probability of a given explanation depends on the quality of such comparisons.

Because of difficulties in analysis and interpretation of these paleopathies, it seems to be legitimate to discuss this category of difficult-to-interpret phenomena in greater detail. Paleopathies with no obvious signs of exogenous physical injury might have been caused by genetic problems, parasitism, ecological/physiological problems, or some combination of these. Tumor-like growths and cysts seem to be most readily explained as results of parasitism. In cases of bacterial infections, tumor-like tissue growths might be expected either to have been checked and then to fade away as ontogeny proceeded or to have spread and, ultimately, to have proved fatal. Examples include the cases of swellings of the conch presented by Keupp (1976) and Hengsbach (1979b).

By contrast, a paleopathy that repeatedly manifests itself in a consistent manner in organisms of a particular taxon may be attributable to genetic causes. However, parasitism cannot be excluded. A specialized parasite can produce highly consistent malformations that occur in the same location in the infected organisms; however, the side of the organism on which the malformation occurs will vary from individual to individual. This latter would not be expected in the case of a genetic cause. Postnatal physiological deviations generally would not show the same consistency in the location of malformation. It is known that parasitic trematodes cause consistent deviations in gastropods (Hengsbach, 1990). In certain instances, marks of injury (e.g., scars) and signs of their repair may be retained for a long time, perhaps even for life (Hölder, 1970, Fig. 9). However, this should be relatively rare, because progressive regeneration would be expected, resulting in the disappearance of the marks of injury.

Irregular, ontogenetically long-lasting paleopathies indicate changes in the size of an infected area over time, such as might reflect the struggle between the infecting population and the host. In the case of symmetropathies, i.e., deviations from bilateral symmetry (Hengsbach, 1991b), a physiological cause is unlikely because one would be expected to lead to extensive pathological areas (such as "**forma cacoptycha**," below).

For those paleopathies that occur relatively frequently, the percentage frequency of their occurrence, both within the taxon and within the biocoenose (community), is important. When dealing with symmetropathies, information about the variability of their intensity and position (for example, with respect to the plane of symmetry of the animal) are of special interest. Even if only a small percentage of the organisms have been affected, teratological causes can not be excluded. Again, such an interpretation may be able to be confirmed by comparison with present-day groups of similar organisms.

If a paleopathy occurs in 100% of the representatives of a taxon, and, additionally, if the variation from one individual organism to another is very slight (e.g., the same side of the organism always is affected, and with the same degree of development), a postnatal cause may well be considered. Obligatory symbiosis, as known in present-day animals, could lead to such a pattern. If there is a correlation with ontogenetic changes (e.g., allometry), however, parasitism is unlikely, because, in cases of parasitism, other frequencies and more variability are to be expected.

If, on the other hand, the frequency of a paleopathy is, for instance, about 70% (i.e., significantly less than 100%), and if the appearance of the paleopathy is consistent, then there is a rather high probability that the paleopathy was produced by a specialist among the parasites. Such a specialist can infect a high percentage of a given population, but this percentage can vary within a broad range. In obligatory symbiosis, on the other hand, all or nearly all individuals in a population are affected. In the case of paleopathies having a genetic origin, consistency of the position of the deviation (for example, on a specific side) would be expected. (Please note that the word "symbiosis" here is used in the restricted sense to connote that the relationship is mutually beneficial to the symbionts.) In symbiosis and parasitosis, there is variability in the position; however, it is questionable that symbiosis can be identified in host organisms at all. Symbionts do not cause tumor-like growths (see Hengsbach, 1979a, 1986a, 1990).

In addition, the presence of similar or equivalent paleopathies in fossils of closely related taxa can be considered as clues for the plausibility of an interpretation. For example, under certain circumstances, it would be entirely possible to find similar parasitoses in closely related groups (Hengsbach, 1981, 1986b, 1990, 1991a).

It is of paramount importance to pay proper heed to logical and biological plausibility. Although this point may appear banal, paleontologists frequently have failed to pay adequate attention to it (see examples given in Hengsbach, 1979a, p. 145; 1979c, p. 89; 1986a, p. 139; and Keupp, 1977, p. 276).

5. Paleopathies in Ammonoids

In order to record and summarize the diverse paleopathies present in ammonoid cephalopods, it is necessary to categorize them. A classification based mainly on the morphological expressions of paleopathies was introduced by Hölder (1956). It features a division into "forma-types," each of which conceptually encompasses a disease, a healing pattern, or both. Each forma-type is named in a manner somewhat analogous to the open nomenclature commonly used by paleontologists and other biologists. The forma-type system makes more or less precise communication about a given paleopathy possible. More than 25 "formae" have been distinguished (Table I).

The forma-type system currently is in common use in Europe. Unfortunately, paleontologists outside Europe have paid scant attention to this scheme of classification, even though non-European contributors to the literature of ammonoid pathology have been rather few. This has led, and still continues to lead, to double designations and parallel descriptions, e.g., those of Landman and Waage (1986) and of Bond and Saunders (1989). The classi-

Table I. Forma Types and Their Supposed Causes

Designation[a]	Author	Date	Supposed cause[b]
"Forma abrupta"	Hölder	1956	i
"Forma alternospinata"	Keupp	1977	i
"Forma aptycha"	Keupp	1977	i, e, p
"Forma bovicornuta"	Keupp	1977	i
"Forma cacoptycha"	Lange	1941	i, e, p
"Forma calcar"	Hölder (first shown by Zieten in 1831–1833)	1956	g, i
"Forma chaotica"	Keupp	1977	i, p
"Forma cicatricocarinata"	Heller	1964	i, (p)
"Forma circumdata"	Hölder (first shown by Martin in 1858)	1956	i
"Forma conclusa"	Rein	1989	p, i
"Forma concreta"	Hengsbach	Herein	p
"Forma disseptata"	Hölder	1956	g, e
"Forma duplicarinata"	Keupp	1976	i
"Forma excentrica"	Hölder	1956	?
"Forma gigantea"	Hengsbach	Herein	p
"Forma inflata"	Keupp	1976	p!
"Forma juxtacarinata"	Hölder	1956	i, p, (g)
"Forma juxtalobata"	Hölder	1956	p, i, (g)
"Forma juxtasulcata"	Geczy	1965	p, i
"Forma mordata"	Hengsbach	Herein	i!
"Forma pexa"	Hölder	1973	i
"Forma pseudocarinata"	Fernandez -Lopez	1987	i
"Forma seccata"	Hölder	1956	i!
"Forma semiverticata"	Hölder	1977	p, i
"Forma substructa"	Hölder	1973	i!
"Forma syncosta"	Hengsbach	1979	g
"Forma undaticarinata"	Heller	1958	i, p
"Forma undatispirata"	Keupp and Ilg	1992	e, p
"Forma verticata"	Hölder	1956	i, p

[a]Terms such as "*circumdata*-like" of Hölder (1970), "*compensation ornamentale*" of Guex (1967), "Morton's Syndrome" of Landman and Waage (1986), and "reduction phenomena" are not included because they are not forma-types, *per se*.

[b]g, genetic; i, mechanical injury; p, parasitosis; e, ecological; !, obvious; (), possibly. Where more than one letter is given, the first is the most likely cause, the second is next, and so on. It is to be stressed that the causes assigned to the forma-types are supposed; in only a few cases is it possible to be sure.

fication scheme of the latter, for example, was based on the reconstructed severity of the injuries.

It would not be very helpful for international communication, were each continent to use its own separate classification system. A lack of homogeneity of terminology and, hence, of understanding would be bound to occur. For any classification system to "work," first and foremost, there must exist general agreement about the morphological pattern of each particular paleopathy. Only then can it be interpreted and compared to other paleopathies.

Because there is no alternative classification scheme that is at once as comprehensive and consistent, I recommend that the forma-type system, as promulgated by Hölder, be used as the basis for common terminology in ammonoid pathology.

However, the forma-types should not be treated as formal taxa as recognized by the International Code of Zoological Nomenclature, nor even by the use of open nomenclature, although this misleading practice is currently customary with some workers. We are not dealing with taxa of ammonoids but, rather, with diseases and injuries that have afflicted the ammonoids. By contrast, the taxa of the affected ammonoids generally are known and definable, or they are described separately. Examples of this false equating of ammonoid taxa with paleopathological expressions are shown in Thiermann (1964), who refers to paleopathies as "taxonomically recognizable units." Even Hölder (1956) described similarly occurring disorders as "taxonomically recognizable forms." Interpretations of this kind are an impediment to the realization that we are dealing with diseases and injuries.

In the following sections, the principal forma-types are discussed briefly, and, to the extent that it appears meaningful and justifiable, their causes are explained.

To avoid confusion of the names of ammonoid taxa with those of forma-types, I have proposed that the latter be designated in a manner different from that mandated by the International Code of Zoological Nomenclature (Hengsbach, 1991b, p. 131). In this chapter, formal taxon names will be given in *italics*, whereas names of formae will be given between quotation marks and in **"boldface."** In circumstances in which it is necessary or advisable to give the name of the person who proposed a particular forma-type, the format will be **"forma seccata"** of Hölder (1956).

In the past, some workers used the word *aegra* as part of the designation for a given paleopathy; for example, **"forma aegra cacoptycha"** of Lange (1941). Because *aegra* is Latin for "sick" or "ill," and all of the forma-types are pathological, this unnecessary word has not been used for many years.

Without doubt, the most widespread paleopathies are what are called, in German, *"Rippenscheitelungen"*; these comprise **"forma verticata"** in Hölder's system (Fig. 1; the German might be translated as "ribbing vertices" or "rib partings"). In the vast majority of cases, these paleopathies were the result of exogenous injury, with localized or puncture damage to the mantle having been the common cause. Keupp (1976) distinguished two types. The

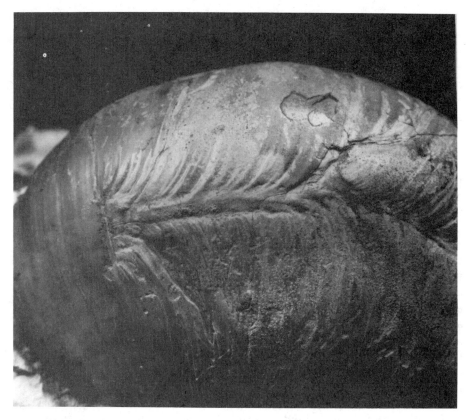

FIGURE 1. "Forma verticata" on a specimen of *Cadoceras* sp. from the Jurassic (lower Callovian) of Gorki (Russia).

first is a simple *Rippenscheitelung* ("ribbing vertex"), which consists of a backward bending of the ribs or growth lines. The second is a bend in the ribs (*Rippenknickung*, in German) that is interrupted by a central furrow, i.e., a scar. In the first instance, the epithelium was only slightly damaged; this led to a bending of the ribs or growth lines as a result of the temporary disorder, but it did not lead to a loss of function of the mantle. In the latter case, however, the injury was more severe and led to a local dysfunction of the mantle, which caused the scar.

Not all cases of **"forma verticata"** are the obvious result of exogenous injury. For *Rippenscheitelungen* that he found located close to the umbilici in southern German dactylioceratids, Keupp (1979) presumed parasitism to have been the cause. He came to this conclusion on the basis of the protected positions of the paleopathies on the shells as well as on the basis of many comparisons to other individuals in the local population. I have designated these cases as **"forma verticata** *sensu* Keupp (1979),"** in contrast to the obviously injury-caused **"forma verticata** *sensu* Hölder (1956)."

FIGURE 2. "**Forma substructa**" on a macroconch of *Jeletzkytes nebrascensis* from the Upper Cretaceous (Maastrichtian) of west-central South Dakota. (From Landman and Waage, 1986, Fig. 6.) Diameter = 115 mm.

Hölder (1977) discussed unilateral *Rippenscheitelungen* ("ribbing vertices") that developed at the umbilicus; because each vertex is incomplete in that it is only one-sided, he designated them "**forma semiverticata.**" (As previously mentioned, Keupp (1979) showed that such paleopathies possibly could have been caused by parasitism.)

Fernandez-Lopez (1987) designated "**forma pseudocarinata**" for a paleopathy that manifests itself in keel-like, raised cicatrices in the ventral area of the conch and, according to H. Keupp (personal communication), also on the flanks. In the opinion of H. Keupp (personal communication), this paleopathy represents a special case of "**forma verticata.**"

Heller (1964) designated cicatrizations of the crenulated keel in amaltheids as "**forma cicatricocarinata.**" Although the epithelium that produced the keel was able to form only a cicatrized keel, a total loss of function did not occur.

A particular form of cicatrization with "ripple structure" was described by Hölder (1973) as **"forma pexa."** These paleopathies exhibit different patterns; Hölder distinguished between ripples that occur over great distances and those that occur only as short scars following injuries. Similar structures also were described by Thiermann (1964, Pl. 1, Fig. 2). It is uncertain whether causes other than mechanical injuries may have been involved; however, most cases appear to be related to such injuries.

Hölder (1973) described a form of paleopathy in which new shell substance substructed the fracture line from behind; i.e., new shell was built out from under the fracture edge. He termed this **"forma substructa"** (Fig. 2). In the opinion of Keupp (e.g., 1977), a discordance of the pattern of ribbing is characteristic for the **"forma substructa"** because, just after the time of the injury, the edge of the mantle withdrew to the point of the injury but maintained production of the sculpture, although not necessarily aligned with the sculpture of the undamaged part of the conch. If only the steinkern is present, a paleopathy of the **"forma-substructa"** type might show no fracture edge but only a local indentation on the steinkern from increased secretion of shell material at the point of the injury. Such increases in shell deposition, resulting in thicker "substructures," also might have resulted in a weakening of the sculpture of the conch (Hengsbach, 1979b).

Hölder (1956) recognized a kind of paleopathy in which there is a longitudinal line of disturbance characterized by interruptions and dislocations of the ribs. He designated this phenomenon **"forma abrupta."** Apparently this type of paleopathy stemmed from an injury of the edge of the mantle that persisted for some considerable portion of subsequent ontogeny and resulted in continuing interruption of the sculpture.

Hölder (1956) named **"forma circumdata"** from its occurrence in *Ammonites circumdatus* Martin, 1858. In this paleopathy (Fig. 3), ribs ordinarily on the flank of the conch extend across the venter of the whorl. Apparently there was a loss of function of the ventral part of the mantle edge, and that portion of the mantle on the flank was stretched over the venter and took over the shell-producing function there.

Instances of **"forma circumdata"** can result in a sculpture pattern characteristic of a different taxon. Some authors have interpreted this phenomenon as the result of mutation and assigned specimens affected to a different genus (Maubeuge, 1949, 1957). However, asymmetry and irregularity generally reveal the pathological nature of the sculpture pattern.

In rare cases, the mutation explanation cannot be excluded. Probably because of this, Hölder (1970) distinguished between true **"forma circumdata,"** in which only the unusual sculpture pattern is present, and *"circumdata*-like," in which evidence of a triggering event also is present.

Keupp (1977) described two of what he considered to be special cases of **"forma circumdata."** In **"forma alternospinata,"** there are alternating nodes. In **"forma bovicornuta,"** there are strongly developed, forwardly oriented horns.

FIGURE 3. "**Forma circumdata**" on a specimen of *Pleuroceras* sp. from the Jurassic (upper Domerian) of Staffelstein (Bavaria, Germany). (From Keupp, 1977, Fig. 57, 6.) Diameter = 54 mm.

Related to "**forma circumdata**" is what Guex (1967) called "*compensation ornamentale*" (which, in English, might be termed "sculptural compensation"). In this paleopathy, the injury was on one side, beside the keel, and the ventral portion of the mantle was not damaged; i.e., it was still able to function. However, it was pulled laterally, toward the injured area, so that the keel and adjacent ornament were dislocated to the damaged side of the venter of the whorl.

Keupp (1977) introduced "**forma chaotica**" as a term for pathological conditions similar to that of the highly deformed ammonoid shell shown by Hölder (1970) and described by him as having "chaotic arietid sculpture" (Fig. 4). The original specimen reported by Hölder (1970, Fig. 12) displays repeated drifting of the keel away from the venter and toward one flank, along with chaotic irregularities in ribbing. Keupp (1977) regarded only the repeated lateral drifting of the keel as characteristic of this forma-type. Hengsbach (1987) emphasized the chaotic character of the sculpture (in a genus in which a keel is not present). Presumably, this forma-type generally was caused by injury, but internal diseases cannot be excluded.

Keupp (1976) introduced the term "**forma duplicarinata**" for a particular paleopathy appearing in the keel area in amaltheids in which there is a doubling of the keel. Apparently, following an injury to the shell-producing epithelium on one side of the keel, but with no involvement of the keel, the mantle edge withdrew in the direction of the point of injury. The stretching of the healthy portion of the mantle to take over the function of the injured

FIGURE 4. "**Forma chaotica**" on a specimen of *Coeloceras pettos* from the Jurassic (Pliensbachian) of Reutlingen (southwestern Germany). (From Hengsbach, 1987, Fig. 1.) Diameter = 36 mm.

part led to a displacement of the keel-producing portion of the epithelium, thereby causing a temporary doubling of the keel. It is improbable that factors other than mechanical injury could have triggered "**forma duplicarinata.**"

Another forma-type, designated "**forma inflata**" by Keupp (1976), can only be explained as caused by parasitism. This forma-type (Fig. 5) is characterized by ontogenetically temporary swellings of the conch with no indications of exogenous injuries. Apparently, a parasitic infestation of the mantle, possibly resulting from its having been laid open by injury, led to its swelling up and, consequently, to the inflation of the conch. Fernandez-Lopez (1987, Pl. 1, Fig. 1) illustrated an ammonoid showing irregularities that seem to be the result of a struggle between the ammonoid host and an infesting parasite. The *Dactylioceras* described by Lehmann (1975) also belongs to this forma-type.

Hölder (1970) used the phrase "juxta-forms" for cases in which there is a separation of medial elements that normally are coincident with one another. There are several different types. In "**forma juxtacarinata**" of Hölder (1956), the keel is dislocated with respect to other structures on the venter of the whorl (Fig. 6). In "**forma juxtasulcata**" of Geczy (1965), the ventral groove is dislocated. In "**forma juxtalobata**" of Hölder (1956), the siphuncle and ventral lobe of the suture (external lobe, in European terminology) are displaced to one side (Fig. 7). The causes of these paleopathies commonly are uncertain.

FIGURE 5. **"Forma inflata."** (a) On a specimen of *Amaltheus* sp. cf. *A. margaritatus* from the Jurassic (lower Domerian) of Tournadous (southern France). (From Hengsbach, 1979b, Fig. 6.) Approximate diameter = 18 mm. (b) On a specimen of *Pleuroceras* sp. from the Jurassic (upper Domerian) of Unterstürming (Bavaria, Germany). (From Keupp, 1977, Fig. 59, 19A.) Diameter = 25 mm.

Not only must exogenous injury be considered but also endogenous diseases (Hölder, 1970, Fig. 9).

Rieber (1963) described a specimen of *Cardioceras* with an asymmetrically situated siphuncle and keel (both **"forma juxtacarinata"** and **"forma juxtalobata"**). This individual shows the lateral displacement of both medial elements after at least three normal whorls and with no indication of injuries.

FIGURE 6. **"Forma juxtacarinata"** on a specimen of *Pleuroceras spinatum* from the Jurassic (upper Domerian) of Staffelstein (Bavaria, Germany). (From Keupp, 1977, Fig. 57, 7.) Diameter = 48 mm.

FIGURE 7. "**Forma juxtalobata**" on a specimen of *Psiloceras plicatulum* from the Lower Jurassic (lower Hettangian) of southwestern Germany. The siphuncle and ventral lobe are located on one ventrolateral flank. (From Hengsbach, 1979a, Fig. 2.) Approximate diameter = 34 mm.

The degree of deviation of the keel and the siphuncle gradually increases adaperturally, with the keel displaying a greater deviation than does the siphuncle. Rieber interpreted this paleopathy as being the result of an infestation by parasites.

I have argued (Hengsbach, 1979a) that at least some of the sutural-asymmetry paleopathies ["**forma juxtalobata** *sensu* Hengsbach (1979a)"], which show no indication of injuries or developmental disorders, were caused by parasitism (Fig. 7). I was able to confirm this later (Hengsbach, 1986a,b). In these cases, I postulated an infestation in young ammonoids shortly after the time of hatching. These parasites were located in the soft body of the animal on the septal mantle at or near the siphuncle; the infestation caused a constant displacement of the siphuncle root (i.e., the beginning of the siphuncle in the soft body) and, hence, of the ventral (external) lobe of the suture. Depending on the exact location and size of the site of infestation, the displacement was to the right or left of the plane of symmetry and to a greater or less extent. Even after the death of the parasite, once the asymmetry had reached a certain level, it no longer was able to be corrected in subsequent growth (Hengsbach, 1986a).

Rein (1989) established **"forma conclusa"** for a kind of pathological condition he found in certain ceratitids. In this paleopathy, a portion of the conch is separated from the rest of the space within the conch by new shell material. The space between these new secretions and the wall of the conch can amount to as much as 25% of the height of the whorl. H. Keupp (personal communication) has found the same phenomenon in Jurassic ammonites. It may be that this paleopathy represents the encapsulation of foreign bodies, such as parasites, or of an injured or diseased area (Hengsbach, 1986a, p. 145). Rein (1989) also gave some consideration to the "encapsulation of a parasite."

Morton (1983) described Scottish graphoceratids that show oscillations of the shell from the plane of symmetry. This paleopathy involves not only sculpture, such as the keel, but also the sutures and, in fact, the whole tube of the phragmocone, all in different ways; however, at the umbilicus, the plane of symmetry is retained. Somewhat later, Landman and Waage (1986) investigated scaphitids that show similar deviations from the plane of symmetry; they named this paleopathy **"Morton's syndrome"** and because of the pattern and course of development and in the absence of any indications of exogenous injury, concluded that these paleopathies were caused by parasitosis. Similar deviations of the conch had been described by Heller (1958) and Keupp (1977) and had been designated by Heller as **"forma undaticarinata"** (Fig. 8b). Even though Heller's and Keupp's descriptions concentrated on the chaotization of the sculpture associated with the paleopathy in question, Heller's name is the elder and should be used. For the same paleopathy but in which no keel is present, I recommend the use of a combination of terms, e.g., **"forma undaticarinata–undaticoncha,"** as was done by Keupp (1977 and later) (Fig. 8a).

Activities of parasites also may lead to shell concretions. House (1960) interpreted regularly arranged pits in the steinkerns of some Devonian goniatites as being the result of pearl-like mounds on the inside of the body chamber that probably were caused by irritation of foreign objects between the shell and mantle. On the steinkerns of two dactylioceratids, Keupp (1986) found similar indentations, along with adhering shell and the corresponding concretions. He assumed that these pearl-like concretions each had developed around an organic nucleus. On the basis of the location, size, and structure of the concretions, Keupp concluded that parasites (trematode larvae) were the likely cause, as they are in similar concretions in present-day molluscs. In the interest of consistency in the type of terminology used, I propose that the kind of paleopathy that involves shell concretions (pearls) be called **"forma concreta."**

Gigantism is one of the paleopathies whose connection to parasitic infestation was recognized relatively early. Böttger (1953a,b) had reported on parasite-caused delay in sexual maturation and the corresponding castration and retardation in growth in the present-day terrestrial snail *Zebrina* Held,

a   b

FIGURE 8. (a) **"Forma undaticarinata-undaticoncha"** on a microconch of *Jeletzkytes nebras-censis* from the Upper Cretaceous (Maastrichtian) of west-central South Dakota. (From Landman and Waage, 1986, Fig. 2, where it is called "Morton's syndrome.") Diameter = 65 mm. (b) **"Forma undaticarinata"** on a specimen of *Lamberticeras* sp. cf. *L. lamberti* from the Middle Jurassic (Callovian) of Villers-sur-Mer (Normandy, France). (From Keupp and Ilg, 1992, Fig. 8.) Diameter = 25 mm.

1837. On the basis of that report, Hölder (1956) suggested, quite reasonably, that cases of parasite-caused gigantism also might have occurred in ammonoids. However, as Tasnadi-Kubacska (1962) stressed, the explanation of large size as caused by parasitism should be restricted to separate individuals in order to exclude large size that resulted from other causes, for example, sexual dimorphism. In the interest of consistency in the type of terminology used, I propose that cases of morbid gigantism be called **"forma gigantea."**

Hölder (1956) established **"forma calcar"** for paleopathies that show, instead of the normally expected two rows of marginal nodes, a single row on the medial line. Long ago, von Zieten (1831–33) showed such a pathological specimen of *Distichoceras* Munier-Chalmas, 1892. From a biological point of view, it is unlikely that a fusion of two rows of marginal nodes exactly on the plane of symmetry of the conch would have resulted from an exogenous injury. Moreover, in the paleopathy originally designated **"forma calcar,"** there is no sign of an obvious cause. However, in many specimens exhibiting so-called "reduction phenomena," there is an "amalgamation keel," which is more or less asymmetrically positioned; that is, it is not exactly on the plane of symmetry. Hölder (1970) referred to such cases, which seem to have been caused by injury, as *"calcar*-like"; they show the relative suppression of one row of protuberances and the displacement of the other row toward the medial plane, but not the fusion of the two rows into one (see Keupp and Ilg, 1992,

Pl. 1, Fig. 15.). Hölder (1970) pointed out that, in the specimen of *Simoceras* shown by Hollmann (1961), the reduction of the two ventral rows of nodes into a single row is a secondary result of a recognizable shell injury. On the other hand, at least some of the taramelliceratids exhibiting "reduction phenomena" described by Hölder (1955a) may have been victims of genetic problems, because the genus involved seems to show a rather high genetic tolerance (Hengsbach, 1979a).

"**Forma syncosta**" was designated by Hengsbach (1979b). It also was meant to be understood in this primarily genetic interpretation. I also pointed out the logical possibility of an atavistic setback in the amaltheids in question. In "**forma syncosta**" there are two independent ribs that have been united in an anomalous manner into a single ventrolateral node. Keupp and Ilg (1992) used "**forma syncosta**" in a different sense from that which was originally intended.

Keupp and Ilg (1992) designated as "**forma undatispirata**" pendulous disturbances of shell growth that affect more or less an entire shell and include extensive exposure of the preceding volution; these could be, for example, a result of encrustations by epizoa. This paleopathy is in contrast to the deviations of the conch described by Morton (1983) and by Landman and Waage (1986), in which symmetry of the shell is retained.

"**Forma undatispirata**" was designated by Keupp and Ilg for forms that seems to be related to "**forma excentrica**," proposed by Hölder (1956) for disturbances in the inner whorls of some widely umbilicate ammonoids, e.g., *Bifericeras* Buckman, 1913 (see Hölder, 1970).

Lange (1941) designated what he called "**forma aegra cacoptycha**" but now is called "**forma cacoptycha**." Lange based this forma-type on specimens in which there is a diminution or partial flattening of the sculpture on one or both flanks of the conch (Fig. 9a). Keupp (1977) designated "**forma aptycha**" as a special case in which there is a total flattening of the sculpture (Fig. 9b). Causes of both of these could be internal diseases, e.g., parasitism or ecological (metabolic–physiological) disorders. But injury-induced "**forma cacoptycha**" also are known (e.g., Keupp, 1977, Fig. 58, 18; Bond and Saunders, 1989, Fig. 1C).

As long ago as 1935, Roll described ammonoid shell fragments that looked as though they had been cut with scissors. He inferred that they had been produced by crabs. Hölder (1955b) agreed with this interpretation and, a year later, designated such phenomena "**forma seccata**." After conducting a paleo-biological analysis of similar shell fragments from the Lias (Lower Jurassic) and Malm (Upper Jurassic), Mehl (1978) came to the conclusion that such shell fragments also could have been the result of attacks by dibranchiate cephalopods (teuthoids).

Mentioned previously were paleopathies in which recognizable bite marks are present; these have been described, for example, by Kauffman and Kesling (1960), Martill (1990), and Keupp (1991). For the sake of uniformity of nomenclature, I propose that these be designated as "**forma mordata**," from the Latin *mordere*, meaning "to bite."

FIGURE 9. (a) **"Forma cacoptycha"** on a specimen of *Amaltheus gloriosus* from the Jurassic (lower Domerian) of Reutlingen (southwestern Germany). The left side is developed normally. (From Hengsbach, 1987, Fig. 2.) Diameter = 27 mm. (b) **"Forma aptycha"** *sensu* Keupp (1977) on a specimen of *Pleuroceras spinatum* from the Jurassic (upper Domerian) of Unterstürming (Bavaria, Germany). (From Keupp, 1977, Fig. 58, 18A.) Diameter = 36 mm.

Hölder (1956) coined the designation **"forma disseptata"** for instances of missing or incomplete septa inside the phragmocone, but in which postmortem destruction was not the cause and in which no signs of exogenous injury are present. A number of such cases have been described in the literature (e.g., Kessler, 1923, 1925; Müller, 1978; and Rein, 1989). Kessler (1925) mentioned that Trueman and Tucher also found these imcomplete septa in Jurassic ammonoids.

6. Grouping the Forma-Types and Evaluating the Groupings

The primary objective of paleopathology is to elucidate the biology of the organisms involved, particularly the causes of the various paleopathies encountered. All paleopathies presently known might be sorted into the following categories:

1. Those caused by mechanical injuries.
2. Those related to parasitosis.
3. Those related to genetics.
4. Those related to the environment.

However, only some of the forma-types can be assigned definitely to any one of these categories. Those paleopathies designated **"forma seccata,"** **"forma substructa,"** and **"forma mordata"** were caused by exogenous injuries. The paleopathy **"forma inflata,"** on the other hand, must have been the result of parasitism.

A broad range of specimens of other forma-types can be attributed to mechanical injuries, but only if evidence of such injuries is present in the specimen at hand, for example, if it exhibits one or more bite marks or definite fractures. In this category would fall many cases of **"forma abrupta," "forma chaotica,", "forma circumdata," "forma duplicarinata," "forma pexa,"** and **"forma verticata."**

Some paleopathies most likely were the result of parasitism (Hengsbach, 1991b): **"forma juxtalobata** *sensu* Hengsbach (1979a)," **"forma juxtacarinata–juxtalobata** *sensu* Rieber (1963)," **"forma verticata** *sensu* Keupp (1979)," including the corresponding **"forma semiverticata"** and certain types of **"forma concreta." "Forma conclusa"** and **"forma chaotica"** also may have been effected by parasitism.

A problem in the classification of forma-types is that they are not all of the same nature. Some may be only different stages or severities of the same actual pathological condition in the living animal. Moreover, not all of the forma-types are defined strictly on morphology. For example, **"forma substructa"** refers to a particular mode of healing, whereas **"forma calcar"** links together specimens that share a particular, strongly defined morphological trait, namely, instead of the two expected rows of nodes, there is but one.

As mentioned previously, certain forma-types seem to be related to one another in that one forma-type can be understood as a more general category that includes others as special cases. Hölder (1956) gave an example in saying that his **"forma verticata"** is a special case of **"forma abrupta."** In the same way, he related **"forma semiverticata"** to **"forma verticata." "Forma cacoptycha"** commonly cannot be distinguished from **"forma aptycha"** readily, because the latter is a special case of the former. Similarly, **"forma bovicornuta"** and **"forma alternospinata"** have been recognized as special cases of **"forma circumdata"** (Keupp, 1977).

Moreover, certain forma-types can be transitional to others. As mentioned previously, Keupp (1977) introduced **"forma chaotica"** for the paleopathy characterized by "chaotic arietid sculpture" reported by Hölder (1970). Keupp characterized this forma-type on the basis of the repeated drifting of the keel toward one flank; however, Hölder's description stressed both keel drift and the chaotic pattern of the ribs (Hengsbach, 1979b). Keupp noticed transitions between **"forma chaotica"** and **"forma circumdata"** and inferred a relationship with **"forma duplicarinata"** (see Keupp, 1977, Fig. 59, 21) and **"forma undaticarinata"** (see Keupp, 1977, Fig. 58, 10).

On the basis of etiology, it is possible to divide known paleopathies into three groups (see Hengsbach, 1991b):

1. Paleopathies that are the result of significant injuries (for example, **"forma seccata," "forma mordata,"** and **"forma substructa"**).
2. Paleopathies that are obviously unrelated to injuries or that never occur with any indications of them (for example, **"forma inflata"** and **"forma syncosta"**).
3. Paleopathies that have been found to occur both after injuries and in otherwise intact shells (for example, **"forma aptycha," "forma cacoptycha," "forma chaotica,"** and **"forma juxtacarinata"**).

A more-detailed classification is not really practicable with the present state of our knowledge. The forma-types are basically different manifestations of healing. The paleopathic expression of various injuries and diseases is the result of physiological phenomena not readily preserved in shells. The reaction of a particular soft tissue to different types of damage and disease can be strikingly different. Different epithelia and other soft tissues may react to the same damage or disease in different manners. In addition, different stages or severities of a disease or injury may lead to different morphological expressions. All of these factors, and more, together yield the diversity of paleopathies that we see in fossils of ammonoid cephalopods.

ACKNOWLEDGMENT. Th. Becker, Freie Universität, Berlin, helped translate the original version of this chapter into English. I gratefully acknowledge his help.

This chapter was composed in German. The transformation into English was materially aided by Frederick Veidt, retired principal of the Fairview German-English Bilingual School, Cincinnati, Ohio. He performed yeoman service, and the editors gratefully acknowledge his help.

References

Bayer, U., 1970, Anomalien bei Ammoniten des Aaleniums und Bajociums und ihre Beziehung zur Lebensweise, *N. Jb. Geol. Paläont. Abh.* **135**(1):19–41.
Bond, P. N., and Saunders, W. B., 1989, Sublethal injury and shell repair in Upper Mississippian ammonoids, *Paleobiology* **15**(4):414–428.
Böttger, C. R., 1953a, Größenwachstum und Geschlechtsreife bei Schnecken und pathologischer Riesenwuchs als Folge einer gestörten Wechselwirkung beider Faktoren, *Zool. Anz.* **17**(*Verh. Dtsch. Zool. Ges.* **46**):468–487.
Böttger, C. R., 1953b, Riesenwuchs der Landschnecke *Zebrina (Zebrina) detrita* (Müller) als Folge parasitärer Kastration, *Arch. Molluskenkd.* **82**:151–152.
Bülow, E. von, 1917, Über einige abnorme Formen bei Ammoniten, *Z. Dtsch. Geol. Ges. B. Mber.* **69**:132–139.
Engel, T., 1894, Über kranke Ammonitenformen im Schwäbischen Jura, *Nova Acta Acad. Caesar. Carol. Leopold* **61**:327–384.
Fernandez-Lopez, S., 1987, Necrocinesis y colonización postmortal en *Bajocisphinctes* (Ammonoidea) de la Cuenca Ibérica; implicaciones paleoecológicos y paleobatimétricas, *Bol. R. Soc. Esp. Hist. Nat. Secc. Geol.* **82**(1–4):151–184.
Geczy, B., 1965, Pathologische jurassische Ammoniten aus dem Bakony-Gebirge, *Ann. Univ. Sci. Budap. Sect. Geol.* **9**:31–37.

Guex, J., 1967, Contribution á l'etude des blessures chez les ammonites, *Bull. Lab. Geol. Mineral, Geophys. Mus. Geol. Univ. Lausanne* **165**:1–16.

Guex, J., 1968, Sur deux conséquences particulière des traumatismes du manteau des ammonites, *Bull. Lab. Geol. Mineral, Geophys. Mus. Geol. Univ. Lausanne* **175**:1–6.

Heller, F., 1958, Gehäusemißbildungen bei Amaltheiden. Ein neuer Fund aus dem fränkischen Jura, *Geol. Bl. Nordost-Bayern Angrenzende Geb.* **8**(2):66–71.

Heller, F., 1964, Neue Fälle von Gehäuse-Mißbildungen bei Amaltheiden, *Paläontol. Z.* **38**(3–4):136–141.

Hengsbach, R., 1979a, Zur Kenntnis der Asymmetrie der Ammoniten-Lobenlinie, *Zool. Beitr. (N.F.)* **25**:107–162.

Hengsbach, R., 1979b, Weitere Anomalien an Amaltheen-Gehäusen (Ammonoidea; Lias), *Senckenb. Lethaea* **60**(1/3):243–251.

Hengsbach, R., 1979c, Zum Problem: Das Selbstverständnis der Paläontologie, *Sitzungsber. Ges. Naturforsch. Freunde Berlin (N.F.)* **19**:81–92.

Hengsbach, R., 1981, Zur Evolution der Grundpläne von Cephalopoda, Ammonoidea und Neoammonoidea. Teil 2. Zur Evolution des Ammoniten-Zweiges der Cephalopoda, *Zool. Beitr. (N.F.)* **27**:223–266.

Hengsbach. R., 1986a, Zur Kenntnis der Sutur-Asymmetrie bei Ammoniten, *Senckenb. Lethaea* **67**:119–149.

Hengsbach, R., 1986b, Ontogenetisches Auftreten und Entwicklung der Sutur-Asymmetrie bei einigen Psilocerataceae (Ammonoidea; Jura), *Senckenb. Lethaea* **67**:323–330.

Hengsbach, R., 1987, Zwei schwer erkrankte Ammoniten aus dem Lias von Reutlingen (Württemberg): *Sitzungsber. Ges. Naturforsch. Freunde Berl. (N.F.)* **27**:183–187.

Hengsbach, R., 1990, Studien zur Paläopathologie der Invertebraten. 1. Die Paläoparasitologie, eine Arbeitsrichtung der Paläobiologie, *Senckenb. Lethaea* **70**:439–461.

Hengsbach, R., 1991a, Studien zur Paläopathologie der Invertebraten. 2. Die Symmetropathie, ein Beitrag zur Erforschung sogenannter Anomalien, *Senckenb. Lethaea* **71**(3/4):339–366.

Hengsbach, R., 1991b, Studien zur Paläopathologie der Invertebraten. III. Parasitismus bei Ammoniten, *Paläontol. Z.* **65**(1/2):127–139.

Hölder, H., 1955a, Die Ammoniten-Gattung *Taramelliceras* im südwestdeutschen Unter- und Mittelmalm, *Palaeontogr. (A)* **106**:37–153.

Hölder, H., 1955b, Belemniten und Ammoniten als Beutetiere, *Aus der Heimat* **63**(5/6):88–92.

Hölder, H., 1956, Über Anomalien an jurassischen Ammoniten, *Paläontol. Z.* **30**(1/2):95–107.

Hölder, H., 1970, Anomalien an Molluskenschalen, insbesondere Ammoniten, und deren Ursachen, *Paläontol. Z.* **44**(3–4):182–195.

Hölder, H., 1973, Miscellanea cephalopodica, *Münster. Forsch. Geol. Paläont.* **29**:39–76.

Hölder, H., 1977, Zwei ungewöhnliche Erscheinungsformen anomaler Jura-Ammoniten der forma aegra verticata, *Paläontol. Z.* **51**:254–257.

Hollmann, R., 1961, *Simoceras (Simoceras) biruncinatum* (Quenstedt, 1845) forma aegra calcar Zieten, 1830 (Ammonoidea) aus dem Untertithon der Sette Comuni (Norditalien), *Mem. Mus. Civ. Stor. Natur. Verona* **9**:267–272.

House, M. R., 1960, Abnormal growths in some Devonian goniatites. *Palaeontology (Lond.)* **3**(2):129–136.

Jessen, W., 1960, Ausgeheilte Schalen-Verletzungen an Goniatiten aus dem Oberkarbon Belgiens, *Bull. Soc. Belge Geol. Paleontol. Hydrol.* **68**:55–58.

Kauffman, E. G., and Kesling, R. V., 1960, An Upper Cretaceous ammonite bitten by a mosasaur, *Contrib. Mus. Paleont. Univ. Mich.* **15**(9):193–248.

Kessler, P., 1923, Beiträge zur Kenntnis der Organisation der fossilen Gehäusecephalopden. II. Die phyletische Entwicklung des Haftmuskels, *Centralbl. Miner.* **1923**:689–702.

Kessler, P., 1925, Über Nautiloideen mit unfertigen Septen, *Paläontol. Z.* **7**:24–31.

Keupp, H., 1976, Neue Beispiele für den Regenerationsmechanismus bei verletzten und kranken Ammoniten, *Paläontol. Z.* **50**(1/2):70–77.

Keupp, H., 1977, Paläopathologische Normen bei Amaltheiden (Ammonoidea) des fränkischen Lias, *Jb. Coburg. Landesst.* **1977**:263–280.

Keupp, H., 1979, Nabelkanten-Präferenz der forma verticata Hölder 1956 bei Dactylioceraten (Ammonoidea, Toarcien), *Paläontol. Z.* **53**(3/4):214–219.

Keupp, H., 1985, Pathologische Ammoniten—Kuriositäten oder Dokumente? Teil 2, *Fossilien* **1985**(1):23–35.

Keupp, H., 1986, Perlen (Schalenkonkretionen) bei Dactylioceraten aus dem fränkischen Lias, *Nat. Mensch (Nuremberg)* **1986**:97–102.

Keupp, H., 1991, Bißmarken oder postmortale Implosionsstrukturen? *Fossilien* **1991**(5):275–280.

Keupp, H., 1992, Wachstumsstörungen bei *Pleuroceras* und anderen Ammonoidea durch Epökie, *Berliner Geowiss. Abh. (E)* **3**:113–119.

Keupp, H., and Ilg, A., 1992, Paläopathologie der Ammonitenfauna aus dem Obercallovium der Normandie und ihre palökologische Interpretation, *Berliner Geowiss. Abh. (E)* **3**:171–189.

Landman, N. H., and Waage, K. M., 1986, Shell abnormalities in scaphitid ammonites, *Lethaia* **19**:211–224.

Lange, W., 1941, Die Ammonitenfauna der *Psiloceras*-Stufe Norddeutschlands, *Palaeontogr. (A)* **93**:1–192.

Lehmann, U., 1975, Über Biologie und Gehäusebau bei *Dactylioceras* (Ammonoidea) aufgrund einer Fraktur-Analyse, *Mitt. Geol.-Paläont. Inst. Univ. Hamburg* **44**:195–206.

Lehmann, U., 1977, *Paläontologisches Wörterbuch*, 2nd ed., Ferdinand Enke, Stuttgart.

Martill, D. M., 1990, Predation on *Kosmoceras* by semionotid fish in the Middle Jurassic lower Oxford clay of England, *Palaeontology (Lond.)* **33**(3):739–742.

Maubeuge, P. L., 1949, Sur quelques échantillons anormaux d'Ammonites jurassiques. Un cas possible de mutation chez les Ammonites, *Arch Inst. Grand-Ducal Luxemb. Sect. Sci. Natur. Phys. Math. (N.S.)* **18**:127–147.

Maubeuge, P. L., 1957, Deux Ammonites nouvelles du Lias moyen de l'Allemagne Septentrionale, *Bull. Soc. Sci. Nancy* **1957**:1–6.

Mehl, J., 1978, Anhäufungen scherbenartiger Fragmente von Ammonitenschalen im süddeutschen Lias und Malm und ihre Deutung als Fraßreste, *Ber. Naturforsch. Ges. Freib. Breisgau* **68**:75–93.

Morton, N., 1983, Pathologically deformed *Graphoceras* (Ammonitina) from the Jurassic of Skye, Scotland, *Palaeontology (Lond.)* **26**(2):443–453.

Müller, A. H., 1978, Über Ceratiten mit fehlenden oder unvollständigen Kammerscheidewänden (Septen) und die Frage nach der Lebensweise der Ammonoidea (Cephalopoda), *Freiberg. Forschungsh. Reihe C* **334**:69–83.

Nicolesco, C. P., 1921, *Étude sur la Dissymetrie de Certaines Ammonites*, Bailliere et Fils, Paris.

Orbigny, A. d', 1842–1849, *Paléontologie Française. Terrains Jurassiques. I. Céphalopodes*, Masson, Paris.

Pompecki, J. F., 1893, Revision der Ammoniten des Schwäbischen Jura, *J. Ver. Vaterl. Naturk. Württ.* **49**:151–248.

Pompecki, J. F., 1894, Über Ammonoideen mit "anormaler Wohnkammer," *J. Ver. Vaterl. Naturk. Württ.* **50**:220–290.

Quenstedt, F. A., 1858, *Der Jura*, Laupp, Tübingen.

Quenstedt, F. A., 1883–1888, *Die Ammoniten des Schwäbischen Jura*. Tome 1–3, Schweizerbart, Stuttgart.

Rein, S., 1989, Über das Regenerationsvermögen der germanischen Ceratiten (Ammonoidea) des Oberen Muschelkalks (Mitteltrias), *Veroff. Naturhist. Mus. Schleusingen* **4**:47–54.

Reynes, P., 1879, *Monographie des Ammonites*, Lias, Paris.

Rieber, H., 1963, Ein *Cardioceras* (Ammonoidea) mit asymmetrischer Lage von Phragmocon und Kiel, *Neues Jahrb. Geol. Paläont. Mh.* **1963**(6):289–294.

Roll, A., 1935, Über Fraßspuren an Ammonitonschalen, *Zbl. Mineral.* **1935**(B):120–124.

Staff, H. von, 1909, Zur Siphonalasymmetrie der Juraammoniten, *Foldt. Kozl.* **39**:489–496.

Stahl, von, 1824, Verzeichnis der Versteinerungen Württembergs, *Correspondenzbl. Württ. Landwirthsch. Ver. Stuttgart* (cited in Hölder, 1956).

Tasnadi-Kubacska, A., 1962, *Paläopathologie*, G. Fischer, Jena.

Theobald, N., 1958, Quelques malformations chez les Ammonites, *Ann. Sci. Univ. Besançon 2e Ser. Geol.* **8**:19–28.

Thiermann, A., 1964, Über verheilte Verletzungen an zwei kretazischen Ammonitengehäusen, *Fortschr. Geol. Rheinl. Westfalen* **7**:27–30.

Vadasz, M. E., 1908, Über eine oberliassische Lytocerasart mit aufgelöster Wohnkammer, *Foldt. Kozl.* **38**:131–136.

Vadasz, M. E., 1909, Über anormale Ammoniten, *Foldt. Kozl.* **39**:215–219.

Zieten, C. H. von, 1831–1833, *Versteinerungen Württembergs*, Schweizerbart, Stuttgart.

Chapter 16

Ammonoid Life and Habitat

GERD E. G. WESTERMANN

GERD E. G. WESTERMANN • Department of Geology, McMaster University, Hamilton, Ontario L8S 4M1, Canada.

Ammonoid Paleobiology, Volume 13 of *Topics in Geobiology*, edited by Neil Landman *et al.*, Plenum Press, New York, 1996.

1. Introduction

Since the last review of Jurassic–Cretaceous ammonoid ecology (Westermann, 1990), much additional work has been done on ammonoid autecology (architecture or macrostructure) as well as on the associations and occurrences of ammonoids in the field (synecology). Important works on Paleozoic through Triassic ammonoids, dispersed in the literature, have not been reviewed previously. Quantitative autecological studies, begun in the mid-1980s, concerned buoyancy and orientation. Electron and light microscopic studies of the shells have also contributed to an understanding of the soft parts. Research on shell fabrication, strength, and hydrodynamics has increased greatly, also contributing to ammonoid autecology. Intraspecific morphological variation has been studied intensively but remains poorly understood ecologically; most authors still fail to consider variation in the functional interpretation of shell shape. Ammonoid synecology was significantly advanced in recent years by the renewed interest in Paleozoic and Mesozoic dysoxic black-shale facies and their relation to eustasy and orbitally enforced cycles. Other recent studies in synecology have emphasized the interrelations among sediment, eustasy, and biofacies. Finally, ammonoid taxonomy has been summarized in *The Ammonoidea* (Special Volume 18, The Systematics Association, 1981).

It recently has been fashionable to assume that the great majority of ammonoids lived close to the epicontinental (epeiric) sea floor, where they obtained their prey, i.e., that they were either "nektobenthic" or "benthopelagic" (= demersal) (e.g., Westermann, 1990). This has been a consequence of the belief that all ammonoids were extremely sluggish and thus unable to pursue prey in midwater. There existed, however, numerous well "streamlined" and usually "keeled" predators, at least some of which were well muscled and, when large, could swim faster than *Nautilus* and with better directional control. Moreover, there existed a planktic food source, now extinct, namely, young ammonoids.

In fact, the majority of ammonoids seem to have been pelagic and divided about equally among swimmers, drifters, and vertical migrants in the cratonic (epeiric and shelf) seas and oceans. Thus, we have come back to the classical belief that ammonoids were generally pelagic. The remainder were demersal. (The term "demersal," commonly applied by biologists to bottom-feeding fish, is used here as a replacement for what has been called by the controversial terms "benthopelagic" and "nektobenthic.") Even the assumption that all ammonoids were euryhaline is now being challenged; a few probably lived in superhaline, others in subhaline waters.

The majority of ammonoids are known from epeiric seas. In contrast to neritic or shelf seas, epeiric seas were far removed from the ocean and more or less surrounded by land. ("Epicontinental" or "epeiric" is used here to refer to a sea located on a continental interior, as opposed to a "neritic" oceanic water mass that happens to lie on the continental shelf.) Their physical and chemical properties differed greatly from shelf seas and oceans, as well as among themselves in "food supply," temperature, turbidity (depth of wave base), clarity (depth of photic zone), oxygenation, and in their more developed stratification that commonly included brackish surface waters. I therefore distinguish three marine biomes, *epeiric*, *neritic*, and *oceanic*.

Note that the epipelagic–mesopelagic boundary, as used here, is considered to have been at about a depth of 240 m; this value is somewhat different from that used by neooceanographers. This is because of the physiology of phragmocone-bearing cephalopods (Westermann, 1990; Hewitt, 1993); 240 m is the lower limit of effective liquid removal from the chambers by osmosis at a 2% salinity differential across the siphuncular membrane (Ward and Westermann, 1985). Mesopelagic ammonoids had to rise regularly to epipelagic depths to compensate for flooding.

The habits and habitats of ammonoids are particularly difficult to understand for three well-known reasons. Ammonoids are extinct, and their softparts are essentially unknown. They lived in the water column, even if demersal, made few traces in or on the sediment, and last, but not least, were prone to postmortem drift. Furthermore, Recent cephalopods are well known to migrate annually over hundreds or thousands of kilometers and, in many cases, to die after spawning in habitats quite different and distant from those in which they spent most of their lives. Much more limited migrations are indicated for many ammonoids by the common separation of juvenile and adult growth phases in the fossil assemblages. Clustered remnants (ammonitellas) of ammonoid eggs have also been found resembling spawning sites. But the evidence suggests that most or all are postmortem accumulations. A mixed aut- and synecological approach is therefore required, together with the accumulation of a multitude of "circumstantial evidence." "Hard" evidence in support of a particular hypothesis is difficult to come by except for morphostructural data from the shell, many of which here are taken directly from the extensive new work by R. Hewitt (Chapter 10, this volume).

The reader might argue rightly that bathymetric estimates based on the calculations of septal strength against ambient pressure for only a single septum or several septa are hardly representative for entire species and even less for genera. Furthermore, even if the data are correct, these estimates are for potential depth limits, not actual local habitat depths. These estimates are, nevertheless, some of the best "hard" data available and, hence, require serious consideration, as do data on shell and aperture orientation, stability, etc. (e.g., Saunders and Swan, 1984; Saunders and Shapiro, 1986).

This chapter first establishes the fundamentals of ammonoid autecology concerning the physical constraints on the living organism as implied by the morphology of the individual shell. We then examine the synecological significance of associations in the field, with special consideration of litho- and biofacies, starting with the deepest, oceanic facies and continuing over ocean-margin and basinal facies to platform facies. Emphasis is given to "restricted" environments in order to examine environmental conditions that limited ammonoid habitats. In this way it is hoped that one of the major pitfalls of ammonoid paleoecology, postmortem drift, can be avoided.

2. Morphological Terms

The terminology used herein follows the Glossary of the *Treatise on Invertebrate Paleontology*, Part L, *Mollusca* 4 (pp. 3–6), but several terms concerning conch shape were either used inconsistently or omitted from the *Treatise* glossary (e.g., discoidal, planulate); and a few new terms are added (planorbicone, discocone, elliptospherocone).

2.1. Shell Shape

Planispirals (Fig. 1) include the following forms:

1. *Spherocone.* Subglobular; adult body chamber remains involute (s. str.), e.g., *Eurycephalites*, or *elliptospherocone*, becoming evolute (egressing), e.g., *Sphaeroceras*.
2. *Discocone.* Involute with ovate whorls, e.g., *Nautilus*.
3. *Cadicone.* Subglobular with open, angular umbilicus, e.g., *Tulites*.
4. *Oxycone.* Involute with subtriangular compressed whorls, e.g., *Oxynoticeras*.
5. *Platycone.* Involute to moderately evolute, with subrectangular compressed whorls, usually keeled, e.g., *Witchellia*.
6. *Planorbicone.* Evolute with subcircular to depressed whorls, e.g., *Stephanoceras*.
7. *Serpenticone.* Very evolute to advolute with subcircular to depressed whorls, longidomic, e.g., *Dactylioceras*.
8. *Leiostraca.* Shells lacking costae, nodes, and spines; usually restricted to "eugeoclinal" facies, i.e., outer shelf and oceanic, e.g., *Phylloceras, Lytoceras, Desmoceras*.
9. *Trachyostraca.* Shells with typical ammonite "ornament," e.g., *Stephanoceras*.
10. *Brevidome.* Body-chamber about one-half whorl.
11. *Mesodome.* Body-chamber about three-fourths whorl.
12. *Longidome.* Body-chamber at least one whorl.

Heteromorphs (Fig. 2) include the following forms:

1. *Orthocone.* Straight; either compressed, e.g., *Baculites*, or circular, e.g., *Sciponoceras*.
2. *Cyrtocone.* Curved with less than one whorl, e.g., *Protancyloceras*.
3. *Gyrocone.* Open spiral with more than one whorl, e.g., *Crioceratites*.
4. *Torticone ("trochospiral").* Helical; either loosely coiled, e.g., *Puebloites*, or closely coiled, e.g., *Turrilites*.

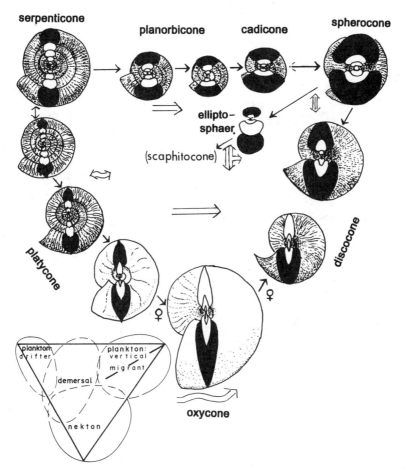

FIGURE 1. The basic planispiral shell shapes, their most common ontogenies (simple arrows), and their principal habitats (small diagram). Plankton includes drifters, such as serpenticones, and vertical migrants (vertical, triple-shafted arrows), mainly spherocones and cadicones; shapes of demersals, such as some planorbicones and platycones, greatly overlap with those of nekton (swimming potential indicated by length and direction of double-shafted arrows), represented by oxycones as well as some platycones and discocones.

    5. *Ancylocone.* Closed or open, planar or helical spiral followed by hook e.g., *Ancyloceras*, *Scaphites*.

    6. *Hamitocone.* Two or more straight shafts; e.g., *Ptychoceras*.

    7. *Vermicone.* Worm-like, seemingly irregular, e.g., *Nipponites*.

2.2. Growth Stages

Based on the shell morphogenesis of Paleozoic (House, 1985; Kant, 1975; Kullmann, 1981; Kant and Kullmann, 1988; Kullmann and Scheuch, 1970, 1972) and Mesozoic ammonoids (Westermann, 1954, 1958; Landman, 1987), the following formal growth stages are distinguished:

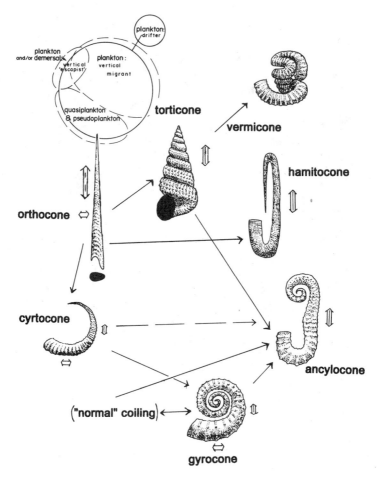

FIGURE 2. The basic heteromorphic shell shapes, their most common ontogenies (simple arrows), and their principal habitats (small diagram). The great majority were megaplanktic, mostly vertical migrants (vertical triple-shafted arrows). Exceptions were: the drifting vermicones; some orthocones and ?torticones that probably were bottom-feeding with rapid vertical escape-swimming (benthic anoxia) or, in the case of large *Baculites*, even nektic (swimming direction, here forward and backward, and speed indicated by double-shafted, double-headed arrows); some small gyrocones that were pseudoplanktic in floating algal mats; and some large cyrtocones that might have been sluggish, demersal swimmers.

1. *Ammonitella.* Embryo, usually subspherical, consists of the subelliptical protoconch or initial chamber and three-fourths to one and a quarter whorls of body chamber (Chapter 11, this volume).
2. *Neanoconch* (as used here). Hatchling/early juvenile, with $2\frac{1}{2}$ to $3\frac{1}{3}$ additional whorls, growing mainly in height, generally planorbiconic and weakly sculpted, diameter 3–5 mm ($\approx$ Hyatt's neanic stage as redefined by Westermann, 1958).

3. *Juvenile stage.* Late juvenile plus adolescent/immature, with several additional whorls, differing in growth parameters, e.g., abrupt growth of width and sculpture, roughly to half adult size.

4. *Adult stage.* Fully grown/mature, with $\frac{1}{2}$ to $1\frac{1}{4}$ or more whorls, generally consisting of final body chamber that changes coiling, cross section, sculpture, and/or peristome; when stage is indistinct (many pre-Permian ammonoids, many Phylloceratina and Lytoceratina), adulthood is established by septal approximation.

3. Evidence from Shell Architecture

3.1. Poise, Stability, and Mobility

3.1.1. Planispirals

In planispiral conchs (Fig. 1), the static orientation of the aperture (apertural angle) is structurally interrelated with stability. Both orientation and stability depend on the spatial distributions of, and relative distance between, the centers of buoyancy and mass. The most critical factor (variable) controlling this is the body chamber angle (angular "length"). In general, this value is between half and one and a quarter whorls (extreme more than two) and is regulated mainly by the whorl-expansion rate (W). The most thoroughly investigated examples are the Carboniferous (Namurian) goniatites and prolecanitids (Saunders and Swan, 1984; Saunders and Shapiro, 1986; Shapiro and Saunders, 1987; Swan and Saunders, 1987). Based on detailed measurements of shell thickness and morphology, these authors calculated body chamber angle ("length"), static orientation of the aperture, and stability under the assumption of neutral buoyancy, including a soft body that could be withdrawn into the body chamber (Fig. 3). With increasing body chamber angle, the orientation of the aperture, measured from the vertical, changed markedly in the floating shell: from 60° in half-whorl body chambers (slightly higher than in *Nautilus* which has an even lower body chamber angle) to 100° in three-fourths- to four-fifths-whorl body chambers. A rapid reversal occurred at angles >300°, where the aperture drops rapidly to 30° at just over one whorl (400°). Significantly, stability decreased continuously to a minimum (10% of *Nautilus*) in shells with the "longest" body chambers, both evolute and involute.

Not all body chamber angles ("lengths") were present in equal frequencies in Paleozoic and Mesozoic ammonoids. Significantly, the frequency peaks (modes) occurred at similar angles in the different ammonoid orders and suborders, indicating adaptive peaks: angles of about one-half whorl (160–180°), three-fourths whorl (260–300°) and one whorl (350–400°) were more abundant than intermediate values (Trueman, 1941; Raup, 1967; Saunders and Shapiro, 1986; Jacobs and Landman, 1993; Figs. 3,4). These three body chamber groupings coincide with the minimal and maximal orientations

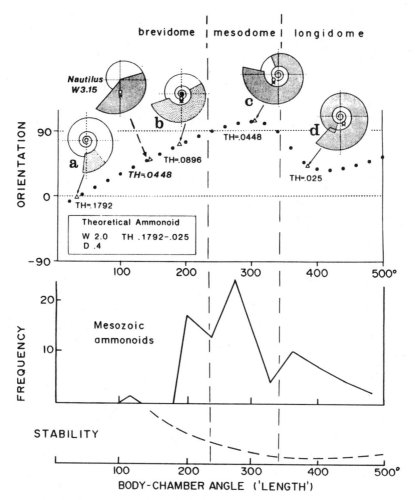

FIGURE 3. Body chamber angle ("length") plotted against aperture orientation in theoretical, moderately evolute ($D = 0.4$) ammonoids with a whorl-expansion rate (W) of 2.0 and with *Nautilus* for comparison (a is hypothetical only). The effects of incrementally decreasing shell thickness (TH) on the neutrally buoyant ammonoids are (1) increased body-chamber angle, (2) change in aperture orientation, and (3) reduced hydrostatic stability, i.e., distance between centers of buoyancy (circle) and mass (cross) (Adapted and modified from Saunders and Shapiro, 1986, Figs. 1,8). The frequency distribution of Mesozoic ammonoids shows three peaks (adapted from Trueman, 1941) that are, respectively, coincident with brevidomes (b), mesodomes (c), and longidomes (d) and interpreted as adaptive optima; for peaks in mid-Carboniferous ammonoids, see Fig. 4.

outlined above and with functional adaptations discussed below; they are here called *brevidome*, *mesodome*, and *longidome*, respectively (Figs. 3, 5).

The location of the hyponome is essential for the hydrodynamic aspects of the ammonoids discussed below. I see little reason to doubt its consistently

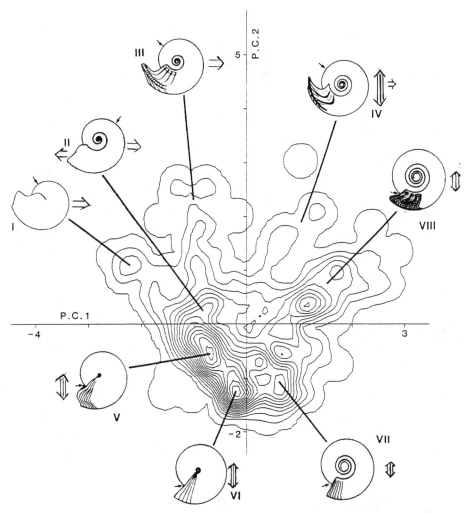

FIGURE 4. Mid-Carboniferous (Namurian) ammonoid morphotypes I–VIII and density contours of 281 species, based on two principal components of variation (20 shell characters). Life orientations (small arrows indicate end of phragmocone, see Fig. 1) and main habits indicated; I–III were presumably active swimmers (double-shafted arrows), and IV–VIII planktic drifters and vertical migrants (triple-shafted arrows). (Adapted from Swan and Saunders, 1987, and Saunders and Shapiro, 1986.)

ventromarginal position (but see Jacobs and Landman, 1993). This is its position in all extant cephalopods except for a very few abyssal forms, where it is absent. The hyponome is tubular in all coleoids (Nesis, 1987) and probably was also in ammonoids.

Intraspecific variation is, of course, closely related to the function of a morphological feature; i.e., variation presumably correlates inversely with the activity or use (realization) of a potential function (see Westermann, 1990).

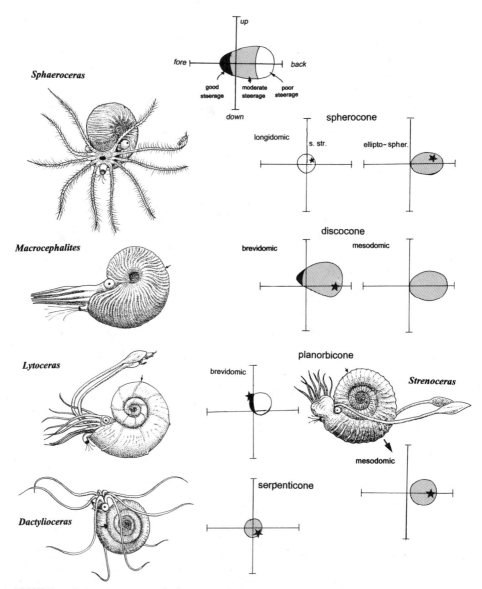

FIGURE 5. Swimming-potential of ammonoids (with soft parts freely reconstructed, see Section 5.3 in text) according to shell shape, aperture orientation, and stability (see Fig. 3). Possible swimming directions and relative velocities are represented as vectors, with the examples indicated by asterisks, and increasing steerage (maneuverability with directional stability) and acceleration by shades of white to black (see Section 3.1.1 in text).

This point nevertheless remains largely misunderstood (e.g., Dagys and Weit-schat, 1993), leading to the false conclusion that shell features that intergraded in one species (or population) with high variation, were similarly nonfunctional in another species with low variation. The interpretations below apply to potential functions in populations marked by low variation.

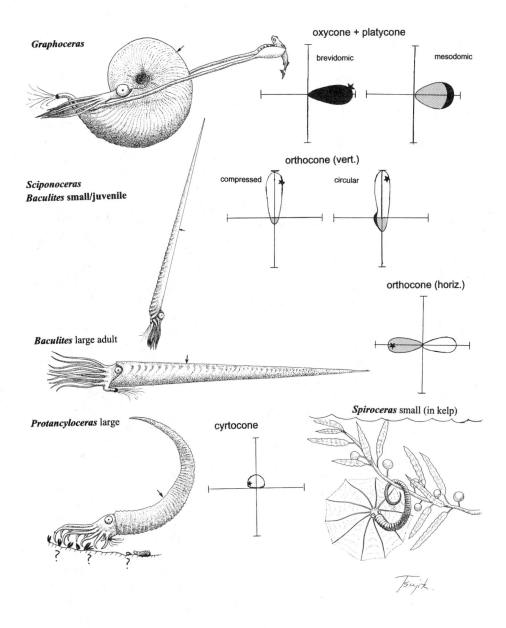

Graphoceras

oxycone + platycone

brevidomic

mesodomic

Sciponoceras
Baculites small/juvenile

orthocone (vert.)

compressed

circular

Baculites large adult

orthocone (horiz.)

Protancyloceras large

cyrtocone

Spiroceras small (in kelp)

Tsujik.

3.1.1a. Brevidomes. Brevidomic shells, with good stability and a lower lateral (ca. 60°) aperture position, were relatively rare in the Paleozoic and mainly confined to the involute Anarcestidae and the evolute Prolecanitida (Fig. 4; morphotype II of Saunders and Swan, 1984) but became abundant in the Mesozoic Ammonitina, especially in the well-"streamlined" oxycones.

Oxycones, at least when strongly compressed with a thickness ratio <0.3 (whorl width/diameter, i.e., length in swimming direction), consistently attained the oxyconic shape well before the hydrodynamically critical size of 5–10 cm; shape rarely tracked ontogenetically the size-related minimum drag and power requirements, which changed from spheroconic to discoconic to oxyconic (Jacobs, 1992a; Chapter 7, this volume). Furthermore, the thickness ratio was often excessively low (ca. 0.2) throughout postneanic growth, and the umbilicus was small or occluded. Two possible adaptations, not necessarily exclusive, are presently proposed for ammonoids with oxyconic shells.

The first is the "added mass" effect (Chapter 7, this volume). Repeated acceleration by funnel propulsion and acceleration reaction forces would have made the thinner shapes with lower mass (including that of "attached" water in the umbilicus) more efficient during brief spurts of relatively rapid swimming. Oxyconic shells would thus imply ambush predation from a stationary midwater position rather than pursuit predation. High stability made them potential carriers of a strong hyponomic propulsion system (Doguzhaeva and Mutvei, 1992; but see Jacobs and Landman, 1993).

The second adaptation is steerage. Oxycones with low morphic variation (not the more inflated forms intergrading with spherocones), ranging in size from a centimeter to over a meter, had improved steerage (maneuverability combined with directional stability) and speed during the pursuit of prey (Bayer, 1982), rather than merely speed as in the discocones (Fig. 5). This would have enabled active but overall slow hunters to capture small crustaceans, e.g., krill of more recent times, ostracodes, or smaller ammonites that used zig-zag escape tactics. Chamberlain (1991) compared them to short, strongly laterally compressed reef fish that substitute maneuverability and steerage for friction drag, although the stiff shells acted quite differently from bending fish. He also reported that *Nautilus* has limited steerage, but it is restricted to bottom feeding (demersal) and presumably does not hunt even small mobile pelagon. Some discocones locally modified the adult body chamber, forming an acuity in the shape and position of a boat's prow to improve steerage, e.g., *Hyattoceras* and *Waagenoceras* (Davis *et al.*, 1969, Pls. 4,5; Davis, 1972, Pls. 16, 20). But brevidomic discocones appear to be rare (Anarcestidae, a few adult Macrocephalitinae; cf. Westermann and Callomon, 1988).

Platycones include highly stable brevidomes as well as moderately stable mesodomes. Platycones are found mainly in the evolute to moderately involute, compressed and smooth shells of the Paleozoic prolecanitids and numerous Mesozoic Ammonitina. "Streamlining" was generally good, notwithstanding the larger umbilicus (Jacobs, 1992a; Elmi, 1993). Steerage (combining directional control with maneuverability) was variable and mostly depended on the shape of the venter; in order of increasing steerage, shape ranged from tabulate (rare) to rounded to sulcate to keeled or fastigate. Platycones were moderately to relatively good swimmers and had better shape than *Nautilus*, which some resembled in poise and stability.

3.1.1b. Mesodomes. Discocones were "streamlined nautilicones" whose shape closely approximated Jacobs's (1992a) paradigm for efficient larger, relatively rapid swimmers, with a thickness ratio of about 0.3 to 0.4 at 1 to 5 cm diameter (a diversity of shapes appears to be hydrodynamically efficient at low and very low velocities, depending on size; O'Dor and Wells, 1990; Jacobs, 1992a). Only a small proportion of Late Paleozoic ammonoids were mesodomic discocones (Fig. 4; morphotype I of Saunders and Swan, 1984) with moderate stability (20% of *Nautilus*). The deep hyponomic sinus lowered the hyponome somewhat from its hypothetical horizontal position to a ca. 60° orientation (8-o'clock position), so that the hyponome was in the correct position for backward swimming, i.e., approximately at the height of the center of buoyancy, preventing rotation (torque) of the shell. In this position, which is higher than that in *Nautilus* and prevents "rocking," the hyponome could not bend backward far enough to produce a horizontal jet. Hence, the animal could not swim forward (Fig. 5). On the other hand, this hyponome position would have been well suited for vertical jet propulsion, both up- and downward, i.e., swimming in all directions except forward. In contrast to *Nautilus* and other brevidomes, these ammonoids could also swim downward, but not forward.

In the Mesozoic, mesodomic discocones (without hyponomic sinus) were prevalent among the Phylloceratina and other "leiostraca" (usually oceanic smooth-shelled ammonoids, e.g., Haploceratidae). They also abound in the Ammonitina, where they tend to be costate, presumably to strengthen the shell, especially the peristome, against predators (Checa and Westermann, 1989; Ward, 1981; Westermann, 1990). Some shells (e.g., *Macrocephalites* macroconchs, Westermann and Callomon, 1988) are particularly well "streamlined" with the required adult modifications (see below) and have very fine, dense costae or lirae (e.g., *Phylloceras*) that could have reduced pressure drag (Chamberlain, 1991). Significantly, other shells (e.g., Cardioceratidae in Callomon, 1985; Ceratitidae in Dagys and Weitschat, 1993) displayed extremely high intraspecific variation, from spheroconic (or subcadiconic) to discoconic to inflated oxyconic, indicating the absence of selection on shell shapes that in other cases were adaptive. These animals displaying high intraspecific variation were presumably planktic (but not all similar morphotypes with small variation, as extrapolated by Dagys and Weitschat, 1993). Much needs to be done, and the examination of large samples is essential.

Mesodomic planorbicones comprised the majority of the Ceratitina and Ammonitina. Depressed and prominently sculpted whorls were poorly "streamlined" and were stable enough for only moderate hyponomic thrust, e.g., in sluggish animals under predator stress (Ward, 1986b). However ovate-compressed whorls with some ribbing provided for both moderate swimming and protection. Very fine ribbing and lirae actually may have reduced drag (Chamberlain, 1991). Some species have exceptional intraspecific variation in shape and sculpture (ornament), usually from finely sculpted and compressed-involute to coarsely sculpted and depressed-evolute morphs, conforming to the Buckman Rule ("Law") of Covariation (Westermann, 1966) (see Section 5). This implies that in these populations, shape and sculpture were nonfunctional.

3.1.1c. Longidomes. Most Late Paleozoic goniatites were spherocones, discocones, or cadicones (morphotypes V–VI of Saunders and Swan, 1984), with "long" body chambers and downward apertural orientation but minimal stability. The actual orientations of the organisms need not have coincided with those calculated from the parameters for liquid-free conchs with the body withdrawn into the body chamber. A modest change in the distribution of cameral liquid or in soft body extension would have markedly altered the poise. In *Nautilus*, with the tenfold stability of these goniatites, the filling of a single chamber rotates the shell by 20° (Hengsbach, 1978; Saunders and Shapiro, 1986). The downward aperture might be interpreted as indicating bottom feeding; however, the clustering around the stability minimum suggests that the longidomic spherocones and serpenticones were adapted to the rotation at will of the pelagic ammonoid, permitting varied feeding modes and slow multidirectional motion.

Longidomic spherocones and discocones less than 5 cm in diameter, even if optimally shaped for swimming (Jacobs, 1992a), could not have been active swimmers. Hydrostatic stability was insufficient for propulsion, even if the apertural orientation could have been regulated statically by cameral liquid or soft body shape, or altered dynamically by directing the hyponome. Consequently, steerage was minimal in any direction with the hyponome in the pushing mode, because any tangential propulsion force would simply have rotated the animal around its center of gravity (Bayer, 1982). However, only the unstable longidomic discocones, rather abundant in the Paleozoic (Fig. 4; morphotypes V and VI of Saunders and Shapiro, 1986), would have fulfilled the requirements for a coleoid-type body that could protrude greatly from the body chamber for efficient propulsion, as postulated by Jacobs and Landman (1993) (see Section 5.2).

Because of the inherent instability of longidomes and the consequent ineffectiveness of thrust, many spheroconic and, especially, discoconic ammonoids were presumably diurnal (and ?annual) vertical migrants, planktic, or/?and demersal. Upward thrust produced by the hyponome was modest, and sinking was achieved at least partly by slightly negative buoyancy, as in *Nautilus* (Ward, 1987).

Serpenticones were already relatively abundant in the Paleozoic (morphotypes IV and VIII of Saunders and Swan, 1984). They differed from the planorbicones significantly in the longidomic body chambers that also made them virtually unable to swim. They ranged from smooth compressed to more commonly costate depressed morphs and were presumably all planktic (passive drifters).

3.1.2. Orthocones, Cyrtocones, and Gyrocones

Late Paleozoic bactritoids (Mapes, 1979), recently placed in the Ammonoidea (House, 1981), lacked sufficient calcareous apical ballast for neutral equilibrium, so they must have had a stable vertical or subvertical poise. The only alternative for achieving neutral equilibrium (and horizontal poise), i.e., water ballast in the adapical chambers, can probably be excluded because the ratio of the length of the phragmocone to that of the body chamber in *Bactrites* was unusually small. Horizontal poise would have required the flooding of over half of the phragmocone and a ratio of 2 or more (Westermann, 1977). The mostly orthoconic bactritids generally had very thinly tapering conchs (i.e., extreme longicones) with circular whorl sections, widely spaced, strong septa permitting habitat depth to 350–400 m (see Chapter 10), and long body-chambers. They were presumably quasiplanktic vertical migrants in quiet waters with few predators.

Cyrtoconic bactritoids generally were more rapidly tapering (breviconic), somewhat compressed (Parabactritidae, Sinubactritidae), and had thinner (weaker) septa that were unevenly and densely spaced. This indicates much shallower habitats in some forms with ontogenetic depth change. Laterally compressed shells suggest very sluggish horizontal mobility of the vertically oriented bactritoid.

Abundant orthoconic to gyroconic heteromorphs were present in the Mesozoic (Figs. 2, 5). Orthocones, with ovate section, had extremely stable vertical orientation (in the general absence of liquid apical ballast; Ward, 1976a). They could move vertically at considerable speed, but backward swimming, including the escape of benthic feeders or saltators (Klinger, 1980) from benthic predators or anoxia, was without directional control. Ovate-compressed forms could, in addition, have moved slowly horizontally. The possible exception were some very large *Baculites* with greatly increased phragmocone/body chamber ratio (from ca. 1 to ca. 2) (see Section 4.4.2), indicating the presence of sufficient apical water ballast for neutral equilibrium that would have permitted quite rapid horizontal forward swimming with good steerage.

Cyrtocones and gyrocones were also stable; their apertures pointed more or less horizontally without the need for apical ballast. Their potential swimming in all directions was restricted by high drag.

3.1.3. Other Heteromorphs

In the scaphitids, the change from planispiral to a hooked shaft is located near the adapertural end of the adult phragmocone. Stability was greatly increased, and the adult aperture directed upward—a combination that cannot be achieved in planispirals (see above); "streamlining" in most adult scaphitids was better vertically than horizontally. This aperture orientation permitted unhindered propulsion upward as well as downward (in contrast to *Nautilus* and most ammonites, in which the arms were in the way of an upward jet; we safely assume a ventromarginal hyponome as discussed above). Prevailing vertical motion is strongly indicated. The latest (Maastrichtian) scaphitids (e.g., *Hoploscaphites*), however, are strongly compressed and densely costate, indicating good horizontal mobility also.

The life orientation of many of the diverse extreme heteromorphs was investigated in experiments with models and by computer simulation; all investigated forms were found to have been neutrally or slightly positively buoyant (Ward, 1976a; Ward and Westermann, 1977; Okamoto, 1984, 1988a,b). Orientation, as evident especially from theoretical morphology (Chapter 8, this volume), was vertical according to the coiling axis and very stable in the high-spired (trochoid) torticones and similar but less stable in the low-spired (helical), sometimes open torticones (Fig. 5). The torticones were also relatively smooth, indicating vertical mobility (Ward, 1976a), either demersal with rapid escape (Klinger, 1980) or planktic; an upturned aperture (*Didymoceras*) would have permitted propulsion for downward swimming (rather than sinking).

Ancylocones, with gyroconic phragmocones, resembled scaphitids in their adult orientation, stability, and directional "streamlining"; some had long spines, presumably for defense against nektic predators, indicating that they were quasiplankton in the photic zone (but see Batt, 1993). The hamitocones, with straight shafts and U-shaped connectives (*Hamites, Ptychoceras, Polyptychoceras*), have alternating stable and unstable growth stages separated by complete "flip-overs" in orientation (Ward, 1976a). They must have been planktic (and the shell external or semiinternal, see below). Vermicones (*Nipponites*) had an even more complicated series of recurrent stages that required repeated "flip-overs" during growth (Ward and Westermann, 1977; Okamoto, 1984, 1988c), and other genera (*Glyptoxoceras, Eubostrychoceras*) had multiphasic growth of entirely different morphs (Ward, 1976a; Okamoto, 1988b). All were planktic and probably mid- to deep epipelagic.

3.2. Adult Modifications and Functional Significance

Adult or mature modifications, so characteristic for Mesozoic ammonites (see Callomon, 1981) but also developed in many Paleozoic taxa (Davis, 1972), have been reviewed repeatedly, especially in discussions relating to sexual dimorphism (Chapter 13, this volume). They are here interpreted as functional changes influencing the last, presumably increasingly specialized and often extended period (sometimes years?) of adult life. In contrast to extant cephalopods, this final growth stage is commonly well differentiated, at least in epicontinental forms. Any combination of features may be affected, often in surprisingly dissimilar ways between the sexes (micro- and macroconchs). Remarkably, this modified stage is usually coincident with the ultimate body chamber, in planispiral forms as well as in many heteromorphs. The adult modifications are here interpreted either to improve or to alter existing juvenile adaptations by the following shifts in functional designs (Figs. 1–3):

1. Improved hydrostatic stability (a) by reducing the body chamber angle ("length") (widespread in most morphogroups) and/or (b) by lowering the center of mass relative to the center of buoyancy. This is achieved by (b') gradual or sudden (elliptical), partial or complete uncoiling [common in scaphitids and ancylocones, both dimorphs of platycones,

discocones, and spherocones (i.e., elliptospherocones) and in platyconic and oxyconic microconchs], or by (b″) modifying the whorl
section by broadening and rounding previously acute or keeled venters
near the end of the body chamber (platyconic and oxyconic macroconchs), or (c) by terminal thickening of the shell (varix) (widespread,
e.g., planorbicones, spherocones).

2. Increased acceleration, speed, and steerage (a) by reducing drag,
   through (a′) narrowing and reshaping the umbilical area (Elmi, 1993)
   (platyconic and oxyconic macroconchs) or (a″) narrowing (compressing) the whorl section (e.g., some spherocones become discocones) or
   (b) by developing an acute or narrowly rounded venter at the "leading
   edge" of the backward-swimming organism (some spherocones) and/or
   (c) by reducing the amplitude of sculpture or by deleting it (many large
   macroconchs of most shapes).
3. Improved protection against predators (a) by coarsening sculpture and
   (b) peristomal constriction and shell thickening (varix) as well as
   protuberances (lappets, rostra, horns), especially in microconchs.
4. Change of propulsion direction by rotation of aperture (a) from vertical
   to horizontal propulsion by reduction of the body chamber angle from
   longidomic to mesodomic (many planorbids and spherocones), or (b)
   from horizontal to vertical propulsion (up and down) by 180° hook
   (ancylocones, scaphitids), and/or by (c) "fine-tuning" of orientation by
   apertural shell thickening (and ? shape of protuberances) for lowering
   of the aperture (widespread).

The following late ontogenetic trends of function, habit, and habitat
changes were prevalent:

1. From planktic (or demersal vertical migrant) to sluggish nektic or
   demersal swimmer—mainly modifications 1a,c, 2, 4a above. Common
   in planorbicones (Stephanoceratidae, Perisphinctaceae), spherocones,
   and discocones (Haloritidae, Lobitidae, Sphaeroceratidae). The listed
   functions as well as shell diameter, itself an important factor for speed
   (Chapter 7, this volume) and protection, imply basic differences in
   adult modes of life between typically differentiated dimorphs.
2. Increased diurnal vertical migration of planktic or demersal forms, with
   the potential for mesopelagic and mesobenthic (>240 m) forays; alternatively to remain at constant depth during vertical water movement—
   mainly modification 1b′ above. Common among heteromorphs
   (Ancyloceratidae, Macroscaphitidae, Scaphitidae) and cadicones and
   spherocones (Tulitidae); may affect mainly microconchs.
3. Increased speed (including acceleration) of nektic or demersal forms;
   also evident from low variation (Tanabe and Shigeta, 1987)—mainly
   modifications 1b″ (i.e., higher stability), 2a′, c above. Common among
   "streamlined" platycones and brevidomic oxycones; mainly macroconchs; some with apparent large muscle scars (but see Jacobs and

Landman, 1993), implying strong funnel action (Chapter 3, this volume)
(Oxynoticeratidae, Hildoceratidae, many Haplocerat0aceae, Platylen-
ticeratidae, Desmoceratidae).

4. From sluggish to mobile, nektic or demersal; improvement of speed by
"streamlining" is similar as in 3, but the adult stage follows a sculpted,
quasiplanktic juvenile stage with high variation (Tanabe and Shigeta,
1987)—modifications as in 3 plus 2c above, restricted to macroconchs.

5. An ontogenetic trend opposite to 4 is prevalent in many of the much
smaller microconchs; they retain the juvenile sculpture, enlarge the
umbilicus, and usually develop drag-producing apertural modifications (Chapter 13, this volume). Common in platycones and oxycones
(Graphoceratidae, Sonniniidae, many Haploceratoaceae).

3.3. Bathymetry

The ammonoid shell wall and septa formed an integral structural "design"
(Chapter 10, this volume). The macrostructural ("architectural") interrelations
between them are not easily discernible in Paleozoic ammonoids but become
obvious in their Mesozoic descendants (Westermann, 1958; Bayer, 1977;
Jacobs, 1992b; Hewitt and Westermann, 1987). There was a great variety of
septal shapes (and septal sutures) in Paleozoic ammonoids, varying in large
part independently of conch shape. Limited functional correlations (covaria-
tions) among septa, sutures, and whorl shape nevertheless existed (for
clymeniids, cf. Korn, 1992) and need to be considered. Overall conch shape
was related to phragmocone wall resistance against ambient pressure, hydrodynamic efficiency, and protection from predators, as discussed above.

3.3.1. Phragmocone Wall

The hydrostatically strongest shells of orthocones and advolute planispirals had circular sections; those of globular conchs had hemicircular sections.
Somewhat weaker are elliptical and ovate whorl sections, which compromise
between hydrodynamic and hydrostatic efficiencies. They lose strength as the
curvature radii of the flanks increase, i.e., the section becomes more compressed for improved "streamlining" and steerage. The well-known amazing
diversity of whorl sections, especially in the Otoceratinae and Ammonitina,
including flat or even concave regions of the phragmocone wall, resulted from
a multitude of adaptations, still poorly understood, that were made structurally possible only by increasingly specialized septa and sutures (Westermann,
1958, 1975; Ward and Westermann, 1985; Jacobs, 1992b; Westermann and
Hewitt, in prep.).

Estimates of the ambient hydrostatic pressure under which the phragmocone failed may be based theoretically on the thickness and curvature of the
shell wall (together with septal spacing, see Hewitt and Westermann, 1987),

but shallow-water ectocochliates required thick shells mainly for protection from predators (Westermann, 1977; Bond and Saunders, 1989).

3.3.2. Septa and Sutures

The most valid strength calculations are based on the septal flutes (Chapter 10, this volume). At any growth stage the last septum withstood the hydrostatic pressure transmitted through the soft body. Although the simplest ammonoid septa, present in the Agoniatitida and Clymeniida, have the shape of part of a torus (e.g., Bactritina, *Cyrtoclymenia*, Korn, 1992) or of part of a cylinder or a combination of the two (e.g., Agoniatitina), the stress in the fluted septa of other orders is more difficult to calculate. The septal flute strength index (a function of maximum septal thickness, minimum whorl radius, and tensile strength), applied earlier to Mesozoic ammonoids (Hewitt and Westermann, 1987; Wang and Westermann, 1993), now has been extended in a more advanced form also to Paleozoic ammonoids, together with recently developed statistical methods to estimate habitat depth limits (Hewitt and Westermann, 1990a; Chapter 10, this volume). This has resulted in markedly reduced estimates for most epicontinental ammonoids.

Habitat depth also is to some degree reflected in the much-discussed grade of complication of the septal sutures (but see, e.g., Saunders, 1995; cf. Hewitt and Westermann, in press); in at least the Ammonitina and Lytoceratina, there probably existed an overall trend of increased "frilling" with depth of habitat. Long ago this was interpreted, mostly correctly, as improving wall support to carry the ambient hydrostatic load (cf. Jacobs, 1992b). This hypothesis was extended to septal shape and, especially, to the spacing between septa that needs to be minimized for strength of the shell wall (Westermann, 1958, 1975). Although septal complication and spacing appear to be useful for bathymetry (the study of depth distribution), much caution is advised, especially in comparing different higher taxa and shell diameters (Ward and Westermann, 1985). The ontogenetic increase of sutural complication, measured by the ratio of curvature length to overall length (span) (Westermann, 1971), is not a linear (exponential or otherwise) function of shell diameter. Therefore, it is also incorrect to use the ratio of complication/span-squared, as applied by Hirano *et al.* (1990) to a Cretaceous ammonoid clade (see Section 5.6.7). Nor has fractal analysis been of much help here, because the resulting line is significantly curved so that no single fractal dimension is valid, as was recently supposed by Boyajian and Lutz (1992). (These authors also misrepresented sutural adaptations by mixing up all ammonoid orders in their analysis of sutural complexity, which they equated with biological complexity; they proceeded to correlate sutural complexity with supposed generic longevity, which they presumed ended by extinction—rather, genera are artificial subdivisions of evolving clades.)

Another, much simpler parameter, proposed by Batt (1989), is the sutural amplitude index, i.e., the ratio of deepest lobe (usually the depth of the lateral lobe L) to total external span. This index is limited, however, to similar whorl sections, because sutural amplitude depends on the curvature of the septal flute axis that it terminates. Amplitude then expresses the degree of convexity of the flute axis: it is highly curved in strong septa and weakly curved in weak septa (Chapter 10, this volume). The current views on the "sutural complexity" problem are presently summarized and discussed by Westermann and Hewitt (in prep).

3.3.3. Septal Necks and Siphuncles

The septal neck is another shell feature probably related to habitat depth (Doguzhaeva, 1988; Tanabe *et al.*, 1993a). Extended necks occur in deep-sea taxa, with high siphuncular strength Index (SiSI) and presumably prevented flooding (Tanabe *et al.*, 1993a; Chapter 10, this volume). The application of the SiSI (Westermann, 1971, 1982) to bathymetry has now been modified to include a constant low tensile strength of connecting ring tissue (Chapters 6 and 10, this volume). The consistent, parallel ontogenetic reduction of SiSI and septal flute strength indicates that the original assumption of Westermann (1971) was correct after all: connecting rings became weaker ontogenetically (in contrast to *Nautilus*).

3.4. Buoyancy and Soft Body Reconstructions

The overwhelming majority of calculations and experiments have shown that postembryonic ammonoids were approximately neutrally buoyant. There are no indications for a snail-like benthic habit as alleged by Ebel (1983).

Buoyancy calculations and experiments have been based on the measurements of shell thickness; the relative volumes of protoconch and body chamber for the complete ammonitella and those of phragmocone and body chamber for the later stages; and on the reasonable assumptions that protoconch and camerae were empty, soft-body volume equaled that of the body chamber (nobody assumed that the soft body could not temporarily protrude from the body chamber as asserted by Jacobs and Landman, 1993), and shell and body densities equaled those in extant cephalopods (Saunders and Swan, 1984; Saunders and Shapiro, 1986; Tanabe and Ohtsuka, 1985; Landman, 1987; Landman and Waage, 1993; Okamoto, 1988a; Shigeta, 1993). Ward (1976a) carried out experiments with scale models of wax with mass added to reproduce the calculated density differentials between phragmocones, body chambers, and sea water in the aquarium. The few erroneous data and/or calculations that resulted in negative buoyancies (Ebel, 1983; Shigeta, 1993) are discussed below and have been rejected (Westermann, 1993).

The shell was external or, exceptionally, semiinternal, with a thin mantle covering the juvenile shell or body chamber only (*Gaudryceras, Ptychoceras*) (Birkelund, 1981; Doguzhaeva and Mutvei, 1989); it was never fully internal and belemnite-like, as suggested by Doguzhaeva and Mutvei (1989, 1993) (see Section 5). The soft parts could usually be withdrawn entirely into the body chamber, as evident from buccal mass remnants (aptychi, radulae) found well back inside the body chamber (Lehmann, 1976; Mapes, 1987; Nixon, 1988). In some forms, soft body extension from the body chamber was constrained, as evident from converging lappets [*Stephanoceras* (*Epalxites*), see Westermann, 1990] or the impingement of the adult aperture on earlier parts of the shell (scaphitid ancylocones). The possible existence of extensively and permanently protruding bodies requires further study, including buoyancy calculations.

The buccal mass (aptychi/mandibles and radulae), crop/stomach, and dietary habits are discussed elsewhere (Chapter 2, this volume) and in Section 5, below.

4. Evidence from Faunal Associations and Biofacies

In this section, associations and biofacies are discussed, proceeding from the oceans to the continental slope, deep structural basins, and continental shelf to epeiric basins that were partly oxygen-depleted and to shallow-water, carbonate platforms.

4.1. Ocean Floor

Little has been left of Jurassic and Early Cretaceous ocean sediments because of subduction, so that even small finds are significant. Evidence for the deepest ammonoid occurrences is based almost exclusively on aptychi from Late Jurassic–Early Cretaceous deep-sea drill cores of the North American Basin in the early North Atlantic and the Magellan Rise in the central Pacific (Jansa *et al.*, 1979; Renz, 1972, 1973, 1978, 1979). The relatively thick calcitic aptychi were preserved, but not chitinous anaptychi or the aragonitic ammonoid shells. The exception is several small, incomplete, almost smooth Aspidoceratidae (?*Physodoceras*, identified as *Aspidoceras* but missing lateral nodes) that presumably were partly oceanic (Hantzpergue in Dommergues *et al.*, 1989), Haploceratidae (or ?Lissoceratidae, i.e., ?*Neolissoceras*), as well as poorly preserved heteromorphs. *Physodoceras* also occurred in shallow epeiric seas, where it probably was at least partly demersal (see 4.5.1). It is conceivable that some or all of the ammonoid shells drifted postmortem into the narrow ocean, especially after damage to the body chamber; in contrast, the heavy aptychi would have fallen out of the body chamber within a day or two of death (Fig. 6).

The following oceanic aptychi were described and assigned to Late Jurassic–Early Cretaceous taxa (Renz, 1972, 1973, 1978, 1979): occasionally abundant *Lamellaptychus* [Haploceratidae, ?Lissoceratidae, ?Oppeliidae]; rare *Punctaptychus* (Haploceratidae), *Laevaptychus* (Aspidoceratidae), and *Praestriaptychus* (Stephanocerataceae and Perisphinctaceae). Their presence is good evidence that the abundant haploceratids were partly also truly pelagic in the ocean, although not necessarily very distant from shore. An epiconti-

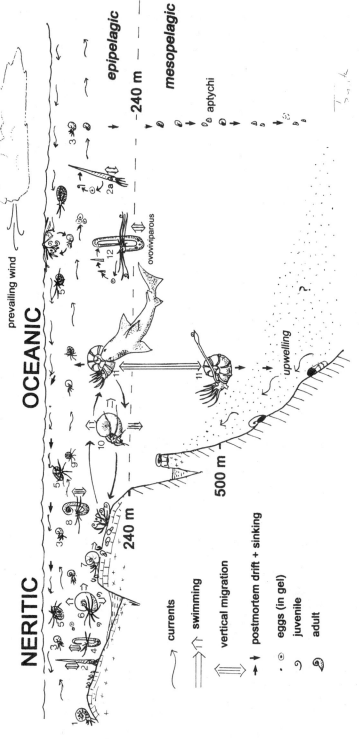

FIGURE 6. Ocean floor and margin. Possible scenario of Jurassic–Cretaceous ammonoid habitats; prevalent habits and several life cycles are indicated. Postmortem sinking occurred mostly at greater depth, especially in mesopelagic habitats, as well as in juveniles and small species. Postmortem ascent, followed by surface drift, happened mostly in shallow habitats and in adults of larger species. (Note also that some dying ammonoids may have ascended to the surface, as is known from extant *Nautilus*, *Spirula*, and some sepiids; R. Hewitt, personal communication, 1993). Ammonitina: 1, *Stephanoceras*; 3, *Haploceras*; 5, *Dactylioceras*; 6, *Macrocephalites*; 7, *Desmoceras*; 9, *Psiloceras*; Ancyloceratina: 2, *Baculites* (a, pelagic form); 4, *Ptychites*; 8, *Ancyloceras*; 12, *Pseudoxybeloceras*; Phylloceratina: 10, *Phylloceras*; Lytoceratina: 11, *Lytoceras*.

nental pelagic habitat was assumed previously for these small, smooth platy-cones and discocones, especially for *Haploceras* (Ziegler, 1967). The very rare *Praestriaptychus* and *Laevaptychus*, however, may have been postmortem drifts, trapped in the meso- to longidomic body chambers of perisphinctids or stephanoceratids.

Abundant isolated aptychi occur in the Jurassic–Cretaceous flysh and radiolarites of the western Carpathian Mountains in Poland (Gasiorowski, 1962) and in the Upper Jurassic of the Betic Cordillera, Spain (J. Sandoval, personal communication, 1993). Again, the haploceratacean aptychi *Lamellaptychus*, *Laevilamellaptychus*, and *Punctaptychus* are by far the most abundant; the others are *Cornaptychus* (Hildoceratidae, Sonniniidae, and Hecticoceratidae) and the aspidoceratid *Laevaptychus*. This suggests an oceanopelagic habitat at least for part of their lives.

There is no evidence in these deposits of the many other taxa believed to have been pelagic, especially oceanopelagic. The Lytoceratida, Phylloceratida, and Desmocerataceae had chitinous anaptychi/mandibles, as did the Dactylioceratidae, Psiloceratidae, Placenticeratidae, and others.

4.2. Ocean Margin

4.2.1. Deep Structural Rises and Basins

In the Jurassic, the well-known affinity of the leiostraca to oceans, distal platforms, and structural rises as well as to deep basins near oceans has been confirmed by many contemporary authors, e.g., Oloriz *et al.* (1991), who distinguished between Platform Ambitus, defined by the absence of the leiostraca Lytoceratina and Phylloceratina, and Basin Ambitus, defined by their abundant presence (Fig. 6).

The classic example of deep distal platforms (structural rises) is the locally developed Mediterranean facies of nodular Ammonitico–Rosso (Fourcade *et al.*, 1991), known especially from southern Poland, Hungary, Italian Alps, and the Betic Cordillera of southern Spain (Geczy, 1982, 1984; Galacz and Horvath, 1985; Clarim *et al.*, 1984; Bruna and Martire, 1985; Cecca, 1992; Oloriz, 1976; Sandoval, 1983). Recognition of hydrodynamic postmortem sorting in condensed assemblages is essential, e.g., Triassic Hallstatt and Han Bulog Limestones (H. Bucher, personal communications, 1993). In the Pliensbachian (Davoei Zone) of Hungary, the proportion of leiostraca among the ammonoids increases with depth, from 66% in cherty beds with silicisponges to 82% in typical Ammonitico–Rosso facies, compared to only 23% in the shallower, bioclastic "Hierlatz facies." Phylloceratina consistently outnumber Lytoceratina, especially in the relatively shallow facies. Among the former, Phylloceratidae outnumber Juraphyllitidae 4:1 to 5:1 in the deepest Ammonitico–Rosso facies; but the proportions are reversed in the shallower "Hierlatz facies," where the mainly epicontinental Juraphyllitidae dominate 3:2. The Ammonitina in this facies consist almost entirely of mobile platycones

(*Fuciniceras, Protogrammoceras*) and planktic serpenticones (*Prodactylio-ceras*). Taxonomic compositions are similar in the Toarcian–Aalenian levels, where leiostraca increase from minimally 50% in marly facies to 80% in typical Ammonitico–Rosso. Phylloceratina again outnumber Lytoceratina, but both occur here together with dactylioceratids and phymatoceratids. The Ammonitina are also dominated by platycones and oxycones (Hildoceratidae, Phymatoceratidae), presumably swimmers, and also include serpenticones (Dactylioceratidae), presumably planktic drifters.

In the mid-Jurassic Ammonitico–Rosso of Hungary, the Ammonitina constitute only 10–40% of the ammonoid fauna, and microconchs are very rare or absent. Again, Phylloceratina greatly outnumber Lytoceratina (2.5:1 to 50:1), independently of the leiostraca/trachyostraca ratio and total abundance. The small, smooth, discoconic *Lissoceras*, another leiostracan, is present in most beds where Lytoceratina comprise more than 5% of leiostraca; and the small, planktic or pseudoplanktic heteromorph *Spiroceras* is present in beds with more than 20% Lytoceratina. Similar conditions existed in Spain, where Ammonitina and Lytoceratina ratios were, in part, higher than in Hungary. This suggests that the lytoceratines as well as *Spiroceras* and probably *Lissoceras* were passively bound to water masses that occasionally invaded the area, whereas the more ubiquitous phylloceratids were more mobile.

The deep, nodular facies of the Rosso Ammonitico Inferiore (Aalenian–Bajocian) of the Alpe Feltrini, Italy, and some localities in the Betic Cordillera have yielded a fauna (not quantified) of diverse Phylloceratina together with some Lytoceratina and the following Ammonitina: oxyconic Leioceratinae and Strigoceratidae; oxyconic to platyconic Sonniniidae, Graphoceratidae, and Hammatoceratidae; discoconic Haploceratidae; elliptospheroconic Hammatoceratidae, Otoitidae, and Cadominitinae; diverse sculpted, planorbiconic as well as serpenticonic Stephanoceratinae; and a few Leptosphinctinae. At least the last two generally have been considered as demersal and are, therefore, of special interest in this pelagic facies. The cosmopolitan Stephanoceratinae are here represented by longidomic serpenticones [*Skirroceras, Kumatostephanus*, and the group of *"Docidoceras" longalvum* (Vacek)] as well as heavily noded brevi- to mesodomic planorbicones and cadicones (e.g., *Teloceras*). The former were presumably planktic and, significantly, occur mainly in the deep, nodular facies; the latter, probably demersal, occur mainly in the shallower, condensed interbeds with hard ground and stromatolites, together with perisphinctids. The overwhelmingly pelagic aspect of the nodular Ammonitico–Rosso is confirmed.

An interesting recurrent community pattern was observed in the deep part of the Atlas Basin of Morocco (Sadki and Elmi, 1991; Fig. 7). The marls-and-carbonates succession of the lower mid-Jurassic (Concava and Discites Standard Zones) includes three repetitions of a rare type of community replacement, i.e., essentially between faunas dominated by either graphoceratids (Ammonitina) or phylloceratids (Phylloceratina). But there are no significant lithofacies associations. The "leiostracan" phylloceratids, found most abun-

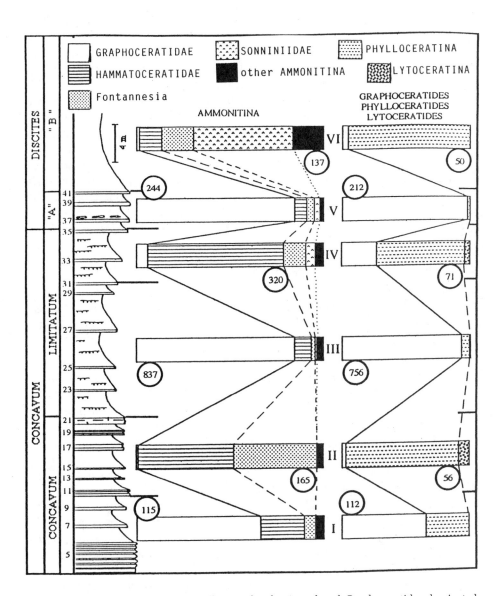

FIGURE 7. Alternation between Phylloceratidae-dominated and Graphoceratidae-dominated ammonoid communities in Aalenian–Bajocian marls of the deep Atlas Basin of Morocco. Phylloceratids are associated with hammatoceratids. It appears that tectonic basin-dynamics repeatedly opened and closed a barrier between the oceanic embayment and the near Mediterranean, permitting the exchange of basinal (cool?) and warm equatorial water masses and their ammonoid faunas. (Reproduced with slight modification from Sadki and Elmi, 1991.)

dantly in the Mediterranean Bioprovince, are associated mainly with hamma-toceratids that are best known from carbonate-platform facies (?demersal) or, in higher beds, with the nektic sonniniids (Ammonitina). The cause of the repeated faunal change was presumably related to basin dynamics, e.g., repeated tectonic opening and closing of barriers between the deep basin and the near Mediterranean ocean with its tropical water masses.

The early Tithonian ammonoids of the pelagic deposits of southern Poland, Hungary, and northern Italy, according to Cecca (1992), were mostly demersal swimmers, from about 100-m depth (*Lytogyroceras, Pseudolissoceras, Hybonoticeras, Virgatosimoceras, Simoceras*) to 400-m depth (*Holcophylloceras, Haploceras*), although the former include several characteristic serpenticones (e.g., *Lytogyroceras*) for which, as well as others, a planktic habitat appears much more likely; only three genera were considered as pelagic, i.e., the gyroconic *Protancyloceras* as epipelagic, *Ptychophylloceras* and *Protetragonites* as mesopelagic, and two, *Aspidoceras* and *Subplanitoides*, as intermediate in habitat.

4.2.2. Distal Shelves and Upper Slopes

4.2.2a. Triassic. In China (Wang and Westermann, 1993), deep outer shelf and upper slope are represented by the following ammonoid "communities" (not in bathymetric order). All or most were pelagic, ranging from planktic serpenticones and spherocones to mobile discocones and oxycones:

1. Bathyal facies in Tibet with the *Owenites* Community (Spathian): mostly smooth spherocones and discocones; elsewhere in shelf facies.
2. Pre-reef-complex slope with the *Paraceratites-Bulogites* and *Balatonites* Communities (Anisian): diverse brachiopods, gastropods, and ammonoids. Dominant were costate–nodose platycones (*Paraceratites, Bulogites, Balatonites*) and discocones with subammonitic sutures (*Beyrichites, Nicomedites, Cuccoceras*); accessories were small serpenticonic ceratites with prominent simple costae (*Paracrochordiceras*) and small, smooth, serpenticonic phylloceratids (*Leiophyllites*).
3. Deep basin with the *Procarnites* Community (Spathian): 73% of the ammonoids are smooth or finely sculpted, and 70% have relatively complicated sutures. Phylloceratinae are represented by serpenticones (*Leiophyllites*) and discocones (*Eophyllites*); among Ceratitina are discocones with subammonitic sutures (*Procarnites*).
4. Deep basin with *Placites–Discophyllites–Arcestes–Cladiscites* Community (Norian) with Hallstatt-type (deep-water) brachiopod assemblage: smooth or finely sculpted discocones, spherocones, and oxycones, with relatively complicated sutures and moderately strong septa and siphuncles (*Cladiscites* with a maximum habitat depth of 80 m, and *Arcestes* at 85–270 m: Chapter 10, this volume).

These Triassic offshore ammonoids from China were mostly, if not entirely, epipelagic. Only *Rhacophyllites* and some *Arcestes* could possibly have lived below 240 m, i.e., have been mesopelagic, in contrast to previous interpretations (Wang and Westermann, 1993).

4.2.2b. Jurassic. Early Tithonian ammonoids from several Mediterranean deep-water, Ammonitico–Rosso facies and from a sub-Mediterranean platform-margin facies illustrate the distributions of the mainly epicontinental Ammonitina (including "trachyostraca," i.e., sculpted forms) versus the mainly oceanic leiostraca (Lytoceratina, Phylloceratina, and some smooth Ammonitina) (Cecca, 1992; and above).

Mediterranean Bioprovince. Basin swells (Umbria and Marche, Italy), 150–200 m, condensed micritic limestone with Perisphinctaceae 35–40%, Lytoceratina and Phylloceratina 35–45%, Haploceratidae 5–40%, Simoceratidae 5–35%, Oppeliidae 8–10%.

Deep structural basin (Rogoznik, Poland), 300–550 m, condensed pink-red ammonite coquina with Haploceratidae 30–70%, Lytoceratina and Phylloceratina 20–25%, Perisphinctaceae 7–30%, orthoconic Bochianitidae 15–20%.

Deep structural basin (Gerecse, Hungary), 400–700 m, Ammonitico–Rosso with Lytoceratina and Phylloceratina 40–60%, Haploceratidae 0–60%, Perisphinctaceae 10–50%, Oppeliidae 1–20%, Simoceratidae 1–20%.

Sub-Mediterranean Bioprovince. Deep shelf margin (SE France), 200–250 m, massive limestone with chert, gray nodular limestone and turbidites with Phylloceratina plus Lytoceratina, Haploceratidae, and Perisphinctaceae in similar proportions.

In conclusion, during the Late Jurassic:

1. In the oceanic area, i.e., the Mediterranean Province (defined by lytoceratid and phylloceratid dominance), Lytoceratina usually outnumber Phylloceratina; and this is reversed on the shelf margin, i.e., the sub-Mediterranean Bioprovince, which is dominated by the Ammonitina (mostly trachyostraca). Both leiostracan suborders are represented by deep-water genera with highly complicated sutures, e.g., *Ptychophylloceras* (high septal and siphuncular strength, Chapter 10, this volume, Table III) and *Protetragonites*, respectively.

2. Except for this leiostracan reversal, the shelf-margin fauna is much like the basin-swell fauna in the westernmost reaches of the Tethys ocean.

3. The smooth-discoconic Haploceratidae (leiostraca), represented mainly by the small *Haploceras* (moderately complicated suture) and *Pseudolissoceras* (simple suture), likely were nektonic in the warm, epicontinental sea as well as in the ocean; all leiostraca tend to become more abundant with increasing water depth, although with much variation.

4. Relative abundance of Haploceratidae tends to be negatively correlated with that of the diverse, oxyconic Oppeliidae (nekton and ?demersal swimmers), the serpenticonic Simoceratidae (*Simoceras*, with simple

suture, in shallow midwater, perhaps also oceanic; Geyssant, 1988), and the compressed-planorbiconic Ataxioceratidae (?demersal swimmers).

5. The first Ancyloceratina heteromorphs, i.e., the orthoconic to cyrtoconic simoceratid *Lytogyroceras* (epiplankton), are recorded in significant numbers only from Rogoznic.

4.2.2c. Cretaceous. The desmerceratacean Ammonitina and some heteromorph Ancyloceratina accompany the Phylloceratina and Lytoceratina in the deep-water habitat. The desmoceratids probably were demersal swimmers on the distal shelf and upper continental slope, i.e., intermediate in habitat depth between characteristic trachyostraca and leiostraca. Occurrences in shallow-shelf facies represent postmortem drift, including the desmoceratids with limpet pits recently given as evidence for a shallow habitat by Kase *et al.* (1994) and rejected by Westermann and Hewitt (1995). This is evident from several analyses of habitat and shell structure (Bulot, 1993; Hewitt and Westermann, 1987; Hirano, 1986; Tanabe, 1977, 1979; Tanabe *et al.*, 1978; Ward, 1976b; Westermann, 1971, 1982, 1990).

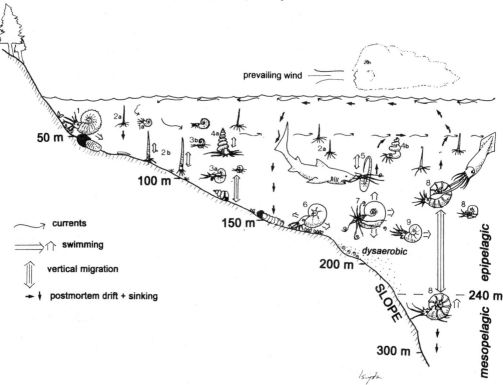

FIGURE 8. Distal shelf and upper slope. Possible scenario of Hokkaido, Japan, during the Late Cretaceous; prevalent habits and postmortem transportation indicated (see text). Ammonitina: 1, Acanthoceratidae and Collignoniceratidae (a, juvenile); 6, Desmoceratidae. Ancyloceratina: 2, Baculitidae; a, *Sciponoceras*; b, *Baculites*; 3, Scaphitidae; a, *Scaphites*; b, *Otoscaphites*; 4, Turrilitidae; 5, Hamitidae. Phylloceratina: 7, Phylloceratidae. Lytoceratina: 8, Lytoceratidae.

Of particular interest here is the Turonian of Hokkaido, Japan (Fig. 8). The well-documented neritic–oceanic gradient is paralleled by a succession of ammonoid biofacies. From shallow to deep, with indication of relative abundance, they are:

1. Above 100 m—almost exclusively trachyostracans; i.e., heavily sculpted spherocones and planorbicones (Acanthoceratidae, Collignoniceratidae) greatly outnumber orthoconic and scaphitoconic heteromorphs (Baculitidae; *Sciponoceras* outnumber Scaphitidae) and few leiostraca.

2. About 100 m to 150–200 m—all morphogroups, i.e., leiostraca with advolute-planorbiconic to platyconic Lytoceratina (Tetragonitidae, Gaudryceratidae) outnumber costate discoconic Desmoceratacea (Puzosiinae outnumber Desmoceratinae), which, in turn, outnumber smooth discoconic Phylloceratina (Phylloceratidae); heteromorphs include orthocones, scaphitocones, hamitocones, torticones, and helicones (Baculitidae, Diplomoceratidae, and Scaphitidae in similar numbers); and trachyostraca as before.

3. Below 150–200 m—leiostraca dominant over heteromorphs; i.e., Lytoceratina (as before) are as abundant as Desmocerataceae (Desmoceratinae outnumber Puzoniinae); Phylloceratina as before; heteromorphs including scaphitocones greatly outnumber orthocones and ancylocones as before.

In conclusion, the small baculitid *Sciponoceras* with circular section occurs in all facies, consistent with a planktic habitat and postmortem drift. Hamitocones, torticones, and helicones are present only in the oceanic facies of 2 and 3, where they were epipelagic, according to the low siphuncular strength (Tanabe, 1979). The widespread association of *Scaphites* and *Otoscaphites*, their greatest abundance in the moderately deep lithofacies 2, and the modest strength of larger juvenile shells suggest that at least the juveniles of both genera (or dimorphs; Tanabe, 1977) were epipelagic. Among the desmoceratids, the fine-ribbed and evolute puzosiines dominate in 2, and the smooth and involute (discoconic) desmoceratines in 3. As in the warm waters of the Jurassic Tethys (Mediterranean Province), the nektic phylloceratids are greatly outnumbered by the planktic lytoceratids. Moreover, there is a consistent contrast in preservation between the leiostracans and all other forms; almost every phylloceratid, tetragonitid (Lytoceratina), and desmoceratid, in different facies, had an incomplete body chamber, suggesting that they were particularly prone to postmortem drift (see above) resulting in taphonomic "telescoping" of the oceanic and neritic faunas. Their body chambers were thin-shelled (for phylloceratids, see Birkelund, 1981) and broken either by predation or during postmortem transportation. The Ammonitina remained largely intact and with their buccal mass in the body chamber, indicating rapid sinking and deposition after death (K. Tanabe, personal communication, 1993).

In the Albian of Poland, shallow and deep facies differed similarly (Marcinowski and Wiedmann, 1985). The shallow-water facies had 99% characteristic trachyostraca (Hoplitidae greatly outnumber Brancoceratidae) and only 1% Desmoceratidae (no Lytoceratina or Phylloceratina). The deep-water facies has 54% varied leiostraca (Desmoceratidae outnumber Lytoceratina, which outnumber Phylloceratina), 34% torticones, ancylocones, and a few scaphitocones (in order of abundance: Turrilitidae, Hamitidae, Anisoceratidae, Scaphitidae), and only 12% trachyostraca (as above). The shallow-water assemblage was mostly mobile demersal; the deep-water assemblage was largely pelagic, including the majority of the planktic heteromorphs, which had an exceptionally wide distribution. The opinion of Marcinowski and Wiedmann (1985) that these heteromorphs were partly benthic (crawlers?) is therefore doubtful. Furthermore, the scaphitid *Eoscaphites* had circular, advolute juvenile whorls followed by a large hook-like adult body chamber, resembling Batt's (1989) planktic morphogroup 16 (see below).

4.2.2d. The Last Ammonoids. The most complete and best-known Maastrichtian–Danian sections containing ammonoids are from continental slope deposits along the sea cliffs of the southeastern Bay of Biscay in the Basque–Cantabrian Basin (Ward, 1990a,b; Ward *et al.*, 1991; Ward and Kennedy, 1993; Wiedmann, 1988a). The great majority of Maastrichtian ammonoid species and all that are known to continue within decimeters of the K/T boundary were oceanic to suboceanic (distal neritic), deep epipelagic to mesopelagic nekton and plankton (Ward, 1987; Wiedmann, 1988a). Evolutionary rates were exceptionally low, and cosmopolitanism high, features characteristic of oceanic deep-water taxa (Ward and Signor, 1983).

From the uppermost few meters of the Cretaceous (upper *Anapachydiscus terminus* Zone), the following seven to nine genera (with eight to ten species) have been documented: *Phylloceras*, *Pseudophyllites*, ?*Zelandites*, ?*Anagaudryceras*, *Pseudokossmaticeras*, *Brahmaites*, *Pachydiscus*, *Anapachydiscus*, and *Diplomoceras*. The Desmocerataceae (60%) had the highest percentage surviving to the end, followed by the Phylloceratidae and Tetragonitidae (together ca. 50%) and, last, the hamitoconic Ancyloceratida (15%). The low number of deep-water heteromorphs surviving to the end is surprising, especially considering their exceptional cosmopolitan distribution; further sampling may still change these data.

This final extinction event resembles the Triassic/Jurassic mass extinction, when oceanic and distal-shelf pelagics were the only survivors. Both contrast with the Permian/Triassic event, when it was the epicontinental, shallow nektonic and demersal taxa that survived, and the oceanic taxa succumbed to extreme anoxia (Wignall and Hallam, 1993).

The fact that the nautilids survived this boundary unscathed indicates their basic difference in habitat. In contrast to the overwhelmingly pelagic last ammonoids that depended on the plankton (see Sec. 5.4.), these nautilids presumably were demersal bottom feeders like *Nautilus*.

4.3. Hypoxic Cratonic Seas

The hypoxic marine conditions and biota throughout the Phanerozoic, especially the Devonian, Carboniferous, and Jurassic, have been studied extensively in the last decade with emphasis on the chemical and biotic conditions in the sediment and seawater (Brett *et al.*, 1991; Jenkyns, 1988; Morris, 1979, 1980; Oschmann, 1991; Savrda and Bottjer, 1991; Tyson and Pearson, 1991; Wilde and Berry, 1984; Wignall, 1987, 1990; Wignall and Hallam, 1991, 1993). Oxygen-poor or dysaerobic ($O_2 = 0.3$–1.0 ml/liter) bottom waters commonly restricted benthic shelly fauna on muddy bottoms to epifaunal rhynchonellid brachiopods and to thin-shelled, generally byssate "flat clams." Whereas some minute "flat clams," e.g., posidoniids, may have been partly or potentially pseudoplanktic in floating algal mats, the majority was almost certainly benthic. They occur in dense populations and dominate low-diversity, inequitable communities for which they were commonly the pioneer species (Kauffman *et al.*, 1992).

The depth sequence of the brachiopod communities has been well worked out (Boucot, 1981). The dysoxic sediment is usually a dark gray shale that is poorly laminated because of bioturbation, usually includes carbonate concretions that may contain pyritic fossils, and is moderately organic-rich. In the Devonian–Carboniferous, the dominant "flat clams" were Posidoniidae and Pectinaceae that characteristically were accompanied by leiorhynchid brachiopods; in the Triassic, the Monotidae (*Daonella, Halobia, Monotis*) and Oxytomidae; in the Jurassic and Cretaceous, the Posidoniidae (*Bositra* or *Posidonia*), Inoceramidae (*Retroceramus, Inoceramus*), Buchiidae (*Buchia, Malayamaorica*), and sometimes Pectinacaceae (*Amussium, Meleagrinella*). Only later Cretaceous inoceramids grew to gigantic size, possibly resulting from a unique chemosymbiosis or greatly extended oxygen-absorption surfaces, adapted to an oxygen-poor or even anoxic environment (Kauffman *et al.*, 1992). Slightly better oxygenation of the mud is indicated by the presence of deposit-feeding nuculid bivalves. In contrast to the Paleozoic black shales, brachiopods were uncommon in this facies during the Mesozoic.

Periodically anaerobic bottom waters ($O_2 < 0.1$ ml/liter) were devoid of benthic biota. The sediment was characteristically organic-rich (kerogen), finely laminated, with disseminated pyrite, and often weathered into "paper shales" with a strictly pelagic and pseudoplanktic fauna, as in the classical *Posidonia* Shales of northwest Europe (see below). This facies is bound to low sedimentation rates, high nutrient input from the surrounding highland causing plankton blooms, and is typically associated with early transgressive phases (Hallam, 1987). Abundant freshwater input is also an important factor in establishing the pycnocline, i.e., halocline, for bottom-water isolation (Van Der Zwaan and Jorissen, 1991). Although haloclines recently have been considered less important than thermoclines for pycnocline establishment, there is new evidence for the presence of the former, including data from ammonoids (below).

4.3.1. Offshore Shelves and Deep Basins

4.3.1a. Jurassic. In the western Mediterranean area, tilting blocks formed numerous basins and subbasins differing in depth and relative isolation (Almeras and Elmi, 1982; Elmi, 1985). The deep-water ammonoid biofacies occur in large basins (F5 of Elmi, 1985) or subsidiary basins (F3).

Large basins of Algeria had benthos of mainly "flat clams" in the Toarcian. Phylloceratids and, more rarely, lytoceratids were dominant; dactylioceratids thrived locally; ancillaries in variable proportions were harpoceratids and hildoceratids. All were presumably pelagic, the mentioned leiostraca deep and the dactylioceratids shallower (see below). Mid-Jurassic (Bajocian–Bathonian) ammonoid faunas were distinctly different at basin center and margin. Basin centers and seamount tops had mainly Phylloceratina and cadiconic to spheroconic Ammonitina (Morphoceratidae), slopes and surrounding platforms, mainly smooth oxycones (Oppeliidae) and evolute, costate platycones (Parkinsoniidae). The ammonoid faunas were presumably pelagic, with the possible exception of most parkinsoniids (?demersal swimmers). In the Bathonian, the 200–400 m deep basin of southeastern France has yielded (in order of dominance) Spirocerataceae of the Ammonitina, Phylloceratina, and Hecticoceratinae of the Ammonitina. Here at least, the discoconic phylloceratines were nektic in deep midwater.

Subsidiary basins were characteristically a few kilometers across and limited by steep seamounts and islands. The Toarcian fauna consisted of dysaerobic benthos ("flat clam" posidoniids) and abundant ammonoids: dominant were small, stunted, as well as young platyconic hildoceratids and oxyconic harpoceratids; ancillaries were phylloceratids and lytoceratids; abundance of the planktic dactylioceratids varied exceptionally, perhaps suggesting swarming (schooling). All were presumably pelagic. Bathonian dysaerobic subsidiary basins (ca. 250 m deep) have yielded (in order of abundance) Phylloceratina and the Ammonitina taxa Perisphinctidae, Spirocerataceae, Oppeliidae, and Hecticoceratinae (Elmi, 1985).

4.3.1b. Cretaceous. In Hokkaido, the Cenomanian black shales of the Yezo Supergroup were deposited in the deep waters of the continental margin (Hirano, 1986, 1993). They yield a few epibenthic inoceramid "flat clams" and (?drifted) trachyostraca (*Calycoceras*) as well as abundant Desmocerataceae (Ammonitina), Lytoceratina, and Ancyloceratina. The desmocerataceans included striate, constricted discocones that were either inflated (*Desmoceras*) (planktic?) or compressed (*Pachydesmoceras*) (nektic); lytoceratines include ubiquitous, round-whorled, evolute planorbicones (*Anagaudryceras*) (planktic) and discocones (*Zelandites*) (nektic); ancyloceratines are represented by torticonic heteromorphs (*Turrilites*) (planktic vertical migrant). All but the last are known to have strong septa and siphuncles indicating deep water. They had oceanic to outer shelf preferred distributions (Westermann, 1990; Chapter 10, this volume, Table II). *Turrilites* is mainly known from epicontinental deposits and has been interpreted as demersal with vertical escape adaptation

(Klinger, 1980) but, in this deep environment at least, was more probably a midwater vertical migrant. Abundant *Desmoceras* were overcome by deep-water benthic anoxia, indicating their demersal habitat and confirming earlier conclusions based on shell structure and distribution (Westermann, 1990; Chapter 10, this volume, Table II).

4.3.2. Offshore Bituminous Limestones

4.3.2a. Devonian. In Morocco and parts of Europe, a diversity of Late Devonian goniatites occurs in the pelagic distal-platform bituminous "cephalopod limestone" that, according to Wendt *et al.* (1984) and Wendt and Eigner (1985), was deposited only 10–20 m deep, and in the nodular limestones at the basin slope, supposedly at a depth of only 50 m, all within the photic zone. In these clear, open, tropical waters, however, the photic zone and storm-wave base may have been twice as deep. Bourrouilh (1981), on the other hand, compared this "Goniatitico Rosso" to the Ammonitico Rosso and assumed a deep-water, slope environment. The macrofauna was dominated by pelagic cephalopods, styliolinids, thin-shelled (planktic) ostracods, and conodonts, with accessory benthos. The discoconic, cheiloceratid goniatites are locally imbricated, indicating storm deposition, thereby casting doubt on their autochthony.

4.3.2b. Triassic. Black, bituminous, laminated limestones and dolomites are well developed in the Middle and Upper Triassic in Europe, western North America, and China. The Gutensteiner Limestone (Anisian) in Steiermark, Austria (Tatzreiter and Vörös, 1991) has shell beds of only the byssate bivalve *Enteropleura* that may have been benthic or pseudoplanktic (H. Bucher, personal communication, 1993) as well as abundant ammonoids. Dominant were the ceratitid *Balatonites* with morphotypes varying from finely to moderately sculpted, involute to evolute platycones. Ancillaries were disco-cones [*Ptychites* *110 m (asterisk indicates maximum habitat depth throughout text, from Chapter 10)]; coarsely costate, involute planorbicones (*Acrocordiceras*) and serpenticones (*Reiflingites*); and smooth platycones with goniatitic suture (*Proavites*). The lithofacies indicates that the ammonoids were all pelagic, i.e., planktic (*Reiflingites*) or sluggish nektonic.

The mid-Triassic Grenzbitumen-Zone of Tessin, Switzerland, consists of bituminous dolomites with interbedded "paper shales," which yield only "flat clams" (*Daonella*) together with abundant marine vertebrates and ammonoids (Rieber, 1973, 1975). Platyconic Ceratitidae were dominant, varying greatly in the usual manner from compressed and finely sculpted to inflated and sculpted morphs, indicating poor mobility (Dagys and Weitschat, 1993). Ancillaries were costate serpenticones (*Celtites*), presumably planktic; and smooth "streamlined" oxycones (*Longobardites* *85 m), discocones (Noritidae, Ptychitidae, *Monophyllites* *95 m), platycones (*Gymnites*), and presumably planktic spherocones (Arcestidae). The analyses of morphotypes support the lithofacies interpretation that the entire ammonoid fauna was pelagic, probably at different depths.

In the Humboldt Range, Nevada, the basinal facies of the Prida Formation consists of Anisian dark gray fossiliferous limestone and interbedded calcareous mudstone, siltstone, and shale (Fossil Hill Member); and of Ladinian dark gray to black, lamellated cherty limestone with increasingly scarce "flat clams" (*Daonella*) and ammonoids. The latter are mostly finely to coarsely sculpted platycones and planorbicones (*Protrachyceras, Frechites*) and small, smooth cadicones (*Thanamites*) (Silberling and Nichols, 1982). Facies and morphotypes suggest that all were sluggish nekton. The famous Fossil Hill fauna, in most anaerobic facies, has yielded the same "flat clams" together with a highly diverse cephalopod fauna of ammonoids, atractitid coleoids, and nautiloids. Facies and fauna resemble the Toad Formation in northeastern British Columbia that is generally dysaerobic and misses lower-latitude elements (McLearn, 1969; H. Bucher, personal communication, 1993; unpublished data from my collection).

Despite differences in facies and paleoclimate (paleolatitude), the ammonoids of both formations include examples of most planispiral morphotypes, although as expected (Tozer, 1981), the mainly low-latitude trachyostraca are more common in the more southerly Fossil Hill fauna: serpenticones, small, smooth to bluntly costate (*Aplococeras*) or midsize with coarse costae (*Pseudodanubites*); planorbicones, costate, evolute with constrictions (*Cuccoceras*) or involute without constrictions (*Acrochordiceras, Czekanowskites* *110 m) or compressed and highly nodose (*Nevadites* *72 m, Protrachyceras*); smooth spherocones (*Proarcestes* *180 m, *Humboldites*); cadicones, smooth to finely costate (*Isculites*), some with platyconic (*Alanites*) or serpenticonic juvenile whorls and fastigate body chamber (*Tropigastrites*); oxycones, entirely smooth and highly compressed (*Longobardites* *86 m, Buddhaites* *70 to ca. 140 m) or very large and with spiral striae (*Sturia*) or moderately sculpted and more inflated (*Eutomoceras* *34 m), some with depressed and strongly sculpted inner whorls (*Lenotropites*); platycones, smooth to moderately sculpted (*Gymnites*, etc.) or coarsely sculpted (*Paraceratites*); and discocones (*Monophyllites, Hollandites* *ca. 72 m, *Gymnotoceras* *72 m, *Parapopanoceras* *150 m).

Intraspecific variation is usually high (Silberling and Nichols, 1982, Pls. 14–17) and follows, as usual, the Buckman rule of covariation (Westermann, 1966), with evolute-inflated forms more coarsely sculpted than involute-compressed forms. The shell structures indicate shallow to moderate depths. Some beds contain all growth stages, ammonitella to adult, in great abundance (e.g., *Gymnotoceras*), indicating mass mortality by rising anoxia. Ammonitellas and neanoconchs also occur in Upper Triassic hypoxic deposits of the Alps (Wiedmann, 1972). The entire assemblage was probably pelagic at different depths, i.e., plankton to sluggish nekton, with the possible exception of some of the coarsely sculpted planorbicones and platycones that might have been demersal swimmers during intervals of relatively good benthic oxygenation.

In the Lower Triassic (upper Spathian) of China, interbedded gray micrite and black mudstone in Anhui (also nodular limestone in Nanjing) have a restricted benthic fauna of "flat clams" (*Claraia* and *Periclaraia*). The pelagic fauna consists of conodonts, reptiles, fish, and ammonoids, i.e., the *Subcolumbites* Community (Wang and Westermann, 1993). Most are platycones or discocones, smooth or simply sculpted (*Procarnites, Subcolumbites*) or smooth serpenticones (*Leiophyllites*) (Wang and Westermann, 1993). All or most ammonoids were presumably sluggish nekton in midwater, not demersal in deep water as assumed previously.

In the latest Triassic (Norian), which was frequently developed in dysoxic "flat clam" facies with monotids, most highly sculpted, evolute morphotypes (presumably shallow midwater and demersal nekton) disappeared, whereas the smooth, involute morphotypes survived (mostly deep-water plankton), and the first heteromorphs appeared (plankton at various depths). The only taxa surviving the Triassic/Jurassic mass extinction event were the discoconic phylloceratids (deep-water nekton) (Tozer, 1981).

4.3.2c. Jurassic. From the Toarcian of the western Mediterranean basin, Almeras and Elmi (1982) and Elmi (1985) described the deep-basinal slope ("hinge zone F2") between seamounts and narrow pelagic basins. Marly, red or gray, nodular carbonates (marly Ammonitico–Rosso) grade upward into condensed facies with "flat clam" coquinas. The scarce macrobenthos consists mostly of brachiopods. Ammonoids are diverse, with abundant nektic platycones (*Hildoceras, Merlaites*) and oxycones (*Harpoceras*), and planktic serpenticones (Dactylioceratidae). Sporadic oceanopelagic phylloceratids and lytoceratids reflect brief connections with the open Tethys.

4.3.3. Epeiric Basins: Black-Shale Facies

In the hypoxic epeiric basins, usually only between 50 and 100 m deep, bottom-water conditions tended to be unstable. This resulted in seasonal alternation between anaerobic and dysaerobic biofacies by the rising and falling H_2S/O_2 boundary, ranging from a few centimeters (exaerobic) or meters (anaerobic) above to a few decimeters below (dysaerobic) the water/sediment interface. This dynamic cycle, named the poikiloaerobic biofacies (Oschmann, 1991), explains the frequent presence of shell accumulations in anoxic (laminated) lithofacies. Annual extended periods (ca. 9 months) of sea floor oxygenation permitted small, fast-growing opportunists to settle and reproduce, e.g., "flat clams"; the shorter intervals (ca. 3 months) of anoxia, as a result of recurrent pycnoclines and weak lateral admixing currents in the small basins, killed off these biota and prevented the establishment of any burrowing infauna, so that lamination prevails. Brief dysaerobic events also remain undetected geochemically in the sediment (Little *et al.*, 1991). Rare, exceptional storms reached the basin floor down to a depth of perhaps 80 m and caused the episodic accumulation of thin but extensive shell beds.

The most recent classification of black-shale biofacies, based on the English epeiric Jurassic, distinguishes six facies (Wignall and Hallam, 1991) with increased oxygenation of the sediment/water interface. All yield ammonites with the exception of Anaerobic Biofacies 1, which contains fish as the only macrofossils, supposedly because the "nektobenthic" ammonites could not survive in the anoxic lower water column. These authors did not consider the reasons for the absence of *pelagic* ammonites; I propose that much of the water column above the oxygen-minimum zone may have had reduced salinity, enabling only euryhaline fish to survive. A very few euryhaline ammonoid taxa probably did exist, e.g., *Ceratites, Placenticeras.* Anaerobic Biofacies 2 has some bedding planes with all size classes of ammonoids, caused by mass killings (e.g., rising anoxia by storms or tectonic events), not postbreeding death. The Lower-Dysaerobic Biofacies 3 (formerly called anaerobic and intermediate) contains a few adult ammonoids and abundant bivalve spat, documenting very short intervals (days) of oxygenation. The Lower-Dysaerobic Biofacies 4, with a low-diversity fauna of small, opportunistic bivalves covering bedding planes, was poikiloaerobic (= seasonally exaerobic). The Upper-Dysaerobic biofacies 5 and 6 represent the common bioturbated gray shales with concretions and pyritic fossils.

4.3.3a. Devonian. The Givetian Hamilton Group of New York and eastern Pennsylvania has been studied thoroughly (e.g., Kammer *et al.*, 1986; Brett *et al.* 1990, 1991) (Fig. 9). Total species richness increased with the relative oxygenation of bottom waters under which the biofacies were deposited: from anaerobic (O_2 <0.1 ml/liter water) in the black, laminated shales (15–20 species with high dominance) to dysaerobic in pyritic dark and midgray shales and mudstones (35–40 species with very high dominance), to aerobic (>1.0 ml/liter water) in calcareous mudstones and limestones (60–100 species with low dominance). The occurrence of aptychi is important evidence for *in situ* deposition of ammonoids (Frye and Feldman, 1991).

The greatest abundance of cephalopods is in the dysaerobic biofacies, with a preponderance of the discoconic, mesodomic goniatite *Tornoceras* (House and Price, 1985), the orthoconic *Bactrites*, and the small orthoceratid "*Michelinoceras*" (Fig. 9). The nautiloids are more abundant in the lower part of this facies, the ammonoids more above, occurring in the midgray mudstones near the aerobic boundary. The sea floor was below the base of all but the most severe storms, but presumably within the photic zone during deposition of the gray mudstone facies, as suggested by algal borings (? 30–50 m). The presence of all size classes and exceptional preservation of the ammonoids suggested to these authors a demersal swimming habit, with ammonoids preying on the epibenthos, i.e., diminutive brachiopods, bivalves, and gastropods. Infaunal nuculid bivalves, which are known for their low oxygen requirement, caused intense bioturbation. The small (?mostly adult) goniatites had smooth, discoconic shells and mesodomic body chambers, indicating that they were active predators. The bactritids were quasiplanktic, vertical migrants (see above). It seems likely that both bactritids and ammonoids (representing different orders) lived in midwater, rather than at the bottom and that the entire fauna was killed by events of rising anoxia. If and when the hypoxic water masses reached the near-surface waters, there was no escape even for the more mobile, larger forms.

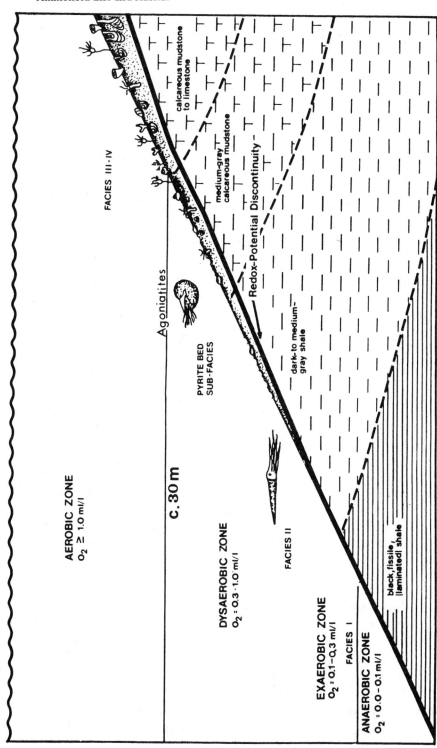

FIGURE 9. Paleozoic black-shale facies. Aerobic–anaerobic ramp model for facies of the Devonian Hamilton Group, with paleoslope highly exaggerated. Solid line in the sediment indicates oxic–anoxic boundary. The ammonoid *Agoniatites* and an orthoconic nautiloid belonged to the dysaerobic biofacies. (Slightly modified from Brett *et al.*, 1991, Fig. 5.)

From the calcareous Cherry Valley and Union Springs Members of the otherwise anoxic Marcellus Formation, in the deepest part of the basin, come the well-preserved Anarcestida *Cabrieroceras, Maenioceras, Subanarcestes,* and *Agoniatites.* Based on the ontogenetic variation of the thin septa, Hewitt (1993; Chapter 10, this volume) has estimated the maximum habitat depth of a large, demersal *Agoniatites* at only 35 m (implosion 78 m), consistent with the reduced recent bathymetric estimates based on benthic invertebrate assemblages (Boucot, 1981; Brett *et al.,* 1991, 1993). The bottom waters at this time were poikiloaerobic (anaerobic and periodically dysaerobic). Significantly, the only Goniatitida species present, *Tornoceras uniangulare* (Conrad), was also the most common in other biofacies. Its mesodomic, discoconic shell is that of a swimmer, and its ubiquitous and recurrent occurrence suggests dependence on distal water masses of considerable habitat depth (146 m; Chapter 10, this volume), rather than on a benthic food source. Hence, a nektic habitat is most likely. Clymeniids occurred in a similar habitat (Schindewolf, 1971; Korn, 1986, 1988, 1992).

For Late Devonian basinal clays of south Devon, England, Goldring (1978) reconstructed a fauna consisting of benthos with blind trilobites and "worms," pseudoplankton with posidoniid "flat clam" bivalves attached to "weed," nekton with conodontophores that fed from the sea floor, and a pelagon with discoconic *Tornoceras* (nekton) and the serpenticonic *Archoceras* (megaplankton) (terms in parentheses here interpreted).

In the Canning Basin, Australia, hypoxic goniatite shales were deposited at more than 60 m of depth, but elsewhere goniatites are found together with stromatolites of the lower photic zone (R. T. Becker, personal communication, 1993). The compressed oxyconic *Beloceras* occurred in shallow subtidal (<25 m deep), back-reef facies, probably the shallowest facies known for Paleozoic ammonoids (unless they were washed over the reef and promptly died). But *Beloceras* is also known from basinal facies (at McWhae Ridge), perhaps 175 m deep, where it was pelagic above 55 m according to its weak septa (Chapter 10, this volume).

The Late Devonian smooth, subtriangular discocones *Manticoceras* and *Carinoceras* had large brevidomic body chambers (R. E. Becker, personal communication, 1992), an exception for goniatites, and weak septa that limited the adults to 50–60 m depth, but juveniles could go twice as deep (Chapter 10, this volume). This suggests that the adults of both were agile hunters in shallow midwater.

4.3.3b. Carboniferous. The North American midcontinental cyclothems have been studied intensively with regard to their marine faunal succession and depth of deposition (Boardman *et al.,* 1984; Kammer *et al.,* 1986; Heckel, 1991; Maynard and Leder, 1992; Becker, 1993) (Fig. 10). From deep to shallow, the following recurrent community sequence has been worked out:

1. *Dunbarella*–ammonoid–radiolarian community—black, laminated, phosphatic shale with poikiloaerobic biofacies at basin center; depth

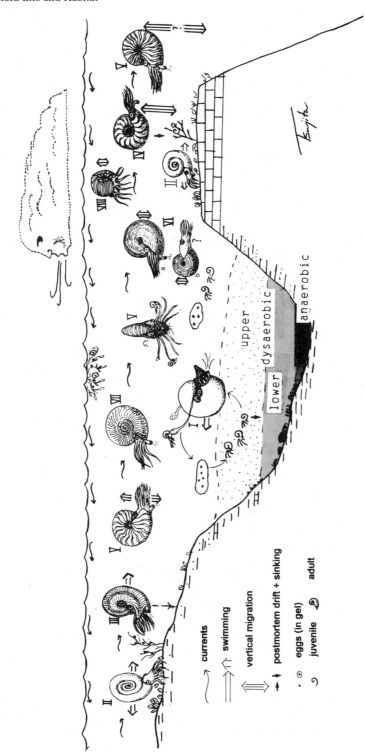

FIGURE 10. Carboniferous (Namurian) scenario of basin and platform facies with the major ammonoid morphotypes I–VIII defined by Saunders and Shapiro (1986) (see text and Fig. 4). VIb shows a hypothetical coleoid-type body temporarily protruding from the body chamber for effective swimming, as postulated by Jacobs and Landman (1993).

ca. 100 m (reduced from the original estimate of 200 m). Dominant were pelagic radiolarians, conodontophores, and epibenthic small "flat clams"; accessories were nautiloids, fish, shark, and the goniatite *Glaphyrites*. This small (?stunted) ammonoid was a longidomic cadicone (Saunders and Shapiro, 1986) with habitat limit of 60–120 m (Chapter 10, this volume). It probably was planktic in deeper, somewhat hypoxic midwater.

2. Divided into (2a) *Sinuata*–juvenile ammonoid–*Anthraconeilo* subcommunity—pyritic black shales with dysaerobic biofacies near basin center, and (2b) *Trepospira*–mature ammonoid–*Anthraconeilo* subcommunity—dark gray, clay-rich shales of uppermost dysaerobic biofacies; depth ca. 50–80 m (reduced from 100 m). Both subcommunities were dominated by similar ammonoid taxa that were very abundant but exclusively neanic (discussed below) in 2a, but mainly adult, less common, and occasionally together with small juveniles in 2b. Boardman *et al.* (1984) tentatively identified 37 genera of the Goniatitida and Prolecanitina. They are here placed into morphogroups, and the percentages of genera in each are: smooth discocones 34%; smooth spherocones 31%; involute cadicones 13%; platycones, smooth 15%; sulcate, and costate 5%; planorbicones 2%. Also present are abundant orthoconic and cyrtoconic (smooth) Bactritida placed in six genera. In conclusion, these assemblages (like other Carboniferous assemblages) consist almost entirely of smooth-shelled genera and mostly of spherocones and discocones, indicating that pelagic habitats were predominant. The presence in 2b of moderately common adults and sporadically massed small juveniles suggests that the adults died "normally," perhaps after spawning in midwater, whereas the small juveniles were overcome by episodically rising anoxia. The abundant presence of mandibles (anaptychi) in body chambers (Mapes, 1987) is evidence that the goniatites have not drifted postmortem. The three shallower communities were probably all less than 25 m deep and devoid of ammonoids. A scenario for Carboniferous (Namurian) habitats is presented in Fig. 10.

Boston and Mapes (1991) postulated that the very young ammonoids were benthic (sessile or vagrant) and were not able to escape the periodic benthic anoxia. They based this on their assumption that the very young goniatites and prolecanitids (neanoconchs and small juveniles?) were unable to empty their chambers as rapidly as required by their high rate of growth and, hence, were negatively buoyant. There is, however, an alternative explanation resembling that for the egg masses discussed below: the neanoconchs were planktic and could not escape brief events of rising anoxia during upwelling or storms. Following death, the small phragmocone was filled with sea water rapidly (owing to the large surface of the siphuncle relative to chamber volume) and sank undamaged to the soft sea floor, with the soft body and mandibles in place. Ammonoids of larger growth stages either (1) lived in higher, photic water masses, being less vulnerable to visual predators than the very small juveniles; (2) could rise rapidly enough to escape the rising anoxia owing to their greater (size-related) swimming velocity; or (3) were able to suffer brief hypoxic events, as can living *Nautilus* (Wells *et al.*, 1992). A combination of all three appears likely.

A large embryonic ammonoid assemblage was described recently from a "core shale" of the Virgilian Oread Formation in Kansas (Mapes *et al.*, 1992; Tanabe *et al.*, 1993b, 1994b). Complete and incomplete ammonitellas as well as a few neanoconchs and small juvenile shells, belonging to at least two taxa of the Goniatitina (*Aristoceras, Vidrioceras*), occur in centimeters-thick and meter-large masses enclosed in lenticular concretions. The matrix is laminated, black mudstone almost devoid of bioturbation, indicating anaerobic to poikiloaerobic bottom conditions. The calculated densities of the well-preserved hatchlings were slightly less than that of sea water (1.012 and 1.003 g/cm^3, respectively) but higher than in Cretaceous ammonoids. Mapes *et al.* (1992), therefore, concluded that the egg masses were floating above the bottom rather than planktic and only slightly transported from nearby spawning sites. It seems to be impossible, however, to distinguish quasiplanktic habitats in midwaters and bottom waters on the basis of slight differences in (calculated) near-neutral buoyancies, and any deviations were probably compensated for by cameral liquid. Furthermore, egg masses had no means of depth control and so were entirely dependent on water masses, which, if sufficiently oxygenated for growth, would not have been strictly stationary.

The oxygen depletion and assemblage composition, however, suggest an alternate interpretation for this unique Kansas occurrence: the eggs, neanic individuals, and small juveniles alike were planktic and quasiplanktic in a moderately deep basin (<100 m); those present in relatively deep waters not far above the dysaerobic zone were overtaken by anoxic upwelling during a storm; the poisoned animals sank to the floor and drifted together short distances before embedding, without being abraded or oriented (being spherical). The larger juveniles and adults lived higher in the water.

Abundant ammonitellas and neanoconchs (*Imitoceras*), bactritids, and conodonts as well as brachiopod and bivalve spat also have been described from the anoxic Exshaw Formation in the earliest Mississippian of Alberta (Pamenter, 1956; Schindewolf, 1959).

A similar bathymetric facies sequence occurred in the cyclic sequence of England (Calver, 1968; Ramsbottom, 1978; Wignall, 1987). Ammonoids are found mainly in the deep, dysaerobic, semipelagic *Gastrioceras*–Pectinoid Facies, i.e., noncalcareous mudstone with small goniatites and some orthocerids, epibenthic bivalves, and radiolarians. The shallower, aerobic, calcareous mudstones contain rare (unidentified) larger ammonoids, nautiloids, and a rich benthic fauna.

In the Pennines, the *Gastrioceras cumbriense* Marine Band represents a sedimentary cycle (Wignall, 1987). The recurrent dysaerobic biofacies has a moderately diverse pectinoid–goniatite assemblage, dominated by the longi-domic discocone *Anthracoceratites*; the anaerobic biofacies has ammonoids and nautiloids, with the dominant goniatite *Cancelloceras* ("*Gastrioceras*"). Accessories are thin-shelled epibenthos; the moderately large, smooth gonia-tite *Homoceratoides*, which is particularly characteristic for the "paper-shales"; and orthoconic (mostly longiconic) and coiled nautiloids. The size–frequency distribution of undamaged *Cancelloceras* shells shows a nor-mal survivorship curve, from complete ammonitellas to adults. The following scenario appears plausible: The midwater or uppermost dysaerobic zone was the habitat of the small planorbiconic *Cancelloceras* with thick-shelled, probably mesodomic, depressed, and costate whorls, indicating that it was sluggish pelagic, perhaps microphagous (Wignall, 1987), and of the larger, discoconic *Homoceratoides*, an active hunter whose prey may have included the abundant *Cancelloceras*.

4.3.3c. Permian. In the Permian Reef Complex of the southern United States, ammonoids are almost restricted to the basin center (Newell *et al.*, 1953; N. Newell, personal communication, 1992). Before reaching maximum depth (ca. 300 m), the basin was inhabited by a mixture of sculptured planorbicones, smooth spherocones and discocones, and oxycones. The oxy-conic *Medlicottia*, the discoconic *Perrinites* and *Pseudogastrioceras*, and the spheroconic *Waagenoceras* could have lived to 100 m depth (Chapter 10, this volume). The deepest facies of the basin consists of gray to black, bituminous, laminated limestone and quartz sandstone and has yielded mostly the minute discocones of *Agathiceras*, *Lilobites*, *Strigogoniatites*, *Timorites*, and *Xenaspis*. Only *Agathiceras* septa have been investigated (Chapter 10, this volume); they were among the strongest in goniatites, permitting a habitat depth of 220–325 m. A benthic fauna was present only toward the basin margin, i.e., leiorhynchid brachiopods and small posidoniid bivalves ("flat clams"). Thus, the basin floor was anaerobic in the center and poikiloaerobic toward the margin. According to Newell *et al.* (1953), the juvenile ammonoids (?also stunted adults) were killed in midwater, well above the bottom, by rising hydrogen sulfide or by anoxia that may have been caused by plankton blooms. Most probably, the entire ammonoid fauna was pelagic. On the reef and shelf edge, drifted shells may abound. For example, *Perrinites* occurs locally in Australia in great quantities within a shallow hypersaline sequence (B. F. Glenister, personal communication, 1993), supporting the hypothesis that deep-water shells had a high potential for postmortem drift (Tanabe, 1979; Westermann, 1990).

The strongly hypoxic Meade Peak Phosphatic Shale Member of the Phos-phoria Formation in Idaho has abundant complete specimens of the metale-goceratid goniatite *Spirolegoceras* (Nassichuk, 1970). The inner whorls are cadiconic and constricted; the outer whorls discoconic and longidomic. This suggests that the adolescent was a planktic drifter and that the adult was mainly a vertical migrant. A specimen of *Pseudogastrioceras* from the same formation could have lived at the moderate depth of 170 m (Chapter 10, this volume).

   4.3.3d. Jurassic. Organic-rich laminated black shales were globally widespread during the earliest Jurassic (early Hettangian) transgression and also during the Sinemurian (Taylor *et al.*, 1983; Jenkyns, 1988). These contain the first Jurassic Ammonitina, the Psilocerataceae, with many smooth or simply costate serpenticones (Psiloceratidae, Echioceratidae), and a few minute spherocones (Cymbitidae), all almost certainly planktic (Tintant *et al.*, 1982), as well as some nektic, shallow-water oxycones (Oxynoticeratidae, see below).

   The Blue Lias of Dorset consists of microcycles that probably were orbitally enforced (House, 1975, 1985, 1993). The biofacies are unfossiliferous anaerobic shale; gray lower-dysaerobic shale with restricted benthic fauna and some crushed ammonites; and upper-dysaerobic bioturbated limestone ("Marston marble") with a low-diversity, abundant bivalve and ammonite fauna. The latter includes minute serpenticonic echioceratids (*Promicroceras*) of all growth stages (in five size classes, according to Trueman, 1941), commonly associated with microgastropods. This resembles the occurrence of the small liparoceratid *Aegoceras* together with these *Trochoturbella* in the clays and suggests a protected habitat in floating plants (R. Hewitt, personal communication, 1993). The considerable shell strength of *Promicroceras*, however, would have permitted it to live in deep-water, benthic algal mats as well. Other limestone beds yield large (40–60 cm) Arietitinae that, because of their prominent sculpture and size, normally would be considered as demersal, especially because they also are found more commonly in nearshore facies. Their serpenticonic, at least partly longidomic whorls (380° for *Vermiceras*, cf. Trueman, 1941) and cosmopolitan distribution suggest a potential planktic habit in the upper, photic part of the water column; their large size and the prominent radial ribs provided protection against predators. This is supported by their weak and irregularly spaced septa with simple sutures (Chapter 10, this volume; Bayer, 1977).

   The Pliensbachian black shales of northern Germany (Hoffmann, 1982) yield a very impoverished benthic macrofauna and abundant small serpenticones with prominent but simple sculpture (Eoderoceratidae and Polymorphitidae) that were probably planktic. Associated genera include evolute microconchs as well as much larger, sometimes highly inflated, spinose, thin-shelled, and variable macroconchs (Liparoceratidae). These larger forms must have moved into shallow water during adolescence, according to their habitat limits (65 m for juveniles and 32 m for adults; Chapter 10, this volume). Ancillaries are subspheroconic, heavily sculpted, and spinose microconchs that belong to large macroconchs with smooth, moderately "streamlined" body chambers (Phricodoceratidae). There are also oxycones with suppressed sculpture (Oxynoticeratidae, Amaltheidae) that were presumably nektic and shallow (*Amaltheus* 25 m). The scarce phylloceratids and lytoceratids were markedly modified compared to their oceanic relatives. The small microconch of *Tragophylloceras* is prominently sculpted, and some species of *Lytoceras* (*L. salebrosum* Pomp.) have dichotomous ribbing resembling that of the Ammonitina. The entire fauna was presumably pelagic.

The lower Toarcian *Posidonia* Shale (Posidonienschiefer, Schistes Cartons, Jet Rock) of northwest Europe and other parts of the world (Jenkyns, 1988) consists of a relatively thin sequence of laminated (anaerobic) shale, marly shale, and bioturbated (dysaerobic) carbonate interbeds. It is famous for its abundant vertebrate and invertebrate fauna (cf. Urlichs *et al.*, 1979; Fig. 11). The carbonate interbeds are tempestites yielding coquinas of "flat clams," belemnites, etc. The epibenthic fauna of the marly shales consists of small inoceramids and oxytomids (*Inoceramus–Steinmannia* Association) as well as posidoniids; that of the laminated shales, only of juvenile posidoniids (*Bositra* Association). Abundant oxyconic Harpoceratinae (with maximum habitat depth *ca. 50 m), platyconic Hildoceratinae (*60+ m), typical serpenticonic Dactylioceratinae, and scarce leiostraca, i.e., discoconic *Phylloceras* and round-whorled planorbiconic *Lytoceras*, range throughout the section. Clusters of harpoceratine fragments probably resulted from squid predation in the upper water column (Mehl, 1978b). Masses of *Dactylioceras* (*ca. 70 m) occur on bedding planes of the posidoniid-rich shales, indicating mass killing by anoxic admixing during storms. Harpoceratids and hildoceratids are especially common in the benthos-poor shales and often contain aptychi, suggesting perhaps a somewhat hypoxic habitat (R. Hewitt, personal communication, 1993). Spinose and more normally coiled dactylioceratids (*Peronoceras et al.*) were rare or absent in this highly bituminous facies, suggesting that they might have been more bottom-dependent than characteristically serpenticonic *Dactylioceras* (Schmidt–Effing, 1972; Loh *et al.*, 1986). Their Pliensbachian ancestors had mostly planorbicones and lived in the warmer and clearer waters of Tethys, probably as sluggish demersals. From there they immigrated into the more northern seas with the Toarcian transgression, following regional disappearance (Hallam, 1987).

The presence of low-salinity surface (photic) waters in the German *Posidonia* Shale seas (Fig. 11) is indicated by the replacement of the stenohaline dinoflagellates and acritarchs by euryhaline prasinophytes and "spheroids" (Loh *et al.*, 1986; Prauss *et al.*, 1991) and by trace-metal enrichments suggesting intense freshwater input, resembling Black Sea sapropels (Brumsack, 1991). In this rather shallow sea, oxygenated waters of normal salinity would have been limited to the mid-water zone, which was periodically reduced to a few meters and occasionally disappeared entirely as a result of rising anoxia and/or a thickened surface layer of low-salinity waters. Only organisms that could survive either slight hypoxia or reduced salinities, e.g., fish and probably a few ammonite species, would have survived under these extreme conditions.

These facies distributions strongly support the notion that the *Dactylioceras* species were planktic (Tintant *et al.*, 1982). If they lived in surface waters as suggested by these authors and supported by dispersal patterns across

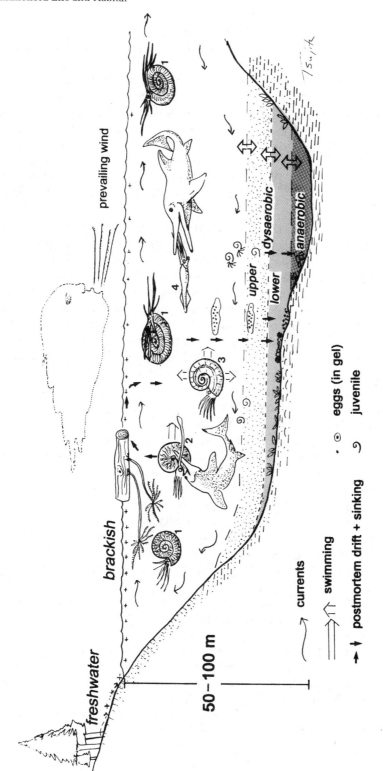

FIGURE 11. Epeiric basin, black-shale facies. Possible scenario of the Early Jurassic *Posidonia* Shale/Jet Rock of northwest Europe. Surface waters were partly brackish, and deeper waters were dysaerobic to anaerobic, changing cyclically (poikiloaerobic). Ammonitina: 1, *Dactylioceras* (planktic drifter); 2, *Harpoceras*; 3, *Hildoceras* (predators). Coleoidea: 4, squid or belemnite. Egg masses in gel and juveniles were also present in lower waters and periodically were killed by rising anoxia: some examples of postmortem sinking and drift are indicated.

shallow shelves (Elmi and Almeras, 1984), then they would have been euryhaline in the *Posidonia* Shale sea, resembling the proposed habitat of the Late Cretaceous *Placenticeras* (see below). More probably, these dactyliocera-tids were drifters in midwater as indicated by their shell and siphuncle strength, which is average for Ammonitina (Chapter 10, this volume). The other common ammonoids of the *Posidonia* Shale, the Harpoceratinae and Hildoceratinae, were, in this facies at least, also pelagic. The "streamlined" shell shape with high stability, the fine costae, and small mandibles (aptychi) found in the stomach or crop of a mature body chamber (Lehmann, 1975; Section 5.4) indicate that especially the harpoceratines were hunters with good steerage and acceleration, preying on small dactylioceratids and other juvenile ammonites in midwater.

The truly global distribution of *Dactylioceras* species in all facies (Schmidt–Effing, 1972) strongly supports their planktic habit in epeiric and neritic (shelf) seas as well as in the oceans. Similar distributions suggest that the Harpoceratinae and Hildoceratinae were nektic in the same waters.

The lower Aalenian Opalinus Clay of north and south Germany is virtually devoid of macrofauna except for Ammonitina: compressed, smooth or weakly costate oxycones with lappeted microconchs and rostrate macroconchs (Leio-ceratinae) and, locally, the small, prominently simple-costate and advolute planorbiconic grammoceratid *Tmetoceras*. It is highly probable that the leio-ceratines were mobile, maneuverable nektic predators, whereas *Tmetoceras* was quasiplanktic, perhaps living in floating algal mats, as supported by its pantropical but sporadic distribution.

The upper Aalenian shales of the North German Basin are developed in dysaerobic "flat clam" (*Retroceramus*) facies. Commonly the only ammonoids found are the platyconic to oxyconic Graphoceratidae, with lappeted and entirely sculpted microconchs and macroconchs that become smoother and "streamlined" at maturity. Bayer and McGhee (1984) claimed that recurrent shallowing trends in basin development were parallelled by the iterative evolution of oxycones, but this has been largely refuted on stratigraphic and taxonomic grounds (Morton, 1984; J. Callomon and G. Dietl, personal com-munication, 1993). Total depth range of graphoceratid habitats was perhaps between 30 and 80–150 m (see Chapter 10, this volume). Curiously, the habits must have differed greatly between the sexes, as in many other Ammonitina. The microconchs were presumably sluggish and depended on their prominent sculpture for protection, whereas the much larger adult macroconchs were potentially mobile predators, especially the oxyconic forms that lived in very shallow, more turbulent waters. Graphoceratids are found in all facies and were possibly pelagic as well as mobile demersal.

On Skye, Scotland, Graphoceratidae occur in two size classes, suggesting different habitats. Smaller juveniles of both sexes are found near shore, together with adult microconchs, whereas larger juveniles occur offshore, together with adult macroconchs (Morton, 1988).

Later Middle Jurassic black, bituminous clays (e.g., "Hamiten Ton") in Germany yield locally minute heteromorphs (Spiroceratinae, Parapatocerati-nae) (Schindewolf, 1963; Dietl, 1973, 1978). An epiplanktic, microphagous habitat is indicated for the orthoconic *Acuariceras, Paracuariceras, Metapa-toceras*, and *Arcuarites* (possibly with uncoiled ammonitella!); the variably cyrtoconic to gyroconic shells of *Spiroceras* and *Parapatoceras* may indicate that these ammonoids lived in a pseudoplanktic manner, with the shell coiled around floating plants (G. Dietl, personal communication). *Spiroceras*, with well-developed jaws and radula (K. Tanabe and N. H. Landman, personal communication, 1993), presumably fed on planktic/pseudoplanktic ostracods and microgastropods that also inhabited the mat. The probably related *Epis-trenoceras*, with advolute (?serpenticonic) shell, however, lived mainly in shallower, aerobic equatorial waters (Sandoval *et al.*, 1990), possibly in benthic algal mats. Similar habitats have been documented for Barremian homeomorphs (Vasicek and Wiedmann, 1994).

The dark-gray clays of the Upper Bathonian Aspidoides Beds at Hildesheim in northern Germany are developed in dysaerobic facies, with a few posidoniids and abundant ammonites (Hahn *et al.*, 1990; Brand and Jordan, 1990). Dominant were the compressed-planorbiconic, costate micro-conchs (and ?juveniles) of the perisphinctid *Choffatia*. Accessories were the oxyconic, smooth oppeliid *Oxycerites*, with complicated sutures and match-ing, strong septa, the elliptospheroconic tulitid *Bullatimorphites* (*Kheraiceras*) with smooth macroconch, and the costate microconch "*Bom-burites*" as well as some macroconchs of *Choffatia*. Facies, cosmopolitan distributions, and shell strength (Chapter 10, this volume) indicate the follow-ing habits for these taxa: deep (100–200 m) for *Oxycerites* and the *Bullatimor-phites* macroconch but shallow (70 m) for its microconch; the perisphinctid microconchs (and possibly also juvenile macroconchs) were sluggish nekton (see also below), but their large macroconchs (more abundant in aerobic facies) were possibly demersal; the tulitids were almost certainly planktic, with the macroconchs becoming diurnal vertical migrants at maturity; and the large oppeliids were predators.

At the center of the back-arc Neuquén Basin, Argentina (Riccardi and Westermann, 1991; Riccardi *et al.*, 1992), the thick, dysoxic black shales of the Los Molles Formation (Bathonian–Callovian) were rapidly deposited (ca. 20 cm/Ka). The invertebrate assemblage characteristically consists of rare inoceramid "flat clams" and of abundant, mainly adult Ammonitina with very low species diversity but high intraspecific variation. Complete body cham-bers are preserved commonly. Adult specimens of the sculpted spheroconic Sphaeroceratidae *Eurycephalites* and *Stehnocephalites* and of the planorbi-conic or serpenticonic Reineckeiidae *Neuqueniceras* are dispersed. Juveniles (about one-third of the adult diameter) are rare; they occur in clusters in calcareous concretions. It appears that the adults died "naturally" and singly, whereas the juveniles suffered mass killings. The young probably lived in deeper waters than the adults, were overcome by periodic upwelling, and their shells drifted together on the sea floor into small, current-produced depressions (C. Tsujita, personal communication, 1993). At about maximum basin depth occur abundant small Phylloceratina, i.e., the oceanomesopelagic *Ptychophylloceras*; Lytoceratina are rare. A very similar, coeval lithofacies and molluscan fauna, here including the elliptospheroconic *Bullatimorphites*, existed in Guerrero, Mexico (Sandoval *et al.*, 1990) and in Atacama province, Chile, where the benthic fauna consisted of a few clustered posidoniids (Westermann and Riccardi, 1979). This ammonite also occurs abundantly in Ammonitico–Rosso facies of Spain and Hungary (Sandoval, 1983; J. Sandoval, personal communication, 1993). These occurrences indicate that these spheroceratids, perisphinctids, tulitids, reineckeiids, and phylloceratids were pelagic throughout life. *Bullatimorphites* was rather deep neritic, as also assumed by Marchand *et al.* (1985), but not oceanomesopelagic as proposed by Tintant *et al.* (1982).

The best-known examples of Upper Jurassic, organic-rich, black shales are the Lower Oxford Clay (Callovian) and the Kimmeridge Clay of England (Cope, 1967, 1974; Callomon, 1985; Duff, 1975; Hudson and Martill, 1991; Morris 1979; Oschmann, 1991; Wignall, 1990; Wignall and Hallam, 1991). The Lower Oxford Clay of Peterborough is developed in characteristic lower-dysaerobic biofacies, with laminated as well as bioturbated shales, some "flat clam" shell beds, and others with semiinfaunal bivalves, some belemnites, and predatory vertebrates. All assemblages have yielded the famous *Kossmoceras* fauna of crushed shells that generally consists of monospecific, adult, strongly dimorphic assemblages with high intraspecific variation. The apertures have usually been damaged, and there are many small shell fragments. Kossmoceratids and, at other levels, perisphinctids had sculpted planorbiconic microconchs and commonly smoother, more "streamlined" macroconchs. The most recent depth estimate is only 30–50 m, even shallower than that of the oxic higher parts of the Oxford Clay. The sedimentation rate was probably rapid, although total accumulation was only moderate, because of innumerable diastems and resuspension events that account for the many thin bivalve and ammonite accumulations, i.e., tempestites (not postbreeding mass death). The abundant small and probably juvenile pectinoids (*Melagrinalla*, *Bositra*) may have lived pseudoplanktic lives in floating algal mats, as is indicated by fibrous algal remnants. A seasonal pycnocline, probably a thermocline, has been documented by oxygen-isotope analysis. Assuming normal salinity throughout the water column, the ammonites lived at 17–20°C (maximum 16–23°C); the epifaunal bivalve *Gryphaea* at 15–18°C; and the infaunal nuculids at 11–18°C. Alternatively, a halocline existed with brackish surface waters. Because the well-oxygenated surface waters had abundant reptiles and, probably, belemnites, the sluggish ammonites may have lived in the deeper, slightly oxygen-deficient (upper-dysaerobic) waters, where the predators could not follow because of their high oxygen metabolism (O'Dor and Wells, 1990). The more serpenticonic and less sculpted kossmoceratids are found in the deepest facies (Marchand *et al.*, 1985), but this does not imply a demersal habitat.

The early Oxfordian Upper Oxford Clay of England and the coeval Renggeri Marls of the eastern Paris Basin and the Jura Mountains (Morris, 1980; Callomon, 1985) contain scarce epibenthos together with scattered pyritized juveniles of mostly microconchiate Cardioceratidae and Perisphinctidae (sex ratios >10 : 1) but only a few adult Oppeliidae. Significantly, an intercalated limestone bed has only adult shells, as in most occurrences. A possible scenario is that the sluggish young cardioceratids and perisphinctids lived in the upper-dysaerobic zone, where they were overcome by rising anoxia caused by storms or earthquakes, whereas the more mobile oppeliids were in the higher, more agitated and aerobic waters and died of old age.

The late Kimmeridgian sea floor of the Kimmeridge Clay in southern England was 50–100 m deep (Oschmann, 1991), and the several interbedded facies resembled that of the *Posidonia* Shale. Bottom-water temperatures of ca. 15°C imply that anoxia of 1–2 months occurred periodically and killed the benthic fauna; yet, many opportunistic bivalves required only 8–9 months to mature. This poikiloaerobic biofacies thus had a seasonal cycle, with roughly $\frac{3}{4}$-year anaerobic and $\frac{1}{4}$-year dysaerobic bottom-water conditions. Virgato-sphinctid and aulacostephanid ammonites (*Pectinatites, Aulacostephanus*) occur in the poikiloaerobic and aerobic facies. The large macroconchs are coarsely costate, mesodomic planorbicones; the microconchs are similar but smaller and in *Pectinatites* have large ventral "horns," which would have protected the hyponome from below but prevented forward swimming. They were both sluggish swimmers, perhaps demersal as well as pelagic, i.e., feeding from the epibenthos during the dysaerobic intervals and from the plankton during the benthic anoxia.

In Milne Land, East Greenland, the late Kimmeridgian Grakloft Member of the Kap Leslie Formation consists of black, finely laminated, carbonaceous, silty shales (anaerobic); very small-scale ripple-lamination occurs at a few levels, and bioturbated horizons (dysaerobic) are present mainly in the middle part. Benthic macrofauna is extremely rare or absent, but crushed juvenile and adult Ammonitina occur in large numbers at several levels, in both the anaerobic and dysaerobic facies (Birkelund and Callomon, 1985). The ammonites belong almost exclusively to the Boreal cardioceratid *Amoeboceras*, and macroconchs greatly outnumber microconchs. Both sexual morphs of the abundant *A. kochi* and *A. elegans* are small (diameters: M, 80–120 mm; m, 15–40 mm), costate and nodose, moderately evolute platycones, with the costae commonly undivided (simple) and the nodes or clavi ventral. They fit perfectly into Batt's (1989) morphogroup 8, which here was almost certainly also pelagic, i.e., sluggish nektic (see Section 4.4.2). *A. decipiens* has several rows of exceptionally prominent nodes and resembles the coarsest *Kossmoceras* from the black shales of the Lower Oxford Clay.

On the Antarctic Peninsula, the thick Nordenskjold Formation was deposited in the extensive and deep Silled Basin (Doyle and Whitham, 1991). The lower Longing Member (Kimmeridgian–Tithonian) consists of organic- and radiolarian-rich, laminated, pelagic muds. The macrofossils of this anaerobic facies are belemnites, crushed ammonites, and pectinid bivalves (*Arctotis*) that grew on 70% of the living ammonites and on driftwood. The small ammonites of the anaerobic facies, which belong to coarsely sculpted, probably serpenticonic perisphinctids (?*Torquatisphinctes*), occur in all size classes (10–110 mm diameter), indicating mass mortality, possibly by storm-induced admixing of anoxic water masses.

Higher in the section, small, smooth, discoconic Haploceratidae (?*Haploceras*) are without epizoans. This suggests that the probable *Torquatisphinctes* was a planktic drifter, perhaps in the upper-dysaerobic zone. The leiostracan *Haploceras* probably swam in the aerobic zone, as assumed previously from its cosmopolitan distribution (see Secs. 4.1 and 4.2). It was immune to drag-producing epizoans because of its smooth surface and/or special periostracum. The Ameghino Member (Tithonian–Berriasian) includes turbidites and bioturbated beds (*Chondrites*), and therefore represents a deep, poikiloaerobic environment. The fauna consists of teleost fish, inoceramid bivalves, and ammonites. Dominant were finely costate virgatosphinctids (*Virgatosphinctes, Lithacoceras*). Accessories were costate, platyconic berriasellids, costate, planorbiconic olcostephanids, as well as leiostracan haploceratids. Curiously, it was again only the perisphinctids that carried a rich epizoan fauna. These perisphinctids are mesodomic, compressed planorbicones to discocones that were presumably moderately mobile and perhaps bottom-feeding (demersal).

4.3.3e. Cretaceous. The Mowry Shale (Albian) of the Western Interior United States includes rare but much discussed hoplitid concretions, from which intergrading series of highly varying, single *Neogastroplites* species (Hoplitidae) were well documented (Reeside and Cobban, 1960; Batt, 1989). The Mowry sea was a cool-temperate, deep Boreal embayment, cut off at this time from the warm, southern waters. The hoplitids in the concretions are all small, with most of their body chambers broken away; they probably are juveniles, and possibly microconchs; the adults occur in the surrounding shale. The accompanying macrofauna consists of much fish debris and rare inoceramids. It appears that the immature, very sluggish hoplitids lived in midwater in the shallow, epeiric sea (with dysaerobic bottom waters) and fell easy prey to large carnivores. The extraordinary intraspecific variation, from inflated and spinose to compressed and weakly costate morphs (Buckman rule of covariation), implies the absence of selection pressure, e.g., on "streamlining" (the ammonites were not adapted to swimming) or defensive ornamentation (periods of heavy predation were presumably too short for a genetic selection in favor of the spinose variants). The single-concretion assemblages, however, need not have originated strictly in the same niche. They could have been collected together by the carnivores or swept together by currents from different microenvironments (Landman and Waage, 1993). Even modest stratic condensation or winnowing cannot be ruled out, as G. Dietl (personal communication, 1993) has suggested for broadly varying fossil "populations" from bedding planes in the Jurassic of Germany.

Dark-gray, bioturbated shales of the Bearpaw Formation (Santonian–Campanian) in southern Alberta and even the bituminous and laminated shales of the coeval Pierre Formation (United States) contain a macrofauna generally limited to the rare "flat clams" *Inoceramus* and abundant ammonites, especially compressed-orthoconic *Baculites* and large, oxyconic *Placenticeras*. Coeval lithofacies with more diverse benthos, indicating better-oxygenated bottom water, also yield Scaphitidae (Landman, 1987; C. Tsujita, personal communication, 1992). Clusters of undamaged neanic *Baculites*, consisting of the septate ammonitella and a body chamber shaft, are found occasionally. Finds of masses of "juvenile" *Baculites* and *Clioscaphites* have also been reported from the Santonian Marias River Shale of Montana as well as from offshore pelagic deposits (Kennedy and Cobban, 1976).

Placenticeras meeki is remarkable in two ways: (1) It had perhaps the highest known incidence of lethal predation, by mosasaur shell-puncturing, that occurred in surface waters (10–20%) (Kauffman, 1990; Hewitt and Westermann, 1990b) as well as by the less obvious but perhaps even more frequent partial removal of the body chamber (with body), as evident from jagged edges in otherwise perfectly preserved shells in southern Alberta. (2) Oxygen and carbon isotope analyses (Morrison, 1986; C. Tsujita, personal communication, 1993) have documented markedly reduced salinities of the surface waters that *P. meeki* inhabited (alternatively, the water temperature was 40–42°C!), contrasting with normal values for the associated *Baculites* and inoceramids as well as for scaphitids (N. Landman, personal communication, 1992). This clearly indicates that the large predator *Placenticeras* lived occasionally in brackish surface waters, much shallower than the 30 m minimum depth of ammonoid habitat assumed previously (e.g., Ziegler, 1967, 1980; Westermann, 1990) and well above wave base. It could dive rapidly to some tens of meters (Chapter 10, this volume) if required, for example, during an approaching storm. The combination of a rather complicated septal suture with a low sutural amplitude index (Batt, 1991), low resistance to hydrostatic pressure, and a high predation rate support the hypothesis that sutural "frilling" also served as elastic wall support against the point loads of predation; single chambers were perforated without phragmocone fracture (Hewitt and Westermann, 1990b).

Baculites is also of special interest. Like *Placenticeras*, it was also abundantly preyed on by mosasaurs as evident from several punctures and elongate depressions; the body chamber of the elongate, fragile shell was usually broken off, producing the very common ragged edges (Kauffman, 1990; C. Tsujita and G. E. G. Westermann, unpublished data). This indicates that *Baculites* was also pelagic in the aerobic, mosasaur-rich upper (but not surface) waters, although it may also have been a potential bottom feeder, being able to dive rapidly in and out of 5–100 m deep dysoxic bottom waters (rather than escaping benthic predators; see Klinger, 1980) (Chapter 10, this volume).

We have found evidence in the shell proportions for an ontogenetic change in shell orientation from vertical to horizontal in several very large *Baculites* specimens [e.g., *B. cuneatus* (Cobban)]; the relative length of phragmocone to body chamber was twice as great in large compared to small specimens. The shell proportions for neutral-equilibrium conditions have been calculated for orthoconic nautiloids; phragmocones with water ballast had to be twice as long as shells without counterweight (Westermann, 1977). Although absence of stability at neutral equilibrium would permit any orientation, this condition was clearly adapted to near-horizontal orientation. Some stability against rolling could have been achieved by trapped gas in dorsal folioles and lobules within the flooded chambers. The much more common small species as well as the juveniles of the giant species were vertical migrants (Klinger, 1980; Batt, 1989; Westermann, 1990), probably mostly but not exclusively in midwater, whereas the giant adults (>0.8–1 m) of *B. cuneatus* became mainly demersal swimmers. This is supported by the exceptional ontogenetic increase in the septal strength of that species (Chapter 10, this volume; C. Tsujita, personal communication, 1993). A pelagic habitat for the host of smaller baculitids is also indicated by the broad species distributions and the endurance of some in deep-sea facies up to the Cretaceous/Tertiary boundary (see Section 4.2.2d).

4.4. Epeiric Basins: Mixed Facies

4.4.1. Jurassic

In lower Bajocian gray shales of northern Germany, two recurrent ammonoid biofacies were distinguished long ago (Westermann, 1954; Geraghty and Westermann, 1994). (1) The Sonniniid Biofacies, mainly in dark-gray shales, has dysaerobic benthic "flat clam" posidoniids together with Sonniniidae that range from costate platycones (some *Sonninia* macroconchs and most microconchs) to smooth oxycones (*Dorsetensia* macroconchs). (2) The Stephanoceratid ("Stephanoid") Biofacies, mainly in silty shales, has an abundant aerobic bivalve fauna, including large, deep-infaunal suspension feeders, together with coarsely sculpted, planorbiconic Stephanoceratinae (*Stephanoceras*, *Teloceras*). The stephanoceratids, also strongly dimorphic (sometimes with high sex ratios that are opposed in related and associated species; Geraghty and Westermann, 1994), were probably sluggish demersal. This is supported by their exceptionally common shell fracture by substrate impact (Bayer, 1970). Most sonniniid macroconchs were well adapted for predation in midwater. Evidence of the pelagic habit of sonniniids comes also from aptychi found in deep-sea deposits (Gasiorowski, 1962). The most common ammonite, *Stephanoceras mutabile* (Quenstedt), lived to about 65–70 m depth (Chapter 10, this volume), indicating that basin depth of the Stephanoceratid Biofacies was 50–65 m and that of the Sonniniid Biofacies was roughly 80–100 m. Although the outcrop with these biofacies was very limited (Alfeld area), abundant subcrop indicates the extensive distribution of these facies through much of the North-German Basin. Significantly, S. Elmi (personal communication, 1994) has observed these facies associations also in Morocco and Algeria. Elsewhere, both "communities" presumably lived at different depths in the same, fully oxygenated sea where they were subsequently buried together.

As in the other lithofacies, most ammonite occurrences are enriched at some levels and impoverished in the larger intervals, even through seemingly monotonous oxic shale facies (see Callomon, 1985; J. H. Callomon, personal communication, 1992–1993), and the shells are almost exclusively adult. The most probable of several possibilities are that (1) the ammonoids schooled; (2) the numerous young were largely eaten by fish and/or larger ammonoids; (3) death followed synchonized spawning, here mostly in the water column (see Section 5); and (4) the empty shells drifted for a few hundred meters to several kilometers at the surface, within the water column, and/or at the floor, during which they were sorted, dispersed, or accumulated before being embedded.

4.4.2. Cretaceous of the Western Interior Seaway

One of the best-known epeiric seas is the Cretaceous Western Interior Seaway of North America that, in its heyday, connected the tropical waters of the Gulf with the Boreal sea. The lithofacies–biofacies interrelations have been worked out in great detail for the Albian–Turonian Greenhorn Cyclothem of the northwestern United States (Kauffman, 1984; Batt, 1989, 1991, 1993; R. J. Batt, personal communication, 1993). The extremely rich ammonoid fauna includes a great diversity of Ammonitina and Ancyloceratina that Batt (1989) placed into 18 morphogroups (Fig. 12). Their habitats are here partly reinterpreted.

- *Group 1.* Heavily sculpted inflated planorbicones with quadrate whorl, large (e.g., adult Acanthoceratidae). Juveniles pelagic; adults sluggish demersal, shallow to moderate depth. (Quadrate whorl section excludes depth greater than ca. 100 m.)
- *Group 2.* Coarsely costate, widely umbilicate spherocones, large (e.g., *Calycoceras*). Demersal or pelagic, somewhat deeper than group 1.
- *Group 3.* Highly spinose, inflated planorbicones and cadicones with quadrate whorl section, midsize (e.g., *Kanabiceras*). Probably quasiplanktic in midwater (whorl section too weak for basinal depth; entered seaway during late and peak transgression).
- *Group 4.* Highly variable, coarsely nodose spherocones to weakly nodose platycones, quadrate-rectangular whorl section (Gastroplitidae, e.g., juvenile and ?microconchiate *Neogastroplites*). Quasiplanktic. (Part of the variation may result from mixing.)

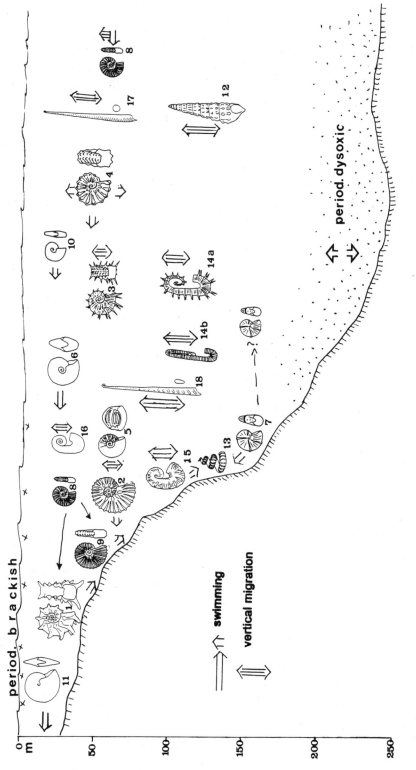

FIGURE 12. The Late Cretaceous morphogroups of Batt (1989) and their principal habits in the Western Interior Seaway of North America, partially revised (for names of morphogroups and examples, see Section 4.4.2 in text). Thin arrows indicate ontogenies; dashed arrow with query, possible alternative habit.

- *Group 5.* Cadicones (and spherocones?) with costate juvenile and smooth adult, suture complicated (e.g., *Fagesia*). Quasiplanktic, adults possibly sluggish demersal, moderately shallow.
- *Group 6.* Discocones with costate juvenile and smooth adult stages, suture generally simple (*Vascoceras*). Nektic, adult especially in shallow water. (Septal amplitude index varied with whorl inflation, see Section 3.3.2; septa of *Vascoceras* from elsewhere indicate only 20 m to 45 m maximum habitat depth; Chapter 10, this volume.)
- *Group 7.* Finely costate, inflated discocones with varices, siphuncular and sutural indices high (Desmoceratidae). Nektic and/or demersal in deep entrance of seaway.
- *Group 8.* Small, costate, evolute platycones and/or compressed serpenticones (e.g., *Neocardioceras*). Planktic in middle or upper midwater.
- *Group 9.* Larger platycones, juvenile (and microconch?) costate but large whorls smooth, suture simple to ceratitic (*Metoicoceras*). Juveniles (and microconchs?) probably sluggish nektic; large adults mobile demersal or nektic; shallow.
- *Group 10.* Very small, almost smooth platycones, suture subammonitic (e.g., *Borissjakoceras*); nektic in upper midwater ("streamlined").
- *Group 11.* Large, mostly smooth oxycones (Placenticeratidae). Mobile swimmers with good steerage, *Placenticeras* in brackish surface water. (See Section 4.3.3e; *Neoptychites* and *Choffaticeras* differ by their discoconic and triangular whorl sections and simplified sutures; distributions suggest nektic habits in offshore midwater.)
- *Group 12.* Tightly coiled, elongate torticones, suture complicated (e.g., *Turrilites*). Pelagic or ?demersal vertical migrant. (Mostly drifted postmortem to near shore. Although restriction to aerobic facies is consistent with demersal, a pelagic habitat is indicated for *Turrilites* by its cosmopolitan occurrence in deep-water and oceanic assemblages.)
- *Group 13.* Loosely coiled torticones (*Puebloites*). Demersal?. ("Benthic crawler" hypotheses need evidence for marked negative buoyancy.)
- *Group 14.* Subdivided into *14A*, ancylocones with gyroconic juvenile stage (e.g., *Allocrioceras*), and *14B*, hamitocones (Ward, 1976a), with several straight juvenile shafts (e.g., *Stomohamites*). Both planktic drifters and vertical migrants in deeper midwater (>80–120 m).
- *Group 15.* Scaphitocones (Scaphitinae); juvenile probably planktic, but adult mostly demersal vertical migrant, cool water. (In seaway restricted to northern part and aerobic facies.) Some compressed forms (*Hoploscaphites*) were nektic.
- *Group 16.* Otoscaphitocones (e.g., *Worthoceras*). Vertical migrant in upper midwater.
- *Group 17.* Small, round orthocones (*Sciponoceras*). Vertical migrant in upper midwater.
- *Group 18.* Large, compressed orthocones (*Baculites*). Quasiplanktic, vertical migrant and potential demersal saltator with sluggish horizontal

mobility; adults of giant species probably horizontal swimmers (Section 4.3.3e). (Vertical mobility would have permitted brief feeding incursions into dysoxic bottom waters.)

In conclusion, the Greenhorn Cyclothem contained the majority of ammonoid morphotypes known from the Cretaceous, except those particular to the absent oceanic Lytoceratina (smooth, round-whorled, advolute planorbicones). The large discoconic shapes of the oceanic Phylloceratina, also missing, are largely represented by the Desmocerataceae (Ammonitina), for which a marginally oceanic habitat is assumed. Compared to the epicratonic taxa of the Triassic and Jurassic, highly sculpted morphotypes are overrepresented as elsewhere in the Cretaceous (Ward and Signor, 1983), as are the newly evolved heteromorphs. Strongly underrepresented are large evolute planorbicones with prominent costae (mostly demersal), discocones (mainly nektic), spherocones and elliptospherocones (mainly planktic), and gyrocones (planktic and pseudoplanktic in algal mats). The habitats of several Greenhorn morphogroups are here interpreted as shallower and more pelagic than suggested previously. Many of these morphogroups had similar adaptations in other taxa and periods, so that a moderately expanded and revised series probably will become very useful in future ecologic analyses of Mesozoic ammonoid assemblages.

4.5. Epeiric Reef Slopes and Carbonate Platforms

4.5.1. Reef Slopes

Only communities of the lower parts of epeiric reef slopes are known to have included ammonoids; those found on upper reef slopes were drifted and trapped there (Ford, 1965). Ramsbottom (1978) reconstructed two lower-reef-slope communities from the Carboniferous of Britain.

In the calcilutites and calcarenites of the Molluscan Community, in- and epifaunal bivalves and the spheroconic *Goniatites* and *Bollandoceras* dominated; crinoids, corals, and bryozoans were accessory, but algae were missing. This epeiric sea floor was relatively hard, below wave base, possibly below the photic zone, but well oxygenated.

The Brachiopod Community differed in the softer substrate without infaunal shelly fauna and in the cephalopod fauna of orthoconic nautiloids.

Perhaps the difference in cephalopod faunas resulted from different (unknown) food preferences (Ramsbottom, 1978). These goniatites were probably planktic and microphagous. The orthoconic nautiloids probably were demersal (Hewitt and Watkins, 1980). (For oceanic reef-complex slopes, see Section 4.2.2a.)

4.5.2. Carbonate Platforms

4.5.2a. Devonian. Condensed pelagic and commonly nodular limestones of Morocco and Europe contained a diverse cephalopod fauna (R. T. Becker, personal communication, 1993). They formed on distal platforms, carbonate banks, and shelf slopes at moderate (but disputed) depth and are commonly bituminous (see Section 4.3.2). Clymeniids showed a morphological depth gradient, with coarse ribbing present in shallow habitats and fine ribbing in deeper habitats (Korn, 1986, 1988, 1992; Section 4.3.3a).

4.5.2b. Carboniferous. Ammonoid-rich carbonate platforms were uncommon in this period and are known mainly from the southern Urals, where there was a great diversity of taxa and morphotypes, with many endemic genera (Saunders and Swan, 1984). The lithofacies of the Namurian Ural sections consists exclusively of structureless fine- to coarse-grained limestone, with bedding developed in crinoidal facies only. Nautiloids as well as ammonoids occur in large accumulations, together with rare other molluscs, solitary corals, brachiopods, and trilobites. The ammonoids vary in size but are usually adult; ammonitellas and immatures are absent. Saunders and Swan (1984) proposed that these shell concentrations were formed by episodic post-mortem transport followed by periods of fine-grained carbonate sedimentation. They concluded that the extraordinary taxonomic diversity indicates warm, shallow-platform deposition, possibly with rich algal association.

The morphotype distribution in the Namurian of the Urals was heavily slanted toward evolute ammonoids, i.e., morphotypes II, VII, and VIII of Saunders and Swan (1984; Saunders and Shapiro, 1986; Swan and Saunders, 1987; Figs. 4, 10). Morphotype II encompasses essentially the Prolecanitida: moderately evolute in the early whorls, compressed, smooth, thick-shelled, brevidomic forms with a small hyponomic sinus, that were highly stable and potentially good, probably demersal swimmers. This morphotype changed little through much of the Carboniferous and Permian, and the prolecanitids were ancestral to the Mesozoic ammonoids. Morphotypes VII and VIII were cosmopolitan but less abundant than morphotype II in all facies (at least juveniles). They were unstable longidomes, including the serpenticonic gastrioceratids *Cancelloceras* and *Phillipsoceras* that were probably planktic (although Swan and Saunders, 1987, attributed a "benthic" habitat to morphotype VIII). The ancillary morphotypes V and VI of the Ural platform were longidomic discocones. They are known from all Namurian marine facies and probably were pelagic, with the potential of occasional bottom feeding.

4.5.2c. Permian. Carbonate-platform facies were widespread at this time, including North America. Some of the Late Permian to earliest Triassic extreme forms, such as the otoceratacean families Araxoceratidae and Otoceratidae (see Glenister and Furnish, 1981), had uniquely "tectiform" whorl sections (subtriangular with concave flanks) that, based on structural grounds (Westermann, 1990), were extremely weak against ambient pressure. The otoceratids lived on shallow carbonate platforms in China and Tibet (see Section 4.5.2d), but the former are found in dark shales with calcareous nodules and greywacke in Coahuila, Mexico (Spinosa *et al.*, 1970), suggesting a pelagic habit. The ancestral Xenodiscaceae included the advolute, compressed *Paraceltites*, which was widespread in tropical-platform limestones and more rarely occurred in shallow-water shale (Spinosa *et al.*, 1975). Its outer whorls, with simple blunt lateral costae, closely resemble planktic Jurassic serpenticones (e.g., Psiloceratidae), but the body chamber was brevidomic (and presumably thick-shelled), giving the organism high stability; the juvenile stage is heavily nodose. This suggests a demersal habitat.

4.5.2d. Triassic. Several platform facies with their characteristic ammonoid "communities" have been distinguished in China (Fig. 13; Wang and Westermann, 1993). From shallow to moderately deep (for basin facies see Section 4.2.2a), these are:

1. Restricted platforms with a *Tirolites–Dinarites* Community, surrounding Early Triassic (Spathian) evaporite basins. Dolomitic limestone and calcareous dolomites had a low-diversity molluscan fauna: euryhaline epifaunal and semiinfaunal bivalves (*Eumorphotis, Unionites, Myophoria*) and small, platyconic to planorbiconic ceratites, commonly with subrectangular whorl section and mostly smooth or with nodes or spines, and sutures goniatitic to subceratitic (Tirolitidae, Dinaritidae). A similar facies in the Middle Triassic of China yields similar nodose platycones (*Progonoceratites*). These ceratites presumably were euryhaline demersal swimmers in very shallow, supersaline waters.

2. Shallow platform with a *Otoceras* Community; latest Permian/earliest Triassic (Griesbachian) transgressive phase. Shallow-water carbonates yield bivalves, brachiopods, ostracods, etc., and some ceratites; *Otoceras* has subtriangular, concave-sided whorls; *Hypophiceras* is platyconic. They were probably shallow-water, demersal swimmers.

3. Offshore platform and shallow basin. (3a) *Ophiceras* Community (Griesbachian); carbonates with some bivalves and ceratites with low diversity; mostly evolute, smooth or finely sculpted; some discocones and cadicones. (3b) *Paranorites* Community (upper Dienerian); brown wackestones with gastropods, bivalves, and discocones (*Paranorites, Clypeoceras*); good swimmers. (3c) *Anasibirites* Association (Smithian) with similar lithofacies, has brachiopods and mostly heavily sculpted planorbicones (Stephanitidae, Sibiritidae, Kashmiritidae); also a few smooth platycones (Prionitidae); mostly sluggish ?demersal. (3d) *Gyronites* Community (lower Dienerian). (3e) *Hedenstroemia* Community (Smithian); light-gray wackestone with bivalves and moderately diverse ceratite fauna that includes more discocones and oxycones with more complicated sutures than in 3a–c; nektic and ?demersal.

In conclusion, the Early Triassic diversification of ammonoids was a global phenomenon after the protracted Permian/Triassic extinction (Fig. 13). Extinction of shallow-water species was caused by extreme regression; the following widespread anoxia led to the extinction of deep-water species during the Griesbachian transgression (Wignall and Hallam, 1993). This resulted in the survival of only a very few shallow-water taxa (Wang and Westermann, 1993). Radiation was slow: Early Triassic ammonoids frequently

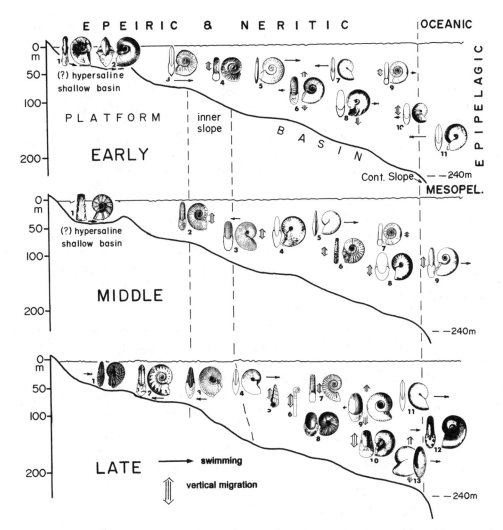

FIGURE 13. Stylized panoramas of Triassic ammonoid habitats (with substantial modifications of the recent reconstructions by Wang and Westermann, 1993, Figs. 10–12). Present interpretations suggest more pelagic and fewer demersal habits, many of them shallower than assumed previously. Thick arrows indicate swimming speed; triple arrows, vertical migration. Early Triassic: 1, *Tirolites*; 2, *Otoceras*; 3, *Inyioites*; 4, *Hellenites*; 5, *Gyronites*; 6, *Anasibirites*; 7, *Hedenstroemia*: 8, *Isculitoides*; 9, *Leiophyllites*; 10, *Paranannites*; 11, *Procarnites*. Middle Triassic: 1, *Ceratites*; 2, *Anolcites*; 3, *Trachyceras*; 4, *Beyrichites*; 5, *Longobardites*; 6, *Balatonites*; 7, *Leiophyllites*; 8, *Ptychites*; 9, *Monophyllites*. Late Triassic: 1, *Tibetites*; 2, *Distichites*; 3, *Acanthinites*; 4, *Discotropites*; 5, *Cochloceras*; 6, *Rhabdoceras*; 7, *Choristoceras*; 8, *Juvavites*; 9, *Tropites*; 10, *Cladiscites*; 11, *Pinacoceras*; 12, *Rhacophyllites*; 13, *Arcestes*.

lived on platforms and above basin slopes, a few also in the basins, probably all above a depth of 120 m. The majority probably were nektic (not demersal as assumed by previous authors) (e.g., *Gyronites, Pseudosageceras, Procarnites, Hedenstroemia*), some demersal (e.g., *Otoceras*), and others planktic (e.g., *Ophiceras, Dieneroceras, Leiophyllites*).

Diversity increased at the end of the Early Triassic (Spathian); some Tirolitidae probably were demersal on shallow platforms that sometimes became hypersaline, whereas the elliptospheroconic *Protropites, Juvenites*, and *Chiotites* presumably were planktic, vertical migrants in offshore basins. In the Middle Triassic, the Ceratitidae became the principal inhabitants of the shallow, epeiric platforms and also may have been euryhaline. In the German Muschelkalk, iterative sutural simplification occurred in several immigrant clades from the deeper Tethys (Urlichs and Mundlos, 1985). Coarse sculpture is also commonly associated with basin-slope habitat, whereas basinal pelagic habitat is usually indicated by smooth or weakly sculptured discocones. A few taxa, e.g., *Monophyllites* and *Psilosturia*, lived on the continental slope of the oceanic region, but, according to their shell strength, were still epipelagic (see Section 4.2.2a; Chapter 10, this volume). In the Late Triassic, radiation affected almost all existing morphs: the proportion of oceanic leiostraca greatly increased, and the first heteromorphic ammonites evolved, i.e., torticones, orthocones, and gyrocones.

4.5.2e. Jurassic. Submarine fissures and caves in Triassic–Jurassic outerneritic limestones of the Mediterranean contain dwarfed ammonoids that were washed into them during periods of nondeposition (Wendt, 1976). This fauna may have lived in benthic algal mats, as indicated by associated microgastropods (Sturani, 1971).

In the Toarcian of the western Mediterranean, tilted platforms have the following fauna (for basins see Section 4.3.1) (Almeras and Elmi, 1982; Elmi, 1985):

1. Distal carbonate platform (open shelf F1). Brachiopods are present only in low-energy carbonates and "flat clam" pectinoids in dysaerobic areas; pelagic ammonoids are abundant, with the nektic *Hildoceras* and planktic dactylioceratid *Collina*.
2. Proximal carbonate platform (shallow shelf F4). Benthic fauna is diverse and equitable; rare Ammonitina are finely to moderately costate platycones that elsewhere are associated with dysoxia and, hence, probably were nektic (*Hildoceras, Pseudogrammoceras*); other taxa are generally common in carbonate facies and presumably were demersal swimmers (*Hammatoceras*).

In the Middle Jurassic of Oregon, Taylor (1982) distinguished several shallow marine depth zones, of which only the relatively deeper ones include Ammonitina.

In the Callovian sequence of the Iberian Chain, Spain, vertical and lateral lithofacies changes were documented together with Ammonitina taxonomic composition (Sequeiros, 1984) (probable habitats here added in parentheses). Cadiconic and elliptospheroconic Tulitidae are most common in calcareous marls (planktic, adults vertical migrants); compressed-planorbiconic Perisphinctidae in limestone (sluggish ?demersal). Oxyconic Oppeliidae are concentrated in marly or high-energy interbeds but are scarce in the carbonates (mobile nekton). The relative abundance of the Perisphinctidae varies inversely with that of the Oppeliidae and discoconic Macrocephalitinae (mobile nekton); the latter dominated in turbulent facies. These perisphinctids belonged to the Pseudoperisphinctinae with moderately "streamlined" adult macroconchs. These presumed females may also have preferred nektic habits rather than demersal, as assumed for most coarsely sculpted perisphinctids.

The advanced bathymetry of Gygi (1986) for some Upper Jurassic epeiric sequences of central Europe has refined the classical work (e.g., Ziegler, 1967, 1980). In a characteristic lower Kimmeridgian succession in northern Switzerland, situated at the transition from a eustatically subsiding platform to a basin, the response of shelly macrofauna to sea floor depth was said to be as follows:

1. Depth <30 m. Mainly benthos (colonial corals, etc.).
2. About 30 m. Benthos (mostly bivalves) outnumber ammonites (20%); Perisphinctidae outnumber Aspidoceratidae.
3. About 40–50 m. Benthos equals ammonites; Perisphinctidae outnumber Aspidoceratidae; few Oppeliidae.
4. About 80 m. Ammonites (80%) and scarce but equitable benthos; Perisphinctidae outnumber Aspidoceratidae and Oppeliidae combined.
5. About 120 m. Ammonites (80%) and scarce benthos of mostly brachiopods; Perisphinctidae equal Aspidoceratidae, Oppeliidae, and Cardioceratidae combined.
6. About 150 m. Benthos almost absent; Oppeliidae outnumber Perisphinctidae, rare Aspidoceratidae and ?Cardioceratidae.

The heavily sculpted aspidoceratid *Paraspidoceras*, Batt's (1989) demersal morphogroup 1, supposedly was demersal at a depth of ca. 100 m in a fauna dominated by Perisphinctidae (45%) and Oppeliidae (39%) (Gygi *et al.*, 1979), i.e., Ziegler's 80- to 100-m assemblage. In Italy, similar aspidoceratids occur in a shallow-water, stromatolitic facies (Clarim *et al.*, 1984) together with coarsely costate perisphinctids that presumably were also demersal.

In light of new strength data for septal failure (Chapter 10, this volume), however, all inferred habitat depths listed above are too deep and need to be reduced by 20–40%.

The famous Tithonian Solnhofen Plate-Limestone of southern Germany has been interpreted as being deposited in a large lagoon behind sponge reefs and with inhospitable benthic conditions (hypersaline and/or dysoxic), so that some authors proposed that at least the benthos and ammonoids were entirely allochthonous, i.e., swept by large waves across the barrier (Barthel *et al.*, 1990). The abundant record of aptychi *in situ* within the body chambers of haploceratid ammonites (Schindewolf, 1958) and even in holostean fish feces (Mehl, 1978a), however, are good evidence that at least the Haploceratidae belonged to the pelagon of the "lagoon."

A modified version of my earlier overview of Jurassic–Cretaceous ammonoid habitats (Westermann, 1990) is presented in Fig. 14.

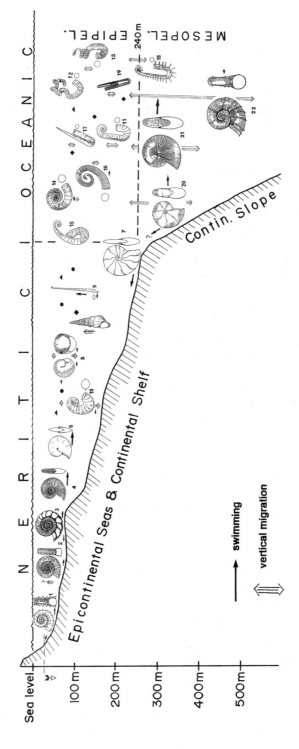

FIGURE 14. Stylized panorama of Jurassic–Cretaceous ammonoid habitats (with substantial modifications of Westermann, 1990, Fig. 7). As for the Triassic (Fig. 13), present interpretations suggest more pelagic, often shallower habitats and more frequent vertical migration. Ammonitina: 1, *Peltoceras*; 2, *Arietites*; 3, *Perisphinctes*; 4, *Harpoceras*; 5, *Sphaeroceras*; 6, *Oxycerites*; 7, *Barremites*. Ancyloceratina: 8, *Turrilites*; 9, *Baculites*; 10, *Scaphites*; 11, *Ancyloceras*; 12, *Nipponites*; 13, *Didymoceras*; 14, *Crioceratites*; 15, *Labeceras*; 16, *Glyptoxoceras*; 17, *Hamulina*; 18, *Anisoceras*; 19, *Pseudoxybeloceras*. Phylloceratina: 20, *Holcophylloceras*; 21, *Phylloceras*. Lytoceratina: 22, *Lytoceras*.

5. Discussion

5.1. Growth Rate and Longevity

Postneanic Mesozoic ammonoids presumably resembled *Nautilus* in the roughly linear rate of growth (Saunders, 1984; Hewitt *et al.*, 1994; Chapter 12, this volume). That is, the ventral length of the coiled shell grew at a more or less constant average rate throughout much of juvenile and adolescent ontogeny (see Section 5.6), notwithstanding short-term interruptions as indicated in many Jurassic–Cretaceous ammonoids by constrictions and megastriae. Maximum growth rates of ammonoid shells probably depended more on chamber growth than on apertural growth. In many if not all ammonoids, septal growth probably took longer than chamber emptying, which decreases roughly linearly with diameter because of its dependance on the ratio of siphuncular surface/cameral volume (Chapter 12, this volume; furthermore, siphuncular radius tended to grow with negative allometry, further decelerating the pumping rate; cf. Westermann, 1971). Because, during the major growth phases, septal thickness tended to remain proportional to septal spacing (chamber length), i.e., the septation retained self-similarity (gnomic growth; cf. Westermann, 1977), cumulative septal thickness per cone or coiling length also remained similar. Constant septal secretion therefore implies a linear growth rate of the phragmocone with limitations set by the condition of neutral buoyancy. Thus, in the common ammonoids that doubled in diameter with each whorl ($W = 2$), the last whorl would have taken about as long to grow as all previous whorls combined, and the chamber (and whorl) production rate decelerated commonly 100 : 1 to as much as 500 : 1 through neanic and juvenile growth (e.g., seven whorls from 1 mm to 128 mm diameter, 128 : 1; eight whorls to 256 mm diameter, 256 : 1; etc.).

The neanic growth rate may or may not have differed greatly from that of the juvenile, especially if there was a major habitat change, e.g., from planktic to demersal. Absolute growth rates depended on a number of endo- and exogenous factors, including shell thickness, connecting ring radius and length, ambient hydrostatic pressure and temperature, food supply and feeding rate, and adult diameter. Significantly, several of these factors combined to slow the growth rate with increasing habitat depth (e.g., increasing pressure reduced the osmotic emptying rate; increasing shell thickness required additional carbonate secretion; decreasing temperature and food supply slowed the metabolism). Accordingly, total life span may have varied greatly, from 1 to 2 years for small, shallow-water ammonoids (<50 mm diameter) to 5 to 10 years for the majority of epeiric and inner-neritic ammonoids, to as much as 50 or 100 years in mesopelagic giants such as the *Lytoceras* of Stevens (1988). But individual rates are extremely hard to come by (Chapter 12, this volume). In general, rates were higher and size larger in shallow platform dwellers than in basin dwellers (Elmi and Benshili, 1987).

5.2. Assumption of Neutral Buoyancy Reexamined

Most calculations and experiments of ammonoid buoyancy that were based on the volumes and estimated (relative) densities of hard and soft tissues as well as on exceptional epizoan loads (attached to the living animal) have supported the hypothesis that all ammonoids were neutrally buoyant at all growth stages (Landman, 1987; Landman and Waage, 1993; Okamoto, 1988a,b; Olivero and Zinsmeister, 1989; Reyment, 1980; Saunders and Swan, 1984; Saunders and Shapiro, 1986; Tanabe and Ohtsuka, 1985; Tanabe *et al.*, 1995; Ward 1976a; review by Westermann, 1993; Chapters 11 and 12, this volume). The quantity of water retained in the shell is unknown, so that calculations of positive buoyancy are irrelevant here. (But note that *Nautilus* is a few grams heavier than sea water: Ward, 1986a). It has to be kept in mind that the sum of uncertainties contained in the density estimates and measurements cannot be reduced to less than about 5% within acceptable confidence limits, besides the hidden potential distortions and volumetric changes from diagenesis. Roughly neutral buoyancy also was estimated for juvenile orthoconic nautiloids (Hewitt, 1988) and was presumably also present in the similar ancestral ammonoids, i.e., the orthoconic to cyrtoconic bactritoids.

There are two notable exceptions to the neutral-buoyancy hypothesis alleging consistent densities higher than sea water for most or all Mesozoic ammonoids. Ebel (1983) concluded and later (Ebel, 1985, 1992) supported his opinion that all (or most) planispiral and heteromorphic ammonoids were benthic crawlers, whereas Shigeta (1993) concluded that the strongly negatively buoyant juveniles and ?adults of all or most Cretaceous ammonoids were demersal or nektic. Shigeta's calculated densities of close to 1.15 g/cm^3 are even higher than those of Ebel and would clearly have prevented any prolonged swimming, demersal or nektic. Both conclusions have recently been rejected (Westermann, 1993).

Ebel's (1983) original density estimates were mostly within acceptable limits of error from neutral buoyancy and based on very rough techniques that tended to overestimate density, i.e., the visual matching of triangles, rectangles, or trapezes to all whorl sections; Ebel applied the same relative shell thickness for all taxa, i.e., the mean with high variance obtained by Westermann (1971) for different families. Furthermore, body chamber angle ("length") was probably overestimated in several extreme cases.

Shigeta (1993) greatly overestimated septal shell volume by erroneously assuming that septal thickness was constant from center to fluted margin (but see Westermann, 1975; Hewitt and Westermann, 1987) and that fluting increased the septal surface by the same factor as the suture lengthened. After correction of the volume proportions for septa (i.e., the septa together comprising 30–50% versus >100%, of the phragmocone wall), overall densities of the later neanic and the juvenile (and adult) stages reduce to about 0.90–0.95 g/cm^3—the same as calculated by Shigeta (1993) for the hatchling (cf. Westermann, 1993). However, even the reduced value of 5% for septal versus

671

cameral volume, as assumed in these revised calculations, is too high for most Ammonitina and Goniatitina. As can be seen in sagittal sections (e.g., Tanabe *et al.*, 1993a, Figs. 2–4), maximum septal thickness tends to be only 2–3% of septal spacing. It is not by accident that septal thickness and spacing covaried in nautiloids and ammonoids (Westermann, 1977; Hewitt and Westermann, unpublished). Any deficiencies between the calculated specific weights of animal and seawater may be accounted for by the unknown amounts of cameral liquid and membranes, which together could have amounted to more than 10% of phragmocone volume, especially in the growing ammonoid.

The only other recent density analysis for adult ammonoids supposedly indicating negative buoyancy of 1.067 g/cm^3, i.e., the Cretaceous heteromorph *Moutoniceras* (Delanoy *et al.*, 1991), is within the 5% error limit set above for meaningful results—surprisingly close considering that the thickness data for shell and septa were obtained from other taxa. The sensitivity of buoyancy values to measurement error (assumed in this case) is easily demonstrated. For example, if one were merely to substitute in the equation of Delanoy *et al.* (1991) the average value of 6% for the septa/total shell proportion (Ward and Westermann, 1977) (or the 4% of *Eubostrychoceras*: Okamoto, 1988a) for the supposed *Diplomoceras* value of 13% (Olivero and Zinsmeister, 1989), the calculated buoyancies would become positive in sea water, i.e., densities approximately 1.0 and 0.98, respectively.

The consistent reduction in body chamber angle ("length") in the adults of planispiral Ammonitina (Westermann, 1971; Lehmann, 1976) indicates that overall density did not increase ontogenetically; this reduction was compensated for by apertural modifications and/or test thickening (Westermann, 1990). In most deep-water heteromorphs (Ward, 1976a) and the oceanomesopelagic lytoceratids and phylloceratids (Birkelund, 1981) as well as in *Nautilus*, however, the body-chamber wall is relatively much thinner than the phragmocone wall (and the body chamber angle presumably not reduced), a promising new bathymetric indicator.

There is little hope of using buoyancy data to distinguish between demersal and pelagic forms and between swimmers and drifters; there is no reason why demersals should have been less buoyant than pelagics. The very slight negative buoyancy of a (demersal) *Nautilus* weighing 1–2 g in sea water (Ward, 1986a), cannot be documented by shell measurements. Pelagic drifters (megaplankton) presumably were exactly neutrally buoyant.

The fact that *Nautilus* does not change its buoyancy by cameral water volume for diurnal/circadian vertical migration (Ward, 1986a, 1987) does not imply that such buoyancy regulation was absent in all ammonoids. *Nautilus* is a demersal deep-water form that requires a thick-walled siphuncle with small diameter (high siphuncular strength index, SiSI; Westermann, 1971, 1982; Chapter 10, this volume). The rapid "pumping" rates required would have been limited to ammonoids (and nautiloids) with large siphuncular radii that are found only among shallow-water forms with small SiSI, such as the Late Cretaceous large ammonite *Placenticeras*. Yet there remains the problem of distinguishing between the two possible major functional adaptations of large siphuncles, i.e., rapid growth (high pumping rate) and rapid buoyancy control for vertical migration (or a combination of the two).

5.3. Soft Body Reconstructions

It is probable that all ammonoids had ten arms as do the extant teuthids (Mehl, 1984; Jacobs and Landman, 1993), but this is by no means certain, and ammonoids without arms have been envisioned by, e.g., Seilacher (1993). The reconstructions in Fig. 5 shows the possible presence of umbrella webs and cirri on the arms as well as the potential secondary function of long tentacles as steering devices during backward swimming.

The spurious recent reconstructions by Ebel (1985, 1992) of ammonites as generalized epibenthic organisms were based on his severely flawed estimates of negative buoyancy, discussed above, and on additional erroneous assumptions of shell growth.

Contrasting with these works are the detailed ultrastructural and morphological studies by Doguzhaeva and Mutvei (1989, 1991, 1992, 1993). They concluded that the shells of the Cretaceous "*Aconeceras*" [*Sinzovia* (Ammonitina); A. C. Riccardi, personal communication, 1991], *Gaudryceras* (Lytoceratina), and *Ptychoceras* (Ancyloceratina) were semiinternal or even belemnite-like internal and that the body commonly was markedly larger than could have been retracted into the shell. The semiinternal condition, with the mantle overlapping part or all of the body chamber, however, differs very significantly from the fully internal condition in which the body was many times the volume of the shell. The semiinternal condition, with the very thin mantle folding back over approximately the last quarter-whorl of the body chamber, was demonstrated by shell duplication for the oxyconic *Sinzovia trautscholdi* (Sinzov). The animal was able to withdraw into its half-whorl (i.e., "normal" brevidomic) body chamber, as the authors themselves demonstrated by their frequent finds of radula and mandibles/aptychi in the body chambers of this species. Ammonitina buccal mass organs *in situ* generally are found about one-third to half-way back in the body chambers in supposed jaw-function position and only rarely in apertural, operculum-function position (Schindewolf, 1958; Lehmann, 1976; Kennedy and Cobban, 1976; Tanabe, 1983; Tanabe and Fukuda, 1987; Mapes, 1987; Nixon, 1988). Shell-wall duplication is also known from the inner four to five whorls (5–6 mm shell diameter) of *Gaudryceras* (Birkelund, 1981), indicating that its (planktic) juvenile shell was exceptionally (?pathologically) covered by mantle.

For the hamitid heteromorph *Ptychoceras*, Doguzhaeva and Mutvei (1989) considered (among other models) an internal shell position in a belemnite-like organism, mainly based on the duplication of the entire shell wall. This hypothesis has caused considerable excitement among ammonitologists but cannot be upheld. Curiously, these authors attempted no buoyancy calculations. The inner, original shell wall had a thickness of about 4% of the whorl radius, as is characteristic for ammonites, but the secondary shell was at least as thick again. Judging from the illustrations, the entire wall thickness amounted to 8% to 10% of the radius. The simplest estimate of (overall) phragmocone density can be done as follows: (1) for the shell wall, multiply the density of nacre, 2.6, by twice the thickness/radius ratio; (2) for all septa (cumulative), multiply 2.6 by the ratio of thickness/cameral length, both measured at septum center; add (1) and (2). In the case of *Ptychoceras*, the estimate for phragmocone density is 0.5–0.6 (unusually high for ammonoids, which usually have values between 0.2 and 0.3). This implies that the phragmocone could buoy up a body-chamber (with body) of only similar volume and relative wall thickness—resembling the present shell. That is, the organism was approximately neutrally buoyant, assuming the model of a semiinternal shell and a soft body of roughly body chamber volume. In contrast, the belemnite-like model of an internal shell within a much larger cephalopod is disproved by the extra buoyancy requirement for a soft body many times the size of the shell. Because soft tissue is about 3% heavier than sea water (1.06 versus 1.03 g/cm^3), this particular *Ptychoceras* phragmocone (using the mean of the density estimate above, 0.55) could buoy up approximately 16 times its own volume of soft tissue $[(1.03 - 0.55)/(1.06 - 1.03) = 16]$, i.e., about the proportion shown in the model *without* the body chamber. It follows that the body could not have been much larger than in the semiinternal model.

There is another, simpler argument against the internal shell model. The hamitid shell, with repeated narrow half-turn bends, could not have grown past these bends; the required rotation would have put the shell repeatedly crosswise, distorting the body beyond reasonable limits (in contrast to the strictly planispiral *Spirula* shell that remains self-similar)—unless one would assume that the shell grew alternately forward and backward.

Doguzhaeva and Mutvei (1991) also documented and confirmed Jordan's (1968) observations of large ventrolateral muscle scars, which they interpreted as funnel muscle attachments, in the brevidomic oxycones *Sinzovia* ("*Aconeceras*"), *Quenstedtoceras*, and *Amaltheus* (but see Jacobs and Landman, 1993, for a different interpretation). This indicates much better swimming-potential than commonly assumed, significantly for ammonoids that also had exceptional stability, steerage, and acceleration.

Based on the valid assumption that longidomic body chambers permitted farther extrusion of the soft body than in the brevidomic *Nautilus* (Jordan, in Westermann, 1990), and with emphasis on the coleoid analogy, Jacobs and Landman (1993) recently advanced new models for mantle cavity contraction in ammonoids: (1) the mantle cavity extended entirely outside the body chamber, where it was free to expand and contract in the usual coleoid manner, or (2) the mantle pump functioned in an apparent combination of the *Nautilus* and coleoid "pumps," within as well as outside the body chamber; either way, the mantle slides effortlessly along the inside of the body chamber.

My main arguments against these highly imaginative hypotheses are that most Mesozoic longidomic shells, i.e., serpenticones, have nearly the worst hydrodynamic and hydrostatic properties for swimmers (see Section 3.1.1c) and that the "piston pump" of model 2 would have been highly inefficient in these shells with narrow, rounded apertures, because the relatively high-velocity incurrent would have opposed the propulsive current in the backward-swimming animal; only in longidomic discocones, rather abundant in the Paleozoic, would such a mechanism possibly be feasible (Fig. 10).

I have suggested that some thinly calcified aptychi/mandibles may have functioned also as louver fans in long mantle cavities (Westermann, 1990). Finally, there is the old suggestion that ammonoids without a hyponomic sinus may have had paired hyponomes (Schmidt, 1930). The fact that all present-day cephalopods have single hyponomes (Nesis, 1987), however, makes this highly unlikely. Largely imaginary soft body reconstructions are here presented in Figs. 5, 6, 8–11, and 15.

5.4. Crop/Stomach Contents, Buccal Organs, and Feeding Strategies

In extant very young cephalopods, the prey consists mainly of mobile minute animals, such as crustaceans, that are pounced upon by rapid "jet" propulsion before being captured either by ejectile tentacles (decapods) or by the arms (octopods) (Nixon, 1988). But arms and tentacles (if present) are little developed at this stage in all extant pelagic cephalopods. A few very young in aquaria have even been observed to eat organic detritus, some even while hanging upside-down from the surface film of the water (S. v. Boletzky, personal communication, 1993). The reduced swimming speed inherent in ectocochliates, however, would have prevented pouncing.

Several possible feeding strategies come to mind for the neanic ammonoids as well as for many juveniles and planktonics:

1. Visual "ambush" feeding on mobile prey with ejectile tentacles.
2. Visual feeding on essentially immobile prey by means of arms, non-ejectile tentacles, or umbrella webs; such prey might have included ammonoid hatchlings and pseudoplanktonic microorganisms, e.g., ostracods, microgastropods in algal mats.
3. Tactile or chemosensory feeding by "pseudoscavenging" of organic particles floating and sinking at all depths.

S. v. Boletzky (personal communication, 1993) rejected strategy 1 for very young ammonoids on the grounds that they probably did not possess the required "highly advanced" ejectile tentacles. Strategy 3, extended to plankton, would have permitted them to live in the aphotic zone, more than 30–50 m deep in epeiric and neritic waters, as well as below 240 m (mesopelagic) in the ocean. Feeding strategy 2 would help to explain the relative scarcity in the fossil record of ammonitellas and smaller neanoconchs. Most feeding strategies including raptorial tentacles are, however, envisioned for more fully grown, larger ammonoids, in accordance with the coleoid paradigm (Jacobs and Landman, 1993).

5.4.1. Evidence of Prey

Stomach and/or crop contents are known only from very few specimens (Lehmann and Weitschat, 1973; Lehmann, 1975, 1976, 1985) but are of eminent importance for the reconstruction of dietary habits. The prey was found fragmented in stomach- or crop-like concentrations in the body chamber, so that postmortem processes such as scavenging of the decomposing ammonoid body and current-caused transportation into the empty shell can be excluded. According to the prey contents, the finds can be summarized as follows:

1. Ostracods. One specimen of Early Triassic *Svalbardiceras* (Meekoceratidae) and one Early Jurassic *Arnioceras* (Arietitidae) contained ostracods (also some foraminiferans). Lehmann (1976) concluded that the shovel-like lower mandibles, i.e., aptychi, so typical of the Ammonitina, could have been used only to scoop up benthic microorganisms in their muddy substrate. Morton and Nixon (1987) modified this into plankton feeding with the "shovel" aiding in shutting the spacious mouth to trap and separate the plankton from the water (but see Westermann, 1990). The "streamlined" *Svalbardiceras* probably would have been able to hunt mobile ostracods in this fashion, but not the serpenticonic, quasiplanktic *Arnioceras*. The habitat of the *Svalbardiceras* prey has now been revised from benthic to pelagic (W. Weitschat, personal communication, 1993). This is in agreement with the anoxic sedimentary facies; i.e., the predator was a nektic carnivore. But the questions of prey capture and precise prey habitat remain unresolved. The shape of most calcified and of many uncalcified aptychi, however, indicates that their main adaptation was as operculum-like structures (see below). Small plankton could have been caught in many ways by specialized arms and/or tentacles, including extended cirri or "umbrella nets." Seilacher's (1993) feeding model, reminiscent of baleen whales, with entirely reduced arms and tentacles, however, is far-fetched. There almost certainly existed since the early Mesozoic, at least, another ostracod habitat that would not have required "scooping" or maneuverability of the predator. The prey ostracodes were neither benthic (shallow-infaunal, epifaunal) nor demersal, but either pseudoplanktic in floating algal mats (W. Braun and G. Hartmann, personal communication, 1993) or planktic in mid-water (R. Watley, personal communication, 1995). Even the most sluggish ammonoids would have been able to gather them with arms and/or simple tentacles

in the floating mats that could have been present even where anoxic soft muds inhibited benthic algae.

2. Very small, mostly young ammonoids. One example is Lower Jurassic *Hildoceras* (Hildoceratidae) with aptychi (mandibles) of juvenile ammonoids; but prey identification even to suborder seems to be impossible. One Late Jurassic *Neochetoceras* (Oppeliidae) has juvenile aptychi (?*Lamellaptychus*) of same or related species. These "streamlined" platycones and oxycones, respectively, were obviously hunters of ammonoids.

3. Crinoids. One specimen of Late Jurassic *Physodoceras* (Aspidoceratidae) has the unique, minute, probably planktic crinoid *Saccocoma*, a short-lived genus of the Tethys. A pelagic habitat for this spheroconic ammonite is also supported by its rare record from the ocean floor (see Sec. 4.1). Milsom (1994) has recently argued that this crinoid was vagrant-benthic, rather than pelagic, with minor swimming potential for escape from benthic predators. It would then appear that this spheroconic ammonite was foraging for *Saccocoma* on the sea floor of the shallow "Plattenkalk" sea.

5.4.2. Buccal Organs

The buccal mass, including radula and mandibles, is discussed elsewhere (Chapter 2, this volume). Many problems of dietary and feeding habits for the diversely shaped buccal organs remain essentially unsolved, even for most extant cephalopods (Nesis, 1986, 1987; Lehmann, 1988). Of particular interest are the aptychus-type mandibles (including anaptychi) present in many but by no means all Mesozoic ammonoids, and the rhynchaptychus-type mandibles of the Lytoceratina (Lehmann *et al.*, 1980). A consensus appears to be that, among the former, the uncalcified anaptychi functioned mainly as feeding organs, i.e., as lower mandibles or, perhaps, an elastic "pump" during suction feeding (Seilacher, 1993); the thickly calcified aptychi (*Lamellaptychus, Laevaptychus, Punctaptychus*) functioned primarily as opercula (Schindewolf, 1958; Lehmann and Kulicki, 1990; Seilacher, 1993). The elastic tissue between the aptychi valves probably served as a folding mechanism to translocate them past the arms, between aperture and buccal mass. These thick (but porous) calcitic plates certainly would have protected against predators, but there are no obvious associations with habitats. Only the lytoceratine mandibles resembled the *Nautilus* jaw, indicating a similar duraphagous diet.

The presence of *Nautilus*-like mandibles in larger hamitid heteromorphs (*Scalarites*) (Chapter 2, this volume) and scaling problems preclude pteropod-like feeding strategies with mucus nets as suggested for similar heteromorphs by Nesis (1986). Similar beak-like, horny upper jaws functioned together with sometimes serrated and horny or partly calcified lower jaws in oxyconic collignoniceratids (*Reesidites*), discoconic desmoceratids (*Damesites, Tragodesmoceroides*), and lytoceratids (*Gaudryceras*) (Tanabe, 1983; Tanabe and Fukuda, 1987). The function of these jaws presumably also resembled that of *Nautilus*, i.e., the cutting up of larger prey so that, besides the micro- and mesophagy documented by crop/stomach contents, macrophagy was also present.

5.4.3. Trophic Scenarios

The few data on prey and buccal organs, together with the results of aut- and synecology discussed above, suggest the following trophic scenarios. The majority of ammonoids belonged to the pelagic food-chain at a number of trophic levels: many small taxa (e.g., longidomic serpenticones and sphero- cones), together with neanic stages and juveniles of many larger taxa, fed directly on the microplankton, perhaps nocturnally with modest vertical migration or at the lower limit of the photic zone. They were eaten by mid-sized ammonoids which, in turn, fell prey to the larger, relatively good swimmers among the ammonoids (many brevidomic discocones and oxy- cones), as well as to vertebrates. Many Ancyloceratinae (e.g., hamitid, ancy- loceratid and vermiconic heteromorphs) were vertical migrants in deep water, including the oceans, where some fed on mesopelagic organisms including sluggish juvenile ammonites, and others caught zooplankton in tentacles modified into umbrella-nets. Only relatively few taxa (among the mesodomes, torticones, orthocones and oxycones) were demersal or potentially demersal, mostly on platforms where they consumed mostly soft-bodied epibenthos, or on the shelf-edge and continental slope where some desmoceratoids and lytoceratids fed, *Nautilus*-like, from the shelly benthos.

5.5. Shell Damage and Predators

Because shell damage and abnormalities are discussed in Chapter 15 (this volume), only a few examples permitting inference on predators are men- tioned here. Best known is the lethal shell perforation by marine reptiles, e.g., in mid-Jurassic *Kosmoceras* (Ward and Hollingworth, 1990) and especially by mosasaurs on Late Cretaceous *Placenticeras* (Kauffman, 1990; Hewitt and Westermann, 1990b) (see Section 4.3.3e). Fish predators have been widely inferred (e.g., Ward, 1981) and documented in Late Jurassic Haploceratidae by aptychi in holostean feces (Mehl, 1978a). Teuthoid predators were sus- pected from clusters of uniformly sized shell fragments of Early Jurassic harpoceratids and Late Jurassic *Gravesia* (Mehl, 1978b). They probably also inflicted most of the peristomal mantle injuries in Jurassic–Cretaceous am- monoids that, unless lethal, caused the well-known shell abnormalities. Kase *et al.* (1994) have recently described limpet home depressions on the shells of what they believe were live Desmoceratids. The evidence, however, strongly favors postmortem infestation (Westermann and Hewitt, 1995).

5.6. Adaptations of the Growth Stages

5.6.1. Ammonitella (Embryo/Egg)

There are two principal spawning strategies among the living cephalopods (Nesis, 1987; Boletzky, 1987; S. v. Boletzky, personal communication, 1993; Hewitt, 1988): (1) some neritic squids and octopus lay large eggs on the oxygenated sea floor (benthic K-type selection strategy); (2) the other neritic and all oceanopelagic taxa shed their minute eggs (<3 mm) continuously for several months, singly or in clusters, into midwater, where they become part of the upper epiplankton or sink to the bottom (planktic r-type selection strategy). Some benthic octopods, however, lay small eggs on the sea floor. With very few exceptions (see below), ammonoid egg size, inferred from the ammonitella, resembled that of the small to smallest eggs of present-day cephalopods. Embryo growth is greatly enhanced by warm water and requires good oxygenation for small eggs. Eggs therefore need full aerobic conditions for their development and hatching, whether attached, directly or indirectly, on the oxic, solid sea floor of epicontinental seas and oceanic rises or planktic, floating in epipelagic midwater or pseudoplanktic in floating algal mats.

Ammonoid embryos, however, differed from squid and octopod embryos by the presence of a high-density shell, the ammonitella. If calcification was indeed its primary state, as most authors have assumed (e.g., Tanabe *et al.,* 1993b), then planktic eggs would have sunk unless they were (1) encased in light gel, singly or in masses; (2) made neutrally buoyant by gradual emptying of the protoconch during embryo growth; or (3) were brooded by a parent. The first possibility would probably have been realized in most taxa. The second probably has to be excluded: the proseptum that fastened the cecum was usually produced only during completion of ammonitella growth (Ward and Bandel, 1987; Tanabe *et al.,* 1993b; Chapter 11, this volume). The third possibility was egg carrying as in some living pelagic octopods, e.g., *Argonauta.* Ovovivipary by the mother ammonoid is supported by the common development of "oversize" female shells (macroconchs, Chapter 13, this volume) that in the scaphitids had bulges interpreted as brood chambers (Landman, 1987). Recently, however, Kulicki and Doguzhaeva (1994) have argued convincingly for an organic primary state of the ammonitella, which was calcified only at the end of the embryonic stage, i.e., not long before hatching, and even then had only part of its final thickness. This would have eliminated the problems related to high egg density in the plankton and permitted large egg masses in modified body-chambers. Figure 15A illustrates several possible life cycles.

Ammonitellas in different stages of development are sometimes found as small clusters in black-shale facies (Schindewolf, 1959; Landman, 1982; Landman and Bandel, 1985; Tanabe *et al.,* 1993b, 1995). These probably originated from midwater planktic egg masses that sank to the sea floor before or after they were overcome by anoxia. It is difficult to visualize the eggs hovering above the sea floor because they lacked depth control (and did not feed). Neither were they laid directly onto the anoxic muds, and the presence of benthic algal mats, to which they might have been attached, is improbable in this facies. Finds of single, complete ammonitellas are best interpreted as hatchlings killed in the plankton by rising anoxia, probably storm or quake induced (Fig. 15B). Dead embryos, hatchlings, and neanoconchs were sometimes concentrated on the sea floor and mixed together by currents.

Complete ammonitellas, sometimes with several septa (hatchlings), and adult shells may form normal death assemblages in limestone beds within black-shale sequences or even in (?hypoxic) Ammonitico–Rosso facies (J. Sandoval, personal communication, 1993). The fact that ammonitellas have rarely been reported from the "normal" ammonoid assemblages in carbonates, sandstone, and shale, however, is probably partly a result of collecting bias as well as diagenetic destruction of the minute fragile aragonitic shell.

5.6.2. Neanoconch (Neanic Stage)

In contrast to egg ecology, that of hatchlings and early juveniles of living cephalopods, sometime erroneously called "larva" (e.g., Nesis, 1987; Boletzky, 1974; S. v. Boletzky, personal communication, 1992), is well known. All eggs of extant cephalopods less than 4 mm in diameter, whether benthic, carried by the mother, or pelagic, develop a planktic hatchling and early juvenile. This pelagic habit is considered to be the phylogenetically "primitive" or original state among the cephalopods, whereas the benthic habit of some octopods and sepiids is clearly a "derived" or secondary development (Boletzky, 1992).

Planispiral shells had variably inflated, planorbiconic to cadiconic neanoconchs that grew mainly in whorl height and had little or no sculpture. The stage began with the hatchling and ended at 3–5 mm diameter. It was almost certainly neutrally buoyant from the beginning, as evident from the constant volume proportions of phragmocone to body chamber and buoyancy calculations (Landman, 1987; Landman and Waage, 1993; Okamoto, 1988a,b; Shigeta, 1993; Tanabe and Ohtsuka, 1985; Tanabe et al., 1995; see Section 5.2). The minute ammonoid hatchling presumably had a functional hyponome and largely resembled the grown animal, as in living cephalopods (Nixon, 1988), but it belonged to the microplankton, presumably with enough propulsion and possibly even buoyancy regulation for vertical migration. The consistently extreme septal and siphuncular strength at this stage (Hewitt, 1988; Chapter 10, this volume) indicates potential life in deep water, below the photic zone. This would have reduced mortality by predators but also would have increased the likelihood for neanoconchs to be overcome by rising anoxia. As in present-day oceans (Boletzky, 1987), the small-sized food required would have been most plentiful in this region, and selection pressure would have been low in this young growth stage. Orthoconic neanoconchs of torticonic and other extreme heteromorphs (cf. Tanabe et al., 1981) were presumably adapted for vertical migration.

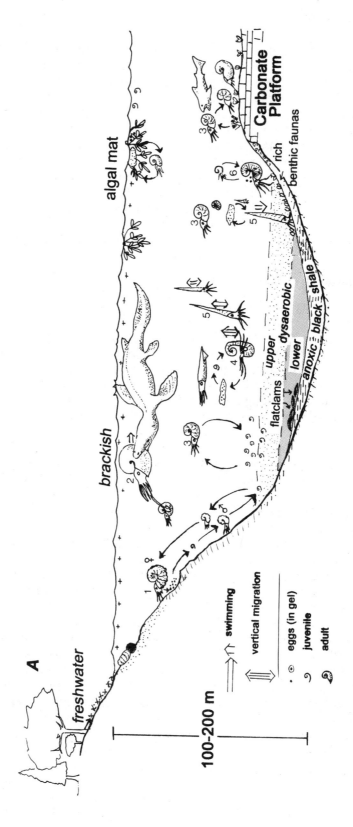

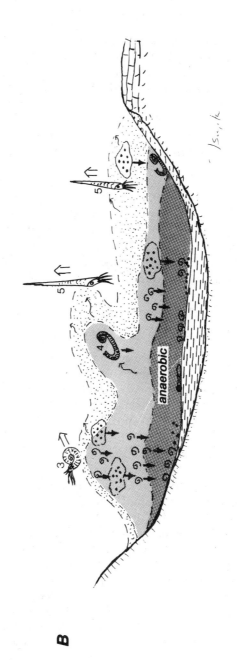

FIGURE 15. Epeiric basin. Possible scenarios of life cycles (A) and death by rising anoxia (B). Eggs may have been enclosed in gel and drifting or attached to aerobic substrate or algae, floating or benthic, or brooded. Ammonitina: 1, Collignoniceratidae; 2, *Placenticeras*; 3, *Acanthoceras*. Ancyloceratidae: 4, *Ancyloceras*; 5, *Baculites*; 6, *Scaphites*.

Neanoconchs are usually found clustered, with or without other growth stages, in hypoxic shales or marls (e.g., Schindewolf, 1959; Landman, 1982; Tanabe *et al.*, 1993b, 1995; Fig. 15B). In the most common ammonitiferous facies, i.e., oxic shales, mudstones, and limestones, however, neanoconchs are usually rare or absent; exceptions are specimens found well dispersed and/or highly localized in deeper-water facies. The most common habitat of the neanic stage probably was the deep midwater of hypoxic, epeiric basins, perhaps just above or below the dysaerobic/aerobic interface. Most presumably fell prey to predatory larger ammonoids, which removed the body from the shell by chemical (Nixon, 1988) or physical means—known crop/stomach contents of ammonoids contain only the remnants of the buccal mass (see Section 5.4.2).

Occasional occurrences of these stages in near-shore sediments (D. Korn, personal communication, 1992) are presumably the result of rare postmortem drift. After death, with or without removal of the soft parts and damage to the shell, the neanoconch generally would have filled up rapidly because of the high ratio of siphuncular surface to chamber volume (Chamberlain *et al.*, 1981), and sunk to the sea floor without marked drift. On the other hand, diagenesis may have destroyed most of the fragile aragonitic shells in these facies.

5.6.3. Juvenile Stage

Juvenile shells also are generally rare in "normal" rich assemblages of the aerobic facies. For example, in the planorbiconic Stephanocerataceae and heteromorphic Scaphitaceae that occur in epeiric gray shales, the common ratio of adult to juvenile shells is more than 1000 : 1, in the microconchs as well as macroconchs (Westermann, 1954; Kennedy and Cobban, 1976; Callomon, 1985; Landman, 1987).

Juvenile Ammonitina occasionally are found in anaerobic and dysaerobic black shales and limestones, sometimes together with neanoconchs, as described above, and adult stages (see below). They were either pelagic or pseudoplanktic. An example for pseudoplankton in floating plant mats was the echioceratids of the Blue Lias (see Section 4.3). The generally higher septal strength of early juveniles (and neanoconchs, see above) compared to later growth stages (Chapter 10, this volume) suggests that the very young pelagics lived deeper. The aphotic and commonly somewhat hypoxic deeper waters would have protected them from vertebrate and coleoid predators. Some small species probably lived in benthic or floating plant mats, depending on the bottom conditions, where they preyed on ostracods, microgastropods, etc. (see Section 5.4.1). Juvenile lytoceratids and phylloceratids have occasionally been found together with herbivorous microgastropods in aerobic, solid-bottom facies at the margin of Tethys. The fauna also included crinoids and a few posidoniid "flat clams." The habitat was almost certainly benthic algal mats on structural uplifts near the ocean (Sturani, 1971; Conti and Fisher, 1981), supporting my earlier life-cycle model for the oceanic lytoceratids and phyl-loceratids (Westermann, 1990).

All longidomic juveniles were presumably pelagic. This probably included all species that were longidomic as adults as well as many planorbiconic species that had a mesodomic adult stage following a longidomic juvenile stage, such as many of the Jurassic Stephanoceratidae and Perisphinctidae (Westermann, 1971; Lehmann, 1976). In several Late Cretaceous hetero-morphs, the outer whorl became "normally" coiled, indicating that the juve-nile was planktic and the adult nektic (cf. Kennedy and Cobban, 1976, Text-Fig. 4). The record suggests that juveniles belonged almost entirely to only two groups, pelagic drifters (plankton) and swimmers (nekton). Morpho-type analysis of Cretaceous Ammonitina also has shown that shell shapes characteristic of pelagics are confined to juveniles of species that develop demersal adult characters, e.g., morphogroup 8 (Batt, 1989) of juvenile *Collignoniceras* developing into adult morphogroup 1, but never in reverse. Other evidence for habitat change from pelagic to demersal comes from shell structure analysis (Chapter 10, this volume): the juvenile septa of *Scaphites* were, exceptionally, weaker (relatively much thinner) than the last 10 septa or so. This confirms the evidence for a habitat change from planktic to demersal based on the occurrence in aerobic facies of adults only (see Section 4.4.2). The other notable exception of adult septal strengthening was in large *Baculites cuneatus* (C. Tsujita, personal communication, 1993), supporting the tentative proposal that they developed from planktic, vertical migrants into demersal swimmers (see Section 4.3.3e).

5.6.4. Adult Stage

In any attempt to attribute special habits to fossil cephalopods, one has to keep in mind the very broad habitats and habits commonly present in the extant coleoids (Nesis, 1987). Neritic and oceanic genera include species living at very different depths and temperatures; many species are pelagic, often over several depth zones, as well as demersal; ontogenetic habitat changes may range from epipelagic, even surface waters, to meso- or bathy-pelagic and sometimes demersal; diurnal vertical migration and annual mi-gration over hundreds or thousands of kilometers may occur, with or without sex segregation; prey is captured in various ways, with arms and/or tentacles that may be strongly modified, including cirri- and finger-like extensions, or with umbrella webs of different sorts. Among the most broadly adapted are species of the eurybathic *Brachioteuthis* (Oegopsina) that migrate several hundred kilometers between shelf and deep ocean and, diurnally, from surface waters to the bottom, preying on fish, crustaceans, and squid. Except for their reduced depth, speed, and, hence, range of migration, ammonoids presumably had similar ecologic diversity.

Table I summarizes the supposed ammonoid habitats. Most bactritids probably were vertical migrants in the epi- and mesopelagic plankton; others

Table I. Facies and Habitats of Selected Adult Ammonoids[a]

	Occurrence							Habitat (Depth, m)
	Oceanic		Epicratonic					
				black shale				
	Oc	AR	DS	off	on	Ox	Pl	
Anarcestida								
Bactritina								
Bactrites					X			Pv 380
Agoniatitina								
Agoniatites						X		Ds 35
Anarcestina								
Anarcestes						X		Ds
Gephuroceratina								
Carinoceras					x			N 70
Manticoceras					x			N 60
Beloceras							x	hyps N 50
Cheiloceratidae			X					N
Clymeniina					x	o		P 170
Goniatitida								
Goniatitina								
Goniatites						X		? 60
Tornoceras				x				Ns 150
Agathiceras				X				P 204
Glaphyrites				x				hypo P 60–125
Antracoceratites				X				N
Cancelloceras				X	x			P
Homoceratoides				x				N
Ammonitida								
Prolecanitina								
Medlicottidae				x				N 105
Otoceratidae						X		Ds 75
Arcestina								
Proarcestes				x				P 180–450
Arcestes		?	o	X				P 100
Ptychitidae				o				N 110
Pinacoceratina								
Pinacoceras								N 80–95
Meekoceras				?				N 50
Buddhaites				o				N 105
Ceratitina								
Ceratitidae						o	X	Ds 35
Beyrichitidae			X					N 50–70
Xenodiscidae							x	Ds
Dinaritidae (incl. Tirolitinae)							x	euh Ds
Sibiritidae							x	?euh N/D
Stephanitidae							x	?euh N/S
Longobarditidae				x				N 85
Acrochordiceratidae				x	o			P
Phylloceratina								
Ussuritidae								
Monophyllites		?		o				N 95
Leiophyllites		?		o				P

Table I. (Continued)

	Occurrence							Habitat (Depth, m)
	Oceanic		Epicratonic					
				black shale				
	Oc	AR	DS	off	on	Ox	Pl	
Phylloceratidae								
Phylloceras	x	X	X	x	o			N/Ds 65–482
Calliphylloceras	x	X		x				N 195–335
Ptychophylloceras	x	x		x				N 340
Holcophylloceras		x		X	o			N/Ds 150
Juraphyllitidae		o		o	o		?	N/Ds
Tragophylloceras				o	o			N
Lytoceratina								
Lytoceratidae [*Lytoc.*]	o	X	X	o	o	o	o	Pv 60–400
Tetragonitidae [*Pseudophyll.*]	?	X		o				N 155
Gaudryceratidae [*Gaudryc.*]			X	o				N
Ammonitina								
Psiloceratidae	(o)		o	x	X	x	o	P
Echioceratidae	(?)		o	X	x	o		P 185
Arietitidae			o	X	X	o		Ds/N/P
Oxynoticeratidae			o	X	x	o		N
Amaltheidae				X	x	o		N 25
Liparoceratidae				X	x			P 30
Dactylioceratidae	o	x		x	X	X	o	P c.70
Hildoceratidae	(?)	x		x	X	x	o	N 50–190
Graphoceratidae		o		x	X	x	X	N c.110
Hammatoceratidae				x	x	X		N/Ds
Sonniniidae				x	X	X		N 80
Strigoceratidae		o		o	o	o	o	N c.100
Oppeliidae	?	o		?	x	x	x	N 35–275
Haploceratidae	x	x	X	o	o	x	x	N/P c.90
Hecticoceratidae				x	x	x	x	N
Otoitidae		o			o	x	x	N/Pv 80–100
Sphaeroceratidae	(?)			o	x	x	o	N/Pv 50–75
Cardioceratidae					X	x	o	P/N 45
Stephanoceratidae								
(serpenticones)	(?)	x			o	x	o	P
(planorbicones)					X	x		Ds 70–85
Kossmoceratidae					X	o	o	P
Parkinsoniidae					x	x	o	Ds/N
Spiroceratidae	(?o)	o		x	x	o	o	P/Pp
Perisphinctidae			o	o	o	x	X	Ds/N/P50–120
Pseudoperisphinctidae					x	o	o	N/P
Tulitidae	(?)				o	x	o	P(v)50–120
Morphoceratidae					o	x	o	Pv
Reineckeiidae								
(planorbicones)					x	x	o	Ds/N
(serpenticones)					x	o	x	P

Table I. (Continued)

	Oceanic		Epicratonic					Habitat (Depth, m)
				black shale				
	Oc	AR	DS	off	on	Ox	Pl	
Aspidoceratidae								
Aspidoceras						o	X	Ds 25
Physodoceras	(?)						o	Ds
Simoceratidae	?		o		o	o	o	P
Aulacostephanidae					x	o	o	N/Ds
Virgatosphinctidae					x	o	o	N/Ds 40–60
Acanthoceratidae						o	X	P/Ds
Gastroplitidae					X	o	o	P
Placenticeratidae					X	o		N 20–80
Vascoceratidae						o	X	D? 20–45
Desmoceratidae								
Desmoceratinae	?		X	x				Ds/N 200–300
Puzosiinae			x					Ds/N?
Pachydiscidae	?		x					Ds/N 70–220
Ancyloceratina								
Ancyloceratidae	?				x	x		Pv
Hamitidae	o		o					Pv
Turrilitidae								
Polyptychoceras	o		o					P 255
Bostrychoceras	o		x			o	o	Pv/Dv 120–155
Baculitidae								
Baculites	?		X	X		o		Pv/Dm 40–120
Sciponoceras	?		o	o	o	o		Pv
Scaphitidae								
Scaphitinae	?		x			X	o	Dv/N
Otoscaphitinae	?		o	o	o	o		Pv
Bochianitidae	?		x					Pv

[a]Abundance: X, dominant; x, abundant; o, ancillary.

Facies: Oc, oceanic; AR, Ammonitico–Rosso; DS, distal shelf/upper slope; off, offshore black shale/limestone; on, onshore black shale; Ox, oxic shale; Pl, platform.

Habitats: hypo, hypoxic; hyps, hypersaline; P, planktic; Pv, planktic vertical migrant; Pp, pseudoplanktic; N, nectic; D, demersal; Ds, demersal swimmer; Dv, demersal vertical swimmer; euh., euryhaline.

(Habitat depth limit and classification from prototype of Table II in Chapter 10, this volume.)

were presumably bottom feeders with vertical mobility. Their very fragile shell indicates that all lived in quiet water well below wave base.

    5.6.4a. Black-Shale Facies. In the Carboniferous cyclothems of the United States (see Section 4.3.3b), the majority of ammonoids were small, smooth spherocones, discocones, and cadicones that lived in about 50–100 m deep seas with hypoxic to anoxic bottomwaters. These dysaerobic and anaerobic biofacies commonly recurred annually, with anoxia dominating (poikiloaerobic), and the surface waters were sometimes brackish. Many of these ammonoids lived during (not necessarily in) all of these facies and, hence, must have been pelagic. Their food was probably mainly plankton and quasiplankton and included at maturity the abundantly present young stages of ammonoids.

    Some black-shale assemblages comprise preadults only, whereas adults with large discoconic shells are generally found together with hatchlings in the interbedded aerobic limestone facies (normal mortality curve). It appears that rising anoxia caused high mortality rates among pelagic embryos (eggs), neanic stages, and juveniles but not among adults. The adults either (1) lived together with the younger animals, just above or even within the upper parts of the dysaerobic zone (Becker, 1993), but they were mobile enough to escape, or were not asphyxiated by, a slowly rising hypoxic water mass (not rapid, storm- or earthquake-induced admixing); or (2) the adults lived separately, higher in the water column, where the hypoxic waters did not reach them. Occasional heavy predation by fish and sharks (Mapes and Hansen, 1984; Bond and Saunders, 1989) as well as by predatory cephalopods including ammonoids, all with high metabolic rates, supports an aerobic habitat of the adults. The larger size would have given the adults better protection against predators and mobility to escape the occasional turbulence of near-surface water.

    Devonian and Permian to Cretaceous hypoxic basins and biofacies adaptations were similar to those of the Carboniferous. Some of the ammonoids in these basins probably were euryoxic or even adapted to slight hypoxia, as recently suggested by Wells *et al.* (1992; M. J. Wells, J. Wells, and R. K. O'Dor, personal communication, 1993) based on aquarium experiments with *Nautilus* and by Becker (1993) on facies distributions. At midwater depth in hypoxic basins, pelagic ammonoids (1) were just below the photic zone in the murky waters and had better chances to escape predators, especially if their habitat was oxygen-deficient so that squids, sharks, and fish (with high metabolic rates) could not follow; (2) they were below the base of common storm waves, which would have tossed about even the relatively mobile swimmers among them; and (3) they would have escaped low-salinity waters, which, during humid times, must have capped some of the epeiric seas.

    5.6.4b. Shallow platforms. A great variety of species lived on warm, rather shallow (ca. 30-50 m) carbonate platforms, but generally only the adults that either migrated there (swimmers) or were drifted there by the "chance" of currents (planktics). The rich fossil record shows that many died there, but generally not after spawning, as has frequently been assumed. There appears to be no record of abundant embryonic (ammonitellas), neanic, or juvenile shells from this facies (although that could in part reflect diagenesis).

Some goniatite and ceratite species on the platforms were probably eury-haline (see Sections 4.5.2a,d). The Devonian *Beloceras* and Triassic *Ceratites*, *Tirolites*, and *Dinarites* lived in or around shallow, hypersaline basins and/or lagoons. Some goniatites (Becker, 1993) and ammonites (*Placenticeras* and perhaps *Dactylioceras*) inhabited the brackish surface waters of epeiric basins during heavy runoff.

Because deep-water shells are particularly prone to postmortem drift, especially when the body chamber was damaged by predation (see above), oceanic phylloceratids and lytoceratids are commonly found as minor elements in shallow-water shelf (not epeiric) assemblages. Good examples are the fragmentary specimens of *Calliphylloceras* sp. recently reported from Kutch, India, which Bardhan *et al.* (1993) attributed to shallow-shelf dwellers because of alleged color patterns. The black bands described from the internal varices (pseudoconstrictions), however, were equivalent to the black, internal, peristomal secretion of adult *Nautilus*, not to its superficial shell banding; color banding would not have restricted *Calliphylloceras* to the shelf in any case. The drifted nature of ammonoids in near-shore deposits is usually quite obvious, e.g., the massed occurrence of the deep-water Permian ammonoid *Perrinites* in a hypersaline sequence of Australia (B. Glenister, personal communication, 1993).

5.6.5. Intraspecific Variation and Buckman Rule of Covariation

The "Buckman Law of Covariation" was named by Westermann (1966) for a consistent but not yet understood phenomenon in Mesozoic Ammonitina first noted by S. S. Buckman a century ago in English Middle Jurassic ammonites. He found in large, morphically intergrading assemblages the common associations of forms with evolute, broad whorls and coarse sculpture together with forms with involute, compressed whorls and fine sculpture. In the classical "typological" approach, the multitude of forms were each given separate species names; more recently, they have been recognized as morphs or formas and variants belonging to a single biospecies with high variation. The thin, carbonate interval that was the source of Buckman's large *Euhoploceras* (Sonniniidae) assemblage has turned out to be stratically condensed from parts of several ammonite zones and/or biohorizons as presently defined (Callomon, 1985). Numerous assemblages, however, have now been carefully collected from single, essentially noncondensed horizons and concretions by numerous authors (e.g., Reeside and Cobban, 1960; Rieber, 1973, 1975; Silberling and Nichols, 1982; Westermann and Callomon, 1988; Dagys and Weitschat, 1993), who have fully verified my original "law" (rule) throughout the Mesozoic and in different facies.

Significantly, some assemblages contain broadly varying species together with narrowly varying species (e.g., Dagys and Weitschat, 1993), indicating marked differences in the selection pressure on, e.g., "streamlining" and defensive sculpture. Hence, variation is itself an important feature signifying the degree of adaptation of the populations. "Streamlining" presumably varied much more in populations of planktic drifters and vertical migrants than in nektic and demersal swimmers that depended on speed, acceleration, and/or steerage for catching prey. H. Bucher (personal communication, 1993) noted that in the Triassic, extreme variation occurred at times of high latitudinal provinciality, with the smoother forms more common in high latitudes and highly sculpted forms in lower latitudes; and that highly varying populations evolved more rapidly than those with low variation. Intricate relationships appear to have existed between sculptural patterns and depth-related factors of the habitat, e.g., temperature. But tropical/subtropical deep water was similarly as cool as temperate shallow water; both tended to produce relatively smooth shells compared to warm shallow-water carbonate platforms. It appears that biogeographic factors influencing habitat, e.g., climate, that can be inferred from the geologic record are basically scaled-up versions of local and regional ecological factors that often remain undetected.

5.6.6. Distribution of Sexual Dimorphs

Dimorphs (Fig. 16) commonly occur in uneven numbers, with ratios up to 100 : 1 for either macroconchs (female shells) or microconchs (male shells), dependent or independent of lithofacies (see Callomon, 1981; Chapter 13, this volume). But great care has to be taken to avoid collecting bias, especially against microconchs (Westermann, 1964; Howarth, 1993). The sex ratios may be opposed even between closely related species in the same assemblage, e.g., *Stephanoceras* (see Section 4.4.1). In these extreme cases, the sexes obviously lived in segregated swarms that were wiped out instantly, perhaps by storm-induced admixing of bottom-water anoxia or upwelling events, as indicated by packed bedding planes of black shales (Callomon, 1985). In cases of dispersed, more-or-less abundant occurrences of adults, usually with both sexes, however, the ammonoids probably died "naturally" of old age. Of particular interest is the relationship of dimorphism to growth stages. In dysoxic-basin facies, juvenile macroconchs are sometimes associated with adult microconchs; and adult macroconchs may be concentrated in carbonate-platform facies. If evidence existed that this happened to the same species in adjoining facies, the conclusion would be obvious: the adult females alone left the hypoxic, cold basinal waters to spawn (and die) on the warm, solid, and oxygenated bottom of the adjoining platform. In most cases and in all kinds of lithofacies, however, the sex ratios are less drastic, and it would not be possible to distinguish the results of single annual episodes from a mixture of several episodes. Furthermore, postmortem drift obviously played an important role, here as in other distributions.

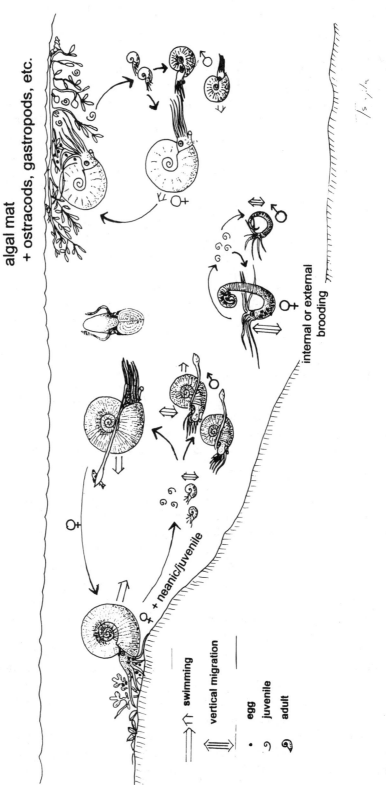

FIGURE 16. Three life-cycle scenarios of dimorphic ammonoid species: nektic with benthic eggs, planktic with planktic eggs, and planktic with pseudoplanktic eggs laid in floating algal mats.

algal mat
+ ostracods, gastropods, etc.

internal or external brooding

+ neanic/juvenile

↗↑ **swimming**

⇔ **vertical migration**

• **egg**

ↄ **juvenile**

಄ **adult**

♀

♂

5.6.7. Habitat Change and Shell Strength in Evolution

Concomitant changes in habitat and shell structure are of particular inter-est. The best examples are the oceanic leiostracans *Phylloceras* and *Lytoceras* that in the Early Jurassic invaded the epicontinental seas of northwestern Europe from Tethys (see Section 4.3.3d). A few adult specimens are known from single horizons in the *Posidonia* Shale of Germany, where they reached giant sizes (>1 m, G. Dietl, personal communication, 1992), just as in deep oceanic habitats of the Pacific (Stevens, 1988). They thrived in the deeper, mainly mesopelagic habitats of the warm and clear waters of Tethys, several hundred kilometers to the south, as well as in some of the surrounding deep subbasins with anoxic bottom waters (Elmi, 1985; see Section 4.3.1). Although it would appear that these specimens were drifted in postmortem, because large size and strong siphuncles favor extensive drift (Tanabe, 1979; Wester-mann, 1990), this was not true, at least in the cases examined; i.e., a shallow epeiric habitat (<100 m) is well reflected in the juvenile and adult shell strength (Chapter 10). The sampled septa of the epeiric Toarcian *Phylloceras heterophyllum* and the Pliensbachian *Lytoceras fimbriatum* would have im-ploded at a depth of about 100–160 m, similar to typically epicontinental Ammonitina, whereas the measured septa of congeneric specimens from oceanic terranes (Peninsula Terrane, Alaska) and continental margins (New Guinea, Argentine Andes) could have withstood depths of 350–600 m.

Interestingly, this marked habitat change did not result in a weakening of the siphuncular tube; the siphuncle strength index (SiSI) of these epeiric specimens does not differ significantly from that of their oceanic close relatives. There is an obvious explanation for this apparent contradiction. Shell growth was accelerated in shallow water, resulting in thinner, more economic, and lighter septa. In present-day *Nautilus*, the pumping rate is inversely proportional to ambient pressure, and septum secretion is termi-nated when the previous chamber is partially emptied (Ward, 1986a, 1987). The connecting rings of the siphuncle were not modified because their volume and mass did not significantly affect buoyancy (neither was their growth rate controlled by ambient water pressure). Relative septal thickness, the major variable relating to septal strength, was therefore either ecologically control-led or evolved rapidly in the new habitat. Both are good evidence for autochthonous populations. Thus, the exceptionally large difference between the septal and siphuncular strengths is indicative of the rapid habitat change from the usual oceano-mesopelagic to the new epeiro-epipelagic habitat.

More widely distributed in the epeiric Lower Jurassic of northwest Europe, but always accessory, are phylloceratids (Juraphyllitidae) (Meister, 1993) and lytoceratids (Alocolytoceratinae, Derolytoceratinae) that evolved from the oceanic forms by developing strong sexual dimorphism, simple sculpture, subrectangular or strongly compressed whorl sections, and relatively weak septa, thus converging on the Ammonitina. They are found in all size classes (G. Dietl, personal communication, 1993), and some (*Pleurolytoceras, Loboly-*

toceras) could have withstood the ambient pressures of the deepest epeiric basins (200–350 m depth) (Chapter 10, this volume, Table II). They occur in different facies and were probably mostly pelagic.

Sutural complication has been used in bathymetry, under the assumption that it improves shell strength against ambient pressure and, hence, varies with habitat depth (Section 3.3.2). A reversal of this general trend has recently been claimed by Hirano *et al.* (1990) for Late Cretaceous desmoceratids. In the *Desmoceras–Tragodesmoceroides* clade, the ratio of sutural length (measured along the meandering curve) to cross-sectional area increased with time, whereas the sampled offshore basin was shallowing. This increase, however, was caused by the phyletic trend of increased whorl overlap; i.e., the internal suture became longer in overall span (measured along the radial saddle envelope or whorl circumference) and length, compared to the external suture. According to my measurements of those (Hirano *et al.*, 1990) illustrations of cross sections, the ratio internal/external sutural span increased from 0.25 in *Tragodesmoceroides subcostatus* to 0.4 in *T. matsumotoi*, and the entire sutural span (dorsum to venter) lengthened by 15%, equaling the supposed phyletic change. The area of cross section also declined, increasing the sutural length/whorl section ratio. Hence, sutural complication remained constant or decreased slightly. To test this hypothesis, sutural length should be measured at equal span or divided by the square of the span (= $\frac{1}{2}$ whorl circumference).

5.7. Dispersal and Biogeographic Distribution

Dispersal of the planktic hatchling and subsequent neanic ammonoid was entirely passive and depended on currents. The water masses would have been chosen by the spawning or egg-carrying (ovoviviparous) mothers and subsequently somewhat controlled vertically by the minute ammonoid (Figs. 6, 10). Compared to the oceans, however, currents were much reduced on broad shelves and especially in epeiric seas. It is therefore reasonable to conclude that (1) dispersal of the neanic animal during the perhaps several weeks' duration (Westermann, 1990) was maximally 1000–2000 km in the open ocean but only a few hundred kilometers in the shelf seas and a few tens of kilometers in smaller, enclosed epeiric seas; (2) the populations in these early planktic stages were genetically much more isolated in the epeiric seas than in the shelf seas and adjoining oceans; (3) the oceanic plankton contained abundant neanic as well as juvenile and even adult animals, which in many ammonoids continued to be pelagic; (4) annual migrations, on the other hand, were limited to tens of kilometers.

Consequently, current transport of pelagic animals could have produced "instant" species dispersal, even across major oceans, and certainly was much more extensive than formerly believed possible for all but the oceanic leiostraca (e.g., Newton, 1988; Wiedmann, 1988b; Westermann, 1990; Smith *et al.*,

1990). Compared to a few weeks of drifting of the neanic animal only, many months to years of potential dispersal were available to ammonoids that had planktic eggs and also pelagic juveniles and, often, adults. (The opposite conclusions by Ebel, 1983, 1985, 1992, and Shigeta, 1993, based on alleged negative buoyancies were rejected above; see Section 5.2.) Indirect dispersal by water currents was certainly much more important for ammonoid dispersal than swimming, and even the best "streamlined" and muscled species may have reached 1–2 km/hr (Jacobs, 1992a). Swimming was effective only for feeding and habitat control, e.g., counteracting currents in demersal habitats.

Because pelagic ammonoids, especially the very young, lived at least several tens of meters below the surface, oceanic dispersal would have resulted from deep, not surface, currents, e.g., equatorial, deep, and commonly rapid countercurrents that cross large oceans in a matter of weeks or, at most, months. Of course, any passive dispersal depended on one overriding (unknown) factor, i.e., food requirements. The dispersal of ammonoids in epeiric habitats was more or less restricted by relative isolation and reduced currents.

This model clearly implies a positive correlation between neritopelagic habits and biogeographic distributions. This is indeed the case, according to a morphotype survey by Ward (1976a). Following the oceanic lytoceratids and phylloceratids, the heteromorphs were most cosmopolitan and least endemic; the "ornamented," "involute inflated," and "evolute inflated" morphogroups were intermediate, and, significantly, the oxycones the least cosmopolitan and most endemic morphogroup. This also explains the problem of many pantropical distributions, some at the species level, at the time of Pangaea: many Late Carboniferous to earliest Jurassic planktic ammonoids, known mainly from epicontinental habitats, presumably crossed even Panthalassa (paleo-Pacific) in this way, with or without "stepping stones." Of course, only relatively few individuals need to have been successful, and the many endemic taxa of single basins or platforms, pelagic or demersal, did not participate in such dispersals.

6. Conclusions

1. In contrast to recent assumptions, the majority of ammonoids were pelagic, especially in the Paleozoic; roughly half of the pelagics were planktic, with or without vertical migration. All neanic and most smaller juvenile ammonoids were quasiplanktic, mainly in aphotic and commonly hypoxic midwater of poorly oxygenated basins; others drifted at intermediate levels in the epiplankton of oceans. At maturity, many, especially the macroconchiate females, became more mobile, i.e., vertical migrants or swimmers, pelagic or demersal. Spawning in hypoxic basins probably occurred in midwater, in floating algal mats, or reproduction was ovoviviparous. Spawning was probably uncommon on aerobic platforms, and then mainly in benthic

algal mats. Spawning in oceans resembled that in hypoxic basins. Adult migrations, presumably for spawning and sometimes of females only, occurred mainly from oceans to platforms and from basin center to basin margin. Many adult pelagic ammonoids were transported by currents, but probably not the preadults that lived in semistagnant, deeper basin waters.

2. A few ammonoids were euryhaline; others probably euryoxic or even adapted to moderate hypoxia that protected them from squid and vertebrate predators.

3. Habitat depths in epicratonic seas were generally shallower than assumed previously, i.e., 30 m to 100 m, possibly 120 m, except for some large, oxyconic ammonites that occasionally came close to the surface, and euryhaline species that lived in shallow lagoons. In pelagics, the juveniles commonly lived deeper than did the adults; the opposite was the case in forms that changed from pelagic to demersal.

4. The principal food was zooplankton, e.g., young ammonoids that abounded in the midwater epipelagon of most seas, planktic crustaceans, and pseudoplanktic ostracodes and microgastropods inhabiting floating algal mats.

5. The dispersal potential of most neritic ammonoids resembled that of the oceanopelagic leiostraca. Because of their many months to years of pelagic life in midwater, they could have been carried long distances by currents, especially deeper water countercurrents, even across major oceans, limited mainly by food resources. Epeiric habitats were much more isolated.

6. Three adaptive peaks are recognized in body chamber angle ("length"): brevidomes (1/2 whorl; stable, "streamlined" swimmers, mostly pelagic); mesodomes (3/4 whorl; moderately stable, sluggish swimmers, commonly demersal); and longidomes (1+ whorl, unstable, pelagic drifters and vertical migrants, possibly with voluntary rotation). Among the brevidomes were the oxycones, with superior steerage and acceleration, and the efficiently swimming discocones. Only a few brevidomic oxycones and discocones potentially could have swum forward; most swam backward. Downward swimming (vs. sinking) was facilitated by upturned apertures, e.g., the adult body chamber hooks of heteromorphs. Note that these potential adaptations were not functional in highly varying populations.

7. The principal predators were among larger oxyconic and, probably, discoconic ammonoids, fish, reptiles, and teuthoids. In hypoxic basins, mass death occurred mainly by upwelling anoxia.

8. Postmortem drift depended on the depth at time of death and on the subsequent ascent. Probability of ascent increased with shell diameter and decreased with depth and siphuncle diameter. Many ammonoids may also have risen to the surface directly preceding death, as in extant ecto- and ?endocochliates. Distance of surface drifting depended on current velocity and on the refilling-rate that was also related to siphuncle diameter. Drifting therefore occurred mostly to larger deep- and open-water adults, shoreward as well as oceanward.

ACKNOWLEDGMENTS. Above all others, I thank Roger Hewitt, Visiting Scientist at McMaster University, for his continued inspiration, help, and constructive criticism. For important information and comments I am grateful to R. Batt of Buffalo, R. T. Becker of Berlin, S. v. Boletzky of Banyuls-sur-Mer, C. Brett of Rochester, H. Bucher of Vancouver, J. H. Callomon of London, E. Cariou of Poitier, F. Cecca of Roma, G. Dietl of Stuttgart, A. Hallam of Birmingham, D. Korn and J. Kullmann of Tübingen, D. Jacobs and N. Landman of New York, R. Mapes of Athens, Ohio, D. Boardman of Stillwater, Oklahoma, L. Martire of Torino, M. Nixon of London, F. Oloriz of Granada, C. F. E. Roper of Fort Pierce, Florida, F. Sandoval of Granada, B. Saunders of Bryn Mawr, K. Tanabe of Tokyo, C. Tsujita of Hamilton, Ontario, P. Ward of Seattle, M. Wells of Cambridge, and W. Weitschat of Hamburg. R. A. Davis of Cincinnati greatly improved the manuscript, and J. Mills Westermann proofread the typescript.

References

Almeras, Y., and Elmi, S., 1982, Fluctuations des peuplements d'ammonites et de brachiopods en liaison avec les variations bathymetriques pendant le Jurassique Inférieur et Moyen en Mediterranée occidentale, *Bol. Soc. Paleont. Ital.* **21**:169–188.

Bardhan, S., Jana, S. K., and Datta, K., 1993, Preserved color patterns of phylloceratid ammonoid from the Jurassic Chari Formation, Kutch, India, and its functional significance, *J. Paleontol.* **67**:140–143.

Barthel, K. W., Swinburne, N. H. M., and Conway Morris, S., 1990, *Solnhofen: A Study in Mesozoic Palaeontology*, Cambridge University Press, Cambridge.

Batt, R. J., 1989, Ammonite shell morphotype distribution in the Western Interior Greenhorn sea and some paleoecological implication, *Palaios* **4**:32–42.

Batt, R. J., 1991, Sutural amplitude of ammonite shells as a paleoenvironmental indicator, *Lethaia* **24**:219–225.

Batt, R. J., 1993, Ammonite shell morphotypes as indicators of oxygenation in ancient epicontinental seas: Example from Late Cretaceous Greenhorn Cyclothem (U.S.A.), *Lethaia* **26**:49–64.

Bayer, U., 1970, Anomalies in Aalenian and Bajocian ammonites as clues to their mode of life, *N. Jb. Geol. Paläont. Abh.* **135**:19–41.

Bayer, U., 1977, Cephalopod septa I. Constructional morphology of the ammonite septum, *N. Jb. Geol. Paläont. Abh.* **154**:290–366.

Bayer, U., 1982, Ammonite maneuverability—a new look at the functions of shell geometry, *N. Jb. Geol. Paläont. Abh.* **164**:154–156.

Bayer, U., and McGhee, G., 1984, Iterative evolution of Middle Jurassic ammonite faunas, *Lethaia* **17**:43–51.

Becker, R. T., 1993, Anoxia, eustatic changes, and Upper Devonian to lowermost Carboniferous global ammonoid diversity, in: *The Ammonoidea: Environment, Ecology and Evolutionary Change*, Systematics Association Spec. Vol. 47 (M. R. House, ed.), Clarendon Press, Oxford, pp. 115–164.

Birkelund, T., 1981, Ammonoid shell structure, in: *The Ammonoidea*, Systematics Association Spec. Vol. 18 (M. R. House and J. Senior, eds.), Academic Press, London, pp. 177–214.

Birkelund, T., and Callomon, J. H., 1985, The Kimmeridgian ammonite faunas of Milne Land, central East Greenland, *Bull. Grønl. Geol. Unders.* **153**:1–56.

Boardman, D. R. II, Mapes, R. H., Yancey, T. E., and Malinky, J. M., 1984, A new model for the depth-related allogenic community succession with North American Pennsylvanian cyclothems and implications on the black shale problem, *Tulsa Geol. Soc. Spec. Publ.* **2**:141–182.

Boletzky, S. v., 1974, The "larvae" of Cephalopoda: A review, *Thalassia Jugosl.* **10**:45–76.

Boletzky, S. v., 1987, Juvenile behaviour, in: *Cephalopod Life Cycles*, Vol. II (P. R. Boyle, ed.), Academic Press, London, pp. 45–60.

Boletzky, S. v., 1992, Evolutionary aspects of development, life style, and reproductive mode in incirrate octopods (Mollusca, Cephalopoda), *Rev. Suisse Zool.* **99**:755–770.

Bond, P. N., and Saunders, W. B., 1989, Sublethal shell repair in Upper Mississippian ammonoids, *Paleobiology* **15**:414–428.

Boston, W. B., and Mapes, R. H., 1991, Ectocochleate cephalopod taphonomy, in: *Processes of Fossilization* (S. K. Donovan, ed.), Bellhaven Press, London, pp. 220–240.

Boucot, A. J., 1981, *Principles of Benthic Marine Paleoecology*, Academic Press, New York.

Bourrouilh, R., 1981, "Orthoceratitico–Rosso" et "Goniatitico–Rosso" facies marqueurs de la naissance et de l'évolution de paleomarges au Paléozoïque, in: *Rosso Ammonitico Symposium Proceedings* (A. Farinacci and S. Elmi, eds.), Tecnoscienza, Roma, pp. 39–58.

Boyajian, G., and Lutz, T., 1992, Evolution of biological complexity and its relation to taxonomic longevity in the Ammonoidea, *Geology* **20**:983–986.

Brand, E., and Jordan, R., 1990, Zur Paläogeographie des Ober-Bathonium (Dogger) im nordwest-deutschen Becke und neue Ergebnisse für den Raum Hildesheim, in: *Zum Ober-Bathonium (Mittlerer Jura) im Raum Hildesheim, Nordwestdeutschland* (R. Jordan, ed.), *Geol. Jahrb. A* **121**:9–20.

Brett, C. E., Miller, K. B., and Baird, G. C., 1990, A temporal hierarchy of paleoecologic processes within a Middle Devonian epeiric sea, in: *Paleocommunity Temporal Dynamics: The Long-term Development of Multispecies Assemblages* (W. Miller, ed.), *Paleontol. Soc. Spec. Publ.* **5**:178–203.

Brett, C. E., Dick, V. B., and Baird, G. C., 1991, Comparative taphonomy and paleoecology of Middle Devonian dark gray and black shale facies from western New York, in: *Dynamic Stratigraphy and Depositional Environments of the Hamilton Group (Middle Devonian) in New York State*, Part II (E. Landing and C. E. Brett, eds.), The State Education Department, Albany, NY.

Brett, C. E., Boucot, A. J., and Jones, B., 1993, Absolute depths of Silurian benthic assemblages, *Lethaia* **26**:25–40.

Brumsack, H.–J., 1991, Inorganic geochemistry of the German 'Posidonia Shale': Palaeoenviron-mental consequences, in: *Modern and Ancient Continental Shelf Anoxia* (R. V. Tyson and T. H. Pearson, eds.), *Geol. Soc. Spec. Publ.* **58**:353–362.

Bruna, G. D., and Martire, L., 1985, La successione giurassica (Pliensbachiano–Kimmeridgiano) delle Alpi Feltrine (Belluno), *Riv. Ital. Paleontol. Stratigr.* **91**:15–62.

Bulot, L. G., 1993, Stratigraphic implications of the relationship between ammonites and facies: Examples taken from the Lower Cretaceous (Valanginian–Hauterivian) of the western Tethys, in: *The Ammonoidea: Environment, Ecology and Evolutionary Change*, Systematics Associa-tion Spec. Vol. 47 (M. R. House, ed.), Clarendon Press, Oxford, pp. 243–266.

Callomon, J. H., 1981, Dimorphism in ammonoids, in: *The Ammonoidea*, Systematics Associa-tion Spec. Vol. 18 (M. R. House and J. R. Senior, eds.), Academic Press, London, pp. 257–273.

Callomon, J. H., 1985, The evolution of the Jurassic ammonite family Cardioceratidae, *Spec. Pap. Palaeontol.* **33**:49–90.

Calver, M. A., 1968, Distribution of Westfalian marine faunas in northern England and adjoining areas, *Yorkshire Geol. Soc. Proc.* **37**:1–72.

Cecca, F., 1992, Ammonite habitats in the Early Tithonian of Western Tethys, *Lethaia* **25**:257–267.

Chamberlain, J. A., Jr., 1991, Cephalopod locomotor design and evolution: The constraints of jet propulsion, in: *Biomechanics and Evolution* (J. M. V. Rayner and R. J. Wooton, eds.), Cambridge University Press, Cambridge, pp. 57–98.

Chamberlain, J. A., Jr., Ward, P. D., and Weaver, J. S., 1981, Postmortem ascent of *Nautilus* shells: Implications for cephalopod paleogeography, *Paleobiology* **7**:494–509.

Checa, A., and Westermann, G. E. G., 1989, Segmental growth in planulate ammonites: Inferences on costae function, *Lethaia* **22**:95–100.

Clarim, P. A., Marini, P., Pastorini, M., and Pavia, G., 1984, Il rosso ammonitico inferiore (Baiociano–Calloviano) nei Monti Lessini settentrionali (Verona), *Riv. Ital. Paleontol. Stratigr.* **90**:15–86.

Conti, M. A., and Fischer, J. C., 1981, Preliminary notes on Aalenian gastropods of Case Canepini (Umbria, Italy), in: *Rosso Ammonitico Symposium Proceedings* (A. Farrinacci and S. Elmi, eds.), Tecnoscienza, Rome, pp. 136–167.

Cope, J. C. W., 1967, The palaeontology and stratigraphy of the lower part of the Upper Kimmeridge Clay of Dorset, *Bull. Br. Mus. (Nat. Hist.) Geol.* **15**(1):1–79.

Cope, J. C. W., 1974, Upper Kimmeridge ammonite faunas of Wash area and a subzonal scheme from the lower part of the Upper Kimmeridgian, *Bull. Geol. Surv. G.B.* **1974**:29–37.

Dagys, A. S., and Weitschat, W., 1993, Extensive intraspecific variation in a Triassic ammonoid from Siberia, *Lethaia* **26**:113–122.

Davis, R. A., 1972, Mature modification and dimorphism in selected Late Paleozoic ammonoids, *Bull. Am. Paleontol.* **62**(272):26–130.

Davis, R. A., Furnish, W. M., and Glenister, B. F., 1969, Mature modification and dimorphism in Late Paleozoic ammonoids, in: *Sexual Dimorphism in Fossil Metazoa and Taxonomic Implications* (G. E. G. Westermann, ed.), Schweizerbart'sche Verlagsbuchhandlung, Stuttgart, pp. 101–110.

Delanoy, G., Magnin, A., Selebran, M., and Selebran, J., 1991, *Moutoniceras nodosum* d'Orbigny, 1850 (Ammonoidea, Ancyloceratina) une très grande ammonite heteromorphe du Barremien inférieur, *Rev. Paleobiol.* **10**:229–245.

Dietl, G., 1973, Middle Jurassic (Dogger) heteromorph ammonites, in: *Atlas of Palaeobiogeography* (A. Hallam, ed.), Elsevier, Amsterdam, pp. 283–285.

Dietl, G., 1978, Die heteromorphen Ammoniten des Dogger (Stratigraphie, Taxonomie, Phylogenie, Okologie), *Stuttg. Beitr. Naturkd. B* **33**:1–97.

Doguzhaeva, L. A., 1988, Siphuncular tubes and septal necks in ammonite evolution, in: *Cephalopods—Present and Past* (J. Wiedmann and J. Kullmann, eds.), Schweizerbart'sche Verlagsbuchhandlung, Stuttgart, pp. 291–302.

Doguzhaeva, L. A., and Mutvei, H., 1989, *Ptychoceras*—a heteromorphic lytoceratid with truncated shell and modified ultrastructure (Mollusca: Ammonoidea), *Palaeontogr. A* **208**:91–121.

Doguzhaeva, L. A., and Mutvei, H., 1991, Organization of the soft body in *Aconeceras* (Ammonitina), interpreted on the basis of shell morphology and muscle scars, *Palaeontogr. A* **218**:17–33.

Doguzhaeva, L. A., and Mutvei, H., 1992, Radula of the Early Cretaceous ammonite *Aconeceras* (Mollusca: Cephalopoda), *Palaeontogr. A* **223**:167–177.

Doguzhaeva, L. A., and Mutvei, H., 1993, Structural features in Cretaceous ammonoids indicative of semiinternal or internal shells, in: *The Ammonoidea: Environment, Ecology and Evolutionary Change*, Systematics Association Spec. Vol. 47 (M. R. House, ed.), Clarendon Press, Oxford, pp. 99–114.

Dommergues, J. L., Cariou, E., Contini, D., Hantzpergue, P., Marchand, D., Meister, C., and Thierry, J., 1989, Homeomorphies et canalisations evolutives: Le rôle de l'ontogenese. Quelques exemples pris chez les ammonites du Jurassic, *Geobios* **22**:5–48.

Doyle, P., and Whitham, A. G., 1991, Palaeoenvironments of the Nordenskjold Formation: An Antarctic Late Jurassic–Early Cretaceous black shale-tuff sequence, in: *Modern and Ancient Continental Shelf Anoxia* (R. V. Tyson and T. H. Pearson, eds.), *Geol. Soc. Spec. Publ.* **58**:397–414.

Duff, K. L., 1975, Palaeoecology of a bituminous shale—the Lower Oxford Clay of central England, *Palaeontology* **18**:443–482.

Ebel, K., 1983, Calculations on the buoyancy of ammonites, *N. Jb. Geol. Paläont. Mh.* **1983**:614–640.

Ebel, K., 1985, Gehausespirale und Septenformen bei Ammoniten unter Annahme vagil benthischer Lebensweise, *Paläontol. Z.* **59**:109–123.

Ebel, K., 1992, Mode of life and soft body shape of heteromorph ammonites, *Lethaia* **25**:179–193.

Elmi, S., 1985, Influences des hauts-fonds sur la composition des peuplements et sur la dispersion des ammonites, in: *La géodynamique des seuils et des hauts-fonds*, Bull. Sect. Scien. Geol. 9:217–228.

Elmi, S., 1993, Area-rule, boundary layer and functional morphology of cephalopod shells (Ammonoids), *Geobios Mém. Spéc.* 15:121–138.

Elmi, S., and Almeras, Y., 1984, Physiography, paleotectonics and paleoenvironment as control of changes in ammonite and brachiopod communities (an example from the Early and Middle Jurassic of western Algeria), *Pelaeogeogr. Palaeoclimatol. Palaeoecol.* 47:347–360.

Elmi, S., and Benshili, K., 1987, Relations entre la structuration tectonique, la composition des peuplements et l'évolution; exemple du Toarcian du Moyen-Atlas meridional (Maroc), *Boll. Soc. Paleontol. Ital.* 26:47–62.

Ford, R. D., 1965, The palaeoecology of the goniatite bed at Cowlow Nick, Castleton, Derbyshire, *Palaeontology* 8:186–191.

Fourcade, E., Azema, J., Cecca, F., Bonneau, M., Peybernes, B., and Dercourt, J., 1991, Essai de réconstitution cartographique de la paléongéographie et des paléoenvironnements de la Téthys au Tithonique supérieur (138 à 135 Ma), *Bull. Soc. Géol. Fr.* 162:1197–1208.

Frye, C. J., and Feldman, R. M., 1991, North American Late Devonian cephalopod aptychi, *Kirtlandia* 49:49–71.

Galacz, A., and Horwath, F., 1985, Sedimentary and structural evolution of the Bakony Mountains (Transdanubian Central Range, Hungary): Paleogeographic implications, *Acta Geol. Hung.* 28:85–100.

Gasiorowski, S. M., 1962, Aptychi from the Dogger, Malm and Neocomian in the western Carpathians and their stratigraphic value, *Stud. Geol. Pol.* 117–165.

Geczy, B., 1982, The Davoi Zone in the Bakony Mountains, Hungary, *Ann. Univ. Sci. Budap. Sect. Geol.* 21:1–11.

Geczy, B., 1984, Provincialism of Jurassic ammonites; examples from Hungarian faunas, *Acta Geol. Hung.* 27:379–389.

Geraghty, M., and Westermann, G. E. G., 1994, Composition and origin of Jurassic ammonite concretions from Alfeld, Germany: A biogenic alternative, *Paläontol. Z.* 68:473–490.

Geyssant, J. R., 1988, Diversity in mode and tempo of evolution within one Tithonian ammonite family, the Simoceratids, in: *Cephalopods—Present and Past* (J. Wiedmann and J. Kullmann, eds.), Schweizerbart'sche Verlagsbuchhandlung, Stuttgart, pp. 79–88.

Glenister, B. F., and Furnish, W. M., 1981, Permian ammonoids, in: *The Ammonoidea*, Systematics Association Spec. Vol. 18 (M. R. House and J. R. Senior, eds.), Academic Press, London, pp. 49–64.

Goldring, R., 1978, Devonian, in: *The Ecology of Fossils* (W. S. McKerrow, ed.), Duckworth, London, pp. 125–145.

Gygi, R. A., 1986, Eustatic sea level changes of the Oxfordian (Late Jurassic) and their effect documented in sediments and fossil assemblages of an epicontinental sea, *Eclogae Geol. Helv.* 79:455–491.

Gygi, R. A., Sadati, S.–M., and Zeiss, A., 1979, Neue Funde von *Paraspidoceras* (Ammonoidea) aus dem Oberen Jura von Mitteleurope—Taxonomie, Ökologie, Stratigraphie, *Eclogae Geol. Helv.* 72:897–952.

Hahn, W., Westermann, G. E. G., and Jordan, R., 1990, Ammonite Fauna of the Upper Bathonian *hodsoni* Zone (Middle Jurassic) at Lechstedt near Hildesheim, Northwest Germany, in: *Zum Ober–Bathonium (Mittlerer Jura) im Raum Hildesheim, Nordwestdeutschland* (R. Jordan, ed.), *Geol. Jahrb. A*, 121:21–64.

Hallam, A., 1987, Radiations and extinctions in relation to environmental change in the marine Lower Jurassic of northwest Europe, *Paleobiology* 13:152–168.

Heckel, P. H., 1991, Thin widespread Pennsylvanian black shales of Midcontinent North America: A record of a cyclic succession of widespread pycnoclines in a fluctuating epeiric sea, in: *Modern and Ancient Continental Shelf Anoxia* (R. V. Tyson and T. H. Pearson, eds.), *Geol. Soc. Spec. Publ.* 58: pp. 259–273.

Hengsbach, R., 1978, Bemerkungen über das Schwimmvermogen der Ammoniten und die Funktion der Septen, *Sitzungsber. Ges. Naturforsch. Freunde Berlin, N.F.*, **18**:105–117.

Hewitt, R. A., 1988, Significance of early septal ontogeny in ammonoids and other ectocochliates, in: *Cephalopods—Present and Past* (J. Wiedmann and J. Kullmann, eds.), Schweizerbart'sche Verlagsbuchhandlung, Stuttgart, pp. 201–214.

Hewitt, R. A., 1993, Relation of shell strength to evolution, in: *The Ammonoidea: Environment, Ecology and Evolutionary Change*, Systematics Association Spec. Vol. 47 (M.R. House, ed.), Clarendon Press, Oxford, pp. 35–56.

Hewitt, R. A., and Watkins, P., 1980, Cephalopod ecology across a late Silurian shelf tract, *N. Jb. Geol. Paläont. Abh.* **160**:96–117.

Hewitt, R. A., and Westermann, G. E. G., 1987, Functions of complexly fluted septa in ammonoid shells. II. Septal evolution and conclusions, *N. Jb. Geol. Paläont. Abh.* **174**:135–169.

Hewitt, R. A., and Westermann, G. E. G., 1990a, *Nautilus* shell strength variance as an indicator of habitat depth limits, *N. Jb. Geol. Paläont. Abh.* **179**:73–97.

Hewitt, R. A., and Westermann, G. E. G., 1990b, Mosasaur tooth marks on the ammonite *Placenticeras* from the Upper Cretaceous Bearpaw Formation of Alberta, *Can. J. Earth Sci.* **27**:469–472.

Hewitt, R. A., Westermann, G. E. G., Checa, A., and Zaborski, P. M., 1994, Growth rates of ammonoids estimated from aptychi, *Geobios Mém. Spéc.* **15**:203–208.

Hirano, H., 1986, Cenomanian and Turonian biostratigraphy of the off-shore facies of the Northern Pacific—an example of the Oyubari area, central Hokkaido, Japan, *Bull. Sci. Eng. Res. Lab. Waseda Univ.* **113**:6–20.

Hirano, H., 1993, Phyletic evolution of desmoceratine ammonoids through the Cenomanian–Turonian oceanic anoxic event, in: *The Ammonoidea: Environment, Ecology and Evolutionary Change*, Systematics Association Spec. Vol. 47 (M. R. House, ed.), Clarendon Press, Oxford, pp. 267–283.

Hirano, H., Okamoto, T., and Hattori, K., 1990, Evolution of some Late Cretaceous desmoceratine ammonoids, *Trans. Proc. Palaeont. Soc. Jpn. N.S.* **157**:382–411.

Hoffmann, K., 1982, Die Stratigraphie, Palaeogeographie und Ammonitenfauna des Unter-Pliensbachium (Carixium, Lias gamma) in Nordwest-Deutschland, *Geol. Jahrb. A* **551**:442.

House, M. R., 1975, Faunas and time in the marine environment, *Yorkshire Geol. Soc. Proc.* **40**(27):45–90.

House, M. R., 1981, On the origin, classification and evolution of the early Ammonoidea, in: *The Ammonoidea*, Systematics Association Spec. Vol. 18 (M. R. House and J. R. Senior, eds.), Academic Press, London, pp. 3–36.

House, M. R., 1985, A new approach to an absolute time scale from measurements of orbital cycles and sedimentary microrhythms, *Nature* **316**:721–725.

House, M. R., 1992, Fluctuations in ammonoid evolution and possible environmental causes, in: *The Ammonoidea: Environment, Ecology and Evolutionary Change*, Systematics Association Spec. Vol. 47 (M. R. House, ed.), Clarendon Press, Oxford, pp. 13–34.

House, M. R., and Price, J. D., 1985, New Late Devonian genera and species of tornoceratid goniatites, *Palaeontology* **28**:159–188.

Howarth, M. K., 1992, The ammonite family Hildoceratidae in the Lower Jurassic of Britain, *Palaeontogr. Soc. Monogr.* Publ. No. 590, vol. 146:1–200.

Hudson, J. D., and Martill, D. M., 1991, The Lower Oxford Clay: Production and preservation of organic matter in the Callovian (Jurassic) of central England, in: *Modern and Ancient Continental Shelf Anoxia* (R. V. Tyson and T. H. Pearson, eds.), Geol. Soc. Spec. Pap. **58**:363–379.

Jacobs, D. K., 1992a, Shape, drag, and power in ammonoid swimming, *Paleobiology* **18**:203–220.

Jacobs, D. K., 1992b, The support of hydrostatic load in cephalopod shells. Adaptive and ontogenetic explanations of shell form and evolution from Hooke 1695 to the present, in: *Evolutionary Biology*, Vol. 26 (M. K. Hecht, B. Wallace, and R. J. MacIntyre, eds.), Plenum Press, New York, pp. 287–349.

Jacobs, D. K., and Landman, N. H., 1993, *Nautilus*—a poor model for the function and behavior of ammonoids? *Lethaia* **26**:101–112.

Jansa, L. F., Emos, P., Tcholke, B. E., Gradstein, F., and Sheridan, R. E., 1979, Mesozoic–Cenozoic sedimentary formations of the North American Basin, western North Atlantic, in: *Deep Drilling Results in the Atlantic Ocean Continental Margins and Paleoenvironment*, Maurice Ewing Series 3 (M. Talman, W. Hay, and W. B. F. Ryan, eds.), American Geophysical Union, Washington, DC, pp. 1–57.

Jenkyns, H., 1988, The early Toarcian (Jurassic) anoxic event. Stratigraphic, sedimentary, and geochemical evidence, *Am. J. Sci.* **288**:101–151.

Jordan, R., 1968, Zur Anatomie mesozoischer Ammoniten nach den Strukturelementen der Gehauseinnenwand, *Geol. Jahrb. Beih.* **77**:1–64.

Kammer, T. W., Brett, C. E., Boardman, D. R. II, and Mapes, R. H., 1986, Ecologic stability of the dysaerobic biofacies during the Late Paleozoic, *Lethaia* **19**:109–121.

Kant, R., 1975, Biometric analysis of ammonoid shells, *Paläontol. Z.* **49**:203–220.

Kant, R., and Kullmann, J., 1988, Changes in conch form in the Paleozoic ammonoids, in: *Cephalopods—Present and Past* (J. Wiedmann and J. Kullmann, eds.), Schweizerbart'sche Verlagsbuchhandlung, Stuttgart, pp. 43–49.

Kase, T., Shigeta, F., and Futakami, M., 1994, Limpet home depressions in Cretaceous ammonites, *Lethaia* **25**:49–58.

Kauffman, E. G., 1984, Paleobiogeography and evolutionary response dynamic in the Cretaceous Western Interior seaway of North America, in: *Jurassic–Cretaceous Biochronology and Paleo-geography of North America* (G. E. G. Westermann, ed.), *Geol. Assoc. Can. Spec. Pap.* **27**:73–306.

Kauffman, E. G., 1990, Mosasaur predation on ammonites during the Cretaceous—an evolutionary history, in: *Evolutionary Paleobiology of Behaviour and Coevolution* (A. J. Boucot, ed.), Elsevier, New York, pp. 184–189.

Kauffman, E. G., Villamil, T., Harries, P. J., and Meyer, C., 1992, The flat clam controversy: Where did they come from? Where did they go? *Paleontol. Soc. Spec. Publ.* **6**:159.

Kennedy, W. J., and Cobban, W. A., 1976, Aspects of ammonite biology, biogeography, and biostratigraphy, *Spec. Pap. Palaeontol.* **17**:1–94.

Klinger, H. C., 1980, Speculations on buoyancy control and ecology in some heteromorph ammonites, in: *The Ammonoidea*, Systematics Association Spec. Vol. 18 (M. R. House and J. R. Senior, eds.), Academic Press, London, pp. 337–355.

Korn, D., 1986, Ammonoid evolution in late Famennian and early Tournaisian, *Ann. Soc. Géol. Belg.* **109**:49–54.

Korn, D., 1988, Oberdevonishe Goniatiten mit dreieckigen Innenwindungen, *N. Jb. Geol. Paläont. Mh.* **1988**(10):605–610.

Korn, D., 1992, Relationship between shell form, septal construction and suture line in clymeniid cephalopods (Ammonoidea; Upper Devonian), *N. Jb. Geol. Paläont. Abh.* **185**:115–130.

Kulicki, C., and Doguzhaeva, L., 1994, Development and calcification of the ammonitella shell, *Acta Palaeontol. Pol.* **39**:17–44.

Kullmann, J., 1981, Carboniferous goniatites, in: *The Ammonoidea*, Systematics Association Spec. Vol. 18 (M. R. House and J. R. Senior, eds.), Academic Press, London, pp. 37–48.

Kullmann, J., and Scheuch, J., 1970, Wachstums-Anderungen in der Ontogenese paläozoischer Ammonoideen, *Lethaia* **3**:397–412.

Kullmann, J., and Scheuch, J., 1972, Absolutes und relatives Wachstum bei Ammonoideen, *Lethaia* **5**:129–146.

Landman, N. H., 1982, Embryonic shells of Baculites, *J. Paleontol.* **56**:1235–1241.

Landman, N. H., 1986, Shell abnormalities in scaphitid ammonites, *Lethaia* **19**:211–224.

Landman, N. H., 1987, Ontogeny of Upper Cretaceous (Turonian–Santonian) scaphitid ammonites from the Western Interior of North America: Systematics, developmental patterns, and life history, *Bull. Am. Mus. Nat. Hist.* **185**:117–241.

Landman, N. H., and Bandel, K., 1985, Internal structures in the early whorls of Mesozoic ammonoids, *Am. Mus. Novit.* **2823**:1–21.

Landman, N. H., and Waage, K. M., 1993, Scaphitid ammonites of the Upper Cretaceous (Maastrichtian) Fox Hills Formation in South Dakota and Wyoming. *Bull. Am. Mus. Nat. Hist.* **215**:1–257.

Lehmann, U., 1975, Über Nahrung und Ernahrungsweise der Ammoniten, *Paläontol. Z.* **49**:187–195.

Lehmann, U., 1976, *Ammoniten, Ihr Leben und Ihre Umwelt,* Enke, Stuttgart.

Lehmann, U., 1985, Zur Anatomie der Ammoniten: Tintenbeutel, Kiemen, Augen, *Paläontol. Z.* **59**:99–108.

Lehmann, U., 1988, On the dietary habits and locomotion of fossil cephalopods, in: *Cephalopods—Present and Past* (J. Wiedmann and J. Kullmann, eds.), Schweizerbart'sche Verlagsbuchhandlung, Stuttgart, pp. 633–640.

Lehmann, U., and Kulicki, 1990, Double function of aptychi (Ammonoidea) as jaw elements and opercula, *Lethaia* **23**:325–331.

Lehmann, U., and Weitschat, W., 1973, Zur Anatomie und Ökologie der Ammoniten. Funde von Kropf und Kiemen, *Paläontol. Z.* **47**:69–76.

Lehmann, U., Tanabe, K., Kanie, Y., and Fukuda, Y., 1980, Über den Kieferapparat der Lytoceraten (Ammonoidea), *Paläontol. Z.* **54**:319–329.

Little, R., Baker, D. R., Leythaeuser, D., and Rullkottner, J., 1991, Keys to the depositional history of the Posidonia Shale (Toarcian) in the Hills Syncline, northern Germany, in: *Modern and Ancient Continental Margin Anoxia* (R. V. Tyson and T. H. Pearson, eds.), *Geol. Soc. Spec. Publ.* **58**:311–333.

Loh, H., Maul, B., Prauss, M., and Riegel, W., 1986, Primary production, marl formation and carbonate species in the Posidonia Shale of NW Germany, *Mitt. Geol.-Paläont. Inst. Univ. Hamburg* **60**:397–421.

Mapes, R. H., 1979, Carboniferous and Permian Bactritoidea (Cephalopoda) in North America, *Univ. Kans. Paleontol. Contrib. Artic.* **64**:1–75.

Mapes, R. H., 1987, Upper Paleozoic cephalopod mandibles: Frequency of occurrence, modes of preservation, and paleoecological implications, *J. Paleontol.* **61**:521–538.

Mapes, R. H., and Hansen, M. C., 1984, Pennsylvanian shark-cephalopod predation: A case study, *Lethaia* **17**:175–183.

Mapes, R. H., Tanabe, K., Landman, N. H., and Faulkner, C. J., 1992, Upper Carboniferous ammonoid shell clusters: Transported accumulations or *in situ* nests? *Paleontol. Soc. Spec. Pub.* **6**:196.

Marchand, D., Thierry, J., and Tintant, H., 1985, Influence des seuls et des hauts-fonds sur la morphology et l'évolution des ammonites, *Inst. Sci. Terre Univ. Dijon Bull. Sect. Sci.* **9**:191–202.

Marcinowski, R., and Wiedmann, J., 1985, The Albian ammonite fauna of Poland and its paleogeographical significance, *Acta Geol. Pol.* **35**:199–218.

Maynard, J. R., and Leder, M. R., 1992, On the periodicity and magnitude of late Carboniferous glacio-eustatic sea-level changes, *J. Geol. Soc. (Lond.)* **149**:303–311.

McLearn, F. H., 1969, Middle Triassic (Anisian) ammonoids from northeastern British Columbia and Ellesmere Island, *Geol. Surv. Can. Bull.* **170**:1–90.

Mehl, J., 1978a, Ein Koprolith mit Ammoniten-Aptychen aus dem Solnhofer Plattenkalk, Jber. Wetterau. *Ges. Naturkunde* **129–130**:85–89.

Mehl, J., 1978b, Anhaufungen scherbenartiger Fragmente von Ammonitenschalen im suddeutschen Lias und Malm und ihre Deutung als Frassreste, *Ber. Naturforsch Ges. Freib. Breisgau* **68**:75–93.

Mehl, J., 1984, Radula and arms of *Michelinoceras* sp. from the Silurian of Bohemia, *Paläontol. Z.* **58**:211–229.

Meister, C., 1993, Parallel evolution in Euboreal and Tethyan Juraphyllitidae: The role of internal and external constraints, *Lethaia* **26**:123–132.

Milson, C. V., 1994, *Saccocoma,* a benthic crinoid from the Jurassic Solnhofen Limestone, Germany, *Palaeontology* **37**:121–130.

Morris, K. A., 1979, A classification of Jurassic marine shale sequences: An example from the Toarcian (Lower Jurassic) of Great Britain, *Palaeogeogr. Palaeoclimatol. Palaeoecol.* **26**:117–126.

Morris, K. A., 1980, Comparison of major regions of organic-rich mud deposition in the British Jurassic, *J. Geol. Soc. (Lond.)* **137**:157–170.

Morrison, J. O., 1986, *Molluscan carbonate geochemistry and paleooceanography of the Late Cretaceous Western Interior Seaway of North America*, unpublished M.Sc. thesis, Brock University, St. Catherines, Ontario.

Morton, N., 1984, Morphological trends in the evolution of some Middle Jurassic ammonites, *Lethaia* **17**:306.

Morton, N., 1988, Segregation and migration patterns in some *Graphoceras* populations (Middle Jurassic), in: *Cephalopods—Present and Past* (J. Wiedmann and J. Kullmann, eds.), Schweizerbart'sche Verlagsbuchhandlung, Stuttgart, pp. 377–385.

Morton, N., and Nixon, M., 1987, Size and function of ammonoid aptychi in comparison with buccal masses in modern cephalopods, *Lethaia* **20**:231–238.

Nassichuk, W. W., 1970, Permian ammonoids from Devon and Melville Islands, Canadian arctic archipelago, *J. Paleontol.* **44**:77–97.

Nesis, K. N., 1986, On the feeding and causes of extinction of certain heteromorph ammonites, *Paleontol. Zh.* **1986**:8–15 (in Russian; Engl. transl. *Paleontol. J.* **20**:5–11).

Nesis, K. N., 1987, *Cephalopods of the World, Squids, Cuttlefishes, Octopuses and Allies*, (transl. from Russian), T.F.H. Publications, Neptune City, NJ.

Newell, N. D., Rigby, J. K., Fisher, A. G., Whitman, A. J., Hickox, J. E., and Bradley, J. I., 1953, *The Permian Reef Complex of the Guadelupe Mountains in Texas and New Mexico*, Freeman & Co., San Francisco.

Newton, C. R., 1988, Significance of "Tethyan" fossils in the American Cordillera, *Science* **242**:385–390.

Nixon, M., 1988, The feeding mechanism and diets of cephalopods—living and fossil, in: *Cephalopods—Present and Past* (J. Wiedmann and J. Kullmann, eds.), Schweizerbart'sche Verlagsbuchhandlung, Stuttgart, pp. 641–652.

O'Dor, R. K., and Wells, M. J., 1990, Performance limits of 'antique' and 'state-of-the-art' cephalopods, *Nautilus* and squid, *Am. Malacol. Union Prog. Abstr. 56 Ann. Meeting*, p. 52.

Okamoto, T., 1984, Theoretical morphology of *Nipponites* (a heteromorph ammonite), in *Fossils* (Kaseki), *Palaeont. Soc. Jpn.* **36**:37–51.

Okamoto, T., 1988a, Analysis of heteromorph ammonoids by differential geometry, *Palaeontology* **31**:35–52.

Okamoto, T., 1988b, Changes in life orientation during the ontogeny of some heteromorph ammonoids, *Palaeontology* **31**:281–294.

Okamoto, T., 1988c, Developmental regulation and morphological saltation in the heteromorph ammonite *Nipponites*, *Paleobiology* **14**:273–286.

Olivero, E. B., and Zinsmeister, W. J., 1989, Large heteromorph ammonites from the Upper Cretaceous of Seymour Island, Antarctica, *J. Paleontol.* **63**:626–635.

Oloriz, F., 1976, *Kimmeridgiano–Tithonico inferior en el sector central de la Cordillera Betica, Zona Subbetica, paleontologia, bioestratigrafia*, Tesis Doct., Universidad de Granada.

Oloriz, F., Marques, B., and Rodriguez-Tovar, F. J., 1991, Eustatism and faunal associations. Examples from the south Iberian margin during the Late Jurassic (Oxfordian–Kimmeridgian), *Eclogae Geol. Helv.* **84**:83–106.

Oschmann, W., 1991, Distribution, dynamics and palaeontology of Kimmeridgian (Upper Jurassic) shelf anoxia in western Europe, in: *Modern and Ancient Continental Shelf Anoxia* (R. V. Tyson and T. H. Pearson, eds.), *Geol. Soc. Spec. Pap.* **58**:381–395.

Pamenter, C. B., 1956, *Imitoceras* from the Exshaw Formation of Alberta, *J. Paleontol.* **30**:965–966.

Prauss, M., Ligouis, B., and Lutterbacher, H., 1991, Organic matter and palynomorphs in the 'Posidonienschiefer' (Toarcian, Lower Jurassic) of southern Germany, in: *Modern and Ancient Continental Shelf Anoxia* (R. V. Tyson and T. H. Pearson, eds.), *Geol. Soc. Spec. Publ.* **58**:335–351.

Ramsbottom, W. H. C., 1978, Carboniferous, in: *The Ecology of Fossils* (W.S. McKerrow, ed.), Duckworth, London, pp. 146–183.

Raup, D. M., 1967, Geometric analysis of shell coiling: Coiling in ammonoids, *J. Paleontol.* **41**:43–65.

Reeside, J. B., and Cobban, W. A., 1960, Studies of the Mowry Shale (Cretaceous) and contemporary formations in the United States and Canada, *U.S. Geol. Surv. Prof. Pap.* **355**:1–126.

Renz, O., 1972, Aptychi (Amnmonoidea) from the Upper Jurassic and Lower Cretaceous of the western North Atlantic (site 105, leg 11, DSDP), in: *Initial Reports DSDP*, No. 11 (C. D. Holister, J. I. Ewing, *et al.*, eds.), U.S. Government Printing Office, Washington, DC, pp. 607–620.

Renz, O., 1973, Two lamellaptychi (Ammonoidea) from the Magellan Rise in the central Pacific, in: *Initial Reports DSDP*, No. 17 (E. L. Winterer and J. L. Hewing, eds.), U.S. Government Printing Office, Washington, DC, pp. 895–901.

Renz, O., 1978, Aptychi (Ammonoidea) from the Early Cretaceous of the Blake–Bahama Basin, leg 44, hole 391c, DSDP, in: *Initial Reports DSDP*, No. 44 (W. E. Benson and R. E. Sheridan, eds.), U.S. Government Printing Office, Washington, DC, pp. 899–909.

Renz, O., 1979, Aptychi (Ammonoidea) and ammonites from the Lower Cretaceous of the western Bermuda Rise, leg 43, site 387, DSDP, in: *Initial Reports DSDP*, No. 43 (B. E. Tucholke and P. R. Vogt, eds.), U.S. Government Printing Office, Washington, DC, pp. 591–597.

Reyment, R. A., 1980, Floating orientations of cephalopod shell models, *Palaeontology* **23**:931–936.

Riccardi, A. C., and Westermann, G. E. G., 1991, Middle Jurassic ammonite fauna of the Argentine–Chilean Andes, III: Bajocian–Callovian Eurycephalitinae, Stephanocerataceae, *Palaeontogr. A* **216**:1–110.

Riccardi, A. C., Gulisano, C. A., Mojica, J., Palacios, O., Schubert, C., and Thomson, M. R. A., 1992, Western South America and Antarctica, in: *The Jurassic of the Circum-Pacific* (G. E. G. Westermann, ed.), Cambridge University Press, New York, pp. 122–161.

Rieber, H., 1973, Ergebnisse paläontologisch-stratigraphischer Untersuchungen in der Grenzbitumenzone (Mittlere Trias) des Monte San Giorgio, *Eclogae Geol. Helv.* **66**:667–685.

Rieber, H., 1975, Der Posidonienschiefer (Oberer Lias) von Holzmaden und die Grenzbitumenzone (Mittlere Trias) des Monte San Giorgio (Kt. Tessin, Schweiz), *Jahresh. Ges. Naturkd. Württemb.* **130**:163–190.

Sadki, D., and Elmi, S., 1991, Fluctuations de la composition des peuplements d'ammonoides en relation avec la dynamique sedimentaire au passage Aalenien–Bajocien dans Haut–Atlas Central Marocain, in *Conference on Aalenian and Bajocian Stratigraphy, Isle of Skye, April, 1991* (N. Morton, ed.), Birkbeck College, University of London, London, pp. 113–122.

Sandoval, J., 1983, *Bioestratigrafia y paleontologia (Stephanocerataceae y Perisphinctaceae) del Bajocense y Bathoniense en las Cordilleras Beticas*, Tesis Doct., Universidad de Granada.

Sandoval, J., Westermann, G. E. G., and Marshall, M. C., 1990, Ammonite fauna, stratigraphy and ecology of Bathonian–Callovian (Jurassic) Tecocoyunca Group, South Mexico, *Palaeontogr. A* **210**:93–149.

Saunders, W. B., 1984, *Nautilus belauensis* growth and longevity: Evidence from marked and recaptured animals, *Science* **224**:990–992.

Saunders, W. B., 1995, The ammonoid suture problem: Relationships between shell septum thickness and suture complexity in Paleozoic ammonoids, *Paleobiology* **21**:343–355.

Saunders, W. B., and Shapiro, E. A., 1986, Calculation and simulation of ammonoid hydrostatics, *Paleobiology* **12**:64–79.

Saunders, W. B., and Swan, R. H., 1984, Morphology and morphologic diversity of mid-Carboniferous (Namurian) ammonoids in time and space, *Paleobiology* **10**:195–228.

Savrda, C. E., and Bottjer, D. J., 1991, Oxygen-related biofacies in marine strata: An overview and update, in: *Modern and Ancient Continental Shelf Anoxia* (R. V. Tyson and T. H. Pearson, eds.), *Geol. Soc. Spec. Pap.* **58**:201–219.

Schindewolf, O. H., 1958, Über Aptychen (Ammonoidea), *Palaeontogr. A* **111**:1–46.

Schindewolf, O. H., 1959, Adolescent cephalopods from the Exshaw Formation of Alberta, *J. Paleontol.* **33**:971–976.

Schindewolf, O. H., 1963, *Acuariceras* und andere heteromorphe Ammoniten aus dem oberen Dogger, *N. Jb. Geol. Paläont. Abh.* **116**:119–148.

Schindewolf, O. H., 1971, Über Clymenien und andere Cephalopoden, *Akad. Wiss. Lit. Abh. Math. Naturwiss. Kl.* **1971**:355–449.

Schmidt, H., 1930, Über die Bewegungsweise der Cephalopoden, *Paläontol. Z.* **12**:194–207.

Schmidt–Effing, R., 1972, Die Dactylioceratidae, eine Ammoniten-Familie des unteren Jura, *Münster. Forsch. Geol. Paläontol.* **25/26**:1–254.

Seilacher, A., 1993, Ammonite aptychi: How to transform a jaw into an operculum, *Am. J. Sci.* **293–A**:20–32.

Sequeiros, L., 1984, Facies y ammonoideos de edad Calloviense al sur de Zaragoza (Cordillera Iberica), *Bol. Geol. Min. (Esp.)* **95**(2):109–115.

Shapiro, E. A., and Saunders, W. B., 1987, *Nautilus* shell hydrostatics, in: *Nautilus* (W. B. Saunders and N. H. Landman, eds.), Plenum Press, New York, pp. 527–545.

Shigeta, Y., 1993, Post-hatching early life history of Cretaceous Ammonoidea, *Lethaia* **26**:23–46.

Silberling, N. J., and Nichols, K. M., 1982, Middle Triassic molluscan fossils of biostratgraphic significance from the Humboldt Range, northwestern Nevada, *U.S. Geol. Surv. Prof. Pap.* **1207**:1–77.

Smith, P. L., Westermann, G. E. G., Stanley, G. D., Jr., and Yancey, T. E., 1990, Paleobiogeography of the ancient Pacific, *Science* **249**:680–683.

Spinosa, C., Furnish, W. M., and Glenister, B. F., 1970, Araxoceratidae, Upper Permian ammonoids, from the Western Hemisphere, *J. Paleontol.* **44**:730–736.

Spinosa, P. L., Furnish, W. M., and Glenister, G. F., 1975, The Xenodiscidae, Permian ceratitoid ammonoids, *J. Paleontol.* **49**:239–283.

Stevens, G. R., 1988, Giant ammonites: A review, in: *Cephalopods—Present and Past* (J. Wiedmann and J. Kullmann, eds.), Schweizerbart'sche Verlagsbuchhandlung, Stuttgart, pp. 141–166.

Sturani, C., 1971, Ammonites and stratigraphy of the "Posidonia alpina" beds of the Venetian Alps, *Mem. Inst. Geol. Min. Univ. Padova* **28**:1–190.

Swan, R. H., and Saunders, W. B., 1987, Function and shape in Late Paleozoic (mid-Carboniferous) ammonoids, *Paleobiology* **13**:297–311.

Tanabe, K., 1977, Functional evolution of *Otoscaphites puerculus* (Jimbo) and *Scaphites planus* (Yabe), Upper Cretaceous ammonites, *Mem. Fac. Sci. Kyushu Univ., D, Geol.* **23**:367–407.

Tanabe, K., 1979, Palaeoecological analysis of ammonoid assemblages in the Turonian *Scaphites* facies of Hokkaido, Japan, *Palaeontology* **22**:609–630.

Tanabe, K., 1983, The jaw apparatus of Cretaceous desmoceratid ammonites, *Palaeontology* **26**:677–689.

Tanabe, K., and Fukuda, Y., 1987, The jaw apparatus of the Cretaceous ammonite *Reesidites*, *Lethaia* **20**:41–48.

Tanabe, K., and Ohtsuka, Y., 1985, Ammonoid early internal shell structure: Its bearing on early life history, *Paleobiology* **11**:310–322.

Tanabe, K., and Shigeta, Y., 1987, Ontogenetic shell variation and streamlining of some Cretaceous ammonites, *Trans. Proc. Palaeont. Soc. Jpn. N.S.* **147**:165–179.

Tanabe, K., Obata, I., and Futakami, M., 1978, Analysis of ammonoid assemblages in the Upper Turonian of the Manji area, central Hokkaido, *Bull. Nat. Sci. Mus. (Tokyo)* C **4**(2):37–62.

Tanabe, K., Obata, I., and Futakami, M., 1981, Early shell morphology in some Upper Cretaceous heteromorph ammonites, *Trans. Proc. Palaeont. Soc. Jpn. N.S.* **124**:215–234.

Tanabe, K., Landman, N. H., and Weitschat, W., 1993a, Septal necks in Mesozoic Ammonoidea: Structure, ontogenetic development and evolution, in: *The Ammonoidea: Environment, Ecology and Evolutionary Change*, Systematics Association Vol. 47 (M.R. House, ed.), Clarendon Press, Oxford, pp. 57–84.

Tanabe, K., Landman, N. H., Mapes, R. H., and Faulkner, C. J., 1993b, Analysis of a Carboniferous embryonic ammonoid assemblage—implications for ammonoid embryology, *Lethaia* **20**:215–224.

Tanabe, K., Shigeta, Y., and Mapes, R. H., 1995, Early life history of Carboniferous ammonoids inferred from analysis of fossil assemblages and shell hydrostatics, *Palaios* **10**:80–86.

Tatzreiter, F., and Vörös, A., 1991, Vergleich der pelsonischen (Anis, Mitteltrias) Ammoniten-faunen von Grossreifling (nordliche Kalkalpen) und Aszofo (Balaton-Gebiet), in: *Jubi-laumsschr. 20 Jahre Geol. Zusammenarbeit Osterreich-Ungarn*, Part 1 (H. Lobitzer and G. Csaszar, eds.), Wien, pp. 247–259.

Taylor, D. G., 1982, Jurassic shallow marine invertebrate depth zones, with exemplification from the Snowshoe Formation, Oregon, *Oregon Geol.* **44**(5):51–58.

Taylor, D. G., Smith, P. L., Laws, R. A., and Guex, J., 1983, The stratigraphy and biofacies trends of the Lower Mesozoic Gabbs and Sunrise Formations, west-central Nevada, *Can. J. Earth Sci.* **20**:1598–1608.

Tintant, H., Marchand, D., and Mouterde, R., 1982, Relations entre les milieux marins et l'évolution des Ammonoides: Les radiations adaptives du Lias, *Bull. Soc. Géol. Fr.* **24**:951–961.

Tozer, E. T., 1981, Triassic Ammonoidea: Geographic and stratigraphic distribution, in: *The Ammonoidea*, Systematics Association Spec. Vol. 18 (M. R. House and J. R. Senior, eds.), Academic Press, London, pp. 397–431.

Trueman, A. E., 1941, The ammonite body–chamber, with special reference to the buoyancy and mode of life of the living ammonite, *Q. J. Geol. Soc. (Lond.)* **384**:339–383.

Tyson, R. V., and Pearson, T. H., 1991, Modern and ancient continental shelf anoxia: An overview, in: *Modern and Ancient Continental Shelf Anoxia* (R. V. Tyson and T. H. Pearson, eds.), *Geol. Soc. Spec. Publ.* **58**:1–26.

Urlichs, M., and Mundlos, R., 1985, Immigration of cephalopods into the German Muschelkalk basin and its influence on the suture lines, in: *Sedimentary and Evolutionary Cycles, Lecture Notes in Earth Sciences* (U. Bayer and A. Seilacher, eds.), Springer, Berlin, pp. 221–236.

Urlichs, M., Wild, R., and Ziegler, B., 1979, Fossilien aus Holzmaden, *Stuttg. Beitr. Naturkd. C* **11**:1–34.

Van Der Zwan, G. J., and Jorissen, F. J., 1991, Biofacial patterns in river-induced anoxia, in: *Modern and Ancient Continental Shelf Anoxia* (R. V. Tyson and T. H. Pearson, eds.), *Geol. Soc. Spec. Publ.* **58**:65–82.

Vasicek, Z., and Wiedmann, J., 1994, The Leptoceratoidinae: Small heteromorph ammonites from the Barremian, *Palaeontology* **37**:203–239.

Wang, Y., and Westermann, G. E. G., 1993, Paleoecology of Triassic ammonoids, *Geobios Mém. Spéc.* **15**:373–392.

Ward, D. J., and Hollingworth, N. T. J., 1990, The first record of a bitten ammonite from the Middle Oxford Clay (Callovian, Middle Jurassic) of Bletchley, Buckinghamshire, *Mesozoic Res.* **2**:153–161.

Ward, P. D., 1976a, *Stratigraphy, Paleoecology and Functional Morphology of Heteromorph Ammonites in the Upper Cretaceous Nanaimo Group, British Columbia and Washington,* unpublished Ph.D. thesis, Department of Geology, McMaster University, Hamilton, Ontario.

Ward, P. D., 1976b, Upper Cretaceous ammonites (Santonian–Campanian) from Orcas Islands, Washington, *J. Paleontol.* **50**:454–461.

Ward, P. D., 1981, Shell sculpture as a defensive adaptation in ammonoids, *Paleobiology* **7**:96–100.

Ward, P. D., 1986a, Rates and processes of compensatory buoyancy change in *Nautilus macrom-phalus*, *Veliger* **1986**:356–368.

Ward, P. D., 1986b, Cretaceous ammonite shell shapes, *Malacologia* **27**:3–28.

Ward, P. D., 1987, *The Natural History of Nautilus*, Allen & Unwin, Boston.

Ward, P. D., 1990a, A review of Maastrichtian ammonite ranges, *Geol. Soc. Am. Spec. Pap.* **247**:519–530.

Ward, P. D., 1990b, The Cretaceous/Tertiary extinctions in the marine realm: A 1990 perspective, *Geol. Soc. Am. Spec. Pap.* **247**:425–432.

Ward, P. D., and Bandel, K., 1987, Life history strategies in fossil cephalopods, in: *Cephalopod Life Cycles*, Academic Press, London, pp. 329–350.

Ward, P. D., and Kennedy, W. J., 1993, Maastrichtian ammonites from the Biscay Region (France, Spain), *Paleontol. Soc. Mem.* **34**:1–58.

Ward, P. D., and Signor, P. W. III, 1983, Evolutionary tempo in Jurassic and Cretaceous ammonites, *Paleobiology* **9**:183–198.

Ward, P. D., and Westermann, G. E. G., 1977, First occurrence, systematics and functional morphology of *Nipponites* from the Americas, *J. Paleontol.* **51**:367–372.

Ward, P. D., and Westermann, G. E. G., 1985, Cephalopod paleoecology, in: *Mollusks, Notes for Short Course* (D. J. Bottjer, C. S. Hickman, and P. D. Ward, organizer; T. W. Broadhead, ed.), *Univ. Tenn. Stud. Geol.* **13**:1–18.

Ward, P. D., Kennedy, W. J., Macleod, K. G., and Mount, J. F., 1991, Ammonite and inoceramid bivalve extinction patterns in Cretaceous/Tertiary boundary sections of the Biscay region (southwestern France, northern Spain), *Geology* **19**:1181–1184.

Wells, M. J., Wells, J., and O'Dor, R. K., 1992, Life at low oxygen tensions: The behaviour and physiology of *Nautilus pompilius* and the biology of extinct forms, *J. Mar. Biol. Assoc. U.K.* **72**:313–328.

Wendt, J., 1976, Submarine Spaltenfullungen, *Zentralbl. Geol. Paläont. Teil II* **1976**:245–251.

Wendt, J., and Eigner, T., 1985, Facies patterns and depositional environments of Paleozoic cephalopod limestones, *Sedim. Geol.* **44**:15–21.

Wendt, J., Eigner, T., and Neugenauer, J., 1984, Cephalopod limestone deposition on a shallow pelagic ridge: The Tafilalt Platform (Upper Devonian, eastern Anti-Atlas, Morocco), *Sedimentology* **31**:601–625.

Westermann, G. E. G., 1954, Monographie der Otoitidae (Ammonoidea), *Geol. Jahrb. Beih.* **15**:1–364.

Westermann, G. E. G., 1958, The significance of septa and sutures in Jurassic ammonite systematics, *Geol. Mag.* **45**:441–455.

Westermann, G. E. G., 1964, Sexual-Dimorphismus bei Ammonoideen und seine Bedeutung für die Taxionomie der Otoitidae, *Palaeontogr. A* **124**:33–73.

Westermann, G. E. G., 1966, Covariation and taxonomy of the Jurassic ammonite *Sonninia adicra* (Waagen), *N. Jb. Geol. Paläontol. Abh.* **124**:289–312.

Westermann, G. E. G., 1971, Form, structure and function of shell and siphuncle in coiled Mesozoic ammonoids, *Life Sci. Contrib. R. Ont. Mus.* **78**:1–39.

Westermann, G. E. G., 1975, A model for origin, function and fabrication of fluted cephalopod septa, *Paläontol. Z.* **49**:235–253.

Westermann, G. E. G., 1977, Form and function of orthoconic cephalopod shells with concave septa, *Paleobiology* **3**:300–321.

Westermann, G. E. G., 1982, The connecting rings of *Nautilus* and Mesozoic ammonoids: Implications for ammonoid bathymetry, *Lethaia* **15**:374–384.

Westermann, G. E. G., 1990, New developments in ecology of Jurassic–Cretaceous ammonoids, in: *Fossili. Evolutione, Ambiente*, Atti II Conv. Int. Pergola 1987 (G. Pallini, F. Cecca, S. Cresta, and M. Santantonio, eds.), Tecnostampa, Ostra Vetere, Italy, pp. 459–478.

Westermann, G. E. G., 1993, On alleged negative buoyancy in ammonoids, *Lethaia* **26**:246.

Westermann, G. E. G., and Callomon, J. H., 1988, The Macrocephalitinae and associated Bathonian and early Callovian (Jurassic) ammonoids of the Sula Islands and New Guinea, *Palaeontogr. A*, **203**:1–90.

Westermann, G. E. G., and Hewitt, R. A., 1995, Do limpet pits indicate that desmoceratacean ammonites lived mainly in surface waters? *Lethaia* **28**:24.

Westermann, G. E. G., and Riccardi, A. C., 1979, Middle Jurassic ammonoid fauna and biochronology of the Argentine-Chilean Andes. Part II: Bajocian Stephanocerataceae, *Palaeontogr. A* **164**:85–118.

Westermann, G. E. G., and Hewitt, R. A. (in prep.), Ammonoid septal fluting, complex sutures, and shell strength: A critique.

Wiedmann, J., 1972, Ammoniten-Nuklei aus Schlammproben der nordalpinen Obertrias—ihre stammesgeschichtliche und stratigraphische Bedeutung, *Mitt. Ges. Geol. Bergbaustud.* **21**:561–622.

Wiedmann, J., 1988a, Ammonite extinction and the Cretaceous–Tertiary Boundary Event, in: *Cephalopods—Present and Past* (J. Wiedmann and J. Kullmann, eds.), Schweizerbart'sche Verlagsbuchhandlung, Stuttgart, pp. 117–140.

Wiedmann, J., 1988b, Plate tectonics, sea level changes, climate and the relationship to ammonite evolution, provincialism, and mode of life, in: *Cephalopods—Present and Past* (J. Wiedmann and J. Kullmann, eds.), Schweizerbart'sche Verlagsbuchhandlung, Stuttgart, pp. 737–765.

Wignall, P. B., 1987, A biofacies analysis of the *Gastrioceras cumbriense* Marine Band (Namurian) of the central Pennines, *Proc. Yorkshire Geol. Soc.* **46**:111–121.

Wignall, P. B., 1990, Observations on the evolution and classification of dysaerobic communities, in: *Paleocommunity Temporal Dynamics: The Long-term Development of Multispecies Assemblies* (W. Miller, ed.), *Paleontol. Soc. Spec. Publ.* 5:99–111.

Wignall, P. B., and Hallam, A., 1991, Biofacies, stratigraphic distribution and depositional models of British onshore Jurassic black shales, in: *Modern and Ancient Continental Shelf Anoxia* (R. V. Tyson and T. H. Pearson, eds.), *Geol. Soc. Spec. Pap.* **58**:291–309.

Wignall, P. B., and Hallam, A., 1993, Griesbachian (earliest Triassic) palaeoenvironmental changes in the Salt Range, Pakistan and southeast China and their bearing on the Permo-Triassic mass extinction, *Palaeogeogr. Palaeoclimatol. Palaeoecol.* **102**:215–237.

Wilde, P., and Berry, W. B. N., 1984, Destabilization of the oceanic density structure and its significance to marine "extinction" events, *Palaeogeogr. Palaeoclimatol. Palaeoecol.* **48**:143–162.

Ziegler, B. 1967, Ammoniten-Ökologie am Beispiel des Oberjura, *Geol. Rundsch.* **56**:439–464.

Ziegler, B., 1980, Ammonoid biostratigraphy and provincialism: Jurassic-Old World, in: *The Ammonoidea*, Systematics Association Spec. Vol. 18 (M. R. House and J. R. Senior, eds.), Academic Press, London, pp. 433–457.

VII

Biostratigraphy and Biogeography

Chapter 17

Paleozoic Ammonoids in Space and Time

R. THOMAS BECKER and JÜRGEN KULLMANN

1. Introduction

Paleozoic ammonoids have attracted much less attention from professional and amateur paleontologists than Mesozoic ammonoids. Because of the Variscan folding in Europe, the classical area of investigation, Devonian and Carboniferous material from Europe is often rather poorly preserved. As a result, few collectors of the 19th and early 20th century have focused their attention on these fossils. In recent decades, however, as our knowledge of

R. THOMAS BECKER • Institut für Paläontologie, Humboldt-Universität Berlin, D-10155 Berlin, Germany.   JÜRGEN KULLMANN • Geologisch-Paläontologisches Institut, Universität Tübingen, D-72076 Tübingen, Germany.

Ammonoid Paleobiology, Volume 13 of *Topics in Geobiology*, edited by Neil Landman *et al.*, Plenum Press, New York, 1996.

Paleozoic ammonoids has expanded, it has become more and more apparent that the evolution and systematics of Paleozoic forms are as complex as those of Mesozoic ones. For example, the number of Devonian genera rose from about 80 in the Treatise of 1957 to a present figure of more than 200.

Paleozoic ammonoids experienced many episodes of radiation and decline. Soon after their first appearance, ammonoids nearly became extinct late in the Givetian, at the Frasnian–Famennian and Devonian–Carboniferous boundaries, and later, at the end of the Permian. Other important extinction events occurred at the end of the Eifelian, in the Namurian, and in the middle Permian. The Carboniferous–Permian transition, in contrast, was relatively smooth and was characterized by increasing diversification.

Because of their high evolutionary rates and broad geographic distributions, Paleozoic ammonoids historically defined "orthostratigraphic" scales. For many intervals, international ammonoid zones allow time resolution on the order of a few hundred thousand years. This potential for inter- and intrabasinal correlation of sedimentological and structural processes has not yet been fully exploited. Future work must also concentrate on the interactions among geological, paleoecological, and evolutionary change.

In this chapter, we provide overviews of the morphology, systematics, geographic distribution, biostratigraphy, and evolution of ammonoids in the Devonian, Carboniferous, and Permian Periods. Sections 1, 2, 3, and 5.1 were written by R. T. Becker, and Sections 4 (excluding bactritids) and 5.2 by J. Kullmann. Other sections were prepared jointly.

2. Pre-Devonian Ammonoids

Currently, it is widely accepted that the uncoiled bactritids form the most primitive group of the Ammonoidea (see, e.g., Erben, 1960, 1964; House, 1981; Teichert, 1988; Becker, 1993a). Bactritids are intermediate between advanced orthocerids (*Sphaerorthoceras, Protobactrites*), with a central to subventral siphuncle, retrochoanitic septal necks, and a spherical protoconch (e.g., Ristedt, 1968), and coiled goniatites. They share with the latter the combination of a marginal (ventral) siphuncle and a narrow ventral lobe of the suture as well as a ventral (hyponomic) sinus at the apertural margin (also visible in growth lines). There is almost a morphological continuum between cyrtoconic bactritids and the earliest gyroconic goniatites (Anetoceratinae: *Kokenia = Borivites = Metabactrites*; see Erben, 1966; Teichert, 1988), which attain just a little more than one full whorl late in ontogeny. Specialized bactritids show such shell morphologies as rectiradiate, biconvex growth lines, regular internal shell thickenings, and bicarinate venters, features that are also present in goniatites but absent in any orthoconic nautiloids. No apomorphy is present in bactritids that justifies equal taxonomic ranking of this group with the whole rest of the Ammonoidea.

The inclusion of the Bactritida in the Ammonoidea extends the range of the Ammonoidea into pre-Devonian time. *Eobactrites* Schindewolf, 1932, from the Ordovician of Scandinavia and Bohemia, which was originally regarded as the oldest bactritid (e.g., Schindewolf, 1933), is much larger than typical representatives of the group. It probably possesses a large embryonic shell (Dzik, 1984) and lacks a distinctive ventral sinus at the apertural margin. Currently, it is regarded as a member of the homeomorphic Baltoceratidae (Ellesmeroceratina) close to *Bactroceras* (Sweet, 1958; Dzik, 1984). Consequently, *Bactrites bohemicus* Ristedt, 1981, from the lower Ludlow of Bohemia is the oldest ammonoid. There is another questionable specimen from the Ludlow of Morocco (Termier and Termier, 1950, Pl. 136, Fig. 31). An alleged Gothlandian *Bactrites* of the Standard Oil collection from Erfoud (Termier and Termier, 1950, Pl. 137, Figs. 27–29) appears to be a Devonian *Lobobactrites*.

3. Devonian Ammonoids

3.1. General Morphology and Major Systematic Categories

3.1.1. Morphology

There is as much morphological variability in Devonian ammonoids as there is in forms from later periods. With the exception of the dorsal siphuncle of clymenids, only some primitive morphological features are restricted to the Devonian. Secondary removal of the siphuncle from its strictly marginal position is present in adults of some bactritids, in Frasnian *Timanoceras* (Gephuroceratidae), and in certain tornoceratids (e.g., *Kirsoceras*), which branched off from the lineage leading to the clymenids.

Early (Zlíchovian) ammonoid evolution witnessed the transformation of orthoconic shells into cyrtoconic, gyroconic, advolute, and finally convolute shapes. Gyroconic coiling of the first whorl (a perforate umbilicus) is last seen in the earliest Eifelian (last *Mimagoniatites*; Chlupáč, 1985). Secondary or adult uncoiling usually occurs only in some Emsian forms, but the last whorl of some Late Devonian species of *Manticoceras* (Gephuroceratidae) and *Wocklumeria* also tends to leave the normal whorl spiral. In *Prolobites* and *Kielcensia* (Wocklumeriidae), the last whorl is somewhat scaphitid-like (Korn, 1984; Czarnocki, 1989). Other extremes are the evolute triangular coiling of *Soliclymenia paradoxa* (Hexaclymeniidae; Schindewolf, 1937) and the tripartition of *Parawocklumeria* and *Mimimitoceras* (Prionoceratinae) by deep regular constrictions.

Serpenticonic shells gave rise to more involute forms. However, this trend was later often reversed, and there are many cases (e.g., Cheiloceratidae and Prionoceratidae; Becker, 1993a; Vöhringer, 1960; Korn, 1994) in which evolute whorls were suddenly introduced into the early ontogeny of completely involute forms (proterogenesis of Schindewolf, 1937). Subdiscoidal forms were the starting point for globular, oxyconic, keeled, or tabulate to bicarinate lineages. Internal shell thickenings are common in certain families (e.g., Cheiloceratidae, Sporadoceratidae).

Ornament is mostly rather simple in comparison with many Mesozoic forms, and most species and genera are unribbed. Spiral ornament is rare and, if present, rather weak. Large, hollow, and triangular flares occur on the tabulate venter of several genera (e.g., *Kosmoclymenia*; Tornoceratinae: *Armatites, Aulatornoceras*; Korn, 1979; Becker, 1993a). The oxyconic "*Falciclymenia*" *excellens* (Rectoclymeniidae) possesses long, hollow, and widely spaced spines on the venter (Bogoslovsky, 1982). In other cases (Platyclymeniidae: *Trigonoclymenia, Spinoclymenia*; Gonioclymeniinae: *Kalloclymenia*), parabolic or marginal nodes end in lateral spines. Growth lines and the apertural margins of Devonian ammonoids have a ventral sinus that sometimes disappears in forms with linear ornament (e.g., *Cycloclymenia*, Phenacoceratidae).

In primitive bactritids and the Mimosphinctidae, the apertural margin is rursiradiate and mostly convex (consisting only of a dorsal salient and a ventral sinus; e.g., Erben, 1964). In derived forms, this pattern is restricted to the embryonic stage. Ornament is introduced at various positions on the protoconch or later on the ammonitella (Erben, 1964). After the second growth constriction, ornament becomes rectiradiate and biconvex. Ribbing and ventrolateral furrows also start immediately after the point of hatching (Erben, 1964; House and Price, 1985). In certain neotenic lineages, the embryonic convex pattern is retained into later ontogeny (e.g., Sobolewiidae, Gephuroceratidae: *Crickites*, CheiloceratACEae). Other forms reduce the juvenile lateral (ocular) sinus secondarily in the adult (e.g., *Phoenixites*, Tornoceratinae).

The simplest sutures consist of only a siphuncular (ventral) lobe, which sometimes disappears with a subventral siphuncular shift (*Devonobactrites*). The formation of a wide lateral lobe in *Lobobactrites* is not simply related to the whorl cross section (Korn, 1992) but is also controlled by the variable curvature of the septum. Later ammonoids also show no simple relationship between whorl cross section and lobation. A dorsal lobe originated in connection with the invention of convolute coiling (e.g., Teicherticeratinae: *Convoluticeras, Taskanites*) as well as with the invention of advolute coiling (*Mimosphinctes*; Kullmann, 1960). In later convolute genera, a dorsal saddle sometimes appears in the middle of the dorsal lobe. Additional sutural elements are formed by the subdivision of all of the three principal lobes (ventral, lateral, dorsal) and intervening saddles.

As a consequence, there is much more variability in sutural ontogeny in the Devonian than in all later periods. The primary suture is trilobate apart from the earliest forms with fewer sutural elements. The tornoceratid origin of clymenids implies that their "lateral" lobe is the midflank adventitious lobe of their ancestors, and the true lateral lobe at the umbilical seam may have been reduced as in contemporaneous prolobitids. Such an interpretation would require, in the future, fundamental revisions of sutural formulas in many clymenid taxa. The most complex septa in the Devonian have more than

50 lobes, but normally there are only four to eight lobes. Slight frilling at the base of lobes occurs in *Ceratobeloceras* (Beloceratidae) and *Devonopronorites*. True ceratitic or even ammonitic sutures are unknown. Angular lobes had already appeared by the early Emsian (dorsal lobe of *Mimagoniatites*; ventral lobes of *Celaeceras*, Auguritidae) but were uncommon until the middle Givetian. Angular saddles are rare exceptions apart from all members of the Beloceratidae.

Other morphological features that are relevant for systematic subdivision are wrinkle layer patterns (Walliser, 1970; House, 1971; House *et al.*, 1985; Korn, 1985), the length of the body chamber, shell thickness (e.g., thick in the Cheiloceratidae and thin in contemporaneous Tornoceratidae), and the presence of strong phosphatic anaptychi in some Acanthoclymeniidae and in the Gephuroceratidae. Septal spacing is distinctive in some taxa [e.g., *Cheiloceras (Raymondiceras) simplex*] but has not yet been systematically studied. Some genera such as *Progonioclymenia* (Hexaclymeniidae) and *Biloclymenia* have very long septal necks forming a continuous tube. Otherwise, the retrochoanitic septal necks are short.

3.1.2. Major Systematic Categories

Devonian ammonoids consist of four orders with five suborders, 14 superfamilies, and 45 families. Subdivisions down to the level of tribes have recently been introduced in a few groups (Bartzsch and Weyer, 1987, 1988; Becker, 1993a).

The Bactritida comprise all ortho- to cyrtoconic (monogeneric Cyrtobactritinae Becker n. subfam.) forms. Devonian members have orthochoanitic septal necks and low apical angles. Variations in septal spacing occur in some species. Secondary siphuncle shifts and loss of the ventral lobe occur in *Devonobactrites*, some *Lobobactrites*, and in a still undescribed new Late Devonian bojobactritid genus. Growth lines of the Bactritidae are rursiradiate convex (Bactritinae) to concavoconvex (Cyrtobactritinae) with a weak lateral sinus in addition to a shallow ventral sinus, as in primitive goniatites, whereas the distinctive Bojobactritidae develop rectiradiate and strongly biconvex apertural margins with dorsal, lateral, and ventral sinuses as in advanced goniatites.

The Agoniatitida embrace all coiled forms without adventitious lobes. Its two suborders differ only by the primary position of the lateral lobe (on the flanks in the Agoniatitina, omnilateral lobe of the Russian terminology, at the umbilical seam in the Anarcestina). The ancestral Mimosphinctidae still have rursiradiate growth lines throughout ontogeny and mostly gyroconic to advolute whorls. Perisphinctid-like ribbing appears in one of the oldest genera (*Mimosphinctes*). Rectiradiate biconvex growth lines characterize the advolute Mimoceratidae and Agoniatitaceae with convolute, very rapidly expanding whorls. A low-diversity early offshoot, the Auguritaceae, subdivide the ventral lobe but retain a perforate umbilicus. Within the Anarcestina, the Anarcestaceae have slowly expanding, depressed whorls and few sutural

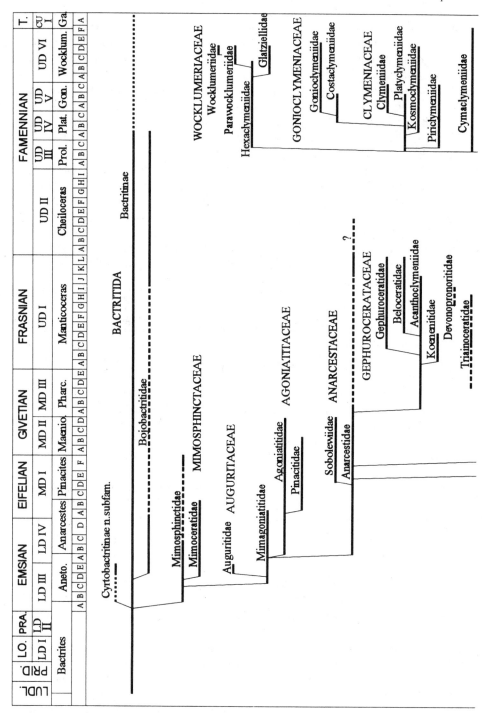

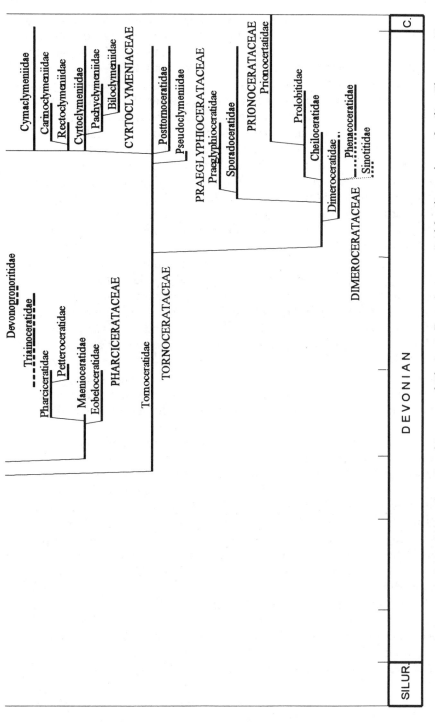

FIGURE 1. Phylogenetic relationships and stratigraphic ranges of Silurian to Devonian ammonoid families and superfamilies. The recognition of polyphyletic groups has led to the dissolution of the former suborders Gephuroceratina and Gonioclymeniina. For details of stratigraphy, see Fig. 2. Abbreviations: LD, Lower Devonian; MD, Middle Devonian; UD, Upper Devonian; CU. Lower Carboniferous.

elements. The ventral lobe is subdivided in two independent lineages (late Givetian Pharcicerataceae and Frasnian Belocerataceae = Gephurocerataceae). In both groups the number of umbilical lobes increases significantly. This is especially true for advanced members of the Pharciceratidae (*Neopharciceras*) and late beloceratids, with about equal numbers of outer umbilical and ventral lobes. The systematic position of the Frasnian Triainoceratidae is still doubtful (Becker and House, 1993). Therefore, the Pharcicerataceae are not yet renamed here as Triainocerataceae as proposed by Bartzsch and Weyer (1988).

The Goniatitida possess at least one adventitious lobe. Devonian representatives belong to the Tornoceratina with undivided ventral lobe. Only in the Praeglyphioceratidae does the position of origin of the adventitious lobes move from the ventral saddle down to the base of the parallel-sided ventral lobe. The Tornocerataceae have short body chambers (less than a whorl), mostly biconvex growth lines, and lack an umbilical lobe. Advanced forms of several lineages (e.g., Posttornoceratidae) subdivide the lateral lobe at the umbilical seam. The Dimerocerataceae (= Cheilocerataceae) differ by their long body chamber and convex growth lines. A shallow lateral sinus appears late in the ontogeny of some forms. In agreement with interpretations by Schmidt (1952), the Prolobitidae and Phenacoceratidae are not combined in a Famennian anarcestid superfamily Prolobitaceae but are assigned to the Dimerocerataceae. In the Praeglyphiocerataceae, a second adventitious (or ventral) and an umbilical lobe are introduced; growth lines may be convex or somewhat biconvex. The Prionocerataceae were the rootstock for all post-Devonian ammonoids, and members of the only Devonian family (Prionoceratidae) have rather invariable sutures with one adventitious and one umbilical lobe each.

The Clymeniida are characterized by a dorsal siphuncle and were derived from Famennian evolute Tornoceratidae (House, 1970). Early forms retain the lingulate dorsal lobe, biconvex growth lines, and the shell form of *Pernoceras* or *Protornoceras*. Rather involute genera and some oxyconic lineages are assigned to the Cyrtoclymeniaceae. Sutures in the brevidomic Clymeniaceae are mostly very simple, and a shallow ventral lobe is seen only in forms with a flattened venter. Several evolute lineages consist of forms with complex ornament. The brevidomic Gonioclymeniaceae were derived via *Nodosoclymenia* from the Platyclymeniidae and retain a true ventral lobe (folded ventral part of the septum) despite the siphuncular migration. They have widely evolute shells and, commonly, much more complex sutures than other clymenids. A maximum of 12 lobes are formed in *Sphenoclymenia*. The Wocklumeriaceae developed from longidomic, evolute ancestors (Hexaclymeniidae) into goniatite-like involute forms with triangular coiling and deep constrictions (Wocklumeriidae, Parawocklumeriidae) or characteristic keels (Glatziellidae).

The phylogenetic relationships among Devonian ammonoid groups down to the family level are illustrated in Fig. 1.

3.2. Biostratigraphy and Evolutionary Events

3.2.1. Principles of Zonation

Early studies of Devonian ammonoid biostratigraphy go back to the late part of the 19th century (e.g., Kayser, 1873; Frech, 1897). From the earliest investigations, it was customary to use successions of genera rather than species to characterize time intervals. Later, because of the work of Wedekind, Schindewolf, and others, proper index species for German zones were established. Several zones were bundled into "Stufen" characterized by the total "life span" of dominating groups, and these "superzones" are still very useful because they define intervals bounded by extinctions and global faunal turnovers (Fig. 2). Hollard (1974) and House (1978) were the first to elucidate independent faunal sequences outside the classical European sections. There are clear differences between different regional zonations; therefore, a succession of nonendemic, that is, widely distributed, species should be used as a standard reference. Interbasinal correlation is, in many cases, facilitated because of the appearance and spreading of genera that locally may be only represented by various allopatric species that evolved in geographical isolation. Genozones for all of the Devonian have recently been established by Becker *et al.* (1993), Becker (1993a), and Becker and House (1994).

3.2.2. Faunal Succession in Time

The succession of ammonoids through the Devonian is shown in Figs. 2 and 3. It is strange that earliest Devonian (Lochkovian and Pragian) cephalopod faunas do not contain bactritids. There is only a single and questionable record from the upper Pragian of Germany (Bender *et al.*, 1974). The appearance of goniatites in the earliest Emsian (Zlíchovian: Yu and Ruan, 1989) was clearly related to a spreading event supported by global transgression (manifested, e.g., in the famous Hunsrück Shale of Germany). Bactritids experienced their first bloom at the same time. It is also obvious that the appearance of goniatites coincided with the decline of graptolites. The initial goniatite radiation has been named the "Basal Zlíchovian Event" by Chlupáč and Kukal (1988).

The *Anetoceras* Stufe comprises most of the lower Emsian (Zlíchovian) and the basalmost part of the upper Emsian (Dalejan). It is defined by the "life span" of members of the Mimosphinctidae such as *Anetoceras*. At the base (LD III-B/C), faunas contain only genera of this family, followed by the appearance of *Mimagoniatites* as the oldest agoniatitacean (LD III-D). In the late Zlíchovian (LD III-E), *Gyroceratites* became abundant, and continuing radiation led to the first diversity peak in early ammonoid history. The Zlíchovian–Dalejan boundary coincided approximately with the phylogenetic transition from *Gyroceratites laevis* to *gracilis*. There was a gradual decline of the Mimosphinctidae, but a significant extinction was linked with the global hypoxic and transgressive Daleje Event at the base of the Dalejan.

series	stages	substages		ammonoid stufen	events
UPPER DEVONIAN (UD)	FAMENNIAN	Wocklumian	UD VI	*Wocklumeria*	Hangenberg
		Dasbergian	UD V	*Gonioclymenia*	Dasberg
		Hembergian	UD IV	*Platyclymenia*	Annulata
			UD III		
		Nehdenian	UD II	*Cheiloceras*	Enkeberg Condroz (Nehden)
	FRASNIAN	Adorfian	UD I	*Manticoceras*	Kellwasser / Middlesex Frasne
MIDDLE DEVONIAN (MD)	GIVETIAN		MD III	*Pharciceras*	Taghanic
			MD II	*Maenioceras*	Kacak
	EIFELIAN		MD I	*Pinacites*	Chotec
LOWER DEVONIAN (LD)	EMSIAN	Dalejan	LD IV	*Anarcestes*	Daleje
		Zlichovian	LD III	*Anetoceras*	Zlichov
				(?bactritids)	
	PRAGIAN		LD II		
	LOCHKOVIAN		LD I		

▩ interregional anoxia

FIGURE 2. Devonian chronostratigraphy, ammonoid Stufen (superzones), and the timing of currently recognized interregional sedimentary and evolutionary events (Chlupác and Kukal, 1988; House, 1985; Becker, 1993b). Abbreviations: LD, Lower Devonian; MD, Middle Devonian; UD, Upper Devonian.

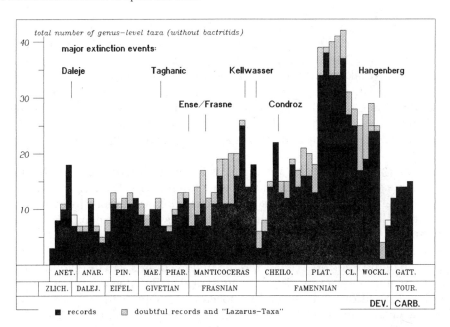

FIGURE 3. Total "generic" (genera plus subgenera) ammonoid diversity in 66 successive faunal intervals from the earliest Emsian (Zlíchovian) to the earliest Carboniferous (early Tournaisian). The notoriously slowly evolving Bactritida are omitted. Goniatite evolution consisted of rapidly fluctuating episodes of diversification and decline interrupted by first-order extinction events at the Frasnian–Famennian (Upper Kellwasser Event) and Devonian–Carboniferous boundaries (Hangenberg Event).

The *Anarcestes* Stufe is defined by the appearance of the earliest Anarcestaceae and ranges from the lower Dalejan into the lowermost Eifelian. Abundant anarcestids were accompanied by early agoniatitids with an imperforate umbilicus and new Mimagoniatitidae (e.g., *Paraphyllites*). *Latanarcestes*, *Sellanarcestes*, and *Anarcestes* appeared in successive zones. The Lower/Middle Devonian boundary is very weakly defined in terms of ammonoid biostratigraphy. Faunas crossing the boundary did not change, apart from the final disappearance of the Mimoceratidae (*Gyroceratites*). Anarcestids with an imperforate umbilicus (*Werneroceras*) appeared at the top of the *Anarcestes* Stufe (MD I-B), and, at the same time, there was a radiation of Agoniatitaceae (*Fidelites, Foordites*).

The base of the *Pinacites* Stufe postdated only slightly the interregional hypoxic Chotec Event, but there was no associated faunal turnover in goniatites. The extremely compressed and oxyconic *Pinacites* was a very characteristic form in the early part of this entirely Eifelian interval, but later, globular (*Subanarcestes*) and cadiconic (*Cabrieroceras*) anarcestids and ribbed agoniatitids (*Agoniatites costulatus* group) dominated. The first tornoceratids (*Parodiceras*) were widespread in the latest Eifelian (MD I-F), but the origin of the Goniatitida is still rather enigmatic because no intermediate forms between anarcestids and tornoceratids are known. In the topmost Eifelian there was again an important interregional hypoxic and transgressive interval (the Kačák Event) causing a bloom of *Cabrieroceras*, *Holzapfeloceras*, and the surprising reappearance of very primitive anetoceratids (*Kokenia*; Holzapfel, 1895; Chlupáč and Turek, 1977), which are morphologically and taxonomically indistinguishable from early Zlíchovian *Borivites* (= *Metabactrites*). Most Anarcestidae and some Agoniatitaceae died out within the event interval.

The marker genus of the *Maenioceras* Stufe appeared immediately after the Kačák extinction event in the topmost Eifelian bed (Becker and House, 1994). Lower and middle Givetian ammonoid zones are defined by evolutionary steps within the Maenioceratidae; these forms had more complex sutures than any earlier goniatite group and had, for the first time, mostly angular lateral lobes. Tornoceratids diversified at the same time. The Taghanic Event at the end of the middle Givetian was the first major break in the evolutionary history of ammonoids (Fig. 3). The Agoniatitidae, Sobolewiidae, holzapfeloceratids, Pinacitidae, and several tornoceratid genera suddenly became extinct. Survivors consisted of a few species of *Tornoceras*, unknown anarcestids that gave rise to *Atlantoceras*, one species of *Maenioceras* (Bensaid, 1974), and an ancestral pharciceratoid (Eobeloceratidae: *Mzerrebites* Becker and House, 1994; see Göddertz, 1987).

Redefinition of the Middle/Upper Devonian series boundary (Klapper *et al.*, 1987) moved the classical *Pharciceras lunulicosta* Zone (Wedekind, 1913) down into the upper Givetian. Correlation with conodonts showed that faunas with the very multilobed Pharciceratidae occurred over a considerable time span, and House (1985), therefore, introduced the *Pharciceras* Stufe. At present, only a regional zonation in southern France has been fully established (House *et al.*, 1985), but North African sequences (Becker and House, 1994) suggest the presence of at least five faunal levels, each characterized by new pharciceratacean genera and first Belocerataceae (Acanthoclymeniidae: *Pseudoprobeloceras*; see Becker and House, 1993).

The redefined Middle/Upper Devonian boundary falls into the youngest levels with Pharciceratacea (especially *Petteroceras*) and abundant *Ponticeras* (Acanthoclymeniidae), but, for convenience, the term *Manticoceras* Stufe is expanded to cover all of the Frasnian. Recently, 12 international divisions have been defined (Becker *et al.*, 1993). Only the upper half of these successive faunas is well represented in the condensed classical German sections. Zones are defined by the appearance of genera of the Koenenitidae, Acanthoclymeniidae, Gephuroceratidae, and Beloceratidae, but the Tornoceratidae radiated as well. There were several small-scale extinction events within the Frasnian, but the Upper Kellwasser Event right at the Frasnian–Famennian boundary was much more drastic (Fig. 3). Apart from a few specimens of *Archoceras*, no member of the Agoniatitida survived into the Famennian. A single tornoceratid species, *Phoenixites frechi*, experienced an enormous opportunistic bloom after the mass extinction; other tornoceratid genera became "Lazarus Taxa" for a long period.

The early Famennian adaptive radiation was characterized by numerous examples of heterochrony (neoteny and proterogenesis as defined by Schindewolf, 1937; Becker, 1986). The Cheiloceratidae, which give the name for the *Cheiloceras* Stufe, and the Tornoceratidae diversified stepwise. The onset of the overall Famennian regression, which has been termed the Condroz Event (Becker, 1993a,b), however, caused a significant extinction soon after the radiation peak. The upper half of the UD II was dominated by involute and smooth sporadoceratids and dimeroceratids. The faunal turnover between the classical do IIα and do IIβ of Wedekind (1908) may be termed the Enkeberg Event (House, 1985; Becker, 1993b).

The succeeding *Platyclymenia* Stufe (German Hembergian, UD III and IV) began at a distinctive level with evolute tornoceratids (*Protornoceras, Pernoceras*) followed by the serpenticonic Pseudoclymeniidae. Subsequently (UD III-C = former do IIIβ), the radiation of the Platyclymeniidae, Cyrtoclymeniaceae, and unusual Prolobitidae produced the major diversity maximum of Devonian ammonoids. The global hypoxic Annulata Event at the base of the UD IV changed ammonoid paleobiogeography (Fig. 4) but had little evolutionary impact. Clymenid diversification continued in the upper part of the classical *Platyclymenia annulata* Zone, which has been only preliminarily subdivided into three generic levels. The transition between the *Platyclymenia* and the succeeding *Gonioclymenia* Stufe needs further study, but the boundary should be drawn with the disappearance of *Platyclymenia* assemblages (Becker, 1992, 1993b). Radiation of Gonioclymeniaceae was typical for the classical do V, but other clymenid families also produced short-ranging marker genera such as *Ornatoclymenia* (Piriclymeniidae).

The terminal Devonian Wocklumeria Stufe is only weakly defined. *Kosmoclymenia (Linguaclymenia)* is the only common genus-level taxon appearing at its base. Prionoceratids (e.g., *Balvia*) experienced a subsequent bloom (UD VI-B). The appearance of the Wocklumeriaceae in the upper half of the German Wocklumian produced another diversity peak. Subsequently, there was a slight gradual clymenid decline (Korn and Kullmann, 1988; Becker, 1993b), but the global Hangenberg Event at the top of the Devonian (Fig. 2) wiped out all remaining clymenids, Praeglyphiocerataceae, the last Tornocerataceae, and practically all species of the Prionocerataceae (Korn, 1994). Two imitoceratid groups were the only ammonoids that managed to survive. Because of complex facies changes during the event interval, details of the mass extinction are still somewhat ambiguous. *Cymaclymenia* survived for a brief period (Korn, 1991). The Devonian–Carboniferous boundary as defined by conodonts falls within the *Acutimitoceras* Genozone ("*A.*" *prorsum* Zone) of the initial postevent adaptive radiation.

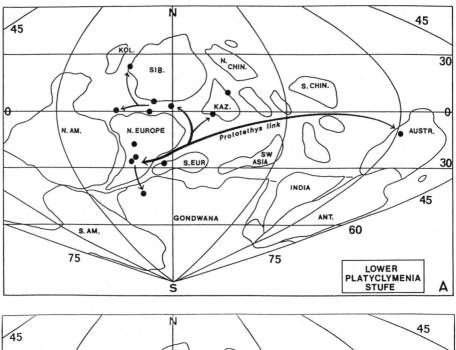

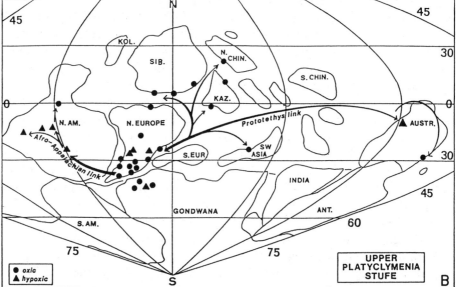

FIGURE 4. Devonian paleogeography (slightly altered from Heckel and Witzke, 1979) and ammonoid paleobiogeography illustrated by two examples from the middle Famennian. (A: Lower *Platyclymenia* Stufe = *Prolobites delphinus* Zone, UD IIIβ; B: Upper *Platyclymenia* Stufe = *Platyclymenia annulata* Zone, early UD IV. The short-term transgressive and hypoxic Annulata Event led to a significant migrational pulse and to a great increase in the number of localities/regions with ammonoid faunas.

3.3. Paleoecologic and Paleogeographic Distribution

The paleoecology and facies control of Devonian ammonoid faunas are still essentially unstudied. Devonian ammonoids were confined to subtropical to tropical seas and to moderately deep epicontinental facies rather than true bathyal facies. The Early Devonian to Eifelian cold-water Malvinokaffric Province of South America and Africa south of the Sahara totally lacked ammonoids. The most diverse assemblages were found at intrabasinal submarine seamounts, where steep slopes provided a range of different depth habitats in a small area. Other rich faunas occurred in wide outer-shelf carbonate platforms (e.g., southern Morocco), ramps (e.g., southern France), hypoxic intrashelf basins, and in the micritic distal slope facies of reef complexes (e.g., Canning Basin of Western Australia). Reef debris limestones, predominantly siliciclastic outer shelves (e.g., New York State, Belgium), and cherty shelf basins (e.g., the Chinese Nandan Facies) showed lower diversity. Specialized species invaded intrareef troughs, episodically drowned backreef areas (e.g., *Beloceras* in the Canning Basin), proximal reef slopes, hemipelagic shallow-water carbonate platforms, and the fine siliciclastic nearshore realm. Single near-shore and shallow-water finds may be related to postmortem drift.

Emsian to Famennian Devonian paleogeography was characterized by sea-level highstand and widely flooded epicratonic areas, producing enormous epicontinental seas unlike seas of the present day. Continents were assembled in a quasi-Pangea configuration with a narrow Prototethys dividing it south of the equator (Fig. 4, reconstruction after Heckel and Witzke, 1979). This situation enabled easy and rapid faunal exchange over large distances. Consequently, there was no significant provincialism in Devonian faunas. There were a number of endemic genera but only a single endemic family (the Devonopronoritidae of the Altai Mountains). The spreading of taxa and the total extent of ammonoid-bearing facies was largely controlled by eustatic movements. Short-term transgressive and often hypoxic episodes such as the Annulata and Hangenberg Events led to the sudden invasion into formerly uninhabited and overly shallow sedimentary provinces (Fig. 4; Becker, 1992, 1993a,c). The synchronous appearance of stratigraphically important but cryptogenic genera (e.g., the earliest tornoceratids, *Cheiloceras, Priono-ceras*) in widely separated sedimentary provinces may be explained by the breaching of geographic barriers by allopatric species during these transgressive episodes. On the other hand, the crossing of deeper basins became viable in peaks of regression.

The fundamentals of Devonian ammonoid paleobiogeography have been established by House (1964, 1973a,b, 1981), who identified three main links of faunal exchange and migration. The North African–European–Urals realm was identified as the major evolutionary center. The trans-Arctic route connected the Urals northward via Novaya Zemlya with the Canadian Rocky Mountains, British Columbia (Wissner and Norris, 1991), and Alaska (House and Blodgett, 1982; Prosh, 1987). The Afro–Appalachian route connected

Morocco and eastern North America. From there, the Transcontinental Arch was occasionally breached during the sea-level highstand, thus explaining the sparse faunas of Nevada, New Mexico, Utah, Montana, and California. The Prototethys route (Fig. 4) allowed faunal exchange among Europe, Turkey, Iran (Walliser, 1966), and the rich occurrences of South China and West Australia. Becker (1993a,c) drew attention to additional branches such as the Appalachian–South American route connecting eastern North America with the poor faunas of Bolivia (House, 1978; Babin *et al.*, 1991) and northwest Argentina (Hünicken *et al.*, 1980). A branch from the southern Urals, Kazakhstan, and Tienshan regions (Central–Northeast Asian route) allowed episodic ammonoid migrations to the Qingling Mountains (Ruan, 1987) and Great Khingan of North China (Chang, 1958) and even further to the Yakutsk (Yatskov, 1992) and Kolyma regions of Northeast Siberia (Bogoslovsky, 1969). A small eastward branch from the Polar Urals led to the still unstudied Famennian Taimyr occurrences.

To reach Queensland and New South Wales, ammonoids had to migrate around Australia because the continent was flooded only around the margin. The few known Tibetan Devonian ammonoids (see Yu and Ruan, 1989) probably arrived from South China. There are some striking examples of goniatite genera that occurred both in North America and in Australia (e.g., *Probeloceras*, *Prochorites*; Becker *et al.*, 1993) but not in the vast area between. Migration along the margins of the gigantic Protopacific, of which little is known, has to be considered. Poorly documented latest Devonian ammonoids from accreted terranes in the Chilean Andes (Breitkreuz, 1986) were probably part of this still unstudied faunal link.

Soon after their origin, primitive goniatites (*Anetoceras* faunas) had already entered all known epicontinental areas with pelagic faunas (Chlupác, 1976). Late Emsian faunas with anarcestids included the oldest goniatites of eastern North America (House, 1978) and Inner Mongolia (Ruan, 1983), but ammonoid facies were already absent from Siberia, Tibet, Australia, and the western United States. A somewhat restricted ammonoid distribution also characterized Eifelian faunas, but there was a remarkable spread of *Pinacites* to more southernly parts of North Africa (Sougy, 1969; Planchon, 1967) and to the Kusnetz Basin. *Cabrieroceras* and some associated forms breached the Transcontinental Arch of North America and reached Nevada (House, 1965), but the genus was completely lacking in Asia.

After the disappearance of cold-water conditions in South America at the end of the Eifelian, goniatites invaded limited areas of Bolivia and Argentina in the early to late Givetian. *Maenioceras* faunas have not yet been found in the western United States but occurred in British Columbia, Virginia, New York, the Appalachians, Europe, and North Africa. Contemporaneous faunas of the Urals and northern (Gansu) and southern (Hunan) China completely lacked Maenioceratidae. This is the only evidence for an episodic and still unexplained provincialism in Devonian ammonoids. Late Givetian *Pharciceras* faunas, in contrast, spread again to the East (Central Kazakhstan, Altai

Mountains, and Yunnan) but were rare in North America. The Urals–trans-Arctic route was obviously no longer open. Frasnian faunas with *Manticoceras* had the widest distribution of all Devonian intervals because of a transgressive maximum causing the drowning of large parts of North America, Northwest Australia, Siberia, and North China. Surprisingly, there are no records from South America or from Africa south of the Sahara.

Early Famennian *Cheiloceras* faunas were almost as widely distributed as middle Frasnian goniatites. They were lacking in most parts of North America apart from New York and the Northwest Territories of Canada but were present in Southeast Australia. The precise age of Argentinean (Hünicken *et al.*, 1980) and Bolivian (Babin *et al.*, 1991) *Sporadoceras* is not yet specified. During the lower part of the *Platyclymenia* Stufe (Fig. 4), with the increasing overall Famennian regression, the ammonoid facies retreated from parts of Europe (e.g., Cantabria; see Kullmann, 1960) North America, many regions of North Africa, China, and Eastern Australia. The trans-Arctic route became defunct for the rest of the Devonian. The paleobiogeographic restriction was reversed by the brief global transgression of the Annulata Event, but migrations did not reach Siberia. Faunas of the *Annulata* Zone have also not been documented from southern China. There are only sporadic western North American clymenid finds, and none from the East that may fall in the *Gonioclymenia* Stufe. By comparison with the *Platyclymenia* Stufe, faunal spreading was again slightly reduced. However, there is good South Chinese evidence of clymenid faunas from this period. The situation remained similar in the *Wocklumeria* Stufe. Records of wocklumeriids have been published from Chile (Breitkreuz, 1986) and Oklahoma (Over, 1992), where no other Devonian ammonoids occurred. The Hangenberg Event allowed a significant but short-term spread of goniatite facies in many regions of North America, parts of Europe (southern Ireland, Moravia), and NE Siberia immediately before the end of the Devonian.

4. Carboniferous Ammonoids

4.1. Major Systematic Categories

Carboniferous Bactritidae do not differ morphologically from Devonian forms, and, therefore, a large number of species have been assigned to *Bactrites* itself. There are claims that *Lobobactrites* was also still present in the Early Carboniferous (e.g., Lai, 1982), but more evidence is needed to exclude the possibility of confusion with compressed, longiconic early parabactritids such as *Aktastioceras* (= *Angustobactrites)* or *Eoparabactrites*, which are characterized by suborthochoanitic to cyrtochoanitic septal necks (see Mapes, 1979). Currently, no bactritids have been described from the late Famennian (UD V to VI) or from the Tournaisian anywhere in the world. Slender Parabactritidae are thought to have given rise to the oldest endocochliate Aulacoceratida during the Early Carboniferous. A Late Carboniferous (Morrowan to Missourian) parabactritid sidebranch consists of the Sinuobactritidae (Mapes, 1979), which are characterized by higher apical angles, smaller embryonic size, a deep constriction separating the protoconch, and strongly inclined septa at early ontogenetic stages.

As mentioned above, representatives of the Goniatitida possess at least one adventitious lobe; the oldest suborder, the Tornoceratina, is characterized by a simple, undivided ventral lobe (Wiedmann and Kullmann, 1981, Fig. 7f; Kullmann and Wiedmann, 1982, Figs. 1,2). Its basic suture consists of five lobes on one side, with the quinquelobate formula EALUI (Russian VLU:ID). [In this chapter, sutural formulas are given in the German suture-symbol terminology of Wedekind (1913, 1916); formulas in parentheses (Russian) correspond to the symbol terminology of Ruzhentsev (1949, 1957). For comparison, see Kullmann and Wiedmann (1970) and Wiedmann and Kullmann (1981).] The Prionoceratidae, which are also quinquelobate, originated in middle Famennian time and persisted until the end of the Early Carboniferous. Only two minor side groups with quinquelobate suture configuration (for details see below), the families Maximitidae (Late Carboniferous) and Pseudohaloritidae (mainly Permian), extended to the end of the Permian.

The ancestral stock of most Paleozoic Goniatitida (Fig. 5) and Prolecanitida was derived from the Prionoceratidae at the beginning of the Tournaisian. Some of them (Karagandocerataceae) developed different modes of subdivision of the ventral lobe, but these disappeared by early to late Tournaisian time. The Karagandoceratidae, *Voehringerites*, and Qiannanitidae–Prodromitidae obviously had a polyphyletic origin within the Prionoceratidae.

The earliest representatives of the suborder Goniatitina, which is also characterized by a subdivided ventral lobe (Wiedmann and Kullmann, 1981, Fig. 7g–i), are recorded from North America and Europe. They are prominently sculptured, as are *Goniocyclus* and other Pericyclaceae, whereas the Muensteroceratidae includes forms with smooth shells.

Three superfamilies descended from the root group of the Goniatitina: Nomismocerataceae, Dimorphocerataceae, and Goniatitaceae. The Nomismocerataceae are confined to forms that have discoidal and serpenticonic shells without ornament or with simple or dichotomous ribbing. The Dimorphocerataceae comprise genera with relatively wide ventral lobes. Among the families within this superfamily, the Girtyoceratidae exhibit well-rounded lateral saddles, the Eogonioloboceratidae show an angular pattern on the ventral and lateral portions of the suture, and the Dimorphoceratidae show denticulate or secondarily subdivided ventral or lateral lobes.

The Goniatitaceae arose at the beginning of the late Viséan. The conch form is discoidal to globular and involute; many species serve as excellent global index fossils. Goniatitidae are characterized by a relatively simple, undivided suture; their descendants Delepinoceratidae and Agathiceratidae exhibit an increase in lobe number through trifurcation of preexisting lobes (Kullmann and Schönenberg, 1975, Fig. 2). The latter family persisted from the earliest

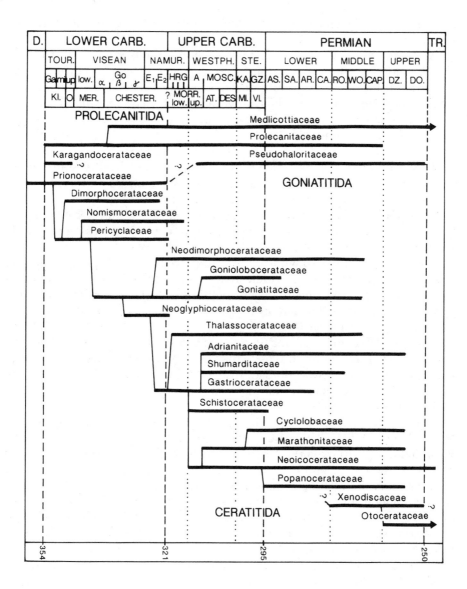

FIGURE 5. Chart showing the phylogeny and range of Carboniferous and Permian superfamilies. Abbreviations: Tour., Tournaisian; mi, middle Tournaisian; up, upper Tournaisian; Namur., Namurian; Westph., Westphalian; A, Westphalian A; Ste., Stephanian; Ga, *Gattendorfia* Stufe; Go, *Goniatites* Stufe; E, *Eumorphoceras* Stufe; H, *Homoceras* Stufe; R, *Reticuloceras* Stufe; G, *Gastrioceras* Stufe; Mosc., Moscovian; KA., Kasimovian; GZ., Gzhelian; AS., Asseilian; SA., Sakmarian; AR., Artinskian; CA., Cathedralian; RO., Roadian; WO., Wordian; CAP., Capitanian; DZ., Dzhulfian; DO., Dorashamian, Changxingian. Subdivision of the Permian according to Glenister (1991).

Namurian until the Wordian without any significant change in shell form and suture.

At the end of the Viséan epoch, the Neoglyphiocerataceae descended from the Goniatitidae; the former flourished during the late Viséan and early Namurian (Chesterian); several index species belong to the Neoglyphiocera- tidae and Cravenoceratidae. The configuration of the suture is distinctive; the rather narrow ventral lobe shows a low median saddle, and the first lateral saddle is broadly rounded.

In the early Namurian, close to the end of the Early Carboniferous, the Gastriocerataceae originated. Members of the family Glaphyritidae are closely linked by transitional forms to the older Cravenoceratidae and have a wide ventral lobe with rather wide prongs that have convex sides. The median saddle is half as high or higher than the height of the entire ventral lobe. The Glaphyritidae usually have smooth shells, and related families differ mainly in the development of sculpture.

A smaller group linked with Dimorphocerataceae or early glaphyritids are the Neodimorphocerataceae, which arose in the later part of the Early Carbon- iferous. In the course of the phylogeny of this group, the ventral lobe widened and became bipartite; the main feature of this group is that the ornamentation consists of bifurcating ribs.

The Late Carboniferous and Permian Thalassocerataceae may also have had their origin in glaphyritids. The Bisatoceratidae, which are thought to be ancestral to the Thalassoceratidae, resemble the Glaphyritidae or Craveno- ceratidae. The Thalassoceratidae experienced a phylogenetic increase in the digitation of the suture.

Among the most prolific goniatite groups of the Late Carboniferous were the Schistocerataceae. This group arose most probably at the beginning of the Westphalian (or Bloydian), but, in contrast to the Thalassoceratidae, the number of lateral sutural elements increased during phylogeny, mainly by trifurcation of the umbilical lobes. Additional features were the widening of the ventral lobe, with a median saddle much higher than half the height of the entire ventral lobe, and the tendency to start coiling in a triangular fashion.

Close to the end of the Carboniferous, several goniatite groups experienced a tremendous increase in sutural elements by trifurcation or digitation of single lobes, as in the Shumarditaceae, the Adrianitaceae, and the Marathoni- taceae. This tendency continued into the Permian.

The Prolecanitida are thought to have descended from the Prionoceratidae. The earliest representative of the Prolecanitida is *Eocanites* from the lower- most Tournaisian, resembling *Acutimitoceras* in whorl shape as well as in the complexity and shape of the suture. The difference is, apparently, that during ontogeny, an adventitious lobe does not develop in *Eocanites*, and the lobes on the flanks are derived through insertion of only umbilical lobes. In *Acutimitoceras*, the lobe in the lateral position is clearly an adventitious one. This mode of lobe insertion was observed by Karpinskiy (1896) and Schinde- wolf (1929), but there are other observations that seem to suggest that more advanced forms may exhibit insertion of an adventitious lobe as well as an increase in the number of umbilical lobes (Schindewolf, 1951; Hodgkinson, 1965; Nassichuk, 1975).

The Prolecanitida are characterized by a suture generally consisting of a series of more or less equal elements on the flanks, ranging from three to more than 30 in number. In the course of phylogeny, the lobes in the lateral position sometimes become subdivided or denticulate. At the end of the Carboniferous, the ventral lobe was transformed in some groups; the ventral prong developed by subdivision of the first lateral saddle. The conch form of the Prolecanitida is, in general, thinly discoidal. It is evolute in early forms but involute in later forms. The surface of the shell is generally smooth; ventrolateral nodes sometimes appear.

4.2. Biostratigraphy, Paleogeographic Distribution, and Evolutionary Trends

4.2.1. Lower Carboniferous Ammonoids

The biostratigraphic subdivision of the Carboniferous is illustrated in Fig. 6. The Lower Carboniferous, roughly equivalent to the Mississippian, has been subdivided into two stages of quite different duration: (1) the Tournaisian (Kinderhookian and Osagean), which is believed to have comprised 8 to 12 million years, and (2) the Viséan and lower part of Namurian A (Meramecian and Chesterian), which is thought to have comprised about 20 million years. [These estimates are based on radiometric data and are derived from a number of sources (Jones, 1988; Harland *et al.*, 1990; Roberts *et al.*, 1991; Claoué-Long *et al.*, 1992; Menning, 1992).]

4.2.1a. Lower Tournaisian. Immediately after the drastic decline of ammonoid diversity at the end of the Devonian, new ammonoid groups proliferated in earliest Tournaisian time (Kullmann, 1981, 1983). Only very few of the evolutionary lineages of the family Prionoceratidae crossed the Devonian–Carboniferous boundary; representatives of *Acutimitoceras*, *Nicimitoceras*, and *Mimimitoceras* are believed to be present in beds below and above that boundary (Korn, 1993).

The base of the *Gattendorfia* Stufe (Ga) has been defined by Heerlen (1935) as the appearance of the earliest species of *Gattendorfia*, *G. subinvoluta* (Prionoceratidae) (Vöhringer, 1960); besides *Gattendorfia*, several well-sculptured representatives of the Prionoceratidae (*Paprothites*, *Pseudarietites*) are used as index fossils for the zonation of the *Gattendorfia* Stufe.

Lower Tournaisian beds contain about 80 species belonging to 13 genera mainly in the order Goniatitida (Figs. 5 and 7). [Estimates of the number of taxa (species and genera) are based on data from the data base GONIAT, version 2.31 (Korn and Kullmann, 1993). Because this data base system is not yet completed, these numbers represent approximate figures. Taxa based on insufficient material or inadequate descriptions are omitted. Representatives

Ma		European Standard			US Standard		Index genera	Gen.	Spec.
295	Upper Carboniferous		Stephanian	Gzhelian	Virgilian		*Shumardites, Uddenites*	40	140
	Upper Carboniferous		Stephanian	Kasimovian	Missourian		*Parashumardites, Prouddenites*	40	140
	Upper Carboniferous		D	upper Moscovian	Desmoinesian			65	160
	Upper Carboniferous		C					65	160
	Upper Carboniferous		B	lower Moscovian	Atokan		*Winslowoceras*	65	160
	Upper Carboniferous		Westphalian A (G_2)	Bashkirian	Morrowan	Bloydian	*Branneroceras*	65	160
	Upper Carboniferous	Namurian C	Yeadonian (G_1)	Bashkirian	Morrowan	Halian	*Cancelloceras*	45	>200
	Upper Carboniferous	Namurian B	Marsdenian (G_2)	Bashkirian	Morrowan	Halian	*Reticuloceras* (R)	45	>200
	Upper Carboniferous	Namurian B	Kinderscoutian (R_1)	Bashkirian	Morrowan	Halian	*Reticuloceras* (R)	45	>200
	Upper Carboniferous	Namurian A	Alportian (H_2) Chokierian (H_1)	Serpukhovian	?		*Homoceras* (H)	20	35
320	Lower Carboniferous	Namurian A	Arnsbergian (E_2)	Serpukhovian	Chesterian		*Eumorphoceras* (E)	>100	>500
	Lower Carboniferous	Namurian A	Pendleian (E_1)	Serpukhovian	Chesterian		*Eumorphoceras* (E)	>100	>500
	Lower Carboniferous		Go gamma 2		Chesterian		*Goniatites* (Go), *Girtyoceras*	>100	>500
	Lower Carboniferous		Upper Viséan (Aprathian, Go)		Meramecian		*Goniatites* (Go), *Girtyoceras*	>100	>500
	Lower Carboniferous	Erdbachian, Pe	Lower Viséan		Osagean		*Ammonellipsites, Merocanites*	33	140
	Lower Carboniferous	Erdbachian, Pe	Upper Tournaisian		Osagean		*Muensteroceras, Pericyclus*	13	60
	Lower Carboniferous	Erdbachian, Pe	Middle Tournaisian		Kinderhookian		*Goniocyclus, Protocanites*	20	60
354	Lower Carboniferous	Balvian	Lower Tournaisian (Ga)		Kinderhookian		*Gattendorfia*	12	80

FIGURE 6. Carboniferous chronostratigraphy and ammonoid biostratigraphy. Numbers of genera and species are approximations (source: GONIAT Database System, version 3.21/3.22, 1993). For abbreviations see Fig. 5.

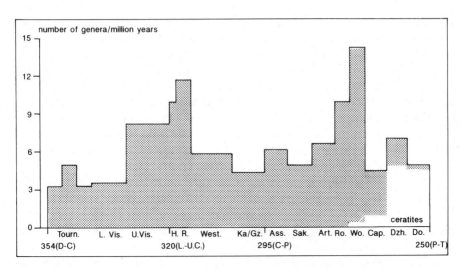

FIGURE 7. Generic ammonoid diversity, as expressed in frequency per millions of years (source: GONIAT Database System, version 3.21/3.22, 1993); durations of time units are estimated. For abbreviations, see Fig. 5; Tourn., Tournaisian; L.Vis., lower Viséan; U.Vis., upper Viséan; West., Westphalian; Ka/Gz., Kasimovian + Gzhelian; Ass., Asselian; Sak., Sakmarian; Art., Artinskian; Ro., Roadian; Wo., Wordian; Cap., Capitanian; Dzh., Dzhulfian; Do., Dorashamian, Changxingian. At the beginning of Wordian: order Ceratitida.

of the order Bactritida are not included.] One genus, *Eocanites*, consisting of about 20 species, is already present in basal Tournaisian beds. Oldest Tournaisian ammonoids have a subdiscoidal, involute, and narrowly umbilicate conch form. There was a tendency, however, to develop discoidal, evolute, and widely umbilicate conch forms, at first in the inner whorls and later in the adult whorls. This tendency had already started before the end of the Devonian but accelerated in early Tournaisian times, leading to widely umbilicate shells (*Gattendorfia*, *Eocanites*). Most species have a smooth shell, but a number of genera develop distinctive sculpture on the flanks and venter (*Gattenpleura*, *Paprothites*, *Pseudarietites*). Except for a few forms (e.g., *Voehringerites*, oldest karagandoceratids; Bartzsch and Weyer, 1988), most early Tournaisian ammonoids possess the typical suture line of the Prionoceratidae with simple lobes and few sutural elements (EALUI; Russian VLUID).

The paleogeographic distribution of ammonoid faunas in the early Tournaisian was mainly restricted to Europe, North Africa, and Asia; ammonoids are known from about 80 localities. Most species occur in different regions of Germany and Austria, but also in northwestern and southern China; only a few forms are recorded from England, Poland, North America, Chile, and Kazakhstan (Becker, 1993c).

4.2.1b. Middle Tournaisian. There are three genera (eight species) of the Prolecanitida and about 15 genera (50 species) of the Goniatitidae in this substage. A major change in sutural configuration occurred in the middle Tournaisian that is best seen in North America (Gordon, 1986) but was also present in the Old World. The distinctive feature of the Goniatitina, starting at the beginning of this substage, was the division of the ventral lobe. Earliest representatives with this sutural pattern are the distinctly ribbed *Goniocyclus* and *Rotopericyclus*. Notable is the appearance of the prionoceratid *Kazakhstania*. The Prodromitidae, characterized by digital lateral lobes, occur mainly in the middle Tournaisian. The Prolecanitida comprise a number of distinct forms indicative of this time interval; *Protocanites* is the first prolecanitid genus with a characteristic V-shaped ventral lobe.

No more than 100 goniatite localities are known from North America, Europe, North Africa, and Asia. Middle Tournaisian strata are best known from Southwest, West, and Central Europe; they are rare in Kazakhstan, South China, and Australia (Becker, 1993c). The main occurrences of this time interval are exposed in North America.

4.2.1c. Upper Tournaisian. About 60 species are known from upper Tournaisian strata, belonging to ten goniatitid and three prolecanitid genera. The beginning of the late Tournaisian was marked by the radiation of the well-known and almost cosmopolitan *Pericyclus* and *Muensteroceras*, producing the root stocks of several different lineages of the Goniatitina.

Only 60 localities of upper Tournaisian beds are recorded, mainly from North America (Indiana, Michigan) and Kazakhstan. A few occurrences are located in Belgium, England, Ireland, and Germany as well as in North Africa, Australia, and South America.

4.2.1d. Lower Viséan (± Meramecian, except upper part). During the Viséan, the diversity of ammonoids increased slowly. The whole time interval is considered to have comprised at least 9 million years; the number of species increased to 140 (30 goniatitid and three prolecanitid genera).

Shortly after the beginning of the early Viséan, the pericyclids disappeared and were replaced by several other superfamilies (nomismoceratids, dimorphoceratids, goniatitids). The muensteroceratid *Beyrichoceras* appeared in the late early Viséan, after an interval that was dominated by the presence of the cosmopolitan genera *Dzhaprakoceras* and *Merocanites* (Riley, 1990a,b). Except for the dimorphoceratids, which in part exhibit subdivided sutural elements, most members of the Goniatitida have a strict sutural plan that follows the formula $(E_1E_mE_1)ALUI$ [Russian $(V_1V_1)LUID$]; the differences in suture lines involve only details of single sutural elements and proportions.

Ammonoids extended their distribution in the early Viséan. They are known not only from many localities all over Europe, North Africa, and the northern Urals but also from Central Asia (Kyrgystan, Iran, and South China), North America (Alaska, Michigan), and rarely in Australia; in all, 250 localities are recorded.

4.2.1e. Upper Viséan (*Goniatites* Stufe), Pendleian, and Arnsbergian (*Eumorphoceras* Stufe) (Chesterian). A remarkable change came in the late Viséan. The appearance of *Girtyoceras* and the radiation of *Goniatites* s. str.

mark the beginning of a series of goniatite zones of rather short duration (Kullmann *et al.*, 1991). Most goniatitid genera were cosmopolitan. At the end of the late Viséan, a world-wide increase in ammonoid diversity took place, which is especially notable in Europe, Asia, and North America. More than 500 species belonging to about 85 goniatitid and 13 prolecanitid genera are recorded almost exclusively from the Northern Hemisphere. A great number of species and genera are used as index fossils for global ammonoid zonation in the *Goniatites* Stufe, Pendleian, and Arnsbergian stages (Kullmann, 1962; Ruzhentsev and Bogoslovskaya, 1971, 1978; Ramsbottom and Saunders, 1985).

The maximum of this faunal changeover was reached in the latest Viséan (in terms of the standard European scale: Go gamma 2 in Germany, V 3c in Great Britain; equivalent in Russia and the southern Urals to Nm1a2; see Ruzhentsev and Bogoslovskaya, 1971), at which time goniatites with triangular inner whorls (*Dombarites, Platygoniatites,* later *Proshumardites, Trigonoshumardites*) became predominant. Neoglyphioceratids with their distinct ornamentation are also useful as index fossils (Korn, 1988). Distinctive cravenoceratids mark the Viséan–Namurian boundary (*Emstites leion* indicates the base of the Namurian), followed by the appearance of the sculptured genus *Eumorphoceras,* the index fossil for the Arnsbergian.

A change in morphological characteristics, especially in the configuration of the suture, appeared in the late Arnsbergian (zone of *Eumorphoceras bisulcatum,* E_2). Goniatitina with a subdivided lateral lobe [A $\rightarrow$ $(A_vA_mA_d)$ $\rightarrow$ $A_vA_mA_d$] [Russian L $\rightarrow$ $(L_2L_1L_2)$ $\rightarrow$ $L_2L_1L_2$], agathiceratids with subdivided ventral and lateral lobes (*Proshumardites* and *Agathiceras*), delepinoceratids (*Delepinoceras*), as well as the conservative Gastrioceratidae (*Syngastrioceras, Glaphyrites,* and related genera) are characteristic of Upper Carboniferous and Lower Permian strata.

Occurrences of ammonoids from upper Viséan, Pendleian, and Arnsbergian rocks are numerous in the northern hemisphere; more than 1500 localities are known so far. Most of them are spread over West, Southwest, Central, and East Europe and North Africa; some goniatite groups are best known from the southern Urals (Russia, Kazakhstan), Central Asia, China, and the midwestern United States; only a few are recorded from Southeast Asia (Thailand) and Australia. Faunal provinces cannot be recognized with certainty.

4.2.2. Upper Carboniferous

At the end of the Early Carboniferous, a severe crisis in ammonoid evolution occurred in which the number of genera and species dropped considerably (Figs. 5 and 7). The first recognizable interval of the Upper Carboniferous, the possibly rather short *Homoceras* Stufe, is not well documented because of the incompleteness of most sections around the Lower–Upper Carboniferous boundary in many parts of the world. The duration of the entire subsystem, which is roughly equivalent to the Pennsylvanian, comprised approximately 25 million years.

4.2.2a. *Homoceras* **Stufe (H, Chokierian + Alportian).** The base of the Upper Carboniferous subsystem is defined by the appearance of the conodont species *Delicognathodus noduliferus*, which is considered to be coeval with the appearance of the Homoceratidae (Saunders and Ramsbottom, 1982). The *Homoceras* Stufe (H) is remarkable because of the rapid changeover of forms; of 85 species from the topmost portion of the *Eumorphoceras* Stufe (E$_{2c4}$), only four are recognized in the *Homoceras* Stufe; about 30 species were new. This shift is also reflected at the generic level: of 32 genera, only nine survived; 11 genera originated during this time interval (Ruzhentsev and Bogoslovskaya, 1978).

The morphology of the new forms belonging to the Homoceratidae was very conservative; conch form, sculpture, and suture line remained almost unchanged. But as in the early Tournaisian prionoceratids and earliest Namurian cravenoceratids, there was a tendency to develop widely umbilicate shells, predominantly starting in the early whorls. In general, the ammonoid faunas were impoverished.

It must be pointed out, however, that the *Homoceras* Stufe is not well documented. It is known from only approximately 100 localities mainly in West, Central, and East Europe, the southern Urals in Russia, Kazakhstan, Tajikistan, and Uzbekistan as well as southern China. No goniatites have been discovered so far in North America; equivalent beds are still doubtful (cf. Grayson *et al.*, 1985).

4.2.2b. Namurian B and C (R$_1$, R$_2$, G$_1$; Halian). After the drastic decline at the beginning of the Late Carboniferous, the middle and late Namurian ammonoids proliferated intensively. Most species of the *Homoceras* Stufe continued into the *Reticuloceras* Stufe (R) and gave rise to a great variety of sculptured forms of worldwide distribution. The general morphological aspect, however, did not change much. The number of species increased to more than 200, belonging to 45 goniatitid and two prolecanitid genera (Ruzhentsev and Bogoslovskaya, 1978).

The time interval under consideration comprises three well-documented goniatite stages, with a high number of short-ranging taxa that provide detailed biostratigraphic resolution: the lower *Reticuloceras* Stufe (R$_1$, Kinderscoutian), the upper *Reticuloceras* Stufe (R$_2$, Marsdenian), and the *Cancelloceras* Stufe (G$_1$, Yeadonian: Nikolaeva and Kullmann, 1995).

Ammonoids from this time interval are recorded from 400 localities, mainly in the Ukraine, Russia, Kazakhstan (southern Urals), South China, Europe, and North Africa but also from North America.

4.2.2c. Westphalian (Bloydian, Atokan, Desmoinesian; upper Bashkirian, Moscovian). During this rather long period of about 10 or 11 million years, more than 160 species were present belonging to 60 goniatitid and four prolecanitid genera. Some groups exhibited a slight increase of sutural elements leading to more complex sutures. Sculptures were very diverse and common in goniatites of Westphalian age.

Approximately 400 localities are known. Beginning with the early Westphalian, the Orulganitidae, an endemic family restricted to Siberia, appeared. Some of these forms persisted through the entire Moscovian; they are not known from other areas. Most localities are recorded from Europe (Germany, Netherlands, Belgium, Russia, Spain), North Africa, and above all from North America, especially the midwestern states (Gordon, 1965); goniatites are extremely rare in South America (Kullmann, 1993) and South Africa.

4.2.2d. Kasimovian + Gzhelian (Missourian + Virgilian). Because of plate tectonic movements during the late Namurian and Westphalian in many parts of the world (e.g., Alleghenian orogenesis in the Appalachian region, Variscan orogenesis in Central and Southwest Europe as well as in the Southern Alps: Kullmann and Loeschke, 1994), the geographic distribution of ammonoids diminished during the Westphalian. Only 115 localities have yielded Late Carboniferous goniatites of the Kasimovian and Gzehlian and its equivalents.

In the Kasimovian and Gzhelian, a major shift in the composition of ammonoid faunas occurred. A number of long-lasting genera persisted and continued into the Permian, but in addition to these, ammonoids with multielement sutures evolved in several separate lineages, not only in the order Goniatitida but also in the Prolecanitida. They represented the forerunners of the Permian ammonoids, which had originated in Late Carboniferous times. These groups are mainly the families Thalassoceratidae, Adrianitidae, Shumarditidae, and the superfamilies Neoicocerataceae, Cyclolobaceae, Marathonitaceae, and Medlicottiaceae.

Only a few occurrences are recorded from Europe. Late Carboniferous ammonoids are best documented in the northern and southern Urals in Russia and Kazakhstan (Ruzhentsev, 1950) and in South China; they are also widespread in North America, especially in the midwestern states and Canada (Nassichuk, 1975).

5. Permian Ammonoids

5.1. Major Systematic Categories

In the Permian there are four orders with one additional suborder and 15 superfamilies of ammonoids. The number of families/subfamilies ranges between 40 and 50 (see, e.g., Glenister and Furnish, 1981).

The Bactritida were still flourishing but consisted only of orthoconic forms similar in shape to those from the Carboniferous. There was an Early Permian (Artinskian) bloom of rather breviconic and cyrtochoanitic members of the Parabactritidae, which have mostly densely spaced and short buoyancy chambers. Secondary reduction of the siphuncular (ventral) lobe occurred in several lineages (*Microbactrites, Tabantaloceras*). Ornament is always simple and convex. Some genera such as *Belemnitomimus* develop apical angles of up to 30°, similar to those in belemnite phragmocones.

Permian Goniatitida were dominated by members of the Goniatitina, but after a long interval of little change, some late Tornoceratina evolved into very specialized, small forms with somewhat complex ornament, spectacular ventrolateral lappets at the mature apertural margin, and ceratitic or ammonitic sutures (e.g., see Frest *et al.*, 1981; Zhou, 1988). These members of the Pseudohaloritaceae = Shouchangocerataceae (sometimes merged with the Prionocerataceae) retain the characteristic (sub)central position of the siphuncle of their Carboniferous ancestors (*Neoaganides* of the Maximitidae). The Permian Goniatitina comprise groups that survived from the Carboniferous, groups with Carboniferous roots that had since diversified, and one new superfamily. The first category includes the Goniatitaceae (with the widespread involute Agathiceratidae displaying tripartite adventitious lobes), the Goniolobocerataceae (*Mescalites*), and the Thalassocerataceae, which have involute shells and whose principal eight lobes and intervening saddles were increasingly incised to form ceratitic or ammonitic sutures. The Gastriocerataceae were still represented by several lineages with rather conservative sutures, but these forms were probably late members of formerly more successful groups.

Five superfamilies that started in the Carboniferous became more characteristic in the Permian. Within the widely umbilicate Neoicocerataceae, the eight-lobed suture was retained (Paragastrioceratidae), or the (primary) lateral lobe at the umbilical seam was increasingly subdivided (Metalegoceratidae). As in Carboniferous goniatite groups, there are forms with a ventral sinus or with a salient in the ornament and at the apertural margin. In the entirely Permian Popanocerataceae, the number of lobes was even further increased (22 to 38), and the lobes were strongly denticulated at the base. Shells are involute (Popanoceratidae) or evolute (Mongoloceratidae) and markedly compressed. The Marathonitaceae showed low diversity (one family) but were rather abundant. Their complex sutures develop both ontogenetically and phylogenetically in a different way than do the sutures of the popanoceratids. A maximum of sutural complication (up to about 56 lobes) is reached in the involute, "ceratitic" (Vidrioceratidae) to "ammonitic" (Cyclolobidae, Hyattoceratidae) Cyclolobaceae. Permian Shumarditaceae belong to the Perrinitidae, which elaborate the general sutural pattern of the superfamily and develop tripartite adventitious and dorsal lobes to complex ammonitic sutures. In the Adrianitaceae, the secondary lobes of the subumbilical region are distinctive, and the shape of the sutural elements is similar to that of the sutural elements in Carboniferous prolecanitids. Many genera of the Adrianitidae have subspherical shells.

Both superfamilies of the Prolecanitida ranged into the Permian. The late Prolecanitaceae include only some advanced genera of the Daraelitidae with a subdivided ventral lobe, ceratitic ventrolateral lobes, and numerous outer umbilical lobes. In *Daraelites* itself, even the first (= median) ventral lobe bears

Stages		Gen.	Spec.
Upper Permian	250 Dorashamian 256	30	>100
	Dzhulfian 262	43	>100
	Capitanian 268	29	>50
	Wordian 272	59	>150
Lower Permian	Roadian 276	38	50
	Artinskian 282	40	150
	Sakmarian 289	34	100
	Asselian 295	37	75

FIGURE 8. Permian chronostratigraphy. Numbers of genera and species are approximations; durations of time units are estimated (source: GONIAT Database System, version 3.21/3.22, 1993).

slight ceratitic incisions. The Medlicottiaceae continued to include narrowly umbilicate and strongly compressed forms. In the ancestral Pronoritidae, the originally simply bifid ventrolateral lobe was increasingly subdivided, and the ventral prong of the latter turned into a rather bizarre, sometimes saw-shaped marginal saddle in Permian Medlicottiidae and later in oxyconic Episageceratidae.

The Ceratitida are thought to have been derived by paedomorphosis from the Daraelitidae (Spinosa *et al.*, 1975) and are characterized by the early ontogenetic elimination of the primary lateral lobe situated at the umbilical seam. The adventitious lobe remains undivided or becomes ceratitic, and numerous umbilical lobes are introduced in the subumbilical region and on the lower flank. Permian members, as far as is documented, still have trilobate primary sutures and can be divided into two superfamilies (or into the Paraceltitina and Otoceratina of Shevyrev and Ermakova, 1979). Compressed, ribbed, and widely evolute forms with few, partly frilled lobes comprise the Xenodiscaceae, which probably gave rise to the main Triassic ceratitid stock. Conspicuously raised umbilical shoulders and tectiform whorl sections are found in the specialized side branch of the Otocerataceae, which also show a corresponding introduction of numerous small lobes in the subumbilical area.

5.2. Biostratigraphy, Paleogeographic Distribution, and Evolutionary Trends

5.2.1. Lower Permian

The changeover of goniatites, which already started in latest Carboniferous Gzhelian times, resulted in the appearance and radiation of three important families in the Gzhelian: (1) the Pseudohaloritidae, the abovementioned aberrant ammonoids with a subcentral to dorsal position of the siphuncle and ceratitic or ammonitic serration of a simple prionoceratid suture line; (2) the Vidrioceratidae, multilobate forms with a denticulation confined to the base of the lobe; and (3) the Medlicottiidae, with modification of the primary lateral lobe. All three groups experienced an increase in their diversity at the beginning of the Permian, but two additional, very important, conservative families appeared: the Pseudoparalegoceratidae, with an unchanged basic goniatitic suture but characterized by conspicuous longitudinal as well as transverse sculpture, and the Metalegoceratidae, with a special mode of increase of the lateral sutural elements: $L \rightarrow L_v L_m L_d \rightarrow L_v L_{mv} L_{mm} L_{md} L_d$ (Russian $U \rightarrow U_2 U_1 U_3 \rightarrow U_2 U_{1.2} U_{1.1} U_{1.2} U_3$).

In Asselian time, about 75 species belonging to 28 goniatitid and nine prolecanitid genera were present (Fig. 7). The Sakmarian included almost 100 species belonging to about 25 goniatitid and nine prolecanitid genera. In the Artinskian, the number of species increased to almost 150, belonging to more than 30 goniatitid and ten prolecanitid genera; this trend continued until the end of the Artinskian Baigendzhinian substage.

Asselian goniatites are known from only 60 localities in Central, Northeast, and Southeast Asia and North America. The distribution of goniatites did not change significantly during Sakmarian times except in Australia; almost 100 Sakmarian localities are known. Artinskian ammonoids became more widespread; over 200 localities yielded goniatites not only in Central Asia and North America (Miller and Furnish, 1940) but also in China, Indonesia (Timor), Oman, Central and South America, Central Europe, and the northern and southern Urals.

5.2.2. Middle Permian

The Roadian was characterized by a decrease of goniatitid and prolecanitid genera and the first appearance of the order Ceratitida. Only 50 species belonging to fewer than 40 genera are known from this probably short time interval. The distribution of Roadian goniatites was mostly restricted to North America, Central Asia, and China; about 100 exposures of rocks of this age are known.

A drastic increase in diversity occurred at the beginning of Wordian time. More than 150 species belonging to about 50 goniatitid, eight prolecanitid, and two ceratitid genera are recorded. The multilobate adrianitids and cyclolobids especially contributed to the diverse aspect of Wordian ammonoid faunas; the ceratitids were rare and missing in most regions. The following Capitanian time interval, however, was characterized by a considerable decrease in the number of forms (Glenister and Furnish, 1981). The number of goniatitid species dropped to about 40 (18 genera); six genera (12 species) represented the order Ceratitida; the number of prolecanitid genera dropped to two (four species).

Wordian exposures are widespread; about 100 localities are known, mainly from North America, China, and Indonesia (Timor); minor occurrences are in Afghanistan, Crimea, Iraq (Kurdistan), Italy (Sicily), Tunisia, and Australia. The Capitanian is less common than the Wordian; only about 70 goniatite localities are known, mainly concentrated in North America and China.

5.2.3. Upper Permian

The proportion of ceratites increased during the Dzhulfian, in which only ten goniatitid genera (21 species) and two prolecanitid genera (two species) existed, but about 30 genera (90 species) of the Ceratitida. In latest Dorashamian (Changxingian), 26 ceratitid genera with more than 100 species are known, but only three goniatitid genera (eight species) are recorded.

The Dzhulfian yielded Goniatitida only at about 40 localities; Dzhulfian faunas were scattered in distribution: Greenland, Azerbaijan, Armenia, Iran (Araxes gorge), India, Pakistan, Madagascar, China, Siberia, and Australia. The uppermost Permian Dorashamian (Changxingian) stage is described from 25 localities in China (Zhao *et al.*, 1981). Outside of China, fewer than 20 localities are recorded, situated in an extremely restricted area in the boundary region of Armenia, Iran, and Azerbaijan. Only a few species of the Goniatitida and Prolecanitida seem to have persisted into the Early Triassic in China.

6. Systematics and Stratigraphic Distribution of Paleozoic Ammonoids

Superorder Ammonoidea
Order Bactritida Shimanskiy, 1951
  Family Bactritidae Hyatt, 1884 (Ludlowian–upper Karnian)
   Subfamily Bactritinae Hyatt, 1884 (Ludlowian–upper Karnian)
   Subfamily Cyrtobactritinae Becker subfam. nov. (lower Emsian)
  Family Bojobactritidae Horny, 1957 (lower Emsian–middle Famennian)
  Family Parabactritidae Shimanskiy, 1951 (Viséan–Permian)
  Family Sinuobactritidae Mapes, 1979 (Morrowan–Missourian)
Order Agoniatitida Ruzhentsev, 1957

Suborder Agoniatitina Ruzhentsev, 1957
  Superfamily Mimocerataceae Steinmann, 1890
    Family Mimosphinctidae Erben, 1953 (lower Emsian–topmost Eifelian)
      Subfamily Anetoceratinae Ruzhentsev, 1957 (lower Emsian–topmost Eifelian)
      Subfamily Teicherticeratinae Bogoslovskiy, 1969 (lower–basal upper Emsian)
      Subfamily Mimosphinctinae Erben, 1953 (lower Emsian)
    Family Mimoceratidae Steinmann, 1890 (lower Emsian–lower Eifelian)
  Superfamily Agoniatitaceae Holzapfel, 1899
    Family Mimagoniatitidae Miller, 1938 (lower Emsian–lower Eifelian)
      Subfamily Mimagoniatitinae Miller, 1938 (lower–upper Emsian)
      Subfamily Parentitinae Bogoslovskiy, 1980 (lower Emsian–basal upper Emsian)
    Family Agoniatitidae Holzapfel, 1899 (upper Emsian–middle Givetian)
    Family Pinacitidae Hyatt, 1900 (lower–upper Eifelian)
  Superfamily Auguritaceae Bogoslovskiy, 1961
    Family Auguritidae Bogoslovskiy, 1961 (lower Emsian)
Suborder Anarcestina Miller and Furnish, 1954
  Superfamily Anarcestaceae Steinmann, 1890
    Family Anarcestidae Steinmann, 1890 (upper Emsian–basal Famennian)
      Subfamily Anarcestinae Steinmann, 1890 (upper Emsian–lower Eifelian)
      Subfamily Werneroceratinae Erben, 1964 (upper Emsian–basal Famennian)
    Family Sobolewiidae House, 1989 (middle Eifelian–middle Givetian)
  Superfamily Pharcicerataceae Hyatt, 1900
    Family Maenioceratidae Bogoslovskiy, 1958 (topmost Eifelian–basal upper Givetian)
    Family Pharciceratidae Hyatt, 1900 (upper Givetian–basal Frasnian)
    ?Family Triainoceratidae Hyatt, 1884 (lower–middle Frasnian)
    Family Petteroceratidae Becker and House, 1993 (topmost Givetian–basal Frasnian)
    Family Eobeloceratidae Becker and House, 1993 (topmost middle–upper Givetian)
  Superfamily Belocerataceae Hyatt, 1884 (= Gephurocerataceae Miller and Furnish, 1954)
    Family Acanthoclymeniidae Schindewolf, 1955 (upper Givetian–middle Frasnian)
    Family Koenenitidae Becker and House, 1993 (lower Frasnian)
    Family Gephuroceratidae Frech, 1897 (lower–topmost Frasnian)

Family Beloceratidae Hyatt, 1884 (middle–topmost Frasnian)
Family Devonopronoritidae Bogoslovskiy, 1958 (middle Frasnian)
Order Clymeniida Wedekind, 1914
Superfamily Cyrtoclymeniaceae Hyatt, 1884
Family Cyrtoclymeniidae Hyatt, 1884 (middle–upper Famennian)
Family Cymaclymeniidae Hyatt, 1884 (middle Famennian–?basal-most Carboniferous)
Family Rectoclymeniidae Schindewolf, 1923 (middle Famennian)
Family Carinoclymeniidae Bogoslovskiy, 1975 (middle–upper Famennian)
Family Pachyclymeniidae Korn, 1992 (middle–upper Famennian)
Family Biloclymeniidae Bogoslovskiy, 1955 (upper Famennian)
Superfamily Clymeniaceae Hyatt, 1884
Family Platyclymeniidae Wedekind, 1914 (middle–upper Famennian)
Family Piriclymeniidae Korn, 1992 (middle–upper Famennian)
Family Clymeniidae Hyatt, 1884 (upper Famennian)
Family Kosmoclymeniidae Korn and Price, 1987 (middle–upper Famennian)
Superfamily Gonioclymeniaceae Hyatt, 1884
Family Costaclymeniidae Schindewolf, 1920 (middle–lower upper Famennian)
Family Gonioclymeniidae Hyatt, 1884 (upper Famennian)
Subfamily Gonioclymeniinae Hyatt, 1884 (upper Famennian)
Subfamily Sellaclymeniinae Schindewolf, 1923 (upper Famennian)
Subfamily Sphenoclymeniinae Korn, 1992 (upper Famennian)
Superfamily Wocklumeriaceae Schindewolf, 1937
Family Hexaclymeniidae Lange, 1929 (middle–upper Famennian)
Family Glatziellidae Schindewolf, 1928 (upper Famennian)
Family Wocklumeriidae Schindewolf, 1937 (upper Famennian)
Family Parawocklumeriidae Schindewolf, 1937 (upper Famennian)
Order Goniatitida Hyatt, 1884
Suborder Tornoceratina Wedekind, 1917
Superfamily Tornocerataceae Arthaber, 1911
Family Tornoceratidae Arthaber, 1911 (upper Eifelian–upper Famennian)
Subfamily Parodiceratinae Petter, 1959 (upper Eifelian–middle Givetian)
Subfamily Tornoceratinae Arthaber, 1911 (lower Givetian–upper Famennian)
Family Pseudoclymeniidae Becker, 1993 (middle Famennian)
Family Posttornoceratidae Bogoslovskiy, 1962 (lower–upper Famennian)
Superfamily Dimerocerataceae Hyatt, 1884
Family Cheiloceratidae Frech, 1897 (lower–middle Famennian)

Family Dimeroceratidae Hyatt, 1884 (lower–middle Famennian)

Family Prolobitidae Wedekind, 1913 (lower–upper Famennian)

Family Phenacoceratidae Wedekind, 1917 (lower–upper Famennian)

Family Sinotitidae Chang, 1960 (lower or middle Famennian)

Superfamily Praeglyphiocerataceae Ruzhentsev, 1957

Family Sporadoceratidae Miller and Furnish, 1957 (lower–topmost Famennian)

Family Praeglyphioceratidae Ruzhentsev, 1957 (lower–middle Famennian)

Superfamily Prionocerataceae Hyatt, 1884

Family Prionoceratidae Hyatt, 1884 (middle Famennian–upper Viséan)

Subfamily Prionoceratinae Hyatt, 1884 (middle Famennian–upper Viséan)

Subfamily Gattendorfiinae Bartzsch and Weyer, 1988 (lower–upper Tournaisian)

Subfamily Pseudarietitinae Bartzsch and Weyer, 1987 (lower–middle Tournaisian)

Superfamily Karagandocerataceae Librovich, 1957

Family Karagandoceratidae Librovich, 1957 (topmost lower–upper Tournaisian; derived from Prionoceratinae)

Family Qiannanitidae Becker, 1993 (Tournaisian; perhaps derived from Pseudarietitinae)

Family Prodromitidae Arthaber, 1911 (middle–upper Tournaisian; probably derived from Qiannanitidae)

Superfamily Pseudohaloritaceae Ruzhentsev, 1957

Family Maximitidae Ruzhentsev, 1960 (Moscovian–upper Changxingian)

Subfamily Maximitinae Ruzhentsev, 1960 (Moscovian–upper Changxingian)

Subfamily Shouchangoceratinae Zhao and Zheng, 1977 (Roadian–Capitanian)

Family Pseudohaloritidae Ruzhentsev, 1957 (Artinskian–Wordian)

Subfamily Pseudohaloritinae Ruzhentsev, 1957 (Artinskian–Roadian)

Subfamily Yinoceratinae Ruzhentsev, 1960 (Roadian–Wordian)

Suborder Goniatitina Hyatt, 1884

Superfamily Pericyclaceae

Family Pericyclidae (middle Tournaisian–lower Viséan)

Family Intoceratidae Kuzina, 1971 (middle Tournaisian–lower Viséan)

Family Muensteroceratidae Librovich, 1957 (middle Tournaisian–middle Viséan)

Superfamily Nomismocerataceae Librovich, 1957

Family Nomismoceratidae Librovich, 1957 (lower Viséan–middle Namurian)

Family Entogonitidae Ruzhentsev and Bogoslovskaia, 1971 (Viséan)

Superfamily Dimorphocerataceae Hyatt, 1884

Family Dimorphoceratidae Hyatt, 1884 (lower Viséan–Yeadonian)

Subfamily Dimorphoceratinae Hyatt, 1884 (lower Viséan–Arnsbergian)

Subfamily Glyphiolobinae Ruzhentsev and Bogoslovskaia, 1969 (upper Viséan–Yeadonian)

Family Berkhoceratidae Librovich, 1957 (upper Viséan–Arnsbergian)

Family Anthracoceratidae Plummer and Scott, 1937 [Namurian–Westphalian (Atokan)]

Family Girtyoceratidae Wedekind, 1918 [upper Tournaisian–upper Westphalian (upper Moscovian)]

Subfamily Sudeticeratinae Kullmann, n.subfam. (upper Tournaisian–Pendleian)

Subfamily Girtyoceratinae Wedekind, 1918 [upper Viséan–upper Westphalian (upper Moscovian)]

Family Eogonioloboceratidae Ruzhentsev and Bogoslovskaia, 1978 [upper Viséan–upper Namurian (Halian)]

Superfamily Goniatitaceae de Haan, 1825

Family Goniatitidae de Haan, 1825 (upper Viséan–Pendleian)

Family Sygambritidae Korn, 1988 (upper Viséan–lower Pendleian)

Family Delepinoceratidae Ruzhentsev, 1957 (upper upper Viséan–Arnsbergian, upper Chesterian)

Family Agathiceratidae Arthaber, 1911 (upper Viséan–Wordian)

Superfamily Neoglyphiocerataceae Plummer and Scott, 1937

Family Neoglyphioceratidae Plummer and Scott, 1937 [upper Viséan–Arnsbergian (upper Chesterian)]

Family Cravenoceratidae Ruzhentsev, 1957 [upper Viséan–upper Arnsbergian (upper Chesterian)]

Subfamily Cravenoceratinae Ruzhentsev, 1957 [upper Viséan–upper Arnsbergian (upper Chesterian)]

Subfamily Nuculoceratinae Ruzhentsev, 1957 (Arnsbergian)

Family Ferganoceratidae Ruzhentsev, 1960 (upper Viséan)

Family Rhymmoceratidae Ruzhentsev and Bogoslovskaia, 1971 (upper Viséan–Pendleian)

Superfamily Neodimorphocerataceae Furnish and Knapp, 1966

Family Ramositidae Ruzhentsev and Bogoslovskaia, 1966 (Arnsbergian–Yeadonian)

Family Neodimorphoceratidae Furnish and Knapp, 1966 (upper Morrowan–Gzhelian)

Superfamily Gastriocerataceae Hyatt, 1884

Family Glaphyritidae Ruzhentsev and Bogoslovskaia, 1971 (Namurian–Artinskian)

Subfamily Glaphyritinae Ruzhentsev and Bogoslovskaia, 1971 (Namurian–Sakmarian)

Subfamily Stenoglaphyritinae Ruzhentsev and Bogoslovskaia, 1971 (Namurian–Yeadonian, ?lower Gzhelian)

Subfamily Somoholitinae Ruzhentsev, 1938 (Moscovian–Artinskian)

Family Homoceratidae Spath, 1934 (Chokierian–Yeadonian)

Subfamily Homoceratinae Spath, 1934 (Chokierian–Yeadonian)

Subfamily Decoritinae Ruzhentsev and Bogoslovskaia, 1975 (Kinderscoutian–Yeadonian)

Family Reticuloceratidae Librovich, 1957 (Kinderscoutian–lower Westphalian, Bashkirian, Bloydian)

Subfamily Surenitinae Ruzhentsev and Bogoslovskaia, 1975 (Kinderscoutian–lower Westphalian, Bashkirian, Morrowan)

Subfamily Reticuloceratinae Librovich, 1957 (Kinderscoutian–Yeadonian)

Family Gastrioceratidae Hyatt, 1884 (Marsdenian–upper Westphalian, Moscovian, ?Kasimovian, Desmoinesian)

Superfamily Thalassocerataceae Hyatt, 1900

Family Bisatoceratidae Miller and Furnish, 1957 (Kinderscoutian, Halian–Missourian, ?Asselian)

Family Thalassoceratidae Hyatt, 1900 (Missourian–Wordian)

Superfamily Schistocerataceae Schmidt, 1929

Family Pseudoparalegoceratidae (lower Westphalian, Bloydian–Virgilian, lower Gzhelian)

Family Schistoceratidae Schmidt, 1929 (lower Westphalian, Bloydian–Gzhelian, ?Sakmarian)

Family Orulganitidae Ruzhentsev, 1965 (lower Westphalian, upper Bashkirian–Kasimovian)

Family Welleritidae Plummer and Scott, 1937 (Bloydian–Desmoinesian)

Subfamily Welleritinae Plummer and Scott, 1937 (Atokan–Desmoinesian)

Subfamily Axinolobinae Ruzhentsev, 1962 (upper Bloydian–Desmoinesian)

Family Christioceratidae Nassichuk and Furnish, 1965 (Atokan)

Superfamily Goniolobocerataceae Spath, 1934

Family Wiedeyoceratidae Ruzhentsev and Bogoslovskaia, 1978 (Halian–Asselian)

Family Gonioloboceratidae Spath, 1934 (Atokan–Asselian)

Superfamily Adrianitaceae Schindewolf, 1931

Family Adrianitidae Schindewolf, 1931 (Desmoinesian–Dzhulfian)

Subfamily Adrianitinae Schindewolf, 1931 (Missourian–Dzhulfian)

Subfamily Dunbaritinae Miller and Furnish, 1957 (Desmoinesian–Virgilian)

Subfamily Hoffmanniinae Mojsisovics, 1888 (Wordian)
Subfamily Texoceratinae Ruzhentsev and Bogoslovskaia, 1978 (Roadian)
Superfamily Cyclolobaceae Zittel, 1895
Family Cyclolobidae Zittel, 1895 (Roadian–Dorashamian)
Subfamily Cyclolobinae Zittel, 1895 (Roadian–Dorashamian)
Subfamily Kufengoceratinae Zhao, 1980 (Roadian–Capitanian)
Family Vidrioceratidae Plummer and Scott, 1937 (Gzhelian–Dorashamian)
Superfamily Marathonitaceae Ruzhentsev, 1938
Family Marathonitidae Ruzhentsev, 1938 (Atokan–Wordian)
Family Hyattoceratidae Miller and Furnish, 1957 (?Sakmarian–Dzhulfian)
Superfamily Neoicocerataceae Hyatt, 1900
Family Neoicoceratidae Hyatt, 1900 (Bloydian–Asselian)
Subfamily Neoicoceratinae Hyatt, 1900 (Bloydian–Asselian)
Subfamily Eupleuroceratinae Ruzhentsev, 1957 (Missourian)
Subfamily Atsabitinae Furnish, 1966 (Artinskian–Wordian)
Family Metalegoceratidae Plummer and Scott, 1937 (Asselian–Wordian)
Subfamily Metalegoceratinae Plummer and Scott, 1937 (Asselian–Wordian)
Subfamily Clinolobinae Miller and Furnish, 1957 (Wordian)
Subfamily Eothinitinae Ruzhentsev, 1956 (Sakmarian–Wordian)
Subfamily Spirolegoceratinae Nassichuk, 1970 (Artinskian–Roadian)
Family Paragastrioceratidae Ruzhentsev, 1951 (Asselian–Triassic, Griesbachian)
Subfamily Paragastrioceratinae Ruzhentsev, 1951 (Asselian–Dorashamian)
Subfamily Pseudogastrioceratinae Furnish, 1966 (Sakmarian–Triassic, Griesbachian)
Superfamily Popanocerataceae Hyatt, 1900
Family Popanoceratidae Hyatt, 1900 (Asselian–Dzhulfian)
Family Mongoloceratidae Ruzhentsev and Bogoslovskaia, 1978 (Wordian–Capitanian)
Superfamily Shumarditaceae Plummer and Scott, 1937
Family Shumarditidae Plummer and Scott, 1937 (Moscovian–Gzhelian)
Family Perrinitidae Miller and Furnish, 1940 (Asselian–Roadian)
Order Prolecanitida Miller and Furnish, 1954
Superfamily Prolecanitaceae Hyatt, 1884
Family Prolecanitidae Hyatt, 1884 (basal Tournaisian–Arnsbergian)
Family Daraelitidae Tchernov, 1907 (Viséan–Wordian)
Superfamily Medlicottiaceae Karpinskiy, 1889

Family Pronoritidae Frech, 1901 (Viséan–Dzhulfian)
Family Medlicottiidae Karpinskiy, 1889 (Desmoinesian–Dzhulfian)
Family Sundaitidae Ruzhentsev, 1957 (Dzhulfian)
Family Episageceratidae Ruzhentsev, 1956 (Capitanian–Nammalian, Lower Triassic)
Order Ceratitida Hyatt, 1884
Superfamily Xenodiscaceae Frech, 1902
Family Xenodiscidae Frech, 1902 (Roadian–Dorashamian, ?Lower Triassic)
Family Dzhulfitidae Shevyrev, 1965 (Dorashamian)
Superfamily Otocerataceae Hyatt, 1900
Family Anderssonoceratidae Ruzhentsev, 1959 (Dzhulfian)
Family Araxoceratidae Ruzhentsev, 1959 (Capitanian–Dzhulfian)
[Family Otoceratidae Hyatt, 1900) (Griesbachian, Lower Triassic)]

References

Babin, C., Rachebeuf, P. R., Le Hérissé, A., and Suarez Riglos, M., 1991, Données nouvelles sur les goniatites du Dévonien de Bolivie, *Geobios* **24**:719–724.
Bartzsch, K., and Weyer, D., 1987, Die unterkarbonische Ammonoidea-Tribus Pseudarietitini, *Abh. Ber. Naturkd. Vorgesch.* **13**:59–68.
Bartzsch, K., and Weyer, D., 1988, Die unterkarbonische Ammonoidea-Subfamilia Karagando-ceratinae, *Freib. Forsch. H.* **C419**:130–142.
Becker, R. T., 1986, Ammonoid evolution before, during and after the "Kellwasser-Event"—review and preliminary new results, in: *Global Bio-Events. A Critical Approach, Lecture Notes on Earth Sciences 8* (O. H. Walliser, ed.), Springer, Berlin, pp. 181–188.
Becker, R. T., 1992, Zur Kenntnis von Hemberg-Stufe und Annulata-Schiefer im Nordsauerland (Oberdevon, Rheinisches Schiefergebirge, GK 4611 Hohenlimburg), *Berl. Geowiss. Abh. (E)* **3**:3–41.
Becker, R. T., 1993a, Stratigraphische Gliederung und Ammonoideen-Faunen im Nehdenium (Oberdevon II) von Europa und Nord-Afrika, *Cour. Forsch.-Inst. Senckenberg* **155**.
Becker, R. T., 1993b, Anoxia, eustatic changes, and Upper Devonian to lowermost Carboniferous global ammonoid diversity, in: *The Ammonoidea: Environment, Ecology, and Evolutionary Change*, Systematics Association Spec. Vol. 47 (M. R. House, ed.), Clarendon Press, Oxford, pp. 115–163.
Becker, R. T., 1993c, Analysis of ammonoid palaeobiogeography in relation to the global Hangenberg (terminal Devonian) and lower Alum Shale (Middle Tournaisian) events, *Ann. Soc. Geol. Belg.* **115**:459–473.
Becker, R. T., and House, M. R., 1993, New early Upper Devonian (Frasnian) goniatite genera and the evolution of the "Gephurocerataceae," *Berl. Geowiss. Abh. (E)* **9**:111–133.
Becker, R. T., and House, M. R., 1994, International Devonian goniatite zonation, Emsian to Givetian, with new records from Morocco, *Cour. Forsch.-Inst. Senckenberg* **169**:79–135.
Becker, R. T., House, M. R., and Kirchgasser, W. T., 1993, Devonian biostratigraphy and timing of facies movements in the Frasnian of the Canning Basin, Western Australia, in: *High Resolution Stratigraphy*, Geological Society Spec. Publ. 70 (E. A. Hailwood and R. B. Kidd, eds.), Geological Society Publishing House, London, pp. 293–321.
Bender, P., Jahnke, H., and Ziegler, W., 1974, Ein Unterdevon-Profil bei Marburg a. d. Lahn, *Notizbl. Hess. Landesanst. Bodenforsch.* **102**:25–45.
Bensaïd, M., 1974, Étude sur des goniatites à la limite du Dévonien moyen et supérieur du Sud Marocain, *Notes Serv. Géol. Maroc* **36**(264):81–140.

Bogoslovsky, B. I., 1969, Devonski Ammonoidei. I. Agoniatity, *Tr. Paleont. Inst. Akad. Nauk SSSR* **124**:1–341 (in Russian).

Bogoslovsky, B. I., 1982, Interesnaya forma priust'yevykh obrazovaniy na rakovine klimeniy, *Dokl. Akad. Nauk SSSR* **364**:1483–1486.

Breitkreuz, C., 1986, Das Palaeozoikum von Nord-Chile, *Geotektonische Forsch.* **70**.

Chang, A. C., 1958, New late Upper Devonian faunas of the Great Khingan and its biological significance, *Acta Palaeontol. Sin.* **8**:180–192 (in Chinese with English summary).

Chlupác, I., 1976, The oldest goniatite faunas and their stratigraphical significance, *Lethaia* **9**:303–315.

Chlupác, I., 1985, Comments on the Lower-Middle Devonian Boundary, *Cour. Forsch.-Inst. Senckenb.* **75**:389–400.

Chlupác, I., and Kukal. Z., 1988, Possible global events and the stratigraphy of the Palaeozoic of the Barrandian (Cambrian–Middle Devonian, Czechoslovakia), *Sbor. Geol. Ved.* **43**:83–146.

Chlupác, I., and Turek. V., 1977, New cephalopods (Ammonoidea, Bactritoidea) from the Devonian of the Barrandian area, Czechoslovakia, *Vestn. Ustred. Ustavu. Geol.* **52**:303–306.

Claoué-Long, J. C., Jones, P. J., Roberts, J., and Maxwell, S., 1992, The numerical age of the Devonian–Carboniferous boundary, *Geol. Mag.* **129**:281–291.

Czarnocki, J., 1989, Klimenie gór swietokrzyskich, *Pr. Panst. Inst. Geol.* **127**.

Dzik, J., 1984, Phylogeny of the Nautiloidea, *Palaeont. Pol.* **45**.

Erben, H. K., 1960, Primitive Ammonoidea aus dem Unterdevon Frankreichs und Deutschlands, *N. Jb. Geol. Paläont. Abh.* **110**:1–128.

Erben, H. K., 1964, Die Evolution der ältesten Ammonoidea. Lieferung 1, *N. Jb. Geol. Paläont. Abh.* **120**:107–212.

Erben, H. K., 1966, Über den Ursprung der Ammonoidea, *Biol. Rev.* **41**:641–658.

Frech, F., 1897, *Lethaea Geognostica. 2 Teil 1. Lethaea Palaeozoica,* Stuttgart.

Frest, T. I., Glenister, B. F., and Furnish. W. M., 1981, Pennsylvanian-Permian cheiloceratacean ammonoid families Matimitidae and Pseudohaloritidae, *J. Paleontol. Suppl. Mem.* **11**:1–296.

Glenister, B. F., 1991, Proposal of Guadalupian as international standard for the Middle Permian Series, *Permophiles* **18**:10–11.

Glenister, B. F., and Furnish, W. M., 1981, Permian ammonoids, in: *The Ammonoidea,* Systematics Association Spec. Vol. 18 (M. R. House and J. R. Senior, eds.), Academic Press, London, pp. 49–64.

Göddertz, B., 1987, Devonische Goniatiten aus SW-Algerien und ihre stratigraphische Einordnung in die Conodonten-Abfolge, *Palaeontogr. Abt. A* **197**(4–6):127–220.

Gordon, M., 1965, Carboniferous cephalopods of Arkansas, *U.S. Geol. Surv. Prof. Pap.* **460**:1–322.

Gordon, M., 1986, Late Kinderhookian (Early Mississippian) ammonoids of the western United States, *J. Paleontol. Suppl. Mem.* **19**.

Grayson, R. C., Davidson, W. T., Westergaard, E. H., Atchley, S. C., Hightower, J. H., Monaghan, P. T., and Pollard, C., 1985, Mississippian–"Pennsylvanian" (Mid-Carboniferous) boundary conodonts from the Rhoda Creek Formation: *Homoceras* equivalent in North America, *Cour. Forsch.-Inst. Senckenb.* **74**:149–180.

Harland, W. B., Armstrong, R. L., Cox, A. V., Craig, L. E., Smith, A. G., and Smith, D. G., 1990, *A Geologic Time Scale 1989,* Cambridge University Press, Cambridge.

Heckel, P. H., and Witzke, B. J., 1979, Devonian world palaeogeography determined from distribution of carbonates and related lithic palaeoclimatic indicators, *Spec. Pap. Palaeontol.* **23**:99–123.

Hodgkinson, K. A., 1965, *The Late Paleozoic ammonoid families Prolecanitidae and Daraelitidae,* unpublished Ph.D. Thesis, University of Iowa, Iowa City.

Hollard, H., 1974, Recherches sur la stratigraphie des formations du Dévonien moyen de l'Emsien supérieur au Frasnien, dans le Sud du Tafilalt et dans le Ma'der (Anti-Atlas oriental), *Notes Serv. Géol. Maroc* **36**:7–68.

Holzapfel, E., 1895, Das obere Mitteldevon (Schichten mit *Stringocephalus Burtini* und *Maeneceras terebratum*) im Rheinischen Gebirge, *Abh. Preuß. Geol. Landesanst. N.F.* **16**.

House, M. R., 1964, Devonian Northern Hemisphere ammonoid distribution and marine links, in: *Problems in Palaeoclimatology* (A. E. M. Nairn, ed.), Interscience Publishers, London, New York, Sydney, pp. 262–269, 299–301.

House, M. R., 1965, Devonian goniatites from Nevada, *N. Jb. Geol. Paläont. Abh.* **122**:337–342.

House, M. R., 1970, On the origin of the clymenid ammonoids, *Palaeontology (Lond.)* **13**:664–676.

House, M. R., 1971, The Goniatite Wrinkle-Layer, *Smithsonian Contr. Paleobiol.* **3**:23–32.

House, M. R., 1973a, An analysis of Devonian goniatite distributions, *Spec. Pap. Palaeontol.* **12**:305–317.

House, M. R., 1973b, Devonian goniatites, in: *Atlas of Palaeobiogeography* (A. Hallam, ed.), Elsevier, Amsterdam, pp. 97–104.

House, M. R., 1978, Devonian ammonoids from the Appalachians and their bearing on international zonation and correlation, *Spec. Pap. Palaeontol.* **21**:1–70.

House, M. R., 1981, On the origin, classification and evolution of the early Ammonoidea, in: *The Ammonoidea*, Systematics Association Spec. Vol. 18 (M. R. House and J. R. Senior, eds.), Academic Press, London, pp. 3–36.

House, M. R., 1985, Correlation of mid-Palaeozoic ammonoid evolutionary events with global sedimentary perturbations, *Nature* **213**:17–22.

House, M. R., and Blodgett, R. B., 1982, The Devonian goniatite genera *Pinacites* and *Foordites* from Alaska, *Can. J. Earth Sci.* **19**:1873–1876.

House, M. R., and Price, J. D., 1985, New Late Devonian genera and species of tornoceratid goniatites, *Paleontology* **28**:159–188.

House, M. R., Kirchgasser, W. T., Price, J. D., and Wade, G., 1985, Goniatites from Frasnian (Upper Devonian) and adjacent strata of the Montagne Noire, *Hercynia* **1985**:1–19.

Hünicken, M., Kullmann, J., and Riglos, M. S., 1980, Consideraciones sobre el devonico Boliviano en base a un nuevo goniatites de la formacion Huamampampa en Campo Redondo, Departemento Chuquisaca, Bolivia, *Bol. Acad. Nac. Cienc. (Cordoba)* **53**:237–253.

Jones, P. J., 1988, Comments on some Australian, British and German isotopic age data for the Carboniferous System, *Newsl. Carbon. Stratigr.* **6**:26–29.

Karpinskiy, A., 1896, Sur l'existence du genre *Prolecanites* en Asie et sur son développement. *Bull. Acad. Imp. Sci. St. Petersb.* **4**:179–194.

Kayser, E., 1873, Studien aus dem Gebiet des rheinischen Devon. IV. Über die Fauna des Nierenkalkes vom Enkeberge und der Schiefer von Nehden bei Brilon, und über die Gliederung des Oberdevon im rheinischen Schiefergebirge, *Z. Deutsch Geol. Ges.* **25**:602–674.

Klapper, G., Feist, R., and House, M. R., 1987, Decision on the boundary stratotype for the Middle/Upper Devonian series boundary, *Episodes* **10**(2):97–101.

Korn, D., 1979, Mediandornen bei *Kosmoclymenia* Schindewolf (Ammonoidea, Cephalopoda), *N. Jb. Geol. Paläont. Mh.* **1979**:399–405.

Korn, D., 1984, *Prolobites aktubensis* Bogoslovskiy—eine devonische Goniatiten-Art (Ammonoidea) mit irregulärem Mundrand, *N. Jb. Geol. Paläont. Mh.* **1984**:66–76.

Korn, D., 1985, Runzelschicht und Ritzstreifung bei Clymenien (Ammonoidea, Cephalopoda), *N. Jb. Geol. Paläont. Mh.* **1985**:533–541.

Korn, D., 1988, Die Goniatiten des Kulmplattenkalkes (Cephalopoda, Ammonoidea; Unterkarbon; Rheinisches (Schiefergebirge), *Geol. Paläont. Westf.* **11**:1–293.

Korn, D., 1991, Three-dimensionally preserved clymeniids from the Hangenberg Black Shale of Drewer (Cephalopoda, Ammonoidea; Devonian-Carboniferous boundary; Rhenish Massif), *N. Jb. Geol. Paläont. Mh.* **1991**:533–563.

Korn, D., 1992, Relationships between shell form, septal construction and suture line in clymeniid cephalopods (Ammonoidea; Upper Devonian), *N. Jb. Geol. Paläont. Abh.* **185**:115–130.

Korn, D., 1993, The ammonoid faunal change near the Devonian–Carboniferous boundary, *Ann. Soc. Geol. Belg.* **115**(2):581–593.

Korn, D., 1994, Devonische und karbonische Prionoceraten (Cephalopoda, Ammonoidea) aus dem Rheinischen Schiefergebirge, *Geol. Paläont. Westf.* **30**.

Korn, D., and Kullmann, J., 1988, Changes in Clymeniid diversity, in: *Cephalopods—Present and Past* (J. Wiedmann and J. Kullmann, eds.), Schweizerbart'sche Verlagsbuchhandlung, Stuttgart, pp. 25–28.

Korn, D., and Kullmann, J., 1993, GONIAT Database System, version 2.31, Tübingen.

Kullmann, J., 1960, Die Ammonoidea des Devon im Kantabrischen Gebirge (Nordspanien), *Akad. Wiss. Lit. Mainz, Abh. Math.-Naturwiss. Kl.* **1960**(7).

Kullmann, J., 1962, Die Goniatiten der Namur-Stufe (Oberkarbon) im Kantabrischen Gebirge, Nordspanien, *Akad. Wiss. Lit. Mainz, Abh. Math.-Naturwiss. Kl.* **1962**(6).

Kullmann, J., 1981, Carboniferous goniatites, in: *The Ammonoidea*, Systematics Association Spec. Vol. 18 (M. R. House and J. R. Senior, eds.), Academic Press, London, pp. 37–48.

Kullmann, J., 1983, Maxima im Tempo der Evolution karbonischer Ammonoideen, *Paläontol. Z.* **57**:231–240.

Kullmann, J., 1993, Paleozoic Ammonoids of Mexico and South America, in *12ième Congr. Int. Stratigr. Géol. Carbon. Perm. Buenos Aires, 1991* **1**:557–562.

Kullmann, J., and Loeschke, J., 1994, Olistholithe in Flysch-Sedimenten der Karawanken: Die Entwicklung eines aktiven Kontinentrandes im Karbon der Südalpen (Paläozoikum von Seeberg und Eisenkappel/Österreich), *N. Jb. Geol. Paläont., Abh.* **194**:15–142.

Kullmann, J., and Schönenberg, R., 1975, Geodynamische und palökologische Entwicklung im Kantabrischen Variszikum (Nordspanien). Ein interdisziplinäres Arbeitskonzept, *N. Jb. Geol. Paläont., Mh.* **1975**:151–166.

Kullmann, J., and Wiedmann, J., 1970, Significance of sutures in phylogeny of Ammonoidea, *Univ. Kans. Paleontol. Contrib. Pap.* **47**:1–32.

Kullmann, J., and Wiedmann, J., 1982, Bedeutung der Rekapitulationsentwicklung in der Paläontologie, *Verh. Naturwiss. Ver. Hamburg (N.F.)* **25**:71–92.

Kullmann, J., Korn, D., and Weyer, D., 1991, Ammonoid zonation of the Lower Carboniferous subsystem, *Cour. Forsch-Inst. Senckenberg* **130**:127–131.

Lai, C., 1982, New materials of Palaeozoic cephalopods from Xizang (Tibet) of China, *Contr. Geol. Qinghai-Xizang (Tibet) Plateau* **7**:1–26.

Mapes, R. H., 1979, Carboniferous and Permian Bactritoidea (Cephalopoda) in North America, *Univ. Kansas Paleontol. Contrib.* **64**:1–75.

Menning, M., 1992, Numerical time scale for the Permian, *Permophiles* **20**:2–5.

Miller, A. K., and Furnish, W. B., 1940, Permian ammonoids of the Guadalupe Mountain region and adjacent areas, *Geol. Soc. Am. Spec. Pap.* **26**:1–242.

Nassichuk, W. W., 1975, Carboniferous ammonoids and stratigraphy in the Canadian Arctic Archipelago, *Geol. Surv. Can. Bull.* **237**:1–240.

Nikolaeva, S., and Kullmann, J., 1995, The Late Namurian Genus *Cancelloceras* (Carboniferous, Ammonoidea) and its distribution, *Paläont. Z.* **69**(3/4):353–376.

Over, J., 1992, Conodonts and the Devonian–Carboniferous boundary in the Upper Woodford Shale, Arbuckle Mountains, South-Central Oklahoma, *J. Paleontol.* **66**:293–311.

Planchon, J.-P., 1967, Observations sur le Dévonien inférieur du Sahara espagnol (Région de Smara), *Mem. Bur. Rech. Géol. Minieres* **33**:321–325.

Prosh, E. C., 1987, A Lower Devonian ammonoid, *Mimagoniatites nearcticus* n. sp., from the Canadian Arctic, *J. Paleontol.* **61**:974–981.

Ramsbottom, W. H. C., and Saunders, W. B., 1985, Evolution and evolutionary biostratigraphy of Carboniferous ammonoids, *J. Paleontol.* **59**:123–139.

Riley, N. J., 1990a, Revision of the *Beyrichoceras* Ammonoid-Biozone (Dinantian), NW Europe, *Newsl. Stratigr.* **21**:149–156.

Riley, N. J., 1990b, A global review of mid-Dinantian ammonoid biostratigraphy, *Cour. Forsch.-Inst. Senckenberg* **130**:133–143.

Ristedt, H., 1968, Zur Revision der Orthoceratidae, *Akad. Wiss. Lit., Abh. Math.-Naturwiss. Kl.* **1968**(4):211–299.

Ristedt, H., 1981, Bactriten aus dem Obersilur Böhmens, *Mitt. Geol.-Paläont. Inst. Univ. Hamburg* **51**:23–26.

Roberts, J., Claoue-Long, J. C., and Jones, P. J., 1991, Calibration of the Carboniferous and Early Permian of the southern New England Orogen by Shrimp Ion Microprobe Zircon Analyses, *Newsl. Carbon. Stratigr.* **9**:15–17.

Ruan, Y.-P., 1983, Discovery of a Devonian ammonoid species from Ejin Banner of western Inner Mongolia, *Acta Palaeont. Sin.* **22**:119–121.

Ruan, Y.-P., 1987, Ammonoids from the upper Middle Devonian in Yiwa Ravine of Tewo, west Qingling Mts., China, in: *Late Silurian–Devonian Strata and Fossils from Luqu-Tewo Area of West Qingling Mountains, China*, 2, Xi'an Institute for Geology Mineral Research, Nanjing Institute of Geology and Paleontology, Academia Sinica, pp. 219–226.

Ruzhentsev, V. E., 1949, Osnornye tipy evolyutsionnykh izmenenii lopastnoi linii verkhuepaleozoiskikh ammonitov, *Tr. Paleont. Inst. Akad. Nauk SSSR* **20**:183–198.

Ruzhentsev, V. E., 1950, Verkhnekamennougolnye ammonity Urala, *Tr. Paleont. Inst. Akad. Nauk SSSR* **29**:1–220.

Ruzhentsev, V. E., 1957, Filogeneticheskaya sistema paleozoiskikh ammonoidei, *Moskov. Obshch. Ispyt. Prirody. Byull. Otdel. Geol.* **32**(2):49–64.

Ruzhentsev, V. E., and Bogoslovskaya, M. F., 1971, Namyurskij etap v evolyutsii ammonoidej. Rannenamyurskie ammonoidei, *Tr. Paleontol. Inst. Akad. Nauk SSSR* **133**:1–382.

Ruzhentsev, V. E., and Bogoslovskaya, M. F., 1978, Namiurskij etap v evolyutsii ammonoidej. Pozdnenamiurskie ammonoidei. *Tr. Paleontol. Inst. Akad. Nauk SSSR* **167**:1–336.

Saunders, W. B., and Ramsbottom, W. H. C., 1982, Mid-Carboniferous biostratigraphy and boundary choices, in: *Biostratigraphic Data for a Mid-Carboniferous Boundary* (W. H. Ramsbottom and B. Owens, eds.), Leeds, pp. 1–5.

Schindewolf, O. H., 1929, Vergleichende Studien zur Phylogenie, Morphogenie und Terminologie der Ammoneenlobenlinie, *Abh. Preuß. Geol. Landesanst., N.F.* **115**:1–102.

Schindewolf, O. H., 1932, Zur Stammesgeschichte der Ammoneen, *Paläontol. Z.* **14**:164–181.

Schindewolf, O. H., 1933, Vergleichende Morphologie und Phylogenie der Anfangskammern tetrabranchiater Cephalopoden, *Abh. Preuß. Geol. Landesanst., N.F.* **148**.

Schindewolf, O. H., 1937, Zur Stratigraphie und Paläontologie der Wocklumer Schichten, *Abh. Preuß. Geol. Landesanst., N.F.* **178**.

Schindewolf, O. H., 1951, Zur Morphogenie und Terminologie der Ammoneen-Lobenlinie, *Paläontol. Z.* **25**:11–34.

Schmidt, H., 1952, *Prolobites* und die Lobenentwicklung bei Goniatiten, *Paläontol. Z.* **26**:205–217.

Shevyrev, A. A., and Ermakova, S. P., 1979, K sistematike tseratitov, *Paleont. Zh.* **1979**(1):52–58.

Sougy, J., 1969, Présence inattendue de *Pinacites jugleri* (Roemer) dans une calcaire situé à la base des siltstones de Tighirt (Couvinien inférieur du Zemmour noir, Mauritanie septentrionale), *Bull. Soc. Géol. Fr.* **11**(7):268–272.

Spinosa, C., Furnish, W. M., and Glenister, B. F., 1975, The Tenodiscidae, Permian ceratitoid ammonoids, *J. Paleontol.* **49**:239–283.

Sweet, W. C., 1958, The Middle Ordovician of the Oslo region, Norway. 10. Nautiloid cephalopods, *Norsk Geol. Tids.* **38**.

Teichert, C., 1988, Main Features of Cephalopod Evolution, in: *The Mollusca, Paleontology and Neontology of Cephalopods*, Vol. 12 (C. M. R. Clarke and E. R. Trueman, eds.), Academic Press, San Diego, pp. 11–79.

Teichert, C., 1990, The Permian–Triassic boundary revisited, *Lect. Notes Earth Hist.* **30**:199–238.

Termier, H., and Termier, G., 1950, Paléontologie Marocaine. II. Invertebres de l'Ere primaire. Fasc. III, Mollusques, *Serv. Géol. Prot. Fr. Maroc Not., Mém.* **78**.

Vöhringer, E., 1960, Die Goniatiten der unterkarbonischen *Gattendorfia*-Stufe im Hönnetal (Sauerland), *Fortschr. Geol. Rheinl. Westfalen* **3**(1):107–196.

Walliser, O. H., 1966, Preliminary notes on Devonian, Lower Carboniferous and Upper Carboniferous goniatites in Iran, *Contr. Pal. East Iran Rep.* **6**:7–24.

Walliser, O. H., 1970, Über die Runzelschicht der Ammonoidea, *Gött. Arb. Geol. Paläont.* **5**:115–126.

Wedekind, R., 1908, Die Cephalopodenfauna des höheren Oberdevon am Enkenberg, *N. Jb. Min. Geol. Paläont., Beil. Bd.* **26**:565–635.

Wedekind, R., 1913, Die Goniatitenkalke des unteren Oberdevon vom Martenberg bei Adorf, *Sitzungsber. Ges. Naturforsch. Freunde Berl.* **1913**:23–77.

Wedekind, R., 1916, Über Lobus, Suturallobus und Inzision, *Centralbl. Miner.* **1916**:185–195.

Wiedmann, J., and Kullmann, J., 1981, Ammonoid sutures in ontogeny and phylogeny, in *The Ammonoidea*, Systematics Association Spec. Vol. 18 (M. R. House and J. R. Senior, eds.), Academic Press, London, pp. 215–255.

Wissner, U. F. G., and Norris, A. W., 1991, Middle Devonian goniatites from the Dunedin and Besa River Formations of Northeastern British Columbia, *Contrib. Can. Paleontol., Geol. Surv. Can. Bull.* **412**:45–79.

Yatskov, S. V., 1992, A new Teicherticeratid member (Ammonoidea) from the Lower Devonian of Yakutia, *Paleontol. Zh.* **1992**:124–128.

Yu, C.-M., and Ruan, Y.-P., 1989, Proposal and comment on the definition of Emsian, *Can. Soc. Petr. Geol. Mem.* **14**(III):179–191.

Zhao Jin-Ke, Sheng Jin-Zhang, Yao Zhao-Qi, Liang Xi-Luo, Chen Chu-Zhen, Rui Lin, and Liao Zhuo-Ting, 1981, The Changhsingian and Permian-Triassic boundary of South China, *Bull. Nanjing Inst. Geol. Palaeont., Acad. Sin.* **2**:1–112.

Zhou, Z., 1988, Several problems on the early Permian ammonoids from South China, *Paleont. Cathayana* **2**:179–209.

Chapter 18

Mesozoic Ammonoids in Space and Time

KEVIN N. PAGE

KEVIN N. PAGE • Earth Science Branch, English Nature, Peterborough PE1 1UA, England.

Ammonoid Paleobiology, Volume 13 of *Topics in Geobiology*, edited by Neil Landman *et al.*,
Plenum Press, New York, 1996.

1. Introduction

Mesozoic ammonoids were diverse and included not only "conventional" ammonites (belonging to the suborders Ammonitina, Haploceratina, and Perisphinctina) but also several other major and well-known groups such as the Triassic ceratites (Suborder Ceratitina) and the dominantly heteromorphic Cretaceous ancyloceratids (Suborder Ancyloceratina). Many of these, and indeed other ammonoid groups, had globally wide distributions, turning up wherever suitable marine environments of appropriate age were available; others, however, were more local in range. Each major group is briefly introduced, and the general features of its distribution in time and space are summarized.

The taxonomic framework followed is similar to that employed by Hewitt (1993) and Page (1993), following Wright (1981) for the Cretaceous, Donovan *et al.* (1981) for the Jurassic (but as modified by Bessenova and Michailova, 1983, 1991), and Tozer (1981a), in part, for the Triassic. Hewitt (1993) and Page (1993) also provided detailed information on the first and last stratigraphic occurrence of taxonomic groups down to family level.

The following section takes a global view of the distribution of ammonoid faunas through the Mesozoic and attempts a reconstruction of the nature and evolution of observed biogeographic provincialism. The basic chronostratigraphic stage framework for the Mesozoic is that used by Harland *et al.* (1989). Nevertheless, two important points need noting. First, the Harland time scale incorporates the Rhaetian as the terminal stage of the Triassic, whereas some Triassic workers regard that stage as little more than equivalent to the last ammonoid zone of the Norian (Wiedmann *et al.*, 1979). Second, the Harland *et al.* time scale employs Tithonian as the terminal Jurassic stage and Berriasian as the basal Cretaceous. Historically, such divisions have been applied only to southern areas (essentially the Tethyan Realm in a biogeographic sense) and not used in faunally distinct northern regions (the Boreal Realm), where the terminal Jurassic stage was the Portlandian or Volgian and the basal Cretaceous, the Ryazanian. The bases of the Portlandian, Volgian, and Tithonian stages are not coincident, but, more importantly, neither is the boundary between the Jurassic and Cretaceous systems in Tethyan and Boreal regions (historically, the Tithonian–Berriasian and Portlandian/Volgian–Ryazanian stage boundaries, respectively); the base of the Tethyan Lower Cretaceous (Berriasian Stage) is probably equivalent to a level within the Boreal Upper Jurassic, approximately the base of the upper Volgian Substage (Wimbledon, 1984; Zeiss, 1986; Cope, 1993).

2. Major Ammonoid Groups in the Mesozoic

Within the Mesozoic, 13 ammonoid suborders are presently recognized, and each is introduced below in order of appearance. Figures 1 and 2

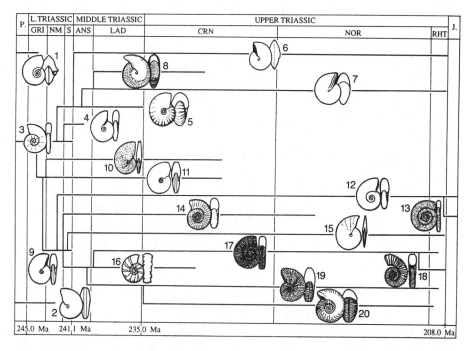

FIGURE 1. A generalized phylogeny for Triassic ammonoid superfamilies (with illustrations of typical genera): 1, Otocerataceae (*Otoceras*); 2, Medlicottiaceae (*Episageceras*); 3, Xenodiscaceae (*Xenodiscus*); 4, Dinaritaceae (*Dinarites*); 5, Ptychitaceae (*Ptychites*); 6, Megaphyllitaceae (*Megaphyllites*); 7, Arcestaceae (*Arcestes*); 8, Lobitaceae (*Lobites*); 9, Noritaceae (*Norites*); 10, Sagecerataceae (*Sageceras*); 11, Meekocerataceae (*Meekoceras*); 12, Phyllocerataceae (*Phylloceras*); 13, Psilocerataceae (*Psiloceras*); 14, Danubitaceae (*Danubites*); 15, Pinacocerataceae (*Pinacoceras*); 16, Ceratitaceae (*Ceratites*); 17, Clydonitaceae (*Clydonites*); 18, Choristocerataceae (*Choristoceras*); 19, Trachycerataceae (*Trachyceras*); 20, Tropitaceae (*Tropites*). Abbreviations as follows: GRI, Griesbachian; NM, Nammalian; S, Spathian; ANS, Anisian; LAD, Ladinian; CRN, Carnian; NOR, Norian; RHT, Rhaetian; P, Permian; J, Jurassic (dates from Harland *et al.*, 1989).

demonstrate the stratigraphic range of the component superfamilies of each suborder, together with a representation of a typical shell morphology.

2.1. Suborder Prolecanitina (Triassic Superfamily: Medlicottiaceae)

The prolecanitines were an exceptionally long-ranging suborder with origins in the Early Carboniferous but surviving until the Early Triassic. They then rapidly disappeared, leaving no descendants, and are, therefore, not a characteristic element of most Triassic ammonoid faunas. Triassic forms characteristically have involute, discoidal, smooth shells with multilobed goniatitic (i.e., simple) sutures.

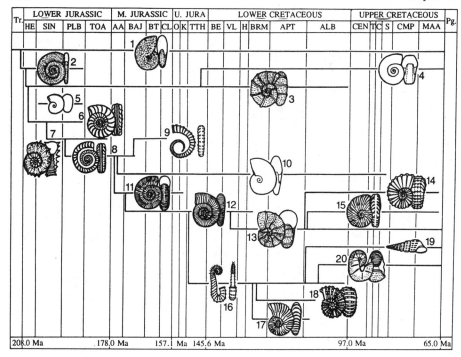

FIGURE 2. A generalized phylogeny for Jurassic and Cretaceous ammonoid superfamilies (with illustrations of typical genera): 1, Phyllocerataceae (*Phylloceras*); 2, Psilocerataceae (*Psiloceras*); 3, Lytocerataceae (*Lytoceras*); 4, Tetragonitaceae (*Tetragonites*); 5, Cymbitaceae (*Cymbites*); 6, Arietitaceae (*Arietites*); 7, Eoderocerataceae (*Eoderoceras*); 8, Hildocerataceae (*Hildoceras*); 9, Spirocerataceae (*Spiroceras*); 10, Haplocerataceae (*Haploceras*); 11, Stephanocerataceae (*Stephanoceras*); 12, Perisphinctaceae (*Perisphinctes*); 13, Desmocerataceae (*Desmoceras*); 14, Acanthocerataceae (*Acanthoceras*); 15, Hoplitaceae (*Hoplites*); 16, Ancylocerataceae (*Ancyloceras*); 17, Deshayesitaceae (*Deshayesites*); 18, Douvilleicerataceae (*Douvilleiceras*); 19, Turrilitaceae (*Turrilites*); 20, Scaphitaceae (*Scaphites*). Abbreviations as follows: Tr, Triassic; HE, Hettangian; SIN, Sinemurian; PLB, Pliensbachian; TOA, Toarcian; AA, Aalenian; BAJ, Bajocian; BT, Bathonian; CL, Callovian; O, Oxfordian; K, Kimmeridgian; TTH, Tithonian; BE, Berriasian; VL, Valanginian; H, Hauterivian; BRM, Barremian; APT, Aptian; ALB, Albian; CEN, Cenomanian; T, Turonian; C, Coniacian; S, Santonian; CMP, Campanian; MAA, Maastrichtian; Pg, Paleogene (dates from Harland *et al.*, 1989).

Their distribution is from the Lower Carboniferous to the Lower Triassic (Nammalian). Triassic forms were widely distributed from Arctic and Western Pacific to Tethyan regions (Tozer, 1981a).

2.2. Suborder Paraceltina (Triassic Superfamilies: Xenodiscaceae, Dinaritaceae)

Of the three ammonoid families to survive the Late Permian mass extinction, the paraceltine Xenodiscidae are by far the most significant. Tozer

(1981a, p. 72) described the family as the "all important rootstock for the majority of Triassic ammonoids" (and indeed for all post-Triassic ammonoids). Of the other two families to survive, the Otoceratidae (Otoceratina) are descendants of Permian Xenodiscidae, and the Medlicottiidae (Medlicottiaceae) represent relatives of the prolecanitine group from which the Paraceltina themselves originated (Glenister and Furnish, 1981).

The Paraceltina were very characteristic of Early Triassic ammonoid faunas; typical shells are evolute and relatively compressed (Xenodiscaceae), but more involute tuberculate forms also occur (some Dinaritaceae). Sutures have few lobes and may be goniatitic or ceratitic (i.e., with serrated lobes).

Their distribution is from the Permian (Roadian) to the Middle Triassic (Middle Anisian). Xenodiscaceae ranged widely from Arctic through Pacific to Tethyan regions, but the Dinaritaceae, in contrast, were dominantly Tethyan (Tozer, 1981a).

2.3. Suborder Otoceratina (Triassic Superfamily: Otocerataceae)

Directly descended from Permian Paraceltina, the Otoceratidae are the only Otoceratina to persist into the Triassic, and although diagnostic of earliest Triassic ammonoid faunas, they did not survive beyond the Griesbachian. Shells are characteristically involute, with a subtriangular section and goniatitic or ceratitic sutures.

Their distribution is from the Permian (Dzhulfian) to Lower Triassic (Griesbachian), as known from Arctic and Tethyan regions, perhaps most common in the former (Tozer, 1981a).

2.4. Suborder Sageceratina (Noritaceae, Sagecerataceae)

The Sageceratina were an important and characteristic Early and Middle Triassic ammonoid group. Noritaceae are variable in morphology, with early forms tending to be compressed and involute with a suture with few lobes, but later groups develop a more involute, even discoidal shape, often with more sutural lobes. This latter type of shell morphology is characteristic of Sagecerataceae, and the resulting homeomorphy of some forms with prolecanitine Medlicottiaceae has previously been used as evidence of evolutionary origins (e.g., Spath, 1934).

Their distribution is Lower Triassic (Griesbachian) to Upper Triassic (Carnian). Although generally widely distributed through Tethyan, Pacific, and Arctic regions, certain families showed a much more restricted distribution; for instance, Aspenitidae (Sagecerataceae) were restricted to Tethyan regions, and Beneckeiidae (Sagecerataceae) only occurred in the Germanic and Sephardic provinces (see Section 3.1; Tozer, 1981a).

2.5. Suborder Arcestina (Ptychitaceae, Megaphyllitaceae, Arcestaceae)

The Arcestina were often abundant in Middle and Late Triassic ammonoid faunas of all but the most environmentally restricted areas (i.e., the Sephardic and Germanic provinces). Although many members of this suborder have relatively primitive goniatitic or ceratitic sutures, the superfamily Arcestaceae is one of the two dominant groups of smooth-shelled Triassic ammonoids with ammonitic (i.e., frilled or complex) sutures (Tozer, 1981a, p. 74). The other group is the Pinacocerataceae (Pinacoceratina).

The origins of the Arcestina are unclear but are assumed to lie within the xenodiscacean (Paraceltina) "rootstock" of Tozer (1981a). The superficial morphological resemblance of the globular to discoidal shells of many groups in the Arcestina to Palaeozoic Goniatitina is remarkable, but sutural studies show this to be a case of convergent evolution and not evidence of origins (Schindewolf, 1968; Tozer, 1981a).

Their distribution is Lower Triassic (Nammalian) to Upper Triassic (Rhaetian) and worldwide (Arctic, Pacific, and Tethyan regions; Tozer, 1981a).

2.6. Suborder Pinacoceratina (Pinacocerataceae, Meekocerataceae)

The suborder includes both relatively evolute and, more characteristically, compressed involute forms with smooth or feebly ribbed shells. Ammonitic sutures are characteristic of the Pinacocerataceae, one of the two dominant, complex-sutured and smooth-shelled ammonoid groups of the Triassic. The origins of the Pinacocerataceae probably lie in the much more simply sutured (ceratitic) Meekocerataceae and, therefore, ultimately in the xenodiscacean "rootstock" (Paraceltina) (Tozer, 1981a).

Their distribution is Lower Triassic (Nammalian) to Upper Triassic (Rhaetian). Pinacocerataceae were widely distributed in the Arctic, Pacific, and Tethyan regions, but Meekocerataceae were largely (and latitudinally?) restricted to Tethyan and Pacific areas.

2.7. Suborder Phylloceratina (Phyllocerataceae)

An exceptionally persistent and morphologically conservative group, Phylloceratina survived from its origins in the Early Triassic until the extinction of all Ammonoidea in the terminal Maastrichtian (there is, in fact, some evidence to suggest that the last ammonoid anywhere could have been a phylloceratid; Wiedmann, 1988, p. 186). The suborder is characterized by sutures with ovoid tips to finely divided lobes (i.e., phylloid sutures).

Early phyllocerataceans, in the Triassic, tended to be relatively evolute and compressed, with no more than fine growth lines as ornament. Similar forms persisted into the Early Jurassic, where some developed very weak ribbing. In the Early Jurassic, a second group evolved, with involute shell morphologies and, frequently, with shell constrictions or flares (the Phylloceratidae). Excepting a few atypical Cretaceous forms with ribs, most Phylloceratidae have little more than fine growth lines as ornament, and all have a characteristically complex phylloid suture.

The Phyllocerataceae are virtually the only ammonoid group to have survived the end-Triassic extinctions, giving rise, just below the Triassic–Jurassic boundary, to all other later Mesozoic ammonoids *via* the earliest "Ammonitina" (Psiloceratidae, as discussed below).

Their distribution is Lower Triassic (Nammalian) to Cretaceous (uppermost Maastrichtian). Phyllocerataceae were widely distributed globally, but many genera occurred primarily in Tethyan regions, where the group was often very abundant. Through much of the Jurassic, for instance, Phylloceratina and the similarly distributed Lytoceratina comprised more than 50% of the ammonoid fauna of the Mediterranean Province (Section 3.2). Although one control on this distribution may have been latitudinal, Phylloceratina were often most abundant in deeper water areas, subject to oceanic influences (e.g., Donovan, 1967; Ziegler, 1981, p. 445; Rawson, 1981, p. 501), and considerably less abundant or absent in neighboring shelf sea deposits. An ecological rather than latitudinal primary control on distribution is supported by the persistent occurrence of Phylloceratina at relatively high latitudes in areas adjacent to open ocean throughout the Jurassic and Cretaceous.

2.8. Suborder Ceratitina (Danubitaceae, Clydonitaceae, Ceratitaceae, Trachycerataceae, Tropitaceae, Choristocerataceae)

The Ceratitina include virtually all the strongly ornamented or tuberculate ammonoids of the Triassic. Shell morphologies and ornament styles are many and varied (Fig. 1), and the suborder includes the only Triassic uncoiled forms (heteromorphs), which belong to the Choristocerataceae (Wiedmann, 1973). Sutures are generally ceratitic but may be goniatitic or ammonitic in certain groups. The origins of the suborder appear to lie in the Meekocerataceae (Pinacoceratina; Tozer, 1981a).

The Ceratitina were very characteristic of Middle and Late Triassic ammonoid faunas, but although many groups were widely distributed, important cases of endemism occurred, especially in the Ceratitaceae; well-known genera such as *Ceratites*, for instance, were entirely restricted to the low-latitude and environmentally restricted Germanic Province (Section 3.1).

Their distribution is Lower Triassic (Spathian) to Upper Triassic (Rhaetian). Most of the included superfamilies had a worldwide distribution (Arctic, Pacific, and Tethyan regions), but some, such as the Clydonitaceae and Tropitaceae, were more characteristic of lower-latitude Tethyan and Pacific areas and were rare in the Arctic. The most characteristic ammonoids of the Germanic and Sephardic provinces were the Ceratitaceae, which included a variety of endemic genera. The Danubitaceae and Trachycerataceae also occurred in the latter provinces but were much rarer than in contemporary Tethyan and Pacific regions.

2.9. Suborder Ammonitina (Psilocerataceae, Arietitaceae, Cymbitaceae, Eoderocerataceae, Hildocerataceae, Spirocerataceae)

Historically, post-Triassic ammonoids have been grouped into four suborders, the Phylloceratina, Lytoceratina, Ammonitina, and Ancyloceratina. Of these groups, the Ammonitina formerly included most "normal" (i.e., nonheteromorph) Jurassic and Cretaceous ammonoids. Based on studies of sutural development, phylogeny, and shell structure, Bessenova and Michailova (1983, 1991) proposed that the suborder Ammonitina should be restricted and proposed two additional suborders, the Haploceratina and the Perisphinctina. The characters used to separate the three "new" suborders are similar to those used to separate Triassic groups of similar rank, and Bessenova and Michailova's scheme is, therefore, accepted here (although with a modified assignment of superfamilies to each suborder).

The first Ammonitina (*sensu* Bessenova and Michailova, 1983, 1991) are conventionally considered to belong to the genus *Psiloceras* (Psilocerataceae), which evolved directly from Triassic Phylloceratina close to the Triassic–Jurassic boundary; forms intermediate between *Rhacophyllites* (Phyllocerataceae) and *Psiloceras* have been illustrated by Guex (1982) associated with late Rhaetian *Choristoceras* (Choristocerataceae) in Nevada. The sutures of *Psiloceras* and its close relatives betray their phylloceratine origins by a tendency to possess phylloid saddles.

Guex (1987, Fig. 1) outlined a detailed scheme for the origin of the earliest Jurassic ammonoid genera and suggested that the early Hettangian radiation of the Ammonitina from *Psiloceras* not only gave rise to later Psilocerataceae but also and separately gave rise to the Lytoceratina via *Pleuroacanthites* (Lytocerataceae), a genus that Guex considered to have also given rise to the Arietitaceae. An alternative source of the Arietitaceae (and, hence, all later Ammonitina) by development of a keel on an evolute ribbed psiloceratid is a more conventional view (e.g., Arkell *et al.*, 1957, p. L235). Guex's suggestion is interesting taxonomically, as it gives the following pattern for the radiation of all post-Triassic ammonoids: Phylloceratina → "Ammonitina" *sensu lato* (Psilocerataceae only) → Lytoceratina → Ammonitina *sensu stricto* → Haploceratina → Perisphinctina → Ancyloceratina.

If this scheme is correct, the Psilocerataceae may have to be assigned to either the Phylloceratina or Lytoceratina or given a suborder of their own (Psiloceratina?) in order to maintain a phylogenetic classification of the Ammonoidea.

The restricted suborder Ammonitina includes a wide range of shell morphologies, from evolute serpenticones to inflated sphaerocones to compressed oxycones. Ornament includes ribs, spines, and ventral keels, in numerous styles and combinations. Shell dimorphism (presumably sexual) is present, but in early groups it is manifested mainly as differences in shell size, and only occasionally are significant differences in the adult morphology of each dimorph recognizable. In later families of the Hildocerataceae, however, dimorphism is well developed, with large forms (macroconchs) having simple apertures and small forms (microconchs) paired lateral lappets (Callomon, 1981). The suborder includes a small short-lived group of Middle Jurassic heteromorphs, the Spirocerataceae, with its origins lying in the Hildocerataceae (Dietl, 1978). Sutures in the suborder are ammonitic but vary in complexity. The primary suture is five-lobed but possibly grades into a four-lobed form (Bessenova and Michailova, 1991, p. 4).

Their distribution is Upper Triassic (uppermost Rhaetian) to Middle Jurassic (middle Callovian). Early Ammonitina were globally widely distributed, even at the generic level. Most forms were characteristic of shelf sea areas and tended to decline in relative abundance, when compared to Phyllo- and Lytoceratina at least, in deeper water regions (especially at low latitudes). A number of provincially restricted species, and sometimes genera, did, however, occur throughout the range of the suborder (most particularly in the late Pliensbachian and Toarcian, when a Boreal–Tethyan realm separation was well developed).

2.10. Suborder Lytoceratina (Lytocerataceae, Tetragonitaceae)

The Lytoceratina, in common with the Phylloceratina, show great evolutionary conservatism throughout their range from the lowest Jurassic to the uppermost Cretaceous. Shell morphologies are always evolute, rarely ribbed, and possess complex, even moss-like, nonphylloid sutures. A characteristic distribution in Tethyan and Pacific areas was also similar.

The first Lytoceratina appeared in the earliest Jurassic but had no obvious Triassic ancestors (Tozer, 1981a, p. 86). Their origin appears to lie in latest Triassic/earliest Jurassic Phylloceratina (Phyllocerataceae) or Ammonitina (Psilocerataceae) as suggested by Guex (1982).

Their distribution is from Lower Jurassic (lower Hettangian) to Upper Cretaceous (uppermost Maastrichtian). The Lytoceratina, as noted above, had a virtually identical distribution pattern to that of the Phylloceratina, the two groups commonly occurring in association in deeper water facies of the Jurassic and Cretaceous. There is, however, some suggestion that latitudinal controls may have been more important for the Lytoceratina, as high-latitude records seem to be less common than for the Phylloceratina.

2.11. Suborder Haploceratina (Haplocerataceae)

The Haploceratina was a long-ranging suborder, with origins in early Middle Jurassic Hildocerataceae. Shell morphologies are dominantly compressed, often tending toward an oxyconic or platyconic shape and with only weak ornament (although keels are often present). Dimorphism is well developed and similar to that of some late Hildocerataceae with simple-apertured macroconchs and lappeted microconchs (Callomon, 1981). Sutures are ammonitic and often very complex in macroconchs. The primary suture is five-lobed.

Their distribution is from Middle Jurassic (Aalenian) to Upper Cretaceous (Santonian). Although a very long-ranging suborder, Haploceratina were consistently and characteristically restricted to Tethyan areas and were usually very rare or absent in Boreal regions.

2.12. Suborder Perisphinctina (Stephanocerataceae, Perisphinctaceae, Desmocerataceae, Hoplitaceae, Acanthocerataceae)

The Perisphinctina were very characteristic of most later Jurassic and Cretaceous ammonoid faunas. The group is morphologically very diverse, with a wide range of ribbing styles and sometimes tubercles. In Jurassic groups, keels are generally absent but reappear in some late Early Cretaceous genera. Dimorphism is well developed in Jurassic genera with lappeted microconchs and smooth-apertured macroconchs (Callomon, 1981). In some Late Cretaceous groups, however, lappets appear to have been lost, and the size difference between dimorphs is often the only obvious distinguishing character. Sutures are characteristically ammonitic and complex in some groups; the primary suture is five-lobed.

Their distribution is from Middle Jurassic (Aalenian) to Upper Cretaceous (Maastrichtian). The Perisphinctina showed some of the most dramatic provincially controlled distributions of any ammonoid group, reflecting, at least in part, the dynamics of Late Jurassic and Early Cretaceous paleogeography. Differentiation was at species, genus, subfamily, or even family level, and the Boreal–Tethyan realm (i.e., North–South) separation was usually the most pronounced.

Although most highly ornamented forms were restricted to shelf sea areas, certain Cretaceous Desmocerataceae (including *Hauericeras*, *Damesites*, *Tragodesmoceroides*, and some Kossmaticeratidae) seem to have consistently cooccurred with Phyllo- and Lytoceratina in deeper water areas (as indicated by Matsumoto, 1973). The smooth, often relatively involute and constricted shells of some of the former are remarkably similar to some of the latter two groups, and presumably this represents similar selection pressures in similar environments.

2.13. Suborder Ancyloceratina (Ancylocerataceae, Douvilleicerataceae, Deshayesitaceae, Turrilitaceae, Scaphitaceae)

The last of the ammonoid suborders to evolve, the Ancyloceratina, include all the well-known heteromorph ammonoids of the Cretaceous and their immediate ancestors in the latest Jurassic. The earliest Ancyloceratceae appeared in the early Tithonian and belong to the loosely coiled genus *Protancyloceras* (Ancyloceratceae); they are completely unrelated to earlier Jurassic heteromorphs (Spirocerataceae; Hildoceratina). The origins of *Protancyloceras* are, however, obscure but are presumed to lie in contemporary Perisphinctaceae (Perisphinctina), perhaps in certain very evolute late Idoceratinae (Perisphinctidae), for instance, late Kimmeridgian *Mesosimoceras* (which has simple ribbing and occasional ventrolateral tubercles, not unlike those of *Protancyloceras*).

The Ancyloceratina later diversified to form a great variety of heteromorphic shell forms ranging from straight to loosely coiled and helicoid to hook-shaped morphologies (Fig. 2; Arkell *et al.*, 1957). Significantly, however, two separate groups, the Douvilleicerataceae and Deshayesitidae, recoiled to form more "conventional" shell forms, but ornamental style and sutural development betray their heteromorphic ancyloceratine ancestry (Wiedmann, 1970). Bessenova and Michailova (1991) included a Suborder Turrilitina within their "Order" Lytoceratida but do not fully discuss the reasons for such an assignment—this proposal is, therefore, not followed here.

Their distribution is from Upper Jurassic (lower Tithonian) to Upper Cretaceous (uppermost Maastrichtian). Although relatively rare in the latest Jurassic and earliest Cretaceous, Ancyloceratina soon came to dominate many Early Cretaceous ammonoid faunas. Many genera were globally widely distributed, although some endemic forms also existed. In the later Cretaceous, a similar distribution pattern existed, but other ammonoid groups, especially certain Perisphinctina, dominated many faunas. There is some evidence of facies control of heteromorph faunas in the Late Cretaceous (Kennedy and Cobban, 1976, p. 44), and important endemic populations of the Scaphitaceae and Baculitidae (Turrilitaceae) were established in the Western Interior of the United States (Section 3.2.3). Other genera and even species were much more widely distributed globally, and several persisted to the end of the Maastrichtian.

3. Ammonoid Provincialism through the Mesozoic

Kennedy and Cobban (1976) discussed at length the factors that may have controlled ammonoid distribution patterns and recognized five basic categories of distributions:

1. Pandemic distributions. Taxa occurred virtually globally and must, therefore, be relatively environmentally tolerant (or migratory?).
2. Latitudinally limited. The diversity of ammonoid faunas, as with other marine organisms, decreased toward the paleopoles; some groups showed specific latitudinal preferences, reflecting in part varying degrees of tolerance of environmental conditions, including temperature.

3. Endemic and provincial distributions. Endemic taxa showed both latitudinal and longitudinal restriction and tended to occur in moderately well-defined faunal provinces. The distinctiveness of such provinces varied from slight differences in generic or specific composition to those characterized by a significant number of endemic taxa. Faunas showing a high degree of endemism were often characterized by low diversities, presumably a direct consequence of the restriction that created the provincial separation in the first place.
4. Disjunct distributions. The local abundance of certain genera or species in widely separated areas can usually be attributed to tectonic separation or nonpreservation and collection failure at intermediate sites. Kennedy and Cobban (1976), however, cited four Cretaceous examples for which they could "detect neither facies, environmental nor palaeogeographic features that might account for their distribution." Looking at Text Fig. 18 of these authors, however, one cannot help getting the impression that at least some of the forms considered are simply scattered records of rare pandemic or longitudinally restricted taxa.
5. Postmortem distribution. The postmortem drift of ammonite shells has generally been compared to that of modern *Nautilus*, where there has been an assumption that shells drift for considerable distances. However, as noted by Callomon (1985, p. 62) and suggested by Kennedy and Cobban (1976), many of the scattered occurrences of *Nautilus* are now known to be directly related to the presence of widely distributed living populations. In this case, postmortem drifting appears to be the exception rather than the rule, and this is also likely to have been true for ammonoids. Cases of postmortem distribution of ammonoids are, therefore, most likely to be identified by isolated and very rare occurrences of otherwise endemic taxa found at significant distances from their principal "home" (that is, where they lived and bred) and are *eudemic sensu* Callomon (1985, p. 63).

Latitidunal and provincial control of ammonoid faunas was common throughout the Mesozoic, greatly restricting the distribution of different taxa at species, genus, family, and higher levels. Pandemic species were, in comparison, rare; disjunct and postmortem distributions were numerically insignificant. In order to understand the broader patterns of ammonoid distribution through the Mesozoic, it is, therefore, necessary to look at the nature and development of faunally distinct areas such as realms and provinces.

Realms are the largest biogeographic areas recognized, and, throughout the Mesozoic, no more than two separate ammonoid realms are distinguished, namely the Tethyan and Boreal realms. The former occupied much of the globe and included both low-latitude and southern high-latitude faunas. The latter occupied a northern high-latitude position and is recognizable through much of the Mesozoic excepting the Late Cretaceous. It is mainly distinguished on the basis of considerably lower faunal diversities, the common presence of endemic ammonoid genera, and the typical absence of many of the families characteristic of contemporary deposits elsewhere in the world within the Tethyan Realm. Occasionally within each realm, and especially within the Tethyan, certain broad areas showed certain faunal similarities, allowing the recognition of *regions* or *subrealms*. The basic "unit" employed, however, when describing geographically distinct faunas is the *province* (see category 3 above). Occasionally, changes in the relative diversity of elements of the faunas and other, more minor faunal differences may lead to the further recognition of *subprovinces*.

Through time, as continents separated and new migration pathways were opened up and others closed, the character and geographic boundaries be-

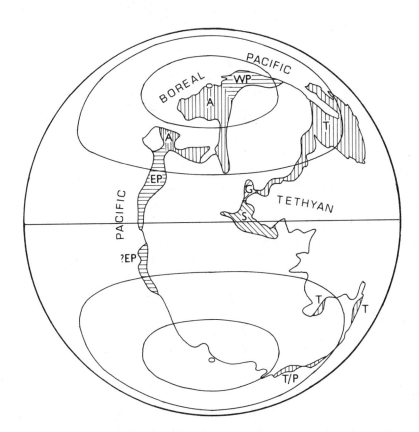

FIGURE 3. Generalized Triassic faunal provinces and realms. Abbreviations as follows: T, Tethyan Realm; A, Boreal Realm, Arctic Province; EP, Pacific Realm, East Pacific Province; WP, Pacific Realm, West Pacific Province; T/P, "mixed" Pacific/Tethyan realm faunas. Continental projection based on Smith and Briden (1977).

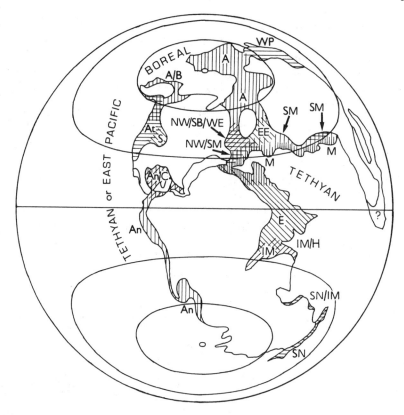

FIGURE 4. Generalized Jurassic faunal provinces and realms. Abbreviations as follows. Tethyan Realm: M, Mediterranean (or West Tethyan) Province; NW, Northwest European Province; SM, Sub-Mediterranean Province; H, Himalayan Province; E, Ethiopian Province; IM, Indo-Malgach Province; SN, Sula–New Guinean Province; C, Cuban Province; WP, West Pacific, Tethyan or East Pacific Realm; At, Athabascan Province; S, Shoshonean Province; An, Andean Province. Boreal Realm: A, Arctic Province; SB, Sub-Boreal Province; WE, West European Province; EE, East European Province; B, Bering Province. Continental projection based on Smith and Briden (1977). Cross-hatching indicates areas where provincial boundaries fluctuated during the Jurassic.

tween different provinces and realms changed dramatically, and, indeed, the actual number and location of faunally recognizable areas varied considerably. The general characters of each of the recognized Mesozoic provinces are summarized below and in Figs. 3–5, which show their generalized global distributions on appropriate continental reconstructions. Nevertheless, because the boundaries and nature of the provincial frameworks changed extensively with time, Fig. 6 is also provided in order to demonstrate the evolution through time of each described province. With each provincial description, a brief selection of key or representative references is provided.

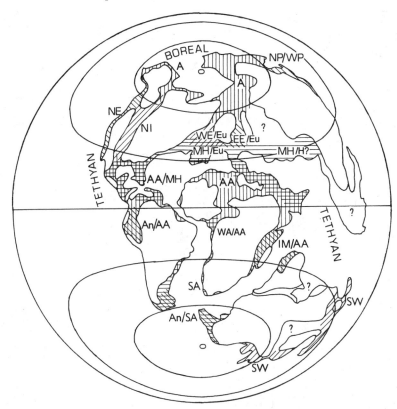

FIGURE 5. Generalized Cretaceous faunal provinces and realms. Abbreviations as follows. Tethyan Realm: MH, Mediterraenean–Himalayan Province; Eu, European Province; An, Andean Province; AA, Afro-Atlantic Province; IM, Indo-Malgach Province; SA, South Atlantic Province; WA, West African (or Angolan) Province; SW, South West Pacific Province; WP, West Pacific Province; NI, North American Interior Province; NE, Northeast Pacific Province (for general distribution of Mediterranean, Sub-Mediterranean, and Himalayan Provinces, see Fig. 4). Boreal Realm: A, Arctic Province; WE, West European Province; EE, East European Province. Continental projection based on Smith and Briden (1977). Cross-hatching indicates areas where provincial boundaries fluctuated during the Cretaceous.

3.1. Triassic Provincialism

The relatively simple continental layout of the Triassic, with a single major North–South continental barrier (Gondwanaland) and narrow shelf sea areas, produced, not surprisingly, a relatively simple pattern of faunal regions and provinces (Figs. 3 and 6). On the west side of the continent, a Pacific Region (or subrealm) is recognized, and on the east, a Tethyan Region. Latitudinally controlled restrictions on faunas produced a third major faunal area, a high-latitude Arctic or Boreal Realm (or subrealm?), although no equivalent Antarctic faunal region is recognizable, at least for ammonoids. Six ammonoid

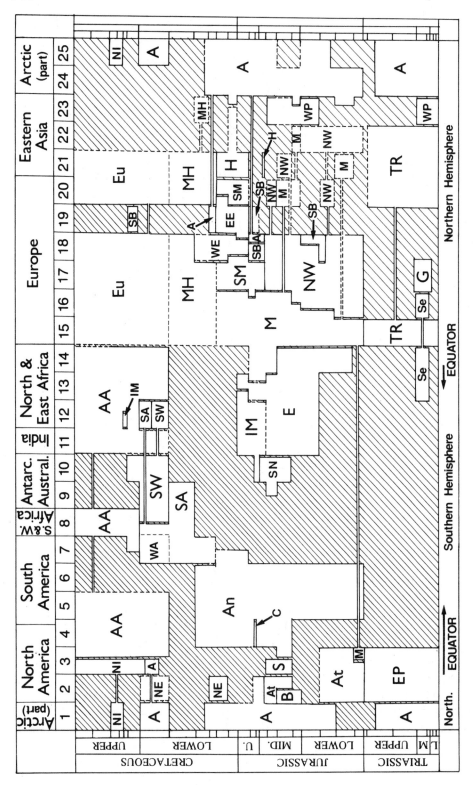

FIGURE 6. The distribution of ammonoid faunal provinces throughout the Mesozoic. Geographic areas are as follows. Arctic (part): 1, Alaska and Arctic Canada; North and Central America: 2, northwest North America and Pacific Coast; 3, southern United States and Western Interior; 4, Central America including Mexico and Caribbean; South America: 5, northwest South America; 6, southwestern South America; 7, southeastern and east-central South America; South and West Africa: 8; Antarctica and Australasia: 9, Antarctic Peninsula, New Zealand; 10, northern and western Australia, New Guinea, Sula Isles; India: 11; North and East Africa and the Middle East: 12, Madagascar, Kenya; 13, Ethiopia, Saudi Arabia, Israel, Turkey, Iran; 14, North Africa; Europe: 15, Balkans, Austria, Italy; 16, southern France, southern Spain; 17, northern Spain, central France, southern Germany; 18, northern France, northern Germany, Britain, Poland; 19, Russian Platform; 20, Caucasus; Southern and Eastern Asia: 21, Pakistan, Himalayas; 22, Southeast Asia; 23, Japan, far eastern former USSR; Arctic (part): 24, Siberia, Arctic of former USSR; 25, Spitzbergen, Greenland. Abbreviations for Provinces, etc.. as follows: A, Arctic Province; EP, East Pacific Province; Se, Sephardic Province; G, Germanic Province; TR, Tethyan Region; WP, West Pacific Province; M, Mediterranean Province; NW, Northwest European Province; SM, Sub-Mediterranean Province; E, Ethiopian Province; IM, Indo-Malgach Province; SN, Sula–New Guinean Province; H, Himalayan Province; An, Andean Province; C, Cuban Province; At, Athabascan Province; S, Shoshonean Province; B, Bering Province; SB, Sub-Boreal Province; WE, West European Province; EE, East European Province; MH, Mediterranean–Himalayan Province; Eu, European Province; SA, South Atlantic Province; WA, West African Province; AA, Afro-Atlantic Province; NI, North American Interior Province; NE, Northeast Pacific ?Province. The absence of an identified province does not necessarily imply the absence of ammonoids. For full details of the sequence of stages within each system, refer to Figs. 1 and 2. Horizontal tie bars link separate parts of the same province that are separated diagrammatically but not, in reality, geographically. Dashed lines indicate uncertainties. Position and relative movement of the equator through the Mesozoic are indicated.

faunal regions and provinces are recognized in the Triassic by Tozer (1981a,b), considerably fewer than are distinguished in the Jurassic. There is, nevertheless, some evidence that certain of these regions, in particular, the Tethyan, could ultimately be subdivisible into additional provinces. Each recognized region or province is discussed further below:

1. Tethyan Realm, Tethyan Subrealm (Austria, Greece, Balkans, Caucasus, Turkey, Himalayas, Pakistan, and Southeast Asia) (T, Fig. 3). The faunas of the Tethys Ocean, with nearly 80% of all known nominal Triassic ammonoid genera (i.e., around 360), were significantly more diverse than those in any other area (undoubtedly linked to the fact that Triassic Tethyan rocks were deposited within around 30° of the paleoequator). Lower diversities in other provinces at higher latitudes are as expected. Cosmopolitan elements that were common in the Tethys, Arctic, and Pacific areas indicate that at certain times throughout the Triassic, but not necessarily all the time, there was communication between all these areas. Nearly half of the ammonoid genera of Tethys were endemic, presumably largely latitudinally limited forms intolerant of the environmental instabilities at higher latitudes (Tozer, 1981a,b), although some of this distinctiveness may actually be a consequence of local centers of endemism (i.e., faunal provinces) within a broader Tethyan Region.

Earliest Triassic faunas were somewhat impoverished, especially when compared to later Permian, largely a result of the end-Permian extinction phase, which left few survivors (Section 2.2). Faunal diversity subsequently increased through the Early Triassic from around nine genera in the Griesbachian to 64 in the Spathian, the latter representing around 75% of all described Early Triassic genera (Kummel, 1973).

Kummel (1973, p. 232) noted that the largest numbers of endemic genera occurred in "embayed regions" along the northern margin of Tethys such as the Mangyslak Peninsula and the Werfen Formation region of Southeast Europe. These and other areas with faunas containing significant numbers of endemics (such as Albania and Chios, Greece; Kummel, 1973) indicate that the generalized "Tethyan Region" (= Subrealm) of Tozer (1981b) is actually divisible into a number of distinct faunal provinces, analogous to those of the Jurassic and Cretaceous. Some of this differentiation may be depth related, as discussed by Wang and Westermann (1993).

Earliest Triassic (Griesbachian) Tethyan Region faunas had characteristic Ophioceratidae (Noritaceae) and Xenodiscaceae. Meekocerataceae often dominated in early Nammalian faunas (Dienerian Substage), but the richer assemblages of the late Nammalian (Smithian Substage) included, in addition, various Sagecerataceae and Ptychitaceae. Many new forms appeared in the Spathian, especially the Dinaritaceae.

Middle Triassic Tethyan faunas were characteristically diverse, with many Ceratitaceae and Danubitaceae in the Anisian and the Trachycerataceae in the Ladinian. The interval saw the initial radiation of the Pinacocerataceae and Arcestaceae, which became important elements of late Middle and Late Triassic faunas. Carnian faunas were still dominated by Trachycerataceae but also included abundant Tropitaceae with various Arcestaceae. The Trachy-ceratataceae persisted into the early to mid-Norian, where they were associated with the Clydonitaceae and Tropitaceae. The Choristocerataceae were typical of the late Norian and Rhaetian, and although some Pinacocerataceae, Arces-taceae, Megaphyllitaceae, and Phyllocerataceae also survived at least as far as the latest Norian, very few persisted into the Rhaetian (Kummel, 1973; Zapfe, 1975, 1978; Guex, 1978; Wiedmann et al., 1979; Tozer, 1981a,b).

2. Tethyan Realm, Pacific Subrealm, East Pacific Province (South Amer-ica, British Columbia, western United States) (EP). Pacific faunas were second in diversity only to those of the Tethys, with just over half of the total number of described Triassic ammonoid genera being recorded. Many of these genera were shared with Tethys, emphasizing that good connections existed (and also making ammonoid-based stratigraphic correlations reasonably good), but a significant number of Pacific forms were endemic; a small number of Arctic genera were also occasionally present (mainly in the Middle Triassic). The general characters are, therefore, those of a midlatitude region with strong Tethyan links but with a decreased diversity because of latitudinal controls. Strong links between East Pacific and Tethyan faunas were particularly characteristic of the Early Triassic, although Griesbachian assemblages were generally absent, as the environments present appear to have been unsuitable for ammonoid life. Occasional Arctic genera occurred in the East Pacific in the early and, particularly, in the mid-Triassic, thereby providing correlative links. Although also generally similar to Tethyan faunas, East Pacific Middle Triassic faunas were less diverse and were dominated by various Ceratitaceae and Trachycerataceae, respectively, in the Anisian and Ladinian.

Abundant Trachycerataceae and Tropitaceae linked Tethyan and East Pacific Carnian faunas, and similarities persisted into the Norian. In the Rhaetian, however, only a few Choristocerataceae survived, but the presence of the first Ammonitina ("Psiloceras," Psilocerataceae) is noteworthy (Silber-ling and Tozer, 1968; Tozer, 1981a,b; Guex, 1982).

3. Tethyan Realm, Pacific Subrealm, West Pacific Province (Japan) (WP). Through much of the Early and Middle Triassic, a West Pacific Province can be distinguished by the presence of a number of endemic species and some genera. Faunal affinities appear to be with the East Pacific Province, but, generally, diversities appear to be considerably lower, presumably in part because of a higher latitude location (Fig. 3). The frequent presence of early Phyllocerataceae (Ussuritidae) appears to be typical (Bando, 1964; Tozer, 1981a,b).

4. Boreal Subrealm, Arctic Province (East Greenland, Canadian Arctic, Spitzbergen, Arctic of former USSR) (A). Arctic faunas were not of high diversity and contained only around a fifth of the total number of Triassic genera. Over half of these were the environmentally tolerant cosmopolitan genera also found in Tethyan and Pacific regions, but nearly a fifth were endemic. The strongest links were with Pacific regions, and a number of noncosmopolitan genera present in Arctic regions were shared with the

Pacific (Tozer, 1981b). The relatively low diversity of Arctic faunas, with a dominance of tolerant cosmopolitan forms, indicates a stressed environment characteristic of high latitude areas. Nevertheless, Triassic Arctic faunas were generally much more diverse than Jurassic and Cretaceous, perhaps suggesting that environmental restriction was less severe early in the Mesozoic.

Early Triassic Arctic faunas were dominated by pandemic Noritaceae, Meekocerataceae, and Ceratitaceae with a small number of endemic genera. Throughout the Middle Triassic, faunal links with the East Pacific were strong, and various endemic genera typified the Arctic Province. Ladinian faunas were relatively impoverished, however, and this pattern persisted into the Late Triassic. For instance, in the Carnian, Trachycerataceae were dominant, but Tropitaceae, which were abundant (and are important for correlation) in the Pacific and Tethys, were very rare. Norian and Rhaetian Arctic faunas were also very rare, possibly in part because of increasingly unfavorable environmental or preservational conditions (Silberling and Tozer, 1968; Tozer, 1981a,b).

5. Tethyan Realm, Germanic Province (Poland, Germany, France) (G) and Sephardic Province (Spain, North Africa, Israel) (S). Triassic ammonoids of the Germanic and Sephardic provinces occur in marine deposits intercalated within continental and evaporitic sequences. Very low-diversity faunas occurred in these areas; only 5% (i.e., around 20) of the total number of Triassic genera have been recorded, and nearly half of these were endemic; correlation with Tethyan areas is, therefore, often difficult.

Environmental conditions, such as abnormal salinities, would have kept diversities low in these areas and presumably led to the origin of specialized genera with a greater tolerance of elevated or depressed salinities. These low-diversity "depauperate" faunas, dominated by *Ceratites* (Ceratitaceae) such as those of the Germanic Muschelkalk facies, provide the popular image of Triassic ammonoid faunas, but, as noted by Tozer (1981b, p. 398), the true variety of Triassic Ammonoidea is much better represented by the rich faunas of the Tethyan Subrealm Halstadt limestones of Austria (as illustrated by Mojsisovics, 1893).

At the end of the Early Triassic, the beginnings of a separate Germanic province were indicated by endemic and very low-diversity faunas with *Beneckeia* (Sagecerataceae). By the mid-Anisian, however, the Germanic and Sephardic provinces were well developed. A few cosmopolitan Ceratitacea and Ptychitaceae in the Anisian of both provinces facilitate the dating of endemic late *Beneckeia* and *Noetlingites* (Danubitaceae).

In Ladinian times, the Germanic Province was virtually entirely isolated from Tethys, and various endemic Ceratitidae developed including *Ceratites* itself. In the Sephardic Province, however, links seem to have been better, and the presence of pandemic *Protrachyceras* (Trachycerataceae) facilitates the dating of endemic *Geranites* (Ceratitidae). Only the Germanic Province appears to have persisted with a few ammonoids into the early Carnian, later

Triassic conditions in the region being completely unsuitable for cephalopods (Parnes, 1975; Tozer, 1981b).

3.2. Jurassic Provincialism

The basic Triassic pattern of ammonoid distributions in Tethyan and Boreal Realms continued into the Jurassic, although Pacific regions were modified by improved communications with other areas. In the Middle Jurassic, however, communication was more restricted, and an East Pacific Subrealm becomes recognizable (Figs. 4 and 6).

Establishment of direct Tethys Ocean–Boreal Sea connections in northern Europe in the late Middle and the Late Jurassic provided more possibilities for interchange between North and South. Faunal distinctions were nevertheless maintained by latitudinal controls, and, frequently, a complex series of essentially latitudinal provincial belts developed (for instance, in the Callovian and Oxfordian) when a high-latitude (Sub-Boreal) to Sub-Mediterranean to Mediterranean (low-latitude) progression developed. Each province had a distinctive fauna, not only in terms of the relative abundance of different taxa but also in the presence or absence of different (and occasionally endemic) genera and subfamilies. A parallel transition from low-latitude diverse faunas to higher-latitude restricted faunas was also occasionally detectable south of the Jurassic equator, toward the Antarctic.

This basic underlying distribution pattern, at different times in the Jurassic, was radically modified by the establishment of further distinct regions or provinces characterized by one or more endemic or very restricted genera or species. Presumably barriers to migration did, in many cases, prevent faunal interchange, but there also seems to be some indication that ecological controls were at work on certain species. The net result is that Jurassic ammonoid provincialism was considerably more complex than that in the Triassic.

The largest and most heterogeneous of the Jurassic ammonoid faunal realms was still the Tethyan, covering the northern and southern shores of the Tethys Ocean, from western Europe eastward to Southeast Asia and Australasia and beyond and around the proto-Pacific Ocean, including the East Pacific seaboard. Tethyan faunas in the Jurassic included both low-latitude high-diversity faunas and Southern Hemisphere high-latitude faunas, which showed a much lower diversity. Strong faunal similarities nevertheless frequently existed between those areas, at least at generic but also often at specific level. In the Middle Jurassic, however, from the late Bajocian to the early Callovian, links between Tethys and the East Pacific were relatively poor, and a separate East Pacific Subrealm can be distinguished (Westermann, 1981; Hillebrandt *et al.*, 1992).

The Jurassic Boreal Realm was the natural "descendant" of the Triassic Arctic Region (or Subrealm) of Tozer (1981a,b). Just as in the Triassic, Jurassic

Boreal faunas were often of low diversity, reflecting environmental stresses at high latitudes, but, in contrast to the Triassic, far fewer truly pandemic taxa occurred in Jurassic Boreal areas. Cited references for each province are simply representative sources, and many other descriptions exist.

1. Tethyan Realm, West Tethyan Subrealm, Mediterranean (or West Tethyan) Province (southern Spain, Italy, Austria, Hungary, Bulgaria, North Africa, etc.) (M, Fig. 4). Many faunas of this province occurred in relatively deep-water regions or at least in areas open to the influence of the Tethys Ocean, and this would account for the characteristically high abundance of Phyllo- and Lytoceratina. Both of these groups, however, occasionally turned up in Southern Hemisphere and Pacific high-latitude deposits, but apparently only in regions open to oceanic influence. Associated Ammonitina were generally almost identical to those of neighboring northwest European or sub-Mediterranean provinces, but there is some evidence of differentiation (and endemism?) in terms of generic or specific composition or relative abundance throughout most of the Jurassic. Mediterranean Province faunas were the most consistently distinctive throughout the Jurassic (Fig. 6; Lange, 1952; Arkell, 1956; Donovan, 1967, 1990; Enay and Geyssant, 1975; Oloriz, 1978; Mangold, 1979; Clari et al., 1984; Cariou et al., 1985; Dommergues et al., 1987; Marchand et al., 1990; Elmi et al., 1995; Meister et al., 1994).

2. Tethyan Realm, West Tethyan Subrealm, Northwest European Province (Britain, Ireland, France, Spain, Germany, Switzerland, etc.) (NW). Essentially an Early to Middle Jurassic phenomenon, the Northwest European Province was typical of Hettangian to early Pliensbachian ("Carixian") times and also Aalenian to Bathonian times, when great faunal uniformity characterized northern Europe. Similar faunas, at generic level at least, occurred in northern North America, but at specific levels there seem to have been some real differences. The diversity of early Northwest European Province faunas was only moderate, but this reflected both the low number of "available" genera early in the development of the Ammonitina and perhaps also the nature of the relatively restricted epicontinental seas of the region. Phyllo- and Lytoceratina were generally rare or absent (except a few species in the Pliensbachian and Toarcian), and Ammonitina alone were characteristic in the Early Jurassic, with Haploceratina and Perisphinctina in the Middle Jurassic.

In the Hettangian, the Psilocerataceae were characteristic but were virtually replaced by the Arietitaceae in the early Sinemurian. In the later Sinemurian, and also in the early Pliensbachian, Eoderocerataceae were locally important (as Eoderoceratidae and Polymorphitidae, respectively). The Hildocerataceae started to become numerically important in Pliensbachian times, and this general pattern persisted until the early Bajocian. Later Bajocian to Bathonian Northwest European faunas were characterized by various Stephanocerataceae, early Perisphinctaceae, and Haplocerataceae (Dean et al., 1961; Mangold et al., 1971; Torrens, 1971; Howarth, 1973; Bloos, 1984; Phelps,

1985; Meister, 1989; Callomon and Chandler, 1990; Page, 1992; Dommergues *et al.*, 1994).

3. Tethyan Realm, West Tethyan Subrealm, Sub-Mediterranean Province (northern Spain, southern and central France, southern Germany, Switzerland, etc.) (SM). Sub-Mediterranean faunas developed a separate character in the Callovian and continued to be distinct throughout much of the remaining Jurassic. Their development parallels that of a separate Sub-Boreal Province within Europe, both provinces being successors to the earlier Northwest European Province. Ammonoid diversities of Sub-Mediterranean faunas were relatively high, but Phylloceratina and Lytoceratina were not important elements. The province essentially occupied the epicontinental seas bordering the northern margins of Tethys. Faunas were generally characterized by a variety of Perisphinctaceae, including Perisphinctidae, Reineckeiidae, Aspidoceratidae, and Ataxioceratidae, some genera of which were virtually restricted to the province; Haplocerataceae were often common associates (Enay *et al.*, 1971b; Gabilly, 1976; Atrops, 1982; Cariou, 1984; Meléndez *et al.*, 1984; Cariou *et al.*, 1985; Meister, 1989: Meléndez, 1989).

4. Tethyan Realm, West Tethyan Subrealm, Ethiopian (or "Arabo-Malgach") Province (North Africa, Ethiopia, Saudi Arabia, Israel, Turkey, Madagascar, western India, etc.) (E). Ethiopian faunas were generally of relatively low diversity when compared to contemporary Sub-Mediterranean assemblages, despite their transpaleoequatorial distribution. This phenomenon was probably related to some environmental control, as parts of the province occupied areas some distance from open ocean, perhaps even at times somewhat restricted or even hypersaline. The presence of some virtually endemic genera and also the location recall the Triassic Sephardic Province.

Ethiopian Province-type faunas were perhaps first recognizable in the Sinemurian, when endemic Arietitaceae developed in Portugal and North Africa (Dommergues *et al.*, 1986; Dommergues and Mouterde, 1987), but it was not until the Toarcian that typical endemic faunas occurred throughout the region with characteristic *Bouleiceras* (Hildoceratidae). The Bajocian to Bathonian faunas of this province were characterized by Haplocerataceae and Stephanocerataceae, and the Tithonian by Perisphinctaceae, usually with endemic genera (Enay *et al.*, 1971a; Howarth, 1973; Ziegler, 1981; Enay and Mangold, 1982, 1984; Zeiss, 1984; Cariou *et al.*, 1985).

5. Tethyan Realm, Indo-Southwest Pacific Subrealm, Indo-Malgach Province (Kenya, Madagascar, northwestern India, Sula Isles) (IM). Faunas of this province were often relatively diverse, recalling Sub-Mediterranean faunas at least to subfamily level. Nevertheless, generic differences often existed, and many Indo-Malgach species were different from Sub-Mediterranean forms. Much of this is not surprising, as the former occurred on the southern margin of Tethys, often at a latitude south of the Jurassic equator equivalent to that of the latter to the north. Latitudinal control of generic distribution is evident, with related species apparently surviving in equivalent positions north and south of the Jurassic equator—some interchange south to north and vice versa

across the equator was apparently feasible (Krishna and Cariou, 1990) but perhaps happened only rarely, the relative isolation allowing taxonomically distinct faunas to develop.

The Indo-Malgach Province was first recognizable in the Callovian with typical and often endemic Perisphinctaceae, Stephanocerataceae, and Haplocerataceae. This general composition persisted into Oxfordian times, but with the extinction of the last Stephanocerataceae (Mayaitinae), only Perisphinctaceae and Haplocerataceae continued into the Kimmerdgian and Tithonian (Krishna, 1984; Krishna and Westermann, 1985; Krishna and Cariou, 1986).

6. Tethyan Realm, Indo-Southwest Pacific Subrealm, Sula–New Guinea Province (Eastern Indonesia, Papua New Guinea) (SN). Although showing some similarities to Indo-Malgach faunas, a number of endemic taxa at generic and specific level occurred in this region during Bajocian, Bathonian, and Callovian times. The province was comparable in latitude to the Sub-Boreal Province of Northwest Europe and perhaps also in faunal diversity, as it apparently had assemblages with relatively few genera when compared with the lower-latitude Indo-Malgach Province. The presence of frequent Phylloceratina was presumably related to the proximity of open ocean, and occasional East Pacific Subrealm taxa suggest that some faunal exchange was possible around the Antarctic. Various Stephanocerataceae, Haplocerataceae, and Persiphinctaceae were present (Westermann, 1981, 1993; Westermann and Callomon, 1988; Hillebrandt *et al.*, 1992; Sukamto and Westermann, 1992).

7. Tethyan Realm, Indo-Southwest Pacific Subrealm, Himalayan Province (Himalayas, Pakistan, Tibet, etc.) (H). At certain times in the Jurassic, particularly during the Callovian and the Tithonian, endemism at both generic and specific levels occurred in this region, indicating a provincial status. The affinities of these faunas were always Tethyan, often with close links to Indo-Malgach assemblages (Enay, 1973; Krishna *et al.*, 1982; Wang, 1988).

8. Tethyan Realm, West Pacific ?Province [Japan, far eastern former USSR, Southeast Asia (part)] (WP). West Pacific faunas occasionally developed a number of endemic species at various times through the Jurassic (for instance, during the Toarcian, Aalenian, and possibly the Oxfordian) but retained a generic base similar to that of contemporaraneous Sub-Mediterranean or Northwest European provinces. The location of these apparently distinct faunas recalls the West Pacific Province of the Triassic (see Section 3.1). Occasionally, restricted circum-Pacific genera also occurred (Westermann, 1981; Sato, 1992).

9. Tethyan Realm (including East Pacific Subrealm), Andean Province (western South America, Chile, Peru, Argentina, Antarctic Peninsula, Mexico) (An). At different times during the history of the Jurassic East Pacific region, sufficiently strong links existed with Tethys that Sub-Mediterranean province faunas showed remarkable similarity, at generic level at least, to those of western South America and western North America. At such times (Hettangian to early Bajocian, mid-Callovian to Tithonian), it is simplest to

consider the latter areas as part of a broad West Tethyan Subrealm. From the late Bajocian and through the Bathonian, however, links were interrupted sufficiently to allow a level of endemism to develop in the East Pacific, which Westermann (1981) considered to be of realm status (although later demoted to a subrealm by Hillebrandt *et al.*, 1992).

The Andean Province was nevertheless more or less recognizable throughout the Jurassic, even when an East Pacific Subrealm was developed. During periods of good Tethyan links, this distinction was primarily at the species level. A few endemic genera did occur, but most others were shared with the Western Tethys, the main connection being via the seaway between the Americas and Africa. During the Middle Jurassic, when these links were interrupted, a number of endemic genera (especially of Stephanocerataceae) developed in a separate East Pacific "Subrealm," and some of these were shared between the Andean Province and North Cordilleran areas. During the Bathonian and Tithonian, northern Andean areas contained a number of European-type Tethyan taxa, which were rare or absent further south. This, combined with some limited endemism, distinguishes the Antofagusta and Neuquén subprovinces of Hillebrandt *et al.* (1992) (Westermann, 1981, 1993; Hillebrandt *et al.*, 1992; Riccardi *et al.*, 1992).

10. Tethyan Realm, Cuban Province (Cuba, ?northern South America) (C). During the Oxfordian, the presence of endemic genera and species (mainly Perisphinctaceae but also including Haplocerataceae) indicates the existence of a separate Cuban Province (Myczynski and Meléndez, 1990; Hillebrandt *et al.*, 1992; Salvador *et al.*, 1992).

11. Tethyan Realm (West Tethyan or East Pacific Subrealm) or Boreal Realm, Athabascan Province (western North America) (At) and Shoshonean Province (Western Interior of the United States) (S). Faunas of the North Cordilleran areas (*sensu* Westermann, 1981) showed both Tethyan and Boreal influences and, from the late Bajocian to the early Callovian, shared characteristic East Pacific Realm genera with the Andean Province. A distinct Athabascan Province was characteristic of much of the Jurassic history of the western United States and Canada, but the Shoshonean Province is primarily known from the late Bajocian to the Oxfordian. The latter was characteristic of the restricted sea of the Western Interior of the United States.

Earliest Jurassic (Hettangian) Athabascan faunas included frequent Phyllo- and Lytoceratina, indicating Mediterranean Province affinities, associated with Psilocerataceae and early Arietitaceae. Sinemurian to Toarcian faunas were broadly similar to those of Northwest Europe, but throughout the Early Jurassic and into the early Middle Jurassic, various endemic species and genera developed. With the development of the East Pacific Subrealm during the Bajocian, Athabascan faunas developed with both endemic and restricted Stephanocerataceae associated with some Hildocerataceae. An increase in Boreal stephanoceratacean influence (Cardioceratidae and Kosmoceratidae), particularly in northern areas, gradually reduced the distinctiveness of the province, and by the Callovian, the Athabascan faunas (and the East Pacific Subrealm) were barely distinguishable.

The Shoshonean Province was characterized by very low diversity, often monospecific faunas with endemic species of more widespread East Pacific and Boreal genera (Westermann, 1981; Taylor *et al.*, 1984; Hillebrandt *et al.*, 1992; Poulton *et al.*, 1992; Smith *et al.*, 1994).

12. Boreal Realm, Arctic (or "Inner Boreal") Province (Scotland, East Greenland, Spitzbergen, Siberia, Canadian Arctic, Alaska) (A) and Bering Province (North Pacific margin; Japan, northwestern North America, northeastern Siberia) (B). Faunas of the Arctic were characterized by some of the lowest diversity of all the Jurassic; some were virtually monospecific, and others consisted of only a couple of genera or species. These faunas were distinct from the earliest Jurassic, although it was not until the Aalenian that endemic genera were significant elements. The Bering Province occupied a marginal area to an inner Arctic (or "Boreal") Province in the Aalenian and early Bajocian but did not extend far into the Boreal Sea. Bering Province faunas were characterized by a small number of endemic genera associated with more typical Arctic Province species.

During the Hettangian and Sinemurian, the Arctic Province was characterized by very low-diversity faunas with some endemic species of otherwise more widespread psiloceratacean and arietitacean genera. From the Pliensbachian onwards, however, endemic or restricted Boreal Realm genera developed, including both Eoderocerataceae and some Hildocerataceae. Late Toarcian to early Bajocian Arctic faunas were frequently virtually monospecific, but from the late Bajocian to the Oxfordian, characteristically Boreal Stephanocerataceae developed (Cardioceratidae and Kosmoceratidae). Latest Jurassic (Kimmeridgian to Tithonian) faunas were composed of Boreal Perisphinctaceae including Aulacostephanidae and Dorsoplanitidae (Howarth, 1973; Sykes and Callomon, 1979; Westermann, 1981; Callomon and Birkelund, 1982; Jeletzky, 1984; Krymholts *et al.*, 1988; Hillebrandt *et al.*, 1992; Poulton *et al.*, 1992; Sey *et al.*, 1992; Smith *et al.*, 1994).

13. Boreal Realm, Sub-Boreal Province (England, northern France, northern Germany, Poland, Russian Platform) (SB). During the early Callovian, opening up of migration pathways south of East Greenland and into Northwest Europe allowed a mingling of previously separated Boreal and Tethyan Realm faunas. The result was an establishment of a province, particularly in the Callovian and Oxfordian, distinguished by the dominance of previously boreally restricted subfamilies and genera, but in association with a number of sub-Mediterranean taxa. Occasionally, endemic species and, more rarely, genera developed. Boreal genera also occurred in the northwestern United States in the Callovian and Oxfordian, and a similar mixing with certain Tethyan-type genera sometimes created a Sub-Boreal style assemblage in that area. However, the apparent absence of many typical European species and the differences in faunal composition suggest that a provincial-level distinctiveness between Europe and North America is probable. An earlier phase of

mixing between Boreal and northwest European genera occurred in the late Pliensbachian and early Toarcian times in northern Europe and created a similar style of provincialism (Dean *et al.*, 1961; Birkelund *et al.*, 1983; Callomon *et al.*, 1989; Page, 1991).

14. Boreal Realm, West European Province (England, northern France) (WE). During the latest Jurassic (Tithonian), reduced exchange between the western and eastern parts of the former Sub-Boreal Province led to the development of ammonite faunas distinguishable largely in terms of the dominance of particular perisphinctacean genera, but apparently also including some more or less endemic species. West European late Tithonian faunas were dominantly of Dorsoplanitidae and characterized a "Portlandian" (local) Stage (Cope, 1967, 1978; Wimbledon and Cope, 1977; Wimbledon, 1984).

15. Sub-Boreal Region, East European Province (Russian Platform, northern Poland, Pechoraland, ?East Greenland) (EE). The East European Province developed in parallel with the West European Province during Tithonian times. Faunas distinguished a "Volgian" (local) Stage; Virgatitidae and Dorsoplanitidae (both Perisphinctaceae) were characteristic, the latter facilitating correlations with the West European Province (Casey, 1973; Wimbledon, 1984; Krymholts *et al.*, 1988).

3.3. Cretaceous Provincialism

During the Early Cretaceous, ammonoid distributions were essentially those of the latest Jurassic (Tithonian/Portlandian/Volgian complex), with a relatively high number of faunally distinct areas or provinces (Figs. 4 and 6). Later in the Cretaceous, however, as continental separation increased considerably, more migrationary pathways were opened up, and endemism was significantly reduced; the hallmark of Late Cretaceous faunas was the virtually pandemic distributions of their genera and even ocasionally species. The main control on early to early Late Cretaceous faunal distributions seems to have been latitude, although even a Boreal Realm for ammonoids ceases to be recognizable after the Cenomanian (although it is recognizable for other groups, such as belemnites; Stevens, 1973). More southern faunas are grouped within a very broad Tethyan Realm (Kauffman, 1979). Virtually the only persistently recognizable province in the Late Cretaceous was that of the Western Interior of the United States and Canada (Figs. 5 and 6). The characteristically high levels of endemism at both specific and generic level in this region are well known (Kennedy and Cobban, 1976; Kauffman, 1984, etc.). Phyllo- and Lytoceratina continued to be characteristic of deeper water, or at least of deep-water-influenced areas, but were often joined in the Cretaceous by the ammonitine Desmocerataceae (Section 2.8).

1. Tethyan Realm, Mediterranean Province (southern France, southern Spain) (equivalent to M, Fig. 4). A Mediterranean Province *sensu stricto* with significant numbers of Phyllo- and Lytoceratina was recognizable during the

earliest Cretaceous (Berriasian to ?Valanginian), but, as noted by Rawson (1981, p. 524), more information is needed on the distribution of these two groups as elements of ammonoid faunas before a Mediterranean/Sub-Mediterranean provincial separation can be established in the later Cretaceous. Rawson consequently adopted a Mediterranean–Himalayan Province to include the previously separate Mediterranean, Sub-Mediterranean, and Himalayan provinces.

Besides Phyllo- and Lytoceratina, Mediterranean Province faunas included typical Neocomitidae and Olcostephanidae (late Perisphinctaceae) in the Berriasian and Valanginian, apparently with some endemism at generic or specific level. The Ancyloceratina (Ancylocerataceae) became significant faunal elements from the Valanginian onward (Hegaret, 1973; Busnardo and Theiuloy, 1979; Rawson, 1981).

2. Tethyan Realm, Sub-Mediterranean Province (southern mid-Europe, Crimea, Caucasus) (SM, Fig. 4). A Mediterranean/Sub-Mediterranean separation is presently demonstrable only in the Berriasian and Valanginian; the usual distinction of the latter province was the relative infrequency of Phyllo- and Lytoceratina, and, in the early Cretaceous, the occasional presence of Boreal Realm genera was noteworthy (for instance, the perisphinctacean Craspeditidae; Thieuloy, 1973). The general composition was otherwise similar to that of the Mediterranean Province (Rawson, 1981; Nikolov, 1982; Zeiss, 1986).

Rawson (1981, p. 509) noted that the Mediterranean Region (Mediterranean and Sub-Mediterranean Provinces) had the richest known Berriasian faunas but that the diversity decreased noticeably both northward and southward. For instance, of 65 nominal genera and subgenera of Berriasian age, 35% occurred in southeast France (Mediterranean Province), 22% on the Russian platform (East European Province), 18% in northern Siberia, and only 8% in the Canadian Arctic (both Arctic Province).

3. Tethyan Realm, Himalayan (or Indo-Pacific) Province (Himalayas, Southeast Asia, ?Japan, ?Australasia) (H, Fig. 4). A small number of endemic genera or subgenera (mainly Neocomitidae) characterized the province during the Berriasian and Valanginian, but the majority of Himalayan ammonites were of "Mediterranean" type (Fatmi, 1972; Rawson, 1981, p. 524; Krishna, 1983).

4. Tethyan Realm, Mediterranean-Himalayan Province (southern Europe to the Himalayas) (MH, in part, Fig. 6). This province was a "descendant" of the Mediterranean-Sub-Mediterranean–Himalayan complex and was the main southern European province of the mid-early Cretaceous (Hauterivian–Barremian). Faunas within the province were diverse in southern areas but tended to decrease in numbers of genera and species northward. Various ammonitine and ancyloceratine genera were characteristic, with some associated Phyllo- and Lytoceratina (although where and in what levels of abundance is not clear). Various Perisphinctaceae persisted into the Hauterivian and Barremian (including late Neocomitidae and Olcostephanidae), but early

Desmocerataceae also became important, as did various Ancylocerataceae (Moullade, 1966; Busnardo, 1965; Rawson, 1981; Bulot and Thieuloy, 1993).

5. Tethyan Realm, European Province (new name) (Europe to the Caucasus) (Eu). The removal of a distinct Boreal Realm by the Aptian and the separation of North Africa from Europe in the Albian led to the development of faunas in the latter area differing from those further south in their relative generic and specific abundances.

As such, the European Province was geographically broadly equivalent to the Northwest European Province of the Early to Middle Jurassic and occupied the shelf seas on the northern edge of the Western Tethys. In the Aptian, cosmopolitan Ancyloceratina, including heteromorphic Ancylocerataceae and the "recoiled" Deshayesitaceae, were typical. By the Albian, late Perisphinctina were once more dominant and included Hoplitaceae and early Acanthocerataceae. The latter were typical from the Cenomanian onwards; Scaphitaceae and Turrilitaceae were usually less frequent, and Desmocerataceae were generally rare (Casey, 1960–1980; Kemper, 1973b; Matsumoto, 1973; Owen, 1973; 1984; Wright, 1979; Wright and Kennedy, 1981, 1984–; Kennedy, 1984, 1985, 1986).

6. Tethyan Realm, South Atlantic Province (new name) (Argentina, Antarctic Peninsula, Australia, South Africa) (SA). The province first became recognizable in the Hauterivian when endemic genera, including Ancyloceratina, started to appear in Argentina, and by the Aptian, the late haploceratacean *Sanmartinoceras* occurred throughout the area; in the Albian, the possible turrilitacean *Labeceras* was characteristic. The province is the closest, in ammonoid terms, to a Cretaceous "southern boreal" or "Antarctic" faunal area, and its development was related to the opening of the South Atlantic and the reduction of marine communications with other areas (Kennedy and Klinger, 1975; Thompson, 1982; Riccardi, 1988; Howlet, 1989; Klinger, 1989).

7. Tethyan Realm, West African (or Angolan) Province (new name) (Angola, Nigeria, eastern Brazil, Madagascar) (WA). The development of endemic Brancoceratidae (Acanthocerataceae) in the Albian distinguishes this province. As with the South Atlantic Province, this was clearly related to the early stages of the opening of the Atlantic—subsequent establishment of marine connections with the North Atlantic and Mediterranean areas removed the distinctiveness of both provinces by the Cenomanian (Howarth, 1965; Reyment and Tait, 1972; Kennedy and Cobban, 1976).

8. Tethyan Realm, Afro-Atlantic Province (new name) (Middle East, Algeria, Niger, Nigeria, Venezuela, Peru, Brazil) (AA). During the Albian, the separation of North Africa and Europe led to the development of faunas differing in relative generic and specific abundances, presumably as a result of latitudinal controls. This general separation appears to have existed at least until the Santonian, but whether the distinction was maintained at all times is unclear, and both northern and southern Tethyan areas had many faunal elements in common (cf. Matsumoto, 1973, p. 124).

The Afro-Atlantic Province occupied not only the southern shelf seas of Tethys but also parts of northern South America, where similar faunas occurred. As the opening of the Atlantic destroyed the distinction of the South Atlantic Province, the latter subsequently developed faunas of a broader Afro-Atlantic Province. Albian Afro-Atlantic faunas were characterized by various virtually endemic Engonoceratidae (Hoplitaceae). By the Cenomanian and into the Turonian, however, cosmopolitan Acanthocerataceae occurred, but associated with various other, more restricted endemic acanthoceratacean genera and species. A similar pattern persisted to the end of the Cretaceous, but with endemism probably mainly at specific level (Freund and Raaks, 1969; Matsumoto, 1973; Owen, 1973; Kennedy and Klinger, 1975; Kennedy and Cobban, 1976; Rawson, 1981; Renz, 1982; Zaborski, 1983, 1985; Courville *et al.*, 1991; Kassab, 1991).

9. Tethyan Realm, Andean Province (Peru, central Argentina, Patagonia, Antarctica) (An). From the Berriasian to the Hauterivian, at least, the development of endemic berriasellid (Neocomitidae) and some olcostephanid genera and species separated this province from the Mediterranean–Himalayan faunas of the central Americas and southern Europe. Notably, Phyllo- and Lytoceratina were apparently absent. As in the Jurassic, the Andean Province of the Early Cretaceous included a northern Antofagusta and a southern Neuquén subprovince (Rawson, 1981, p. 524; Riccardi, 1988).

10. Boreal Realm or Tethyan Realm, North American Interior Province (Western Interior of the United States and Canada) (NI). Probably the best known of all Cretaceous ammonoid provinces, the North American Interior Province developed endemic ammonitine and ancyloceratine faunas at generic and specific levels in an inland sea with restricted connections to both Tethyan and Boreal areas (Kauffman, 1984). The province became distinguishable in the late Albian, when endemic and essentially Boreal Gastroplitinae (Hoplitaceae) appeared, and remained distinct until the disappearance of ammonoids in the latest Maastrichtian. Even within the province, there is some evidence of latitudinal control, with increased numbers of Tethyan genera such as Acanthocerataceae in more southerly areas.

Cenomanian faunas were dominated by cosmopolitan Acanthocerataceae, but some endemic genera and species of this and other groups developed. By the Turonian, distinctive endemic faunas of Scaphitidae (Scaphitaceae) and Baculitidae (Turrilitaceae) became characteristic and distinguish the province throughout the remaining Late Cretaceous. Associated were typical cosmopolitan Late Cretaceous acanthoceratacean and hoplitacean genera (Jeletzky, 1971; Matsumoto, 1973; Kennedy and Cobban, 1976; Kauffman, 1984; Kauffman *et al.*, 1978; Caldwell, 1975).

11. Tethyan Realm, Southwest Pacific ?Province (new name) (northern Australia, New Guinea, New Zealand) (SW). Occasionally in the Cretaceous, for instance, in the Cenomanian, the occurrence of apparently endemic genera or species (including certain Hoplitaceae or Acanthocerataceae) suggests some provincial separation of this region. Associated were typical subpan-

demic Desmocerataceae, Acanthocerataceae, and Ancyloceratina (Matsumoto, 1973).

12. Tethyan Realm, Indo-Malgach ?Province (Madagascar, southern India, South East Africa) (IM). There is some evidence of the development of endemic vascoceratids (Acanthocerataceae) in this region, in the Turonian at least, but faunas generally had a Tethyan Realm or, more specifically, an Afro-Atlantic Province affinity (Collignon, 1965; Matsumoto, 1973).

13. Boreal Realm, Arctic Province (= Siberia–North American Region of Rawson, 1981; northern North America, Greenland, Siberia) (A). Cretaceous Arctic faunas were of low diversity and tended to be characterized by only a very small number of taxa, some of which also occurred in neighboring sub-Boreal areas. This province was most distinctive from Berriasian to earliest Barremian times and then again in the later Albian. Subsequent records of ammonites in the Arctic appear to be only stray occurrences of forms that occurred abundantly elsewhere.

Berriasian to early Barremian Boreal faunas were characterized by various largely endemic Craspeditidae (Perisphinctaceae), with some species restricted to the Arctic Province. Arctic faunas were not distinctive in the Aptian, but the development of endemic *Gastroplites* in the Albian defined the province. In the Cenomanian, *Schloenbachia* (Hoplitaceae), although present within European Province faunas, persisted further northward onto the Russian Platform, where other typical groups were absent or rare, but a clear provincial separation is lacking (Jeletzky, 1971, 1973, 1984; Saks, 1972; Balan, 1973; Owen, 1973; Rawson, 1981).

14. Boreal Realm, Sub-Boreal Region, West European Province (eastern England, northern Germany, Poland) (WE). Berriasian western European Boreal ammonite faunas are known only from eastern England and are of Arctic type. Transgression in the Valanginian, however, brought Arctic faunas across northern Europe, and the development of endemic ammonitine and ancyloceratine genera and species in western Europe separated a West European Province from an East European Province (analogous to the development of these two provinces during the earlier Tithonian). The former is recognizable at least until the earliest Barremian, and faunas included various occurrences of more typically Mediterranean–Himalayan genera.

Berriasian faunas of the West European Province contained only Craspeditidae, but, by the Valanginian, various Tethyan immigrants such as Neocomitidae and Olcostephanidae were frequent, and Ancyloceratidae became important in the Hauterivian (and included endemic forms) (Saks, 1972; Casey, 1973; Kemper, 1973a; Rawson, 1973, 1975, 1981; Kemper et al., 1981).

15. Boreal Realm, Sub-Boreal Region, East European Province (Russian Platform) (EE). Faunas of this area showed a dominance of Arctic–type ammonites (Craspeditidae) but also included endemic genera and species mixed with a few taxa of Mediterranean–Himalayan affinities (such as Neocomitidae). The latter were much less common than in the West European Province but increased in numbers southward toward the Caucasus and

Crimea. The East European Province was recognizable primarily during the Berriasian and the Valanginian, but the subsequent dominance of Arctic forms in the Hauterivian meant that it then could no longer be distinguished (Sazunova, 1961; Saks, 1972; Saks and Shulgina, 1973; Rawson, 1981).

16. Tethyan Realm, West Pacific ?Province (Japan, ?far east of former USSR) (WP). The presence of a few endemic taxa including the turrilitacean *Nipponites* in the Coniacian and Santonian suggests that some faunal differentiation, perhaps at a provincial level, had occurred. Associated are typical circum-Pacific taxa including Desmocerataceae and Tetragonataceae (Matsumoto, 1973; Matsumoto *et al.*, 1978; Pergament, 1977).

17. Tethyan or Boreal Realm, Northeast Pacific ?Province (new name) (Pacific coast of North America) (NE). Some evidence of endemism, at specific and occasionally generic level, suggests that a Northeast Pacific Province can be recognized occasionally at certain levels in the Cretaceous. Examples include certain endemic Craspeditidae (Perisphinctina) in the Valanginian and Hauterivian and a few restricted Desmocerataceae and Hoplitaceae in the Albian (Casey, 1954; Imlay, 1960; Jeletzky, 1971, 1973; Rawson, 1981). In the latest Cretaceous, faunas of Turrilitaceae, Desmocerataceae and late Hoplitaceae (Placenticeratidae) apparently showed significant differences from those of the Western Interior, at least in terms of faunal composition, that a provincial-level separation is possible (Popenoe *et al.*, 1960; Ward and Haggart, 1981).

3.4. Evolution of Ammonoid Faunal Provinces through the Mesozoic

As with any form of faunal distribution observed in a predominantly nektic or nektobenthic group, that of ammonoids is controlled simply by latitude and barriers to migrations. Throughout the Mesozoic, plate tectonics continuously modified the second control and hence created the array of provinces now observed. Figure 6 attempts to demonstrate the origins, distribution, evolution, and even extinction and rebirth of such provinces throughout the Mesozoic.

It is only really in the latest Cretaceous (from the Coniacian onwards), when ammonoid faunas were dominated by virtually pandemic or at least widely distributed genera, that distinct provincialism appears to have lingered on in only one restricted area, namely, the North American Interior. Whether this was entirely the result of a virtual removal of all other barriers to movement or whether it represents the demise of all but the most environmentally tolerant ammonoid groups is unclear. The latter, at least, persisted in moderate diversity to very near to the end of the Cretaceous, and the last ammonoid to disappear may even have been a representative of the most long-ranging and, therefore, presumably most environmentally tolerant of all Mesozoic ammonoid groups, a phylloceratid (Section 2.7; Wiedmann, 1988).

References

Arkell, W. J., 1956, *Jurassic Geology of the World*, Oliver and Boyd, Edinburgh.

Arkell, W. J., Kummel, B., and Wright, C. W., 1957, Mesozoic Ammonoidea, in: *Treatise on Invertebrate Palaeontology*, Part L, *Mollusca 4* (R. C. Moore, ed.), Geological Society of America and University of Kansas Press, Lawrence, KS, L80–L471.

Atrops, F., 1982, La sous famille des Ataxioceratinae (Ammonitina) dans le Kimméridgien inférieur du Sud-Est de la France. Systématique, évolution, chronostratigraphie des genres *Orthosphinctes* et *Ataxioceras*, *Doc. Lab. Géol. Fac. Sci. Lyon* **83**.

Balan, T. M., 1973, Schloenbachiidae from the Cenomanian of the southwestern Russian platform, in: *Palaeontology and Stratigraphy of the Mesozoic of the Southern Outskirts of the Russian Platform* (L. P. Kopaevich, ed.), Ministry of the People's Education of the Moldavian SSR, Kishinev, pp. 67–79, (in Russian).

Bando, Y., 1964, The Triassic stratigraphy and ammonite fauna of Japan, *Sci. Rep. Tokyo Univ.* (*Geol.*) **36**:1–137.

Bessenova, N. V., and Michailova, I. A., 1983, The evolution of the Jurassic–Cretaceous ammonoids, *Dokl. Akad. Nauk. SSSR* **269**:733–737 (in Russian).

Bessenova, N. V., and Michailova, I. A., 1991, Higher taxa of Jurassic and Cretaceous Ammonitida, *Paleontol. J.* **25**:1–19.

Birkelund, T., Callomon, J. H., Clausen, C. K., Nøhr Hansen, N., and Salinas, I., 1983, The Lower Kimmeridgian Clay at Westbury, Wiltshire, England, *Proc. Geol. Assoc.* **94**:289–309.

Bloos, G., 1984, On Lower Lias ammonite stratigraphy—present state and possibilities of revision, in: *International Symposium on Jurassic Stratigraphy. I.* (O. Michelsen and A. Zeiss, eds.), Geological Survey of Denmark, Copenhagen, pp. 146–157.

Bulot, L. G., and Thieuloy, J. P., 1993, Le Cadre stratigraphique du Valanginien supérieur et de l'Hauterivian du sud-est de la France: Définition des biochronozones et caractérisation de nouveaux biohorizons, *Geol. Alp.* **68**:13–56.

Busnardo, R., 1965, Rapport sur l'étage Barrémien, *Mem. Bur. Res. Géol. Min. Fr.* **34**:161–169.

Busnardo, R., and Theiuloy, J. P., 1979, Hypostratotype mésogéen de l'étage Valanginien (sud-est de la France), Busnardo, R., Theiuloy, J. P., and Moullade, M., (eds.) *Les Stratotypes Français* **6**:58–68, 127–134.

Caldwell, W. G. E. (ed.), 1975, The Cretaceous System in the Western Interior of North America, *Geol. Assoc. Can. Spec. Pap.* **13**:31–54.

Callomon, J. H., 1981, Dimorphism in ammonoids, in: *The Ammonoidea*, Systematics Association Spec. Vol. 18 (M. R. House and J. R. Senior, eds.), Academic Press, London, pp. 257–273.

Callomon, J. H., 1985, The evolution of the Jurassic ammonite family Cardioceratidae, *Spec. Pap. Palaeontol.* **33**:49–90.

Callomon, J. H., and Birkelund, T., 1982, The ammonite zones of the Boreal Volgian (Upper Jurassic) in East Greenland, in: *Arctic Geology and Geophysics* (A. F. Embry and H. R. Balkwill, eds.), *Mem. Can. Soc. Pet. Geol.* **8**:349–369.

Callomon, J. H., and Chandler, R. B., 1990, A review of the ammonite horizons of the Aalenian and Lower Bajocian stages in the Middle Jurassic of southern England, in: *Proceedings of Meeting on Bajocian Stratigraphy* (G. Cresta and G. Pavia, eds.), *Mem. Descr. Cart. Geol. Ital.* **40**:85–111.

Callomon, J. H., Dietl, G., and Page, K. N., 1989, On the ammonite faunal horizons and standard zonation of the Lower Callovian stage in Europe, in: *2nd International Symposium on Jurassic Stratigraphy, Lisbon, 1988* (B. Rocha, ed.), Universidade Nova de Lisboa, pp. 359–376.

Cariou, E., 1984, Biostratigraphic subdivision of the Callovian Stage in the Subtethyan Province of ammonites: Correlations with the Sub-Boreal Zonal Scheme, in: *3rd International Symposium on Jurassic Stratigraphy*, (O. Michelsen and A. Zeiss, eds.), Universidade Nova de Lisboa, pp. 316–326.

Cariou, E., Contini, D., Dommergues, J.-L., Enay, R., Geyssant, J. R., Mangold, C., and Thierry, J., 1985, Biogéographie des ammonites et évolution structurale de la Téthys au cours du Jurassique, *Bull. Soc. Géol. Fr.* **8**:679–697.

Casey, R., 1954, New genera and subgenera of Lower Cretaceous ammonites, *J. Wash. Acad. Sci.* **44**:106–115.

Casey, R., 1960–1980, A monograph of the Ammonoidea of the Lower Greensand, *Palaeontogr. Soc. Monogr. (Lond.).*

Casey, R., 1973, The ammonite succession at the Jurassic–Cretaceous boundary in eastern England, in: *The Boreal Lower Cretaceous* (R. Casey and P. F. Rawson, eds.), *Geol. J. Spec. Issue* **5**:193–266.

Clari, P. A., Marini, P., Pastorini, M., and Pavia, G., 1984, Il Rosso Ammonitico Inferiore (Baiociana–Calloviana) nei Monti Lessini settentrionali (Verona), *Riv. It. Paleont. Strat.* **90**:15–86.

Collignon, M., 1965, *Atlas des Fossiles Caractéristiques de Madagascar. 12. Turonien*, Service Géologique, Tananarive.

Cope, J. C. W., 1967, The palaeontology and stratigraphy of the lower part of the Upper Kimmeridge Clay of Dorset, *Bull. Br. Mus. (Nat. Hist.) Geol.* **15**:1–17.

Cope, J. C. W., 1978, The ammonite faunas and stratigraphy of the upper part of the Upper Kimmeridge Clay of Dorset, *Palaeontology* **21**:469–533.

Cope, J. C. W., 1993, The Bolonian Stage. An old answer to an old problem, *Newsl. Stratigr.* **28**:151–156.

Courville, P., Meister, C., Lang, J., Mathey, B., and Thierry, J., 1991, Les corrélations en Téthys occidentale et l'hypothèse de la liason Téthys-Atlantique Sud: Intérêt des faunes d'ammonites du Cénomanien supérieur-Turonien moyen basal du Niger et du Nigeria (Afrique de l'Ouest), *C.R. Acad. Sci. (Paris)* **313**:1039–1042.

Dean, W. D., Donovan, D. T., and Howarth, M. K., 1961, The Liassic ammonite zones and subzones of the Northwest European Province, *Bull. Br. Mus. (Nat. Hist.) Geol.* **4**:435–505.

Dietl, G., 1978, Die heteromorphen Ammoniten des Dogger, *Stuttg. Beitr. Naturkd. B* **33**.

Dommergues, J.-L., and Mouterde, R., 1987, The endemic trends of Liassic ammonite faunas of Portugal as the result of the opening up of a narrow epicontinental sea, *Palaeogeogr. Palaeoclimatol. Palaeoecol.* **58**:129–137.

Dommergues, J.-L., Faure, P., and Deybernes, B., 1986, Le Lotharingien inférieur du Djebel Ouest (Tunisie): Description d'ammonites nouvelles (Asteroceratinae, Arieticeratinae), *C.R. Acad. Sci. (Paris)* **302**:1111–1116.

Dommergues, J.-L., Marchand, D., and Thierry, J., 1987, Biogéographie des ammonites jurassiques et reconstitution palinspastique de la Tethys, *Geodin. Acta (Paris)* **1**:273–281.

Dommergues, J.-L., Page, K. N., and Meister, C., 1994, A detailed correlation of Upper Sinemurian (Lower Jurassic) ammonite biohorizons between Burgundy (France) and Britain, *Newsl. Stratigr.* **30**:61–73.

Donovan, D. T., 1967, The geographical distribution of Lower Jurassic ammonites in Europe and adjacent areas, in: *Aspects of Tethyan Biogeography*, Systematics Association No. 7 (C. G. Adams and D. V. Ager, eds.), Academic Press, London, pp. 111–134.

Donovan, D. T., 1990, Sinemurian and Pliensbachian ammonite faunas of central Italy, in: *Atti del Secondo Convegno Internazionale,. Fossili, Evoluzione, Ambiente, Pergola, 1987* (G. Pallini, F. Cecca, S. Cresta, and M. Santantonio, eds.), Tecnostampa, Roma, pp. 253–262.

Donovan, D. T., Callomon, J. H., and Howarth, M. K., 1981, Classification of the Jurassic Ammonitina, in: *The Ammonoidea*, Systematics Association Spec. Vol. 18 (M. R. House and J. R. Senior eds.), Academic Press, London, pp. 101–155.

Elmi, S., Gabilly, J., Mouterde, R., Rocha, R., and Pulleau, L., 1995, L'Étage Toarcien de l'Europe et de la Téthys: Divisions et corrélations, *Geobios* **17**:149–160.

Enay, R., 1973, Upper Jurassic (Tithonian) ammonites, in: *Atlas of Palaebiogeography* (A. Hallam, ed.), Elsevier, Amsterdam, pp. 292–397.

Enay, R., and Geyssant, J. R., 1975, Faunes tithoniques de chaines bétiques (Espagne méridionale), in: *Colloque sur la Limite Jurassique-*Crétacé, *Lyon-Neuchâtel, 1973, Mém. Bur. Rech. Géol. Min. Fr.* **86**:34–55.

Enay, R., and Mangold, C., 1982, Dynamique, biogéographique et l'évolution des faunes d'ammonites du Jurassique, *Bull. Soc. Géol. Fr.* **24**:1025–1046.

Enay, R., and Mangold, C., 1984, The ammonite succession from Toarcian to Kimmeridgian in Saudi Arabia: Correlation with European faunas, in: *International Symposium on Jurassic Stratigraphy* (O. Michelsen and A. Zeiss, eds.), Geological Survey of Denmark, Copenhagen, pp. 642–651.

Enay, R., Martin, C., Monod, O., and Thieuloy, J. P., 1971a, Jurassique supérieur à ammonites (Kimmeridgian–Tithonique) dans l'autochtone du Tauras de Beysehir (Turquie méridionale), *Ann. Inst. Geol. Publ. Hung.* **54**:397–422.

Enay, R., Tintant, H., and Cariou, E., 1971b, Les faunes Oxfordiennes d'Europe meridionale, essai de zonation, in: *Colloque sur la Jurassique, Luxembourg, 1967, Mém. Bur. Rech.* Géol. *Min. Fr.* **25**:635–664.

Fatmi, A. N., 1972, Stratigraphy of the Jurassic and Lower Cretaceous rocks and Jurassic ammonites from northern areas of West Pakistan, *Bull. Br.. Mus (Nat. Hist.) Geol.* **20**:299–380.

Freund, R., and Raab, M., 1969, Lower Turonian ammonites from Israel, *Spec. Pap. Palaeontol.* **4**.

Gabilly, J., 1976, Le Toarcien à Thouars et dans le Centre-Ouest de la France: Biostratigraphique et evolution de la faune (Harpoceratinae, Hildoceratinae), *Stratotypes Français* **3**.

Glenister, B. F., and Furnish, W. M., 1981, Permian Ammonoids, in: *The Ammonoidea*, Systematics Association Spec. Vol. 18 (M. R. House and J. R. Senior, eds.), Academic Press, London, pp. 49–64.

Guex, J., 1978, Le Trias inférieur des Salt Ranges (Pakistan), *Eclogae Geol. Helv.* **71**:105–141.

Guex, J., 1982, Relations entre le genre *Psiloceras* et les Phylloceratida au voisinage de la limite Trias–Jurasique, *Bull. Geol. Lausanne* **260**:47–51.

Guex, J., 1987, Sur la phylogenèse des ammonites du Lias inférieur, *Bull. Geol. Lausanne* **292**:455–469.

Harland, W. B., Armstrong, R. L., Cox, A. V., Craig, L. E., Smith, A. G., and Smith, D. G., 1989, *A Geologic Time Scale 1989*, Cambridge University Press, Cambridge.

Hegaret, G. le, 1973, Le Berriasien du sud-est de la France, *Doc. Lab. Géol. Fac. Sci. Lyon* **43**.

Hewitt, R., 1993, Mollusca, Cephalopoda (Pre-Jurassic Ammonoidea), in: *Fossil Record 2* (M. J. Benton, ed.), Chapman and Hall, London, pp. 189–212.

Hillebrandt, A., Westermann, G. E. G., Callomon, J. H., and Detterman, R. L., 1992, Biogeography. 24. Ammonites of the circum-Pacific, in: *The Jurassic of the Circum-Pacific* (G. E. G. Westermann, ed.), Cambridge University Press, Cambridge, pp. 342–359.

Howarth, M. K., 1965, Cretaceous ammonites and nautiloids from Angola, *Bull. Br.. Mus. (Nat. Hist.) Geol.* **10**:237–412.

Howarth, M. K., 1973, Lower Jurassic (Pliensbachian and Toarcian) ammonites, in: *Atlas of Palaeobiogeography* (A. Hallam, ed.), Elsevier, Amsterdam, pp. 275–282.

Howlet, P. J., 1989, Late Jurassic–Early Cretaceous cephalopods of eastern Alexander Island, Antarctica, *Spec. Pap. Palaeontol.* **41**.

Imlay, R. W., 1960, Ammonites of Early Cretaceous age (Valanginian and Hauterivian) from the Pacific coast states, *Prof. Pap. U.S. Geol. Surv.* **647**:1–59.

Jeletzky, J. A., 1971, Marine Cretaceous biotic provinces of western and arctic Canada: Illustrated by a detailed study of ammonites, *Geol. Surv. Can. Pap.* **70**(22):1–92.

Jeletzky, J. A., 1973, Biochronology of the marine boreal latest Jurassic, Berriasian and Valanginian in Canada, in: *The Boreal Lower Cretaceous* (R. Casey and P. F. Rawson, eds.), *Geol. J. Spec. Issue*, 41–80.

Jeletzky, J. A., 1984, Jurassic–Cretaceous boundary beds of western and Arctic Canada and the problem of Tithonian–Berriasian stages in the Boreal Realm, in: *Jurassic–Cretaceous Biochronology and Palaeogeography of North America* (G. E. G. Westermann, ed.), *Geol. Assoc. Can. Spec. Pap.* **27**:175–256.

Kassab, A. S., 1991, Cenomanian–Coniacian biostratigraphy of the northern eastern desert of Egypt, based on ammonites, *Newsl. Stratigr.* **25**:25–35.

Kauffman, E. G., 1979, Biogeography and biostratigraphy: Cretaceous, in: *Treatise of Invertebrate Palaeontology* Part A, *Introduction* (R. A. Robinson and C. Teichert, eds.), Geological Society of America and University of Kansas Press, Lawrence, KS.

Kauffman, E. G., 1984, Palaeobiography and evolutionary response dynamic in the Cretaceous. Western Interior Seaway of North America, in: *Jurassic–Cretaceous Biochronology and Palaeogeography of North America* (G. E. G. Westermann, ed.), *Geol. Assoc. Can. Spec. Pap.* **27**:273–306.

Kauffman, E. G., Cobban, W. A., and Eicher, D. L., 1978, Albian through lower Coniacian strata; biostratigraphy and principal events, in Western Interior states, in: *Evénements de la partie moyenne du Crétacé* (R. A. Reyment and G. Thomel, eds.), *Ann. Mus. Hist. Nat. Nice* **4**(34), 1–52.

Kemper, E., 1973a, The Valanginian and Hauterivian stages in northwest Germany, in: *The Boreal Lower Cretaceous* (R. Casey and P. F. Rawson, eds.), *Geol. J. Spec. Issue* **5**:327–344.

Kemper, E., 1973b, The Aptian and Albian stages in northwest Germany, in: *The Boreal Lower Cretaceous* (R. Casey and P. F. Rawson, eds.), *Geol. J. Spec. Issue* **5**:345–360.

Kemper, E., Rawson, P. F., and Thieuloy, J.–P., 1981, Ammonites of Tethyan ancestry in the early Lower Cretaceous of North West Europe, *Palaeontology* **24**:251–311.

Kennedy, W. J., 1984, Ammonite faunas and the standard zones of the Cenomanian to Maastrichtian Stages in their type areas, with some proposals for the definition of the stage boundaries by ammonites, *Bull. Geol. Soc. Den.* **33**:147–161.

Kennedy, W. J., 1985, Ammonite faunas of the Coniacian, Santonian and Campanian stages in the Aquitaine Basin, *Geol. Mediterran.* **10**:103–113.

Kennedy, W. J., 1986, Campanian and Maastrichtian ammonites from Northern Aquitaine, France, *Spec. Pap. Palaeontol.* **36**.

Kennedy, W. J., and Cobban, W. A., 1976, Aspects of ammonite biology, biogeography and biostratigraphy, *Spec. Pap. Palaeontol.* **17**.

Kennedy, W. J., and Klinger, H. C., 1975, Cretaceous faunas from Zululand and Natal, South Africa: Introduction, Stratigraphy, *Bull. Br. Mus. (Nat. Hist.) Geol.* **25**:263–315.

Klinger, H. C., 1989, The ammonite subfamily Labeceratinae Spath 1925: Systematics, phylogeny, dimorphism and distribution, *Ann. S. Afr. Mus.* **98**:189–219.

Krishna, J., 1983, Callovian–Albian ammonoid stratigraphy and paleobiogeography in the Indian subcontinent with special reference to the Tethys Himalaya, *Himalayan Geol.* **2**:43–72.

Krishna, J., 1984, Current status of the Jurassic stratigraphy of Kachchh, western India, in: *International Symposium on Jurassic Stratigraphy* (O. Michelsen and A. Zeiss, eds.), Geological Survey of Denmark, Copenhagen, pp. 732–742.

Krishna, J., and Cariou, E., 1986, The Callovian of western India: New data on the biostratigraphy, biogeography of the ammonites and correlation with western Tethys (Submediterranean Province), *Newsl. Stratigr.* **17**:1–8.

Krishna, J., and Cariou, E., 1990, Ammonoid faunal exchanges during lower Callovian between the Indo-East African and Submediterranean Provinces: Implications for long distance east–west correlations, *Newsl. Stratigr.* **23**:109–122.

Krishna, J., and Westermann, G. E. G., 1985, Progress report on the Middle Jurassic ammonite zones of Kachchh, western India, *Newsl. Stratigr.* **14**:1–11.

Krishna, J., Kumar, V. S., and Singh, I. B., 1982, Ammonoid stratigraphy of the Spiti Shale (Upper Jurassic), Tethys, Himalaya, India, *N. Jb. Geol. Paläontol. Mh.* **1982**(10):580–592.

Krymholts, C. Y., Mesezhnikov, M. S., and Westermann, G. E. G. (eds.), 1988, The Jurassic ammonite zones of the Soviet Union, *Geol. Soc. Am. Spec. Pap.* **223**.

Kummel, B., 1973, Lower Trassic (Scythian) Molluscs, in: *Atlas of Palaebiogeography* (A. Hallam, ed.), Elsevier, Amsterdam, pp. 223–233.

Lange, W., 1952, Der Untere Lias Fonsjoch (östliches karwendelgebirge) und seine Ammoniten fauna, *Palaeontogr. A* **102**:49–162.

Mangold, C., 1979, Le Bathonien de l'est du Subbétique (Espagne du Sud), *Cuad. Geol.* **10**.

Mangold, C., Elmi, S., and Gabilly, J., 1971, Les faunes du Bathonien dans la moitié sud de la France: Essai de zonation et de correlations, in: *Colloque sur la Jurassique, Luxembourg, 1967, Mém. Bur. Rech. Géol. Min. Fr.* **75**:103–132.

Marchand, D., Fortwengler, D., Dardeau, C., Gracianksi, P.-C. de, and Jacquin, T., 1990, Les peuplements d'ammonites du Bathonien Supérieur à l'Oxfordien moyen dans les Baronnies (Basin du sud-est France): Comparisons avec la plate-forme Nord Européenne, *Bull. Centres Rech. Explor.—Prod. Elf. Aquitaine* **14**:465–479.

Matsumoto, T., 1973, Cretaceous Ammonoidea, in: *Atlas of Palaeobiogeography* (A. Hallam, ed.), Elsevier, Amsterdam, pp. 421–429.

Matsumoto, T., Okada, H., Harano, H., and Tanabe, K., 1978, Mid-Cretaceous zonation in Japan, in: *Evénements de la Partie Moyenne du Crétacé* (R. A. Reyment and G. Thomel, eds.), *Ann. Mus. Nat. Hist. Nice* **4**(33):1–6.

Meister, C., 1989, *Les Ammonites du Domerian des Causses (France)*, Éditions C.N.R.S., France.

Meister, C., Blau, J., and Böhm, F., 1994, Ammonite biostratigraphy of the Pliensbachian Stage in the Upper Austroalpine Jurassic, *Eclogae Geol. Helv.* **87**:139–155.

Meléndez, G., 1989, *El Oxfordiensé en el Sector Central de la Cordillera Ibérica. I. Bioestratigrafia. II. Paleontologia (Perisphinctidae, Ammonoidea)*, Institución "Fernando el Catolico," Instituto de Estudios Turolenses, Zaragoza.

Meléndez, G., Sequeiros, L., and Brochwicz-Lewinski, W., 1984, Tentative biostratigraphic subdivision for the Oxfordian of the Submediterranean Province on the basis of perisphinctids, in: *International Symposium on Jurassic Stratigraphy* (O. Michelsen and A. Zeiss, eds.), Geological Survey of Denmark, Copenhagen, pp. 482–501.

Mojsisovics, E. V., 1893, Die Cephalopoden der Hallstätter Kalk, *Abh. Geol. Reichsanst. Wien* **10**:1–835.

Moullade, M., 1966, Étude stratigraphique et micro–paléontologique du Crétacé inférieur de la "fosse vocontienne," *Doc. Lab. Géol. Fac. Sci. Lyon* **15**.

Myczynski, R., and Meléndez, G., 1990, On the current state of progress of the studies on Oxfordian ammonites from western Cuba, *1st Oxfordian Meeting, Zaragoza, 1988, Publ. Sepaz.* **2**:185–189.

Nikolov, T. G., 1982, *Les ammonites de la famille Berriasellidae Spath 1922*, Académie Bulgare des Sciences, University of Sofia, Sofia.

Oloriz, F., 1978, *Kimmeridgiense-Tithonico Inferior en el Sector Central de las Cordilleras Béticas (Zona Subbetica), Paleontologia, Bioestratigrafia*, Thesis Doctoral, Universidad de Granada.

Owen, H. G., 1973, Ammonite faunal provinces in the middle and upper Albian and their palaeogeographical significance, in: *The Boreal Lower Cretaceous* (R. Casey and P. F. Rawson, eds.), *Geol. J. Spec. Issue* **5**:145–154.

Owen, H. G., 1984, Albian stage and substage boundaries, *Bull. Geol. Soc. Denmark* **33**:183–189.

Page, K. N., 1991, Ammonites, in: *Fossils of the Oxford Clay* (D. M. Martill and J. D. Hudson, eds.), *Palaeontological Association Field Guide to Fossils 4*, Paleontological Association, London, pp. 86–143.

Page, K. N., 1992, The sequence of ammonite correlated horizons on the British Sinemurian (Lower Jurassic), *Newsl. Stratigr.* **27**:129–156.

Page, K. N., 1993, Mollusca: Cephalopoda (Ammonoidea: Phylloceratina, Lytoceratina, Ammonitina, Ancyloceratina), in: *Fossil Record 2* (M. J. Benton, ed.), Chapman and Hall, London, pp. 213–228.

Parnes, A., 1975, Middle Triassic ammonite biostratigraphy in Israel, *Geol. Surv. Isr. Bull.* **66**:1–35.

Pergament, M. A., 1977, Stratigraphy and correlation of mid-Cretaceous of the USSR Pacific regions, *Palaeont. Soc. Jpn. Spec. Pap.* **21**:85–95.

Phelps, M., 1985, A refined ammonite biostratigraphy for the Middle and Upper Carixian (Ibex and Davoei zones, Lower Jurassic) in Northwest Europe and stratigraphical details of the Carixian–Domerian boundary, *Geobios* **18**:321–362.

Popenoe, W. T., Imlay, R. W., and Murphy, M. A., 1960, Correlation of the Cretaceous formations of the Pacific Coast (United States and northwestern Mexico), *Geol. Soc. Am. Bull.* **71**:1491–1540.

Poulton, T. P., Detherman, R. L., Hall, R. L., Jones, D. L., Peterson, J. A., Smith, P., Taylor, D. G., Tipper, H. W., and Westermann, G. E. G., 1992, Regional geology and stratigraphy. 4. Western Canada and United States, in: *The Jurassic of the Circum-Pacific* (G. E. G. Westermann, ed.), Cambridge University Press, Cambridge, pp. 29–92.

Rawson, P. F., 1973, Lower Cretaceous (Ryazanian–Barremian) marine connections and cephalopod migrations between the Tethyan and Boreal Realms, in: *The Boreal Lower Cretaceous* (R. Casey and P. F. Rawson, eds.), *Geol. J. Spec. Issue* **5**:131–144.

Rawson, P. F., 1975, Lower Cretaceous ammonites from northeast England: The Hauterivian heteromorph *Aegocrioceras*, *Bull. Br. Mus. (Nat. Hist.) Geol.* **26**:129–159.

Rawson, P. F., 1981, Early Cretaceous ammonite biostratigraphy and biogeography, in: *The Ammonoidea*, Systematics Asociation Spec. Vol. 18 (M. R. House and J. R. Senior, eds.), Academic Press, London, pp. 499–529.

Renz, O., 1982, *The Cretaceous Ammonites of Venezuela*, Birkhäuser Verlag, Basel.

Reyment, R. A., and Tait, E. A., 1972, Biostratigraphical dating of the early history of the South Atlantic Ocean, *Phil. Trans. R. Soc. Lond. B* **264**:55–95.

Riccardi, A. C., 1988, The Cretaceous system of southern South America, *Geol. Soc. Am. Mem.* **168**:1–161.

Riccardi, A. C., Gulisano, C. A., Mojica, J., Palacios, O., Schubert, C., and Thomson, M. R. A., 1992, Regional geology and stratigraphy. 6. Western South America and Antarctica, in: *The Jurassic of the Circum-Pacific* (G. E. G. Westermann, ed.), Cambridge University Press, Cambridge, pp. 122–161.

Saks, V. N. (ed.), 1972, *The Jurassic–Cretaceous Boundary and the Berriasian Stage in the Boreal Realm, Akademiya Nauk SSSR*, Novosibirsk (in Russian; translation published 1975 by Keter Publishing House, Jerusalem).

Saks, V. N., and Shulgina, N. I., 1973, Correlation of the Jurassic Cretaceous boundary beds in the Boreal Realm, in: *The Boreal Lower Cretaceous* (R. Casey and P. F. Rawson, eds.), *Geol. J. Spec. Issue* **5**:387–392.

Salvador, A., Westermann, G. E. G., Olóriz, F., Gordon, M. B., and Gusky, H.-J., 1992, Regional geology and stratigraphy. 5. Meso America, in: *The Jurassic of the Circum-Pacific* (G. E. G. Westermann, ed.), Cambridge University Press, Cambridge, pp. 93–121.

Sato, T., 1992, Regional geology and stratigraphy. 9. Southeast Asia and Japan, in: *The Jurassic of the Circum–Pacific* (G. E. G. Westermann, ed.), Cambridge University Press, Cambridge, pp. 194–213.

Sazunova, I. G., 1961, Standardised stratigraphical scheme of lower Cretaceous deposits of the Russian Platform, *Tr. Vses. Neft. Nauchno-Issled. Geol. Razv. Inst.* **29**:5–28 (in Russian).

Schindewolf, O. H., 1968, Studien zur Stammesgeschichte der Ammoniten, Lief. 1–7, *Akad. Wiss. Lit. Abh. Math.-Naturwiss. Kl. Mainz* **1968**(3):731–901.

Sey, I. I., Repin, Y. S., Kalachera, E. D., Okuneva, T. M., Paraketsor, K. V., and Polubotko, I. V., 1992, Regional geology and stratigraphy. 11. Eastern Russia, in: *The Jurassic of the Circum–Pacific* (G. E. G. Watermann, ed.), Cambridge University Press, Cambridge, pp. 225–246.

Silberling, N. J., and Tozer, E. T., 1968, Biostratigraphic classification of the Marine Triassic in North America, *Geol. Soc. Am. Spec. Pap.* **110**.

Smith, A. G., and Briden, J. C., 1977, *Mesozoic and Cenozoic Palaeocontinental Maps*, Cambridge University Press, Cambridge.

Smith, P. L., Beyers, J. M., Carter, E. S., Jakobs, G. K., Pálfy, J., Pessagno, E., and Tipper, H. W., 1994, Jurassic taxa ranges and correlation charts for the circum-Pacific. 5. North America (G. E. G. Westermann and A. C. Riccardi, eds.), *Newsl. Stratigr.* **31**:33–70.

Spath, C. F., 1934, *Catalogue of the Fossil Cephalopoda in the British Museum (Natural History). Part IV. The Ammonoidea of the Trias*, British Museum (Natural History), London.

Stevens, G. R., 1973, Cretaceous belemnites, in: *Atlas of Palaeobiogeography* (A. Hallam, ed.), Elsevier, Amsterdam, pp. 385–401.

Sukamto, R., and Westermann, G. E. G., 1992, Regional geology and stratigraphy. 8. Indonesia and Papua New Guinea, in: *The Jurassic of the Circum-Pacific* (G. E. G. Westermann, ed.), Cambridge University Press, Cambridge, pp. 180–183.

Sykes, R. M., and Callomon, J. H., 1979, The *Amoeboceras* zonation of the Boreal Upper Oxfordian, *Palaeontology* **22**:112–121.

Taylor, D. G., Callomon, J. H., Hall, R., Smith, R. L., Tipper, H. W., and Westermann, G. E. G., 1984, Jurassic ammonite biogeography of western North America: The tectonic implications, in: *Jurassic–Cretaceous Biochronology and Palaeogeography of North America* (G. E. G. Westermann, ed.), *Geol. Assoc. Can. Spec. Pap.* **27**:121–142.

Thieuloy, J. P., 1973, The occurrence and distribution of boreal ammonites from the Neocomian of southeast France (Tethyan Province), in: *The Boreal Lower Cretaceous* (R. Casey and P. F. Rawson, eds.), *Geol. J. Spec. Issue* **5**:289–302.

Thompson, M. R. A., 1982, A comparison of the ammonite faunas of the Antarctic Peninsula and Magallanes Basin, *J. Geol. Soc. (Lond.)* **139**:763–770.

Torrens, H. S., 1971, Standard zones of the Bathonian, in: *Colloque sur la Jurassique, Luxembourg 1967*, *Mém. Bur. Rech. Géol. Min. Fr.* **75**:581–604.

Tozer, E. T., 1981a, Triassic Ammonoidea: Classification, evolution and relationship with Permian and Jurassic forms, in: *The Ammonoidea*, Systematics Association Spec. Vol. 18 (M. R. House and J. R. Senior, eds.), Academic Press, London, pp. 66–100.

Tozer, E. G., 1981b, Triassic ammonoids: Geographic and stratigraphic distribution, in: *The Ammonoidea*, Systematics Association Spec. Vol. 18 (M. R. House and J. R. Senior, eds.), Academic Press, London, pp. 397–451.

Wang, Y.–G., 1988, Jurassic taxa ranges and correlation charts for the circum-Pacific. 2. China (G. E. G. Westermann and A. C. Riccardi, eds.), *Newsl. Stratigr.* **19**:95–130.

Wang, Y., and Westermann, G. E. G., 1993, Paleoecology of Triassic ammonoids, *Geobios* **15**:373–392.

Ward, P. D., and Haggart, J., 1981, The Upper Cretaceous (Campanian) ammonite and inoceramid bivalve succession at Sand Creek, Colusa County, California, and its implications for establishment of an Upper Cretaceous Great Valley Sequence ammonite zonation, *Newsl. Stratigr.* **10**:140–147.

Westermann, G. E. G., 1981, Ammonite biochronology and biogeography of the circum-Pacific Middle Jurassic, in: *The Ammonoidea*, Systematics Association Spec. Vol. 18 (M. R. House and J. R. Senior, eds.), Academic Press, London, pp. 66–100.

Westermann, G. E. G., 1993, Global bio-events in mid-Jurassic ammonites controlled by seaways, in: *The Ammonoidea: Environment, Ecology, and Evolutionary Change*, Systematics Association Spec. Vol. 47 (M. R. House, ed.), Clarendon Press, Oxford, pp. 187–226.

Westermann, G. E. G., and Callomon, J. H., 1988, The Macrocephalitinae and associated Bathonian and Early Callovian (Jurassic) ammonoids of the Sula Islands and New Guinea, *Palaeontogr. Abt. A* **203**:1–90.

Wiedmann, J., 1970, Über den Ursprung der Neoammonoideen—das Problem einer Typogense, *Eclogae Geol. Helv.* **63**:923–1020.

Wiedmann, J., 1973, Upper Triassic heteromorph ammonites, in: *Atlas of Palaeobiogeography* (A. Hallam, ed.), Elsevier, Amsterdam, pp. 235–249.

Wiedmann, J., 1988, Ammonoid extinction and the "Cretaceous–Tertiary boundary event," in: *Cephalopods—Present and Past* (J. Wiedmann and J. Kullman, eds.), Schweizerbart'sche Verlagsbuchhandlung, Stuttgart, pp. 117–140.

Wiedmann, J., Frabricus, F., Kristyn, C., Reitner, J., and Urlichs, M., 1979, Über Umfgang und Stellung des Rhaet, *Newsl. Stratigr.* **8**:133–152.

Wimbledon, W., 1984, The Portlandian, the terminal Jurassic Stage in the Boreal Realm, in: *International Symposium on Jurassic Stratigraphy* (O. Michelsen and A. Zeiss, eds.), Geological Survey of Denmark, Copenhagen, pp. 534–549.

Wimbledon, W. A., and Cope, J. C. W., 1977, The ammonite faunas of the English Portland Beds and the zones of the Portlandian Stage, *J. Geol. Soc.(Lond.)* **135**:183–190.

Wright, C. W., 1979, The ammonites of the English Chalk Rock (Upper Turonian), *Bull. Br. Mus. (Nat. Hist.) Geol.* **31**:341–382.

Wright, C. W., 1981, Cretaceous Ammonoidea, in: *The Ammonoidea*, Systematics Association Spec. Vol. 18 (M. R. House and J. R. Senior, eds.), Academic Press, London, pp. 157–174.

Wright, C. W., and Kennedy, W. J., 1981, The Ammonoidea of the Plenus Marls and the Middle Chalk, *Palaeontogr. Soc. Monogr. Lond.* **134**:1–148.

Wright, C. W., and Kennedy, W. J., 1984–, The Ammonoidea of the Lower Chalk, *Palaeontogr. Soc. Monogr. Lond.* Part I, 1984, **137**:1–126; Part 2, 1987, **139**:127–218; Part 3, 1990, **144**:219–294.

Zaborski, P. M. P., 1983, Campano-Maastrichtian ammonites, correlation and palaeogeography in Nigeria, *J. Afr. Earth Sci.* **1**:59–63.

Zaborski, P. M. P., 1985, Upper Cretaceous ammonites from the Calaban region, southeast Nigeria, *Bull. Br. Mus. (Nat. Hist.) Geol.* **39**:1–72.

Zapfe, U. (ed.), 1975, Die stratigraphie der alpin-Mediterraenean Triassic, *Ost. Akad. Wiss. Schrift. Erdwiss. Komm.* **2**.

Zapfe, U. (ed.), 1978, Beiträge zur Biostratigraphie der Tethys-Trias, *Ost. Akad. Wiss. Schrift. Erdwiss. Komm.* **4**.

Zeiss, A., 1984, Contributions to the biostratigraphy of the Jurassic System in Ethiopia, in: *International Symposium on Jurassic Stratigraphy* (O. Michelson and A. Zeiss, eds.), Geological Survey of Denmark, Copenhagen, pp. 572–581.

Zeiss, A., 1986, Comments on a tentative correlation chart for the most important marine provinces at the Jurassic/Cretaceous boundary, *Acta Geol. Hung.* **29**:27–30.

Ziegler, B., 1981, Ammonoid biostratigraphy and provincialism: Jurassic—Old World, in: *The Ammonoidea*, Systematics Association Spec. Vol. 18 (M. R. House and J. R. Senior, eds.), Academic Press, London, pp. 433–457.

Chapter 19

Crises in Ammonoid Evolution

JOST WIEDMANN[†] and JÜRGEN KULLMANN

1. Paleozoic Ammonoids

During the Devonian, Carboniferous, and Permian periods, several severe changes occurred in the configuration of the tectonic plates, resulting in major changes in the areas of oceans and epicontinental seaways. The relatively rapid rate at which ammonoids evolved allows detailed information on changes in the environmental conditions and geological events that influenced the habitats of these animals (House, 1985a,b, 1989; Kullmann, 1983, 1985).

Periods of mass extinction have been discussed intensively, first by Schindewolf (1954) and Newell (1967) and then in detailed ammonoid studies by House (1983, 1989). Fluctuations of sea level in connection with anoxic oceanic overturns are regarded as major factors causing drastic changes in

[†]Deceased.

JOST WIEDMANN and JÜRGEN KULLMANN • Geologisch-Paläontologisches Institut, Universität Tübingen, D-72076 Tübingen, Germany.

Ammonoid Paleobiology, Volume 13 of *Topics in Geobiology*, edited by Neil Landman *et al.*, Plenum Press, New York, 1996.

Table I. Number of Ammonoid Species (Bactritida Omitted) in the Paleozoic[a]

	Rad. Age	Myr	Spec.	Spec/Myr
Lower (part) and Middle Devon.	395–368	27	317	11.7
Frasnian	368–363.3	5	155	31.0
Famennian	363.3–354	10	660	66.0
Lower Carboniferous	354–319	35	959	27.4
Upper Carboniferous	319–295	24	512	21.3
Lower Permian	295–272	23	359	15.0
Upper Permian	272–250	22	442	20.0
Devonian (part)	395–354	41	1115	27.2
Carboniferous	354–295	59	1464	24.8
Permian	295–250	45	788	17,5
	395–250	~145	~3700	~25

[a]Rad. Age, estimated radiometric age, based on data derived from a number of sources (Claoué-Long *et al.*, 1992; Harland *et al.*, 1990; Jones, 1988; Menning, 1992; Roberts *et al.*, 1991); Myr, estimated duration in millions of years; Spec., approximate number of species based on data from the data base GONIAT, Tübingen (Korn and Kullmann, 1993); Spec/Myr, number of species per million years.

ammonoid diversity (House, 1989; Becker, 1993a,b). Global geotectonic influences, i.e., intervals of tectonic unrest in various parts of the world, are also thought to be reflected by changes in diversity (Kullmann, 1985; Korn and Kullmann, 1988). However, the fact that major diversity fluctuations seem to have occurred more frequently and more intensely during the Devonian than during the Carboniferous and the Permian, which were characterized by widespread geotectonic activities (Becker and Kullmann, Chapter 17, this volume), is in direct contrast to the idea that plate tectonic movements were the major source of perturbations in the ammonoid record (Kullmann, 1994). In addition, a number of as yet unidentified environmental, tectonic, and genetic factors (Ziegler and Lane, 1987) may also have contributed to changes in ammonoid diversity.

The overall diversity of Paleozoic ammonoids is reflected by the number of recognized species during the major time units of the Devonian, Carboniferous, and Permian (Table I). In this chapter, the number of Paleozoic taxa (species and genera) is based on data from the GONIAT data base, Tübingen, version 2.40 (Kullmann *et al.*, 1993; Korn and Kullmann, 1993; Korn *et al.*, 1994). Because this data base is not yet completed, these numbers represent approximations of the real numbers of taxa; taxa based on insufficient material or inadequate descriptions are omitted. Representatives of the Bactritida are extremely rare and insufficiently known and are, therefore, not included in this study.

1.1. Devonian

Ammonoids appeared suddenly in the upper Lower Devonian (lower Emsian, lower Zlichovian) with several lineages. The most important group was the Mimosphinctidae, loosely coiled ammonoids with a perforate umbilicus and a suture consisting of two or three lobes. Several species exhibited rather complicated ornament. The Auguritidae, known from only two localities in Central Europe and the Northern Urals, are the only forms with complicated sutures on the ventral side.

The earliest ammonoids (see Chapter 17, this volume, Fig. 2), about 50 species belonging to 13 or 14 genera, seem to have been widespread in the Northern Hemisphere. They are known from nearly 50 localities: Czech Republic, France, Germany, Russia (Northern Urals, Siberia), Spain, China, Middle Asia, and North America (United States, Canada).

During the uppermost part of the Zlichovian, the number of species seems to have decreased. It is not known, however, if there was a drastic decline of species immediately at the top of the Zlichovian. House (1985a, 1989) suggested that a special extinction event ("Daleje event") occurred at that time, which was related to a widespread facies change to black mudstones.

Ammonoid diversity was low at the base of the upper Emsian (Dalejan) but began to increase again in the middle part of the upper Emsian in the *serotinus* conodont zone. The early Zlichovian Mimagoniatitidae gave rise to two important groups, the Anarcestidae and Agoniatitidae, both of which proliferated in late Early Devonian and Middle Devonian. During Dalejan time, about 30 species were present and are recorded from more than 50 localities in Europe, North Africa, Central Asia, China, and North America.

In Eifelian and early Givetian times, the number of species increased to 120–130, and more than 30 genera can be counted in this time interval of about 10 to 15 million years. Slight changes in the composition of ammonoid faunas (extinction of *Pinacites*, appearance of *Tornoceras*) at the end of the Eifelian and beginning of the Givetian have led to the inference that a minor extinction event called the "Kacák event" occurred at this time (House, 1985a).

A total changeover in ammonoid succession was marked by the extinction of the index genus *Maenioceras* and the families Anarcestidae, Agoniatitidae, and Pinacitidae in the middle Givetian (middle *varcus* conodont zone); this is assumed to be connected with the Taghanic event. Except for *Tornoceras* and *Protornoceras*, no other genus is recorded as crossing this boundary; 12 genera with more than 30 species became extinct.

Following this drastic extinction event, the new stock of ammonoids included ones with complex sutures. Conch form as well as shell surface features varied considerably. Almost 20 new genera with 75 species are recorded at the start of the late Givetian and Frasnian intervals. The guide genus for the upper Givetian is *Pharciceras* (with about 20 species); this genus is restricted to the upper Givetian (*Pharciceras* Stufe); 11 genera with more than 120 species became extinct at the end Givetian at what is called the "Frasnes event" (House, 1985a).

Only five genera survived; about 20 genera and 70 species originated at the beginning of Frasnian time (*Manticoceras* Stufe; House, 1985a). One hundred thirty species (in 22 genera) are recorded from this interval. Among Frasnian goniatites, several genera with multilobate sutures were present, e.g., the Beloceratidae; variation in sculpture was minor.

At the top of the *Manticoceras* Stufe, the "Kellwasser event" apparently diminished the goniatite faunas; more than 40 species became extinct, and only four genera survived into the uppermost Frasnian *Crickites* zone (I delta). However, 70 species were still present.

The lower Famennian *Cheiloceras* Stufe is characterized by Tornoceratina with convex growth lines; the number of genera increased again to about 35, with 150 species. In general, the diversity in conch shape and sutural differentiation was relatively low. At the end of this time interval most genera and species disappeared. Because of an apparently short gap in one of the most complete sections (Enkeberg, Sauerland, Germany), an extinction event called the "Enkeberg event" has been postulated (House, 1985a).

The *Prolobites* Stufe that followed is marked by the appearance of the Clymeniida, which show migration of the siphuncle to a dorsal position during early ontogeny. The number of genera and species increased again to 45 genera (20 clymeniids, 25 tornoceratids) with about 180 species; these were the most diverse ammonoid faunas that existed in the Devonian.

The *Platyclymenia* Stufe is characterized by clymeniids that exhibit a conspicuous ornamentation and sculpture; index fossils include representatives of *Platyclymenia*. The diversity of the clymeniids stagnated; the tornoceratids decreased considerably (20 genera of clymeniids and 14 genera of tornoceratids, about 100 species in total). During this time span, black shales were deposited in at least parts of the Rheinische Schiefergebirge, known as the "annulata event." It is not known if this event had any global influence on the overall evolution of the ammonoids.

The last stage of Famennian ammonoid evolution (*Clymenia* and *Wocklumeria* Stufen) witnessed a tremendous diversification in conch form and sutural configuration in the clymeniids as well as in the tornoceratids but less in the prionoceratids. Clymeniid diversity increased to over 40 genera (10 genera of Goniatitida; a total of 275 species) and reached a second maximum in the uppermost Famennian *Wocklumeria* Stufe (Korn and Kullmann, 1988). The peak of this development was reached shortly before the final decline and extinction of the Clymeniida. In the last zone with diverse clymeniid faunas, the "upper *paradoxa* zone," the number of genera dropped to about 20 clymeniid and six goniatitid genera, and no more than about 100 species were present. The last representative of the Clymeniida occurred just below the Devonian–Carboniferous boundary; only a few (perhaps two) goniatitid lineages crossed that boundary. House (1985a) regarded this crisis as influenced partly by euxinic black shale facies of the Hangenberg Schiefer ("Hangenberg event").

1.2. Carboniferous (Mississippian, Pennsylvanian)

After the drastic decline at the end of the Devonian, goniatites proliferated again in the early Tournaisian (see Chapter 17, this volume, Figs. 5 and 6). In the lower Tournaisian, 80 species (12 genera) were present, and during the Lower Carboniferous (Mississippian), the diversity of ammonoids increased almost continuously. This is especially true for the Viséan; in early Viséan time, 140 species (more than 30 genera) can be recognized. At the end of the late Viséan a maximum of more than 500 species (almost 100 genera) was reached. It must be taken into account, however, that these intervals are of extremely long duration; the early Viséan is considered to span 9 million, and the late Viséan and early Namurian, 12 million years.

The absolute peak in the diversity of Paleozoic ammonoids was reached in the uppermost portion of the upper Viséan (middle and upper Chesterian). Several new features appeared, such as a change in the mode of coiling of the inner whorls (triangular coiling of the inner whorls), an increase in the size of the umbilicus of the inner whorls, and an increase in sutural elements and mode of sutural development. Most taxa of species and generic rank lasted only a short time (Kullmann, 1985).

Almost at the top of the Lower Carboniferous, most of the new goniatite species and genera of the *Eumorphoceras* Stufe (E) disappeared again; of 85 species (32 genera) in the uppermost zone (E_{2c4}) of the Lower Carboniferous, only four species seem to have survived the Lower–Upper Carboniferous boundary ("E_2 crisis"). Nine genera occurred in beds below and above this boundary.

At the beginning of the Upper Carboniferous, no goniatites are known from the New World, Africa, and Antarctica. The few lineages that survived the E_2 crisis produced a new stock of goniatites; in the *Homoceras* Stufe 30 species (11 genera) were new.

Beginning with the Bashkirian (Kinderscoutian, Halian, *Reticuloceras* Stufe), the ammonoids are again known from many places in the northern hemisphere. The radiation of different stocks which had begun in the *Homoceras* Stufe continued during the following interval. In the Kinderscoutian (early Halian) about 140 species (almost 40 genera) were present, but this number decreased considerably in the Marsdenian (middle Halian) to 50 species (about 25 genera). Seventy-five ammonoid species (more than 30 genera) were present in the late Halian interval, the Yeadonian (G_1). Half of them originated in this interval. More than 50 species became extinct at the end of the Yeadonian (Halian–Bloydian boundary).

The next step in ammonoid evolution was at the beginning of Westphalian time (late Bashkirian, Bloydian). The number of ammonoid species reached

100 (about 40 genera); most of the Reticuloceratidae, with many short-ranging index species, disappeared and were replaced by the Gastrioceratidae and Pseudoparalegoceratidae. The trend of diversifying sutural elements continued slowly. The most remarkable change was the onset of provincialism among goniatites. The Orulganitidae were restricted exclusively to Siberia; the separation of this area seemed to persist until the Early Permian (Sakmarian).

   Starting with the Moscovian (Atokan and Desmoinesian), a period of low diversity began. In the Moscovian, which is thought to have spanned 9 million years, about 85 species (40 genera) were present; a relatively high number of long-ranging genera (e.g., *Agathiceras*, *Glaphyrites*, *Syngastrioceras*, *Somoholites*, *Boesites*, and *Stenopronorites*) predominated. But, in addition, ammonoids with complicated sutures (*Aktubites* of the Shumarditidae, *Wellerites* and *Winslowoceras* of the Welleritidae, and *Marathonites* of the Marathonitidae) appeared for the first time in the Goniatitida. The Prolecanitida persisted nearly unchanged from the late Early Carboniferous. In the Kasimovian (late Desmoinesian) diversity continued to decline (30 species, about 20 genera), but the appearance of the Adrianitidae (*Emilites*) and Thalassoceratidae (*Prothalassoceras*) added two new groups of multilobate ammonoids. At the same time, the first Medlicottiidae (*Prouddenites*) appeared, exhibiting broad complicated ventral prongs of the lateral lobe. In the Gzhelian, this group proliferated greatly.

1.3. Permian

   The trend to complication of the suture and the addition of sutural elements became widespread at the beginning of the Permian, not only in the Goniatitida but also in the Prolecanitida. In addition, two major groups, which were derived from mid-Carboniferous neoicocerataceans, persisted with long-ranging genera throughout the entire Permian: the Metalegoceratidae and the Paragastrioceratidae. The latter group maintained the basic suture of the Goniatitina $(E_1E_mE_1)$ALUI [Russian: (V_1V_1)L:UI][*] until the beginning of the Triassic. The changeover at the Carboniferous–Permian boundary is remarkable (see Chapter 17, this volume, Figs. 5–7); in the Gzhelian, about 115 species (almost 40 genera) were present, whereas in the Asselian, 75 species (37 genera) were present. Only 12 Gzhelian genera survived. In the Early Permian, however, the diversity of Goniatitida and Prolecanitida increased continuously until the middle of the Permian (Glenister and Furnish, 1981).

   In the Roadian, the uppermost stage of the Lower Permian, a distinct changeover began, apparently connected with the appearance of the Ceratitida. In this interval, the numbers of genera in the Goniatitida, Ceratitida, and

---

[*]For suture terminology, see Chapter 17, this volume, Section 4.1.

Prolecanitida were about 30, 1, and 4, respectively (a total of more than 50 species). The peak diversity was reached in the following interval, the lower part of the Upper Permian Wordian stage; more than 150 species were present, and the numbers of goniatitid, ceratitid, and prolecanitid genera were about 50, 2, and 8, respectively. In the Capitanian (a total of almost 60 species), the number of goniatitid genera dropped to 10, the number of ceratitid genera increased to six, and only the number of prolecanitid genera remained constant. In the Dzhulfian stage (more than 110 species), the Ceratitida became predominant. More than 110 species were present belonging to ten goniatitid, 30 ceratitid, and two prolecanitid genera. The uppermost part of the Upper Permian (Dorashamian, Changxingian) yielded 26 ceratitid genera; of the Goniatitida, only one genus, *Pseudogastrioceras*, persisted until the Lower Triassic Griesbachian (Teichert, 1990). No prolecanitid genus was recorded, but some must have existed; *Episageceras*, the last member of the Medlicottidae, occurred in the Lower Triassic Griesbachian.

The main crisis at the end of the Permian affected the Goniatitida and Prolecanitida. Apparently only one genus of each order survived the Permian–Triassic boundary, but both became extinct in the Early Triassic. The Ceratitida, however, gave rise to the otoceratids at the beginning of the Triassic.

2. Mesozoic Ammonoids

The state of taxonomy of Mesozoic ammonoids is based on a number of diverse studies, which makes it difficult to arrive at any fixed conclusion. Statistical evaluation of radiations and extinctions as has been done for the Paleozoic is, for the time being, possible only for the Triassic. The synthetic paper of Tozer (1981) especially allows calculations on the generic and stage levels. Nothing comparable is available for the Jurassic and Cretaceous, at least since the publication of the ammonoid *Treatise* (Arkell *et al.*, 1957). This disadvantage is, however, of minor importance because no spectacular crises can be detected within the Jurassic and Cretaceous Periods nor at their mutual boundary. This is, indeed, different in the Triassic Period.

2.1. Triassic

A considerable amount of time was involved in recovering from the Permo–Triassic boundary crisis (Fig. 1). Late in the Nammalian, the first maximum diversity for the Lower Triassic was achieved, resulting in a total of 51 ceratitid genera and two genera of phylloceratids. A second maximum can be recognized at the top of the Spathian with a total of 61 ceratitid and four phylloceratid genera. But during the crisis that immediately followed, 48 ceratitid and three phylloceratid genera became extinct. This is the most severe event during the Triassic. Nine families did not cross the Lower to Middle Triassic boundary, i.e., the Xenodiscidae and the Xenodiscaceae as a

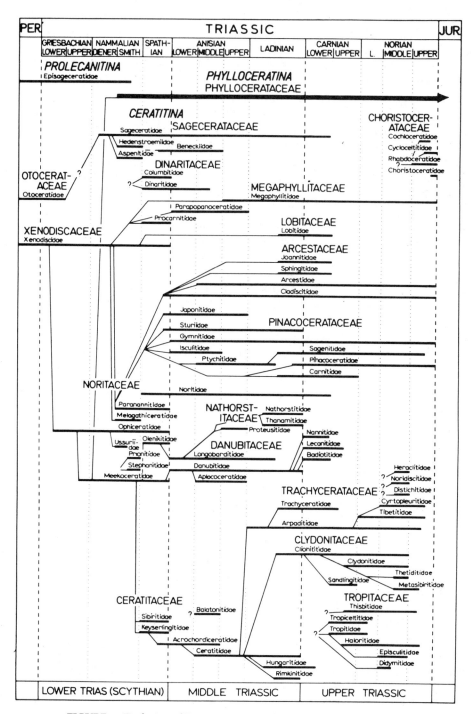

FIGURE 1. Evolution of Triassic ammonoid families (after Tozer, 1981).

whole: the Paranannitidae, Ophiceratidae, Olenikitidae, and Meekoceratidae of the Noritaceae; the Sibiritidae and Keyserlingitidae of the Ceratitaceae; the Columbitidae of the Dinaritaceae; and the Procarnitidae of the Megaphyllitaceae.

It is worth mentioning that there is a pronounced extinction peak at the Lower–Middle Triassic boundary as far as families are concerned according to the computation of Sepkoski (1986, Fig. 2). There is also a pronounced minimum at the same time when genera are counted (Sepkoski, 1986, Fig. 3).

Triassic ammonoids underwent their second severe crisis at the Anisian–Ladinian boundary. At this time, 40 of 61 ceratitid genera and two of four phylloceratid genera disappeared. This event involved four families, i.e., the Aplococeratidae of the Danubitaceae, the Proteusitidae of the Nathorstitaceae, the Parapopanoceratidae of the Megaphyllitaceae, and, finally, the Japonitidae of the Pinacocerataceae. No major crisis has been detected at this boundary according to Raup and Sepkoski (1984) and Sepkoski (1986).

Not as severe as the two previous crises is the Ladinian–Carnian extinction that followed. During late Ladinian time, ceratites and phylloceratids flourished again with a total of, respectively, 48 and three genera. The subsequent extinction at the boundary is characterized less by the disappearance of genera (ten) than by extinction at the family level. The Danubitidae of the Danubitaceae disappeared as well as the Ceratitidae and Rimkinitidae, together with most of the Ceratitaceae, the Nathorstitidae, and Thanamitidae, with all of the Nathorstaceae, and, finally, the Sturiidae of the Pinacocerataceae.

Another minor extinction event can be detected in the Carnian near or at the Jul–Tuval boundary. During the Jul, 39 ceratitid and five phylloceratid genera were present, of which eight and five, respectively, did not cross the boundary. Again, the number of families that became extinct was considerable. The Badiotitidae of the Danubitaceae nearly disappeared; the Noritidae of the Noritaceae, the Lobitidae of the Lobitaceae, and the Joannitidae and Sphingitidae of the Arcestaceae became extinct. Of the two extinctions of marine fossil genera detected by Sepkoski (1986) for the Triassic, the older one seems to correlate with this intra-Carnian crisis.

After some recovery, the diversity of ceratites increased for the last time to a total of 50 genera during the late Alaun. During Sevat and Rhaetian times, the final decline of the Ceratitina was continuous or at least stepwise. At the Alaun–Sevat boundary, e.g., 14 of these genera became extinct; on the family level, this "event" was again much more pronounced. Many families did not cross this boundary: the Heraclitidae; the Noridiscitidae and Distichitidae of the Trachycerataceae; the Clionitidae, Clydonitidae, and Thetiditidae of the Clydonitaceae; and the Episculitidae of the Tropitaceae. It is this crisis that may correlate with the "upper Norian extinction" recorded by Sepkoski (1986).

Only 19 genera of the Ceratitina and Phylloceratina persisted into the Rhaet, during which time they gradually disappeared. These included six genera of the Choristocerataceae, four genera of the Arcestidae and Cladisci-

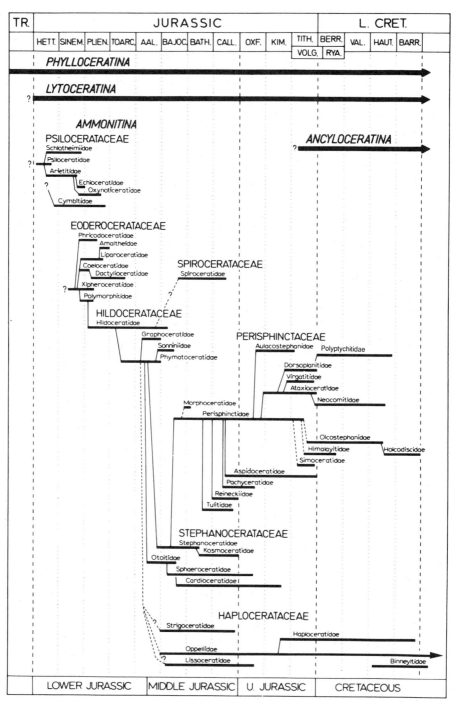

FIGURE 2. Evolution of Jurassic and Lower Cretaceous ammonoid families (after Donovan *et al.*, 1981).

tidae (Arcestaceae), one genus of the Megaphyllitidae and, thus, the Megaphyllitidaceae, and three genera of the Triassic phylloceratids, i.e., the Ussuritidae and Dicophyllitidae.

The picture given of the turnover of ammonoids at the Triassic–Jurassic boundary (Wiedmann, 1970, 1973a) is still fully valid. At present, no ammonoid genus or species is known that was common to both the Triassic and the Jurassic, although one must have existed. But despite the total disappearance of the highly flourishing Triassic Ceratitina, no sudden extinction can be recognized at this boundary.

2.2. Jurassic

Judging from a modern classification of the Jurassic ammonites, e.g., Donovan *et al.* (1981), it is very difficult to recognize any evolutionary crisis comparable to those previously discussed (Fig. 2). Significant extinctions have been described by Donovan (1988) within and at the end of the Sinemurian, when the Echioceratidae disappeared and the psiloceratids were replaced by the eoderoceratids. But the total rate of extinction was not spectacular.

A post-Pliensbachian extinction can only be inferred from other molluscs and various marine invertebrates (Hallam, 1986, 1987; Sepkoski 1986). The disappearance of the Amaltheidae should not be overemphasized, even though it coincides with the 26-million-year cyclicity described by Raup and Sepkoski (1984). In addition, the "lower Bajocian extinction" of Sepkoski (1986) is not well documented in ammonoids; instead, a stepwise extinction can be observed, first of the Hildoceratidae, then of the Sonniniidae, and, finally, during the mid-Bajocian, of the Otoitidae. A mid-Jurassic extinction peak at the generic level, as shown by House (1985b), may therefore be an artifact resulting from the relatively broad time slice considered.

No "upper Tithonian extinction" (Sepkoski, 1986) can be observed in ammonoids. Instead, two minor extinctions occurred near the Jurassic–Cretaceous boundary. At or near the boundary, five perisphinctid families disappeared, i.e., the Ataxioceratidae, Dorsoplanitidae, Virgatitidae, Aspidoceratidae, and Simoceratidae. The Perisphinctaceae persisted as a whole, however, well into the Cretaceous, as did the Haplocerataceae. The previously published picture of this boundary (Wiedmann, 1968, 1975) needs no major change.

2.3. Cretaceous

A second minor extinction event near the Jurassic–Cretaceous boundary coincided with the Berriasian–Valanginian boundary (Fig. 3). At this time, the perisphinctids Himalayitidae, Berriasellinae, Spiticeratinae, and Craspeditinae disappeared.

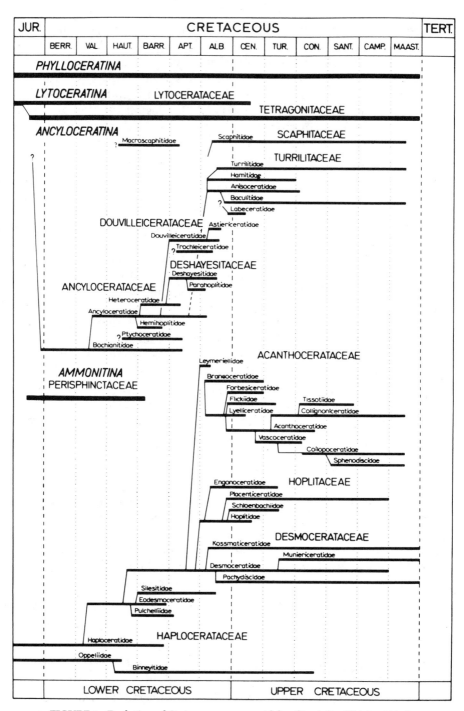

FIGURE 3. Evolution of Cretaceous ammonoid families (after Wright, 1981).

During the whole Cretaceous, however, no spectacular time of ammonoid extinction can be detected at the family level. The change in diversity during the Cretaceous perfectly mirrored changes in global sea level. Both increased continuously through the Lower Cretaceous towards the Cenomanian sea-level maximum and decreased thereafter except for a short-term sea-level rise during the lower Campanian.

It is interesting to compare our data for Cretaceous extinctions with those of Sepkoski (1986, Fig. 3). His early Aptian and late Cenomanian events, which fit perfectly into a 26-million-year cyclicity, do not involve ammonoids. The reason is that Sepkoski's "synthetic" way of calculating marine inverte-brate diversity lumps together totally different events, such as the global regression during the early Aptian, which coincides with the drowning of Tethyan carbonate platforms (Schlager, 1981; Föllmi *et al.*, 1994), the global black-shale event at the close of the Cenomanian (Arthur *et al.*, 1990), and the really sudden marine surface plankton extinction event caused by a cosmic impact at the Cretaceous–Tertiary boundary (Alvarez *et al.*, 1980). Not only are the causes of all these extinctions totally different, but so are the fossil groups affected. Uncritical statistical analyses do not separate out the affected biotopes, which is an essential requirement for understanding any extinction event (see Wiedmann, 1969, Fig. 23); the resulting cyclicity is thus a mere artifact.

What can be said today about the ultimate ammonoid extinction at the close of the Cretaceous? Much effort has been spent during the last decade to explain how and why ammonoids disappeared. From our knowledge of the continuous boundary sections in the Bay of Biscay and in southwest France (Wiedmann, 1969, 1986, 1988a,b; Ward and Wiedmann, 1983; Ward, 1988; Hancock and Kennedy, 1993; Hancock *et al.*, 1993), we can summarize that it was very similar to the Triassic–Jurassic boundary, a long period of decline with a maximum number of extinctions long before the end of the Cretaceous. Of about 100 ammonoid genera that lived during the sea-level highstand in the lower Campanian, fewer than 50 persisted into the middle Maastrichtian. At the two most complete boundary sections, Zumaya and Hendaye (Wied-mann, 1986, 1988a,b; Ward, 1988), just eight genera are recorded 15 m ($\approx$150,000 years) below the boundary and the well-documented iridium layer, i.e., *Neophylloceras, Gaudryceras, Saghalinites, Pseudophyllites, Vertebrites, Anapachydiscus, Pachydiscus,* and *Diplomoceras* with a total of ten species. In the last 0.5 m below the boundary at Zumaya, *Neophylloceras ramosum,* and at Hendaye, *Gaudryceras* sp., *Pachydiscus gollevillensis,* and *P.* sp., have been collected by the author (Fig. 4).

The ultimate extinction of ammonoids was, therefore, a continuous and long-lasting decline that can be traced over several million years and was comparable to the major extinctions at other Mesozoic system boundaries. This extinction was unrelated to the K–T boundary event itself, which affected the oceanic surface plankton and was of either cosmic or volcanic origin. It is worth mentioning that evolutionary rates in ammonoids decreased during the

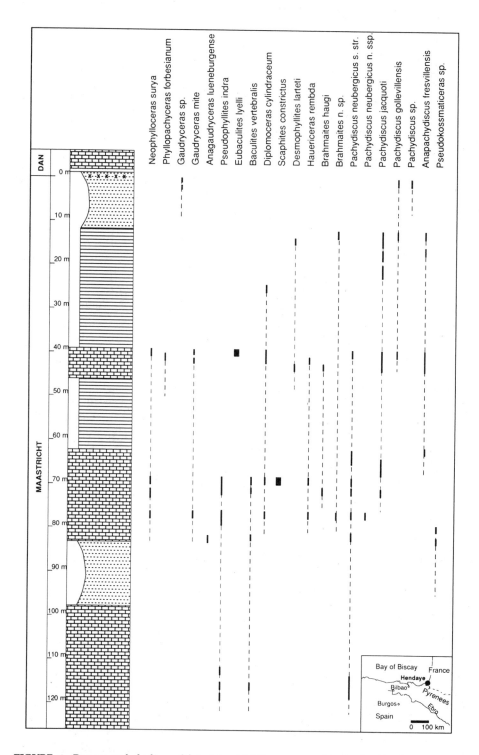

FIGURE 4. Ranges and decline of late Maastrichtian ammonoids in the Hendaye section, southwestern France.

Late Cretaceous (Wiedmann, 1969); for example, the Maastrichtian species had a much longer time range than the Turonian ones. It is also noteworthy that deep-water phylloceratids and lytoceratids were predominant among the last ammonoids. Two other important mollusc groups occupying the same habitat during the Cretaceous, i.e., belemnites and inoceramids, followed the same extinction pattern, the latter disappearing even somewhat earlier than the ammonoids (Wiedmann, 1988a,b).

2.4. Do Mesozoic Extinctions Have a Common Cause?

On the basis of the detailed knowledge we now have of all the Mesozoic system boundaries, which coincide with and are even based on major ammonoid extinctions, we can conclude that the pattern of ammonoid extinction was perfectly comparable, if not identical, at all these boundaries. In all these cases, the decline was continuous, and no sudden extinctions occurred. In most cases, the new forms that replaced the old ones developed long before the boundary, i.e., the Xenodiscaceae and Otocerataceae in the Permian (Spinosa *et al.*, 1975; Ruzhentsev, 1959), the Phyllocerataceae and, eventually, the Pinacocerataceae in the Triassic (Wiedmann, 1970, 1973a; Tozer, 1971, 1981), and the Ancyloceratina and Haplocerataceae in the Late Jurassic (Wiedmann, 1973b; Ziegler, 1974). Were the sexlobate Cretaceous Tetragonitaceae the prospective stock for Tertiary ammonoids, or were ammonoids genetically "exhausted" at the end of the Cretaceous? Both speculations have some support, but the final conclusion remains difficult. The evolutionary rate of ammonoids continuously decreased during the Late Cretaceous. However, the heteromorphic shapes of the Ancyloceratina do not indicate any genetic "exhaustion" ("Typolyse," Schindewolf, 1945), a theory that was completely refuted by Wiedmann (1969).

The Mesozoic deep-water stocks, i.e., the phylloceratids and lytoceratids, were not only well represented among the last few ammonoids, they generally exhibited the best survival rates; phylloceratids were the only form that unquestionably crossed the Triassic–Jurassic boundary. This may indicate that water depth was a factor in ammonite extinction. The pattern of major extinctions also suggests that explanations should be based on processes that operated over long time periods.

Few direct relations can be observed between plate tectonics and the diversity of ammonoids during the Mesozoic (Wiedmann, 1988b) except for (1) the breakup of Pangaea at the Paleozoic–Mesozoic boundary, (2) changing patterns of endemic versus cosmopolitan distributions, and (3) the presumed dependence of larger sea-level changes on pulses of ocean floor spreading.

To relate the course of ammonoid diversity to major sea-level changes, i.e., extinctions with regressions and radiations with transgressions, is by no means a new idea. It was put forward as a general explanation by Moore (1950) and Newell (1952) and has been discussed in the context of Mesozoic am-

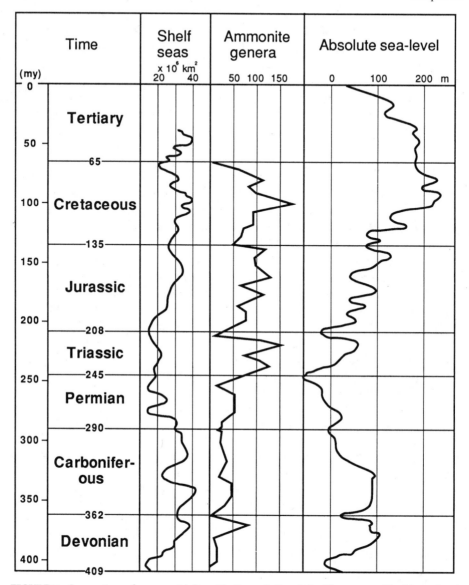

FIGURE 5. Comparison of ammonoid diversity through time (after House, 1985b) with sea-level fluctuations of shelf seas (after Sliter, 1976; Yanshin, 1973) and long-term eustatic sea-level changes (after Haq *et al.*, 1987; House, 1989).

monoids by Wiedmann (1969, 1973a, 1988c) and Hallam (1986, 1987) and in the context of Paleozoic ammonoids by House (1985a,b, 1989). The comparison of the pattern of ammonoid diversity through time with global sea-level changes or with first-order cycles in sequence stratigraphy (Fig. 5) shows such a striking similarity that the obvious conclusion is that they were related.

There is no need to involve a major cosmic impact to explain the final decline of ammonoids.

ACKNOWLEDGMENTS. This is publication no. 14 of the series "Database System GONIAT Tübingen." We thank D. Korn (GONIAT Tübingen) for valuable suggestions concerning data retrieval in the data base GONIAT as well as for helpful comments about the problems of Carboniferous time units. J. Lehmann and M. Wippich are thanked for help with Cretaceous literature. The referees (N. Landman and S. Klofak) provided thoughtful reviews that improved this paper.

References

Alvarez, L. W., Alvarez, W., Asaro, G., and Michel, H. V., 1980, Extraterrestrial cause for the Cretaceous–Tertiary extinction, *Science* **208**:1095–1108.

Arkell, W. J., Furnish, W. M., Kummel, B., Miller, A. K., Moore, R. C., Schindewolf, O. H., Sylvester-Bradley, P. C., and Wright, C. W., 1957, *Treatise on Invertebrate Paleontology*, Part L, *Mollusca 4*, Geological Society of America and University of Kansas Press, Lawrence, KS.

Arthur, M. A., Brumsack, H. J., Jenkyns, H. C., and Schlanger, S. O., 1990, Stratigraphy, geochemistry, and palaeoceanography of carbon-rich Cretaceous sequences, in: *Cretaceous Resources, Events and Rhythms* (R. N. Ginsburg and B. Beaudoin, eds.), Kluwer Academic Publisher, Norwell, MA, pp. 75–119.

Becker, R. T., 1993a, Analysis of ammonoid palaeobiogeography in relation to the global Hangenberg (terminal Devonian) and Lower Alum Shale (Middle Tournaisian) events, *Ann. Soc. Geol. Belg.* **115**:459–473.

Becker, R. T., 1993b, Anoxia, eustatic changes, and Upper Devonian to lowermost Carboniferous global ammonoid diversity, in: *The Ammonoidea: Environment, Ecology, and Evolutionary Change*, Systematics Association Spec. Vol. 47 (M. R. House, ed.), Clarendon Press, Oxford, pp. 115–163.

Claoué-Long, J. C., Jones, P. J., Roberts, J., and Maxwell, S., 1992, The numerical age of the Devonian–Carboniferous boundary, *Geol. Mag.* **129**:281–291.

Donovan, D. T., 1988, Evolution of the Arietitidae and their descendants, *Cah. Inst. Cath. Lyon, Ser. Sci.* **1**:123–138.

Donovan, D. T., Callomon, J. H., and Howarth, M. K., 1981, Classification of Jurassic Ammonitina, in: *The Ammonoidea*, Systematics Association Spec. Vol. 18 (M. R. House and J. R. Senior, eds.), Academic Press, London, pp. 101–155.

Föllmi, K. B., Weisser, H., Bisping, M., and Funic, H., 1994, Phosphogenesis, carbon-isotope stratigraphy and carbonate-platform evolution along the Lower Cretaceous northern Tethyan margin, *Geol. Soc. Am. Bull.* **106**:729–746.

Glenister, B. F., and Furnish, W. M., 1981, Permian Ammonoids, in: *The Ammonoidea*, Systematics Association Spec. Vol. 18 (M. R. House and J. R. Senior, eds.), Academic Press, London, pp. 49–64.

Hallam, A., 1986, The Pliensbachian and Tithonian extinction events, *Nature* **319**:765–768.

Hallam, A., 1987, Radiations and extinctions in relation to environmental change in the marine Lower Jurassic of northwest Europe, *Paleobiology* **13**:152–168.

Hancock, J. M., and Kennedy, W. J., 1993, The high Cretaceous ammonite fauna from Tercis, Landes, France, *Bull. Inst. R. Sci. Nat. Belg. Sci. Terre* **63**:149–170.

Hancock, J. M., Peake, N. B., Burnett, J., Dhondt, A. V., Kennedy, W. J., and Stokes, R. B., 1993, High Cretaceous biostratigraphy at Tercis, south-west France, *Bull. Inst. R. Sci. Nat. Belg. Sci. Terre* **63**:133–148.

Haq, B. U., Hardenbol, J., and Vail, P., 1987, Chronology of fluctuating sea level since the Triassic, *Science* **235**:1156–1167.

Harland, W. B., Armstrong, R. L., Cox, A. V., Craig, L. E., Smith, A. G., and Smith, D. G., 1990, *A Geologic Time Scale 1989*, Cambridge University Press, Cambridge.

House, M. R., 1983, Devonian eustatic events, *Proc. Ussher Soc.* **5**:396–405.

House, M. R., 1985a, Correlation of mid-Paleozoic ammonoid evolutionary events with global sedimentary perturbations, *Nature* **313**:17–22.

House, M. R., 1985b, The ammonoid time-scale and ammonoid evolution *Geol. Soc. Lond. Mem.* **10**:273–283.

House, M. R., 1989, Ammonoid extinction events, in: The Chronology of the Geological Record (E. J. Snelling, ed.), *Phil. Trans. R. Soc. Lond. [Biol.]* **325**:307–326.

Jones, P. J., 1988, Comments on some Australian, British and German isotopic age data for the Carboniferous System, *Newsl. Carbon Stratigr.* **6**:26–29.

Korn, D., and Kullmann, J., 1988, Changes in clymeniid diversity, in: *Cephalopods—Present and Past* (J. Wiedmann and J. Kullmann, eds.), Schweizerbart'sche Verlagsbuchhandlung, Stuttgart, pp. 25–28.

Korn, D., and Kullmann, J., 1993, GONIAT Database System, version 2.40, Tübingen.

Korn, D., Kullmann, J., Kullmann, P. S., and Petersen, M. S., 1994, GONIAT, a computer retrieval system for Paleozoic ammonoids, *J. Paleontol.* **68**:1257–1263.

Kullmann, J., 1983, Maxima im Tempo der Evolution karbonischer Ammonoideen, *Paläontol. Z.* **57**:231–240.

Kullmann, J., 1985, Drastic changes in Carboniferous ammonoid rates of evolution, in: *Sedimentary and Evolutionary Cycles* (U. Bayer and A. Seilacher, eds.), *Lect. Notes Earth Sci.* **1**:35–47.

Kullmann, J., 1994, Diversity fluctuations in ammonoid evolution from Devonian to mid-Carboniferous, *Cour. Forschungsinst. Senckenb.* **169**:137–141.

Kullmann, J., Korn, D., Kullmann, P. S., and Petersen, M. S., 1993, The Database System GONIAT—a tool for research on systematics and evolution of Paleozoic ammonoids, *Geobios Mém. Spec.* **15**:239–245.

Menning, M., 1992, Numerical time scale for the Permian, *Permophiles* **20**:2–5.

Moore, R. C., 1950, Evolution of the Crinoidea in relation to major paleogeographic changes in Earth history, *Rep. 18th Sess. Intern. Geol. Cong., Great Britain 1948* **12**:27–53.

Newell, N. D., 1952, Periodicity in invertebrate evolution, *J. Paleontol.* **26**:371–385.

Newell, N. D., 1967, Revolutions in the history of life, *Geol. Soc. Am. Spec. Pap.* **89**:63–91.

Raup, D. M., and Sepkoski, J. J., 1984, Periodicity of extinctions in the geologic past, *Proc. Natl. Acad. Sci. U.S.A.* **81**:801–805.

Roberts, J., Claoué-Long, J. C., and Jones, P. J., 1991, Calibration of the Carboniferous and Early Permian of the Southern New England Orogen by Shrimp Ion Microprobe Zircon Analyses, *Newsl. Carbon. Stratigr.* **9**:15–17.

Ruzhentsev, V. E., 1959, Klassifikatsia nadsemeistva Otocerataceae, *Paleontol. Zh.* **1959**(2):56–67.

Schindewolf, O. H., 1945, Darwinismus oder Typostrophismus? *Magy. Biol. Kutato Intezet. Munkai* **16**:104–177.

Schindewolf, O. H., 1954, Über die möglichen Ursachen der grossen erdgeschichtlichen Faunenschnitte, *N. Jb. Geol. Paläont. Mh.* **1954**:457–465.

Schlager, W., 1981, The paradox of drowned reefs and carbonate platforms, *Geol. Soc. Am. Bull.* **92**:197–211.

Sepkoski, J. J., 1986, Global bioevents and the question of periodicity, in: *Global Bio-Events* (O. H. Walliser, ed.), *Lect. Notes Earth Sci.* **8**:47–61.

Sliter, W. V., 1976, Cretaceous foraminifers from the southwestern Atlantic Ocean, Leg 36, Deep Sea Drilling Project, *Init. Reps. DSDP* **36**:519–537.

Spinosa, C., Furnish, W. M., and Glenister, B. F., 1975, The Xenodiscidae, Permian ceratoid ammonoids, *J. Paleontol.* **49**:239–283.

Teichert, C., 1990, The Permian–Triassic boundary revisited, *Lect. Notes Earth Hist.* **30**:199–238.

Tozer, E. T., 1971, One, two or three connecting links between Triassic and Jurassic ammonoids? *Nature* **232**:565–566.

Tozer, E. T., 1981, Triassic Ammonoidea: Classification, evolution, and relationship with Permian and Jurassic forms, in: *The Ammonoidea*, Systematics Association Spec. Vol. 18 (M. R. House and J. R. Senior, eds.), Academic Press, London, pp. 46–100.

Ward, P. D., 1988, Maastrichtian ammonite and inoceramid ranges from Bay of Biscay Cretaceous–Tertiary Boundary sections, in: *Palaeontology and Evolution: Extinction Events* (M. A. Lamolda, E. G. Kauffman, and O. H. Walliser, eds.), *Rev. Esp. Paleont.* núm. extraord., pp. 119–126.

Ward, P. D., and Wiedmann, J., 1983, The Maastrichtian ammonite succession at Zumaya, Spain, *Symposium Cretaceous Stage Boundaries, Abstracts*, pp. 203–208.

Wiedmann, J., 1968, Das Problem stratigraphischer Grenzziehung und die Jura/Kreide-Grenze, *Eclogae Geol. Helv.* **61**:321–386.

Wiedmann, J., 1969, The heteromorphs and ammonoid extinction, *Biol. Rev.* **44**:563–602.

Wiedmann, J., 1970, Über den Ursprung der Neoammonoideen—Das Problem der Typogenese, *Eclogae Geol. Helv.* **63**:923–1020.

Wiedmann, J., 1973a, Evolution or revolution of ammonoids at Mesozoic System boundaries, *Biol. Rev.* **48**:159–194.

Wiedmann, J., 1973b, Ancyloceratina (Ammonoidea) at the Jurassic/Cretaceous boundary, in: *Atlas of Paleobiogeography* (A. Hallam, ed.), Elsevier Scientific, Amsterdam, pp. 309–316.

Wiedmann, J., 1975, The Jurassic-Cretaceous boundary as one of the Mesozoic System boundaries, in *Coll. Int. Lim. Jur.–Crét., Lyon and Neuchatel 1973*, *Mém. Bur. Rech. Geol. Min.* **86**:14–22.

Wiedmann, J., 1986, Macro-invertebrates and the Cretaceous–Tertiary boundary, in: *Global Bio-Events* (O. H. Walliser, ed.), *Lect. Notes Earth Sci.* **8**:397–409.

Wiedmann, J., 1988a, Ammonoid extinction and the "Cretaceous–Tertiary boundary, event," in: *Cephalopods—Present and Past* (J. Wiedmann and J. Kullmann, eds.), Schweizerbart'sche Verlagsbuchhandlung, Stuttgart, pp. 117–140.

Wiedmann, J., 1988b, Plate tectonics, sea level changes, climate and the relationship to ammonite evolution, provincialism, and mode of life, in: *Cephalopods—Present and Past* (J. Wiedmann and J. Kullmann, eds.), Schweizerbart'sche Verlagsbuchhandlung, Stuttgart, pp. 737–765.

Wiedmann, J., 1988c, The Basque coastal sections of the K/T boundary: A key to understanding "mass extinction" in the fossil record, in: *Palaeontology and Evolution: Extinction Events* (M. A. Lamolda, E. G. Kauffman, and O. H. Walliser, eds.), *Rev. Esp. Paleont.* núm. extraord., pp. 127–140.

Wright, C. W., 1981, Cretaceous Ammonoidea, in: *The Ammonoidea*, Systematics Association Spec. Vol. 18 (M. R. House and J. R. Senior, eds.), Academic Press, London, pp. 157–174.

Yanshin, A. L., 1973, On so-called world transgressions and regressions, *Byull. Mosk. Ova. Ispyt. Prir. Otd. Geol.* **48**(2):9–44 (in Russian).

Ziegler, B., 1974, Über Dimorphismus und Verwandtschaftsbeziehungen bei Oppelien des oberen Juras, *Stuttg. Beitr. Naturkd. Ser. B* **11**:1–43.

Ziegler, W., and Lane, H. R., 1987, Cycles in conodont evolution from Devonian to mid-Carboniferous, in: *Palaeobiology of Conodonts* (R. C. Aldridge, ed.), British Micropalaeontological Society, London, pp. 147–163.

Chapter 20

Ammonoid Extinction

PETER WARD

1. Introduction

With a range of over 300 million years, the ammonoids represent one of the most successful groups of organisms in all of earth history. Their very success, however, makes the nature of their demise especially curious. In this short review I examine questions pertaining to extinction in ammonoids.

There have been several recent excellent discussions about ammonoid extinction or, perhaps more accurately, the various "outages," "crises" (Teichert, 1985), or "fluctuations" (House, 1993) of this long-lived group. Other recent summaries include the work of Ward (1983), Becker (1993), Westermann, (1990; Chapter 16, this volume), and Wiedmann and Kullman (Chapter 19, this volume). My goal here is to differentiate the various factors that may have been involved in diversity changes within the group. As most of my personal research has dealt with Cretaceous ammonoid faunas, I concentrate on this part of ammonoid history.

PETER WARD • Department of Geological Sciences, University of Washington, Seattle, Washington 98195.

Ammonoid Paleobiology, Volume 13 of *Topics in Geobiology*, edited by Neil Landman *et al.*, Plenum Press, New York, 1996.

2. Factors Affecting Ammonoid Diversity

The initial evolution of the ammonoids from orthoconic ancestors in the Silurian Period involved coiling of the shell, development of a marginal siphuncle, folding and fluting of the septa, and, perhaps most importantly of all (with regard to future evolutionary tempo of the group), a fundamental change in reproductive strategy. The ammonoids, throughout their long reign, hatched at a small size (usually less than or equal to 1 mm in diameter) and presumably spent some period of time freely floating in the plankton after hatching. In contrast, most nautiloids apparently assumed a benthic mode of life immediately after hatching. From the Devonian onward, the ammonoids showed origination and extinction rates far exceeding other groups of mollusks; I believe this difference in evolutionary tempo (at least for the nautiloids) is related to differences in reproductive strategy.

The actual factors leading to the characteristically high extinction rates of ammonoids can be differentiated into biological and (nonbiological) environmental factors. The biological factors included aspects of competition, predation, and habitat depth, especially as the last was related to growth rate and buoyancy control. Nonbiological environmental factors included the global events so elegantly and recently discussed by House (1993), including small-scale events (anoxia, transgression, cooling), medium-scale (global) events (such as offlap and onlap, bolide impact), and the large-scale events that House viewed as some highly lethal combinations of the former. Each of these is described below.

2.1. Biological Factors

2.1.1. Competition

Direct paleontological evidence of biological factors is rare in the fossil record. Hence, conclusions about these biological factors most often depend on inference. Although evidence of predation (and even the ancient predator) can sometimes be identified by the nature of shell breaks on cephalopod fossils, the identify of competitors is at best a guess.

Almost all investigators have agreed that ammonoids were mobile carnivores. (Virtually the sole opponent of this view has been Klaus Ebel, who has argued that ammonoids were entirely benthic and were incapable of swimming above the bottom: Ebel, 1985, 1992.) The phragmocone system of (presumably) neutral buoyancy attainment, adjudged as such based on the numerous studies of *Nautilus* and *Sepia* buoyancy attainment and use, has left us a picture of variably rapid swimming creatures of predatory aspect. As such, their most obvious competitors would have been nonammonoid cephalopods (the nautiloids and coleoids) and the various fish.

One of the few methods of elucidating competitive effects comes from analysis of either morphological or habitat change of one particular group following the evolution or appearance of a second group. Within the cephalopods, it is clear that the evolution of the ammonoids in the Devonian created widespread and apparently catastrophic effects on the Devonian nautiloids. This can be seen in two ways. First, soon after the first evolution of the ammonoids, the number of nautiloids relative to ammonoids markedly diminished. This occurred both in species-level diversity (Dzik, 1984) as well as in relative numbers of individuals (W. B. Saunders, personal communication). Second, the very nature of nautiloid shell shapes (and their relative abundance to one another) changed as well. Pre-Devonian nautiloid shells belonged to a wide spectrum of types, including planispirals, orthocones, cyrtocones, and torticones. Following the Devonian, however, most nautiloids were planispirally coiled with a minor number of orthocones (Fig. 1). From the Triassic onward, all nautiloids were planispirally coiled. Ward (1980, 1987) showed that Jurassic and Cretaceous nautiloids utilized a spectrum of planispiral shell shapes that were generally more involute and had higher whorl-expansion rates than the majority of ammonoid shells of those periods (higher W, lower D, and higher S, using the parameters of Raup, 1967). Following the extinction of the ammonoids at the end of the Cretaceous, many Tertiary nautiloid species assumed shell shapes that reoccupied portions of the ammonoids' WDS morphospace (Ward, 1980).

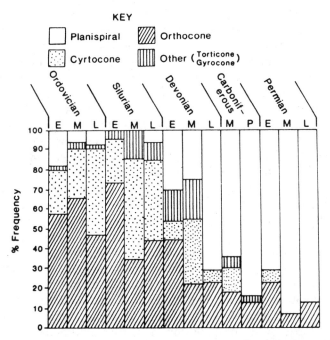

FIGURE 1. Chambered cephalopod shell shape categories (from Ward, 1987, Fig. 7.4).

The other cephalopod group that may have competed with the ammonoids was the Coleoidea. Many workers have suggested that ammonoids were more "coleoid-like" than "nautiloid-like," and this theme recently was reiterated in detail by Jacobs and Landman (1993). With the loss of an external, calcareous shell, the Mesozoic coleoids were probably more rapid and maneuverable than were the majority of ammonoids and may indeed have competed with ammonoids for food. There is no direct evidence for this, however.

The final group that may have competed with the ammonoids for food or living space were various fish groups. Because all present-day cephalopods and most marine fish are predators of one sort or another, it is reasonable to assume that the ammonoids interacted competitively with various fish, just as modern coleoids do today with bony fish (Packard, 1972). Again, however, there is no direct evidence of this, and whether these interactions (if they occurred at all) had any effect on ammonoid extinction can only be left to supposition.

2.1.2. Predation

Adult-cephalopods today are preyed upon mainly by cetaceans, fish, and other cephalopods (and humans); the young fall prey to a far wider spectrum of creatures, including many invertebrates. The predators on ammonoids are poorly known, with the only direct evidence that can be reliably related to past predators coming from bite or tooth marks, such as those in Cretaceous ammonites ascribed to mosasaurs (Chapter 15, this volume). There is abundant evidence, however, suggesting that predation by shell-breaking predators commonly occurred, for break marks are common on Jurassic and Cretaceous ammonites and are known from earlier forms as well. The evolution of shell-breaking (durophagous) predators has been one of the dominant themes of the Mesozoic, according to Vermeij (1977), with teleost fish, sharks, skates, rays, gastropods, and a variety of crustaceans all having evolved the ability to attack and penetrate previously impenetrable external shells. Vermeij has pointed out that this phenomenon, which he dubbed the Marine Mesozoic Revolution, intensified during the Jurassic and Cretaceous Periods.

By the very nature of the design, ectocochliate cephalopods are especially vulnerable to shell breakage of any sort, not necessarily just by action by predators. By analogy with *Nautilus* and *Sepia*, we can be fairly sure that ammonoids maintained neutral buoyancy in the sea; that is, they had neither a positive nor a negative weight if weighed in sea water. This balance is produced by the lower-density phragmocone offsetting the higher (than sea water) density of the shell and soft parts in the body chamber. The external shell around the body chamber is by far the densest part of this system, and, thus, removal of even small amounts of this shell material creates a great buoyancy imbalance. In experiments with present-day *Nautilus*, Ward (1986a, 1987) showed that removal of as little as 5 g of shell material from a mature *Nautilus* will trigger a compensatory response in the buoyancy system, whereas removal of between 7 and 10 g will cause the animal to rise to the surface uncontrollably, an action generally lethal to the *Nautilus*. Shell breakage, then, is doubly dangerous to an ectocochliate cephalopod: breakage of shell not only exposes the more vulnerable soft parts but, at the same time, creates a dangerous buoyancy imbalance. Because of this, we might expect to see some sort of evolutionary response to the danger of shell breakage in the history of the group. Just such a response was indeed present, namely, the addition of shell ornament.

Ward (1981) showed that degree of ornament did increase in the ammonoids over the course of their evolutionary history. Coarse ornament, which appears to be defensive in nature, was virtually unknown during the Paleozoic but became increasingly common during the Mesozoic, especially during the early and middle Cretaceous (Fig. 2). Eventually, however, we could expect that the penalty exacted by increasing ornament (lowered buoyancy and maneuverability) would offset gains produced by greater protection. The acme of coarse ornament was reached during the Albian and Cenomanian Stages of the Cretaceous, and thereafter it diminished. I have hypothesized that the ammonite design essentially became obsolete in shallow-shelf communities, places where ammonites had dominated for 300 million years (Ward, 1987). The response was exploitation of new niches, specifically the midwater niche still favored and dominated by cephalopods today. Design changes during the Cretaceous, especially the evolution of heteromorphic or nonplanispiral forms, may have been to morphologies adapted to a passive midwater existence in a habitat devoid of shell-breaking predators (Ward, 1986b, 1987).

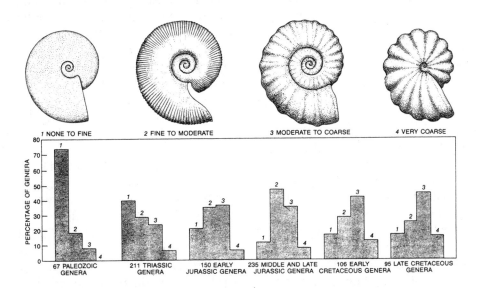

FIGURE 2. Ammonoid shell ornament categories (from Ward, 1983).

I have argued that Cretaceous ammonites exploited both mid- and deep-water benthic habitats during the latter part of the Cretaceous Period, two ecological regions not commonly inhabited by pre-Cretaceous ammonoids. This type of habitat adjustment may also have been adopted by the nautiloids. Throughout much of the Cretaceous Period, and perhaps during most of their Paleozoic and Mesozoic existence as well, the nautiloids were essentially shallow-water creatures; even up to the early part of the Late Cretaceous, nautiloids such as *Eutrephoceras* and *Cymatoceras* are most commonly found in shallow-water deposits. This was probably as much for physiological as ecological (resource) reasons: because of the nature of the phragmocone buoyancy system, growth rates are inversely related to ambient pressure and, hence, water depth. Emptying of chambers in phragmocone-bearing cephalopods is accomplished by an osmotic pump (Denton and Gilpin-Brown, 1966). This system, however, becomes increasingly inefficient with increasing depth and ceases to function (in *Nautilus*) at depths of about 300 m (Ward, 1987). Because of this, the time necessary for growth to maturity may be as much as 15 years or more at the 200 to 300 m average habitat depths favored by all living *Nautilus* species, whereas full growth to maturity in aquarium-maintained specimens of the same species can be accomplished in as little as 2 to 3 years (Ward, 1987). Seemingly, the deep-water habitat now inhabited by the last nautiloids exacts a terrible penalty, because their slow rate of growth keeps them in the dangerous juvenile growth stages for inordinately long periods of time.

I have hypothesized that *Nautilus* occupies its current deep fore-reef-slope habitat not because it is a good place to live but because the plethora of visual, shell-breaking predators have forced them out of the shallow-shelf habitat where the first cephalopods evolved and where the majority of the chambered cephalopods (including most ammonoids) lived, at least until the middle part of the Cretaceous. After that, predation pressure from the shell-breaking predators forced the ammonoids (and nautiloids) into less favorable habitats: among the ammonoids, it was off the shelf and up into the water column, and among the nautiloids, it was into deeper-water shelf habitats. The nightly vertical migration observed in *Nautilus* (Carlson *et al.*, 1984; Ward *et al.*, 1984; Ward, 1987) is probably also an antipredator adaptation.

By the later ages of the Cretaceous, the ammonoids, ever the stuff of great evolutionary novelty and plasticity, had remade themselves: no longer shallow-shelf creatures, almost half of their numbers had colonized the midwater, whereas the rest were either rapid swimmers (the neoceratites) or deep-shelf forms (Ward, 1986b). In my view, these survivors pulled off one of the great evolutionary changeovers in the history of life, and (in my estimation) they would be here still but for an unprecedented catastrophe, the extraordinary Chicxulub Impact event that ended the Cretaceous world and more than half of its species.

2.2. Environmental Factors

Environmental perturbations affecting the course of ammonoid evolution have been summarized recently and concisely by House (1993), and I will not repeat his excellent discussion. Suffice it to say that it is clear that ammonite diversity was clearly affected by sea-level change and by oceanic anoxic events. Because many ammonoids appeared to inhabit epeiric seas, it is perhaps no surprise that short-term sea-level drops (thus eliminating large areas of such regions) adversely affected ammonoid diversity. Similarly, global anoxia, with its devastating effects on the biota of shallow-shelf regions, was also a major ammonoid killer.

The two major external factors so widely cited as causes of local ammonoid extinction are regression and anoxia. It is instructive to consider the taphonomic consequences of each. Regression reduced the amount of preserved rock material and should have altered the sample of ammonoids noticeably. It has been widely shown that the surface area of rock available of a given age is one of the most important factors in determining sampled species richness (see Signor, 1985, for a summary). Moreover, reduction of habitat area during a regression would lead to extinctions because of species area effects. However, the amount of diversity decrease that is related not to actual extinction but to inadequate sampling is to be determined. In a situation where regression empties shallow seas, ammonoid faunas may have been restricted to narrow continental margins, which themselves are both less areally extensive (than epeiric seas) and more susceptible to tectonic deformation than continental interiors; the perceived reduction in ammonoid diversity following regression thus may be more apparent than real.

Surely more lethal to ammonoids than regressions were periods of anoxia. There is now well-documented literature on global anoxic events, and there is good comprehension of the relationship between these and ammonoid extinctions. House (1993) and Becker (1993) showed that Paleozoic anoxic events led directly to decline and, in some cases, catastrophic decline in ammonoid abundance. Other cases are known from the Mesozoic, with the Cenomanian–Turonian event, accompanied by worldwide decrease in ammonite abundance, one of the best-known examples.

3. The Final Decline and Extinction of the Ammonoids: Diversity Analyses and Long-Term Cretaceous Decline

The final extinction of the ammonoids has been long discussed and documented. Most workers ascribe it to some long, multicausal series of events, with the final extinction only capping a long period of decline. This point of view was summarized most recently by House (1993, p. 28): "But, as is well known, the group [ammonoids] had been in decline. . . ." This statement echoes other, virtually identical declarations on the subject (Wiedmann, 1986; Teichert, 1985; Hallam, 1987). House used family-level data to support his claim (House, 1993, Fig. 2.4). He showed a Cretaceous maximum of 23 families in the Cenomanian Age. This number dropped to 15 families by the Santonian Age and then to 14 during the Campanian Age. It should be noted, however, that the 14 families found during the Campanian Age are not significantly different from the average number of families found throughout the Cretaceous, based on House's own data (from which I have computed a mean of 14.6 families per interval for the 35 intervals compiled by House). The Campanian Age, therefore, was not aberrant for the Cretaceous and indeed showed a higher number of families than did the stages of the Jurassic Period, with a mean number of 9.2 families for the 34 intervals tabulated by House.

I interpret the drop in families between the middle part of the Cretaceous and the Campanian Age as being caused largely by the disappearance of the heavily armored shelf ammonites. These forms were virtually gone from the record by Campanian time. Their disappearance, however, did little to alter world ammonoid diversity when viewed in the context of the entire Jurassic and Cretaceous. Many new forms arose during this interval (particularly among the heteromorphs). However, these new taxa arose within already existing families. House (1993) makes much of the lack of new families evolving during the last stages of the Cretaceous; in terms of the degree of evolutionary innovation, however, the end of the Cretaceous was one of the most evolutionarily fecund in the history of the group rather than a sterile period as suggested by House.

And what of the Maastrichtian Age, the last of the Cretaceous? If long-term decline was the norm, we should see a steady reduction in ammonoid families and genera during this interval, and, indeed, a seemingly catastrophic decline is apparent in House's diagram. However, the numbers shown by House do not match currently known and documented Maastrichtian ammonoid diversity, much of which was discovered by intensive collecting during the last decade.

The last known episode of global anoxia in the seas is documented for the end of the Campanian (Kaiho, 1994), and this appears, as all such episodes did, to have caused extinction and diversity reduction among ammonoids. For instance, House (1993, Fig. 2.2) showed a nearly 50% extinction rate of families at the end of the Campanian, whereas an ammonoid generic data set compiled by J. J. Sepkoski and published by Collom (1986) showed a 40% generic extinction rate at the end of the Campanian. During the Maastrichtian, however, virtually no extinction took place until the very end of the age (Collom, 1986). My analysis of a generic data set culled from taxonomic literature (Ward, 1990) yielded similar results: there was little extinction during the Maastrichtian. Thirty-five ammonite genera are known from the lower Maastrichtian, and 27 from the Upper. This finding is even more impressive because sampling is skewed in favor of the Lower Maastrichtian by the larger number of lower than upper Maastrichtian localities.

In summary, following an end-Campanian anoxic event, which did reduce world ammonoid diversity, ammonoid numbers stayed approximately constant until the end of the Maastrichtian. The ammonites were then suddenly extinguished from the face of the earth.

The final extinction of the ammonoids was undoubtedly brought about by the effects following the impact of the Chicxulub Comet. Leaving a crater 300 km in diameter (Sharpton *et al.*, 1994), this impact event was by far the largest of the Phanerozoic and has left the largest recognizable impact crater on the earth of any age. The virtual extinction of the plankton following this impact was probably the actual killing mechanism of the ammonites, a view recently subscribed to as well by Kennedy (1993). As both Landman (1984) and I (Ward, 1983, 1987) have hypothesized earlier, the survival of the nautiloids may have been in part from the characteristic reproductive strategy of producing large eggs at depth. Perhaps all adult nautiloids were exterminated soon after the impact, but their deep eggs provided the seeds of nautiloid survival. The ammonites, hatching directly into the plankton, were sudden victims.

The suddenness of ammonite extinction at the various K–T boundary sites has now been documented, and the supposed survival or lowered extinction rate at the high-latitude sites (Keller, 1994; Zinsmeister and Feldman, 1994) may not be as pronounced as previously thought. Recent statistical analyses of the Seymour Island sites (Marshall, 1995) are entirely compatible with a sudden extinction.

The pattern of ammonoid disappearance at the K–T boundary is seemingly mimicked at the Triassic–Jurassic boundary. This event may have been brought about by impact as well (although a far smaller impact) and should provide a new frontier for ammonoid undertakers.

ACKNOWLEDGMENTS. This effort was considerably strengthened by the critiques of the editors.

References

Becker, R. T., 1993, Anoxia, eustatic changes, and Upper Devonian to lowermost Carboniferous global ammonoid diversity, in: *The Ammonoidea: Environment, Ecology and Evolutionary Change*, Systematics Association Spec. Vol. 47 (M. R. House, ed.), Clarendon Press, Oxford, pp. 115–164.

Carlson, B., McKibben, J., and de Gruy, M., 1984, Telemetric investigation of vertical migration of *Nautilus* in Palau, *Pac. Sci.* **38**:183–188.

Collom, C., 1986, Extinction rates in Cretaceous ammonites, in: *Global Bio-Events* (O. H. Walliser, ed.), *Lect. Notes Earth Sci.* **8**:249–258.

Denton, E., and Gilpin-Brown, J., 1966, On the buoyancy of the pearly Nautilus, *J. Mar. Biol. Assn. U.K.* **46**:723–759.

Dzik, J., 1984, Phylogeny of the Nautiloidea, *Acta Palaeontol. Pol.* **45**:1–219.

Ebel, K., 1985, Gehäusespirale und Septenformen bei Ammoniten unter Annahme vagil benthsicher Lebensweise, *Paläont. Z.* **59**:109–123.

Ebel, K., 1992, Mode of life and soft body shape of heteromorph ammonites, *Lethaia* **25**:179–193.

Hallam, A., 1987, End Cretaceous mass extinction event; argument for terrestrial causation, *Science* **238**:1237–1242.

House, M. R., 1993, Fluctuations in ammonoid evolution and possible environmental controls, in: *The Ammonoidea: Environment, Ecology and Evolutionary Change*, Systematics Association Spec. Vol. 47 (M. House, ed.), Clarendon Press, Oxford, pp. 13–34.

Jacobs, D., and Landman, N., 1993, *Nautilus*—a poor model for the function and behavior of ammonoids? *Lethaia* **26**:101–111.

Kaiho, K., 1994, Planktonic and benthic foraminiferal extinction events during the last 100 m.y., *Palaeogeogr. Palaeoclimatol. Palaeoecol.* **111**:45–71.

Keller, G., 1994, Effects of the KT boundary event: Mass extinction restricted to low latitudes, *L.P.I. Contrib.* **825**:57–58.

Kennedy, W., 1993, Ammonite faunas of the European Maastrichtian; diversity and extinction, in: *The Ammonoidea: Environment, Ecology and Evolutionary Change*, Systematics Association Spec. Vol. 47 (M. R. House, ed.), Clarendon Press, Oxford, pp. 285–326.

Landman, N., 1984, Not to be or to be? *Nat. Hist.* **93**:34–41.

Marshall, C., 1995, Distinguishing between sudden and gradual extinctions in the fossil record: Predicting the position of the Cretaceous-Tertiary iridium anomaly using the ammonite fossil record on Seymour Island, Antarctica, *Geology* **23**(8):731–734.

Packard, A., 1972, Cephalopods and fish, the limits of convergence, *Biol. Rev.* **47**:241–307.

Raup, D., 1967, Geometric analysis of shell coiling: Coiling in ammonoids, *J. Paleontol.* **41**:43–65.

Sharpton, V., Marin, L., and Schuraytz, B., 1994, The Chicxulub multiring basin: Evaluation of geophysical data, well logs, and drill core samples, *L.P.I. Contrib.* **825**:108–110.

Signor, P., 1985, Real and apparent trends in species richness through time, in: *Phanerozoic Diversity Patterns* (J. Valentine, ed.), Princeton University Press, Princeton, pp. 129–141.

Teichert, C., 1985, Crises in cephalopod evolution, in: *Mollusks—Notes for a short course*, (D. Bottjer, C. Hickman, and P. Ward, eds.), *Univ. Tenn. Dept. Geol. Sci. Stud. Geol.* **13**:202–214.

Vermeij, G., 1977, The Mesozoic marine revolution: Evidence from snails, predators, and grazers. *Paleobiology* **3**:245–258.

Ward, P., 1980, Comparative shell shape distributions in Jurassic–Cretaceous ammonites and Jurassic–Tertiary nautilids, *Paleobiology* **6**(1):32–43.

Ward, P., 1981, Shell sculpture as a defensive adaptation in ammonoids, *Paleobiology* **7**(1):96–100.

Ward, P., 1983, The extinction of the ammonites, *Sci. Am.* **249**:136–147.

Ward, P., 1986a, Rates and processes of compensatory buoyancy change in *Nautilus macromphalus*, *Veliger* **28**:356–368.

Ward, P., 1986b, Cretaceous ammonite shell shapes, *Malacologia* **27**:3–28.

Ward, P., 1987, *The Natural History of Nautilus*, Allen & Unwin, London.

Ward, P., 1990, A review of Maastrichtian ammonite ranges, *Geol. Soc. Am. Spec. Pap.* **247**:519–530.

Ward, P., Carlson, B., Weekly, M., and Brumbaugh, B., 1984, Remote telemetry of daily vertical and horizontal movement in *Nautilus*, *Nature* **309**(5965):248–250.

Westermann, G. E., 1990, New developments in ecology of Jurassic–Cretaceous ammonoids, in: *Atti del Secondo Convegno Internazionale, Fossili, Evoluzione, Ambiente, Pergola, 1987* (G. Pallini, F. Cecca, S. Cresta, and M. Santantonio, eds.), Tectnostampa, Ostra Vetere, Italy, pp. 459–478.

Wiedmann, J., 1986, Macro-invertebrates and the Cretaceous–Tertiary boundary, in: *Global Bio-Events* (O. H. Walliser, ed.), *Lect. Notes Earth Sci.* **8**:397–409.

Zinsmeister, W., and Feldmann, R., 1994, Antarctica, the forgotten stepchild: A view from the high southern latitudes, *L.P.I. Contrib.* **825**:134–135.

GLOSSARY

This is a selection of terms used in this volume. Most of them are morphological; for additional such terms and illustrations, the reader is referred to the *Treatise on Invertebrate Paleontology* (Arkell *et al.*, 1957, Arkell, 1957). For definitions of the forma-type terms used in discussions of ammonoid pathology, the reader is referred to Chapter 15 (this volume).

Acceleration reaction: The force required to accelerate added mass. This force is likely to be an important component of an ammonoid's locomotor energy costs, particularly at small size. It is especially important in unsteady swimming modes that characterize most cephalopods.

Adapertural (= adoral): Toward the aperture.

Adapical: Away from the aperture.

Added mass: The mass of liquid entrained in an ammonoid's boundary layer and wake. Because this fluid must be accelerated, it results in an additional cost associated with acceleration, e.g., acceleration reaction.

Adoral: Same as adapertural.

Adult stage: The ontogenetic stage that begins with the onset of sexual maturity, which, in ammonoids, is believed to be demarcated by a number of morphological changes in the shell, most of which affect the final body chamber and last few septa. (compare: neanic stage, gerontic stage)

Advolute: Refers to an evolute, planispirally coiled shell in which the dorsum of one whorl just touches the venter of the preceding whorl. (compare: evolute, involute)

Aerobic (= oxic): Having 1 liter of dissolved oxygen or more per cubic meter of water. (compare: anoxic, dysoxic, hypoxic)

Ammonitella: The embryonic shell of an ammonoid; also used to describe the embryonic animal as a whole, including the shell. (compare: initial chamber [= protoconch], neanoconch)

Amphichoanitic septal neck: A septal neck that is projected both adapically and adorally from the septum. (compare: modified retrochoanitic, prochoanitic, retrochoanitic)

Anaerobic: Same as anoxic.

Anaptychus: A single plate found in the body chamber of some ammonoids that is organic in composition. It is generally considered to have been the lower jaw of the animal, although some workers are of the opinion that it may have served as an operculum. (compare: aptychus)

Ancylocone, ancyloconic: Refers to a shell consisting of a closed or open planispiral or helical part followed by a straight shaft and hook, e.g., *Ancyloceras*. (see: coiled)

Angular deposit: A deposit between the proximal part of a septal neck and the cuff in prochoanitic septal necks.

Angusteradulata: The taxon consisting of the Ammonoidea and Coleoidea characterized by the possession of a narrow radula with seven teeth in each transverse row. (compare: Lateradulata)

Annulation: One of a series of strong ridges or plications extending across the shell at an angle of approximately 90° to the direction of growth. (see: ornament)

Anoxic (= anaerobic): Having a level of dissolved oxygen below 0.1 liter per cubic meter of water. (compare: aerobic, anoxic, dysoxic; see: exaerobic, poikiloaerobic)

Anticlastic curvature (= negative Gaussian curvature): A surface of double curvature with the orthogonal radii of curvature on opposite sides of the surface, e.g., cooling tower morphology. (see synclastic curvature)

Antidimorph: In a species in which sexual dimorphism is present, one of the two opposite members of the dimorphic pair.

Apical: Refers to the area of the apex or beginning of the shell.

Approximation: See septal approximation.

Aptychus: One of a pair of calcareous plates found in the aperture or body chamber of some ammonoids. These structures have been variously interpreted as opercula and as jaws. (compare: anaptychus)

Auxiliary deposit: A spherulitic–prismatic deposit on the adoral side of a septal neck.

Axial radius of curvature: The radius of curvature in the axial plane of a fold.

Basal lamellae: The first lamellae in a cuff or auxiliary deposit, which differ in their degree of calcification from the other lamellae.

Bathypelagic: Pelagic below a depth of 1000 m. (see: pelagic)

Beak: See jaw.

Beccublast: One of the columnar cells that secrete the tissues of the jaw.

Bending moment and stress: The development of a stress gradient from compression on the concave surface to tension on the adjacent convex surface of a beam–like shell element. A central plane ("neutral axis" of zero bending stress) separates these two equal layers of the shell.

Benthic: Refers to organisms that live in association with the bottom. (compare: demersal, pelagic)

Benthic feeding: See demersal.

Benthopelagic: See demersal.

Biome: A major ecological "zone." On land, biomes are characterized by climatic conditions (temperature, precipitation, etc.); in the ocean, they are characterized by water depth, temperature, salinity, etc.

Body chamber (= living chamber): That portion of the shell of an ammonoid not subdivided into camerae; during life, the bulk of the body of the animal occupied the body chamber. (compare: phragmocone)

Body chamber angle (= body chamber "length"): The angular "length" of the body chamber measured from the septal neck (medial saddle of the external lobe) of the ultimate septum to the peristome, excluding lappets and rostra; defined as the angle of rotation of the generating curve around the coiling axis. (see: brevidomic, mesodomic, and longidomic)

Body chamber "length": Same as body chamber angle.

Brevicone, breviconic: Refers to a short, stout orthocone or cyrtocone. (compare: longicone)

Brevidomic (= microdomic): Having a body chamber of approximately one-half whorl or less. (compare: mesodomic and longidomic)

Buckling: The sudden compressive failure of a structure caused by the inadequate flexural stiffness of its design and the inadequate stiffness of the material of which it is built. The structure bends and then breaks in tension without being subjected to an increased load.

Buoyancy: The upward force exerted by a fluid on an immersed object. Total buoyancy is equal to the weight of the fluid displaced by the object. What is called "neutral buoyancy" exists when an object's buoyancy exactly balances its weight; this is the condition that most ammonoids are thought to have approximated, which allowed them to float without rising or sinking in the water column.

Cadicone, cadiconic: Refers to a subglobular shell with a deep umbilicus, e.g., *Cadoceras*. (see: coiled)

Caecum: The bulb-like beginning of the siphuncle.

Calcification front: The zone along which calcification of the primary organic shell of the ammonitella occurs.

Callus: Same as umbilical plug.

Camerae: The chambers into which the phragmocone is subdivided by septa.

Cameral membranes: Thin, organic membranes that cover the adapical and adoral surfaces of septa and the outer surfaces of septal necks and the siphuncular tubes in ammonoids; homologous with the brown membranes (pellicle) within camerae of *Nautilus, Spirula,* and *Sepia.* (see: pseudoseptum)

Center of buoyancy: The point through which the line of action of the buoyant forces generated by an immersed object passes. It generally is equivalent to the center of gravity of the water displaced by the object.

Center of gravity (= center of mass): The point through which the line of action of an object's weight passes. In most ammonoids, the center of gravity is determined by the mass distribution of shell material and body tissues.

Center of mass: Same as center of gravity.

Central conchiolin accumulations: Same as central organic accumulations.

Central organic accumulations (= central conchiolin accumulations): Relatively long elements of conchiolin oriented perpendicular to the lamination in the nacreous layer of ammonoids and occupying the central portions of the stacks of nacreous tablets.

Coefficient of drag: A dimensionless number that relates the magnitude of the drag force to the shape and speed of an object and to the density and viscosity of the fluid through which the object moves. It commonly is considered an index of the hydrodynamic efficiency of an object, i.e., of the efficiency with which thrust is converted to velocity.

Coiled: Refers to any shell that is so curved as to have at least one full whorl. A tremendous variety of coiling is exhibited within the Ammonoidea. Various kinds are listed separately in this glossary: ancyloconic, advolute, cadiconic, discoconic, elliptospheroconic, evolute, gyroconic, hamitoconic, heteromorphic, involute, "normal" coiling, planispiral, planorbiconic, playtyconic, scaphitoconic, serpenticonic, spheroconic, torticonic, vermiconic.

Columnar nacre: Same as stack nacre.

Compressed: Refers to a whorl cross section that is higher than wide. (compare: depressed)

Conchorhynch: See jaw.

Connecting ring (= siphuncular tube): The tube that extends from the adoral side of one septal neck to the adapical side of the septal neck of the adjacent septum.

Constriction: A radial or subradial groove on the shell; it may appear as a complete flexure in the shell wall or only manifest itself on the surface. (see: ornament)

Costa: Same as rib.

Cost of transportation: The amount of energy required for the movement of a unit of mass a unit of distance. This metric permits comparison of efficiency of transportation among various organisms and at various swimming speeds.

Cuff: A spherulitic–prismatic deposit that commonly occurs on the adapical sides of prochoanitic septal necks and of type B modified retrochoanitic septal necks.

Cusp: A projection on a radular tooth.

Cyrtocone, cyrtoconic: Refers to a shell that is curved but is not coiled (i.e., it is less than one whorl), e.g., *Cyrtoceras*. (compare: orthoconic)

D: See fractal dimension.

Decoupling: A possible mode of chamber emptying in which the liquid reservoir within a camera becomes isolated from the siphuncle during the emptying process.

Demersal: Swimming or floating above the sea floor and dependent on it as a food source; hitherto, this way of life has been designated variously as benthopelagic, nektobenthic, or epibenthic.

Demersal saltator: An organism that swims mainly vertically, presumably to escape from benthic predators or anoxia.

Depressed: Refers to a whorl cross section that is wider than high. (compare: compressed)

Direct development: A mode of early ontogenetic development in which there is no larval stage. (compare: indirect development)

Discocone, discoconic: Refers to an involute, planispiral shell with compressed, ovate whorls, e.g., *Nautilus*. (see: coiled)

Dorsal wall: In normally coiled ammonoids, the wall bounded by the left and right umbilical seams and superimposed on the ventral wall of the preceding whorl.

Drag bands: Spiral imprints appearing as grooves or color lines on the steinkerns of some ammonoids; they currently are interpreted as traces showing the movement of the rear of the soft body within the shell.

Drag force: The force resisting steady motion of an object through a fluid. It is produced by friction generated as the fluid slides over the surface of the object and also by pressure inequalities set up around the object. Drag force depends on the object's shape and speed and on the density and viscosity of the fluid.

Dysaerobic: Same as dysoxic.

Dysoxic (= dysaerobic): Having a level of dissolved oxygen between 0.1 and 1.0 liter per cubic meter of water. (compare: aerobic, anoxic, hypoxic; see: exaerobic, poikiloaerobic)

Elastic: Describes a material that reacts immediately to the application of a stress, for only as long as it is present, and upon its removal, immediately reverts back to the prestressed state.

Elastic constants: See Young's modulus (E) and Poisson's ratio.

Elliptospherocone, elliptospheroconic: Refers to a planispiral, subglobular shell in which the adult body chamber partially uncoils, e.g., *Sphaeroceras.* (see: coiled)

Encrusting layer: A layer secreted in some normally coiled ammonoids that covers the shell from the outside, except along the dorsal part of the whorl.

Epeiric: Pertaining to inland seas, i.e., those within continents. (compare: epicratonic)

Epeiropelagic: Pelagic within an epeiric sea. (see: pelagic)

Epibenthic: See demersal.

Epicontinental: Same as epicratonic.

Epicratonic (= epicontinental): Pertaining to both epeiric and neritic environments and biomes.

Epinektic: Nektic above a depth of 240 m. (see Chapter 16, this volume)

Epipelagic: Pelagic above a depth of 240 m. (see: pelagic; see also Chapter 16, this volume)

Epiplanktic: Planktic above a depth of 240 m. (see Chapter 16, this volume)

Evolute: Refers to a coiled shell in which the whorls overlap little or not at all, so that there is a wide umbilicus. Evolute is a relative term; some workers would distinguish between advolute and evolute. (see: coiled; compare: advolute, gyroconic, involute).

Exaerobic: Refers to the boundary between anoxic and dysoxic biofacies, more or less coinciding with the bottom/sea-water interface. (compare: poikiloaerobic)

External septum: Same as "last septum."

External suture: That portion of the suture of a coiled shell that is visible outside of the umbilical seam.

Factor of safety: The ratio, e.g., 2/1, by which the breaking load or ultimate stress exceeds the working load (i.e., pressure) or stress at the maximum habitat depth.

Fixed coordinate system: An artificially introduced framework to express the position of a given point. The origin and axes of such a system are fixed in place. Examples include Cartesian coordinate and polar coordinate systems.

Flange: The dorsal lip of the initial chamber.

Flank: The portion of the whorl between the dorsum and the venter in a coiled shell.

Foliole: A minor element of a saddle of a suture. (compare: lobule)

Foliole-flute: A minor, marginal anticlastic corrugation of an ammonoid septum, characteristically producing an adorally convex semicircle (foliole) on the adjacent suture line. The foliole-flute axis is, therefore, adorally concave; it would have generated transverse tensile stresses during direct loading of the "last septum" by water pressure. (see: saddle-flute; compare: lobule-flute)

Fractals: Geometric objects displaying invariance over a large range of scales; broccoli is an example of a fractal object.

Fractal dimension (D): A single parameter characteristic of a fractal structure. The fractal dimension is usually a fractional number that varies with the degree of complexity of the structure.

Frenet frame: A local coordinate system used to describe the shape of a curve in space. This frame is composed of the unit length of the tangent, principal normal, and binormal vectors, which all intersect each other perpendicularly. (see: standardized moving frame)

Froude efficiency: The energy used in accelerating the jet (i.e., generating propulsive force) relative to energy used to drive the animal forward (i.e., overcoming drag). Froude efficiency is maximized when a large volume of propellent (in this case, water) is accelerated to low velocities to produce thrust; it is minimized when small volumes of water are accelerated to high velocity.

Generating curve: A hypothetical curve that moves and expands in space during a simulation of shell growth. This curve is usually circular or elliptical and moves in a direction normal to the plane containing it. A generating curve would be analogous to the whorl cross section of an actual coiled shell.

Gerontic stage: The stage following the adult stage (synonymous with old age, senility, and senescence); in many present-day cephalopods, this stage is characterized by a deterioration leading to death; it is unlikely that this stage was preserved in the morphology of the ammonoid shell.

Gyrocone, gyroconic: Refers to a planispirally coiled shell in which the whorls do not touch one another, e.g., *Gyroceras*. (see: coiled; compare: orthoconic, cyrtoconic)

Habitat depth limit: The maximum depth visited at least once during the growth of each chamber of an ammonoid during a particular stage of its ontogeny.

Hamitocone, hamitoconic: Refers to a shell with two or more straight shafts, e.g., *Hamitoceras*. (see: coiled)

Height of suture: Same as sutural amplitude.

Heliconic: Same as torticonic.

Heteromorph, heteromorphic: Refers to any ammonoid shell that is not planispirally coiled with successive whorls in contact with one another; includes, for example, ancyloconic, hamitoconic, scaphitoconic, torticonic, etc. (compare: "normal" coiling)

Hoop stress: The stress developed parallel to the surface of a thin shell and normal to the plane of rotation of the axial radius of curvature.

Hydrodynamic stability: See stability.

Hydrostatic stability: See stability.

Hypoxic: Having a level of dissolved oxygen below normal. (compare: aerobic, anoxic, dysoxic)

Implosion depth: The depth at which fragmentation of the weakest part of the shell occurs as the result of the difference between the external hydrostatic pressure and the pressure within the phragmocone. The implosion depth also can be defined, more specifically, as the mean implosion depth of the septa of a particular species at a particular stage of ontogeny.

Index of sutural complexity (ISC): The total length of a suture divided by the whorl circumference. (see: sutural amplitude index)

Indirect development: Mode of early ontogenetic development in which there is a larval stage. (compare: direct development)

Initial chamber (= protoconch): The portion of the ammonitella up to the proseptum.

Inner lamella of the jaw: The inner layer of the ammonoid jaw.

Intercrystalline membranes: Organic membranes separating adjacent hexagonal plates in a single lamella within the nacreous layer of an ammonoid.

Interlamellar membranes: Organic membranes separating the lamellae within the nacreous layer of an ammonoid.

Internal height: In involute shells, the distance from the middorsum of a whorl to the midventer of the same whorl measured in the plane of bilateral symmetry. (compare: whorl height)

Inverse septal thickness (IST): The maximum, central septal thickness divided by one-quarter the distance between opposite ends of the largest saddle-flute on a septum.

Inverted suture: A suture in which all of the elements are supposed to have changed their polarity from adorally to adapically directed. (compare: pseudoinverted suture)

Involute: Refers to a planispirally coiled shell in which the whorls overlap. Involute is a relative term; in general, the term is used for shells in which the whorls overlap to a high degree so that the umbilicus is relatively narrow. (see coiled; compare: advolute, evolute)

Isomorphic shell-form: Refers to a shell form produced by a constant mode of growth; includes orthoconic, planispiral, and torticonic forms. Note that mature modifications exhibited by a given ammonoid may include a departure from an isomorphic shell-form.

Iteroparous: A type of reproduction in which an organism has many reproductive cycles during its lifetime. (compare: semelparous)

Jaw: One of two structures at the entrance to the buccal cavity of a cephalopod; together, they resemble the beak of a bird, but the tip of the upper jaw fits within that of the lower jaw. In most present-day cephalopods, the jaws are organic in composition; however, in *Nautilus*, calcareous parts are present. Some fossil cephalopods are known to have possessed jaw structures similar to those of *Nautilus*; these are known as rhyncholites and conchorhynchs. Some workers are convinced that aptychi and anaptychi are jaws or parts of jaws.

Juvenile stage: The ontogenetic stage between neanic and adult; it is sometimes divided into early, middle, and late, mainly on the basis of shell size.

Keel: A ventral ridge located on the plane of bilateral symmetry of a planispirally coiled shell. (see: ornament)

Lappet: One of a pair of adapertural projections of the peristome that are lateral or ventrolateral in position; generally one occurs on each side, but in some instances there are two pairs of lappets. (compare: rostrum)

Larval stage: In animals with indirect development, the stage between hatching and metamorphosis.

"Last septum" (= external septum): The adoralmost septum of an ammonoid shell; it separates the external hydrostatic pressure (as transmitted through the body) from the lower pressure within the most recently formed chamber of the phragmocone.

Lateradulata: Those forms, mainly Nautiloidea, characterized by the possession of a wide radula with nine teeth in each transverse row. (compare: Angusteradulata)

Leiostraca: A general term used for members of those ammonoid suborders in which shells commonly show little or no ornament, e.g., the Lytoceratina and Phylloceratina. (compare: trachyostraca)

Life orientation: Same as life position.

Life position (= life orientation): The orientation assumed by a living ammonoid when resting in the water column. It is the orientation that occurs when the centers of buoyancy and gravity of the animal are aligned vertically (with the center of buoyancy positioned above the center of gravity), and when the buoyancy and weight forces act in opposite directions along the same line of action. Life position commonly is measured as the angle made by the peristome of the shell with the vertical, in lateral view.

Lira (pl., lirae): A minor ridge on the shell surface; lirae can be spiral (normal to the aperture) or transverse (parallel to the aperture). (see: ornament)

Living chamber: Same as body chamber.

Lobe: An adorally concave, major element of a suture. (compare: saddle)

Lobe-flute: A major, anticlastic corrugation of an ammonoid septum producing a pair of lobes on opposite sides of the suture. The lobe-flute axis is, therefore, adorally convex and characteristically has a parabolic profile. (compare: saddle-flute)

Lobule: A minor element of a lobe of a suture. (compare: foliole)

Lobule-flute: A minor, marginal, anticlastic corrugation of an ammonoid septum, characteristically producing V-shaped "tie points" between the folioles. The lobule-flute axis joins the wall at a lower angle than do the adjacent foliole-flute axes and is either short and straight or adorally convex. (compare: foliole-flute)

Local coordinate system: A coordinate system used to express a local condition within a larger coordinate system. Unlike a fixed coordinate system, this system can move in space. The Frenet frame is an example.

Longicone, longiconic: Refers to a slowly tapering orthocone or gyrocone with a very acute apical angle. (compare: brevicone)

Longidomic (= macrodomic): Having a long body chamber, i.e., one or more whorls long. (compare: brevidomic, mesodomic)

Macroconch: The larger of the antidimorphs in an ammonoid species that exhibits sexual dimorphism; it generally has been considered to be the female. (compare: megaconch, microconch, miniconch)

Macrodomic: Same as longidomic.

Mature modification: A change in the ammonoid shell at the approach of sexual maturity. Common mature modifications include a constriction of the aperture, a change in coiling, a change in whorl cross section, and septal approximation.

Maximum growth point: A theoretical point that signifies the point of maximum displacement on a generating curve; this point coincides with the outermost margin of a tubular shell.

Medial cross section: A longitudinal section along the plane of symmetry of a bilaterally symmetrical conch.

Megaconch: Some workers have recognized three morphs in an ammonoid species rather than two; the third form has been called a megaconch because it is larger than either the macroconch or the microconch.

Megastria (pl., megastriae): A distinctive thick line that extends continuously around the flanks and venter of an ammonoid shell; it represents a pause in growth. (see: ornament)

Membrane stresss: A tensile or shear stress developed parallel to the external surface of a shell; it does not vary by more than 10% within the various layers of the shell below any one point on the surface.

Mesocone: The major cone of the central, rhachidian tooth of a radula.

Mesodomic: Having a body chamber of intermediate length, i.e., about three-fourths whorl long. (compare: brevidomic and longidomic)

Mesopelagic (includes mesoplanktic and mesonektic): Pelagic below 240 m (to about 1000 m). (see: pelagic; see also Chapter 16, this volume).

Microconch: The smaller of the antidimorphs in an ammonoid species that exhibits dimorphism; it generally has been considered to be the male. (compare: macroconch, megaconch, miniconch)

Microdomic: Same as brevidomic.

Miniconch: At least one worker has reported three morphs in an ammonoid species in which the third form is smaller than either the macroconch or the microconch. The smallest morph has been called a miniconch.

Modified retrochoanitic septal neck: A septal neck that is projected adapically from the septum on the ventral side but both adapically and adorally (type A) or only adorally (type B) from the septum on the dorsal side. (compare: amphichoanitic, prochoanitic, retrochoanitic)

Mohr diagram: A graphic representation of tensile and compressive stresses (and their maximum values, termed strengths) as circles; invented by Otto Mohr around 1880.

Nacre: Aragonite with a laminated microstructure; nacre forms the bulk of the ammonoid shell and pearls.

Neanic stage: The ontogenetic stage of an ammonoid immediately after hatching and before the juvenile stage.

Neanoconch: The shell of an ammonoid during the neanic stage of ontogeny (and, by extension, the animal as a whole); the shell extends adaperturally up to approximately two whorls beyond the ammonitella, corresponding to a diameter of 3–5 mm. (compare: ammonitella)

Negative Gaussian curvature: Same as anticlastic curvature.

Nektic (= nektonic): Refers to pelagic organisms that swim well enough to be able to move independently of currents. (compare: planktic)

Nektobenthic: See demersal.

Nepionic constriction: Same as primary constriction.

Neritic: Refers to the environment and biome of the seas of the continental shelves, in general, the shallow waters along coasts. (compare: epeiric, oceanic)

Neritopelagic: Pelagic within the neritic biome. (see: pelagic)

Neutral buoyancy: See buoyancy.

"Normal" coiling: The coiling exhibited by a planispirally coiled ammonoid shell in which the whorls overlap or just touch one another. (see: coiled; compare: heteromorphic)

Oceanic: Refers to the environment and biome of the ocean beyond the continental shelf. (compare: epeiric, neritic)

Oceanomesopelagic: Oceanopelagic between depths of about 240 m and 1000 m. (see: pelagic; see also: Chapter 16, this volume)

Oceanopelagic: Pelagic within the oceanic biome. (see: pelagic)

Ornament (= ornamentation): A term applied to various ribs, plications, grooves, etc., that may be observed in ammonoid shells. Although the term "ornament" implies that these features serve no function, at least some certainly served to strengthen the shell, and some record events in the ontogeny of the animal. Many terms have been proposed for the various kinds of ornament, e.g., annulation, constriction, keel, lira, megastria, rib, varix; for their orientation, e.g., prorsiradiate, rectiradiate, rursiradiate; and for their configuration.

Orthocone, orthoconic: Refers to a shell that is straight. (compare: coiled, cyrtoconic)

Orthogonal stress ratio: The ratio of membrane stresses at any one point on the external shell surface (including the "last septum").

Outer lamella of the jaw: The outer layer of the ammonoid jaw.

Oxic: Same as aerobic.

Oxycone, oxyconic: Refers to an involute shell with strongly compressed, subtriangular whorls and a sharp venter, e.g., *Placenticeras*. (see: coiled)

Pelagic: Floating or swimming in seawater, but not near the bottom (i.e., independent of the bottom). Pelagic organisms may be divided on the basis of their mode of life into floating (planktic) and swimming (nektic) organisms. They may be divided on the basis of water depth into epipelagic, mesopelagic, and bathypelagic. They may be divided on the basis of whether they live in the open ocean, above the continental shelf, or in inland seas into oceanopelagic, neritopelagic, and epeiropelagic, respectively. Moreover, the different schemes of division can be combined to produce such terms as oceanomesopelagic. (compare: benthic)

Periostracum: The external, organic layer of the shell present in most molluscs.

Periostracal groove: The external groove at the edge of the mantle in molluscs; it is the site of secretion of the periostracum.

Peristome: The apertural edge of an ammonoid shell.

Phi: See theta (θ) angle.

Phragmocone: That portion of the shell of an ammonoid subdivided into camerae. (compare: body chamber)

Planispiral: Coiled in a single plane. (see: coiled)

Planktic (= planktonic): Pelagic but without much mobility. This includes passive floaters (e.g., some microorganisms) and weak swimmers (e.g., neanic ammonoids, presumably); organisms that can move vertically, for example, by changing buoyancy, but not laterally could be included here also. (compare: nektic, vertical migrant)

Planorbicone, planorbiconic: Refers to an evolute shell with a subcircular or depressed whorl cross section, e.g., *Stephanoceras*. (see: coiled)

Platycone, platyconic: Refers to an involute to moderately evolute, planispirally coiled shell with a subrectangular, compressed whorl cross section and, generally, a keel, e.g., *Collignoniceras*. (see: coiled)

Poikiloaerobic: Refers to conditions on the sea floor that shift seasonally between dysoxic and anoxic.

Poisson's ratio: The ratio of the small transverse strain to the large axial strain caused by axial stress. The nacreous shell material of *Nautilus* has a Poisson's ratio of about 0.314.

Positive Gaussian curvature: Same as synclastic curvature.

Primary constriction (= nepionic constriction): The groove in the shell wall at the end of the ammonitella; this groove is especially well expressed along the venter.

Primary varix: The thickening of nacre at the end of the ammonitella associated with the primary constriction.

Prochoanitic septal neck: A septal neck that is projected adorally from the septum. (compare: amphichoanitic, modified retrochoanitic, retrochoanitic)

Prorsiradiate: Slanting forward; a prorsiradiate rib, for example, is slanted so that it is more adapertural at the venter of the whorl than it is at the dorsum. (see: ornament)

Proseptum: The first septum in an ammonoid; it closes off the initial chamber.

Prosiphon: The attachment band between the caecum and the inside surface of the initial chamber.

Protoconch: Same as initial chamber.

Pseudohexagonal trilling: Refers to cyclic twins in orthorhombic carbonates, e.g., aragonite.

Pseudoinverted suture: A suture in which the inversion is only apparent. (compare: inverted suture)

Pseudoseptum (pl., pseudosepta): An intracameral organic membrane with a topology similar to that of the septum. (see: cameral membranes)

Pseudosuture: The line of attachment of a pseudoseptum to the ammonoid shell wall; it is usually preserved as a faint groove on the steinkern.

Radial: Refers to an orientation parallel to a transverse plane radiating outward from the center of coiling of a planispiral shell, and, hence, in a given portion of a whorl, refers to orientations parallel to former apertural positions and to septa in that portion of the whorl.

Radial conchiolin membranes: Same as radial organic membranes.

Radial organic membranes (= radial conchiolin membranes): Organic membranes that separate adjacent sectors in a single hexagonal plate of stack nacre.

Radius of curvature: The distance from the middle of a curved shell surface to a locus defined by the normals to the tangents to the shell surface in a single plane.

Radula: The ribbon between the upper and lower jaws bearing numerous transverse rows of teeth.

Rectiradiate: Oriented perpendicular to the direction of growth, i.e., directly transverse a whorl. (see: ornament)

Retrochoanitic septal neck: A septal neck that is projected adapically from the septum. (compare: amphichoanitic, modified retrochoanitic, prochoanitic)

Reynolds number: A dimensionless number describing flow state. Reynolds number compares the magnitude of the inertial (pressure) forces to viscous (frictional) forces generated when an animal moves through the water; it is defined as linear dimension times speed divided by the kinematic viscosity. For ammonoids, Reynolds numbers are thought to have ranged from 10 for small animals moving slowly to 5×10^5 for large animals moving more rapidly.

Rhachidian tooth: The central tooth of the transverse row of teeth of a radula.

Rhyncholite: See jaw. (The name "rhyncholite" sometimes is spelled "rhyncolite", which is the same spelling as a generic name [Teichert *et al.*, 1964].)

Rib (= costa): A radial or subradial ridge on the shell. Strictly speaking, a rib affects only the exterior of the shell, whereas a plication or corrugation affects both the shell exterior and the shell interior. Thus, a rib would not show on the internal mold, but a plication would. In more casual usage, any radial or subradial ridge or plication is called a rib. (see: ornament; compare: lira)

Rib index: In a heteromorphic ammonoid, the number of ribs in a distance equal to the whorl height at the midpoint of the interval over which the ribs are counted.

Rostrum: (1) In an ammonoid shell as a whole, a ventrally positioned adapertural projection of the peristome. (compare: lappet) (2) In an ammonoid jaw, the anterior tip.

Rule-defined simulation: A computer simulation of shell growth in which certain rules of growth are specified.

Runzelschicht: Same as wrinkle layer.

Rursiradiate: Slanted backward; a rursiradiate rib, for example, is slanted so that it is more adapertural at the dorsum of the whorl than it is at the venter. (see: ornament)

Saddle: An adorally convex, major element of a suture; the term originated from the similarity of the associated saddle-flute to the saddle of a horse. Unfortunately, engineers have since used this term for any anticlastic fluted structure, whereas ammonoid taxonomists have tended to restrict this term to the two-dimensional suture line. (compare: lobe)

Saddle-flute: A saddle-like major anticlastic corrugation of the septal surface. Each saddle-flute has an adorally concave, relatively flat, tensile axis that may or may not connect saddles on opposite sides of the septum. Smaller saddle-flutes grade into marginal foliole-flutes. (see: foliole-flute; compare: lobe-flute)

Safety factor: Same as factor of safety.

Scaphitocone, scaphitoconic: Refers to an ancylocone with an involute phragmocone and a hooked body chamber, e.g., *Scaphites*. (see: coiled)

Semelparous: A type of reproduction in which an organism has only one reproductive cycle and then dies. (compare: iteroparous)

Septal angle: The angle between two adjacent septa measured in the medial plane.

Septal approximation: In general, septal spacing increases adaperturally; however, in certain circumstances, for example, as sexual maturity is approached, the septa may become more closely spaced than they had been. This septal crowding is termed septal approximation.

Septal differentiation: The progressive complication of the septal mantle through ontogeny.

Septal flute strength index (SFSI): A measure of the strength of a septum to the hydrostatic pressure transmitted through the body.

Septal neck: Collar-shaped projection of the septum at the point of connection with the siphuncular tube. (see: amphichoanitic, modified retrochoanitic, prochoanitic, retrochoanitic)

Septal-neck/siphuncular complex (= siphuncle): A tube of alternating rigid and flexible rings characteristic of the chambered shells of cephalopods. It permits the removal of liquid from the chambers of the phragmocone by means of an osmotic pump mechanism.

Septal recess: A tunnel-like structure formed by a backward folding of the middorsal septal surface and terminating at the mouth of its equivalent on the preceding septum; it is characteristic of some lytoceratid ammonoids.

Serpenticone, serpenticonic: Refers to very evolute or advolute shells with subcircular or depressed whorls, e.g., *Dactylioceras*. (see: coiled)

Siphuncle: (1) Same as septal-neck/siphuncular complex. (2) The strand of soft tissue that occupied the septal-neck/siphuncular complex during life.

Siphuncular strength index (SiSI): A measure of the strength of the connecting rings to the hydrostatic pressure transmitted through the body.

Siphuncular membranes: Intracameral membranes connected with the siphuncular tube. (see: cameral membranes)

Siphuncular tube: Same as connecting ring.

Spherocone, spheroconic: Refers to a planispiral, subglobular shell with an involute body chamber, e.g., *Eurycephalites*. (see: coiled; compare: elliptospherocone)

Squirming instability: The lateral buckling of an axially restrained tube by internal pressure.

Stability: The moment (force multiplied by the moment arm) acting to deflect an ammonoid from its orientation in the water. There are two types of stability: hydrostatic stability and hydrodynamic stability. Hydrostatic stability applies to an ammonoid sitting motionless in the water and is defined as the ammonoid's weight, including both shell and the soft body, times the distance between its centers of buoyancy and gravity. It commonly is viewed as the stability of the neutrally buoyant organism against rotation caused by the propulsive force of water expelled through the hyponome. Hydrodynamic stability applies to swimming ammonoids and is defined by a complex interaction of hydrostatic stability plus moments deriving from thrust, drag, and other hydrodynamic forces produced by the animal's motion.

Stack nacre (= columnar nacre): A certain type of nacreous layer present in ammonoids, nautiloids, and gastropods in which hexagonal aragonitic plates are piled on top of one another, forming a so-called stack.

Standardized moving frame: A local coordinate system used to describe a tubular shell. The tangent, principal normal, and binormal vectors in this frame enlarge with growth in conjunction with the radius of the tube.

Steerage: Potential directional control during swimming, which characteristically was backward in ammonoids.

Stress: The internal forces exerted across an imaginary plane of separation within each element of the shell wall. In the case of shear stress, the plane of separation is parallel to these forces, and, in the case of tensile or compressive stress, it lies normal to the forces.

Stretch pathology: Refers to a portion of a shell characterized by a decrease in whorl width and height, an attenuation of ornament, and a thinning of the shell wall; it is common in macroconchs of some Late Cretaceous scaphitid ammonoids.

Subglobular: Refers to an ammonoid shell that is almost globular in shape.

Sutural amplitude (= height of suture): Maximum depth of the suture measured along the spiral.

Sutural amplitude index: The ratio of the depth of the deepest lobe (usually *L*) to the total external span of the suture, times 10. In planspiral shells, the total external span of the suture is measured along the shell surface from the umbilical seam to the midventer. This ratio can only be used to compare similar whorl cross sections. (see: index of sutural complexity)

Synclastic curvature (= positive Gaussian curvature): A surface of double curvature with the orthogonal radii of curvature on the same side of the surface; domes, for example, exhibit synclastic curvature. (compare: anticlastic curvature)

Theta angle: The larger portion of the angle of rotation of the radius of curvature of an anticlastic surface, the orthogonal, smaller angle being termed phi. Theta is more commonly used to describe the total rotation angle of planispiral ammonoids.

Thickness ratio: The maximum width of an ammonoid shell relative to shell diameter. For most ammonoids, thickness ratio is a major factor in determining drag force.

Thrust: The propulsive force generated by the action of an ammonoid's locomotor musculature. It is the force that drives the animal through the water. In steady motion, thrust generally is equal to drag, but, in the pulsed motion characteristic of cephalopods, thrust and drag are not necessarily equal at all times. During forward movement, the line of action of propulsive force is in the animal's direction of motion, but, to change direction, thrust is directed off–center.

Tie point: A discrete attachment point of the septal mantle, presumably located originally at the tip of each adapically directed fold of the suture. (see: lobule–flute)

Toroidal whorl: A portion of a coiled shell with the shape of a doughnut and a whorl expansion rate (*W*) approximately equal to 1.

Torticone, torticonic (= heliconic, trochoceroid, trochoconic, trochospiral): Refers to a shell that is coiled not in a single plane, e.g., *Turrilites*; the threads of a screw are torticonic. (see: coiled; compare: planispiral)

Torus radius: The third radius of curvature of the relatively unfluted central region of an ammonoid septum, lying in the plane of the generating curve and defined by a radial pattern of fluting on each septum.

Trachyostraca: A general term used for members of those ammonoid suborders in which shells commonly show some sort of ornament, e.g., the Ammonitina. (compare: leiostraca)

Translocation: The forward movement of the soft body from one septum to the next.

Trochoceroid, trochoconic, trochospiral: Same as torticonic.

Umbilical plug (= callus): The mineralized and organic material that fills the umbilicus in some ammonoids and nautiloids.

Varix (pl., varices): A thickening of the shell wall commonly associated with a constriction. (see: ornament)

Vault: An arched roof or series of arches supporting their own weight by acting as a beam and thrusting outward into lateral supports.

Vermicone, vermiconic: Refers to a seemingly irregular shell, resembling a twisted tube worm but, in reality, following a complex growth program, e.g., *Nipponites*. (see: coiled)

Vertical migrant: An organism that makes regular vertical movements through the water column, for example, an animal that moves from epipelagic to mesopelagic depths and back again on a daily (circadian) basis. This might be accomplished by changes in buoyancy, swimming, or both.

Viscoelastic deformation: The property of a material that increasingly deforms the longer a force is applied. Once the stress is removed, recovery is gradual but is only partial, such that some of the deformation is permanent. Any measurement of strain in a viscoelastic material is time dependent. (compare: elastic)

Viscous fingering: A growth pattern created by the displacement of the interphase between two liquids with different viscosities.

Wall proper of the initial chamber: The calcified primary wall of the initial chamber.

Weibull modulus: A statistical measure of the variance of the strength of a brittle material, invented by W. Weibull in 1938. It is equal to 1.27 times the mean divided by one standard deviation.

Whorl expansion rate (W): The increase in the distance from the center of coiling of a shell to the venter of a given whorl over the distance of the previous half-whorl; as defined by Raup, $W = (d/e)^2$ in planispiral shells, where d and e are the distances from the center of coiling of the shell to the venter one-half whorl apart.

Whorl height: The distance from the most dorsal point of a whorl to the midventer of the same whorl, measured in a line parallel to the plane of bilateral symmetry; in a gyroconic or advolute shell, the most dorsal point most likely will be in the plane of bilateral symmetry; in an involute shell, on the other hand, the most dorsal point might be at the umbilical seam or on the umbilical wall, depending on the shape of the whorl cross section. (compare: internal height)

Working circumference: The distance around the whole whorl surface in a radial plane.

Working stress: The maximum stress applied to shell material during life, for example, at the habitat depth limit.

Wrinkle layer (= *Runzelschicht*): The layer of the dorsal wall in "normally" coiled ammonoids that is characterized by the presence of wrinkles, which may resemble the ridges and grooves of human fingerprints. (compare: encrusting layer)

Young's modulus: The axial stress divided by the elastic strain in the same direction. Young's modulus approximately equals 45,800 MPa for the nacreous material in a *Nautilus* shell when the stress is parallel to the surface and the nacreous lamination.

Zonal indices: Abbreviations of zonal names based on abbreviations of chronostratigraphic units (e.g., MD II, Middle Devonian II = Lower Givetian; CU, Lower [German: "*Unter-*"] Carboniferous) and abbreviations of marker genera (e.g., Go = *Goniatites*).

References

Arkell, W. J., 1957. Introduction to Mesozoic Ammonoidea. in: *Treatise on Invertebrate Paleontology, Part L, Mollusca 4* (Moore, R. C., ed.), Geological Society of America and University of Kansas Press, New York and Lawrence, Kansas, pp. L81–L129.

Arkell, W. J., Kummell, B., Miller, A. K., and Wright, C. W. 1957. Morphological Terms Applied to Ammonoidea. in: *Treatise on Invertebrate Paleontology, Part L, Mollusca 4* (Moore, R. C., ed.), Geological Society of America and University of Kansas Press, New York and Lawrence, Kansas, pp. L2–L6.

Teichert, C., R. C. Moore, and Zeller, D. E. Nodine. 1964. Rhyncholites. in: *Treatise on Invertebrate Paleontology, Part K, Mollusca 3* (Moore, R. C., ed.), Geological Society of America and University of Kansas Press, New York and Lawrence, Kansas, pp. K467–K484.

Index

Like most indexes, this one is nowhere near as thorough as it might have been; the tremendous number of items that might have been included would have made it completely unwieldy. People have not been indexed; those mentioned in each chapter are listed in the bibliography at the end of the chapter. Items that obviously would be expected to be in a particular chapter are not indexed for that chapter. For items that appear many times throughout a chapter or a section, only the first occurrence in that chapter or section is indexed, with the page number followed by "ff". Certain items appear so many times in so many chapters throughout the book that to have listed them in the index would have taken up tens or hundreds of entries each; hence, things like ammonoid, Ammonoidea, Paleozoic, Mesozoic, Jurassic, Cretaceous, and *Nautilus* are not indexed as such.